Marriages and Families

MAKING CHOICES IN A DIVERSE SOCIETY

Sixth Edition

Mary Ann Lamanna

University of Nebraska at Omaha

Agnes Riedmann

California State University, Stanislaus

WADSWORTH PUBLISHING COMPANY

I(T)P® An International Thomson Publishing Company

Belmont, CA • Albany, NY • Bonn • Boston • Cincinnati • Detroit • Johannesburg • London • Madrid • Melbourne • Mexico City • New York • Paris • San Francisco • Singapore • Tokyo • Toronto • Washington

SOCIOLOGY EDITOR: Eve Howard

ASSISTANT EDITOR: Susan Shook

EDITORIAL ASSISTANT: Deirdre McGill

PRODUCTION: Robin Lockwood & Associates

PRINT BUYER: Karen Hunt

PERMISSIONS EDITOR: Jeanne Bosschart

COPY EDITOR: Jennifer Gordon

PHOTO RESEARCH: Linda Rill

TECHNICAL ILLUSTRATOR: Hans & Cassady

DESIGNER: Cuttriss & Hambleton

COVER DESIGNER: Cuttriss & Hambleton

COMPOSITOR: GTS Graphics, Inc.

PRINTER: Quebecor Printing/Hawkins

COVER PHOTOGRAPH: ©David Hanover, Tony Stone Images

Printed in the United States of America
1 2 3 4 5 6 7 8 9 10

For more information, contact Wadsworth Publishing Company, 10 Davis Drive, Belmont, California 94002, USA

International Thomson Publishing Europe
Berkshire House 168-173
High Holborn
London, WC1V 7AA, England

Thomas Nelson Australia
102 Dodds Street
South Melbourne 3205
Victoria, Australia

Nelson Canada
1120 Birchmount Road
Scarborough, Ontario
Canada M1K 5G4

International Thomson Publishing GmbH
Königswinterer Strasse 418
53227 Bonn, Germany

International Thomson Editores
Campos Eliseos 385, Piso 7
Col. Polanco
11560 México D.F. México

International Thomson Publishing Asia
221 Henderson Road
#05-10 Henderson Building
Singapore 0315

International Thomson Publishing Japan
Hirakawacho Kyowa Building, 3F
2-2-1 Hirakawacho
Chiyoda-ku, Tokyo 102, Japan

International Thomson Publishing Southern Africa
Building 18, Constantia Park
240 Old Pretoria Road
Halfway House, 1685 South Africa

Library of Congress Cataloging-in-Publication Data
Lamanna, Mary Ann.
    Marriages and families: making choices in a diverse society/
  Mary Ann Lamanna, Agnes Riedmann.—6th ed.
        p.   cm.
    Includes bibliographical references and index.
    ISBN 0-534-50553-8
    1. Marriage.   2. Family.   3. Remarriage.   4. Single people.
    5. Choice (Psychology)   I. Riedmann, Agnes Czerwinski. II. Title.
    HQ734.L23   1997
    306.8—dc20   96-18509

To our families, especially
Bill, Beth, Chris, and Natalie
Larry, Valerie, Sam, Janice, and Simon

About the Authors

MARY ANN LAMANNA is Professor of Sociology at the University of Nebraska at Omaha. She received her bachelor's degree from Washington University (St. Louis), her master's degree from the University of North Carolina, Chapel Hill, and her doctorate from the University of Notre Dame. Her teaching and research interests are in family, gender, and reproduction, especially law and public policy in these areas. Current research projects concern teenage pregnancy, Durkheim's sociology of the family, and the sociological aspects of transplant medicine. She has two children, Larry, 32, and Valerie, 28.

AGNES RIEDMANN is Assistant Professor of Sociology at California State University, Stanislaus. She attended Clarke College, Dubuque, and received her bachelor's degree from Creighton University, her master's degree in sociology from the University of Nebraska at Omaha, and her doctorate from the University of Nebraska. Her professional areas of interest are theory and family. She received a one-year Fulbright award for demographic research at Australian National University, Canberra. She is the author of *Science That Colonizes: A Critique of Fertility Studies in Africa,* Temple University Press, 1993. Current research projects concern adult sibling relationships and the sociology of demography. She has two children, Beth, 31, and Bill, 29, and a granddaughter, Natalie, 3.

Brief Contents

Detailed Contents

PART  *Becoming Partners*   *81*

P A R T   *III*   *Defining Your Marriage and Family*   *179*

260-263

PART *IV*   *Experiencing Family Commitment*   *303*

P A R T *V*   *Family Change and Crises*   *433*

Preface

As we complete our work on the sixth edition of *Marriages and Families,* we look back over five earlier editions. Together, these represent twenty years spent observing the contemporary American family. Not only has the family changed during this time but so has sociology's interpretation of it. Even though our observation of change is not always encouraging, it *is* gratifying to find ourselves a part of the enterprise of learning about the family and to share that knowledge with students.

Our own perspective on the family has developed and changed during this period as well. We have studied demography and history and have come to pay more attention to social structure in our analysis. We continue to affirm the power of families to direct the course of their lives. But the American social milieu seems less promising today than it did when we began this book. Consequently, we now give more attention to policies needed to provide support for today's families: working parents, poor and minority families, single parents, remarried families, gay and lesbian couples, and other nontraditional families—in short, virtually all families, as the once-standard nuclear family becomes less and less prevalent (though it remains a cultural ideal and common family experience). Marriage and family values continue to be important in contemporary American life. Our students come to a marriage and family course because family life is important to them. Our aim now, as in the first edition, is to help students question assumptions and to reconcile conflicting ideas and values throughout their lives. We enjoy and benefit from the contact we've had with faculty and students who have used this book. Their enthusiasm and criticism have stimulated many changes in the book's content. To know that a supportive audience is interested in our approach to the study of the family has enabled us to continue our work over a long period of time.

The Book's Themes

Several themes are interwoven throughout this text: People are influenced by the society around them as they make choices; social conditions change in ways that may impede or support family life; there is an interplay between individual families and the larger society; and individuals make family-related choices throughout adulthood.

MAKING CHOICES THROUGHOUT LIFE   The process of creating and maintaining marriages and families requires many personal choices, and people continue to make decisions, even "big" ones, throughout their lives.

PERSONAL CHOICE AND SOCIAL LIFE   Tension frequently exists between the individual and the social environment. Many personal troubles result from societal influences, values, or assumptions, inadequate societal support for family goals, or conflict between family values and individual values. By understanding some of these possible sources of tension and conflict, individuals can perceive their personal troubles more clearly and can work constructively toward solutions. They may choose to form or join groups to achieve family or individual goals. This may include involvement in the political process to shape state or federal social policy. The accumulated decisions of individuals and families may also shape the social environment.

A CHANGING SOCIETY   In the past, people tended to emphasize the dutiful performance of social roles in marriage and in the family structure. Today people are more apt to view marriages as committed relationships in which they expect to find companionship and intimacy. This book examines the implications of this shift in perspective.

Individualism, economic pressure, social diversity, and decreasing marital permanence are features of the social context in which personal decision making takes place today. As fewer social guidelines seem fixed, personal decision making becomes even more challenging.

The Themes Throughout the Life Course

The book's themes are introduced in the prologue, and they reappear throughout the text. We developed these themes by looking at the interplay between findings in the social sciences and experiences of the people around us. Ideas for topics arose from the needs and concerns we perceived. We observed many changes in the roles people play and in the ways they relate to each other. Neither the "old" nor the "new" roles and relationships seemed to us as stereotyped or as free of ambivalence and conflicts as is often indicated in books and articles. The attitudes, behavior, and relationships of real people have a complexity that we have tried to portray in this book.

Interwoven with these themes is the concept of the life course—the idea that adults may change through reevaluation and restructuring throughout their lives. This emphasis on the life course creates a comprehensive picture of marriages and families and enables this book to cover many topics that are new to marriage and family texts. This book makes these points:

- Many people reexamine the decisions they have made about marriage and family not once or twice but throughout their lives.

- People's personal problems and their interaction with the social environment change as they and their marriages and families grow older.

- People reevaluate their relationships and their expectations for relationships as they and their marriages and families mature.

- Because marriage and family forms are more flexible today, people may change the style of their marriage and family throughout their lives.

Marriages and Families Making Choices

Making decisions about one's marriage and family, either knowledgeably or by default, begins in early adulthood and lasts into old age. People choose whether they will adhere to traditional beliefs, values, and attitudes about gender roles or will adopt more androgynous roles and relationships. They may clarify their values about sex and become more knowledgeable and comfortable with their sexual choices.

Women and men may choose to remain single or to marry, and they have the option today of staying single longer before marrying. Single people make choices about their lives, ranging from decisions to engage in sex for recreation to decisions to abstain from sex altogether. In the courtship process people choose between the more formal custom of dating and the less formal "getting together."

Once individuals choose their partners, they have to decide how they are going to structure their marriages and families. Will the partners be legally married? Will theirs be a dual-career marriage? Will they plan periods in which just the husband or just the wife works interspersed with times in which both work? Will they have children? Will they use the new reproductive technology to become parents? Will other family members live with them—parents, for example? They will make these decisions not once but over and over during their lifetimes.

Within the marital relationship, couples choose how they will deal with conflicts. Will they try to ignore conflicts and risk devitalized relationships? Will they vent their anger in hostile, alienating, or physically violent ways? Or will they practice bonding ways of communicating and fighting—ways that emphasize sharing and can deepen intimacy?

How will the partners distribute power in the marriage? Will they work toward a no-power relationship in which the individual is more concerned with helping and supporting the other than with gaining a power advantage? How will the partners allocate work responsibilities in the home? What value will they place on their sexual lives together? Throughout their experience family members continually face decisions about how to balance each one's need for individuality with the need for togetherness.

Parents also have choices. In raising their children, they can assume the role of martyr or police officer, for example, or they can simply present themselves as human beings who have more experience than their youngsters and who are concerned about developing supportive, mutually cooperative relationships.

Many spouses face decisions about whether to divorce. They weigh the pros and cons, asking themselves which is the better alternative: living together as they are or separating? Even when a couple decides to divorce, there are choices to make: Will they try to cooperate as much as possible or insist on blame and revenge? What living and economic support arrangements will work best for themselves and their children? How will they handle the legal process?

The majority of divorced individuals eventually face decisions about remarriage. And in the absence of cultural models, they choose how they will define step-relationships.

When families meet crises—and every family will face *some* crises—members must make additional decisions. Will they view each crisis as a challenge to be met, or will they blame each other? What resources can they use to handle the crisis?

An emphasis on knowledgeable decision making does not mean that individuals can completely control their lives. People can influence but never directly determine how those around them behave or feel about them. Partners cannot control each other's changes over time, nor can they avoid all accidents, illnesses, unemployment, deaths, or even divorces. Society-wide conditions may create unavoidable crises for individual families.

Families *can* control how they respond to such crises, however. Their responses will meet their own needs better when they refuse to react automatically and choose instead to act as a consequence of knowledgeable decision making.

Key Changes in This Edition

As marriages and families have evolved over the last twenty years, so too has this text. Its new subtitle, *Making Choices in a Diverse Society*, illustrates the vast changes that have taken place over the last decade. With its thorough updating and inclusion of current research, plus its new emphasis on students' being able to make choices in an exceedingly *diverse* society, this book has become an unparalleled resource for gaining insights into today's marriages and families.

Over the past five editions, we have had two goals in mind for student readers: first, to help them better understand themselves and their family situations; and second, to make students more conscious of the personal decisions that they will make throughout their lives and of the societal influences that affect those decisions. These two goals continue and are reflected in the four themes of this text, described earlier.

The decision-making theme is, in fact, enhanced, in this edition. We highlighted the theme of making choices by introducing a group of theme boxes ("As We Make Choices . . .") throughout the text. These boxes emphasize human agency and are designed to help students through those moments of crucial decisions. Many of these are boxes from earlier editions, ones which our reviewers especially liked and particularly asked us not to delete.

Moreover, we have added a third goal for this edition: to help students better appreciate the variety and diversity among families today. In order to accomplish this third goal we have presented the latest research and statistical information on diverse family forms, lesbian and gay male families, and on families of diverse races and ethnicities. We consciously integrated these new materials throughout the textbook, always with an eye toward avoiding stereotypical, simplistic generalizations and, instead, explaining data in sociological and sociohistorical context. Besides integrating information on ethnic diversity throughout the text proper, we have added an entirely new series of boxes entitled, "A Closer Look at Family Diversity." These boxes focus on the latest theory or research as it applies to some area of family diversity, from Hmong families in Minneapolis to sexuality between lesbian partners.

In addition (and as users have come to expect), we have thoroughly updated the text's research base and statistics, emphasizing cutting-edge research that addresses the diversity of marriages and families as well as all other topics. In accordance with this, users will notice as many as twelve entirely new tables and twelve entirely new figures. Revised tables and figures have been updated with the latest available statistics.

As our reviewers requested, we have included more discussion of how to apply an understanding of theory and methods to current research on families. And we have reviewed every sentence, paragraph, and section for readability, eliminated wordiness and made explanations and points clearer where appropriate.

The material formerly covered in the Managing Family Resources chapter has been moved, putting the budget material into an appendix and adding policy material in appropriate places throughout the text. Meanwhile, we have kept an applied perspective in the text, with its emphasis on decision making in today's increasingly diverse society.

Because reviewers felt that the material in the former Prologue was too important not to be designated a regular chapter, we expanded the Prologue and renamed it Chapter 1, "Family Commitments: Making Choices in a Changing Society." The former box, "Some Facts about Families in the United States Today," a favorite of our users, now appears in Chapter 1. This chapter now defines the concept *family* and sets the stage for the text, introducing the themes and goals of the text, introducing the controversy over when families are in "decline" or simply "change," and giving statistics that show briefly what the situation of families in the United States is today—statistics that will be expanded upon throughout the textbook. Chapter 1 also includes a new "Closer Look at Family Diversity" box on family ties and immigration.

Chapter 2 discusses theoretical perspectives on families, perspectives that are integrated throughout the text wherever possible, and adds a new section on the family ecology perspective that includes family policy discussion taken from our previous edition's Chapter 13. Box 2.1 is updated, and there is new, focused discussion on feminist perspectives and a new "Closer Look at Family Diversity" box on studying ethnic minority families.

The focus of Chapter 3 has changed from gender stereotypes and gender scripts to gender identities and how these are negotiated and maintained. There are new sections on masculinities and femininities; revised, updated discussion of gendered socialization in schools, including a new section on African American, Latina, and Asian American girls in middle school; and new material on how men change. The section on the men's movement has been revised, and a new "Closer Look at Family Diversity" box explores Chicana feminism.

Chapter 5 underwent a major revision. It now begins with a section on sexual orientation, the entire chapter having been reframed with much new material in the section on sex and society and a new section, "Evolutionary Psychologists—Once again: Is Anatomy Destiny?" that reinforces and extends the discussion in Chapter 3. A new section, "An Historical Perspective," presents material on heterosexism and homophobia; another new section, "An Interactionist Perspective: Negotiating Cultural Messages," introduces this concept and relates back to the subject's introduction in Chapter 2. In response to our reviewers' requests that we further discuss research methods wherever possible, we have added a new section, "How do We Know What We Do?" that explains surveys in general and surveys about sex in particular. The section, "Platform and Pulpit," now begins with material on Masters and Johnson's research (formerly in an appendix), and there is new, updated material on the New Christian Right.

In Chapter 6, the "Changing Attitudes Toward Marriage and Singlehood" section has been extensively revised and updated; material dealing with single women as compared with married women and single men as compared with married men has been rewritten, and there is a new "Closer Look at Family Diversity" box on African American men's friendship.

In Chapter 7, the table on the ratio of men and women by age and race has been revised to include figures for the following ethnic groups: Asian/Pacific Islander, American Indian/Eskimo/Aleut, African American, Hispanic, and non-Hispanic white. The concept of complementary needs, along with complimentary needs theory, have been added. There is a new "Closer Look at Family Diversity" box on early marriage in a Hmong cohort.

Chapter 8, "Marriage: A Unique Relationship," has been a substantial revision. We believe it is the best chapter on this topic that we have ever written. The chapter begins with a new section on marriage and kinship that includes new concepts such as dominant dyad, consanguineous relationship, and la familia. The chapter also reframes the discussion on intrinsic versus utilitarian marriages. The section on extramarital sex has been totally rewritten, and a new section on the controversy over whether same-sex couples should legally marry is now included.

Chapter 9 includes a new discussion of family systems theory that has been incorporated into the text proper rather than appropriated to a box, a change of chapter focus from just resolving conflicts to implementing positive communication techniques, and a new "Closer Look at Family Diversity" box on religion's role in rural African American marriages.

In Chapter 10, we reinstated French and Raven's six bases of power, as requested by our reviewers, and added a new "Closer Look at Family Diversity" box on immigrant children, speaking English, and family power. We added material on family violence and updated that section thoroughly as well as added a new section on abuse among lesbian and gay male couples.

Material throughout Chapter 11 has been updated, while technical information on high tech fertility from the 5th edition has been moved from this chapter, as requested by our reviewers, and reformulated as an appendix (Appendix E).

In Chapter 12, new "parents and children" sections relate specifically to African Americans, Native Americans, Mexican Americans, and Asian Americans. There is a newly revised section on raising minority children in a racist and discriminatory society, as well as a new "Closer Look at Family Diversity" box in which an Asian student talks about discrimination he encountered. Le Master's five parenting styles have been taken out of their 5th-edition box and incorporated into the text proper.

Chapter 13 has a new section on race, ethnicity, and housework, and a new "Closer Look at Family Diversity" box on race/ethnicity, immigration, social class, and childcare.

In Chapter 14 there is a new "Closer Look at Family Diversity" box on how low-income, African American mothers care for children with sickle-cell disease, along with a new section that specifically addresses the "sandwich" generation, as requested by our reviewers.

In Chapter 15, there is a new section, "Other Factors Associated with Divorce," that points to such demographic factors as marrying young and premarital childbearing; the section on the economic consequences of divorce has been revised; there are new sections that explore the issue of no-fault divorce laws and if they're responsible for ex-wives' and their children's poverty; and there is a new section on survey versus clinical findings regarding children of divorce. There is a new section on the possible reasons for the negative effects of divorce on children, adult children of divorced parents, and intergenerational relationships, and the stability of marriages for adult children of divorced parents.

Chapter 16 has been retitled "Remarriages" in the plural to demonstrate that remarriages are not all alike; and a section has been added to show the diversity of remarriage types, from those without any children to those with step and biological children. This chapter has a new case study, "My (Remarried) Family," along with a new section on remarriage and children's living arrangements.

The Afterword (formerly the Epilogue) now includes an entirely new section on family values. This section is designed to be especially text-student interactive: Students are asked to think critically and answer questions about what they think family values are and the relationship between these and social policy.

All the appendices have been revised and updated as needed. Two new appendices have been added from material previously in the text proper: Appendix E: High-Tech Fertility, and Appendix I: Managing a Family Budget.

The Summary sections of each chapter have been revised, as have all Study Questions sections at the end of each chapter.

Special Features of Marriages and Families

- **A broad, up-to-date research base.** The content is based on extensive research. More than 2,500 sources are cited, including many recent studies.

- **Interview case studies.** Agnes Riedmann talked with individuals of all ages about their experiences in marriages and families. These interviews appear as boxed excerpts, balancing and expanding topics presented in the chapters. We hope that the presentation of individuals' stories in their own words will help students to see their own lives more clearly and will encourage them to discuss and reevaluate their own attitudes and values.

- **Enrichment material.** Boxed material on making choices and on family diversity, drawn from classic research, new studies, popular social science, and newspaper and journal articles supports and expands the chapter content.

- **Pedagogical aids.** An outline of topics to be covered introduces each chapter. A large number of charts and diagrams present current data in easily understood form. End-of-chapter study aids include a summary, a list of key terms, study questions, and annotated suggested readings. A comprehensive glossary defines and illustrates important terms.

Ancillary Materials Available

The Instructor's Resource Manual with Test Items (written by David Treybig, Baldwin-Wallace College) includes for each chapter lecture outlines and suggestions, student activities, chapter review sheets, and video suggestions. Each chapter also includes multiple-choice, true-false, and completion questions with page references.

The Study Guide (also written by David Treybig of Baldwin-Wallace College) contains for each chapter an overview and summary, points to ponder, key terms, key research studies and concepts, study tips, and test items (that include true/false, short-answer, multiple choice and essay questions with an answer key).

New to this edition, the *Marriages and Families* World Wide Web Home Page offers materials for students and instructors, including links to Internet resources relevant to the study of marriages and families, exercises for students using the World Wide Web, Instructor's Resource Manual material, and more. Access the Lamanna/Riedmann Web site through the Wadsworth URL: http://www.wadsworth.com/wadsworth.html.

In addition, new videotapes selected to reflect the diversity of marriages and families covered within the textbook will be available to adopters; contact your ITP sales representative for details.

The Small Group Activities and Collaborative Exercises Workbook consists of suggestions and resources for student activities appropriate for small groups and collaborative learning environments.

Wadsworth Marriage and Family Videotape is available upon adoption. This features four 30-minute segments from the Intelecom telecourse series, "Portrait of a Family," which was developed to support this text. Topics include gender roles, communication, conflict and divorce.

Transparencies contain forty two-color acetates that include charts and graphs from the text.

Testing now includes both computerized testing and new online testing for adopters.

Acknowledgments

This book is a result of a joint effort on our part; neither of us could have conceptualized or written it alone. We want to thank some of the many people who helped us.

Looking back on the long life of this book, we again acknowledge Steve Rutter for his original vision of the project and his faith in us. We also want to thank Sheryl Fullerton and Serina Beauparlant, who saw us through earlier editions as editors and ongoing friends.

As has been true of our past editions, the people at Wadsworth Publishing Company have been professionally competent and a pleasure to work with. We are especially grateful to Eve Howard, our editor, who worked with us diligently on this edition. We value the skill and patience of Robin Lockwood, production manager, who shepherded our book through a complex production. Hal Humphrey served as project editor. Thanks also go to copy editor Jennifer Gordon for a wonderful job, to Linda Rill for photo research, and to Jeanne Bosschart and Linda Rill for permissions. We are also grateful to editorial assistants Deirdre McGill and Carrie Kahn, as well as to former assistant editor at Wadsworth, Susan Shook. Our cover and interior design are the work of Stuart Cuttriss and Jane Hambleton. Our illustrations were done by Hans & Cassady, and our indexing, by Katherine Stimson. Thanks to everyone.

Agnes Riedmann especially acknowledges her mother, Ann Langley Czerwinski, Ph.D., who helped her significantly with past editions.

Sam Walker has contributed to each edition of this book through his enthusiasm and encouragement for

Mary Ann's work on the project. Larry and Valerie Lamanna have enlarged their mother's perspective on the family by bringing her into personal contact with other family worlds—those beyond the everyday experience of family life among the social scientists!

Reviewers gave us many helpful suggestions for revising the book, and although we have not incorporated all of their suggestions, we considered them all carefully and used many. The review process made a substantial contribution to the revision. Peter Stein's work as a thorough, informed, and supportive reviewer throughout the various editions has been an especially important contribution. Sixth edition reviewers include:

Ellen M. Bjelland, University of Missouri-Columbia
Sampson Lee Blair, Arizona State University
Stephen Ciafullo, Central Missouri State University
Peter J. Collier, Portland State University
Laura Workman Eells, Wichita State University
Henry W. Fischer, III, Millersville University
Craig Forsyth, The University of Southwestern
    Louisiana
Deborah Jacobvitz, The University of Texas-Austin
Karen Seccombe, University of Florida
Glenna Spitze, State University of New York, Albany
Larry Stearley, Charles Stewart Mott Community
    College
Janine A. Watts, University of Minnesota, Duluth
David West, College of San Mateo
Frank R. Williams, The University of Arizona
Sue Marie Wright, Eastern Washington University
Fifth edition reviewers include:
Alexa Albert, University of Rhode Island
Jason L. Blank, Rhode Island College
Barbara Carson, Ball State University
Donna Christenson, University of Arizona
Stephen Ciafullo, Central Missouri State University

Leilani Clar, Grossmont College
Marilyn Coleman, University of Missouri
Peter J. Collier, Portland State University
John Deaton, Northland Pioneer College
Laura Workman Eells, Wichita State University
Mary Ann Fitzpatrick, University of Wisconsin,
    Madison
Jerry Gale, University of Georgia
Stephen Hall, Appalachian State University
Gary L. Hansen, University of Kentucky
Jane Hood, University of New Mexico
Helen Hoover, Arizona State University
Ira J. Hutchison, University of North Carolina,
    Charlotte
Mary Dellmann Jenkins, Kent State University
Joan Jurich, Purdue University
James Koval, California State University, Long Beach
Jay Mancini, Virginia Polytechnic Institute and State
    University
William Marsiglio, University of Florida
Marilyn McCubbin, University of Wisconsin,
    Madison
Jack Peterson, Mesa Community College
Barbara Risman, North Carolina State University
Kay Sears, University of Central Oklahoma
Peter J. Stein, William Paterson College
Janine Watts, University of Minnesota
Julie M. Wigen, University of Delaware
Mari S. Wilhelm, University of Arizona
Kersti Yllo, Wheaton College
Students and faculty members who tell us of their interest in the book are a special inspiration.

To all of the people who gave their time and gave of themselves—interviewees, students, our families and friends—many thanks.

Mary Ann Lamanna
Agnes Riedmann

A Student's Guide to Learning

As you start out on this exploration of marriages and families, you no doubt hope to gain some insights into this social institution that is a major element in shaping our lives. For instance, what is a functional family? For that matter, what is a family? How do so many different interpretations of marriage and family work within one society? Where is this institution called family headed and how will all these rapid changes affect you?

This book will help you gain both a better understanding of yourself and your own personal family situation, as well as a deeper understanding and broader perspective of the societal diversity that exists outside—and maybe even within—your family structure. Through the text's focus on decision making, you'll learn to become more aware of the personal decisions you must make throughout your life and how to make educated choices based on knowledge. You'll also discover how society influences your decision making.

As marriages and families have evolved over the last twenty years, so too has this text. Its new subtitle, *Making Choices in a Diverse Society,* illustrates the vast changes that have taken place over the last decade. With its thorough updating and inclusion of current research, plus its new emphasis on your being able to make choices in an exceedingly *diverse* society, this book is an unparalleled resource for gaining insights into today's marriages and families.

Take a few minutes to look over the next few pages. This Guide to Learning is a quick run-through of the text's major elements—all developed to help you to understand the changing nature of families today and to make informed choices in an increasingly complex and fluid society.

Making Choices . . .

The Decision Making Perspective

What's really behind the decisions you make? This diagram of the cycle of knowledgeable decision making is a great visual aid in helping you gain a better understanding of the decision making process and how social forces affect the personal choices you make. The diagram also illustrates one of this book's important themes: decision making within the context of a diverse society.

(From Chapter One: Family Commitments: Making Choices in a Changing Society)

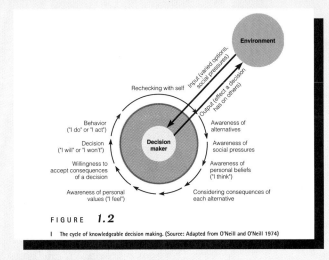

FIGURE **1.2**

I The cycle of knowledgeable decision making. (Source: Adapted from O'Neill and O'Neill 1974)

Communicating About Sex

Negotiating our own sexuality and sexual expression in face-to-face relationships can be tricky. (Maybe that's why cybersex is so popular!) Sexual expression carries the promise of "transcending the limits of the individual and relating intimately... to other persons" (Francoeur 1987, p. 531). At the same time, sharing our sexuality makes us tremendously vulnerable. Both women and men need clear and positive feedback to know what brings their partners the greatest joy and fulfillment. One popular seminar leader on relationship communication and sex recommends that couples take a half hour every so often, particularly when they are not feeling negative about sex, and talk about it.

Here are some questions to stimulate an informative conversation:

- What do you like about having sex with me?
- How did you feel when I did _____?
- Would you like more sex?
- About how much sex would you like each week (or month, or day)?
- Would you like for us to spend more (or less) time when we have sex?
- Is there something specific you would like me to do in the next month during sex?
- Is there a new way you would like me to touch you? If yes, would you show me?

- Is there anything new you would like me to try?
- Is there anything you would like to try that we've never done?
- Is there anything I used to do that you would like me to do more of or again?

Note that a conversation like this needs to be mutual and reciprocal: *Both* partners ask and answer these questions—and they accept each other's answers without judgment either of themselves or of their partner.

What makes it difficult to talk about our sexual desires is that we don't want to hurt our partner's feelings and, conversely, we would hate to hear that we are disappointing our partner in any way at all. Meanwhile, we don't want to feel pressured to do anything that would not feel comfortable to us. So when answering these questions, it is important that you make it clear that you are not demanding more. The dialogue works best if its only purpose is to share information.

Then, too, you should not do things that do not feel right to you. But if your partner wants something that doesn't seem important to you or seems unpleasant, you might consider keeping an open mind. And when your partner does not seem open to things you favor, try to accept that at least for now. "A secret of great sex is to build on the strengths you have and not focus on the problems or what you are missing" (Gray 1995, pp. 53–55).

Many New "As We Make Choices" Boxes to Help You in Those Moments of Crucial Decisions

This box, "As We Make Choices: Communicating About Sex," provides you with some useful questions that can help you turn what might have been awkward, stilted, self-conscious conversation about sex into a sincere, open, meaningful dialogue.

(From Chapter Five: Our Sexual Selves)

. . . in a Diverse Society

NEW! "A Closer Look at Family Diversity" Boxes Will Get You Thinking About the Many Diversity Issues Facing America Today.

In this box, "A Closer Look at Family Diversity: Family Ties and Immigration," you'll gain a different perspective on immigration and learn how it's powered predominantly by family bonds. These "Closer Look…" boxes can be found in 11 of the text's 16 chapters and amplify the text's emphasis on diversity.

(From Chapter One: Family Commitments: Making Choices in a Changing Society)

1.2 BOX   A CLOSER LOOK AT FAMILY DIVERSITY

Family Ties and Immigration

Today there is more racial and ethnic diversity among American families than ever before, and much of this diversity results from legal immigration. Interestingly, a great deal of this immigration is spurred by family ties.

Before 1965, immigrants were mainly European. The 1924 Immigration Act had established country of origin as the principal criterion for relocating to the United States, setting quotas favoring immigrants from northern and western Europe. Federal reforms in 1965 changed the criteria for obtaining an immigrant visa. The new policy gave priority to people with needed job skills or to people with family members already living here. In 1993, 55 percent of immigrants arrived here because they had a family member living in the United States. About half of those were reuniting with spouses, children, or parents who were United States citizens.

This policy change drastically changed the composition of U.S. immigration. Since 1965, Asia and Latin America, not Europe, have been the major sending regions. During the 1980s, Asian and Central and South American countries, taken together, sent 85 percent of all immigrants to the United States. The top five countries of origin were Mexico, the Philippines, China, Korea, and Vietnam.

Why do immigrants choose to come here? For the most part, immigrants leave a poorer country for a richer one. In their nation of origin, families may calculate how each member can best contribute to the family's status and/or income. This household economics of migration can encourage a rural Mexican family, for instance, to send a daughter to Mexico City to be a secretary and a son to Los Angeles to be a day laborer. Typically, the children send money home so that the family can try

ment, or a television set. As children establish themselves in the host country, they begin to send for their relatives, so that gradually ethnic kin or community networks develop regionally. Every major U.S. city today is truly multicultural, characterized by disparate ethnic neighborhoods.

The increase in immigrant families has led policymakers and others to consider many issues. For one thing, the number of American children who speak a language other than English has increased markedly. The 1990 census found a total of 6.3 million schoolchildren in the United States who spoke languages other than English at home. Some school districts use bilingual education to teach these students more effectively, while others teach English as a second language, and still others offer no special help at all. How to best educate immigrant children—many of them already U.S. citizens and many more on the path toward naturalization—is a matter of public debate.

Another issue is whether, on balance, immigrant families are an asset or a liability to the nation's economy. Many studies are underway in this regard, and the verdict is still out. But it appears that immigrant families pay more in payroll, property, and sales taxes than they cost the government in public services. However, controversy rages because most immigrant family tax dollars go to the federal government while the costs of their schooling or health care are largely paid for by local governments. This is the situation in Los Angeles County, for example (Martin and Midgley 1994). Whatever our views on immigration, immigrants are generally responsible family members doing what they can to improve their lives as well as the lives of their relatives.

A Rock-Solid Research Base . . .

This latest edition of **Marriages and Families** is filled with a wealth of statistics and research references that are the latest and most solid data available. As you read through the many topics on marriages and families and the data that supports the material, you'll gain new insights into the forces shaping our society and your own daily life.

Some 500 new sources have been added to this edition, including references from *The American Sociological Review*, *The American Journal of Sociology*, and *The Journal of Marriage and the Family*, as well as periodicals such as *The New York Times*, *The Wall Street Journal*, and *Newsweek*.

Children learn much about gender roles from their parents, whether they are taught consciously or unconsciously. Parents may model roles and reinforce expectations of appropriate behavior. On the other hand, children also internalize messages from available cultural influences and materials surrounding them.

Topics of High Interest Backed by Recent Research

This discussion of girls and boys in families is richly supported by current research references. The text's hundreds of current references give you an informed and analytical view of the status of marriages and families today.
(From Chapter Three: Our Gendered Identities)

playing with other-sex toys (Lytton and Romney 1991).

Boys have toys that develop spatial ability and creative construction; girls have toys that encourage social skills (Lips 1995). Toys considered appropriate for boys encourage physical activity and independent play, whereas "girls' toys" elicit closer physical proximity and more talk between child and caregiver (Caldera, Huston, and O'Brien 1989). As children get older, Patrick is more likely to have a computer than is Patricia (Nelson and Cooper 1990). And Patrick may join the Boy Scouts, an organization originated to "make big men of little boys."

Beginning when children are about 5 and increasing through adolescence, parents allocate household chores—both the amounts and kinds—to their children differentially, that is, according to their sex. Patri-

cia will more likely be assigned cooking and laundry tasks; Patrick, painting and mowing (Burns and Homel 1989; McHale et al. 1990).[6] Because typically girls' chores are daily and boys' sporadic, girls spend more time doing them—a fact that "may convey a message about male privilege" (Basow 1992, p. 131). Some parents unconsciously allow (even expect) sons to be inconsiderate, rude, and interrupting while their sisters are reminded to be a little quieter and less argumentative (Orenstein 1994, pp. 46–47).

6. The number of girls and boys in a family apparently influence chores are distributed. Families with all girls, for example, may assign traditional male-child tasks to girls than do families with one boy. And the age and gender of sibs are important. Resea found that when an older sibling is of the same sex, play acti to be gender-stereotyped; if the older sibling is of the other s sex play is common (Brody and Steelman 1985).

Girls and Boys in the Family

Most studies on gender socialization beliefs and practices have been conducted on non-Hispanic white families. One study of black families indicated that both sons and daughters are socialized toward independence, employment, and child care (Hale-Benson 1986). Still, because most of the research presented in this section has focused on middle-class whites, findings may or may not apply to other racial/ethnic or class categories.

At least among middle-class whites, parents rear female and male infants differently. From the 1970s on, parents have reported treating their sons and daughters similarly (Maccoby and Jacklin 1974; Antill 1987), but differential socialization exists and typically is subtle and not deliberate (Shapiro 1990). Even parents who support nonsexist child rearing for their daughters are often concerned if their sons are not aggressive or competitive "enough"—or are "too" sensitive (Pleck 1992).

Research shows that parents handle infant sons more roughly and respond more quickly to crying baby girls (Lips 1995). As a toddler, Patricia will probably have a doll; Patrick, a truck. A study of 120 babies' and toddlers' rooms found that girls had more dolls, fictional characters, children's furniture, and the color pink; boys had more sports equipment, tools, toy vehicles, and the colors blue, red, and white (Pomerleau et al. 1990). And most parents, especially fathers, discourage their children, especially sons, from

CHAPTER 3   Our Gendered Identi

Parents also model different behaviors. As Chodorow notes, most children are cared for by mothers primarily. This situation may convey to children the idea that child care (and nurturance in general) is "women's work." Meanwhile, fathers may appear to be involved with "more important" and prestigious things (Burns and Homel 1989).

Gender socialization in early childhood begins a continuing process whereby females and males are channeled into separate spheres of skills and interests (Lips 1995). Although relations in the family provide early feedback and help shape a child's developing identity, play and peer groups become important as children try out identities and adult behaviors. We turn our attention now to research on differences in boys' and girls' play.

. . . Comprehensive and timely

Throughout this book, you will find comprehensive coverage of important and timely topics such as this provocative discussion of same sex marriages in Chapter 8. Sensitive, complete coverage of issues like these are important to your understanding of the many controversial issues related to marriage and the family.

... same-sex couples marry? The question is controversial, even among gay men and lesbians.

...ist Craig Dean and his lover, Patrick Gill, ...he Equal Rights Marriage Fund after they ...nied a marriage license in 1991. At about the ...e, after being denied a marriage license in ...Patrick Lagon, a graphic artist in his 30s, and ...Mellio, a chef in his late 40s, sued to gain ...legal recognition of their sixteen-year rela-... Two lesbian couples joined the suit. Lagon ...d that both he and Mellio were raised to ex-...ll in love and marry "like anybody else" ...a 1994). The men talk proudly of their "sim-...ogether, characterized by TV, pet dogs, a ...rden, and weekend drives (Gross 1994). ...ponse to the suit, in 1993 the Hawaii ... Court voted 3-2 that refusal to recognize ... marriages may violate sex discrimination ...mong other things, the court said that defining

marriage as necessarily between one man and one woman was "circular and unpersuasive" reasoning. The court sent the case back to a lower court where the state was required to prove that it had a "compelling interest" in denying same-sex couples the right to marry.

As a result, in their 1994 legislative session, Hawaii's lawmakers sought to clarify the state's marriage statute. They considered, on the one hand, a state constitutional amendment that would limit marriage to heterosexuals and, on the other, a broad domestic partnership act (see Box 6.2) that would appease same-sex couples without giving them the right to marry legally. Legislators finally agreed on a bill restating that marriage is meant for "one man and one woman" but also creating a commission to propose remedies for any discrimination against same-sex

...gest that lifelong fidelity is far from universal; on the other hand, specific acts of infidelity may not be routine or continual. The contrasting data on extramarital sex suggest that today, as in the past, many spouses are torn between lifelong commitment to a sexually exclusive marriage and desire for outside sexual relationships. For the majority of mates, however, the decision to marry involves the promise to forgo other sexual partners. If we interpret fidelity to mean a primary commitment to one's partner and the relationship, then maintaining emotional intimacy becomes essential to being faithful. Without continued intimacy and self-disclosure, partners may not remain in love or keep a central place in each other's lives. Emotional commitment, in this view, is the essential element of primariness, as much of outside sexual fidelity per se. This ideal of primariness can make marriage relationships more satisfying.

To sum up, marriage entails making ongoing choices about permanence and primariness. In the next section, we'll examine whether same-sex couples should marry legally.

Should Same-Sex Couples Marry?

Many gay male and lesbian couples live together in long-term, committed relationships (Allen and Demo 1995). Partners may exchange vows or

rings or both. Couples may publicly declare their commitment in ceremonies among friends or in some congregations and churches, such as the Unitarian Universalist Association or the Metropolitan Community Church, the latter expressly dedicated to serving the gay community. Catholics have access to a union ceremony designed by Dignity, a support association for gay and lesbian Catholics. In a 1993 gay rights march in Washington, DC, 1,500 same-sex couples participated in a mass wedding with ministers and rice (Salholz 1993). None of these couples is legally married, however. By law, dating to a U.S. Supreme Court decision in 1974, marriage is defined as a union between one man and one woman; hence same-sex marriage is not legally recognized in the United States.

Meanwhile, some cities and counties have passed *domestic partner* laws (defined in Box 6.2). According to this concept, unmarried couples may register their partnership and then receive some of the legal benefits of marriage, such as joint health or auto insurance or bereavement leave, for example. Besides practical benefits, registering as domestic partners has emotional significance for some same-sex couples, who do so as a way of publicly expressing their commitment (Ames 1992). But registering as domestic partners lacks the deep symbolism of marriage. "Gays and lesbians were raised in the same culture [as] everyone else," notes gay historian Eric Marcus. "When they settle down they want gold bands [and] legal documents" (in Salholz 1993).

Attitudes toward gay rights have generally become more liberal. Nevertheless, a 1993 poll by the *Washington Post* found that 53 percent of Americans oppose homosexual relationships in general and 70 percent are against same-sex marriage (Salholz 1993). For religious fundamentalists and other conservative groups, the case against gay marriage is unambiguous: Marriage is intrinsically straight. From this point of view, the move to legalize marriage is an "attempt to deconstruct traditional morality." As the posters proclaim, "God made Adam and Eve, not Adam and Steve" (Salholz 1993; Gross 1994).

Legal Marriage and Hawaii

While several western European nations are beginning to consider it, the only countries that legally recognize gay male and lesbian marriages are Denmark and Norway (Singer and Deschamps 1994; Wyman 1994). This situation may be about to change, however. U.S.

couples that might result from denying them the legal right to marry (Gross 1994). The outcome will be either legalization of same-sex marriage by Hawaii's Supreme Court or a broad and far-reaching domestic partnership act—and perhaps both (Gross 1994). At the time of this writing, many observers think that Hawaii is about to legalize same-sex marriage.

By itself, any court ruling in Hawaii would not affect other states. But many same-sex couples from throughout the United States could marry legally in Hawaii, then seek recognition of their marital status in their home states. States usually recognize one another's legal decisions, according to the *principle of reciprocity.* The only exception is when one state claims a violation of local public policy, which is expected to be the explanation for not recognizing gay marriage in the twenty-three states where sodomy remains a criminal act (see Chapter 5). Within the next several years the issue is likely to find its way to the U.S. Supreme Court (Gross 1994).[1]

Views of Gay Rights Activists on Same-Sex Marriage

Lesbians and gay men who favor legalized same-sex marriage argue that legal marriage yields economic and other practical advantages. Marrieds can lower their taxes by filing a joint return; if one spouse dies or is disabled, the other is entitled to Social Security benefits; legal partners can inherit from one another without a will; spouses are immune from subpoenas requiring testimony against each other. Some companies offer health insurance to an employee's legal spouse and dependents. And when one or both partners suffer from AIDS or other serious illness, the protections of marriage could prove highly beneficial (Stoddard 1989; Wyman 1994).

Moreover, unlike heterosexual couples, homosexual couples receive little social support for continuing long-term relationships. On average, relationships for both gay males and lesbians last two years to three years, and a pattern of serial monogamy exists (Harry

1983). As Mary Mendola, lesbian author of the *Mendola Report* (1980), explains:

The major difference separating us as heterosexual and lesbian couples: our lesbian marriage had none of the support systems Mr. and Mrs. Next Door enjoyed. I had not had a bridal shower or a bachelor party, depending on how one looks at it. . . . Aunts and uncles did not come to visit and admire our home. We never received anniversary cards. As trivial as these things may seem, they represent something vitally important: heterosexual couples are encouraged to stay together. . . . Lesbian and homosexual couples have no such support systems. Rather than being encouraged to stay together, we are conditioned to believe there is no future for us as couples. (p. 4)

Being able to marry legally could help to change this situation.

Furthermore, denying lesbians and gay men the right to marry implies that they are not as good or valuable as heterosexuals. The right to marry legally could remedy this. Some lesbians and gay men believe that legal marriage would be their "most important civil rights victory" yet; related successes would follow. For instance, laws banning same-sex sodomy (see Chapter 5) would very likely be repealed or declared unconstitutional. "And social custom—already in flux—would undoubtedly change" (Wyman 1994).

Gay and lesbian opponents of legal same-sex marriage, such as gay rights attorney Paula Ettelbrick, object to mimicking a traditionally patriarchal institution based on property rights and institutionalized husband–wife roles. As one lesbian explained,

Within my home I feel married to Frances, but I don't consider us "married." The marriage part is still very heterosexual to me. One of the reasons I don't like to associate with marriage is because heterosexual marriage seems to be in trouble. It's like booking passage on the *Titanic.* Frances is my life partner; that's how I'm accustomed to thinking of her. (in Sherman 1992, pp. 189–90)

More generally, opponents object to giving the state power to regulate primary adult relationships. They stress that legalizing same-sex unions would further stigmatize any sex outside marriage, with unmarried lesbians and gay men facing heightened discrimination (Ettelbrick 1989).

1. As we go to press, Congress is considering the Defense of Marriage Act, a proposed federal statute declaring marriage to be a "legal union of one man and one woman," denying gay couples many of the civil advantages of marriage, and also relieving states of the obligation to grant "full faith and credit" to marriages performed in another state. How the U.S. Supreme Court's ruling of May 20, 1996, which struck down a Colorado challenge to local gay rights ordinances, will impact the proposed Defense of Marriage Act is "anybody's guess" (Vicki Haddock, "Activists Hail Shift by U.S. Justices," San Francisco Examiner, May 21, 1996, pp. A-10).

A Host of Helpful End-of-Chapters Aids . . .

Every chapter ends with these review materials, such as the ones shown here from *Chapter Two: Exploring the Family*. These will help you determine how well you've absorbed the chapter material.

In Sum
This section summarizes the major points of the chapter.

Key Terms
The key terms listed at chapter's end offer a good check for your comprehension of the chapter's main ideas.

Study Questions
This series of questions offers exercises that foster critical thinking and introspection.

Suggested Readings
If you want to read more on the chapter's main topics, this list of suggested readings is the ideal starting point. It offers numerous suggestions for both recently published books, as well as in-print works that are classics in the study of marriages and families. Each listing is accompanied by a capsule description/critique.

---

4. How is your family structured? That is, is it a traditional nuclear family, or does it represent the diversity of American families? How would you apply the current debate over family values to your own family experience?

5. Review the techniques of scientific investigation, and discuss why science is often considered a better way to gain knowledge than is personal experience alone. When might this not be the case?

Suggested Readings

Acock, Alan C. 1994. *Family Diversity and Well-Being.* Thousand Oaks, CA: Sage. A social scientific analysis of family diversity in the United States today that employs the family ecology perspective.

Babbie, Earl. 1995. *The Practice of Social Research.* 7th ed. Belmont, CA: Wadsworth. Respected textbook in sociological methods. Recommended for those who would like additional information or clarification.

Barber, Kristine M. and Katherine R. Allen. 1992. *Women and Families: Feminist Reconstructions.* New York: Guilford. A good example of scholarship on the family from a feminist perspective, this work examines women's intimate relationships, both straight and lesbian, and argues that a critique of hierarchical gender relations must be central to family studies.

Boss, Pauline G., William J. Doherty, Ralph LaRossa, Walter R. Schumm, and Suzanne K. Steinmetz, eds. 1993. *Sourcebook of Family Theories and Methods: A Contextual Approach.* New York: Plenum. Don't let the size of this book scare you off! It is filled with very readable essays on all the theoretical perspectives and methods described in this chapter and more.

DaVanzo, Julie. 1993. *American Families: Trends and Policy Issues.* Santa Monica, CA: Rand. Current data on today's families, together with policy implications and suggestions. The book integrates the family ecology and feminist theoretical perspectives.

DuBois, W. E. B. 1970. *The Negro American Family.* Cambridge, MA: MIT Press. Originally published in 1908, this is one of the first careful studies of the African American family in the United States, done by an early and important African American scholar. Coming from what would later be called the family ecology perspective, DuBois argued that black family instability was a function of limited economic opportunity and adjustment to urban living.

Duvall, Evelyn M. and Brent C. Miller. 1985. *Marriages and Family Development.* 6th ed. New York: Harper & Row. Family development textbook whose senior author is one of the pioneers of this framework.

Hill, Robert B., Andrew Billingsley, Eleanor Engram, Michelene R. Malson, Roger H. Rubin, Carol B. Stack, James B. Stewart, and James E. Teele. 1993. *Research on the African-American Family: A Holistic Perspective.* Westport, CT: Auburn House. This is a well-written work by a cadre of family scholars on the current status of the African American family. The authors give particular attention to the inadequacy of theoretical and research perspectives guiding social scientists and policymakers as they examine black families.

Lee, Joann Faung Jean. 1991. *Asian American Experiences in the United States: Oral Histories of First to Fourth Generation Americans.* Jefferson, NC: McFarland. Illustrates the oral history method of family research and also gives a rich contextual view of Asian American families.

Levy, Frank S. and Richard C. Michel. 1991. *The Economic Future of American Families: Income and Wealth Trends.* Washington, DC: Urban Institute Press. An examination of the growing gap between wealthy and poor families in the United States today with implications for the future, should this trend continue.

Maddock, J. W., M. J. Hogan, A. L. Antonove, and M. S. Matskovsky, eds. 1993. *Perestroika and Family Life: Post-USSR and US Perspectives.* New York: Guilford. Not all cross-cultural analyses deal with nonindustrialized societies. This work uses a globally comparative approach to examine families in a changing society, with emphasis on the late 1980s and early 1990s.

McAdoo, Harriette Pipes. 1993. *Family Eth[...] [...] Diversity.* Newbury Park, CA: Sage. An e[...] nic diversity among American families w[...] avoiding theoretical or methodological b[...] is explored from the cultural variant mo[...]

Stack, Carol. 1974. *All Our Kin: Strategies f[...]* York: Harper & Row. An example of pa[...] tion research that conveys what the expe[...] plifies the cultural variant approach to r[...] minority families, and shows how new th[...] tives can emerge to better describe ethni[...]

Zill, Nicholas. 1994. *Running in Place: Hou[...] Are Faring in a Changing Economy and a [...] ciety.* Washington, DC: Child Trends. Zi[...] stricting economy, coupled with failure t[...] families require socio-economic support [...] public policy, makes for stress—and othe[...] today's families.

Zimmerman, Shirley L. 1992. *Family Polici[...] Being: The Role of Political Culture.* Newl[...] Sage. An example of a work in the family[...] this book illustrates how the political and[...] in which families exist influences what g[...]

How do we know what families are like? We can call upon personal opinion and experience for the beginning of an answer to this question. But everyone's personal experience is limited. Scientific investigation—with its various methodological techniques—is designed to provide a more effective way of gathering knowledge about the family.

Key Terms

agreement reality	heterosexism
case studies	identity
conflict perspective	interactionist perspective
cultural deviant	kin scripts framework
cultural equivalent	longitudinal studies
cultural variant	monogamy
exchange theory	naturalistic observation
experiential reality	polygamy
experiment	scientific investigation
extended family	self-concept
family development perspective	self-sufficient economic unit
family ecology perspective	social institutions
family life cycle	structure-functional perspective
family policy	surveys
family systems theory	theoretical perspective
feminist perspective	

Study Questions

1. Choose one of the major theoretical perspectives on the family and try to explain how you might use it to understand something about life in your family.

2. Choose a magazine photo and analyze its content from a structure-functional perspective. (*Hint:* Together the people in the photo can constitute the group under analysis, with each person meeting certain of the group's needs, or functional requisites.) Then analyze it from one or more other perspectives. How do your insights differ?

3. Why is the family a major social institution? Does your family fulfill each of the functions identified in the text? How?

Marriages and Families

PART One

A New Look at Familiar Worlds

We all have families of some sort. But what *is* a family? How do we study families?

In Chapter 1 we consider definitions of the family in light of contemporary family patterns. We learn that U.S. families are increasingly diverse—both in form and due to differences in social class, race/ethnicity, and sexual orientation. In Chapter 2 we see that what we define as a family depends partly on our initial conceptual framework or theory of the family. We review the methodology of family study, together with some difficulties encountered in the study of families.

Gender identities and the social definition of gender are central to family life. They intersect with the important dimensions of race/ethnicity and class discussed in Chapter 1.

Chapter 3 examines Americans' changing views about gender roles.

But change doesn't come about all at once; each of us probably carries some traditional ideas about gender roles, along with some newer ones. Chapter 3 points out how combining these two different sets of ideas can affect individual women and men. Throughout the remainder of the text we'll touch on how combining traditional and modern ideas affects relationships, marriages, and families. And we'll see that how partners combine these ideas is influenced by their social class, race, ethnicity, and sexual orientation. There is no one kind of family; that's why we titled this book *Marriages and Families*—in the plural.

CHAPTER 1

Family Commitments: Making Choices in a Changing Society

Human nature is not a machine to be built after a model, and set to do exactly the work prescribed for it, but a tree, which requires to grow and develop itself on all sides, according to the tendency of the inward forces which make it a living thing.

JOHN STUART MILL, 1859

We are neither totally victims of systems nor totally captains of our own souls.

CATHERINE CHILMAN

Today's Americans are cautious and apprehensive—and at the same time creative and hopeful—about marriages and families. On the one hand, most of us still hope to experience ongoing happiness in committed unions and families. We hold onto the idea that a family can provide us with a "haven in a heartless world" (Lasch 1977). Yet, we have reason to wonder about the chances of finding such happiness. For one thing, the high divorce rate has called into question the stability of marriage. Beginning in the 1980s, sociologists began to speak of "the reconceptualization of marriage as a nonpermanent relationship" (Safilios-Rothschild 1983, p. 306); they claimed that the family had been so shaken up that "all bets are off" (Wiley 1985, p. 2).

Meanwhile, we remain hopeful about family commitment because families are central to society. Families are commissioned with the pivotal tasks of raising children and providing continuing intimacy, affection, and companionship to members. But hoping alone won't make enduring or emotionally satisfying families, although faith and dedication to the possibility of committed happiness and stability would seem an es-

sential precondition. Maintaining a family requires both commitment and knowledge of what you're doing. That theme is a good part of what this book is about. We will return to it later in this chapter and throughout this text. Right now, though, we need to explore what a family is.

Defining Family

What is a family? In everyday conversation we make assumptions about what families are or should be. Traditionally, both law and social science specified that the family consisted of people related by blood, marriage, or adoption. Some definitions of the family specified a common household, economic interdependency, and sexual and reproductive relations (Murdoch 1949).

In their modern classic, *The Family: From Institution to Companionship* (1945), Burgess and Locke defined the family as "a group of persons united by the ties of marriage, blood, or adoption; constituting a single household; interacting and communicating with each other in their respective social roles (husband and wife, mother and father, son and daughter, brother and sister); and creating and maintaining a common culture" (p. 8). This definition goes beyond earlier ones to talk about family relationships and interaction. Burgess and Locke saw the family as a **primary group**—a term coined by sociologist Charles Horton Cooley (1909, p. 23) more than eighty years ago to describe any group in which there is a close, face-to-face relationship.[1] In a primary group, people communicate with one another as whole human beings. They laugh and cry together, they share experiences, and they quarrel, too, because that's part of being close. Primary groups can give each of us the feeling of being accepted and liked for what we are.

Burgess and Locke gave us a definition of the family that described some family relationships. But their view of family interaction was more limited than our current ones. It assumed that interaction occurred primarily in the context of traditional (heterosexual, married-couple) social roles, rather than emphasizing spontaneity, individuality, and intimacy. Social scientists today are still studying the family's responsibility

---

1. Another example of a primary group relationship is a close friendship. A **secondary group**, in contrast, is characterized by more distant, practical relationships, as, for example, in a professional organization or business association.

Families are commissioned with the pivotal tasks of raising children and providing members with ongoing intimacy, affection, and companionship.

in performing necessary social roles, such as child rearing. But many social scientists (including us) reject Burgess and Locke's limited definition of family. Family members are not necessarily bound by legal marriage, by blood, or by adoption.

Burgess and Locke also specified that family members "constitute a household." We would expand this definition to include, for example, commuter couples, noncustodial parents, parents with adult children living elsewhere, extended kin such as aunts and uncles, and adult siblings and stepsiblings. In other words, the term *family* does not necessarily include only partners, parents, and any children in one household. Throughout much of this century, the traditional **nuclear fam-**

ily (husband, wife, and children in one household) has been considered the **modern family.** The progressively increasing family diversity that we see now has led some scholars to refer to today's family as the **postmodern family.** Indeed, a few postmodern theorists (e.g., Stacey 1990; Edwards 1991) now argue that the concept of "family" no longer has any objective meaning of any sort.

This liberalization of the definition of family has affected others besides social scientists: Since June 1977, federal regulations have permitted unmarried low-income heterosexual and homosexual couples to qualify as families and live in public housing. Recently, some cities and states have adopted similar definitions,

A primary group is a small group marked by close, face-to-face relationships. Group members share experiences, express emotions, and, in the ideal case, know they are accepted and valued. In many ways, teams, friends, and families are similar primary groups: Joys are celebrated spontaneously, tempers can flare quickly, and expression is often physical.

voluntarily or through court challenge to certain laws (Gutis 1989b). Furthermore, to be a family not *all* the roles mentioned—husband, wife, children—need be included. Single-parent households and childless unions are families. And roles may be nontraditional: Social scientists no longer assume that a family has a male breadwinner and a female homemaker; and dual-career and reversed-role (working wife, househusband) combinations are also families.

As families have become less traditional (see Box 1.1, "Some Facts About Families in the United States Today"), the legal definition of a family has become much more flexible and nonspecific and is not limited to people linked by legal marriage, blood, or adoption. Law, government agencies, and to some extent private bureaucracies such as insurance companies must make decisions about what a family is. If zoning laws, rental practices, employee privileges, and insurance policies cover families, decisions must be made about which

groups of people can be considered a family. In recent years, the courts have determined that unmarried heterosexual or gay or lesbian couples, elderly people and their caregivers, handicapped people who live together, and even coresident groups of students constitute families.

In defining family, judges have used criteria of common residence and economic interdependency and the more intangible qualities of stability and commitment (Robson and Valentine 1990; Scanzoni and Marsiglio 1993). According to Judge Vito J. Titone of the New York Court of Appeals, the definition of the family "should not rest on fictitious legal distinctions or genetic history, but instead should find its foundation in the reality of family life." Rather, "it is the totality of the relationship as evidenced by the dedication, caring and self-sacrifice of the parties which should, in the final analysis, control" (Gutis 1989c, p. C-6).

BOX 1.1

Some Facts About Families in the United States Today

What do U.S. marriages and families look like today? Statistics can't tell the whole story, but they are an important beginning.

1. *The racial/ethnic distribution of the population, and so, of families, is changing.* This is due to lower fertility among non-Hispanic white women than among minority women and the increased immigration of Asians, Latinos, and some Central American and African blacks (Martin and Midgley 1994). If present immigration rates and birthrates continue, non-Hispanic whites are projected to be 68 percent of the population in 2010, with Hispanics making up 14 percent, African Americans 13 percent, Asians 5 percent, and Native Americans and others just under 1 percent (U.S. Bureau of the Census 1995, Table 19). Compare this to 1980, when non-Hispanic whites were 80 percent of the population, with African Americans constituting 12 percent, Hispanics 6 percent, and Asians and others 2 percent (Saluter 1994).

2. *More and more Americans are living alone*—25 percent of households in 1993 contained only one person, compared with 17 percent in 1970. As a result, the average size of a U.S. household dropped from 3.14 people in 1970 to 2.67 in 1994 (U.S. Bureau of the Census 1995, Tables 66 and 68). However, this drop also reflects the fact that, while this varies by ethnicity, families are generally smaller due to declining fertility.

3. *People have been postponing marriage in recent years.* Median age at first marriage, after half a century of decline, has been rising since the early sixties. The median age at first marriage in 1993—24.5 for women and 26.5 for men—is comparable to that at the beginning of the twentieth century and quite different from that in 1960 (20.3 for women and 22.8 for men).

   In 1993, the proportion of women ages 30–34 who had never married (19 percent) was more than three times as high as in 1970 (6 percent) (Saluter 1994, Tables B and C).

4. *Unmarried-couple households increased by more than six times over the past twenty-five years.* In 1970, there was 1 unmarried-couple household for every 100 married couples. By 1993, there were 6 unmarried-couple households for every 100 married couples (Saluter 1994, Table D). About 5 percent of couples living together are unmarried.[a] The proportion of people who have *ever* lived with an unmarried partner is higher. According to a 1988 survey, about one-third of U.S. women under age 44 have lived with an unmarried male partner at some point in their lives (London 1991).

5. *The rate of births to unmarried mothers is high.* The rate rose to a peak in the mid-sixties, dropped in the seventies, then rose again to previously unseen levels. In 1993, 31 percent of all births in the United States were to unmarried mothers (with 68 percent of African American births occurring outside marriage) (Ventura et al 1995). Of all babies born outside marriage in 1992, 59 percent were to white women (U.S. Bureau of the Census 1995), Table 94). Teen unwed birthrates remain high, but the proportion of teens in the population has declined. Consequently, unwed childbearing is now highest, both in numbers and rate, among women in their 20s.

6. *The total fertility rate* (which represents the number of children born to a hypothetical typical woman during her childbearing years) *has generally declined:* It was 2.07 in 1993, as compared to 2.9 in 1965 and 3.6 in 1955, the height of the baby boom. At least among non-Hispanic whites, women have delayed childbearing since the early 1970s so that an increasing number are having a first child in their 30s or even 40s (U.S. Bureau of the Census 1989b, Table 87; "Births to Women" 1992; U.S. Bureau of the Census 1995, Tables 91 and 99). As a result, the birthrate for 1994 (the latest year for which statistics were available at this printing) was the lowest since 1978 (Singh et al. 1995).

(continued)

(continued)

7. *Half of all family households today contain no children under age 18* (U.S. Bureau of the Census 1995, Table 66). These are couples who have not yet had children or who do not plan to, and couples whose children are older.

8. *More and more lesbians and gay men are establishing families with children.* It is estimated that in the United States between 3 and 5 million gay men or lesbian parents have produced about 8 million children. The vast majority were born in heterosexual marriages before a parent defined her- or himself as gay (Rothblum and Cole 1989; Harry 1990).

9. *More mothers are employed.* In 1994, 59 percent of all U.S. women were in the labor force (compared with 75 percent of all men). The proportion of women in the labor force is up from 38 percent in 1960. Three-quarters of married mothers with children ages 6–17 were labor force participants in 1994, compared with 39 percent in 1960. Among married mothers with children under age 6, 62 percent were in the labor force in 1994, compared with 19 percent in 1960 (U.S. Bureau of the Census 1995, Tables 628 and 638).

10. *Young adults are increasingly likely to be living with their parents.* For 1994, the Census Bureau reported that 60 percent of men and 46 percent of women ages 18–24 were still—or once again—living with their parents. These percentages were up from 52 percent for men and 35 percent for women in 1960 (U.S. Bureau of the Census 1988, Table A-6, 1995, Table 63). This trend reflects both increased economic difficulties and changes in marital patterns (delayed marriage and high divorce rates) (Goldscheider and Goldscheider 1994).

11. *The rate of divorce has doubled since 1965,* peaking in 1979 and 1981, then dropping slightly (U.S. Bureau of the Census 1995, Table 87). Projections are that in the future about half of all marriages will last a lifetime (Bumpass, Raley, and Sweet 1995).

One result is that the proportion of the population currently married has decreased. The percentage of all U.S. men over 18 who are married fell from 78 percent in 1970 to 62 percent in 1994. The proportion for African American men dropped from 67 to 46 percent. Proportions for Hispanics just about match those for all men (U.S. Bureau of the Census 1995, Table 58).

12. As a result of divorce and unmarried parenthood, *more than one-fourth of all U.S. children under 18* (21 percent of white, 32 percent of Hispanic, and 57 percent of black children) *lived with just one parent in 1993* (Saluter 1994, Table F). Compare this with 1960, when the overall proportion was 9 percent. According to one prediction, 70 percent of white children born in 1980 will spend some time living in single-parent families by the time they reach age 17; the projection is 94 percent for African American children (Popenoe 1993, p. 531).

13. *Remarriages make up an increasingly large proportion of all marriages.* In 1988, 46 percent of all marriages were remarriages for one or both partners, as compared to 31 percent in 1970 (U.S. Bureau of the Census 1995, Table 143).

14. Because people are living longer and fertility has declined, *the proportion of the population over 65 has increased by 40 percent since 1960*—from 9 percent in 1960 to 13 percent in 1995. About 5 percent of our population today is over age 75, and that figure is projected to reach nearly 13 percent by 2050 (U.S. Bureau of the Census 1995, Table 17).

Generally, the elderly maintain their own homes as long as possible. Forty-three percent of people over 75 are husbands and wives in their own households (70 percent of men, 26 percent of women). Forty percent of those over 75 live alone (21 percent of men, 52 percent of women); 15 percent live with other relatives, mostly children. Just 2 percent live with nonrelatives; this includes nursing homes (U.S. Bureau of the Census 1995, Table 62).

15. *The proportion of middle-income families in the United States has dropped since 1970, while the proportion of families with low or high incomes has increased.* In 1970, 44 percent of U.S. families were earning between $25,000 and $50,000 annually. By 1993, that proportion has dropped to 33 percent. In 1970, 8.2 percent of all U.S. families were earning under $10,000 annually; by 1993, the proportion had risen to 9.6 percent. Seventeen percent of all U.S. families earned between $50,000 and $74,999 in 1970; in 1993, the figure was 19 percent. And the proportion of families earning $75,000 or more doubled between 1970 and 1992—from 7.5 to 15.5 percent. (All figures are given in 1993 dollars to take inflation into account and to indicate comparability in buying power.) This trend is similar for all races, although white families are far less likely to earn under $10,000 annually than are African Americans or Hispanics (U.S. Bureau of the Census 1995, Table 731).

16. *The proportion of the population below the poverty line has generally risen over the past twenty-five years, with* poverty unevenly distributed across racial/ethnic groups. Poverty rates have increased for all Americans, including non-Hispanic whites (although they still have remarkably lower rates when compared with other racial/ethnic groups). Between 1960 and 1969, the poverty rate was halved from 22 to 11 percent. But the 1993 poverty rate of 15 percent was the highest since 1966. In 1993, 39 million Americans lived below the poverty line of $14,763 for a family of four. Twelve percent of non-Hispanic whites lived below the poverty line, compared with 33 percent of African Americans and 31 percent of Hispanics (U.S. Bureau of the Census 1995, Table 744, 1994b, p. 3).

[a]This is only a rough estimate for two reasons. First, the Census Bureau cannot really tell if two unrelated people living together are an "unmarried couple" in our sense. The census category that is the source of the data is specified as two adults, not related and of opposite sex. It could also include roommates, dependents and their caregivers, and so on. But the Census Bureau assumes that most are heterosexual couples living together. Second, because the census category specifies "opposite sex," by definition it does not include gay or lesbian couples.

Some employers have proactively sought to redefine "family." The city of San Francisco has granted family privileges to *domestic partners,* defined as "two people who have chosen to share one another's lives in an intimate and committed relationship. The two must live together and be jointly responsible for basic living expenses; neither may be married to anyone else" (Bishop 1989).

In this text, we have adopted a definition of the family that combines elements of some of the definitions discussed here: A **family** is any sexually expressive or parent–child or other kin relationship in which people—usually related by ancestry, marriage, or adoption—(1) form an economic unit and care for any young, (2) consider their identity to be significantly attached to the group, and (3) are committed to maintaining that group over time. This definition combines some practical and objective criteria with a more social–psychological sense of family identity. Ultimately, there is no one correct answer to the question "What is a family?" We hope our definition and the others presented here will stimulate your thoughts and discussion.

To a significant extent, the diversity that we see in families today is a result, over time, of people making personal choices. We turn now to a discussion of such choices.

The Freedom and Pressures of Choosing

This text is different from others you may have read. It is not intended specifically to prepare you for a particular occupation. Instead, it has three other goals: first, to help you understand yourself and your family situations; second, to help you appreciate the variety and diversity among families today; and third, to make you more conscious of the personal decisions you must make throughout your life and of the societal influences that affect those decisions. As families have become less rigidly structured, people have made fewer choices "once and for all." Many people reexamine their decisions about family throughout the course of their lives, continually reassessing and reevaluating their relationships. Thus, choice is an important emphasis of this book.

The best way to make decisions about our personal lives is to make them knowledgeably. It helps to know something about all the alternatives; it also helps to know what kinds of social pressures affect our decisions. As we'll see, people are influenced by the beliefs and values of their society. In a very real way, we and our personal decisions and attitudes are products of our environment.

But in just as real a way, people can influence society. If you don't agree with something in this text, you can write us; your opinion may influence ours. Similarly, every time you participate in class discussions, you help to shape the content of the class. Individuals create social change by continually offering new insights to their groups.[2] Sometimes this occurs in conversation with others. Sometimes affecting social change requires forming social organizations and becoming politically involved. Sometimes creating social change involves many people living their lives according to their principles, even when these fly in the face of accepted group or cultural norms.

We can apply this view to the phenomenon of living together, or cohabitation. Thirty years ago, it was widely accepted that unmarried couples who lived together were immoral. But in the seventies, college students challenged university restrictions on cohabitation, and many people—students and non-students, young and old—chose to live together. As cohabitation rates increased, societal attitudes changed. Consequently, it is now easier for people to choose this option. While we are influenced by the society around us, we are also free to influence it. And we do that every time we make a choice.

Moreover, if people are to shape the kinds of family living they want, they must not limit their attention to their own marriages and families. Making knowledgeable decisions about one's family increasingly means getting involved in national and local political campaigns, finding out what candidates have in mind for families, and writing and phoning government representatives once they are in office. One's role as a family member, as much as one's role as a citizen, has come to require participation in public policy decisions. Although no family policy, of course, can guarantee "ideal" families, such a policy may contribute to a good foundation for family life.

Personal Troubles and Societal Influences

People's private lives are affected by what is happening in the society around them. Many of their personal troubles are shared by others, and these troubles often reflect societal influences.[3] When a family breadwinner cannot find work, for example, the cause may not lie in his or her lack of ambition but rather in the economy's inability to provide a job (see Figure 1.1). This text assumes that people need to understand themselves (and their problems) in the context of the larger society. Individuals' choices depend largely on the alternatives that exist in their social environment and on cultural values and attitudes toward those alternatives.

Moreover, our ability to make good choices in response to changing alternatives depends to a degree on how quickly societal changes occur. Demographer Paul Glick notes that when changes come suddenly, as they have over the past three decades, the problem of adjustment is "especially great" and leads to increasing signs of personal and family stress.

Glick may be most concerned for children. More of them today than even a few decades ago "have never known the benefits of lifelong dependable ties between themselves and their parents" (Glick, 1988, p. 871). Individual choices are shaped by the social context in which they are made, so we'll examine some social factors that influence adult decisions.

Social Influences and Personal Choices

Social factors influence people's personal choices in two ways. First, it is always easier to make the common choice; norms about appropriate behavior make the alternatives psychologically and socially difficult. For example, in the 1950s and early 1960s people tended to marry earlier than they do now. With the median age at marriage at about 20 for women and 22 for men then (compared with about 24 for women and 26 for men now), it was more difficult to remain single after graduation, and women in their last year of college sometimes became panicky enough

2. We have drawn this theme from Peter Berger and Thomas Luckmann, *The Social Construction of Reality* (1966). According to these social scientists, people *externalize* their own ideas, impressions, opinions, and ways of doing things (that is, they voice them or act them out). In the process of externalization, things may come to seem real, to become part of assumed common knowledge. At the same time, individuals often *internalize* externalized impressions or points of view. They thus begin to believe that commonly held opinions are true.

3. This theme is drawn from C. Wright Mills, *The Sociological Imagination*. In Mills' words, people must begin to grasp the "problems of biography, of history and of their interactions within a society" ([1959] 1973, p. 6).

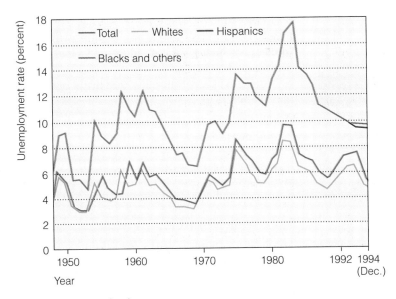

FIGURE **1.1**

Unemployment rates in the United States, 1948–December 1994. These rates make it clear that it was more difficult for individuals to find a job in 1984—or today—than in 1953. As you can see, the rate for "blacks and others" has consistently been about twice that for whites. According to December 1994 figures, African Americans could expect more difficulties finding a job than either Hispanics or whites. These difficulties result from societal influences, not lack of personal motivation, because only people actively seeking work are included in unemployment figures. Considering this situation, to what extent can personal problems be viewed apart from their socioeconomic context? (Sources: Data from Economic Report of the President 1989, Table B-39; U.S. Bureau of Labor Statistics, 1995a)

to marry men they did not really care for. Now, staying single longer is much more acceptable in our society and, consequently, a more comfortable choice for individuals. Second, social factors limit people's options. For example, American society has never offered legal polygamy. We are presently watching how state and federal courts and legislatures may affect individuals' options concerning gay male or lesbian unions, and abortion.

Historical Events

Specific historical events, such as war, depression, inflation, and social change, affect options, choices, and the everyday lives of families. For example, the Vietnam War may have played a part in the rise in the American divorce rate in the sixties and seventies (Norton and Moorman 1987).

The work of sociologist Glen Elder (1977) and demographer Richard Easterlin (1987) reminds us that family life has been a different experience in the Great Depression, in the optimistic fifties, in the tumultuous sixties, and in the economically constricted eighties and nineties. In the Depression years, couples delayed marriage and parenthood and had fewer children than they wanted. During World War II, married women were encouraged to get defense jobs, and they went to work in factories for the first time. Day-care centers were popular.

In the affluent fifties, young people could better afford to get married. They had large families cared for by stay-at-home mothers. Divorce rates slowed their long-term increase (Easterlin 1987). The expanding economy provided a sound basis for this family life. Today, one person can no longer earn a "family wage." More and more wives seek employment, regardless of their preference to be employed or not. Our constricting economy limits options for housing, hiring household help, and visiting geographically distant relatives, among other things. More and more people who might prefer to live alone, but cannot afford to, find

This family portrait is of a gay male couple and their relatives. The portrait reminds us of two ideas: First, individuals are freer to make personal choices today than in the past. These men can be fairly open about choosing a same-sex relationship. Second, social factors influence individuals' personal choices. One way they do so is by limiting our options. When this photo was taken, this couple could not marry legally in any state, although Hawaii has been considering legalizing same-sex marriages.

themselves returning home to live with parents (Goldscheider and Goldscheider 1994).

Historical change involves not only specific events but also short- or long-term change in basic indicators of human life. As recently as a hundred years ago, one-third of our population died before reaching adulthood. Presently, three-fourths of the U.S. population live to be 65, and many reach 85 and older. Among the consequences of this increased longevity are more

years invested in education, longer marriages for those who do not divorce, a longer period during which parents and children are both adults, and a long retirement.

We will increasingly be influenced by the age structure of our society, as increasing numbers of elderly people must be cared for by a smaller group of middle-aged and young adults. This impact will be felt not only economically but also at the level of personal care for kin. The smaller family sizes of today will result in fewer adults who can share intergenerational family responsibilities in the future (Treas 1995).

Race and Ethnicity

The historical moment shapes people's options and decisions, but so does their place within our culturally diverse society. Marriage and family patterns vary among racial and ethnic groups. For example, Hispanic fertility rates are the highest of all the major racial/ethnic groups (U.S. Bureau of the Census 1995, Table 92). This trend probably has to do with the strong influence of the Catholic religion on Hispanic culture. Native American traditions, formalized in federal law, encourage more communal (as opposed to strictly parental) rights and responsibilities for children ("Navajo Tribal Court" 1988). As a third example, Asian Americans may be more strongly influenced by their parents in choices about marriage than are other Americans.

The growth of immigration in recent decades should increase the impact of ethnicity on family life because new immigrants retain more of their ethnic culture than do immigrants who have been in the United States longer. Foreign-born women have half again as many children as do native-born Americans (Martin and Midgley 1994), and marriages of recent immigrants seem less egalitarian than do those of couples of similar ethnic background whose families have been in the United States longer (Pozzetta 1991). The concept of "choice" itself may be alien or incompatible in cultures in which family life is experienced as carrying on tradition and accepting the decisions of family heads.

The impact of racial discrimination and economic disadvantage on African American families is indicated by the fact that blacks are more than twice as likely as whites to suffer the death of an infant, and black children are significantly less likely than whites to finish high school or go on to college. A 1992 national poll

Many social factors condition people's options and choices. One such factor is an individual's place within our culturally diverse society. This Navaho child is being taught to shear sheep by her grandparents. Race and ethnicity influence and shape people's options, decisions, marriages and family life.

of African Americans taken by *Newsweek* found that 51 percent of respondents felt that over the past ten years the quality of life for blacks had gotten worse (Morganthau 1992).

In 1993, 78 percent of white children and 36 percent of African American children under age 18 were living with two parents (Saluter 1994, Table F). Our most recent analysis tells us that the difference in the proportions of two-parent families among African Americans and whites is not new. Throughout the twentieth century, the percentage of black children with at least one absent parent has been about twice that for whites. Especially over the past three decades, however, race differences in family structure have grown (Ruggles 1994). Experts strongly disagree on the cause of the increase in female-headed households, but one reason has been a decline in the number of black husbands who can help support a wife and chil-

dren. The elimination of entry-level positions paying a family wage is responsible for low levels of black employment and income, which in turn preclude marriage or doom it from the start (Wilson 1987). Then too, African American women may be less dependent on marriage for economic survival. But the pattern may also "reflect a difference in social norms between blacks and whites, which could have developed either through the experience of slavery or could have its roots in differences between European and African cultures" (Ruggles 1994, pp. 147, 148; see also McDaniel 1994).

People's racial/ethnic heritage affects their options and decisions. Box 1.2, "Family Ties and Immigration," looks at these issues with regard to recent immigrants to the United States. In taking account of racial and ethnic differences, it is important not to rely on assumptions about differences that may not exist.

Family Ties and Immigration

Today there is more racial and ethnic diversity among American families than ever before, and much of this diversity results from legal immigration. Interestingly, a great deal of this immigration is spurred by family ties.

Before 1965, immigrants were mainly European. The 1924 Immigration Act had established country of origin as the principal criterion for relocating to the United States, setting quotas favoring immigrants from northern and western Europe. Federal reforms in 1965 changed the criteria for obtaining an immigrant visa. The new policy gave priority to people with needed job skills or to people with family members already living here. In 1993, 55 percent of immigrants arrived here because they had a family member living in the United States. About half of those were reuniting with spouses, children, or parents who were United States citizens.

This policy change drastically changed the composition of U.S. immigration. Since 1965, Asia and Latin America, not Europe, have been the major sending regions. During the 1980s, Asian and Central and South American countries, taken together, sent 85 percent of all immigrants to the United States. The top five countries of origin were Mexico, the Philippines, China, Korea, and Vietnam.

Why do immigrants choose to come here? For the most part, immigrants leave a poorer country for a richer one. In their nation of origin, families may calculate how each member can best contribute to the family's status and/or income. This household economics of migration can encourage a rural Mexican family, for instance, to send a daughter to Mexico City to be a secretary and a son to Los Angeles to be a day laborer. Typically, the children send money home so that the family can try planting a new kind of crop, for example, or buy farm equip-

ment, or a television set. As children establish themselves in the host country, they begin to send for their relatives, so that gradually ethnic kin or community networks develop regionally. Every major U.S. city today is truly multicultural, characterized by disparate ethnic neighborhoods.

The increase in immigrant families has led policymakers and others to consider many issues. For one thing, the number of American children who speak a language other than English has increased markedly. The 1990 census found a total of 6.3 million schoolchildren in the United States who spoke languages other than English at home. Some school districts use bilingual education to teach these students more effectively, while others teach English as a second language, and still others offer no special help at all. How to best educate immigrant children—many of them already U.S. citizens and many more on the path toward naturalization—is a matter of public debate.

Another issue is whether, on balance, immigrant families are an asset or a liability to the nation's economy. Many studies are underway in this regard, and the verdict is still out. But it appears that immigrant families pay more in payroll, property, and sales taxes than they cost the government in public services. However, controversy rages because most immigrant family tax dollars go to the federal government while the costs of their schooling or health care are largely paid for by local governments. This is the situation in Los Angeles County, for example (Martin and Midgley 1994). Whatever our views on immigration, immigrants are generally responsible family members doing what they can to improve their lives as well as the lives of their relatives.

Social Class

Social class or status may be as important as race or ethnicity in affecting people's choices. Clearly, the distribution of wealth in the United States is skewed. In 1992 the top 20 percent of U.S. families received 47 percent of the nation's total income; the poorest 20 percent of Americans received just 4 percent. These distributions have shown little change since World

War II (U.S. Bureau of the Census 1993). Money may not buy happiness, but it does afford a myriad of options, from sufficient and nutritious food to eat to comfortable residences to better health care to keeping in touch with family and friends through the Internet to education at prestigious universities to vacations, household help, and family counseling.

Besides distinguishing families on the basis of income and/or assets, social scientists have often distinguished between white- and blue-collar workers. White-collar workers include professionals, clerical workers, salespeople, and so forth who have traditionally worn white shirts to work. Working-class people, or blue-collar workers, are employed as mechanics, truckers, police officers, machine operators, and factory workers—jobs typically requiring uniforms.

Social scientists do not agree on whether blue- and white-collar workers have become increasingly alike in their values and attitudes in recent decades. But white- and blue-collar employees may continue to look at life differently even at similar income levels. Regarding marriage, for example, working-class couples tend to emphasize values associated with parenthood and job stability, whereas white-collar partners are more inclined to value companionship, self-expression, and communication. Middle-class parents value self-direction and initiative in children, whereas parents in working-class families stress obedience and conformity (Luster, Rhoades, and Haas 1989).

Religion, Region, and Rural–Urban Residence

Other social-group memberships can influence family life in ways too complex to analyze here. The family heritage of major religious groups (D'Antonio and Aldous 1983) is a significant influence. Catholics, for example, appear to have shifted from the traditional church teachings to modern conceptualizations of the family and sexuality (D'Antonio et al. 1989). But studies still find contemporary Catholics to be different from Protestants in their more communal, rather than individualistic, worldview (Greeley 1989). And Mormon encouragement of large families is reflected in the distinctly high fertility of the state of Utah (U.S. National Center for Health Statistics 1994, Table 1).

Regional family differences have not been widely researched, and we run the risk of arriving at false generalizations. One study, however, used General Social Survey data to explore regional differences in attitudes about corporal punishment of children. Results showed that a majority of adults throughout the United States favored spanking children, but the proportion was significantly higher (86 percent) among Southerners (Flynn 1994).

Age Expectations

In the same way that historical events and social location influence choices, so do **age expectations.** When individuals shop for clothes, they sometimes wonder whether they are too old or too young to wear something they like. They are at least unconsciously aware that society views some attitudes and behaviors as appropriate for a given age and others as inappropriate. Despite growing diversity and broader time frames, there is still a sense of a "right time" to have children or to "slow down" or retire, for example.

Individuals are aware of their own life timing in relation to social expectations. Accordingly, they view themselves as early, late, or on time in family, occupational, and educational events, and such age expectations influence people's personal choices. For example, some spouses admit that their choice to marry was influenced by the feeling that it was time they were married.

Being early, late, or on time can also affect our options. For example, individuals choose from the largest pools of eligibles when looking for marriage partners when they are "on time." And women find it biologically easier to have children before their late 30s.

In sum, social factors such as historical events, social location, and age expectations limit people's options. (We will discuss the impact of socially defined gender roles in Chapter 3.) Even though individuals may be limited in their choices by social conditions in ways that they do not realize, becoming conscious of social influences permits a more knowledgeable choice. It also minimizes self-blame for what may be a socially structured lack of options and can inspire collective effort to alter social conditions.

Let's look more closely at two forms of decision making—choosing by default and choosing knowledgeably—along with the consequences of each.

Making Choices

All people make choices, even when they are not aware of it. One effect of taking a course in marriage and the family may be to make you more aware

Midlife changes can be both exhilarating and intimidating, as these college graduates have probably found. Certainly, the decision of a middle-aged adult to earn a college degree involves many emotional and practical changes. But by making knowledgeable choices—by weighing alternatives, considering consequences, clarifying values and goals, and continual rechecking—personal decisions and changes can be positive and dynamic.

of when choices are available and of how a decision may be related to subsequent options and choices, so that you can make decisions more knowledgeably.

Choosing by Default

Unconscious decisions are called **choosing by default.** Choices made by default are ones people make when they are not aware of all the alternatives or when they pursue the proverbial path of least resistance.

If you're taking this class, for example, but you're unaware that a class in modern dance (which you would have preferred) is meeting at the same time, you have chosen *not* to take the class in modern dance, but you have done so by default because you didn't find out about all the alternatives when you registered.

Another kind of decision by default occurs when people pursue a course of action primarily because it

seems the easiest thing to do. Many times college students choose their courses or even their majors by default. They arrive at registration only to find that the classes they had planned to take are closed. So they register for something they hadn't planned on, do pretty well, and continue in it just because that seems easier than rearranging their curriculum programs.

Many decisions concerning marriages and families are also made by default. Spouses may focus on career success, for example, to the neglect of their relationship simply because this is what society expects of them.

Although most of us have made at least some decisions by default, almost everyone can recall having the opposite experience: **choosing knowledgeably.**

Choosing Knowledgeably

Today, society offers many options. People can stay single or marry; they can choose to live together outside legal marriage; they can form communes or family-like ties with others; they can decide to divorce or to stay married. One important component of choosing knowledgeably is recognizing as many options or alternatives as possible. This text is designed in part to help you do that.

A second component in making knowledgeable choices is recognizing the social pressures that may influence personal choices. Some of these pressures are economic, whereas others relate to cultural norms that have been taken for granted. Sometimes people decide that they agree with socially accepted or prescribed behavior. They concur in the teachings of their religion, for example. Other times, though, people decide that they strongly disagree with socially prescribed beliefs, values, and standards. Whether they agree with such standards or not, once people recognize the force of social pressures they are free to choose whether to act in accordance with them.

An important aspect of making knowledgeable choices is considering the consequences of each alternative rather than just gravitating toward the one that initially seems most attractive. A couple deciding whether to move so that one partner can be promoted, for example, may want to list the consequences, both positive and negative. In the positive column, that partner may have a higher position and earn more money, and the region to which they would move may have a friendlier climate. In the negative column, the other spouse may have to give up or disrupt his or her career, and both may have to leave relatives. Listing positive and negative consequences of alternatives—either mentally or on paper—helps one see the larger picture and thus make a more knowledgeable decision.

An element in this process is personally clarifying your own values. In recent years, counselors and others have given considerable attention to values clarification—becoming aware of your values and choosing to act consistently with those values. Society today holds up contradictory sets of values. For example, there are varied standards of nonmarital sex ranging from abstinence to sex for recreation only, without personal affection. Contradictory values can cause people to feel ambivalent about what they want for themselves.

Clarifying one's values involves cutting through this ambivalence in order to decide which of several orientations are more strongly valued. To do this, it is important to respect the so-called gut factor—the emotional dimension of decision making. Besides rationally considering alternatives, people have subjective, often almost visceral feelings about what feels right or wrong, good or bad. Respecting one's feelings is an important part of making the right decision.

One other component of decision making should be mentioned, and that is rechecking. Once a choice is made and a person acts on it, the process is not necessarily complete. People constantly recheck their decisions, as Figure 1.2 suggests, throughout the entire decision-making cycle, testing these decisions against their own subsequent feelings and against any changes in the social environment.

An assumption underlying this discussion has been that individuals cannot have everything. Every time people make an important decision or commitment, they rule out alternatives—at least for the time being, and perhaps permanently. People cannot simultaneously have the relative freedom of a child-free union and the gratification that often accompanies parenthood.

In some respects, though, people can focus on some goals and values during one part of their lives, then turn their attention to different ones at other times. As we will see in the next section, choices throughout life often are made according to this pattern.

Personal Choices and Change

Three or four decades ago, we used to think of adults as people who entered adulthood in their early 20s, found work, married, had children, and continued on the same track until the end of the life course. That view has changed. Today we view adulthood as a time with potential for continued personal development, growth, and change.

The Life Spiral

The **life spiral** is a concept that takes into account a significant feature of the contemporary life course: Not everyone moves into and out of family life cycle stages "on time." The term *spiral* conveys a sense of "the incorporation of traditional and alternative roles in the life course" (Etzkowitz and Stein 1978, p. 434) and the variations in a person's role commitments over the life course.

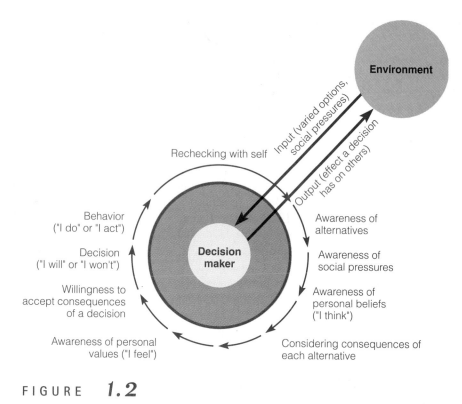

FIGURE **1.2**

| The cycle of knowledgeable decision making. (Source: Adapted from O'Neill and O'Neill 1974)

The life spiral model of adult change is meant to accommodate the variety of lifestyle choices people make today. While some adults follow fairly traditional patterns, others shift from traditional to alternative patterns, and vice versa. People marry, divorce, and then remain single, for example. Or they might live alone or in communal groups or with friends and marry later. Another pattern involves people choosing and remaining in alternative lifestyles; an example would be unmarried, cohabiting parenthood. Although Etzkowitz and Stein do not make this connection, the life spiral idea leaves room for the observation that parents may be influenced by their children and come to share the values of a later generation.

We have been discussing changes experienced by individuals as they move through the life course. But marriage and other domestic partnerships involve *two or more* adults. Family living becomes a situation in which multiple life courses must be coordinated. Moreover, life in American families reflects a tension in American culture between family solidarity and individual freedom.

A Family of Individuals

Family values (familism) permeate American culture. These values, such as family togetherness, stability, and loyalty, focus on the family as a whole. Many of us have an image of the ideal family in which members spend considerable time together, enjoying one another's company. Togetherness, in other words, represents an important part of family life. For many of us, the family is a major source of stability; we are disturbed when it is disrupted. Then, too, the group most deserving our loyalty, we believe, is often our family. Most of us who are married today vowed publicly to stay with our partners as long as we live. We expect our partners, parents, children, and even our more distant relatives to remain loyal to the family unit. Because of familism, home is sometimes considered "the place where you can scratch anywhere you itch."

While families are composed of individuals, each seeking self-fulfillment and a unique identity, they also offer a place to learn and express togetherness, stability, and loyalty. Families also perform a special archival function: events, rituals, and histories are created and preserved and, in turn, become intrinsic parts of each individual.

Families as a Place to Belong

Whether families are traditional or newer in form (such as communes or cohabiting partnerships), they create a place to belong in at least two ways. First, families create *boundaries,* both physical and psychological, between themselves and the rest of the world. Whether in multiple- or single-family dwellings, families in virtually all cultures mark off some physical space that is private and theirs alone (Boulding 1976). Family members determine "what kinds of things are allowed to enter the family space and under what conditions and what kinds of items are simply not permitted admission" (Kantor and Lehr 1975, p. 68).

With an idea of how the external world resembles and differs from the family interior, family members screen off certain aspects of the larger, outside culture. They put up fences so that they can barbeque or sunbathe in privacy, for example, and they prevent certain people, books, pictures, words, and topics of conversation from entering the family interior. Family members may patrol one another's activities, so that a child or a spouse is accompanied by another family member when leaving the family home, in order to provide safety from the real or imagined hazards of the outside world.

A second way families create a place to belong is by performing an **archival family function.** That is, families create, store, preserve, and pass on particular objects, events, or rituals that members consider relevant to their personal identities and to maintaining the family as a unique existential reality or group. The archives contain a variety of symbols:

> These are snapshots of happy times, posed and unposed (apparently never snapshots of sad times); family movies of celebrations or rites of passage or vacations; . . . artifacts from infancy or childhood transmuted into relics of lost and sweet identities; symbols of recognition and achievement such as diplomas; . . . pointed anecdotes about infancy or youth which reinforce a particular identity as the reckless one, the always helpful one, or the unlucky one ("Remember the time you accidentally broke two of Mr. Jones's porch windows!"); and various . . . symbols which function almost like debts binding one to past relationships with the implication of potential future obligations ("Here is part of the plaster cast from your left leg which you broke in sixth grade, when I had to care for you at home for six weeks—remember that when I am unable to walk around by myself anymore"). (Weigert and Hastings 1977, p. 1174)

But just as family values permeate American society, so also do **individualistic (self-fulfillment) values.** Expressing our individuality within the context of a family requires us to negotiate innumerable day-to-day issues. How much privacy can each person be allowed at home? What things and places in the family dwelling belong just to one particular individual? What family activities should be scheduled, how often, and when? What outside friendships and activities can a family member have?

Family Rhythms of Separateness and Togetherness

In every family, members regulate personal privacy (Kantor and Lehr 1975). Members come together for family rituals, such as playing games or watching television, but they also need and want to spend time alone. Being shortchanged on privacy is associated with irritability, weariness, family violence, poor family relationships, and emotional distance from one's spouse.

Time represents another important dimension of family life. Each family member has personal feelings about timing. Staying at a party until the wee hours might feel good to one spouse, but the other might prefer to call it quits earlier. Working out a common rhythm for dialogue, sexual expression, and joint activities can be difficult for spouses, for example, when a "night person" marries a "morning person."

In all intimate relationships, partners alternately move toward each other, then back away to reestablish a sense of individuality or separateness. The needs of different individuals vary, of course, and so do the temporary or permanent balances that couples and families strike between togetherness and individuality.

"Family life in any form has both costs and benefits. . . . Belonging to any group involves loss of personal freedom" (Chilman 1978). The juxtaposition of opposing values[4]—familism and individualism—creates in society and in ourselves tension that we must resolve. Where are our solutions leading us? Sociologist Lynn White (1987), among others, believes that "the pendulum between freedom and constraint has swung so far toward freedom that it has altered the family in basic and undesirable ways" (p. 470). Family sociologist David Popenoe (1993) similarly warns that self-interest has led to "family decline" over the past three decades, the consequences of which—particularly for children—are "cause for alarm" (p. 539).

Not every family expert completely agrees, however; John Edwards (1987) argues that the situation may not be this bleak. He suggests that we may be putting too much emphasis on the weakening of the

---

4. Contradictory family and individual values are part of a general "value cleavage" growing out of changes in our society. Philip Slater in *The Pursuit of Loneliness* (1976) contrasts the "old" culture, which stresses social forms, with the "new" emphasis on personal expression. Ralph Turner (1976) points to a growing emphasis on acts of self-expression, as opposed to past self-realization through dutiful performance of accepted social roles.

In a world of demographic, cultural, and political changes, our views of family structure and life cycle have begun to broaden. For example, today there are more single-parent families, gay partners and parents, remarried families, and families in which adult children care for their aging parents. Whatever their form, families can remain a center of love and support.

nuclear family without recognizing that other structural arrangements may be equally capable of rearing children well. Similarly, sociologist Judith Stacey (1993) sees the concept of "family decline" as unfairly value-laden and argues that there are many ways to define the word *family* besides that of husband, wife, and children. Moreover, according to Stacey, "Short of exhorting or coercing people to enter or remain in unequal, hostile marriages, family decline critics offer few social proposals to address children's pressing needs" (p. 547).

Providing something of a middle ground, sociologist Norval Glenn states his position:

I believe that marriage should be satisfying to the spouses. . . . [However,] I do not give the spouses' needs and desires priority over the needs of their children or over the social need for the proper specialization of children. Second, I do not believe that each spouse's giving priority to his or her own needs, deemphasizing duty and obligation in mar-

riage, and giving up the ideal of marital permanence will maximize the happiness and satisfaction of American adults. (Glenn 1987a, pp. 350–51)

Glenn (1993) makes a plea for objective assessment of the notion of family decline through research.

The Life Spiral and Partners

Shifts in the balance of individuality and familism have meant that family members have become less predictable than in the past. The course of family living results in large part from decisions and choices two *individual* adults make, both moving in their own ways and at their own paces through their own lives. Assuming that partners' respective beliefs, values, and behaviors mesh fairly well in marriage, any change in either spouse is likely to adversely affect the fit. One consequence of ongoing developmental change in two individuals is that a marriage is no longer as likely to be permanent, as we'll see in Chapter 15. If both husband and wife change little over the years, their match

Cartoon by Lynn Johnston, from Roz Warren [Ed.] *Men Are from Detroit, Women Are from Paris,* p. 91. Reprinted by permission of Universal Press Syndicate.

Marriages and Families: Four Themes

Throughout this book, we will develop the following four themes.

1. Personal decisions must be made throughout the life course. Decision making is a trade-off; once we choose an option, we discard alternatives. No one can have everything. Thus, the best way to make choices is knowledgeably.

2. People are influenced by the society around them. Cultural beliefs and values influence our attitudes and decisions. And societal or structural conditions can limit or expand our options.

3. We live in a changing society, characterized by increased ethnic, economic, and family diversity; by increased tension between family and individualistic values; and by decreased marital and family permanence. This can make personal decision making not only more difficult than in the past but also more important.

4. Personal decision making feeds back into society and changes it. Making personal decisions in order to effect family-related social change can mean choosing to become politically involved. But we affect our social environment every time we choose, so it's best to choose according to our thought-through principles whether or not these agree with what's accepted.

In Sum

This chapter introduced the subject matter for this course and presented the four themes that this text develops. The chapter began by addressing the challenge of defining the term *family*. In Box 1.1, "Some Facts About Families in the United States Today," we have pointed to much evidence that we live in a changing society, characterized by increasing ethnic, economic, and family diversity. Family diversity has progressed to the point that there is no typical family form today.

It is now widely recognized that change and development continue throughout adult life. People make choices, either actively and knowledgeably or by default, that determine the courses of their lives. These choices are influenced by a number of factors, includ-

can remain mutually satisfying. But if one or both change considerably over time, the danger exists that they will grow apart instead of together. A challenge for contemporary relationships is to integrate divergent personal change into the relationship.

How can partners make it through such changes and still stay together? Two guidelines may be helpful. The first is for people to take responsibility for their own past choices and decisions rather than blaming previous "mistakes" on their mates. The second is for individuals to be aware that married life is far more complex than the traditional image commonly portrayed. It helps to recognize that a changing spouse may be difficult to live with for a while. A relationship needs to be flexible enough to allow for each partner's individual changes—to allow family members some degree of freedom. At the same time, we must remind ourselves of the benefits of family living.

We have defined the term *family* and discussed decision making and diversity in the context of family living. We can now state explicitly the four themes of this text.

ing age expectations, race and ethnicity, religion, social class, gender, and historical events. People must make choices or decisions throughout their life courses, and those choices and decisions simultaneously are limited by social structure and are causes for change in that structure.

Marriages and families are composed of separate, unique individuals. That uniqueness stems partly from the fact that human beings are able to make choices. They have creativity and free will: Nothing they think or do is totally programmed.

At the same time, all of the individuals in a particular society share some things. They speak the same language and have some common attitudes about work, education, and marriages and families. Moreover, within a socially diverse society such as ours, many individuals are part of a racial, ethnic, or religious community or social class that has a distinct family heritage.

Our culture values both familism and individualism. Whether individualism has gone too far and led to an alarming family decline is a matter of debate. Even though families fill the important function of providing members a place to belong, finding personal freedom within families is an ongoing, negotiated process.

Adults change, and because they do, marriages and families are not static: Every time one individual in a relationship changes, the relationship changes, however subtly. Throughout this text we will discuss some creative ways in which partners can alter their relationship in order to meet their changing needs.

Remembering, then, that no two people are exactly alike—and that in many ways every adult continues to change—we begin our study by looking at choices in the context of individual development and social change.

Key Terms

age expectations
archival family function
choosing by default
choosing knowledgeably
family
family values (familism)
individualistic (self-
    fulfillment) values

life spiral
modern family
nuclear family
postmodern family
primary group
secondary group

Study Questions

1. Has marriage in the United States been reconceptualized as a nonpermanent relationship, as sociologist Constantina Safilios-Rothschild says? If not, give examples to support your point. (Are your examples from personal experience or scientific investigation?) If you agree with Safilios-Rothschild, explain how you think this has happened, giving examples.

2. Without looking to find ours, write your definition of the family. Now compare yours to ours. How are the two similar? How are they different? Does your definition have advantages over ours? If so, what are they?

3. What important changes in family patterns do you see today?

4. In what ways does personal decision making involve pressures? In what ways does it involve freedom? Give concrete examples.

5. This chapter gave the example of a person's being unemployed as a personal problem that can result from societal influences. What are some other examples of this concept? In general, do the beliefs in our society reflect this concept—or do they put responsibility for personal problems on the individual? Give examples to support your position.

6. When might a personal problem *not* result from societal influences?

7. What are some of the social factors that influence people's choices? How do these factors operate? Does the fact that social factors limit people's options mean that important life decisions are predetermined? Why or why not?

8. What are the components of knowledgeable decision making? How does knowledgeable choosing differ from choosing by default? How might the consequences be different? Give some examples of the two forms of decision making.

9. What is the archival function of families? How important is it?

10. Do you agree with sociologist David Popenoe that "family decline" today is cause for alarm? Why or why not? How might this issue be objectively researched and assessed?

11. Do you want your family life to be similar to or different from that of your parents? In what ways?

Suggested Readings

Ahlburg, Dennis A. and Carol J. DeVita. 1992. *New Realities of the American Family*. Washington, DC: Population Reference Bureau. With statistics and discussion, this work explores the social, economic, and demographic trends that have contributed to the changing structure of American families, describes various types of families prevalent today, and projects their numbers into the future. Forty pages long, it is a handy source of useful information on the family.

Coontz, Stephanie. 1992. *The Way We Never Were: American Families and the Nostalgia Trap*. New York: Basic Books. Social historian Coontz gives data and evidence to argue that the happy traditional family style that we imagine existed in the past is not historically accurate. Rather, the idea is a socially constructed image that developed retrospectively given particular social and political interests.

Easterlin, Richard. 1987. *Birth and Fortune*. 2d ed. rev. Chicago: University of Chicago Press. Data-based analysis of how cohort size (the number of people born at the same time) affects economic opportunity, chance of marrying, chance of divorcing, crime rate, and mental health, among other phenomena.

Edmonds, Susan. 1993. *Native Peoples of North America: Diversity and Development*. New York: Cambridge University Press. A general analysis of Native Americans with some exploration of Native American families.

Goldscheider, Frances K. and Linda J. Waite. 1991. *New Families, No Families?: The Transformation of the American Home*. Berkeley: University of California Press. A highly acclaimed work destined to become a modern classic, this book examines whether the family is in decline or simply changing.

Gubrium, Jaber B. and James Holstein. 1990. *What Is Family?* Mountain View, CA: Mayfield. Important book on the family that demonstrates how family members act to construct and define a family. Focuses on families coping with the issues of aging and health care.

Hoobler, Dorothy and Thomas Hoobler. 1994. *The Mexican American Family Album*. New York: Oxford University Press. Appreciative description of Mexican American families in the United States today.

Ingoldsby, Bron B. and Suzanna Smith. 1995. *Families in Multicultural Perspective*. New York: Guilford. An analysis of U.S. families that takes into account the growing diversity in our society and culture today.

Kitano, Harry H. L. and Roger Daniels. 1995. *Asian Americans: Emerging Minorities*. 2d ed. Englewood Cliffs, NJ: Prentice-Hall. Reviews the literature on Asian Americans and discusses various socio-cultural aspects of their lives in the United States, including family traditions and change.

Pozzetta, George E., ed. 1991. *Immigrant Family Patterns: Demography, Fertility, Housing, Kinship, and Urban Life*. New York: Garland. A collection of readings that examine immigrant families as they strive to retain their traditions and at the same time grapple with issues of acculturation.

Rubin, Lillian B. 1994. *Families on the Fault Line: America's Working Class Speaks About the Family, the Economy, Race, and Ethnicity*. New York: HarperCollins. This book is based on interviews Rubin conducted with 162 working-class and lower middle-class families of various ethnic and racial backgrounds during a period in which the economy has contracted and shifted. The author examines the detrimental impact and stress this has had on family life.

Spanier, Graham B. 1989. "Bequeathing Family Continuity." *Journal of Marriage and the Family* 51 (1): 3–13. This was Spanier's presidential address to the National Council on Family Relations. He asks, and begins to answer, two questions: (1) Are the current changes confronting American families indicators of pathology, deterioration, and instability? (2) How is it possible and by what mechanisms do families with considerable dysfunction transmit to succeeding generations a commitment to the concept of the family?

Staples, Robert, ed. 1995. *The Black Family: Essays and Studies*, 5th ed. 1994. Belmont, CA: Wadsworth. This collection of very current readings, put together by a leading authority on African American families, explores trends and issues surrounding the black family today.

Staples, Robert and Leanor Boulin Johnson. 1993. *Black Families at the Crossroads: Challenges and Prospects*. San Francisco: Jossey-Bass. This book explores the challenges that black families face today within a socio-historical context and focuses on how black history influences today's African American family life.

Stone, Elizabeth. 1988. *Black Sheep and Kissing Cousins: How Our Family Stories Shape Us*. New York: Penguin. The shaping of family identity and family "rules" through the medium of tales told about the family's history.

Weston, Kath. 1991. *Families We Choose: Lesbians, Gays, Kinship*. New York: Columbia University Press. The premise of this work is we create our own families; families are those persons whom we actively choose to include as members. This theme is specifically applied to lesbians and gay men as they create families through involved decision making.

Williams, Norma. 1990. *The Mexican American Family: Tradition and Change*. Dix Hills, NY: General Hall. Sociological description and analysis of Mexican American families in the United States today.

Exploring the Family

There is nothing so practical as a good theory.

KURT LEWIN

When we begin a course on the family, we are eager to have our questions answered, including:

- "What's a good family?"
- "How do I make that happen?"
- "Whom should I choose?"
- "What do I need to know to be a good parent?"

Perhaps the most important question of all is:

- "What's happening to the family today?"

Box 1.1, "Some Facts About Families in the United States Today," outlines some trends in family life, some changes and developments you may have learned about from the media or deduced from observing life around you. But what these trends mean—how to interpret them—is not always so easy. There are many visions of the family, and what an observer reads into the data depends partly on his or her perspective.

For some, these trends mean that the family is "declining" or "vanishing," whereas for others it is simply "changing." From politicians, we hear about family trends with words such as *conservative* or *liberal* attached. But in forming their interpretations, social scientists use the more formal vocabulary of social theory and research methodology to characterize marriage and family patterns.

This chapter invites you to share this way of seeing families. We look at some theoretical perspectives that shape our thinking about families and at the knotty problem of studying a phenomenon as close to our hearts as family life.

Theoretical Perspectives on the Family

Theoretical perspectives are ways of viewing reality, lenses through which analysts organize and interpret what they see. A theoretical perspective leads researchers to identify those aspects of families that are of interest to them. There are several different theoretical perspectives on the family. We shall see that what is significant about families varies from one perspective to the next. Sometimes, the perspectives complement one another and appear together in a single piece of research. In other instances, the perspectives compete, a situation that can lead analysts and policymakers to heated debate. All this can be frustrating to students who grope for the one "correct" answer. Instead, it is useful to think of a theoretical perspective as a point of view. As we move around an object and see it from different angles, we have a better grasp of what it is than if we look at it from a single fixed position.

In this chapter, we describe seven theoretical perspectives on the family: family ecology, family development, structure-functionalism, the interactionist perspective, exchange theory, family systems theory, and conflict/feminist perspectives. Each of these perspectives will be further explored in subsequent chapters. In actual research, two or more of these perspectives may come together. We will see that each perspective not only illuminates our understanding in its own way but also emerged in its own social and historical context.

The Family Ecology Perspective

The **family ecology perspective** explores how a family influences and is influenced by the environments that surround it. Every family is embedded in "a set of nested structures, each inside the next, like a set of Russian dolls" (Bronfenbrenner 1979, in Bubolz and Sontag 1993, p. 423). The neighborhood in which a family lives includes components of these environments (see Figure 2.1). All parts of the model are interrelated and influence one another. We use the family ecology perspective throughout this book when we stress that society does not determine family members' behavior but does present limitations and constraints, as well as possibilities and opportunities, for families.

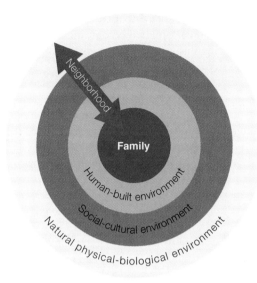

FIGURE **2.1**

The family ecology perspective. The family is embedded in the natural physical–biological, social–cultural, and human-built environment. (Adapted from Bubolz and Sontag 1993, p. 432)

This perspective emerged in the latter part of the nineteenth century, a period marked by social concern about the health and welfare of families. After losing ground to the family development and structure-functional perspectives (discussed below), the family ecology model resurfaced in the 1960s with increased societal awareness of the interdependence between families and their political and economic environments (Bubolz and Sontag 1993). Head Start programs, for instance, part of President Lyndon Johnson's War on Poverty of the 1960s, were largely propelled by family ecology. In our current troublesome economy, the family ecology model is again prominent in research and in political discussion and debate. Furthermore, today's family ecologists, in an increasingly global society, stress the interdependence of all the world's families—not only with each other but also with our fragile physical-biological environment (Bubolz and Sontag 1993). While crucial, this last focus is beyond the scope of this text. Here our interest centers on families enmeshed in their socio-cultural environments.

The ecology model leads researchers to investigate how various socio-cultural environments impact families. Put another way, the family ecology model is concerned with *family policy*—how to influence the effects that circumstances in the broader society have on families. For instance, family ecologists would point to poverty as a real problem for many U.S. families (see item 16 in Box 1.1). More and more Americans worry about making ends meet: How will we support ourselves, find comfortable housing, educate (or feed) our children, get affordable health care, finance our old age? Meanwhile, economic and political developments over at least the last fifteen years have resulted in declining and inadequate government resources and spending for social services.

FAMILY POLICY   In a narrow sense, **family policy** is all the procedures, regulations, attitudes, and goals of government that affect families. Family ecologists might point out that the United States provides fewer services to families than does any other industrialized nation; western Europe offers many examples of a successful partnership between government and families in the interests of family support.

In addition to noting society's impact on families, the ecology perspective encourages researchers and policymakers to investigate what families might do to create environments that improve their quality of life. How can families become activists for the kinds of changes they want? This, of course, is a thorny issue. In 1980, President Carter summoned ordinary citizens to the White House Conference on Families. His goal was to get us all involved in shaping family policy. It quickly became clear that although almost everyone endorsed the idea of a nationally stated family policy, Americans were divided on what that policy should be. Some argued that only heterosexual, nuclear families should be encouraged, whereas others believed in supporting a variety of families—single-parent or gay or lesbian families, for example. Indeed, the diversity of family lifestyles in the United States makes it extremely difficult to develop a national family policy that would satisfy all, or even most, of us.

Then too, more government help to families would be costly, and hence sure to be opposed in this era of an enormous federal deficit. Many Americans, imbued with individualistic values and facing economic difficulties, do not want to support costly family programs that they perceive to benefit others. On the other hand, the estimated costs of not having family programs may be higher. Disadvantaged children may eventually cost society more in unemployment compensation and incarceration expenses, for example, than would preventive investments in support of these children and their families.

A strength of the family ecology perspective is that it sensitizes us to significant politico-economic and socio-cultural issues that may not be addressed in other theories. A weakness is that the perspective is so broad and inclusive that virtually nothing is left out. As a result, it can be difficult for analysts to focus on specific causes for particular circumstances or problems.

The Family Development Perspective

Whereas family ecology analyzes the family and society (and even the physical-biological environment) as interdependent parts of a whole, **family development perspective** emphasizes the family itself as its unit of analysis. The concept of the *family life cycle* is central here. Typical stages of family life are marked off by (1) the addition or subtraction of family members (through birth, death, and leaving home), (2) the various stages the children go through, and (3) changes in the family's connection with other social institutions (retirement from work, for example). These stages of family development are termed the **family life cycle.** They succeed one another in an orderly progression and have their requisite developmental tasks. For example, newly established couples must make homes of their own. If the developmental tasks of one stage are not successfully completed, adjustment in the next stage will be more difficult.

Various versions of the stages of the family life cycle have been offered, but there is some convergence on a seven-stage model (Duvall 1957; Aldous 1978; Rodgers and White 1993). In this view, the *newly established couple* stage comes to an end through a dramatic change occasioned by the *arrival of the first baby.* Experts disagree on whether the transition to the child-bearing family can be termed a "crisis," but research shows that it does change couple dynamics, leisure patterns, sexual frequency, and labor-force participation.

Entry of the oldest child into school brings about changes in family life reflecting the need for parents of schoolchildren to coordinate schedules with another social institution (education). Here parents are faced with the task of helping their children meet the school's expectations. *Parents of secondary-school children* may be dealing with more complex problems in-

volving adolescent sexual activity or drug and alcohol abuse, for example. Children become increasingly expensive during this stage, while anticipated college costs add to parents' financial pressures.

Parents of young adults help their offspring enter into the adult world of employment and their own family formation. Later, *families in the middle years* return to a couple focus with (if they are fortunate!) the time and money to pursue leisure activities. Still later, aging families must adjust to *retirement* and perhaps health crises or debilitating chronic illness. *Death of a spouse* marks the end of the family life cycle.

The family development perspective emerged and prospered from the 1930s through the 1950s—an era in which the "natural" family was nuclear, two monogamous heterosexual parents and their children. Accordingly, the model assumed that family life follow certain conventional patterns: Couples marry, and marriage precedes parenthood; families are nuclear and reside independently from other relatives; all families have children; parents remain together for a lifetime. Of course today—and, in fact, in the past as well—many of us do not proceed so predictably along these well-marked paths.

Furthermore, critics have noted the white, middle-class bias of the family life cycle perspective (Hogan and Astone 1986). Due to economic, ethnic, and cultural differences, two families in the same life cycle stage can still be very different from one another in many respects. For these reasons, the perspective is less popular now than it was once. Meanwhile, family development theorists continue to see the model as useful (Mattessich and Hill 1987, p. 445), and some have modified it to recognize racial/ethnic and other social variations, such as child-free unions, single parenthood, or divorce (Rodgers and White 1993). In sum, family development theory can sensitize us to important family transitions and challenges. But its usefulness is limited by the implied assumptions that families are essentially similar, that they share a fairly traditional way of life, and that they change according to conventional time-of-life transitions.

The Structure–Functional Perspective

The **structure-functional perspective** sees the family as a social institution that performs certain essential functions for society. When social scientists use the term *institution,* they are not referring to a university, a

If a society is to persist beyond one generation, it is necessary that adults not only bear children but feed, clothe, and shelter them during their long years of dependency. Doing all this is an important function of families. Despite the fact that they are homeless, these parents are obviously trying to fulfill that function.

hospital, or a prison. In the social sciences, the term has an abstract meaning: **Social institutions** are patterned and predictable ways of thinking and behaving—beliefs, values, attitudes, and norms—that are organized around vital aspects of group life and serve essential social functions. In the structure-functional perspective (as well as others), the family is the institution, or structure commissioned to perform some very basic social functions. In preindustrial or traditional societies, the family structure was extended to involve whole kinship groups and performed most societal functions. In industrial or modern societies, the appropriate family structure is nuclear (husband, wife, children) and has lost many functions (Goode 1963). Nevertheless, in contemporary society, the family remains principally accountable for at least three important functions: to raise children responsibly, to provide economic support, and to give emotional security.

FAMILY FUNCTION 1: TO RAISE CHILDREN RESPONSIBLY   If a society is to persist beyond one generation, it is necessary that adults not only bear children but feed, clothe, and shelter them during their long years of dependency. Furthermore, a society needs new members who are properly trained in the ways of the culture and who will be dependable members of the group. All this requires that children be responsibly raised. Virtually every society places this essential task on the shoulders of families—either extended or nuclear. Accordingly, a related family function has traditionally been to control its members' sexual activity. Although there are several reasons for the social control of sexual activity, the most important one is to ensure that reproduction takes place under circumstances that guarantee the responsible care and socialization of children. Even though between one-quarter and one-third of U.S. births today are to single women, the universally approved locus of

In today's impersonal world, the family has grown more important as a source of emotional support. Family members cannot fill all of one another's emotional needs, but committed family relationships can and do offer important emotional security.

reproduction remains the married-couple family. But as Case Study 2.1, "The Family as a Child-Rearing Institution," illustrates, in the United States today the child-rearing function is often performed by divorced or never-married parents.

FAMILY FUNCTION 2: TO PROVIDE ECONOMIC SUPPORT   A second family function involves providing economic support. For much of history, the family was primarily an economic rather than an emotional unit. The family was the unit of economic production in societies with agriculture, craft, and earlier industrial or commercial economies (Shorter 1975; Stone 1980). Although the modern family is no longer a **self-sufficient economic unit,** virtually every family engages in activities aimed at providing for such practical needs as food, clothing, and shelter.

Family economic functions now consist of earning a living outside the home, pooling resources, and making consumption decisions together. In assisting one another economically, family members create some sense of physical security. For example, mates can offer each other a kind of unemployment insurance. They reciprocally ensure that if one is not employed—because of long illness, the inability to find work, the wish to change jobs, or to stay home with children—the other will support both of them. And family members care for each other in additional practical ways, such as nursing and transportation during illness. Chapter 13 explores how individuals integrate work and family today.

FAMILY FUNCTION 3: TO GIVE EMOTIONAL SECURITY   In today's world, the family has grown more important as a source of emotional security (Berger, Berger, and Kellner 1973; Popenoe 1993). This is not to say that families can solve all our longings for affection, companionship, and intimacy. They

The Family as a Child-Rearing Institution

Jo Ann is 38 and has been divorced for six years. She has five children. Gary, 19 and her oldest, lives with his father, Richard. At the time of this interview, Jo Ann and Richard and their children had recently begun family counseling. The purpose, Jo Ann explained, was to create for their children a more cooperative and supportive atmosphere. Jo Ann and Richard do not want to renew an intimate relationship, but they and their children are still in many ways a family filling traditional family functions.

We've been going to family counseling about twice a month now. The whole family goes—all five kids, Richard, and me. The counselor wants to have a videotaping session. He says it would help us to gain insights into how we act together. The two older girls didn't want any part of it, but the rest of us decided it might be really good for Joey to see how he acts.

[Joey's] the reason we're going in the first place. At the first counseling sessions, he sat with his coat over his head....

Joey's always been a problem. He's used to getting his own way. Some people want all the attention. They will do anything to get it. I guess I never knew how to deal with this.... He drives us nuts at home. He calls me and the girls names.... He was disrupting class and yelling at the teacher. And finally they expelled him.... Joey gets anger and frustration built up in him....

So I took him to a psychologist, and the psychologist said he'd like the whole family to come in, including Richard. Well, Richard still lives in this city and sees all the kids, so I asked him about it. And he said okay....

In between sessions, the counselor wants us to have family conferences with the seven of us together. One day I called Richard and asked him over for supper. In the back of my mind I thought maybe we could get this family conferencing started.

Well, after dinner Joey, our 11-year-old, started acting up. So I went for a walk with him. We must have walked a mile and a half, and Joey was angry the whole time. He told me I never listen to him; I never spent time with him. Then he started telling me about how he was mad at his dad because his dad won't listen to him.

He said his dad tells all these dumb jokes that are just so old, but he just keeps telling them and telling them. So when he got home, I saw Richard was still there, and I asked Joey, "Would you like to have a family conference? Maybe tell your dad some of the things that are bothering you?" And he said, "Could we?" ...

[During the conference] Joey talked first. Then everybody had a turn to say something. There was one time I was afraid it was going to get out of hand. Everybody was interrupting everybody else. But the counselor had told me you have to set up ground rules. This is where we learned that we've got some neat kids because when I said "Let somebody else talk," everybody did! So it went real well.... And then finally Richard said, "I think it's time for us to come to a conclusion." I said, "Well, you're right."

How might this divorced family differ from the same family before divorce? How is it the same? Even though Jo Ann and Richard's family is no longer intact, what functions does it continue to perform for its members? For society?

cannot. (Frequently, in fact, the family situation itself is a source of stress, as we'll discuss in Chapters 9 and 10.) Neither is it true that family members or intimate partners never experience loneliness—nor that they can fill all of one another's emotional needs. But families and committed intimate relationships can offer important emotional security. The broader family network—grandparents, grown sisters and brothers,

The broader family network, or extended family—grandparents, aunts and uncles—is often an important source of security. Extended families meet most needs in a traditional society—economic and material needs and reproduction and child care, for example—and they have strong bonds. In urban societies, specialized institutions, such as factories, schools, and public agencies, often meet practical needs. But extended families may also help each other materially in urban society, especially during crises.

aunts and uncles—is often an important source of emotional security. Family can mean having a place where you can be yourself, even sometimes your worst self, and still belong.

In anthropology, the structure-functional perspective calls attention to cross-cultural variation in family structure. Box 2.1, "Marriages and Families Across Cultures," gives some examples. As it dominated family sociology in the United States during the 1950s, however, the structure-functional perspective emphasized the heterosexual nuclear family as "normal," "good," or "functional" (Parsons and Bales 1955). Furthermore, the perspective gave much attention to arguing the functionality of specialized gender roles: the "instrumental" husband-father who makes the living and wields authority, and the "expressive" wife-mother whose main function is to enhance emotional relations at home (Parsons and Bales 1955). (Gender roles and issues are discussed in detail in Chapter 3.) Generally shared values would lead family members to perform these roles. Deviance from them would likely result in "family disorganization" as measured by divorce and juvenile delinquency. These views fit nicely in a post–World War II society, characterized by an expanding economy and widespread attitudes specifying husbands as providers and wives as homemakers. While many women had been gainfully employed during World War II as well as during the preceding Depression years, a return to "normal" after the war meant an emphasis on traditional husband–wife roles (Kingsbury and Scanzoni 1993).

Much of the current debate over family "decline" can be better understood once one has a grasp of the structure-functional perspective. Those who argue that the family is in decline generally come from this perspective and see the nuclear family as *the* structure for performing necessary social functions (Popenoe 1993). Their opponents refuse to view the nuclear family as "normal," "natural," or best. Both sides would agree, however, that many families in our contemporary society do depart from the nuclear family structure. As you can see in Figure 2.2, only 26 percent of

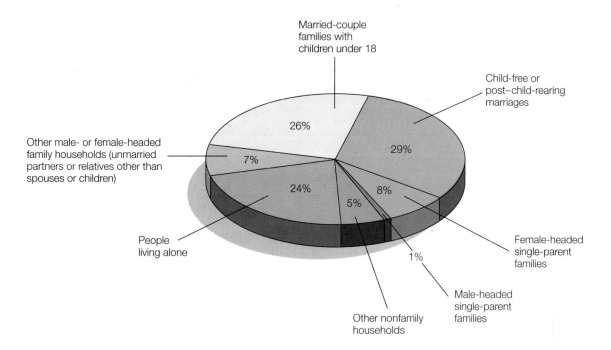

Married-couple families with children under 18 — 26%

Child-free or post–child-rearing marriages — 29%

Other male- or female-headed family households (unmarried partners or relatives other than spouses or children) — 7%

People living alone — 24%

Other nonfamily households — 5%

Male-headed single-parent families — 1%

Female-headed single-parent families — 8%

FIGURE 2.2

I The many kinds of American households, 1994. (Source: U.S. Bureau of the Census 1995, Table 66)

American households in 1994 took the form of the typical nuclear family (married-couple families with children under 18).

The structure-functional perspective has been criticized for giving us an image of smoothly working families characterized by shared values while overlooking such issues as power disadvantages, spousal or parent–child conflict, and even family violence. The perspective has been further criticized because it generally fails to recognize that what is functional for one group or category of people may not be so for others. For instance, structure-functionalism does not take into consideration racial/ethnic or class variation in family structures. Nonetheless, virtually all social scientists assume the one basic premise underlying structure-functionalism: that families comprise an important social institution (however they might be defined or structured) with essential social functions to be performed.

The Interactionist Perspective

Unlike the three theoretical orientations that we have already described, the **interactionist perspective** looks within families, at internal family dynamics. This point of view explores the interaction of family members—that is, the back and forth talk, gestures, and actions that go on in families. Members respond to what other members say (verbally or nonverbally) and do. These interchanges take on a reality of their own; they construct or create a family. Put another way, something called "family" emerges from the relationships and interactions among family members. Unlike structure-functionalism, which posits a standard family form, the interactionist perspective refuses to identify a "natural" family structure. The family is not a stock social unit but the creation of its participants as they spontaneously relate to one another.

Based on the work of Charles Cooley (1909) and George Herbert Mead (1934), the interactionist perspective was the principal theoretical orientation in sociology during the 1920s and early 1930s when the family studies field was establishing itself as a legitimate social science. The orientation remains a popular and fruitful one (LaRossa and Reitzes 1993). Interactionists are keenly interested in the two concepts,

Marriages and Families Across Cultures

Some features of family life that we take for granted do not necessarily occur in all cultures. This box describes cross-cultural variation in family life. Some of the societies whose practices are reported here are small, whereas others are from among the world's major nations. Nevertheless, they illustrate the wide range of family practices societies consider normal. How many of the following six statements have you assumed to be characteristic of what is normal or universal?

1. *Marriage is monogamous.* When we think of marriage we think of **monogamy,** the sexually exclusive union of one man and one woman. Actually, only 20 percent of the world's cultures insist on monogamous marriage. Many societies permit **polygamy** (a person's having more than one spouse) and practice both polygamy and monogamy. In some societies adultery is accepted, though often limited to certain classes of partners.

2. *Young people select their own marriage partners.* Few traditional societies allow young people to choose partners without the approval of parents or other relatives. Often, though not always, the child's preference is deferred to, at least somewhat.

3. *Love is universal and provides the basis for marriage.* Many Americans believe that marriage should be the outcome of falling in love, but love is not considered a reason for marriage in all cultures. The notion of romantic love seems to be a phenomenon peculiar to Western culture or to societies influenced by Western values.

4. *The nuclear family is a separate economic unit.* We assume that members of a nuclear family—husband, wife, and children—share income, property, and expenses. They buy necessities together as a unit. In our society, one or more adults work to support the nuclear family. Other relatives or adult children are usually not included in this economic pool. In other societies, the economic unit is the extended

"self" and "identity." They assume that an individual develops concepts of self (**self-concept:** the basic feelings people have about themselves, their abilities, and their worth) and **identity** (a sense of inner sameness developed by individuals throughout their lives) through social interaction. The self, in turn, is able to assess and assign meaning and value to ongoing family activities. Families shape the identities and self-concepts of all their members, including adults. Family identities and traditions emerge through interaction as the family defines itself with the growth of relationships and the creation of rituals (Bossard and Boll 1943; Fiese et al. 1993).

Thinking about families this way leads interactionists to investigate questions like the following: How do two separate individuals interact in a marriage or otherwise committed partnership to fashion a couple identity (Berger and Kellner 1970)? Conversely, how is the couple identity dismantled through divorce (Vaughan 1986)? How do families define the appropriateness of feelings (Hochschild 1983)? How do family members communicate intimacy? How are family roles constructed and learned? By what processes exactly do parents socialize children—and children, their parents? What mechanisms underlie family power dynamics? By what process do family members arrive at more or less shared goals, beliefs, values, and norms (LaRossa and Reitzes 1993)? If, as structure-functionalists assume, family members share common values, interactionists would want to know how this situation develops.

How do geography, race/ethnicity, class, gender, or age relate to family dynamics? For example, how do American families differ from Asian or African fami-

family. In an **extended family,** three or more generations, including, perhaps, relatives besides parents and children, share work and resources. In still other cases, the wife and husband may each be economically self-sufficient. Among the Hopi Indians, for example, it is the wife who owns farmland, orchards, livestock, and water holes. They are not joint property.

5. *The best way to raise children is with two married parents in the home.* Some researchers argue that societies with either extended or polygamous family forms may even do a better job of rearing children because more adults are involved.

6. *Premarital conception is never condoned.* Not all societies condemn premarital intercourse or conception. In some societies, a couple doesn't marry or isn't formally considered to be married until after the woman becomes pregnant.

In what ways are expectations concerning marriage in other cultures more—and less—realistic than ours? Would you like to live where you couldn't choose your own marriage partner? What possible conflicts do you think might arise when families immigrate to the United States from cultures, like the ones described here, that are very different from ours?

For additional examples of cross-cultural variations in family life, see Stuart A. Queen, Robert W. Habenstein, and Jill Quadagno (eds.), *The Family in Various Cultures,* 5th ed. (New York: Harper & Row, 1985); and Bron B. Ingoldsby and Suzanna Smith, *Families in Multicultural Perspective* (New York: The Guilford Press, 1995).

Marriage and family practices and forms differ from culture to culture, although the ceremony and significance of marriage is important to most peoples. The Hindu marriage ceremony, like this one performed in New York, adheres to its own distinct rites. In a multicultural and pluralistic society like the United States, ethnic groups retain many of their traditions.

lies? How is family interaction distinctive in a Muslim society that gives legal authority to *sharia,* the Islamic code? As another example within our own society, how are rural families in poverty dissimilar to their urban counterparts? How do families change—or avoid changing—in the wake of general social upheaval or more personal crisis (LaRossa and Reitzes 1993)?

An often-voiced criticism of the interactionist perspective is that (similar to family ecology) interactionism makes intuitive sense but is difficult to test empirically. A related criticism is that, because it is qualitative and relatively subjective, the research connected with the interactional perspective lacks rigor. Perhaps a more serious criticism is that interactionism overestimates the power of individuals to create their own realities, ignoring the extent to which humans inhabit a world *not* of their own making.

Exchange Theory

Exchange theory grew out of the application of an economic perspective to social relationships, beginning around 1960 and flourishing during the 1960s and 1970s. This orientation focuses on how individuals' various personal resources affect their relative positions in families or other groups. The basic premise here is that people use their resources to bargain and secure advantage in relationships. Exchange theorists pay attention to the exchange of rewards and costs between participants in a relationship or family unit. Such transactions form and stabilize a relationship or group.

Relationships based on equal or equitable (fair, if not actually equal) exchanges thrive, whereas those in which the exchange balance is one-sided eventually dissolve. Whether an unhappy couple divorces or remains married can be analyzed in exchange terms.

Exchange theory must fight the human tendency to see family relationships in far more romantic and emotional terms. Yet, dating relationships, marriage and other committed partnerships, divorce, and even parent–child relationships show signs of being influenced by the relative assets of the parties. Money is power, and the children of wealthier parents are more likely to share their parents' values. Marriages tend to take place between people of equal status (see Chapter 7). Decision making within a marriage, as well as decisions to divorce, are affected by the relative resources of the spouses. People without resources or alternatives to a relationship defer to the preferences of others and are less likely to leave it (see Chapter 10).

Family Systems Theory

Family systems theory, like interactionism, is an umbrella term for a wide range of specific theories. Growing out of psychotherapy, this theoretical framework looks at the family as a whole. Even though they are like interactionism in being concerned with the interaction among family members, systems theories have some unique concepts and propositions.

Systems theory uses the model of—a system! Like an organic system (the body), a mechanical system, or especially like a cybernetic system (a computer), the parts of a family make a whole that is more than the sum of the parts (Laing 1971).

A family functions regularly in a certain way; emotional expression and behavior of family members tend to persist. Put another way, systems tend toward equilibrium. Like a computer program directing a space vehicle, information about behavior provides feedback to the system, which then adjusts itself. Change in the external environment or in one of the internal parts sets in motion a process to restore equilibrium. What this means in a family is that there is pressure on a changing family member to revert to his or her original behavior within the family system. For change to occur, the family system as a whole must change. Indeed, that is the goal of family therapy based on systems theory.

Without therapeutic intervention, families may replicate problem behaviors over the generations. Similarly, it might be unrealistic for a spouse to attempt to resolve marital difficulties by leaving the family (by divorcing). An individual may recreate or enter a family system similar to the one she or he left. Then too, family systems can survive divorce.

These folks are waiting patiently for medical attention in a neighborhood clinic. How might scholars from different theoretical orientations see this photograph? Family ecologists might remark on the quality of the facilities and how this affects family health and relations. Scholars from the family development perspective would likely note that this woman is in the child-rearing stage of her family life cycle. Structure-functionalists would be quick to note the child-raising (and, perhaps, expressive) function(s) that this woman is performing for society. Interactionists would be more inclined to explore the mother's body language: What is she saying nonverbally to the child on her lap? What is he saying to her? Exchange theorists might speculate about this woman's personal power and resources relative to others in her family. Family system theorists might point out that this mother and child are part of a family system: Should one leave or become seriously and chronically ill, for example, the roles and relationships in the entire family would change and adapt as a result. Feminist theorists might point out that typically it is mothers, not fathers, who are primarily responsible for their children's health—and ask why.

Systems theory and family therapy overlap. But social scientists have moved systems theory away from its therapeutic origins to use it in a more general analysis of families. They have been especially interested in how family systems handle information, deal with problems, respond to crises, and regulate contact with the outside world. Boundaries of the family, as well as closeness or distance of family members from one another, are important issues.

Criticism of the family systems perspective relates to its nonspecificity: So the family is a system; then what? In working concretely with families in therapy, it can be very useful to both therapists and clients. If family members come to understand how their family system operates, they can use this knowledge to achieve desired goals.

Systems theory does not take note of social structure (the class system or race/ethnicity), but presumes instead that families are families the world over. Transactions with a particular family's external world of work, school, religious affiliation, and extended family may be addressed, but the structure of economic opportunity and other features of the larger society are not analyzed. Systems theory tends to diffuse responsibility for conflict, by attributing dysfunction to the system. This makes it difficult to extend social support to victimized family members while establishing legal accountability for others, as in incest or other domestic violence (Stewart 1984).

But systems theory often gives family members insight into the effects of their behavior and is a good analytic tool. It can make visible the hidden benefits or costs of certain family patterns. For example, doctors were puzzled by the fact that death rates were higher among kidney dialysis patients with supportive families. Family systems theorists attributed the higher rates to the unspoken desire of the patients to lift the burden of care from the close-knit family they loved (Reiss, Gonzalez, and Kramer 1986).

The Conflict/Feminist Perspectives

We like to think of families as being supportive. We do not like to think of cost-conscious family behavior (exchange theory) or of dysfunctional families (systems theory). For years, sociologists talked about how traditional family roles were functional for society, ignoring in the process the politics of gender. Interactionism ignored social conflict in the larger society to concentrate on the inner workings of families.

The conflict/feminist perspectives bring latent family and social conflict out into the open. A first way of thinking about the **conflict perspective** is that its position is the opposite of functional theory: Not all of a family's practices are good; not all family behaviors contribute to family well-being; what is good for one family member is not necessarily good for another. Family interaction can include domestic violence as well as holiday rituals—sometimes both on the same day.

In the 1960s and subsequently, the application of the conflict perspective to families permitted us to see some things about families that had been overlooked before. Social scientists discovered child abuse, wife abuse, marital rape, husband abuse, elder abuse, child sexual abuse, parent abuse, and sibling abuse, for instance.[1]

The conflict perspective calls attention to unequal power within groups or larger societies. This orientation traces its intellectual roots to Karl Marx, who analyzed disparate power in terms of wealth, or capital. Marx viewed industrialized workers, because they were required to sell their own labor, as fundamentally exploited by industry-owning capitalists. When applied to families (e.g., Engels 1942 [1884]), the conflict perspective has emphasized the sex–gender system. Feminist theories are conflict theories. Although there are many variations within the **feminist perspective,** this theory focuses on how male dominance in the family and society is oppressive to women.

Unlike the perspectives already described, which emerged primarily among academic researchers, feminist theories developed from political and social movements over the past thirty years. As a result, the feminist perspective is fairly unique: Its mission is using knowledge to confront and end the oppression of women and related patterns of subordination based on social class, race/ethnicity, age, or sexual orientation. Feminist theorizing has contributed to political action regarding families in the following ways, among others:

1. Changes in policies that economically weaken households headed by women (for example, efforts to end gender and race discrimination in wages)

---

1. Actually, a physician "discovered" child abuse. He published an article on "battered child syndrome" based on the hidden injuries to children revealed by X rays (Kempe et al. 1962). Social scientists then pursued their interest in child abuse and other forms of domestic violence. The discovery of child abuse was thus somewhat like Columbus' discovery of America in that each was always there but had not been noticed by authorities or academics.

Studying Ethnic Minority Families

Just a few decades ago, white, middle-class (and male) scholars conducted the preponderance of theorizing and research on families. But beginning about thirty years ago, men and women from diverse ethnic backgrounds began to study families scientifically. One of the first contributions they made was to point out how limited, and hence biased, our theoretical and research perspectives have been. In this chapter, for example, we note the white, middle-class bias in family development theory. (And it becomes obvious from our text discussion that the structure-functional perspective can be seen as biased against women.)

To begin to understand this question of bias, we might think of theory and research on ethnic minority families as falling into one of three frameworks: cultural equivalent, cultural deviant, and cultural variant (Allen 1978). The **cultural equivalent** approach emphasizes those features that minority families have in common with mainstream white families. An example would be the finding that middle-class black parents treat their children much the same way as do middle-class white parents (Taylor et al. 1991). The **cultural deviant** approach views the qualities that distinguish minority families from mainstream

2. Changes in laws that reinforce the privileges of men and of heterosexual nuclear families versus other family types (for example, divorce laws that disadvantage women economically or laws that exclude nontraditionally defined families of economic and legal supports offered to marrieds)

3. Efforts to stop sexual harassment and sexual and physical violence against women and children

4. Advances in securing women's reproductive freedom (for example, through abortion rights)

5. New recognition and support for women's unpaid work, by involving men more fully in housework and child care and by efforts to fund quality day care and paid parental leaves

6. Transformations in family therapy so that counselors recognize the reality of gender inequality in family life and treat women's concerns with respect (Goldner 1993)

Conflict theory is difficult to accept for those in privileged categories. For some social scientists, it is too political, too value-laden, too tied to advocacy for social change. One form of conflict theory, critical theory, has radical social change as a specific goal. For those who find radical change to be unlikely, conflict

theory is too utopian. They would prefer an analysis that helps families adapt to incremental change and cope effectively with current problems. And for some scholars the categories selected in conflict analysis are too vague and ahistorical. For example, patriarchy, found in all societies from primitive to postindustrial, seems to lose its meaning as an analytic category. Is there really no difference between America in the 1990s and ancient Rome, where husbands had life-and-death power over wives?

There Is No Typical Family

It is important to emphasize as we close this section that diversity exists not only among family theorists but also among families themselves. Despite its functions, the structure of the family institution is remarkably flexible. Until recently, Judeo-Christian tradition, the law, and societal attitudes converged in a fairly common expectation about what form the American family should take. Over the past three decades, however, the predominance of this pattern has disappeared. *There is no typical American family today.*

Our response to this change in American families is the stuff of talk radio, serious political debate, and academic analysis. We, the authors, have worked to balance in this text an appreciation for flexibility and

families as negative or pathological. An example would be analysis that laments the high incidence of female-headed families among Hispanics (22 percent) when compared with non-Hispanic whites (13 percent) ("We the American Hispanics" 1993).

The **cultural variant** approach calls for making culturally and contextually relevant interpretations of minority family lives. A classic example is Carol Stack's (1974) participant observation in a black community; her resulting book, *All Our Kin,* emphasized previously ignored strengths in extended African American families. In the first two approaches, white mainstream families are considered the standard against which "other" or "less valid" families are compared, either favorably or unfavorably—a situation conducive to bias. In the third and preferred approach, minority families are studied on their own terms, and comparisons are made *within* those groups (for example, comparing urban Native American parenting styles with those of American Indians on reservations).

Besides pointing out bias in theory and research, some scholars have introduced new, more culturally relevant, theoretical concepts. One of these is the **kin scripts framework** for studying ethnic minority families. This theoretical framework includes three culturally relevant family concepts: kin-work, kin-time, and kin-scription. *Kin-work* is the labor families must accomplish in order to survive. *Kin-time* refers to family norms concerning the temporal and sequential nature of such transitions as parenthood, marriage, and grandparenthood. *Kin-scription* is the active recruitment of family members to do kin-work. Besides helping to make family theory and research less biased and more relevant to minority families, thinking like this is potentially an example of reverse theorizing. That is, although the kin scripts framework was derived from research on black extended families, the concepts might be used to study nonminority mainstream families as well (Dilworth-Anderson, Burton, and Johnson 1993).

diversity in family structure and relations—and for freedom of choice—with the increased concern of many social scientists about what they see as diminished marital and child-rearing commitment. Still, in varied circumstances, chosen or not, we find that families are made of people: committed people who are married or in otherwise committed intimate relationships, people trying to raise their children or contribute to the rearing of others' children, and people providing material and emotional support to each other in relationships they identify as "family."

Studying Families

The great variation in family forms and the variety of social settings for family life mean that few of us can rely on firsthand experience alone in studying the family. Although we "know" about the family because we have lived in one, our **experiential reality**—beliefs we have about the family—may not be accurate. We may also be misled by media images and common sense—what everybody knows. Everybody knows, for example, that the American family is a nuclear one—or else that it is dying. This **agreement reality**—what members of a society agree is true—may misrepresent the actual experience of families.

We turn now to consider the difficulties inherent in studying the family and then to a presentation of various methods used by social scientists. Although imperfect, the methods of scientific inquiry can bring us a clearer knowledge of the family than either personal experience or speculation based on media images. Scientific methods represent a form of agreement reality that sets special standards for the acceptance of statements about the family.

The Blinders of Personal Experience

Most people grow up in some form of family and know something about what marriages and families are. But while personal experience provides us with information, it can also act as blinders. We assume that our own family is normal or typical. If you grew up in a large family, for example, in which a grandparent or an aunt or uncle shared your home, you probably assumed (for a short time at least) that everyone had a big family. Perceptions like this are usually outgrown at an early age, but there may be more subtle, not-yet-apparent differences between your family experiences and those of others. For instance, the members of your family may spend a lot of time alone, perhaps reading, whereas in other families it may be cause for alarm if a

Traditional Muslim families follow the Islamic code of sharia. In this photo of a Saudi family, the daughters' Western dress and uncovered hair suggest that cultural beliefs and norms concerning women's dress and behavior have been privately negotiated within this family—a conclusion that reflects the interactional perspective.

family member—adult or child—is not talking with others around the kitchen table.

Personal experience, then, can make us believe that most people's family lives are similar to our own when often this is not the case. We may be very committed to the view of family life shaped by our experiences and our own choices: "The search for regularities and generalized understanding is not a trivial intellectual exercise. It critically affects our personal lives" (Babbie 1992, p. 25).

In looking at marriage and family customs around the world (see Box 2.1), we can easily see the error of assuming that all marriage and family practices are like our own. But not only do the traditional American assumptions about family life not hold true in other places, they also frequently don't even describe our own society. **Heterosexism,** the tendency to see heterosexual or straight families as the standard, can lead us to ignore lesbian or gay male families. Black, His-

panic, and Asian families; Jewish, Protestant, Catholic, Mormon, Islamic, Buddhist, and nonreligious families; upper-class, middle-class, and lower-class families all represent some subtle (often not-so-subtle) differences in lifestyle. However, the tendency to use the most familiar yardstick for measuring things is a strong one; even social scientists fall victim to it. (See Box 2.2 on pages 40–41, "Studying Ethnic Minority Families," for a look at how social scientists have begun to address the problems of cultural bias.)

Scientific Investigation: Removing Blinders

Seeing beyond our personal experience involves learning what kinds of families other people are experiencing, and with what consequences. To do this, we rely on data gathered systematically from many sources through techniques of **scientific investigation,** from which it often is possible to generalize. The tech-

Personal experience gives us some information about families but it can also fit us with "blinders": We don't see what's going on in others' families. This extended African American family is celebrating Kwanzaa, derived from African tradition and observed at the end of December. An estimated 10 million black Americans now celebrate Kwanzaa as a ritual of family, roots and community.

niques—surveys, laboratory observation and experiments, naturalistic observation, case studies, longitudinal studies, and historical and cross-cultural data—will be referred to throughout this text, so we will briefly describe them now.

SURVEYS   **Surveys** are part of our everyday experience. When conducting scientific surveys, researchers either engage in face-to-face or telephone interviews or distribute questionnaires to be answered and returned. Questions are often structured so that after a statement, such as "I like to go places with my partner," the respondent has answers from which to choose. The possible responses might be: Always, Usually, Sometimes, Not very often, Never. Researchers spend much energy and time wording such *close-ended* questions so that, as much as possible, all respondents will interpret questions in the same way.

Interview questions can also be *open-ended*. For example, the question might be, "How do you feel about going places with your partner?" or "Tell me

about going places with your partner." Many social scientists and respondents alike believe that open-ended questions give those answering more opportunity to express how they really feel or what they actually believe than do more structured survey questions, which require people to choose from a predetermined set of responses.

Once the returns are in, survey responses are tallied and statistically analyzed, usually with computers. After the survey data have been analyzed, the scientists conducting the research begin to draw conclusions about the respondents' attitudes and feelings. Then the scientists must decide to whom their conclusions are applicable. Do they apply *only* to those whom they interviewed directly, for example, or to other people similar to the respondents? In order to ensure, moreover, that their conclusions can be *generalized* (applied to people other than those directly questioned), survey researchers do their best to ensure that their respondents constitute a *representative sample* of the people they intend to draw conclusions about. Popular magazine surveys, for example, are seldom representative of the total American public. A survey on attitudes about premarital sex in *Cosmopolitan* or *Playboy* will likely yield different findings than one in *Family Circle* or *Reader's Digest*. It would not be scientifically accurate to generalize from a set of conclusions drawn from a *Vanity Fair* survey about how American adults feel about premarital sex. In the same way, results from a survey in which all respondents are white, middle-class college students cannot be considered representative of Americans in general. Researchers and political pollsters may instead use random samples, in which households or individuals are randomly selected from a comprehensive list (see Babbie 1995 for a more detailed discussion). A random sample is considered to be representative of the population from which it is drawn. A national random sample of approximately 1,500 people can validly represent the U.S. population.

Survey research has certain advantages over other inquiry techniques. The main advantage is uniformity. Presumably, all respondents are asked exactly the same questions in the same way. Also, surveys are relatively efficient means of gathering large amounts of information. And, provided the sample is designed to be accurately representative, conclusions drawn from that information can be applied to a large number of people.

Surveys have disadvantages, too. Because they ask uniform or standardized questions, surveys may miss points respondents consider important. Surveys neither tell us about the context in which a question is answered nor guarantee that in a real-life situation a person will act in a manner consistent with the answer given to the interviewer (Babbie 1995).

Other disadvantages of surveys result from respondents' tendency to say what they think they *should* say, rather than what they in fact believe. Social scientists refer to this problem as the tendency of respondents to give *normative answers*. If asked whether or how often physical abuse occurs in the home, for example, those who often engage in family violence might be especially reluctant to say so.

Another disadvantage of surveys, closely akin to the tendency of respondents to say what they think they should, is the tendency of respondents to forget or to reinterpret what happened in the past. (Because of this, social scientists recognize the value of longitudinal studies—studies in which the same group of re-

spondents is surveyed or interviewed intermittently over a period of years.) For this reason, most researchers agree that asking about attitudes or events that occurred in the past seldom yields valid results. Another disadvantage of surveys, again closely related to respondents' tendency to give normative answers, is a problem specific to interviewer surveys. Depending on the sex, age, race, and style of the person questioning you, you might tend, even without knowing it, to give a certain response. Scientists call this tendency *interview effects of an interviewer*. Certain characteristics of the interviewer—age, sex, race, clothing style, hair length—may tend to elicit a certain kind of response. A 45-year-old male, for example, asked about his experience concerning sexual impotence, can be expected to reply differently to a male interviewer of 20 than he would to one of 60—or to a female of any age. Perhaps the most significant disadvantage of every research survey is that what respondents say may not accurately reflect what they do.

LABORATORY OBSERVATION AND EXPERIMENTS

Because of the relative ease of conducting surveys, the flexibility of the format, and the availability of samples (even a classroom sample can provide useful information), surveys have been the primary source of information about family living. Other techniques are also used, however. In a laboratory observation or **experiment,** behaviors are carefully monitored or measured under controlled conditions. These methods are particularly useful in measuring physiological changes associated with anger, fear, sexual response (as discussed in Chapter 5), or behavior that is difficult to report verbally. In family problem solving, for example, families may be asked to discuss a hypothetical case or to play a game, while their behavior is observed and recorded.

In an experiment, subjects from a pool of similar participants will be randomly assigned to groups (experimental and control groups) that will be given different experiences (treatments). Families whose child is undergoing a bone marrow transplant may be asked to participate in an experiment to determine how they can best be helped to cope with the situation. One group of families may be assigned to a support group in which the expression of feelings, even negative ones, is encouraged (Experimental Group 1). Another set of families may be assigned to a group in which the emphasis is on providing factual information about transplantation and maintaining a positive, cheerful out-look (Experimental Group 2). A third group of families may receive no special intervention (Control Group).

If at the conclusion of the experiment the groups differ in attitudes and behavior according to some explicit measures of coping behavior, mental health, and family functioning, then this outcome is presumed to be a result of the experimental treatment. Put another way, because no other differences are presumed to exist among the randomly assigned groups, the results of the experiment provide evidence of the effects of the therapeutic interventions.

A true experiment has these features of random assignment and experimental manipulation of the important variable. Laboratory observation, on the other hand, simply means that behavior is *observed* in a laboratory setting, but it does not involve random assignment or experimental manipulation of a variable.

The experiment just described takes place in a field (real-life) setting, but experiments are often conducted in a laboratory setting because the researcher has more control over what will happen. He or she has more chance to plan the activities, measure the results, determine who is involved, and eliminate outside influences. In the previous example, families might have obtained additional personal counseling on their own, which would affect the results, or they may have attended the group only infrequently.

Experiments, like surveys, have both advantages and disadvantages. One advantage of experiments is that social scientists can observe human behavior directly, rather than depending, as they do in surveys, on what respondents *tell* them about what they think or do. The experimenter can control the experience of the subjects and can ensure, to some extent, the initial similarity of subjects in the two groups. A disadvantage of this research technique is that the behaviors being observed often take place in an artificial situation, and whether an artificial or simulated testing situation is analogous to real life is virtually always debatable. A family asked to solve a hypothetical problem through group discussion may behave very differently in a formal laboratory experiment than they would at home around the kitchen table discussing a real problem.

The fact that in social research the subject pool is often drawn from college classrooms and is not very representative of the general population is another limitation. Sometimes "volunteer" participants are re-ally draftees or subjects who have responded to financial incentives; in such cases the authenticity of results is often in question. And volunteer subjects may be very different from their peers who decide not to participate.

NATURALISTIC OBSERVATION   Of course, many aspects of human behavior and interaction just don't lend themselves to study in laboratory settings, so social scientists use another technique in an attempt to overcome virtually all artificiality (or as much as possible). In **naturalistic observation,** the researcher lives with a family or social group or spends extensive time with family or group members, carefully recording their activities, conversations, gestures, and other aspects of everyday life. Sometimes the researcher is a member of the family. The researcher attempts to discern family interrelationships and communication patterns and to draw implications and conclusions from them for understanding family behavior in general.

The principal advantage of naturalistic observation is that it allows us to view family behavior as it actually happens in its own natural—as opposed to artificial—setting. The most significant disadvantage of this tool is that findings and conclusions may be highly subjective. That is, what is recorded, analyzed, and assumed to be accurate depends on what one or very few observers think is significant. Another drawback to naturalistic observation is that it requires enormous amounts of time to observe only a few families. And these families may not be representative of family living in general. Perhaps because of these disadvantages, relatively few studies use this technique.

Still, Fred Davis' (1991) study of the families of polio victims, first published in 1963, remains an important piece of observational research that provides useful insights into family dynamics. It may become even more relevant because as medical science has enabled more and more children to survive serious illness, they and their families will live with crisis and chronic disease for an extended period.

An especially important use of naturalistic observation has been in research on racial/ethnic communities or in other settings not easily accessible to survey or experimental research. Studies such as anthropologist Carol Stack's (1974) participant observation of families in a lower-class black community and sociologist Judith Stacey's (1990) study of families in Silicon Valley give us the context in which families live their lives.

They often reveal that families play an active role in using the resources of their social environment to shape their destinies.

CLINICIANS' CASE STUDIES   A fourth way that we get information about families is from **case studies** compiled by clinicians—psychologists, psychiatrists, marriage counselors, and social workers who counsel people with marital and family problems. As they see individuals, couples, or whole families over a period of time, these counselors become acquainted with communication patterns and other interactions within families. Clinicians offer us knowledge about family behavior and attitudes by describing cases to us or by telling us about their conclusions based on a series of cases.

The advantages of case studies are the vivid detail and realistic flavor that enable us to experience vicariously the family life of others. The insights of clinicians can be helpful.

But case studies also have weaknesses. There is always a subjective or personal element in the way the clinician views the family. Inevitably, any one person has a limited viewpoint. Clinicians' professional training may also lead them to over- or underemphasize certain aspects of family life and to see family behavior in a certain way. For example, as a group, psychiatrists used to assume that the assertiveness or career interests of women caused the development of marital and sexual problems.

Furthermore, people who present themselves for counseling may differ in important ways from those who do not. Most obviously, they may have more problems. For example, throughout the fifties psychiatrists reported that gays and lesbians in therapy had many emotional difficulties. Subsequent studies of gay males not in therapy concluded that gays were no more likely to have mental health problems than were heterosexuals.

LONGITUDINAL STUDIES   **Longitudinal studies** provide long-term information about individuals or groups, as a researcher or research group conducts follow-up investigations (by means of interviews or questionnaires), most often for several years after the initial study. Observational or experimental studies could be repeated, but this is rarely done.

Booth and White's ongoing research, in which adults who were married at the time of the first survey in 1980 were reinterviewed three more times during

the 1980s and 1990s, is a good example of longitudinal research. The researchers traced the demographic and relationship patterns affecting marital quality and stability, divorce, and remarriage (Booth, Johnson, and White 1984; White and Booth 1985a, b; White, Booth, and Edwards 1986; Booth and Johnson 1988; White and Keith 1990; White and Booth 1991).

A study in which the same community, though not precisely the same individuals, is resurveyed would also be considered a longitudinal study. The city of "Middletown" (Muncie, Indiana) was restudied by a research team (Caplow et al. 1982) some fifty years after the initial community study (Lynd and Lynd 1929).

A difficulty encountered in longitudinal studies, besides the almost prohibitive cost, is the frequent loss of subjects due to death, emigration, or loss of interest. Social change occurring over a long period of time can make it difficult to ascertain what, precisely, has influenced family change. Yet cross-sectional data (one-time comparison of different groups) cannot show change in the same individuals over time.

HISTORICAL AND CROSS-CULTURAL DATA
The Middletown study of family life reached back into the nineteenth century through the use of historical records, a research approach that is becoming more and more common in the study of the family. Some interesting work by social historians in France (Ariès 1962) and in England (Laslett 1971) began to attract the interest of sociologists about thirty years ago, affecting current generations of scholars and teachers.

Historical research, whether done by historians or sociologists, has had a powerful influence on the study of the family. Zelizer's (1985) study of insurance documents and other historical materials conveys the changing status of the child from economic asset to emotional asset. Linda Gordon's (1988) research using social agency files from the early twentieth century reveals how lower- and working-class women used social agencies to cope with domestic violence, child sexual abuse, and other family problems.

Drawbacks to historical research are the unevenness and unavailability of data. Scholars must rely on only those data to which they have access. Typically, the upper classes, who had both the leisure and resources to record their activities, are overrepresented. But historical scholars have been very creative. Hanawalt (1986) constructed a rich picture of the medieval family from an examination of death records. Demographic and economic data and legal records, among the most reliable sources, are especially useful for analyses of the family institution (see Glendon 1989). Scholars are on less solid footing in describing intimate family matters because they must rely on materials such as individuals' diaries, which may not be representative of the period.

Sociologists, especially those who place more emphasis on cross-cultural comparison than we are able to do in this text, continue to look to anthropological fieldwork for information on family life and structure in societies in both developed and developing nations. Many make use of anthropological data that have been compiled in a systematic way, such as the Human Relations Area Files, in which various features of a multitude of societies are described in a standard format. Since the former Soviet Union fairly recently became open to social scientists from around the world, scholars have analyzed families in Russia (Maddock et al. 1993).

The Application of Scientific Techniques

All research tools represent a compromise. Each has its special strengths and weaknesses. The strengths of one research tool, however, can make up for the weaknesses of another. Findings that result from direct observations, for example, supplement survey reports in an important way. Whereas the former allow scientists to observe actual behavior among a limited number of people, surveys provide information about attitudes and reported behavior of a vast number of people. To get around the drawbacks of each technique, social scientists may combine two or more tools in their research. Ideally, a number of scientists examine one topic by several different methods. In general, the scientific conclusions in this text result from many studies and from various and complementary research tools. Despite the drawbacks and occasional blinders, the total body of information available from sociological, psychological, and counseling literature provides a reasonably accurate portrayal of marriage and family life today.

In Sum

Different theoretical perspectives—family ecology, family development, structural-functional, interactionist, exchange, family systems, and conflict/feminist—illuminate various features of families. Just as there is no one correct family theory, there is no typical American family today.

Structure-functionalism leads us to see families as a social institution commissioned to perform basic functions for society—responsible child rearing, economic support, and emotional security.

How do we know what families are like? We can call upon personal opinion and experience for the beginning of an answer to this question. But everyone's personal experience is limited. Scientific investigation—with its various methodological techniques—is designed to provide a more effective way of gathering knowledge about the family.

Key Terms

agreement reality	heterosexism
case studies	identity
conflict perspective	interactionist perspective
cultural deviant	kin scripts framework
cultural equivalent	longitudinal studies
cultural variant	monogamy
exchange theory	naturalistic observation
experiential reality	polygamy
experiment	scientific investigation
extended family	self-concept
family development	self-sufficient economic
perspective	unit
family ecology perspective	social institutions
family life cycle	structure-functional
family policy	perspective
family systems theory	surveys
feminist perspective	theoretical perspective

Study Questions

1. Choose one of the major theoretical perspectives on the family and try to explain how you might use it to understand something about life in your family.

2. Choose a magazine photo and analyze its content from a structure-functional perspective. (*Hint:* Together the people in the photo can constitute the group under analysis, with each person meeting certain of the group's needs, or functional requisites.) Then analyze it from one or more other perspectives. How do your insights differ?

3. Why is the family a major social institution? Does your family fulfill each of the functions identified in the text? How?

4. How is your family structured? That is, is it a traditional nuclear family, or does it represent the diversity of American families? How would you apply the current debate over family values to your own family experience?

5. Review the techniques of scientific investigation, and discuss why science is often considered a better way to gain knowledge than is personal experience alone. When might this not be the case?

Suggested Readings

Acock, Alan C. 1994. *Family Diversity and Well-Being*. Thousand Oaks, CA: Sage. A social scientific analysis of family diversity in the United States today that employs the family ecology perspective.

Babbie, Earl. 1995. *The Practice of Social Research*. 7th ed. Belmont, CA: Wadsworth. Respected textbook in sociological methods. Recommended for those who would like additional information or clarification.

Barber, Kristine M. and Katherine R. Allen. 1992. *Women and Families: Feminist Reconstructions*. New York: Guilford. A good example of scholarship on the family from a feminist perspective, this work examines women's intimate relationships, both straight and lesbian, and argues that a critique of hierarchical gender relations must be central to family studies.

Boss, Pauline G., William J. Doherty, Ralph LaRossa, Walter R. Schumm, and Suzanne K. Steinmetz, eds. 1993. *Sourcebook of Family Theories and Methods: A Contextual Approach*. New York: Plenum. Don't let the size of this book scare you off! It is filled with very readable essays on all the theoretical perspectives and methods described in this chapter and more.

DaVanzo, Julie. 1993. *American Families: Trends and Policy Issues*. Santa Monica, CA: Rand. Current data on today's families, together with policy implications and suggestions. The book integrates the family ecology and feminist theoretical perspectives.

DuBois, W. E. B. 1970. *The Negro American Family*. Cambridge, MA: MIT Press. Originally published in 1908, this is one of the first careful studies of the African American family in the United States, done by an early and important African American scholar. Coming from what would later be called the family ecology perspective, DuBois argued that black family instability was a function of limited economic opportunity and adjustment to urban living.

Duvall, Evelyn M. and Brent C. Miller. 1985. *Marriages and Family Development*. 6th ed. New York: Harper & Row. Family development textbook whose senior author is one of the pioneers of this framework.

Hill, Robert B., Andrew Billingsley, Eleanor Engram, Michelene R. Malson, Roger H. Rubin, Carol B. Stack, James B. Stewart, and James E. Teele. 1993. *Research on the African-American Family: A Holistic Perspective*. Westport, CT: Auburn House. This is a well-written work by a cadre of family scholars on the current status of the African American family. The authors give particular attention to the inadequacy of theoretical and research perspectives guiding social scientists and policymakers as they examine black families.

Lee, Joann Faung Jean. 1991. *Asian American Experiences in the United States: Oral Histories of First to Fourth Generation Americans*. Jefferson, NC: McFarland. Illustrates the oral history method of family research and also gives a rich contextual view of Asian American families.

Levy, Frank S. and Richard C. Michel. 1991. *The Economic Future of American Families: Income and Wealth Trends*. Washington, DC: Urban Institute Press. An examination of the growing gap between wealthy and poor families in the United States today with implications for the future, should this trend continue.

Maddock, J. W., M. J. Hogan, A. L. Antonove, and M. S. Matskovsky, eds. 1993. *Peristroika and Family Life: Post-USSR and US Perspectives*. New York: Guilford. Not all cross-cultural analyses deal with nonindustrialized societies. This work uses a globally comparative approach to examine families in a changing society, with emphasis on the late 1980s and early 1990s.

McAdoo, Harriette Pipes. 1993. *Family Ethnicity: Strength in Diversity*. Newbury Park, CA: Sage. An examination of ethnic diversity among American families with an emphasis on avoiding theoretical or methodological bias so that ethnicity is explored from the cultural variant model.

Stack, Carol. 1974. *All Our Kin: Strategies for Survival*. New York: Harper & Row. An example of participant observation research that conveys what the experience is like, exemplifies the cultural variant approach to research on ethnic minority families, and shows how new theoretical perspectives can emerge to better describe ethnic minority families.

Zill, Nicholas. 1994. *Running in Place: How American Families Are Faring in a Changing Economy and an Individualistic Society*. Washington, DC: Child Trends. Zill argues that a constricting economy, coupled with failure to recognize that families require socio-economic support endorsed through public policy, makes for stress—and often disruption—in today's families.

Zimmerman, Shirley L. 1992. *Family Policies and Family Well-Being: The Role of Political Culture*. Newbury Park, CA: Sage. An example of a work in the family ecology tradition, this book illustrates how the political and economic context in which families exist influences what goes on inside them.

Our Gendered Identities

An 80-year-old woman with a Ph.D. in pharmacology (coauthor Agnes Riedmann's mother) tells this story, one of her earliest childhood memories. She is about three years old. She is seated on a stool at the foot of her father's chair. He is reading to her, his legs crossed. She works and works to cross her legs just the way his are. Finally she manages it and is delighted. Her father looks up from his book and says, "You mustn't do that." "Do what?" she asks. "Cross your legs," he says. "But *your* legs are crossed," she says. "Girls and women don't cross their legs," he replies.

Hearing this story today, we are struck with how gender expectations have changed. When we realize that this event occurred before 1920, when U.S. women were first granted voting rights, we are doubly cognizant of change. But we may also become aware that living in our society remains a different experience for males and females. Although gender expectations have changed and continue to change, they have not done so completely.

Gender influences virtually every aspect of people's lives and relationships. Put another way, we are all **gendered.** In this chapter we will examine various aspects of gender, especially those that more directly affect committed relationships and families. In doing so, we'll consider personality traits and cultural scripts typically associated with masculinity and femininity. We'll describe male dominance, discuss the possible influence of biology, and examine the socialization process, the study of which helps us discover whether people learn to behave as either females or males, or whether they are born that way. We'll discuss the lives of adults as they select from options available to them and speculate about what the future is likely to hold.

Gendered Identities

We are using the term *gender* rather than *sex* for an important reason. The word **sex** refers only to male or female anatomy and physiology. Sex includes the different chromosomal, hormonal, and anatomical components of males and females that are present at birth. We use the term **gender** (or **gender role**) far more broadly—to describe attitudes and behaviors expected of and associated with the two sexes. Gender, as distinguished from sex, involves socially constructed roles regarding what it means to be masculine or feminine.

The first two chapters pointed out that our personal decisions, attitudes, and behaviors are influenced by our social environment, which we in turn influence and shape. The interactionist perspective (Chapter 2) encourages us to recognize that our culture and the social arrangements surrounding us are neither God-given nor natural but rather are socially constructed. Hence gender roles, or attitudes and behaviors, are socially constructed as well. People internalize others' expectations regarding gender; by acting accordingly, they reinforce those expectations for themselves, for others around them, and for all who will follow them in subsequent generations. People who refuse to act accordingly precipitate change.

Let's examine the predominant gender expectations in our culture.

Gender Expectations

You can probably think of some characteristics typically associated with being feminine or masculine. Stereotypically masculine people are often thought to have *agentic* (from the root word *agent*) or **instrumental character traits**—those that enable them to accomplish difficult tasks or goals. A relative absence of agency characterizes our expectations of women, who are thought to embody *communal* or **expressive character traits:** warmth, sensitivity, concern about others' needs, and the ability to express tender feelings.

Talcott Parsons, a prominent structure-functionalist in the 1950s, helped to establish these gender expectations when he wrote

The masculine personality tends more to the predominance of instrumental interests, needs, and functions, presumably in whatever social system both sexes are involved, while the feminine personality tends more to the primacy of expressive inter-

Traditional stereotypes of children define boys as aggressive and competitive, and girls as sensitive and concerned for others. Real behavior is far more varied than these stereotypes, however.

ests, needs, and functions. We would expect, by and large, that other things being equal, men would assume technical, executive, and "judicial" roles, women more supportive, integrative, and "tension-managing" roles. (Parsons and Bales 1955, p. 101)

Gender Expectations and Diversity

This view of men as instrumental and women as expressive is based primarily on people's images of white, middle-class heterosexuals. But subcultural variations exist. For example, African American males and females may be more similar to each other in terms of expressiveness and competence than are non-Hispanic whites (Blee and Tickamyer 1995). Compared with white men, black men are viewed as more emotionally expressive and less competitive. Compared with white women, black women are viewed as less passive and

less dependent. Latinas (Vazquez-Nuttall, Romero-Garcia, and De Leon 1987) and Asian women (Chow 1985) are stereotyped as being more submissive than non-Hispanic white women.

In addition to these racial/ethnic components, there are age, class, and sexual-orientation differences in gender expectations. We may think of elderly men as less aggressive, for example. Working-class women are expected and thought to be more hostile, inconsiderate, and irresponsible than middle-class white women (Cazenave 1984). Gays are stereotyped as possessing feminine traits, whereas lesbians are stereotyped as possessing masculine ones (Kite and Deaux 1987).

Cultural Messages

The particular ways in which men are expected to show agency and women expressiveness are embedded in the culture around us. Let's examine some of our cultural messages about masculinity and femininity in turn.

MASCULINITIES   In writing about men and gender, we need to state the obvious: Men are not all alike. Recognizing this, scholars have begun to analyze **masculinities** in the plural, rather than the singular—a recent and subtle change meant to promote our appreciation for the differences among men. Anthropologist David Gilmore (1990), having examined expectations for men cross-culturally, argues that what is common among the world's concepts of "masculinity"—and what separates these from cultural messages regarding women—is that a man must somehow prove that he is a "real man" (versus "no man at all") whereas a woman is allowed to take gender for granted.

How do men go about demonstrating their manhood? Twenty years ago, sociologists Deborah David and Robert Brannon (1976) pointed to four masculine "scripts" that our culture provides as guidelines. The first was no "sissy" stuff, according to which men are expected to distance themselves from anything considered feminine. In a second cultural message, a man should be occupationally or financially successful, a "big wheel." Third, a man is expected to be confident and self-reliant, even tough—a "sturdy oak." A fourth cultural message emphasizes adventure, sometimes coupled with violence and or the need to outwit, humiliate, and defeat. Adult men "give 'em hell" or "kick ass" in barroom brawls, contact sports, and war. If a male finds that legitimate avenues to occupational success are blocked to him because of, for example, social class or racial/ethnic status, he might "make it" through subcultural standards, such as physical aggression, rapping, or striking a "cool pose." The latter involves dress and postures manifesting fearlessness and detachment, adapted by racial/ethnic minority males for emotional survival in a discriminatory and hostile society (Majors and Billson 1992). During the 1980s, another cultural message emerged and was lauded by many (including us) as the preferred option for men. According to this message, the "new" or "liberated" male is emotionally sensitive and expressive, valuing tenderness and equal relationships with women (Kimmel 1989). These divergent cultural messages—coupled with the male's need to prove his manhood—contribute to men's ambivalence and con-

fusion in today's changing society (Gilmore 1990; Gerson 1993), a point explored later in this chapter. Case Study 3.1, "An Inside View of Masculinity," illustrates various masculine cultural messages.

FEMININITIES   The pivotal expectation in **femininities** for a woman requires her to offer emotional support. Traditionally, the ideal woman was physically attractive, not too competitive, a good listener, and adaptable. Considered fortunate if she had a man in her life, she acted as his helpmate, facilitating and cheering his accomplishments. In addition to caring for a man, a woman has been expected to be a good mother and put her family's and children's needs before her own. The "strong black woman" cultural message combines assertiveness, independence, employment, and child care (Basow 1992, p. 132).

A feminine expectation that has emerged over the past twenty years is the "professional woman"—independent, ambitious, self-confident. This cultural message may combine with the traditional one to form the "superwoman" message, according to which a good wife and/or mother also efficiently attains career success or supports her children by herself. An emerging female expectation is the "satisfied single"—a woman (either lesbian or heterosexual and usually employed) who is quite happy not to have a serious relationship with a male.

A theme that has wound through cultural images of femininity is the good girl/bad girl or virgin/whore theme. Women have traditionally been stereotyped as either sexually conservative or sluts.

To What Extent Do Individual Women and Men Follow Cultural Expectations?

It is one thing to recognize cultural messages but another to choose to live accordingly. Consequently, we can ask to what extent actual men and women exhibit gender-expected behaviors. The first thing we need to recognize here is that gender traits are sometimes *thought* to be opposite and mutually exclusive; that is, we may think that a person cannot be both masculine and feminine. But this is not the case. Indeed, fairly recent descriptions by college students of "typical" men and women overlap considerably (De Lisi and Soundranayagam 1990).

For one thing, acting according to cultural expectations may be situational. The same woman may speak

An Inside View of Masculinity

The following is from an essay written several years ago by a male student of marriage and family in his 20s. He spent six years in the Marine Corps and served in Vietnam. Here he looks back at his childhood and the pressures he felt to conform to the traditional male role. Ironically, as you will see, it was his Marine Corps experience during the 1960s that led him to view the traditional masculine stereotype as more myth than reality.

Growing up to be a man has been confusing and frustrating. From the time I was very young until now, I had to live with the idea of how things should be, not how they are.

As a young boy, I was brought up in the traditional male role, with constant pressure on what I should be, not what I am. In grammar school, I began to believe that power was everything. No matter what you said, if you could back it up with power, you were right. So I went out to prove I was powerful. There were triumphs and defeats, but I did get my point across—that I was a force to be reckoned with.

What I eventually achieved was splitting up the people I knew into two groups: people I could push around, and people who could push me around....

I rallied all the kids on my block to attack the kids on the next block. We made swords out of wood, used garbage cans as shields, and went into battle. We threw rocks. What a thrill it was to lead the troops to battle....

As I got older, I did other things to prove my masculinity. I would hitch a ride on a freight train and climb the highest trees, all to prove I was daring and brave. I would do things that I would never do without someone watching, all the time never letting anyone know I was afraid of the Ferris wheel. I couldn't tolerate being called a "sissy" or a "fraidy cat." That just wasn't what they called a man.

In high school I participated in the roughest sports. Being on the football team was proof in itself that I had to be a man....

I wrestled because the people who wrestled would tease the basketball players about playing a sissy sport. I ran the 880 in track because it was referred to as a long, grueling sprint. Wrestling built up the muscles, football was a definite man's sport, and track demonstrated endurance. All of these served as my identity. I even took woodshop one year, not because I was interested in working with wood, but I felt that a man should know how to do it.

There were certain things I liked to do, but I would do them privately until I could feel out the crowd to see if they would fit into the acceptable standards of being a man. One was art. I became very accomplished as an oil painter, but I would only paint the things that men would be interested in—paintings of football scenes and the like....

I remember the first time I felt I was in love. The girl and I were both 14. I met her at a Little League baseball game. She was watching her brother, and I was watching her. I used to walk to her house to see her. I was playing the he-man game. I was so afraid of saying the things that I felt for fear she would think I was a sissy. I also felt that I had never been really close to anyone before, and I didn't know how to act. At the time I didn't realize that girls were going to play such a large part of proving my manliness. We ran around for about six months. I never even kissed her—something I never overlooked in later relationships....

Now I feel that maybe at the time I actually resented girls. When I was younger, men didn't get mushy. When I was asked as a small boy if I was going to get married, I would promptly say no! I would chase little girls with worms or anything I felt they wouldn't like. All my friends did the same because there was a feeling of not getting too close to [girls], for fear they would try to get you to play house....

When I got out of high school, I made my most drastic move ever to prove my manliness. I joined the Marines. What could be

(continued)

(continued)

more manly than that? Everyone knows the Marine Corps builds men, and I felt I could use more building.

It wasn't until I got in the Marine Corps that I realized that I could get close to men. In boot camp and in Nam I found out these people felt the same way I did. They had the same self-doubts and were trying to satisfy the same needs. It wasn't until this time that I found out that other people are not all living up to the myth. I got answers and true, gut feelings from these guys, and I found out there was nothing wrong with me. We had all been victims of having to live up to a standard that never existed in the first place.

How does this essay illustrate both masculine and feminine cultural messages? As a little boy, this young man saw getting too close to girls as dangerous because girls "try to get you to play house." Does this situation have a parallel for older male–female dating relationships? What limits did the writer impose on himself by trying to live up to masculine expectations? How does his story illustrate men's changing and renegotiating their traditional attitudes and roles?

forcefully when in a job interview and demurely when on a date, for example. A man who suffers pain without wincing on a football field may cry in a less public or less competitive situation. Then too, beginning in the 1970s (Bem 1975, 1981; Swim 1994) researchers have found that about half of American individuals actually see themselves as having both agentic/instrumental and communal/expressive traits.

We can visualize the extent to which females and males actually differ on a trait as two overlapping normal distribution curves (see Figure 3.1). For example, although the majority of men are taller than the majority of women, the area of overlap in men's and women's heights is considerable. Furthermore, the shaded area in Figure 3.1 indicates that some men are shorter than some women and vice versa. It is also true

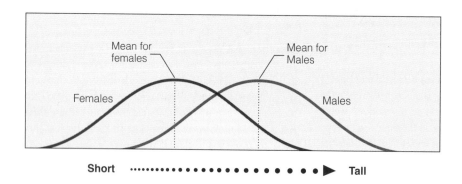

F I G U R E **3.1**

How females and males differ on one trait, height, conceptualized as overlapping normal distribution curves.

Cartoon by Annie Gibbons, from Roz Warren [Ed.], *Men Are from Detroit, Women Are from Paris*, p. 100. Crossing Press. Reprinted by permission of Annie Gibbons.

that differences among women or among men ("within-group variation") are usually greater than the differences between men and women ("between-group variation").

Although some research finds women more concerned about others' well-being than men and men more competitive than women (Beutel and Marini 1995), generally the gendered expectations we've discussed fit an overlapping pattern (Basow 1992). An exception is male dominance, although this may be changing.

Male Dominance

On an interpersonal level, **male dominance** describes a situation in which the male(s) in a dyad or group assume authority over the female(s). On the societal level, male dominance is the assignment to men of greater control and influence over society's institutions. Sociologist Clyde Franklin II cites both mixed-gender play situations in which boys take over playground equipment "whenever they want" and

men harassing women on jobs as evidence that male dominance "seems ubiquitous" (1988, p. 30).

Research shows that on an interpersonal level, males in groups tend to dominate verbally. Men talk louder and longer and interrupt other speakers and control conversational topics more than women do. Also, females restrict themselves more in claiming personal space, smile more when smiling is not related to happiness, and touch others less in groups but are touched more. Although these findings may reflect personality differences, they also indicate male–female status differences (Henley and Freeman 1995). On an institutional level, male dominance is evident in politics, religion, and the economy.

MALE DOMINANCE IN POLITICS Before 1992 there had never been more than two women among our 100 U.S. senators. (Now California has two female U.S. senators, a national first.) Between 1947 and 1991, the proportion of women in Congress fluctuated between about 2 and 6 percent. In 1968,

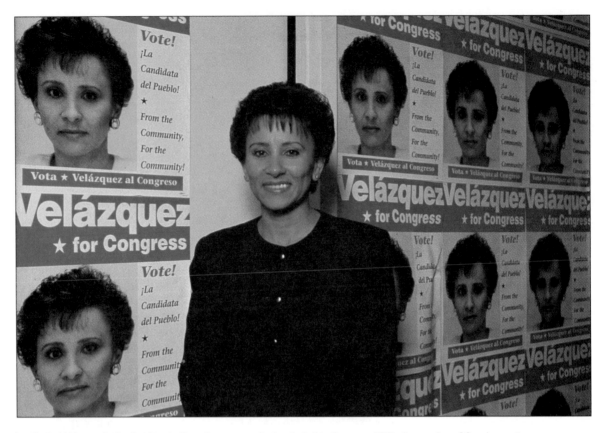

Nydia Velázquez is the first Puerto Rican–born woman to be elected to Congress. While the number of female senators and members of the House of Representatives has increased in recent years, women remain a minority in positions of political power.

Shirley Chisholm was the first African American woman elected to Congress; until 1992, only seven other black women, one Asian American woman, and one Latina had served in the House (Smolowe 1992). Women gained a toehold in the 1992 congressional elections. The number of female senators went from two to seven and the number of representatives from 29 to 48. Placing women in positions of political power seems to have resulted in some changes. It was California Senator Barbara Boxer, for instance, who first called for Senator Bob Packwood's ultimate resignation due to sexual harassment charges (Yoachum and Freedberg 1995). But women "on the Hill" are still an obvious minority and have more difficulty than men gaining access to power positions, such as working on congressional committees (Riordan and Kirchhoff 1995).

MALE DOMINANCE IN RELIGION   Religion evidences male dominance as well. Although most U.S. congregations have more female than male participants, men more often hold positions of authority, with women performing secretarial and housekeeping chores. Gallup polls show that two-thirds of U.S. Catholics believe women priests "would be a good thing," but only 10 percent of U.S. bishops agree (Reese 1992). The Plan of Action document, resulting from the 1995 United Nations Fourth World Conference on Women, in Beijing, declares women's rights as human rights to be actively endorsed by the world's governments. But the document's insistence of women's right to control their bodies and sexuality angered the Vatican, Moslem Iran, and various Christian fundamentalist groups in the United States (Chen 1995).

Although more females than males attend churches, men hold more positions of authority in the clergy and on church boards. As more women enter the clergy, however, gender balance may shift within churches.

The effects of personal religious involvement on women's daily lives are complex. On the one hand, the phenomenal growth of fundamentalist Islam religions, evangelical Protestantism, and the similar charismatic renewal in the Catholic church have fostered a traditional family ideal of male headship and a corresponding rejection of feminist-inspired redefinitions of family roles. On the other hand, some feminist evangelicals mingle the two movements as other evangelical and charismatic women redefine male family authority to foster greater male involvement in the household, more open communication, and shared decision making (Stacey 1990). There is also a feminist movement among Arab women, who seek to combine their religio-cultural heritage with equal rights for females (Tucker 1993).

MALE DOMINANCE IN ECONOMICS   Although the situation is changing, men as a category have been and continue to be dominant economically. Nearly twice as many employed women as men earn less than $300 weekly (U.S. Bureau of Labor Statistics 1994a, Table B). In 1993, employed women earned 77 percent of what employed men did. Table 3.1 presents

the ratio of women's to men's median weekly earnings for 1983 and 1993.

As you can see from Table 3.1, the differences in women's and men's earnings vary by age, race, and occupation. Younger women earn a larger proportion of what men do, although in no age category are earnings equal. Black and Hispanic ratios are higher than those of non-Hispanic whites, primarily because black and Hispanic men have much lower earnings than do white men. For instance, white men still run corporate America: 92 percent of corporate officers and 88 percent of corporate directors were white men in 1993 (Galen 1994). Although racism blocks the path to management for minority men, both racism and sexism block the path for minority women.

Even in the same occupational categories, women earn less than men. For instance, in 1993 female professionals made 77 percent of what their male counterparts did. Women's earnings in managerial occupations were 67 percent of men's. To some extent, these ratios reflect the type of professional and management jobs that women hold. Despite equal proportions of college graduates and master's degrees, men's and women's employment remains segmented into

TABLE 3.1

Ratio of Women's to Men's Median Weekly Earnings, 1983 and 1993

	1983	1993
Age		
Total, 16 years+	67	77
16–24 years	88	95
25–54 years	66	75
25–34 years	73	83
35–44 years	61	73
45–54 years	59	67
55 years+	62	68
Race		
White	66	76
Black	79	89
Hispanic	–	89
Occupation		
Executive, administrative and managerial	64	67
Professional specialty	73	77
Technical, sales, and administrative support	64	70
Service occupations	68	74
Precision production, craft, and repair	66	67
Operators and laborers	66	72

NOTE: Weekly earnings are annual averages. Ratios are expressed as percentages.

Source: Adapted from U.S. Bureau of Labor Statistics 1994a, Table A.

dual labor markets, with women in a narrower range of jobs offering fewer benefits and advancement opportunities. Women are underrepresented in many higher paying management occupations, such as financial managers and marketing. Moreover, women's earnings in specific and comparable management fields are often below men's as well.

Table 3.1 also shows that the earnings gap between men and women narrowed between 1983 and 1993. Unfortunately for both sexes, this narrowing is not due so much to rising wages for women as to falling wages for men. Since the 1970s highly paid industrial production has declined, white-collar corporate structures have grown leaner and meaner, and the labor force has become increasingly characterized by lower paying, less secure jobs. Men's average wages fell by 5 percent between 1979 and 1989 and by another 2 percent between 1989 and 1991. Except for men in the upper 10 percent of the wage ladder, male workers across the wage hierarchy have suffered declining or stagnant earnings (Gerson 1993, pp. 268–69). The influence of this decline for men on their relationships and families will be addressed later in this chapter and periodically throughout this text.

We have been discussing male dominance in the United States today. Cross-culturally and historically, it appears that virtually all societies have been characterized by some degree of male dominance. This leads us to ask whether the cause might be genetic. Put another way, is what Sigmund Freud once proposed true—that "anatomy is destiny"?

Is Anatomy Destiny?

We need to ask which aspects of gender-associated behavior depend on physiology and which have people learned. For example, about 80 percent of adult American females shave their body hair (Basow 1991). Is this behavior genetic in females, or learned? More seriously, are males biologically destined to be more agentic, instrumental, and dominant than females? Or do men and women learn and enforce their different behaviors and statuses?

Genetics-Based Arguments

Advocates of the genetic basis for gender-related differences invoke various arguments to support their contentions. Some find "proof" for inborn differences in religious works such as the Bible; various religions see aspects of male dominance as sacred and ordained by God. In science, arguments for genetics as the basis of gender differences have been provided by sociobiologists (Udry 1994) and by ethologists, primatologists who study human beings as an evolved animal species (Tiger 1969; Goldberg 1973). Having studied baboons primarily, these authors found males to be dominant and argue that *Homo sapiens* inherited this condition through evolution. Attributing masculine instrumentality and dominance to testosterone, characteristically present at higher levels in males, they

| Wilma Mankiller, Chief of the Cherokee Nation, shows that anatomy is not necessarily destiny.

hold that the prehistoric male's greater size and musculature let him better perform instrumental tasks, such as hunting and defense. Because these roles were functional to the species, the argument continues, they were the basis of male political and familial controls that became genetically established in humans.

In the 1980s, much of this "baboons-with-briefcases" reasoning (Sperling 1991) was scientifically debunked. Newer data on nonhuman primates have challenged these conclusions as socially constructed myths—as "politics by other means" (Haraway 1989)—or as biased science at best.[1] Although male baboons are dominant and very aggressive, not all monkeys behave that way. Males are much less dominant and aggressive among chimpanzees than among

baboons, and the former may be more closely related to humans (McGrew 1981). Female dominance seems typical of lemurs, and gibbons manifest no sex difference in dominance (Hardy 1981).

Most social scientists refuse to impute animal behavior to that of humans. Humans share ancestry with the apes, but we did not descend directly from them; the two families have evolved separately for millions of years. Moreover, gender-expected behaviors have "clearly followed no logical pattern based on biological differences. Even though men possessed the requisite skills, for example, they were much less likely than women to bottle-feed babies" (Huber 1989, p. 111).

Society-Based Arguments

Sociologists stress that stratification and the division of labor in society (how necessary tasks are divided by

1. As Cynthia Fuchs Epstein (1988) points out, "biological explanations have been used to support inequality between the sexes, as they have been used to support inequality between the races and other dominant and subordinate groups" (p. 51).

gender, age, and social class) shape gender roles. In addition, cultural ideas enforce gender stratification.

FORAGING AND HOE SOCIETIES

In both foraging (hunting and gathering) and hoe societies, food production was relatively compatible with pregnancy, childbirth, and breast-feeding; women thus played an important part. Among foragers, for example, 60 to 80 percent of the food comes from gathering activities, performed predominantly by women, and both sexes may hunt small game (Linton 1971). With women fully participating economically, males are less dominant than in agricultural or industrial societies (Chafetz 1988).

Native Americans, members of what were once hunting and gathering and hoe cultures, have a complex heritage that varies by tribe but may include a matrilineal tradition in which women owned, and may still own, houses, tools, and land. But Native American women's political power declined with the spread of Europeans into their territories and the subsequent reorganization of Indian life by federal legislation in the 1920s. Recently, Native American women have begun to regain their power: A woman serves as chief of the Cherokee Nation, and Navajo women now hold more local offices than they have since 1920. Many Native American women are politically active in the Pan-Indian movement, which strives to maintain or regain traditional tribal culture (Almquist 1995).

AGRICULTURAL SOCIETIES

Agricultural (agrarian) societies, based on plow cultivation, the domestication of animals, and laying claim to land, developed about 5,000 years ago. Plow agriculture requires greater physical strength and full-time labor, conditions less compatible with pregnancy and nursing (Basow 1992, p. 108). Although women continued to contribute significantly to the family economic enterprise (Huber 1986), men may have taken over plow and other heavy work, thereby making women's productive labor less visible. About this time, patriarchy—a societal organization based on the supremacy of fathers and inheritance through the male line—became firmly established. As it became possible to accumulate wealth through large landholdings, concern with property inheritance—and hence with legitimacy of offspring—increased the social control exerted over women, particularly in non-European areas.

INDUSTRIAL SOCIETIES

With industrialization, beginning about 200 years ago in Europe, economic production gradually shifted from agriculture to mechanized production of manufactured goods.[2] The status of women declined further as industrialization separated work from home and family life, transferring work traditionally done by women (such as clothing production) from homes to factories. Owners organized factories in such a way that industrial work could not easily be combined with domestic tasks and the supervision of children. White women no longer contributed directly to economic production; their indirect contribution to the economy through domestic support and reproduction of the labor force became virtually invisible.

In the middle and upper classes, an ideology of "separate spheres" arose to support this separation of men's and women's roles. Men came to be seen as possessing the instrumental traits referred to earlier, as being more comfortable than women with the competition and harshness of the world outside the home. Middle- and upper-class white women were "angels of the house" whose delicacy and passivity were appropriate to their more sheltered lives.[3] White women's

2. Men who had been peasants, serfs, or slaves became urban wage workers. Men's responses to changed work lives produced the socialist and labor movements of nineteenth-century Europe. Sociologist Joan Huber (1986) labels those movements "men's movements" because, as responses to changed forms of economic participation, they parallel the later women's movements.

3. "While law and public opinion idealized motherhood and enforced protection of white women's bodies, the opposite held true for black women's," notes historian Evelyn Brooks Higginbotham (1992, p. 257). In fact, during the nineteenth century southern state courts ruled slave women outside the statutory rubric "woman." Higginbotham explains by citing the 1855 case of *State of Missouri v. Celia*. "Celia was fourteen years old when purchased by a successful farmer, Robert Newsome. During the five years of his ownership, Newsome habitually forced her into sexual intercourse. At age nineteen she had borne a child by him and was expecting another. In June 1855, while pregnant and ill, Celia defended herself against attempted rape by her master. Her testimony reveals that she warned him she would hurt him if he continued to abuse her while sick. When her threats would not deter his advances, she hit him over the head with a stick, immediately killing him.... [S]he then burned his body in the fireplace and the next morning spread his ashes on the pathway. Celia was apprehended and tried for first-degree murder. Her counsel sought to lower the charge of first degree to murder in self-defense, arguing that Celia had a right to resist her master's sexual advances, especially because of the imminent danger to her health. A slave master's economic and property rights, the defense contended, did not include rape. The defense rested its case on Missouri statutes that protected women from attempts to ravish, rape, or defile. The language of these particular statutes explicitly used the term 'any woman,' while other unrelated Missouri statutes explicitly used terms such as 'white female' and 'slave' or 'negro' in their criminal codes. The question centered

submissiveness and emotional sensitivity, the expressive qualities described earlier, supposedly enabled them both to provide a haven for their wage-earner husbands and to bring up innocent children (Cancian 1987).[4]

These breadwinner and housewife roles applied mainly to middle- and upper-class white men's and women's work—and for only a brief period, from the late nineteenth century to the mid-twentieth century. Most immigrant, black, and working- and lower-class women did not have the housewife option but instead needed employment in domestic service, factories, or in home-based activities such as piecework performed for money. Nevertheless, although the idealized portrait of men, women, work, and personality is changing, we have retained remnants of it to the present.

We have seen that sociobiologists and ethologists stress genetics, whereas most social scientists stress the importance of society in the formation of gender roles and expectations and stratification. Very probably, biology and society interact to create **gender-linked characteristics and roles.** For example, men's greater average physical strength, a result of higher testosterone levels, may have resulted in force and/or threats of force to effect some degree of male dominance in virtually all societies. But, as we have seen, the various forms of economic organization either mitigated or exaggerated male dominance.

The Interaction of Culture and Biology

Strong support exists for the proposition that gender-linked characteristics result from the interaction of cultural learning and biological hormones.[5] Some of the most striking research in this regard is that of John Money and Anke Ehrhardt at Johns Hopkins University Hospital in Baltimore.

Money and Ehrhardt (1974) studied **hermaphrodites,** people whose genitalia cannot be clearly identified at birth as either female or male. Their research supported arguments for both socialization and genetics as influences on behavior. For the argument for socialization, Money and Ehrhardt found that the most crucial factor in these infants' ultimate self-perception as girls or boys, and later as women or men, was their **assignment** (at birth or shortly after, hermaphrodites are often arbitrarily assigned a sex identity and are then treated accordingly). Hermaphrodites assigned *male* grow up to think and behave as men, whereas those assigned *female* grow up to think and behave as women.

Money and Ehrhardt also found evidence for genetic influence. Androgynized females (female infants who had been exposed prenatally to a greater amount of the male sex hormone androgen through drugs taken by the mother) were much more apt than normal girls to be tomboys. Taller and stronger than average, they were more likely to be athletic, play with boys, assert themselves, compete, and reject feminine adornment. They tended to be uninterested in either dolls or babies. As adult women, they were oriented much more toward careers than marriage. Such masculinization, the researchers concluded, probably results from the action of male sex hormones on the developing fetus. But it is possible that parents' awareness of their daughters' exposure to masculine hormones may have affected the parents' behavior toward their daughters, so that Money and Ehrhardt's findings cannot be presumed to be solely biological in origin. Furthermore, their larger body size may have exposed androgynized females to socialization experiences more common to boys.

AGGRESSIVENESS   More than twenty years ago, psychologists Eleanor Maccoby and Carol Jacklin (1974) reviewed all of the research on sex differences through the early 1970s. The only difference between girls and boys that clearly seemed to Maccoby and

---

on her womanhood. The court found Celia guilty. . . . [She] was hanged in December 1855 after the birth of her child" (Higginbotham 1992, pp. 257–58).

4. We should note that some women were able to use the moral authority given to them by a separate spheres ideology to lead antislavery and temperance crusades and suffrage movements. During this era some women advocated far more radical changes in society and the family (Rossi 1973).

5. **Hormones** are chemical substances secreted into the bloodstream by the endocrine glands; they influence the activities of cells, tissues, and body organs. Sex hormones are secreted by male or female gonads, or sex glands. The primary male sex hormone is testosterone, produced in the male testes. Females secrete testosterone also, but in smaller

amounts. The primary female hormones are estrogen and progesterone, secreted by the female ovaries.

Sex hormones influence sexual dimorphism: sex-related differences in body structure and size, muscle development, fat distribution, hair growth, and voice quality. Various researchers report findings that testosterone levels correlate positively with tendencies toward physical and verbal aggression, although not necessarily with desire for achievement or competitiveness (Bardwick 1971; Goleman 1990a).

Jacklin to be biologically based was aggressiveness. The term *aggressiveness* here refers to physical or verbal hostility and attempts to injure another, not simply competitiveness. The consistently reported sex difference in aggression may or may not be genetically based (Hyde and Plant 1995).

More recent studies have investigated the relationship between testosterone levels (often determined from saliva samples) and gender-related behaviors. In one study (Baucom and Besch 1985), eighty-four female students completed masculinity and femininity scales (questionnaires identifying one's personality traits as either mainly instrumental or mainly expressive). The researchers found that women with "the stereotypic feminine personality" had relatively low levels of testosterone and that those with more stereotypically masculine traits had somewhat higher levels. Other research among males (Booth and Dabbs 1992) suggests that those with more testosterone may be less likely to marry and more likely to divorce. When married, they may experience a lower quality of spousal interaction, are more likely to report hitting or throwing things at their wives, and may be more likely to have extramarital sex.

But what's happening in one's environment can influence hormone secretion levels. For example, hormone levels regulating a new mother's ability to produce milk are influenced by her culturally learned attitudes toward nursing (Bardwick 1971, p. 80). And when a husband batters his wife—at least partly because he lives in a society and a culture that condone violence—the action may stimulate increased secretions of testosterone (which may in turn result in a greater tendency to act aggressively in the future). We might conclude that even though evidence for an intrinsic sex difference in aggressiveness continues to build, society and socialization still play a significant—if not a major—role.

VERBAL AND SPATIAL SKILLS   Many researchers have found girls to be better at verbal skills and boys to be better at math and visual-spatial tasks (Hedges and Nowell 1995). The suggestion associated with such findings is that these differences are genetic. Today some research on brain lateralization supports that conclusion (for example, see Gorman 1992; Wade 1994). Brain lateralization refers to the relative dominance and the synchronization of the two hemispheres of the brain. Some scientists have argued that male

and female brains differ due to greater amounts of testosterone secreted by a male fetus. Different sides of the brain may be dominant in males and females, or males and females may differ in the degree to which the two brain halves work together. Overall, however, lateralization studies have produced conflicting evidence concerning sex differences or their connection to various cognitive activities such as verbal tasks or spatial relations.

Moreover, girls' consistently observed disadvantage in math is arguably a consequence of social expectations and opportunities (Marecek 1995). Test scores in this and other areas appear to be converging. A recent analysis of research done through 1988 found girls and boys to be equivalent in mathematical ability, with comparable performance through middle school. Although boys performed better in high school and college, this difference was limited to the college-bound population rather than all high school students, and it has declined over the years (Hyde, Fennema, and Lamon 1990).

These facts point to what sociologist Alice Rossi (1984) has called the interactive influence of both "nature" (genetics) and "nurture" (learning) on sex-linked attitudes and behavior. In Rossi's words,

> It makes no sense to view biology and social experience as separate domains contesting for election as "primary causes." Biological processes unfold in a cultural context, and are themselves malleable, not stable and inevitable. So too, cultural processes take place within and through the biological organism; they do not take place in a biological vacuum. (p. 10)

Beginning at birth, however, and throughout their lives, males and females learn and negotiate sex-appropriate attitudes and behavior. It's important to recognize that cultural learning can either exaggerate (as seems to be the case in our society) or minimize whatever genetic tendencies exist.

Gender and Socialization

Societal attitudes influence how we behave. As people in a given society learn to talk, think, and feel, they **internalize** cultural attitudes; that is, they make the attitudes their own. Besides attitudes, people internalize cultural expectations about how to behave. The process by which society influences members to internalize attitudes and expectations is called **socialization.** The socialization process is an important concept in the interactional theoretical perspective (Chapter 1). Interactionists point out that individuals do not automatically absorb, but rather negotiate, cultural attitudes and roles. Nevertheless, in various ways society encourages people to adhere, often unconsciously, to culturally acceptable gender roles. We'll examine in detail how language, family, and school function in gender socialization.

The Power of Cultural Images

Our cultural images in language and in the media convey the gendered expectations described earlier in this chapter. You are no doubt aware of the many—and sometimes controversial—efforts to make our language less gender-oriented over the past two decades. A recent edition of *Webster's College Dictionary,* for example, defines new words such as "waitron" (gender-neutral for waiter) and "womyn," meant to avoid the perception of sexism in the word *men.* But our language also continues to accentuate male–female differences rather than similarities (Adams and Ware 1995). Soon after birth, most infants receive either a masculine or a feminine name. From that day on, gender identity is stamped on the individual so thoroughly that people who want to avoid gender identification (for instance, through the mail or in phone-book listings) must replace their first names with initials.

Besides first names, titles, adjectives, nouns, and verbs remind people that males and females differ—and in stereotypic ways. A new mother may be told that she has either a "*lovely* girl" or a "*sturdy* boy." We have "women doctors" and "male nurses"; we have actors and actresses. In her book about the differences in masculine and feminine communication styles, linguist Deborah Tannen writes:

> If I wrote, "After delivering the acceptance speech, the candidate fainted," you would know I was talking about a woman. Men do not faint; they pass out. And these terms have vastly different connotations that both reflect and affect our images of women and men. *Fainting* conjures up a frail figure crumpling into [rescuing arms]. . . . Passing out suggests a straightforward fall to the floor. (Tannen 1990, pp. 241–42)

The media promote gender stereotypes as well. Children's programming more often depicts boys than girls in dominant, agentic roles. Beginning in 1991, all Saturday morning children's programs deliberately began to feature dominant males as lead characters (Carter 1991). This was a deliberate marketing decision by television executives based on the finding that girls will watch shows with either male or female lead characters, but boys will watch only shows with male leads. On music videos such as those on MTV, females are likely to be shown trying to get a man's attention. Some videos broadcast shockingly violent misogynist (hatred of women) messages.

The proportion of lead characters on prime-time television who are working women has increased since about 1985 (Atkin, Moorman, and Lin 1991). Nevertheless, most shows featuring a female professional focus on family issues, not work issues. And white men continue to outnumber women as leads, especially in adventure shows. In TV commercials, men predominate by about nine to one as the authoritative narrators or voice-overs, even when the products are aimed at women. Daytime ads, targeted for women, portray men in family and/or dominant roles, but weekend ads, aimed more toward men, emphasize male escape from home and family (Craig 1992; Kilbourne 1994).

Cultural images in the media and language help to socialize individuals to expected gender attitudes and roles. They do so by portraying what is "normal" and by influencing socializing agents, such as our parents, peers, and teachers. We turn now to a brief examination of socialization theories.

Theories of Socialization

How do cultural ideas about gender get incorporated into personality and behavior? We don't have a definitive answer to that question. There are a number of competing theories of gender socialization, each with some supporting evidence.

SOCIAL LEARNING THEORY   Much of what we have just described fits a **social learning theory** in which children learn gender roles as they are taught by parents, schools, and the media.

As children grow older, toys, talk of future careers or marriages, and admonitions about "sissies" and "ladies" communicate parents' ideas about appropriate behavior for boys and girls. The rewards and punishments, however subtle, that parents assign to gender roles are sometimes seen to be the key to behavior patterns. Other theorists emphasize the significance of parents as models of masculine and feminine behavior, although researchers have found little association between children's personalities and parents' characteristics. Fathers seem to have stronger expectations for gender-appropriate behavior than mothers (Losh-Hesselbart 1987; Anderson 1988).

SELF-IDENTIFICATION THEORY   Some psychologists think that what comes first is not rules about what boys and girls should do but rather the child's awareness of being a boy or a girl. In this **self-identification theory,** at least by age 3, children categorize themselves. They then identify behaviors in their families, in the media, or elsewhere appropriate to their sex, and they adopt those behaviors. In effect, children socialize themselves from available cultural materials. This perspective, developed by psychologist Lawrence Kohlberg (1966), can account for the fact that boys without a father in the home may be just as masculine as boys in intact families (Vander Zanden 1981; Anderson 1988).

CHODOROW'S THEORY OF GENDER   Sociologist Nancy Chodorow (1978) has constructed a **theory of gender** that combines psychoanalytic ideas about identification of children with parents with an awareness of what those parents' social roles are in our society.

According to Chodorow, infants develop a "primary identification" with the person primarily responsible for their early care. Later, as young children, they must distinguish between "self" and "other." Put another way, developing children must learn to differentiate psychologically and emotionally between themselves and their primary caregiver.

Cross-culturally and historically, children's primary caregivers are virtually always female. Both daughters and sons make their primary identification with a female; as a result, the task of separation is more difficult for a boy. Because a daughter is developing a gender identity similar to her principal caregiver's (that is, an identity similar to her mother's), she can readily model her mother's behavior. And her feeling of oneness with her caregiver is not sexually threatening. But a boy cannot model his mother's behavior and also develop a culturally consistent gender identity. He learns instead that he is "not female." Also, he must suppress "feelings of overwhelming love, attachment, and dependence on his mother that are charged by the sexual

current between mother and son" (Thurman 1982, p. 35).

Chodorow attributes commonly held beliefs about the differences between men and women to this divergence in the early socialization experiences of boys and girls. Boys are disappointed and angry at the necessary but abrupt and emotionally charged detachment from their mother. Gradually, however, they come to value their relatively absent fathers as models of agency, independence, and "the superiority of masculine . . . prerogatives" (Thurman 1982, p. 35). Conversely, "relatedness" or expressiveness is allowed and fostered among girls.

As with other theories we are discussing, research does not always support Chodorow. Because gender socialization theories are abstracted from empirical observations, each seems plausible. We'll turn now to some empirical findings regarding gender socialization.

Girls and Boys in the Family

Most studies on gender socialization beliefs and practices have been conducted on non-Hispanic white families. One study of black families indicated that both sons and daughters are socialized toward independence, employment, and child care (Hale-Benson 1986). Still, because most of the research presented in this section has focused on middle-class whites, findings may or may not apply to other racial/ethnic or class categories.

At least among middle-class whites, parents rear female and male infants differently. From the 1970s on, parents have reported treating their sons and daughters similarly (Maccoby and Jacklin 1974; Antill 1987), but differential socialization exists and typically is subtle and not deliberate (Shapiro 1990). Even parents who support nonsexist child rearing for their daughters are often concerned if their sons are not aggressive or competitive "enough"—or are "too" sensitive (Pleck 1992).

Research shows that parents handle infant sons more roughly and respond more quickly to crying baby girls (Lips 1995). As a toddler, Patricia will probably have a doll; Patrick, a truck. A study of 120 babies' and toddlers' rooms found that girls had more dolls, fictional characters, children's furniture, and the color pink; boys had more sports equipment, tools, toy vehicles, and the colors blue, red, and white (Pomerleau et al. 1990). And most parents, especially fathers, discourage their children, especially sons, from

Children learn much about gender roles from their parents, whether they are taught consciously or unconsciously. Parents may model roles and reinforce expectations of appropriate behavior. On the other hand, children also internalize messages from available cultural influences and materials surrounding them.

playing with other-sex toys (Lytton and Romney 1991).

Boys have toys that develop spatial ability and creative construction; girls have toys that encourage social skills (Lips 1995). Toys considered appropriate for boys encourage physical activity and independent play, whereas "girls' toys" elicit closer physical proximity and more talk between child and caregiver (Caldera, Huston, and O'Brien 1989). As children get older, Patrick is more likely to have a computer than is Patricia (Nelson and Cooper 1990). And Patrick may join the Boy Scouts, an organization originated to "make big men of little boys."

Beginning when children are about 5 and increasing through adolescence, parents allocate household chores—both the amounts and kinds—to their children differentially, that is, according to their sex. Patricia will more likely be assigned cooking and laundry tasks; Patrick, painting and mowing (Burns and Homel 1989; McHale et al. 1990).[6] Because typically girls' chores are daily and boys' sporadic, girls spend more time doing them—a fact that "may convey a message about male privilege" (Basow 1992, p. 131). Some parents unconsciously allow (even expect) sons to be inconsiderate, rude, and interrupting while their sisters are reminded to be a little quieter and less argumentative (Orenstein 1994, pp. 46–47).

6. The number of girls and boys in a family apparently influences how chores are distributed. Families with all girls, for example, more readily assign traditional male-child tasks to girls than do families with at least one boy. And the age and gender of sibs are important. Research has also found that when an older sibling is of the same sex, play activities tend to be gender-stereotyped; if the older sibling is of the other sex, cross-sex play is common (Brody and Steelman 1985).

Parents also model different behaviors. As Chodorow notes, most children are cared for by mothers primarily. This situation may convey to children the idea that child care (and nurturance in general) is "women's work." Meanwhile, fathers may appear to be involved with "more important" and prestigious things (Burns and Homel 1989).

Gender socialization in early childhood begins a continuing process whereby females and males are channeled into separate spheres of skills and interests (Lips 1995). Although relations in the family provide early feedback and help shape a child's developing identity, play and peer groups become important as children try out identities and adult behaviors. We turn our attention now to research on differences in boys' and girls' play.

Play and Games

Girls play in one-to-one relationships or small groups of twosomes and threesomes; their play[7] is relatively cooperative, emphasizes turn taking, requires little competition, and has relatively few rules. In "feminine" games like jump rope or hopscotch, the goal is skill rather than winning (Basow 1992). Boys more often play in fairly large groups, characterized by more fighting and attempts to effect a hierarchical pecking order. From preschool through adolescence, children who play according to traditional gender roles are more popular with their peers; this is more true for boys (Martin 1989). Sex segregation at play and leisure begins in preschool and intensifies in elementary school. When boys and girls do interact on school playgrounds, their games (for example, "girls chase the boys" or "chase and kiss") emphasize the differences

7. The role of play is an important concept in the interactionist perspective, particularly symbolic interaction theory, developed by philosopher/social psychologist George Herbert Mead (1934). This theory argues that gender identity, like other aspects of self-identity, is developed by and anchored in social relations. One's self-concept, including one's sense of masculinity and femininity and adult gender roles, depends on the responses of significant others (emotionally important people such as parents). A boy or a girl begins to see himself or herself as competent or as delicate and nurturing, depending on the comments of parents and others.

Those others may be other children at play. Play, in Mead's theory, is not idle time but rather a significant vehicle through which children develop appropriate conceptions of adult roles, as well as images of themselves. His theory distinguishes two stages: "play," one-to-one role playing, and "game," in which children must take multiple positions and roles into account, as in a baseball game with two teams and many positions. (A third stage is that of the "generalized other," in which the norms of the society as a whole are taken into account.)

and cultural boundaries between them (Thorne 1992).

Furthermore, from about second grade, boys' play begins to incorporate the cultural message, "no sissy stuff"—as in "girls have cooties." "In societies where men's higher status is obvious and differential treatment of male and female children pronounced, boys' struggle to disassociate themselves from females is particularly strong" (Whiting and Edwards 1988, p. 12). The tendency to degrade whatever is feminine increases into adolescence—and for some males, into adulthood—when males are expected to put "the guys" first in their priorities. Peer status for adolescent girls, in contrast, more often rests on being popular with boys (Basow 1992, p. 139). These differences in childhood play and games work to teach boys and girls divergent attitudes, skills, and gendered identities. The process is reinforced in schools.

Socialization in Schools

School organization, classroom teachers, and textbooks all convey the message that boys are more important than girls. Let's consider each element of the educational system in turn.

SCHOOL ORGANIZATION   In 1990, about 87 percent of elementary school teachers were women, and between two-thirds and one-half of junior and senior high teachers were women. But the vast majority of principals were men, ranging from 82 percent in elementary school to 98 percent in senior high. About 90 percent of college and university presidents were and are men (Basow 1992, p. 153). With men in positions of authority (coordinators, principals, superintendents) and women in positions of service (teacher's aides, secretaries), school organization itself models male dominance.

TEACHERS   Research shows that, at least among white children, teachers pay more attention to males than to females, and males tend to dominate classroom environments from nursery school through college (Lips 1995). Compared to girls, boys are more likely to receive a teacher's attention, to call out in class, to demand help or attention from the teacher, to be seen as a model student, or to be praised by teachers. Researchers who observed more than 100 fourth-, sixth-, and eighth-grade classes over a three-year pe-

Sex segregation at play and leisure begins in preschool and intensifies in elementary school. This girl, unlike her mother, plays soccer—but on an all-girls team.

riod found that boys consistently and clearly dominated classrooms. Teachers called on and encouraged boys more often than girls. If a girl gave an incorrect answer, the teacher was likely to call on another student; but if a boy was incorrect, the teacher was more likely to encourage him to learn by helping him discover his error and correct it (Lips 1995).

In subtle ways teachers may reinforce the idea that males and females are more different than similar. There are times when boys and girls interact together relatively comfortably—in the school band, for example. But especially in elementary schools, many cross-sexual interaction rituals are based on and reaffirm boundaries and differences between girls and boys. Sociologist Barrie Thorne (1992) calls these rituals **borderwork.** Between 1976 and 1981, Thorne spent eleven months doing naturalistic or participant observation (see Chapter 2) in two elementary schools, one in California and one in Michigan. She found several

types of borderwork both on playgrounds (for example, "boys chase the girls") and in the classroom. In classrooms, for instance, teachers often pitted girls and boys against each other in spelling bees or math contests. On one occasion, the teacher wrote two score-keeping columns on the board: "Beastly Boys" and "Gossipy Girls." In thinking about this research, remember that Thorne's participant observation was done some fifteen years ago. It would be interesting to research the number of teachers who use designations like those today. However, it seems likely that borderwork still persists in schools.

AFRICAN AMERICAN GIRLS, LATINAS, AND ASIAN AMERICAN GIRLS IN MIDDLE SCHOOL

More recently, journalist Peggy Orenstein (1994) spent one year observing in two California middle schools, one mostly white and middle class and the other predominantly African American and Hispanic and of lower socioeconomic status. Orenstein found that in both schools girls were subtly encouraged to be quiet and nonassertive while boys were rewarded for boisterous and even aggressive behaviors. Furthermore, African American girls were louder and less unassuming than non-Hispanic white girls when they began school—and some continued along this path. In fact, they called out in the classroom as often as boys did. But Orenstein noted that teachers' reactions differed. The participation, and even antics, of white boys in the classroom were considered inevitable and rewarded with extra attention and instruction, while the assertiveness of African American girls was defined as "menacing, something that, for the sake of order in the classroom, must be squelched" (p. 181). Orenstein further found that Latinas, along with Asian American girls, had special difficulty being heard or even noticed. Probably socialized into quiet demeanor at home and often having language difficulties, these girls' scholastic or leadership abilities largely went unseen. In some cases, their classroom teachers did not even know who the girls were when Orenstein mentioned their names.

WOMEN IN COLLEGE   Some women, especially those returning to college and working in sexist corporate environments, find the college environment liberating. But compared to men, college and university women tend to be "marginalized" (forced to operate on the fringes of power and involvement). Like teachers in grade school and high school, college instructors pay more attention to male students than to female students (Sadker and Sadker 1986). With the exception of women's studies, females are relatively absent from college texts as well (Peterson and Kroner 1992).

Even though female faculty members serve as role models for female students, a minority of college teachers are women, especially in some fields like physics or engineering. Moreover, men's harassment of campus women is fairly common (Paludi 1990). Some of the unwanted sexual attention comes from male faculty members, and other sources of this problem exist as well. Sexual harassment and rape on campus may be associated with both fraternity and sorority membership (Copenhaver and Grauerholz 1991). Among male students, attitudes of dominance are associated with fraternity membership (Kalof and Cargill 1991). And because they have even fewer role models and suffer from both racism and sexism, black and other minority women have additional problems at predominantly white colleges and universities (Guy-Sheftall and Bell-Scott 1989; Nieves-Squires 1991).

Biology and socialization interact throughout infancy, childhood, and adolescence to produce the adult women and men we see around us. But socialization continues throughout adulthood as we negotiate and learn new roles—or as those already learned are renegotiated and sometimes, though assuredly not always, reinforced. The varied opportunities we encounter as adults influence the adult roles we play. In the next section we examine various aspects of gender in adults' lives today.

Gender in Adult Lives

Old patterns can be difficult to alter. This is true for women and men, both of whom may be torn between what they have learned to value from the past and newer ideas. But new roles promise some rewards.

Gender and Stress

Traditional gender roles result in some characteristic stresses. Oriented to others at the expense of self, many women have felt depressed, bored, empty, dissatisfied with life, inadequate, and excessively guilty. Suicide attempts are more common among women than men (although men are more likely to succeed in suicide efforts and therefore have a higher suicide rate). Women also experience higher rates of mental illness and eating disorders and are particularly prone to depression (Basow 1992; Bower 1995). Traditional sex roles are hazardous to women's health.

Traditional gender expectations are hazardous to men as well. Overemphasis on productivity, competition, and achievement creates anxiety or emotional stress, which may contribute to men's shorter life expectancy. Undue, physically dangerous behavior and violence shorten men's lives as well.

The masculine role not only requires that men undergo pressure to "prove it" but also encourages them to discount or ignore their anxiety and physical symptoms of stress. Men also learn to hide emotions of vulnerability, tenderness, and warmth when they are in public. When hiding tender feelings, men cannot share their inner selves. Hence, they block avenues to intimacy, isolating themselves. Isolation, like the lack of assertiveness or autonomy, limits people's potential to feel positive about themselves, and it can engender stress.

Gender and Personal Change

Sometimes in response to these stresses—and also in response to available options—adults reconsider earlier choices regarding gender roles. For example, a small proportion of men choose to be full-time fathers and/or househusbands. More may effect more subtle changes, such as breaking through previously learned isolating habits to form more intimate friendships.

That choices in adult development and social roles are contingent on opportunity and, to some degree, chance, is illustrated by Kathleen Gerson's research on women's and men's role choices. Gerson (1985) studied sixty-three women who were between ages 27 and 37 when interviewed in 1978–79. Extensive and repeated interviews enabled Gerson to construct life histories and to trace gender-related decisions made in adulthood, particularly decisions about work and motherhood.

The women fell into four groups: Two groups of women made role choices early in life and held to them. The other two groups are more interesting for our purposes here. These women had entered young adulthood with clear expectations for their lives. One set planned a rather conventional marriage and family; work would be an interim activity. The other group gave high priority to careers, regardless of any hopes for marriage and children.

But their lives turned out to be quite different from their expectations. Sometimes women who had planned careers met obstacles, often overt or covert discrimination. They became dissatisfied and pessimistic about their chances of realizing their original goals. Or they fell into very satisfying relationships and may also have found children more enjoyable than they had expected. These circumstances combined to channel their movements into a more traditional family lifestyle. Here is one example:

Vicki was never especially oriented toward motherhood. Instead, since she was old enough to know who the police were, she wanted to be a policewoman. . . . Forced to take the best job she could find after high school, Vicki became a secretary-clerk. She also took the qualifying exam for police work and passed with high marks. No jobs were available, however. . . . In the meantime, she met and married Joe.

. . . She ultimately grew to hate working, for it usually involved taking orders from bosses she did not respect. . . . After the birth of her first child, Vicki discovered that staying at home to rear a child was more important than her succession of boring, deadend jobs. By her mid-thirties, she was a full-time mother of two. Today she has given up hope of becoming a policewoman, but in return for this sacrifice she feels she has gained the secure home life she never knew as a child. (Gerson 1985, pp. 18–19)

Vicki's story illustrates a point we made earlier: Individuals can choose only options that are available in their society. At the time Vicki sought a job as a policewoman, probably few women were being hired, so she was not able to have the career she wanted. Other women may have been discouraged from working by the lack of child care or inadequate maternity leave policies. In recent years, women have been hired as policewomen in large numbers, partly as a consequence of the courts' support for affirmative action hiring practices; this change indicates that it is possible to create wanted options through changes in public policy. Making choices about family life may include political activity directed toward creating those choices.

Some of the women in Gerson's study who planned to marry and start families did not develop permanent relationships or were unable to have children. Or they chanced into career opportunities, particularly as women began to be included in formerly male-dominated positions. Or both. These women found themselves very involved in their careers and perhaps not married or not wanting to become a parent.

Several years later, Gerson (1993) turned her attention to men's changing lives. She interviewed 138 men, mostly in their 30s, and found that, like women, they were reexamining choices made when they were younger. Subsequent changes often involved the family roles they played, a point we will return to throughout this text. But some of these men (like the student in Case Study 3.1) had also changed their attitudes toward masculinity in general. In one example, Carlos, a Mexican American social worker, recalled his earlier days:

"In addition to the usual fun and friends, we guarantee that by the end of the summer she won't take any crap from males."

When I was in high school, I was more of a traditional Hispanic male—sort of macho. At least, I was playing with that idea. A relationship started that was more of a traditional relationship. I expected that person to give to me more than I gave them. It was almost like a fetch-me type of relationship. I think if we had lived together, she would have cooked, cleaned the house, raised the kids.

A growing sense of discomfort led Carlos to later reject this cultural message:

I felt that I wouldn't want to be treated that way, and I shouldn't treat someone else that way. I saw there could be an abuse of the traditional male role, and also I saw limitations in that type of relationship. The woman is limited within the family, and the man gets locked into an image I didn't enjoy. At that point, I pretty much decided that the type of person I wanted to be did not match with the traditional Hispanic role model. (Gerson 1993, p. 159)

The import of Gerson's two studies for our understanding of change over the course of life is that child-hood socialization and early goals do not necessarily predict adult lifestyles. The adult life course is determined by the interaction of the individual's goals, values, abilities, and motivation with the opportunities that present themselves. Throughout adulthood, individuals make a series of choices about their lives, and different individuals respond differently to those options that are the product of a particular time. The development of adult life is a process of ongoing decision and choice, early choices shaping later ones, within the possibilities of a particular historical time.

Both men and women may rethink gender roles in adulthood. Today that rethinking takes place in a society changed somewhat by the women's movement.

Options and the Women's Movement

The separate spheres ideology of gender retained its power into the fifties and early sixties, as media glorification of housewife and breadwinner roles made them seem natural despite the reality of increased women's employment. But contradictions between what women were actually doing and the roles prescribed for them became increasingly apparent. Among whites, higher levels of education for women left college-educated women with a significant gap between their abilities and the housewife role assigned to them (Friedan 1963; Huber 1973). Employed women chafed at the unequal pay and working conditions in which they labored and began to think that their interest lay in increasing equal opportunity. Further, the civil rights movement of the 1960s provided a model of activism. Economic change precipitated a social movement, the second wave of the women's movement, that challenged the heretofore accepted traditional roles in favor of increasing gender equality. (The first wave of the women's movement in the United States occurred from about the last half of the nineteenth century until about 1920, when women obtained the right to vote.)

Women vary in their attitudes toward the women's movement. Some white women deplore the rise of feminism and encourage traditional marriage and motherhood as the best path to females' self-fulfillment (Marshall 1995). As a group, African American women are torn with regard to the women's movement. In one poll, in fact, black women were more likely (85 percent) than white (64 percent) or Hispanic women (76 percent) to agree with the statement that "the United States continues to need a strong women's movement to push for changes that benefit women" (Cowan 1989a). Still, many racial and ethnic minority women view the contemporary feminist movement as a white middle-class creation, with origins in the historical experience of educated upper middle-class homemakers (Friedan 1963) and New Left activists. Some women of color and white working-class women find the women's movement irrelevant to the extent that it focuses on psychological oppression or on professional women's opportunities. Black women have always labored in the productive economy under duress or out of financial necessity and did not experience the enforced delicacy of women in the Victorian period. Nor were they ever housewives, so much of the feminist critique of that role seems irrelevant to African American women (Terrelonge 1995).

There is an increasing number of important black feminist scholars. Nevertheless, black women may be ambivalent about committing themselves to feminist goals that lead to conflict with black men, given the disadvantage the latter have experienced. One response to the obvious disadvantage of black men in particular,

and blacks in general, is to treat gender discrimination as less pressing than racial discrimination, particularly in view of the image of black women as "strong" (see Higginbotham 1992). A sometimes concurrent theme is that "black men must regain a leadership position and take control of their households" (Ransford and Miller 1983, p. 49). How women of color are to participate in the women's movement is problematic, although viewed as desirable by feminist organizations and by black feminists (Almquist 1995; White 1995).

Despite its divisions, the women's movement has caused changes in society and, consequently, in people's options and decisions. A 1989 Yankelovich telephone poll of 1,000 women across the country taken for *Time* magazine found that 94 percent said the women's movement had helped females become more independent; 86 percent agreed that the movement has given women more control over their lives; and 82 percent said the movement "is still improving the lives of women" (Wallis 1989, p. 85). (Thirty-five percent of the women polled said the movement "looks down on women who do not have jobs," and 24 percent saw the movement as "antifamily.") Box 3.1 presents Chicana feminism.

Despite highly publicized expanding options for women over the past two decades, both men and women who choose nontraditional roles can experience discrimination and negative sanctioning. Men, for example, often face prejudice when they take jobs traditionally considered women's, such as day-care workers. Because the pay and prestige are so low, people assume male day-care workers might be child molesters (Campbell 1991). In professions that are traditionally men's, such as politics, the military[8] (Schmitt 1992; Swartz 1992; Warner 1992), police work, law, and medicine, men may harass women who challenge boundaries. Taken together, separate studies of state judicial systems have found "a pattern of sexual harassment by judges: They offer to drop charges in exchange for sexual favors, make salacious remarks to employees and litigators in court or behind closed doors, and they abuse female law clerks." A

---

8. Female military cadets at Texas A & M University report being threatened with knives, raped, and otherwise assaulted by their male classmates (Swartz 1992). Between 1989 and 1992 nearly 1,600 women reported having been raped on U.S. military installations around the world; more than half of these assaults took place on army bases. Using 1988 Pentagon survey figures, an estimated 60,000 female veterans may have been raped or assaulted while serving in the military (Warner 1992).

Chicana Feminism

Chicanas are U.S. "women of color" of Mexican heritage. They share a distinct Chicano/Mexican culture with Mexican American men, or Chicanos (Segura and Pesquera 1995). Some Chicanas have immigrated fairly recently. Others trace their roots on U.S. soil to before the 1846–48 U.S.–Mexico war, when today's southwestern United States belonged to Mexico.

Like other women, Chicanas feel the effects of patriarchal national, community, and family organization. But unlike non-Hispanic white women, Chicanas' membership in a subordinated racial/ethnic group and concentration among the poor and working class, along with their gender, have resulted in triple oppression. In some ways, such as their experiences with racial/ethnic discrimination and their relatively low wages, Chicanas are more like Mexican American men, who are also subordinated, than they are like non-Hispanic white women. Generally Chicanas support women's economic issues, such as equal employment and day care, while showing less support for abortion rights than do Anglo women. Chicanas see the latter as a family issue and are no doubt also influenced by their long affiliation with the Catholic church, a pro-life organization.

Various strains of feminism exist among Chicanas. Like the current mainstream women's movement, contemporary Chicana feminism emerged during the 1960s and 1970s. At that time, various civil rights activities by people of color, including the Chicano movement, opposed power and privilege on the basis of race/ethnicity. And the feminist movement was challenging existing power relations on the basis of gender, or patriarchy. Many Chicanas, however, felt that their particular concerns were ignored or trivialized within both the Chicano and the women's movements. The Chicano movement, organized

Massachusetts task force concluded that such misconduct has effectively isolated female attorneys, causing them and sometimes their clients not only to "feel unwelcome" [in the legal profession] but also "to doubt their own abilities and effectiveness" (Hayes, 1991b).

Negative sanctions (punishments for breaking norms) can also effectively close off opportunities. For example, a Minnesota state trooper, Cheryl Turner, left the force after years of sexual harassment and took a poorer-paying job as a clerk in a convenience store. Among other abuses, Turner had consistently been called "T.T.T.T." (for Tiny Tits Trooper Turner) and "generally humiliated and dehumanized" (McCarthy 1991, p. 28). In another example, in 1991 Dr. Frances Conley, charging "gender insensitivity" on the part of male colleagues, resigned from Stanford University's medical school, where she had been a professor for sixteen years. Examples of "gender insensitivity," she said,

seem trivial, but they are real, and they do affect a person who has a professional life. If I am in an op-

erating room, I have to be in control of the team that is working with me. That control is established because people respect who I am and what I can do. If a man walks into the operating room and says, "How's it going, honey?" what happens to my control? (in L'Hommedieu 1991, p. 43)

Some observers point to a backlash against women's continuing gains in self-awareness and political and economic power. They argue that, on average, "life is no better for today's woman"; the idea that it is largely due to media features that falsely give the impression of trends ("trend stories"), when really these stories are describing situations that are statistically rare (Faludi 1992).

According to Gloria Steinem, founder of *Ms.* magazine, the women's movement has drawn public attention to issues but has yet to change behavior significantly. "The consciousness raising, which is the first stage of any revolution, is pretty much complete. . . . We realize that [women's position] is political, and it

around the ideal of *la familia* ("the family"), placed high value on family solidarity, with individual family members' needs and desires subsumed to the collective good. Chicanas' critiques of unequal gender relations in *la familia* often met with hostility. Chicana feminists were called *vendidas* (sell-outs), *aqabachadas* (white-identified), or *malinche* (betrayer). Chicana feminists argued that they were none of these things—that to say a woman had to be on one side (against racial oppression) or on the other (against patriarchal oppression) was "a bunch of bull" (Nieto-Gomez, quoted in Segura and Pesquera 1995, p. 621). For Chicanas, the struggle against male domination was central to their movement for liberation.

At the same time, Chicana feminists took exception to how mainstream feminists regarded oppression as based almost exclusively on gender. Furthermore, Chicanas felt excluded from mainstream feminist organizations and scholarship. And they resented what they defined as "maternal chauvinism" among mainstream feminists—the idea that white feminists can comprehend, analyze, and devise the best solutions to Chicanas' concerns.

As a result, Chicanas formed grass-roots community organizations of their own. These offer Chicanas social services such as job training, community-based alternatives to juvenile incarceration, and bilingual child development centers. In 1974, Chi-

canas established the Mexican American Women's National Association (MANA) with headquarters in Washington, DC. From its beginning, MANA's goal has been "striving for parity between Chicanas and Chicanos as they continue their joint struggle for equality ... creating a national awareness of the presence and concerns of Chicanas and the active sharing of its Mexican American heritage."

Like mainstream feminists today, Chicana feminists do not all see things in exactly the same ways. One group, *Chicana liberal feminists*, advocates strategies that empower Chicanas within existing social institutions. Their proposals and activities range from personal support to affirmative action initiatives. A second category, *Chicana cultural nationalist feminism,* includes women who identify themselves as feminists but who are simultaneously committed to preserving traditional Mexican American culture. This form of Chicana feminism downplays how cultural traditions often uphold patriarchy, however, and has difficulty reconciling a critique of gender relations among Mexican Americans with the overall preservation of Chicano culture. Finally, *Chicana insurgent feminism* argues that real liberation for Chicanas is not possible without a radical restructuring of society and advocates revolutionary change to end all forms of oppression (Segura and Pesquera 1995).

can be changed." But it could take decades before belief in women's equality leads to substantive action ("Gloria: . . ." 1992). Similarly, author Betty Friedan, sometimes credited with sparking the second wave of the women's movement, sees the trend toward equality for women as "an endless process. It may never be completed. It's an evolution" ("Women's Summit . . ." 1992).

The Men's Movement

As the women's movement encouraged changes in gendered cultural expectations and social organization, some men responded by initiating the men's movement. The first National Conference on Men and Masculinity was held in 1975 and has been held almost annually ever since. The focus of the men's movement is on changes men want in their lives and how best to get them. One goal has been to give men a forum—in consciousness-raising groups, in men's studies college courses, and, increasingly, on the Internet—in which to air their feelings about gender.

Kimmel (1995) divides today's men's movement into three fairly distinct camps: antifeminists, profeminists, and masculinists. *Antifeminists* believe the women's movement has caused the collapse of the natural order that guaranteed male dominance, and they work to reverse this trend. As an example, the National Organization for Men (NOM) opposes feminism, which it claims is "designed to denigrate men, exempt women from the draft and to encourage the disintegration of the family" (Kimmel 1995, p. 564). Some antifeminist responses emphasize men's rights, especially fathers' rights, and no guilt in relationships with women. According to Mark Kann, men's self-interest may reasonably lead to an antifeminist response:

I would suggest as a rule of thumb that men's immediate self-interest rarely coincides with feminist opposition to patriarchy. Consider that men need money and leisure to carry out their experiments in self-fulfillment. Is it not their immediate interest to monopolize the few jobs that promise affluence and autonomy by continuing to deny women equal access to them? Further, men need social space or freedom from constraints for their experiments. Why should they commit themselves to those aspects of feminism that reduce men's social space? It is one thing to try out the joys of parenting, for example, but quite another to assume sacrificial responsibility for the pains of parenting. Is it not men's immediate self-interest to strengthen the cultural presumption that women are the prime parents and thus the ones who must diaper, chauffeur, tend middle-of-the-night illnesses, launder, and so forth? (Kann 1986, p. 32)

Profeminists support feminists in their disdain for patriarchy. They analyze men's problems as stemming from a patriarchal system that privileges white heterosexual men while forcing all males into restrictive gender roles. In 1983, profeminist men formed the National Organization for Changing Men (changed in 1990 to the National Organization for Men Against Sexism, or NOMAS), whose purposes are to transcend gender stereotypes while supporting women's and gays' struggles for respect and equality (Doyle 1989). There are also many regional profeminist men's organizations, such as RAVEN (Rape and Violence End Now) in St. Louis.

The newer *masculinists,* who emerged in the early 1990s, tend not to focus on patriarchy as problematic (although they might agree that it is). Instead, masculinists analyze their own gendered expectations and behaviors. They work to develop a positive image of masculinity, one combining strength with tenderness. The path to this is through therapy, consciousness-raising groups, and rituals. The latter encourage men to release the "wild man" from within the socialized or "civilized" man. Through rituals men are to get in touch with their inner feelings and heal the buried rage and grief caused by the oppressive nature of corporate culture, the psychological and/or physical absence of their fathers, and men's general isolation due to a learned reluctance to share their feelings. Robert Bly's *Iron John* (1990) is a prominent example of the ideas of this camp. Retreats ("wildman gatherings") encourage men to overcome the barriers to intimate friendships with each other (Kimmel 1995).

Gloria Steinem (1992) suggests the more inclusive term "Wild Child" and encourages both women and

men to release the child within by getting in touch with buried feelings of rage and resentment. Steinem argues that both women and men, having been socialized as children to limiting gender roles, experience grief over the loss of "a part of myself" (p. 160).[9] Finding and developing that other part of oneself is discussed next.

Ambivalence, Confusion—and Hope

In the 1970s, feminist social scientists typically proposed androgyny as an answer to the stresses and inequalities that result from traditional masculine and feminine expectations and conditions. **Androgyny** (formed from the Greek words *andro,* meaning "male," and *gyne,* meaning "female") is the social and psychological condition by which individuals think, feel, and behave both instrumentally and expressively (Bem 1975). In other words, androgynous persons evidence the positive qualities traditionally associated with both masculine and feminine roles. More recently, feminists have disagreed about androgyny as a model for women. Probably the majority continue to emphasize equal treatment of men and women and to encourage nonsexist child rearing that would produce similarity in personality between men and women. Others seek to acknowledge biological differences and celebrate a "women's culture" that mitigates individual ambition to emphasize communitarian values and women's nurturing capacity (Mitchell and Oakley 1986; Fox-Genovese 1991). Nevertheless, the fact remains that in a modern, complex society such as ours, people need to be assertive and self-reliant and also to depend on one another for intimacy and emotional support (Helgeson 1994). Allowing people both to develop their talents fully and to be emotionally expressive can greatly expand the range of behaviors and possibilities open to everyone.

---

9. In Steinem's words, "The more I talked to men as well as women, the more it seemed that inner feelings of incompleteness, emptiness, self-doubt, and self-hatred were the same, no matter who experienced them, and even if they were expressed in culturally opposite ways. I don't mean to gloss over the difficulties of equalizing power, even when there is the will to do so: to the overvalued and defensive, the urge to control and dominate others may be as organic as a mollusk's shell; and to the undervalued and resentful, the power to destroy the self (and others who resemble the self) may be the only power there is. But at both extremes—as well as in the more subtle areas between, where most of us struggle every day—people seemed to stop punishing others or themselves only when they gained some faith in their own unique, intrinsic worth" (1992, p. 5).

Drawing by Wm. Hamilton; © 1991 The New Yorker Magazine, Inc.

"I know this is going to sound completely crazy and off the wall—but do you, by any chance, know how to iron?"

Ironically, however, that very expansion in the range of people's opportunities has led to new ambivalences, or mixed feelings and conflicts, both within ourselves and between men and women. Although women's attitudes and behaviors began to change appreciatively more than twenty-five years ago, women continue to experience a great deal of ambivalence and conflict. Stay-at-home moms may feel stigmatized for not being employed, for instance. Others, who are employed, may wish they could stay home full-time with their families. Moreover, a wife's career success may push her into renegotiating gender boundaries at home. Today's women are typically expected both to assume primary responsibility for nurturing others and also to pursue their own achievement—goals that can conflict. This conflict is more than psychological. It is in part a consequence of our society's failure to provide support for all employed mothers in the form of adequate maternity (and paternity!) leave or day care, for example. More than women in many other industrialized societies, such as those in western Europe, American women continue to deal individually with problems of pregnancy, recovery from childbirth, and early child care as best they can. Adequate job

performance, let alone career achievement, is difficult under such conditions regardless of a woman's ability. These points are further developed in later chapters.

Currently, however, it is men's ambivalence and confusion (even anger) that receives the most attention. For one thing, husbands are still expected to succeed as principal family breadwinners—even in an economy that makes this increasingly difficult. Declining economic opportunities, coupled with criticisms of patriarchy sparked by the women's movement, can lead a man to feel unfairly picked on. Recently a male student in author Agnes Riedmann's family course wrote this, for example: "Why is it that when the male 'good provider' fails to achieve his duties, he is considered a failure, but if the woman (mother) fails in completing her duties or job, she and others will push the blame on the father? It seems no matter what, the male will be downgraded for failures of his family!" A second example comes from a man's letter to the syndicated column, "Working Woman":

Women don't need encouragement to stand up for themselves. They're all too strong, assertive, outspoken and ready for an argument. Their husbands, on the other hand, are all too anxious, apologetic, and eager to please. . . . Not a single man in this country does enough around the house, judging by your columns. . . . I work for a Fortune 500 company, furthermore, where every man I know spends a great deal of time and energy watching every word he says because if he isn't on guard every minute, he can be accused of everything from sexism to sexual harassment. (Scott 1992)

Not all men are angry, of course. Nonetheless, today's society offers them copious opportunities for ambivalence and confusion (Gerson 1993). Currently, varied cultural messages no longer provide clear-cut guidelines for men's behavior. The "new" man is expected to succeed economically and to value relationships and emotional openness. Although women want men to be sensitive and emotionally expressive, they also want them to be self-assured and confident. And they may want male leadership to relieve them of minor, or even major, choices. Taken together, today's "masculinities" are not only multifaceted but also paradoxical (Gerson 1993).

Furthermore, today's men, like women, find it increasingly difficult to have it all. If women find it difficult to combine a sustained work career with motherhood, men face a conflict between maintaining their privileges and enjoying supportive relationships. In the face of societal changes and new options, some men will choose to opt out of family living altogether, a topic discussed in Chapters 6 and 7. But others "will find that equality and sharing offer compensations to offset their attendant loss of power and privilege" (Gerson 1993, p. 274).

In Sum

Roles of men and women have changed over time, but living in our society remains a different experience for women and for men. Gendered messages and social organization influence people's behavior, attitudes, and options. Women tend to be seen as more expressive, relationship-oriented, and "communal"; men are considered more instrumental or agentic.

Stereotypes of African American men and women are more similar to each other than are those of other Americans. Generally, traditional masculine expectations require that men be confident, self-reliant, and occupationally successful and engage in "no sissy stuff." During the 1980s the "new male" (or "liberated male") cultural message emerged, according to which men are expected to value tenderness and equal relationships with women. Traditional feminine expectations involve a woman's being a man's helpmate and a "good mother." An emergent feminine role is the successful "professional woman"; when coupled with the more traditional ones, this results in the "superwoman."

The extent to which men and women differ from one another and follow these cultural messages can be visualized as two overlapping normal distribution curves. Means differ according to cultural expectations, but within-group variation is usually greater than between-group variation. An exception is male dominance, evident in politics, in religion, and (although this is changing for many men) in the economy.

Biology interacts with culture to produce human behavior, and the two influences are difficult to separate. Sociologists give greater attention to the socialization process, for which there are several theoretical explanations. Overall, however, sociologists stress socialization in the family, in play and games, and in school as encouraging gendered attitudes and behavior.

Turning our attention to the actual lives of adults, we find women and men negotiating gendered expectations and making choices in a context of change at work and in relationships. Change brings mixed responses, depending on class, racial/ethnic membership, religion, or other social indicators. New cultural ideals are far from realization, and efforts to create lives balancing love and work involve conflict and struggle.

Key Terms

androgyny	expressive character traits
assignment	femininities
borderwork	gender
Chodorow's theory of gender	gendered
	gender role

hermaphrodites
hormones
instrumental character
traits
internalize
male dominance

masculinities
self-identification theory
sex
socialization
social learning theory

Study Questions

1. What are some characteristics generally associated with males in our society? What traits are associated with females? How do these affect our expectations about the ways men and women behave?

2. How are our role expectations for men and women related to changes in the larger society?

3. What evidence is used to demonstrate the influence of heredity on gender-linked behavior? To show the influence of the cultural environment? Describe the research and findings of Money and Ehrhardt on this issue.

4. How are people socialized to adhere, often unconsciously, to culturally acceptable gender roles? Can you give some examples of how you were socialized into your gender role?

5. Discuss Chodorow's theory of gender. How does it help to explain how and why men tend to develop autonomy and women a capacity for relatedness?

6. How have traditional gender roles been stressful for men and women? How have changing roles altered the stresses on men and women?

7. Women and men may renegotiate and change their gendered attitudes and behaviors as they progress through life. What evidence do you see of this in your own or others' lives?

8. Describe the women's movement and the men's movement. What evidence is there of a backlash against the women's movement?

9. Explain the concept of "androgyny" as an alternative to traditional gender roles. Why might people in our society feel ambivalent about androgyny?

10. What do we mean by "masculinities" and "femininities" today?

Suggested Readings

Bly, Robert. 1990. *Iron John: A Book About Men,* Reading, MA: Addison-Wesley. The much publicized bible of the latest branch in the promasculinist camp of the men's movement.

Davis, Angela Y. 1990. *Women, Culture, and Politics.* New York: Vintage. Most recent book by an important black leader.

Faludi, Susan. 1991. *Backlash: The Undeclared War Against American Women.* New York: Crown. A vigorous response to resistance to the women's movement for equality. Discusses many of the academic and media controversies touched on in this text.

Freeman, Jo, ed. 1995. *Women: A Feminist Perspective.* 5th ed. Mountain View, CA: Mayfield. Classic feminist reader touching many aspects of women's lives.

Gerson, Kathleen. 1993. *No Man's Land: Men's Changing Commitments to Family and Work.* New York: HarperCollins, Basic Books. Readable and convincing analysis of the changing options and expectations for men in today's constricting economy, along with the various choices they make.

Giddings, Paula. 1984. *When and Where I Enter: The Impact of Black Women on Race and Sex in America.* New York: Morrow. History of black women in America.

Gilmore, David D. 1990. *Manhood in the Making: Cultural Concepts of Masculinity.* New Haven, CT: Yale University Press. A cross-cultural analysis of what it means to be a "real man" in various cultures around the world.

Gordon, Richard A. 1990. *Anorexia and Bulimia: Anatomy of a Social Epidemic.* Cambridge, MA: Basil Blackwell. An analysis of these eating disorders as culturally influenced and at least partly resulting from anxiety related to a changing female role.

Green, Rayna. 1980. "Native American Women." *Signs* 6:248–67.

Grimes, Ronald L. 1995. *Marrying and Burying: Rites of Passage in a Man's Life.* Boulder, CO: Westview. An interesting and readable study and analysis of men's life transitions.

Kimmel, Michael S. 1995. *Manhood in America: A Cultural History.* Berkeley: University of California Press. This exploration of how manhood has been historically defined in U.S. culture argues that it was not until about the 1890s that manhood became masculinity—something that had to be proven through sports, fraternities, and fashion.

Kimmel, Michael S. and Michael A. Messner, eds. 1992. *Men's Lives.* 2d ed. New York: Macmillan. Important readings on men and male roles.

Majors, Richard G. and Janet M. Billson. 1992. *Cool Pose: The Dilemmas of Black Manhood in America.* Lexington, MA: Heath.

Messner, Michael A. 1992. *Power at Play: Sports and the Problem of Masculinity.* Boston: Beacon. Sociologist Messner interviewed thirty male former athletes and used these data to speculate about how our "jock culture" creates barriers to intimacy for men. He concludes that boys and men, as well as girls and women, can find camaraderie and pleasure in sports, but the rules of the game must change first.

Orenstein, Peggy. 1994. *SchoolGirls: Young Women, Self-Esteem, and the Confidence Gap.* New York: Doubleday. A very good book, written in association with the American Association of University Women, on young women in today's middle schools. Based on naturalistic observation in two California schools.

Sidel, Ruth. 1990. *On Her Own: Growing Up in the Shadow of the American Dream.* New York: Viking. A study of women in their 20s, who fall into three categories we might think of as gendered roles: (1) women who are pursuing the traditionally male professional/managerial career path, (2) women who would like to recreate the traditional housewife role, and (3) women who seem to be drifting. Sidel points out the contradictions between dreams and reality for many young women today.

Smith, Barbara, ed. 1983. *Home Girls: A Black Feminist Anthology.* New York: Kitchen Table-Women of Color Press. A collection of essays by feminists of color, including "Black Macho and Black Feminism" by Linda C. Powell.

Tavris, Carol. 1992. *The Mismeasure of Woman.* New York: Simon & Schuster. A review of research and a critique of common misconceptions and stereotypes of women, both old and new. Chapter 1 includes a critical review of sociobiological perspectives on gender.

Thorne, Barrie and Marilyn Yalom, eds. 1992. *Rethinking the Family: Some Feminist Questions.* Rev. ed. Boston: Northeastern University Press. This book raises and addresses thorny questions, such as why women in industrialized societies will continue to choose to have children when their parenting is seldom or meagerly supported by advanced capitalistic societies.

Tsuchida, Nobuya, ed. 1982. *Asian and Pacific American Experiences: Women's Perspectives.* Minneapolis: Asian/Pacific American Learning Resource Center and General College, University of Minnesota. Family living experiences from Asian women's points of view.

Becoming Partners

Individuals form relationships and fall in love. The various facets of choosing partners and falling in love—or not falling in love—are addressed in Part Two. Falling in love—indeed, the word *love* itself—has vastly different meanings to different people. In Chapter 4 you'll see what the authors (backed, we think, by counseling psychologists) mean by love. Loving is a caring, responsible, and sharing relationship involving deep feelings. Chapter 5 describes the sexual aspects of ourselves and of our partnerships. A sexual relationship can be a profound way to communicate love because partners literally and symbolically shed their protective coverings.

But not everyone chooses to fall in love, and among those who do, not all couples' intimate relationships lead to marriage. Chapter 6 discusses being single. The distinction between single and married is less sharp today than it used to be, as legally single individuals can be living together in a relationship or in other kinds of families. Advantages and drawbacks of being single are discussed. Singles are compared with married men and women in terms of loneliness and life satisfaction.

Even though more people are single today than in the fifties and sixties, the majority will choose a marriage partner at least once. Chapter 7 examines the process of choosing a spouse and influences on that process. It also shows how one goal in becoming partners—gaining a loved one's commitment to getting married—can conflict with the essence of intimacy—honest self-disclosure.

Loving Ourselves and Others

Americans love love. We marry and remarry for love. Because it involves the will, "love implies choice. We do not have to love. We choose to love." But when asked what love is, most of us have trouble answering. Love *is* difficult to define; in attempting to do so, we "toy with mystery" (Peck 1978).

In this chapter we'll discuss the need for loving in today's society and describe one writer's view of some styles love takes. We'll examine what love is (and isn't). We'll explore the idea that self-esteem, or loving oneself, is a prerequisite to loving others. We begin by looking at what love means in impersonal, modern society.

Personal Ties in an Impersonal Society

We say that modern society is impersonal because so much of the time we are encouraged to think and behave in ways that deny our emotional need to be cared for and to care for others (Bellah et al. 1985). An impersonal society exaggerates the rational aspect of human beings and tends to ignore people's feelings and their need for affection and human contact.[1] We expect people to leave their friends and relatives to move to job-determined new locations because it is an economically efficient way of organizing production and management. Salespeople and service employees (flight attendants, for example) may be required to display certain emotions and repress others as part of their job (Hochschild 1983).

In the context of this impersonality, most people search for at least one caring person with whom to share their private time. Some find love and nourish it;

some don't discover loving at all. But we all need it. Physicians point out that loving enhances physical health, and many psychologists insist that loving is essential for emotional survival. It helps confirm a person's sense of individual worth.

What Is Love?

Love exists between parents and children, and between people of the same and opposite sexes. Love may or may not involve sexuality. When it does involve sexuality, love can be heterosexual or homosexual. Psychoanalyst Rollo May defines love as "a delight in the presence of the other person and an affirming of his [or her] value and development as much as one's own" (1975, p. 116). We include this definition here because we like it and think it complements ours. But for our purposes we will define love in more detail. **Love** is a deep and vital emotion resulting from significant need satisfaction, coupled with a caring for and acceptance of the beloved and resulting in an intimate relationship. We'll discuss each part of this definition.

Love Is a Deep and Vital Emotion

An **emotion** is a strong feeling, arising without conscious mental or rational effort, that motivates an individual to behave in certain ways. Loving parents, for example, are motivated to see what is wrong if their child begins to cry. Anger, reverence, and fear are other emotions that also evoke certain behaviors. When people get angry, for instance, they may feel like screaming or throwing things. Emotions are sometimes difficult for social scientists to come to grips with. Yet, even the most careful of researchers are able to find the poetry of love in their psychological data. The part of the brain associated with fantasy rather than rationality is most closely connected with emotional response and spontaneous behavior toward others (McClelland 1986).

Love Satisfies Legitimate Personal Needs

Human beings need recognition and affection, and a second element of love is that it fills this basic need. Loving enables people to fulfill their needs for nurturance, creativity, and self-revelation.

It's all very well to state that a person's emotional needs can be fulfilled by love, but what kind of—and how many—needs can we expect to be satisfied? Psy-

---

1. This impersonality of modern society has been a principal concern of sociologists (see Durkheim 1893; Weber 1948; Simmel 1950; Berger, Berger, and Kellner 1973) since sociology first appeared as a distinct discipline in the Western world, roughly at the time of the Industrial Revolution.

Love is a deep and vital emotion, coupled with a caring for and acceptance of the beloved, and resulting in an intimate relationship.

chologists stress that love cannot fulfill all needs, and they distinguish between legitimate and illegitimate needs.

LEGITIMATE NEEDS   Sometimes called "being needs," **legitimate needs** arise in the present rather than out of deficits accumulated in the past (Crosby 1991, pp. 50–51). People who have a healthy self-concept expect emotional support and understanding, companionship, and often sexual sharing from their partners. But they do not expect their partners to make them feel lovable or worthwhile; they already take those things for granted. Achieving a sense of individuality and personal identity is a step that best precedes relationship formation (Vannoy 1991). People's legitimate need in loving becomes the desire to share themselves with loved ones to enrich their lives (Maslow 1943).

ILLEGITIMATE NEEDS   Sometimes called "deficiency needs," **illegitimate needs** arise from feelings of self-doubt, unworthiness, and inadequacy. People who feel deficient often count on others to convince them they are worthwhile (Crosby 1991, pp. 50–54). "If you are not eternally showing me that you live for me," they seem to say, "then I feel like nothing" (Satir

1972, p. 136). They strive to borrow security from others.

To expect others to fill such needs is asking the impossible: No amount of loving will convince a person that he or she is worthwhile or lovable if that person doesn't already believe it. Hence, illegitimate needs for affection are insatiable. Love does satisfy legitimate needs, however.

Love Involves Caring and Acceptance

A third element of love is the acceptance of partners for themselves and "not for their ability to change themselves or to meet another's requirements to play a role" (Dahms 1976, p. 100). People are free to be themselves in a loving relationship, to expose their feelings, frailties, and strengths.

Related to this acceptance is caring: the concern a person has for the partner's growth. Rollo May defines this caring as a state "in which something does *matter*. . . . [It is] the source of eros, the source of human tenderness. It is a state composed of the recognition of another; a fellow human being like one's self; of identification of one's self with the pain or joy of the other" (1969, p. 289). Ideally, lovers support and encourage each other's personal growth. If a loved one wants to join a softball league or spend some evenings

Committed partners work on their relationship, express themselves authentically, and have fun together. Committed love can also give partners the confidence to step outside their own relationship and share their lives with others in the community.

studying, for example, he or she is encouraged to do so. At the same time, partners respond to each other's needs for affection by recognizing budding feelings of insecurity or jealousy in each other and trying to be reassuring: "Just because I want to play softball doesn't mean I don't still love you," they explain; or "I love you, but I want to study now; let's get ice cream later."

Do Men and Women Care Differently?

Note also that experiencing feelings and expressing them may not be entirely the same thing. Sociologists have observed that in our society women verbally express feelings of love more than men do. Various writers have expanded Chodorow's theory of gender (see Chapter 3) to illustrate how girls' and boys' differing socialization experiences affect male–female relationships later in life. Men are encouraged to develop independence and to separate emotionally from their mothers, and by extension to give autonomy priority over intimacy. They bring a sharper sense of self to relationships, but at the expense of interdependence.

Research indicates that women today do not believe that they should be more self-sacrificing in relationships (Heiss 1991). But their socialization has been directed more strongly to attachment than to autonomy. Women have been raised to be more aware of both their feelings and ways of communicating them. Men tend to be more often baffled by questions about inner feelings.

During the twentieth century, seeking emotional satisfaction in marriage has been part of a general trend toward self-development and the cultivation of emotional intimacy for both sexes. Before the nineteenth century, men's and women's domestic activities involved economic production, not personal intimacy. With the development of separate gender spheres, love and *feeling* became the domain of women, whereas *work* was seen to be the appropriate masculine mode. Men, consequently, felt estranged from the "female world of love and ritual" (Smith-Rosenberg 1975), the social world of women that is anchored in the family and close friendships between women. We have come to see men as inadequately equipped for emotional relatedness (Cancian 1987).

Women tend to view sex as one of several means for communicating an already established emotional closeness. Men, on the other hand, tend to view sex as *the* emotional communication (Rubin 1983; Cancian 1985; Critelli, Myers, and Loos 1986; Vannoy 1991). Sociologist Francesca Cancian maintains that men are equally loving, but that love is expressed in our society on feminine terms and women are the verbal sex. Nonverbal expressions of love such as men may make

Love involves intimacy, the capacity to share one's inner self with someone else and to commit oneself to that person despite some personal sacrifices. But love—and commitment—aren't meant to be all work. Love needs to feel supportive and fun, at least sometimes, as well.

through doing favors, reducing their partners' burdens, and so on, are not credited. This is especially true for middle-class as opposed to working-class people.

Cancian argues that in our society women, not men, are made to feel primarily responsible for love's endurance or success. This situation results in *both* partners feeling manipulated and powerless. When there are problems in a relationship, the woman is likely to propose discussing them. But for a man, "talking about the relationship, like she wants, feels like taking a test that she made up and he will fail." Probably he won't refuse to talk outright because he, like she, has been taught that verbal intimacy is good. So his typical response is withdrawal or passive aggression, behavior that will "probably make her feel helpless and controlled" (Cancian 1985, p. 260). "The consequences of love would be more positive if love were the responsibility of men as well as women and if love were defined more broadly to include instrumental help as well as emotional expression" (p. 262).

Cancian also argues that a more balanced view of how love is to be expressed—one that includes masculine as well as feminine elements—would find men equally loving and emotionally profound. Less abstractly, a lover could consider the possibility that the partner expresses love differently and accept such differences, or else negotiate change openly. It is possible that as gender roles change generally, men and women will develop more balanced capacities for autonomy and intimacy, "the very capacities necessary for sustaining the loving relationships on which marriages now depend" (Vannoy 1991, p. 262).

Love and Intimacy: Commitment to Sharing

Love involves **intimacy,** the capacity to share one's inner self with someone else and to commit oneself to that person despite some personal sacrifices. We'll look more closely at two elements of that definition: first, the **commitment** involved in intimacy; and second,

Sharon and Gary: Discovering Love After Twenty-Five Years

Married at ages 16 and 18, respectively, Sharon and Gary were together nearly twenty-five years and had four children before separating for ten months. After the separation they got back together, publicly restating their wedding vows in a religious ceremony. Here they talk about what commitment to loving means to them.

INTERVIEWER: How did you guys get back together?

SHARON: Connection. For me it was that I had to be connected. I just need someone regular to check in with, to have dinner with, to care where you are and what you are doing.

GARY: For me it was just love. For a long time I didn't love Sharon after we got married, and then I grew to love her very much. Then when we separated—it was both of our idea to separate—I think once she moved out it was an empty place. I knew that I just needed her back. I needed to have her there to share my love with her. It kinda took her moving out to really find that, to determine that was what I wanted.

SHARON: I think our troubles started when we got married. I was pregnant, our parents didn't approve, and we both thought we were doing the right thing. We tried for years to do the right things, and it wasn't quite right. A lot of things

we never talked about and buried, they got deep... he told me he didn't love me when we first got married.... I got married because I loved him like crazy and I knew I could make it work. I tried really hard. I did all the right things. I thought if we had a child it would tie us together.

GARY: I got married because it was the honorable thing to do. I don't regret being married. I don't regret being married as young as I was.... It took a few years and I ended up loving her.

INTERVIEWER: You got to know Sharon?

GARY: I don't know if I've ever got to know Sharon. Probably the last year I have got to know Sharon more than I had up until then.

INTERVIEWER: Why's that?

GARY: That's because we didn't say nothing. We could argue and neither one of us would tell why we were mad or what the problem was.

SHARON: We didn't argue that much either.

GARY: No.

SHARON: Just glaring, that kind of stuff.... We were young, and I used to cry for his attention. And when I cried, he left. A few of those and you think, "What good is this doing me?" So I quit crying.

---

the experience of sharing intimacy. Then we'll look at the triangular theory of love, which tries to put it all together.

COMMITMENT   In love, committing oneself to another person involves the determination to develop a relationship "where experiences cover many areas of personality; where problems are worked through; where conflict is expected and seen as a normal part of the growth process; and where there is an expectation that the relationship is basically viable and worth-

while" (Altman and Taylor 1973, pp. 184–87). Case Study 4.1, "Sharon and Gary: Discovering Love After Twenty-Five Years," illustrates commitment.

Committed lovers have fun together; they also share more tedious times. They express themselves freely and authentically. Committed partners do not see problems or disagreements as indications that their relationship is over. They view their relationship as worth keeping, and they work to maintain it in spite of difficulties. Commitment is characterized by this willingness to work through problems and conflicts as

GARY: Well, if I had known what else to do, I probably would have done that. Today I give her a hug and try to hold her. Back then, you know, laughter was easier.

SHARON: I don't remember if I ever told him what I needed then or not. I don't think so. I just wanted comfort and understanding, I guess. I think I usually cried over something—lack of money or fears.

GARY: We had plenty of all that.

SHARON: I found out with the separation, though, finally I realized that Gary loved me. The turning point was when Gary made a commitment to help me get into the apartment. It made me look at him differently. It was the commitment; it was obvious that he cared—and not for his own personal thing. He wasn't just saying to get out. He was saying, "I want you to be well." He came to help me do that.

INTERVIEWER: What did he do?

SHARON: Painted, laid carpet and ripped wall paper, those things.

GARY: I was doing that because I cared about her. It surprised me that it was the turning point because I was helping her get out. We had set this date that she had to be moved, but where she was moving to—it was impossible to move in there at that date. So we set another week, and we cleaned and got the place ready for her. That was the time that she decided that I cared for her.

SHARON: I remember it was a Sunday that I first got into the apartment where I was going to move to. It was pretty awful. I called Gary and said could I have another week, and he came over and brought breakfast....

INTERVIEWER: Why did you guys have a public ceremony when you got back together?

GARY: I think one of the reasons that I wanted it was because there were a lot of our friends who knew what had gone on between us. I just wanted them to be part of our getting back together.... It was very touching to me, very emotional. We never had anything like this. We got married by the justice of the peace—very cold, very. I don't know who was the witness now, just some person in the courthouse at the time.... That wasn't emotional; it was cold. It was just: "Let's get this over with so people can start counting the months." And hoping that the baby would come up a month late or something.

SHARON: This time we invited our whole church and our whole square-dance club and our whole group of people that we both work with.... We called our parents and asked for their blessing.

GARY: What they wanted to do was to make believe that we had already done this 25 years ago.

SHARON: But it was important that they came this time.

GARY: [At this ceremony] I felt a lot of love. We had it taped. I've watched the tape lots of times already. In the tape I see it. Sharon's trying to hold this thing together. She doesn't want to cry. She is just trying to hold a stiff lip, you know, and keep this thing going. You can see every once in a while, when I would break down and cry, she would try to stiffen up for the whole bunch.

SHARON: Sometimes I do that too well.

How does Gary and Sharon's story illustrate the idea of commitment in loving? How does it illustrate Francesca Cancian's (1985) view that men love differently than women do and that a more balanced view of what loving is would find men equally loving?

opposed to calling it quits when problems arise. In this view, commitment need not include a vow to stay together exclusively for life or even for a certain period of time. But it does imply that love involves effort: Committed partners "regularly, routinely, and predictably attend to each other and their relationship no matter how they feel" (Peck 1978, p. 118).

PSYCHIC AND SEXUAL INTIMACY   Besides commitment, intimacy involves sharing. This sharing can take place on two often overlapping planes. At one

level is **sexual intimacy.** In popular terminology, people who have a sexual relationship are "intimate" with each other. At another level is **psychic intimacy:** people sharing their minds and feelings. This is the sense in which we use the term in this book. Although sexual intimacy can either result from or lead to psychic intimacy, the two concepts are not synonymous. Strangers and people who like each other can enjoy sexual intimacy. Those who share with and accept each other experience psychic intimacy. They engage in the "work of attention": making the effort to set

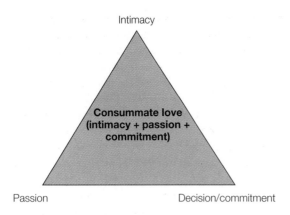

Intimacy

Consummate love (intimacy + passion + commitment)

Passion   Decision/commitment

FIGURE *4.1*

The three components of love: triangular theory. (Source: Adapted from Sternberg 1988, p. 121)

aside existing preoccupations in order to listen to each other (Peck 1978, pp. 120–21). Research on married couples indicates that partners who more openly express feelings of love to each other score higher on measures of marital adjustment (Davidson, Balswick, and Halverson 1983).

THE TRIANGULAR THEORY OF LOVE Psychological research expands upon these notions but presents a more complex picture. Psychologist Robert Sternberg believes that the qualities most important to a lasting relationship are not so visible in the early stages. In his research on relationships varying in length from one month to thirty-six years, he found three components of love: intimacy, passion, and commitment.

According to **Sternberg's triangular theory of love, intimacy** "refers to close, connected, and bonded feelings in a loving relationship. It includes feelings that create the experience of warmth in a loving relationship . . . [such as] experiencing happiness with the loved one; . . . sharing one's self and one's possessions with the loved one; receiving . . . and giving emotional support to the loved one; [and] having intimate communication with the loved one." **Passion** "refers to the drives that lead to romance, physical attraction, sexual consummation, and the like in a loving relationship." **Commitment**—actually, the "decision/commitment component of love" consists of "two aspects, one short term and one long term. The short-term one is the de-

cision that one loves someone. The long-term aspect is the commitment to maintain that love" (Sternberg 1988, pp. 120–21).

Sternberg used these three dimensions to create a typology of "love" (see Figure 4.1). (The word is in quotes because many types of love would not meet our definition of love for romantic partners.) **Consummate love,** composed of all three components, is "complete love, . . . a kind of love toward which many of us strive, especially in romantic relationships. Attaining consummate love can be difficult, but keeping it is even harder. We do not seek consummate love in all our loving relationships or even in most of them. Rather, we tend to reserve it for those loves that mean the most to us and that we want to make as nearly complete as possible" (Sternberg 1988, p. 129).

The three components of consummate love develop at different times, as love grows and changes. "Passion is the quickest to develop, and the quickest to fade. . . . Intimacy develops more slowly, and commitment more gradually still" (Sternberg, quoted in Goleman 1985). Passion peaks early in the relationship but continues at a stable lower level and is important to the long-term maintenance of the relationship. Intimacy, which includes understanding each other's needs, listening and supporting each other, and sharing common values, becomes increasingly important. In its most emotional form, intimacy may not always be visible, but it emerges when the relationship is interrupted—for example, by travel, sickness, or death. Commitment is essential, but a commitment without intimacy and passion is hollow. In other words, all these elements of love are important. Because these components each develop at a different rate and so exist in various combinations of intensity, no relationship is stable but rather will be always changing (Sternberg 1988).

The triangular theory of love has been explored further in research (Chojnack and Walsh 1990). Sternberg developed a scale measuring the three dimensions. Researchers have found a good deal of overlap in measurements of the three dimensions, and modifications of the scale have improved it. Social scientists find the theory conceptually appealing, and reviews of a number of theories of love and their measures identify similar elements. Generally, commitment has been found to be the factor that is most predictive of happiness in relationships. Initial research finds passion to decline over time, as Sternberg theorizes, but only for women. Intimacy did not decline in longer relationships (Hendrick and Hendrick 1989; Acker and Davis 1992).

Of course, the triangular theory of love is not the only way of looking at love. One interesting typology developed by social scientist John Alan Lee looks at the wide individual variation in love styles.

Six Love Styles

Loving relationships can take many forms or personalities, just as the individuals in a relationship can.

John Alan Lee classified six love styles, initially based on interviews with 120 respondents, half of them male and half female. All were heterosexual, white, and of Canadian or English descent. Lee subsequently applied his typology to an analysis of gay relationships (Lee 1981). Although not all subsequent research has found all six dimensions, this typology of love styles has withstood the test of time and has proven to be more than hypothetical (Borello and Thompson 1990).

Love styles are distinctive characteristics or personalities that loving or lovelike relationships can take. The word *lovelike* is included in this definition because not all love styles are genuine loving as we have defined it. Moreover, people do not necessarily confine themselves to one style or another; they may incorporate different aspects of several styles into their relationships. In any case, these love styles tell us that people can love passionately, quietly, pragmatically, playfully, self-sacrificingly, and crazily.

EROS **Eros** (pronounced "AIR-ohs") is a Greek word meaning "love"; it forms the root of our word *erotic.* This love style is characterized by intense emotional attachment and powerful sexual feelings or desires. When erotic couples establish sustained relationships, these are characterized by continued active interest in sexual and emotional fulfillment, plus the development of intellectual rapport. Romeo and Juliet, who fell in love on first meeting, were erotic lovers; so are the older couple you know who seem endlessly fascinated by each other.

STORGE **Storge** ("STOR-gay") is an affectionate, companionate style of loving. This love style focuses on deepening mutual commitment, respect, friendship over time, and common goals. Whereas eros emphasizes emotional intensity and sexual passion, storge

does not. Sexual intimacy comes about as partners develop increasing understanding of one another. The storgic lover's basic attitude to his or her partner is one of familiarity: "I've known you a long time, seen you in many moods" (Lee 1973, p. 87); my partner is my friend.

Storgic lovers are usually seen to have a lot in common. These lovers, who are also friends, may not understand the sudden attraction of the erotic couple who seem such a surprising combination.

PRAGMA **Pragma** ("PRAG-mah") is the root word for *pragmatic.* Pragmatic love emphasizes the practical element in human relationships, particularly in marriages. Pragmatic love involves rational assessment of a potential partner's assets and liabilities. Here a relationship provides a practical base for both economic and emotional security. The arranged marriages of royalty or the pairing of the children of business partners or rivals are visible examples of pragma. But so is the person who decides very rationally to get married to a suitable partner because it's time.

AGAPE **Agape** ("ah-GAH-pay") is a Greek word meaning "love feast." Agape emphasizes unselfish concern for the beloved's needs even when that means some personal sacrifice. Often called *altruistic love,* agape emphasizes nurturing others with little conscious desire for return other than the intrinsic satisfaction of having loved and cared for someone else. The sexual component of love seems less important in agape.

LUDUS **Ludus** ("LEWD-us") focuses on love as play or fun. Ludus emphasizes the recreational aspects of sexuality and the enjoyment of many sexual partners rather than searching for one serious relationship. Of course, ludic flirtation and playful sexuality may be part of a more committed relationship based on one of the other love styles.

MANIA **Mania,** a Greek word, designates a wild or violent mental disorder, an obsession or craze. Mania rests on strong sexual attraction and emotional intensity, as does eros. It differs from eros, however, in that manic partners are extremely jealous and moody, and their need for attention and affection is insatiable. Manic lovers alternate between euphoria and depression. The slightest lack of response from the love

Agape is a love style that emphasizes unselfish concern for another's needs. Often called altruistic love, agape emphasizes nurturing. This love style exists between partners and also between and among other family members.

partner causes anxiety and resentment; any small sign of warmth evokes enormous relief. It is probably safe to say that manic lovers lack self-esteem, discussed later in this chapter.

We may learn of manic love in the newspaper or on TV when a relationship ends violently. But more humdrum versions of mania exist. Many of us have experienced the would-be or former lover who just won't quit or have been appalled at our own jealousy or obsessiveness in love.

These six love styles represent different ways people can feel about and behave toward each other in love-like relationships. In real life a relationship is never entirely one style, and the same relationship may be characterized, at different times, by features of several styles. Lovers can be erotic or pragmatic. Loving can assume qualities of quiet understanding and respect,

along with playfulness. We turn now to an examination of two things love isn't.

Two Things Love Isn't

Love is not inordinate self-sacrifice. And loving is not the continual attempt to get others to feel or do what we want them to—although each of these ideas is frequently mistaken for love. Next we'll examine these misconceptions in detail.

Martyring

Love isn't martyring. **Martyring** involves maintaining relationships by giving others more than is received in return. Martyrs usually have good intentions; they believe that loving involves doing unselfishly for others without voicing their own needs in return. Consequently, they seldom feel that they receive genuine affection. Martyrs do more for the relationship—or feel that they do—than do their partners. Martyrs may

- Offer to do things for others that others can, should, or would prefer to do for themselves

- Be reluctant to suggest what they would like (concerning recreation or entertainment, for example) and leave decisions to others

- Allow others to be constantly late for engagements and never protest directly

- Work on helping loved ones develop talents and interests while neglecting their own

- Be sensitive to others' feelings and problems while hiding their own disappointments and hurts

Although it sounds noble, there's a catch to martyring. Aware that they're not receiving as much as they're giving, martyrs grow angry, even though they seldom express their anger directly. (In Chapter 9 we will discuss how unexpressed anger can damage a loving relationship.) Psychologists tell us that martyrs often think "it is better to be wanted as a victim than to not be wanted at all" (Szasz 1976, p. 62). The reluctance of martyrs to express even legitimate needs is damaging to a relationship, for it prevents openness and intimacy.

Martyring has other negative consequences. Social psychologists have been researching the concept of equity, the balance of rewards and costs to the partners in a relationship. In love relationships and marriage, as well as in other relationships, people seem most comfortable when things are fair or equitable—that is, when partners are reasonably well balanced in terms of what they are giving to and getting from the relationship. (We will discuss this theoretical perspective, termed *exchange theory*, in Chapter 7.)

Manipulating

Manipulators follow this maxim: If I can get him (or her) to do what I want done, then I'll be sure he (or she) loves me. **Manipulating** means seeking to control the feelings, attitudes, and behavior of your partner or partners in underhanded ways rather than by assertively stating your case. Manipulating is not the same thing as love. Manipulators may

- Ask others to do things for them that they could do for themselves, and generally expect to be waited on

- Assume that others will (if they "really" love them) be happy to do whatever the manipulators choose, not only regarding recreation, for example, but also in more important matters

- Be consistently late for engagements ("if he [or she] will wait patiently for me, he [or she] loves me")

- Want others to help them develop their interests and talents but seldom think of reciprocating

Manipulators, like martyrs, do not believe that they are lovable or that others can really love them; that is why they feel a continual need to test their partner. Aware that they are exploiting others, habitual manipulators often experience guilt. They try to relieve this guilt by minimizing or finding fault with their loved one's complaints. "You don't really love me," they may accuse. Manipulating, along with the guilt that often accompanies it, can destroy a relationship.

You may already have noticed that martyring and manipulating complement each other. Martyrs and manipulators are often attracted to each other, forming what family counselor John Crosby (1991) calls **symbiotic relationships** in which each partner depends on the other for a sense of self-worth. Often, such symbiotic relationships are quite stable. Should one symbiotic partner learn to stand on his or her own two feet, however, the relationship is less likely to last.

Manipulating and martyring are both sometimes mistaken for love. But they are not love, for one simple reason: Both relationships share a refusal to accept oneself or one's partner realistically—a quality we'll examine in greater detail now. The next section examines a necessary prerequisite for loving: self-esteem.

Self-Esteem as a Prerequisite to Loving

I'm more interested," writes Leo Buscaglia, "in who is a loving person. . . . I believe that probably the most important thing is that this loving person is a person who loves him [or her] self" (1982, p. 9).

According to many psychologists, high self-esteem is a prerequisite for loving others because, among other reasons, loving requires self-disclosure—sharing who each of us is and how we feel. Individuals who are higher in self-esteem experience romantic love more frequently and report richer love experiences (Dion and Dion 1988).

Drawing by Ziegler; © 1991 The New Yorker Magazine, Inc.

Self-esteem is part of a person's self-concept; it involves feelings people have about their own worth. Social psychologist Stanley Coopersmith defines self-esteem as an evaluation a person makes and maintains of her- or himself. "It expresses an attitude of approval or disapproval, and indicates the extent to which the individual believes himself [or herself] to be capable, significant, successful, and worthy" (Coopersmith 1967, pp. 4–5). More recently, psychologist Nathaniel Branden has defined self-esteem as "the disposition to experience oneself as competent to cope with the basic challenges of life and as worthy of happiness" (1994, p. 21).

Branden (1994), as well as other contemporary psychologists, argues that self-esteem is a *consequence,* not simply an uncontrollable condition. Because self-esteem is a consequence of all the little things we do, we can enhance it purposefully. Accordingly, Branden identifies six "pillars" of self-esteem:

1. The practice of living consciously, or aware of what's going on
2. The practice of self-acceptance
3. The practice of self-responsibility
4. The practice of self-assertiveness
5. The practice of living purposefully
6. The practice of personal integrity

You can evaluate your own self-esteem by answering questions in Box 4.1, "A Self-Esteem Checklist." Box 4.2, "Learning to Love Yourself More," gives suggestions for enhancing self-esteem.

Self-Love Versus Narcissism

In defining self-esteem we should also explain a related concept (actually, a synonym in this book's context), that of *self-love.* People commonly confuse self-love with conceit—a self-centered, selfish outlook. But psychologists point out that self-love and narcissism are opposites. Self-love is the same as high self-esteem, and it enhances a person's capacity to love others. **Narcissism** is concerned chiefly or only with oneself, without regard for the well-being of others. Narcissism results from low self-esteem. Disregarding others' needs results from preoccupation with one's own feelings of insecurity, together with the desire to compensate for such feelings.

Self-love—or "honoring the self"—is a habit worth cultivating. To honor the self is to

think independently, to live by our own mind, and to have the courage of our own perceptions and judgments. . . .

know not only what we think but also what we feel, what we want, need, desire, suffer over, are

A Self-Esteem Checklist

In this brief self-test you can broadly evaluate your own level of self-esteem.

- Do you believe strongly in certain values and principles, so that you are willing to defend them?

- Do you act on your own best judgment, without regretting your actions if others disapprove?

- Do you avoid worrying about what is coming tomorrow or fussing over yesterday's or today's mistakes?

- Do you have confidence in your general ability to deal with problems, even in the face of failures and setbacks?

- Do you feel generally equal—neither inferior nor superior—to others?

- Do you take it more or less for granted that other people are interested in you and value you?

- Do you accept praise without pretense of false modesty and accept compliments without feeling guilty?

- Do you resist the efforts of others to dominate you, especially your peers?

- Do you accept the idea—and admit to others—that you are capable of feeling a wide range of impulses and desires, ranging from anger to love, sadness to happiness, resentment to acceptance? (It does not follow, however, that you will act on all these feelings and desires.)

- Do you genuinely enjoy yourself in a wide range of activities, including work, play, creative self-expression, companionship, and just plain loafing?

- Do you sense and consider the needs of others?

If your answers to most of these questions are "yes" or "usually," you probably have high self-esteem.

Sources: Hamachek 1971, pp. 248–51; Hamachek 1992, especially pp. 355–6; see also Branden 1994.

---

frightened or angered by—and to accept our right to experience such feelings. . . .

preserve an attitude of self-acceptance—which means to accept what we are, without self-possession or self-castigation, without any pretense aimed at deceiving either ourselves or anyone else. . . .

live authentically, to speak and act from our innermost convictions and feelings. . . .

refuse to accept unearned guilt, and to do our best to correct such guilt as we may have earned. . . .

be in love with our own life, in love with our possibilities for growth and for experiencing joy, in love with the process of discovery and exploring our distinctively human potentialities.

Thus we can begin to see that to honor the self is to practice *selfishness* in the highest, noblest, and least understood sense of that word. (Branden 1980, pp. 3–4)

Whether people honor themselves in this sense—as opposed to being narcissistic—affects their personal relationships.

Self-Esteem and Personal Relationships

Research indicates that self-esteem has a lot to do with the way people respond to others. On a broader social level, people with high self-esteem feel that they have less difficulty in making friends, are more apt to express their opinions, are less sensitive to criticism, and are generally less preoccupied with themselves (Coopersmith 1967, pp. 48–71). In personal relationships,

Learning to Love Yourself More

We all feel inadequate at times. What can you do with feelings of inadequacy besides worry about them? People can work in many ways to improve their self-esteem. For example, you may choose to

- Pursue satisfying and useful occupations that realistically reflect your strengths and interests ("Young children make me nervous, but I'd like to teach, so I'll consider secondary or higher education")

- Work to develop the skills and interests you have rather than fret about those you don't

- Try being more honest and open with other people

- Make efforts to appreciate the good things you have rather than focus on the more negative things in your life

- Avoid excessive daydreaming and fantasy living ("Boy, things would be different if only I were a little taller," or "If only I had not gotten married so young, things would be better")

- "Keep on keeping on"—even when you are discouraged, realizing that one little step at a time in the right direction is often good enough

- Reevaluate the standards by which you have learned to think of yourself as inadequate—you try to be satisfied to be smart *enough*, slender *enough*, or successful *enough* rather than continue to impose unrealistic standards on yourself

- Relax ("If I feel this way, it means this is a human way to feel, and everybody else has probably felt this way too at one time or another")

- Decide to be your own good friend, complimenting yourself when you do things well enough and not criticizing yourself too harshly

- Try to lighten up on yourself, remembering that you don't have to be perfect ("Easy does it").

Sources: Hamachek 1971, 1992.

---

people with very low self-esteem often experience a persistent and insatiable need for affection. As psychologist Erich Fromm puts it, "If I am attached to another person because I cannot stand on my own two feet, he or she may be a life saver, but the relationship is not one of love" (Fromm 1956, p. 112).[2] We will explore the difference between loving and dependency in the next section.

Self-esteem affects love relationships in other ways. People with high self-esteem are more responsive to

2. Fromm chastises Americans for their emphasis on wanting to *be loved* rather than on learning *to love*. Many of our ways to be loved or make ourselves lovable, he writes, "are the same as those used to make oneself successful, 'to win friends and influence people.' As a matter of fact, what most people in our culture mean by being lovable is essentially a mixture between being popular and having sex appeal" (Fromm 1956, p. 2).

praise, whereas men and women with low self-esteem "are forever on the alert for criticism . . . and remember it long afterward." People high in self-esteem are better at picking up signs of interest from other people and responding to them, whereas people low in self-esteem often miss such cues and, in general, are "set for rejection" (Walster and Walster 1978, pp. 54–55).

Emotional Interdependence

Besides self-esteem, or self-love, a quality necessary for loving is the ability to be emotionally interdependent. Interdependence is different from both dependence and independence. **Dependence** involves reliance on another or others for continual support or assurance, coupled with subordination—being easily influenced or controlled by those who are so greatly needed. **In-**

| Love is a process of discovery, which involves continual exploration, commitment, and sharing.

dependence, on the other hand, involves self-reliance and self-sufficiency and may imply that the individual functions in isolation from others. It emphasizes separation from others.

Loving is different from both dependence and independence as we have defined them. Loving is **interdependence,** a relationship in which people with high esteem make strong commitments to each other. Therapist John Crosby has distinguished between A-frame (dependent), H-frame (independent), and M-frame (interdependent) relationships. **A-frame relationships** are symbolized by the capital letter A: Partners have a strong couple identity but little individual self-esteem. They think of themselves as a unit rather than as separate individuals. Like the long lines in the letter A, they lean on one another. The relationship is structured so that "if one lets go, the other falls" (Crosby 1991, p. 55). And that is exactly what happens—at least for a while—when one partner outgrows his or her dependency in a martyr–manipulator relationship.

H-frame relationships are structured like a capital H: Partners stand virtually alone, each self-sufficient and neither influenced much by the other. There is little or no couple identity and little emotionality: "If one lets go, the other hardly feels a thing" (Crosby 1991, p. 55). Research supports the notion that high levels of individualism are associated with a more detached style of loving (Dion and Dion 1991).

M-frame relationships rest on interdependence: Each partner has high self-esteem (unlike in the A-frame relationship), and partners experience loving as a deep emotion (unlike in the H-frame relationship). The relationship involves mutual influence and emotional support. M-frame relationships exhibit a meaningful couple identity: "If one lets go, the other feels a loss but recovers balance" (Crosby 1991, p. 55).

Acceptance of Self and Others

Loving partners need to have another quality besides self-love and interdependence—that of acceptance or empathy. Love "requires the ability to have empathy with the other person, to appreciate and affirm his [or her] potentialities" (May 1975, p. 116).

Each partner must try to understand and accept how the other perceives situations and people. If a loved one tells you he or she dislikes a friend of yours, for example, the accepting response is not "That's impossible!" but "Tell me why." Accepting relationships rest on unconditional positive regard. This doesn't mean that you, too, have to decide to dislike your friend. But even when loved ones do not share or condone specific attitudes and behavior, they accept each other as people (Dahms 1976, pp. 100–101). This, in fact, is the hallmark of a committed relationship.

Only people with high self-esteem can accept others as people. Because they accept their own feelings, anxieties, and frailties, they do not fear seeing these emotions in others. In other words, the more that people can accept themselves the more they can accept others. We will now discuss how love happens.

Love as a Discovery

Love is discovered, not just found. The words *discover* and *find* have similar meanings. But to discover involves a process, whereas to find refers to a singular act. One definition of *discover,* for example, is to reveal or expose through exploration; *finding* more often means to attain or succeed in reaching. Loving is a process of discovery. It is something people must do—and keep doing—rather than just a feeling they come upon. In Erich Fromm's words, love is "an activity, not a passive affect; it is a 'standing in,' not a 'falling for'" (1956, p. 22).

Discovering Love

It's fine to say that love is a process of discovery. But that doesn't answer the question of how love happens. Don't people fall in love when least expecting it? Do they find they're in love after knowing each other a while? Or do they work at learning to love each other? The answer to all these questions is "sometimes."

Some people know what kind of social characteristics they want in a partner before they find one, as Chapter 7 explores further. He or she must be the right age, for example, and have the appropriate occupation, area of residence, or education. These people often find partners by joining organizations in which they think it most likely to meet fitting mates.

Other people report that they fell in love on an ac-cidental first meeting. Typically, their partners also fit well-defined images of those whom they could love, but these images emphasize physical attractiveness and potential for emotional and intellectual rapport.

Still others do not have any clear image or specific demands regarding potential love partners. Not actively looking for love, they choose activities for their own sake rather than as a means to meet a partner. In these activities, however, they meet other people with the same interests.

People can find love in any of these ways and often in any combination. Psychologists warn, however, that romantically inclined individuals who insist on waiting for their ideal lover to come around the next corner may wait forever—and miss opportunities to love real people. George Bach and Ronald Deutsch counseled many single adults who hoped to find love. Often, they reported, clients ask where to meet potential partners. "Ironically, the questioner is [frequently] in the presence of some of the people he is looking for" (Bach and Deutsch 1970, p. 31). These psychologists pointed out, too, that waiting for accidental meetings to occur can be a barrier to discovering love. A more realistic alternative is simply to introduce oneself to others who seem appealing.

Once a person meets the right partner, love can begin to develop. To describe this process of development, social scientist Ira Reiss has proposed what he calls the "wheel theory of love."

The Wheel of Love

According to Reiss' theory, there are four stages in the development of love, which he sees as a circular process—a **wheel of love**—capable of continuing indefinitely. The four stages—rapport, self-revelation, mutual dependency, and personality need fulfillment—are shown in Figure 4.2, and they describe the span from attraction to love.

RAPPORT   Feelings of rapport rest on mutual trust and respect. People vary in their ability to gain rapport with others. Some feel at ease with a variety of people; others find it hard to relax with most people and have difficulty understanding others.

One factor that can make people more likely to establish rapport is similarity of background—social class, religion, and so forth, as Chapter 7 will discuss. The outside circle in Figure 4.2 is meant to convey this point. But rapport can also be established between

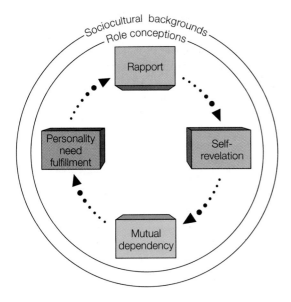

Sociocultural backgrounds
Role conceptions

Rapport

Self-revelation

Mutual dependency

Personality need fulfillment

FIGURE 4.2

Reiss' wheel theory of the development of love. (Source: Reiss and Lee 1988, p. 103)

people of different backgrounds, who may perceive one another as an interesting contrast to themselves or see qualities in one another that they admire (Reiss and Lee 1988).

SELF-REVELATION   Self-revelation, or **self-disclosure,** involves gradually sharing intimate information about oneself. People have internalized different views regarding how much self-revelation is proper. Men, for example, are socialized to be less self-disclosing than women. The middle circle, Role Conceptions, signifies that ideas about gender-appropriate behaviors influence how partners respond to each other's self-revelations and other activities. And as we saw earlier in this chapter, people with low self-esteem will likely find it difficult to share themselves fully.

Even for people with high self-esteem, however, loving produces anxiety. They fear their love won't be returned. They worry about being exploited. They are afraid of becoming too dependent or of being depended on too much. One way of dealing with these anxieties is, ironically, to let others see us as we really are and to share our motives, beliefs, and feelings.

Altman and Taylor (1973) see the self as having penetrable layers (see Figure 4.3). People's "outer lay-

ers" are easily accessible to many others. We are willing to share with the general public such information as our middle name, for example, or our religion. But the inner layers are progressively more private: These are revealed to fewer and fewer people and only after knowing them over time. As the figure shows, the self can be visualized as being opened in the shape of a wedge, gradually penetrating to the center. At the outer layers the symbolic wedge makes a wide space, allowing room for many others. As the wedge penetrates, the space gradually narrows, allowing increasingly fewer people into the center. As reciprocal self-revelation continues, an intimate relationship may develop while a couple progresses to the third stage in the wheel of love: developing interdependence or mutual dependency.

MUTUAL DEPENDENCY   In this stage of a relationship, the two people desire to spend more time together and thereby develop the kind of interdependence or, in Reiss' terminology, mutual dependency, described in the discussion of M-frame relationships. Partners develop habits that require the presence of both partners. Consequently, they begin to depend on or need each other. For example, a woman may begin to need her partner as an audience for her jokes. Or watching evening television or videos may begin to seem lonely without the other person because enjoyment now depends not only on the program but on sharing it with the other. Interdependency leads to the fourth stage: a degree of mutual personality need fulfillment.

PERSONALITY NEED FULFILLMENT   As their relationship develops, two people find they satisfy a majority of each other's emotional needs. As personality needs are satisfied, greater rapport is developed, which leads to deeper self-revelation, more mutually dependent habits, and still greater need satisfaction. Reiss uses the term *personality need fulfillment* to describe the stage of a relationship in which a stable pattern of emotional exchange and mutual support has developed. The relationship meets basic human needs, both practical and emotional.

Returning to Reiss' image of this four-stage process as a wheel, then, the wheel turns indefinitely in a lasting, deep relationship. Or the wheel may turn only a few times in a passing romance. Finally, the wheel can reverse itself and turn in the other direction. As Reiss

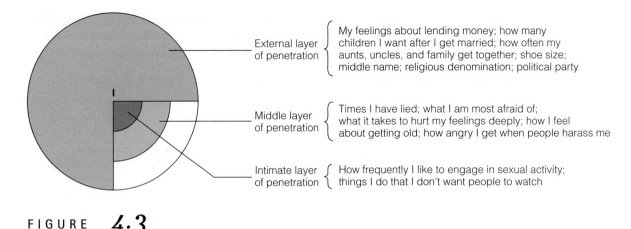

External layer of penetration } My feelings about lending money; how many children I want after I get married; how often my aunts, uncles, and family get together; shoe size; middle name; religious denomination; political party

Middle layer of penetration } Times I have lied; what I am most afraid of; what it takes to hurt my feelings deeply; how I feel about getting old; how angry I get when people harass me

Intimate layer of penetration } How frequently I like to engage in sexual activity; things I do that I don't want people to watch

FIGURE **4.3**

❙ Social penetration. (Source: Adapted from Altman and Haythorn 1965, pp. 411–26)

explains: "If one reduced the amount of self-revelation through an argument, . . . that would affect the dependency and need-fulfillment processes, which would in turn weaken the rapport process, which would in turn lower the revelation level even further" (1988).

Keeping Love

The wheel theory suggests that once people fall in love, they may not necessarily stay in love. Relationships can "keep turning," or they may slow down or reverse themselves. We need to point out that sometimes this reversal, and eventual break up, is a good thing. "Perhaps the hardest part of a relationship is knowing when to salvage things and when not to" (Sternberg 1988, p. 242). Chapter 7 explores this issue more fully. But here we need to point out that being committed is not always noble—as in cases of physical violence or consistent verbal abuse, for example.

How do people stay in love? Keeping love requires continual discovery of both oneself and one's partner through mutual self-disclosure. As partners penetrate deeper toward the center of each other's personalities, they continually discover the remarkable and unique.

Love, then, is a continual process. Partners need to keep on sharing their thoughts, feelings, troubles, and joys with each other, an effort that will receive much more of our attention later in this text.

In a marriage relationship a vital dimension of this sharing is sexual intimacy. In Chapter 5 we'll turn to the physical expressions of love.

In Sum

In an impersonal society, love provides an important source of fulfillment and intimacy. Genuine loving in our competitive society is rare and difficult to learn (May 1975, p. 115). Our culture's emphasis on self-reliance as a central virtue ignores the fact that all of us are interdependent. We rely on parents, spouses or partners, other relatives, and friends far more than our culture encourages us to recognize (Cancian 1985, pp. 261–62). Loving is one form of interdependence.

Despite its importance, love is often misunderstood. It should not be confused with martyring or manipulating. There are many contemporary love styles that indicate the range that lovelike relationships—not necessarily love—can take. John Lee lists six love types: eros, or passionate love; storge, or friendship love; pragma, or pragmatic love; agape, or altruistic love; ludus, or love play; and mania, or possessive love.

We *can* learn to love, even if it's difficult. A first step is knowing what love and loving are. Love is a deep and vital emotion resulting from significant need satisfaction, coupled with caring for and acceptance of the beloved, and resulting in an intimate relationship. Loving is a caring, responsible, and sharing relationship involving deep feelings, and it is a commitment to intimacy. Intimacy involves disclosing one's inner feelings, a process that is always emotionally risky. For this reason, the people who are most capable of intimate loving are those who first love themselves or have high self-esteem (self-love). Loving also takes the ability to be emotionally interdependent, an acceptance of oneself as well as a sense of empathy, and a willingness to let down barriers set up for self-preservation.

People discover love; they don't simply find it. The term *discovering* implies a process, and to develop and maintain a loving relationship requires self-disclosure. This mutual self-disclosure requires time and trust.

Key Terms

A-frame relationships	love styles
agape	ludus
commitment	mania
commitment (Sternberg's theory)	manipulating
	martyring
consummate love	M-frame relationships
dependence	narcissism
emotion	passion (Sternberg's theory)
eros	pragma
H-frame relationships	psychic intimacy
illegitimate needs	self-disclosure
independence	self-esteem
interdependence	sexual intimacy
intimacy	Sternberg's triangular theory of love
intimacy (Sternberg's theory)	
	storge
legitimate needs	symbiotic relationships
love	wheel of love

Study Questions

1. Do you agree that loving is essential for emotional survival today? Why or why not?

2. What kinds of needs can a love relationship satisfy? What needs can never be satisfied by a love relationship?

3. Discuss the characteristics necessary for a fully loving relationship. Do you agree or disagree with what the text says? Which characteristics seem most important to you as you observe the world around you?

4. Sternberg offers a very specific triangular theory of love. What are its components? Are they useful concepts in analyzing any love experience(s) you have had?

5. What are some misconceptions of love? Why is each of them not love?

6. What is the difference between self-love and narcis-

sism? How is each related to a person's capacity to love others?

7. Describe A-frame, H-frame, and M-frame relationships.

8. Describe Reiss' wheel theory of love. Compare it with your idea of how love develops.

9. How do the two outer circles (Role Conceptions and Sociocultural Background) move Reiss' theory from a purely social-psychological one to a theory with commonalities with the family ecology perspective discussed in Chapter 2?

Suggested Readings

Berger, Peter L., Brigitte Berger, and Hansfried Kellner. 1973. *The Homeless Mind: Modernization and Consciousness.* New York: Random House. This theoretical work examines modern consciousness as it is shaped by an impersonal society in which people feel little sense of being rooted in communities.

Berne, Eric. 1964. *Games People Play: The Psychology of Human Relationships.* Secaucus, NJ: Castle. This classic book examines how people may play games in their relationships in order to avoid reality or genuine intimacy.

Branden, Nathaniel. 1994. *Six Pillars of Self-Esteem.* New York: Bantam. A best-seller by a long-recognized authority in this field, this book both explains self-esteem (in a way that won't make you feel worse!) and outlines practices to improve yours.

Burns, David D., M.D. 1989. *The Feeling Good Handbook.* New York: Penguin. Recommended by counselors and therapists, this workbook by a psychiatrist can help you fight unreasonable negative thoughts that undermine self-esteem and lead to anxiety and/or depression.

Cancian, Francesca. 1987. *Love in America: Gender and Self-Development.* New York: Cambridge University Press. A look at how Americans' ideas have changed; a consideration of feminine love, masculine love, and androgynous love.

Crosby, John F. 1991. *Illusion and Disillusion: The Self in Love and Marriage.* 4th ed. Belmont, CA: Wadsworth. Professor Crosby has been a minister, is a marriage and family therapist, and teaches marriage and family courses at the University of Kentucky. "None of the above professional experience," he writes, "can really be separate from my personal quest for authentic self-hood as a husband and father" (p. xiii). The book is scholarly but personal and is about the "entangled joys and sorrows of the illusion and disillusion process" of women, men, and love.

Derlega, Valerian J., Sandra Metts, Sandra Petronio, and Stephen T. Margulis. 1993. *Self-Disclosure.* Newbury

Park, CA: Sage. This book reviews the self-disclosure literature to date within the context of close relationships.

Fromm, Erich. 1956. *The Art of Loving.* New York: Harper & Row. A classic, this book talks about love as an active choice to care for another person.

Peck, M. Scott, M.D. 1978. *The Road Less Traveled: A New Psychology of Love, Traditional Values and Spiritual Growth.* New York: Simon & Schuster. More contemporary than *The Art of Loving* (Fromm), this popular work speaks of love as active, willful, disciplined, committed, and attentive.

Powell, John. 1969. *Why Am I Afraid to Tell You Who I Am?* Niles, IL: Argus Communications.

Power, John S. J. 1969. *Why Am I Afraid to Tell You Who I Am?* Niles, IL: Argus. A book about self-understanding and the development of intimacy.

Sternberg, Robert J. and Michael L. Barnes, eds. 1988. *The Psychology of Love.* New Haven, CT: Yale University Press. Academic work presenting current social science research on love; contains chapters by virtually all the important researchers in the field.

Our Sexual Selves

I have a fancy body.

MR. ROGERS, "MR. ROGERS' NEIGHBORHOOD"

The real issue isn't making love; it's feeling loved.

WILLIAM H. MASTERS AND VIRGINIA E. JOHNSON, THE PLEASURE BOND

From infancy to old age, people are sexual beings. Sexuality has a lot to do with the way we think about ourselves and how we relate to others. It goes without saying that sex plays a vital role in marriages and other relationships. Despite the pleasure it can give, sexuality can be one of the most baffling aspects of ourselves. In our society sex is both exaggerated and repressed; rarely is it treated naturally. For this reason, people's sexual selves may be sources of ambivalence, even discomfort. Finding mutually satisfying ways of expressing our sexuality can be a challenge.

This chapter and Appendixes A through G discuss the sexual and reproductive aspects of people's lives. Here we will define sexual orientation and examine various cultural messages regarding men's and women's sexuality. We will discuss sex as a pleasure bond that requires open, honest, and supportive communication, then look at the role sex plays throughout marriage. We will consider the impact of HIV/AIDS on relationships. Finally, we will examine ways that politics and religion have combined to influence sexual expression in our society today.

Before we begin, we want to point out our society's tendency to reinforce the differences between women and men and to ignore the common feelings, problems, and joys that make us all human. The truth is, men and women aren't really so different. As Appendix A illustrates, many physiological parts of the male and female genital systems are either alike or directly analogous. Furthermore, as Appendix B illustrates, the patterns of sexual response are very similar in men and women. Space limits our ability to present much de-

Self-disclosure and physical pleasure are key qualities in building sexually intimate relationships. Tenderness is a form of sexual expression, valued not just as a prelude to sex but as an end in itself.

tail on the various possibilities of sexual expression (kissing, fondling, cuddling, even holding hands); but we want to point out that intercourse, or coitus, is not the only mode of sexual relating.

Sexual Orientation

Sexual orientation refers to whether an individual prefers a partner of the same or opposite sex. **Heterosexuals** are attracted to opposite-sex partners and **homosexuals** to same-sex partners. Everyday terms are *straight* (heterosexual), *gay,* and *lesbian.* Technically, the term *gay* is synonymous with homosexual and refers to males or females. But often "gay" (or "gay male") is used in reference to men; "lesbian" is used to refer to gay women. The Commission on Lesbian and Gay Concerns of the American Psychological Association (1991) prefers the words *gay male* or *gay man* and *lesbian* to *homosexual* because it thinks that the latter term may perpetuate negative stereotypes. All these terms designate one's choice of sex partner only, not general masculinity or femininity or other aspects of personality.

We tend to think of sexual orientation as a dichotomy: One is either gay or straight. Actually, sexual orientation may be a continuum. Freud and many contemporary psychologists and biologists maintain that humans are inherently bisexual; that is, we all may have the latent physiological and emotional structures necessary for responding sexually to either sex. From the interactionist point of view (Chapter 2), the very concepts "bisexual," "heterosexual," and "homosexual" are social inventions. They emerged during the nineteenth century when a new category of scientists (sexologists) first began to examine sexual behavior (Weeks 1985). Put another way, conceptual categories—along with the notion of sexual orientation itself—were created by social interaction, not by nature. Hence, developing a sexual orientation may be influenced by our tendency to think in dichotomous terms: Individuals may sort themselves into the categories and behave accordingly. In time, social pressures to view oneself as either straight or gay may inhibit latent bisexuality (Beach 1977a; Gagnon 1977).

The Development of Sexual Orientation

We know little about how an individual's sexual orientation develops. The origins of both heterosexual and gay identities remain a mystery. The existence of a fairly constant proportion of gays in virtually every society—in societies that treat homosexuality permissively as well as those that treat it harshly—suggests a biological imperative (Ford and Beach 1971; Bell, Weinberg, and Hammersmith 1981). Some anatomical and genetics research has found a possible relationship between physiology and sexual orientation (Gelman 1992; LeVay and Hamer 1994). However, no clear genetic differences between heterosexuals and gays have been conclusively established (Byrne 1994). It may not be clear just how sexual orientation develops, but nonetheless we each make choices regarding it and many other aspects of sexuality. As we have seen, we negotiate decisions within the parameters established by society. The next section takes a look at how sexuality and society interface.

Sex and Society

As with gender roles (Chapter 3), the way people think and feel about sex has a lot to do with the messages society gives them. Cultural messages give us legitimate reasons for having sex, as well as who should take the sexual initiative, how long a sexual encounter should last, how important it is to experience orgasm (see Appendix B), what positions are acceptable, and whether it is appropriate to masturbate, among other things. Saying this may already have made you aware that in this chapter we, the authors, take primarily an **interactionist perspective on human sexuality:** Women and men negotiate and are influenced by the "sexual scripts" (Laumann et al. 1994) they learn from society. There is a competing point of view, however, called evolutionary psychology, and we will examine it now.

Evolutionary Psychologists—Once Again: Is Anatomy Destiny?

Chapter 3 asked whether gendered behavior is due mainly to biological factors (genetics). *No!* The question can also be raised regarding human sexuality. **Evolutionary psychology** argues that humans have an evolutionary biological origin that affects their sexual relations. This perspective has its roots in Charles Darwin's *The Origin of the Species,* first published in 1859. Darwin proposed that all species evolve according to the principle **survival of the fittest:** Only the stronger, more intelligent, and adaptable of a species survive to reproduce, a process whereby the entire species is strengthened and prospers over time.

According to evolutionary psychologists, humans—like the species from which they evolved—are designed for the purpose of transmitting their biological hereditary material (genes or chromosomes) to the next generation. They do this in the most efficient ways possible. Human sexual attitudes and behaviors are best understood in this context. Like males in other primate species, men are inclined toward casual sex with many partners while women are more discriminating. (After all, male gorillas keep a harem, while female gorillas do not.) In this view, the basic source for this difference is as follows: A woman, regardless of how many sex partners she has, can generally have only one offspring a year. Compared with men, a woman's anatomy limits her ability to pass on her genes, and so she must be more discriminating when choosing a chromosomal partner. For a man, each new mate offers a real chance for carrying on his genetic material into the future. Evolutionary psychologists also assume, and attribute to biology, men's association of sex with power. Their evidence for this is that among baboons, the higher a male ranks in the social hierarchy, the more sex he has (Wright 1994).

Even proponents of the biosocial perspective, of which evolutionary psychology is a part, recognize that biosocial influences are *both* genetic and social in character (Troost and Filsinger 1993). As we concluded in our discussion of gender in Chapter 3, at best, genetics has an interactive influence with society on human behavior. In contrast to evolutionary psychology, the social-historical perspective shows us that in the United States (and elsewhere), societal messages about sex have changed over time.

An Historical Perspective

From colonial times until the nineteenth century, the purpose of sex in America was reproduction; the so-

"Morning on the Cape" by Leon Kroll.

ciocultural attitudes were patriarchal. A new definition of sexuality emerged in the nineteenth century and has flourished in the twentieth. Sex became significant as a means of couple communication and intimacy (D'Emilio and Freedman 1988). We will explore the cultural messages of several historical periods in more detail.

EARLY AMERICA: PATRIARCHAL SEX   In a patriarchal society, descent, succession, and inheritance are traced through the male genetic line, and the socioeconomic system is male dominated. **Patriarchal sexuality** is characterized by many beliefs, values, attitudes, and behaviors developed to protect the male line of descent. Exclusive sexual possession by a man of a woman in monogamous marriage ensures that her children will be legitimately his, and men control women's sexuality. Sex is defined as a physiological activity, valued for its procreative potential. Men are thought to be born with an urgent sex drive, while women are naturally sexually passive; orgasm is expected for men but not for women. Unmarried men

and husbands whose wives do not meet their sexual needs can gratify those needs outside marriage. Sex outside marriage is wrong for women, however; only tainted or loose women do it.

THE TWENTIETH CENTURY: EXPRESSIVE SEXUALITY.   A different sexual message has emerged as the result of several societal changes, including the decreasing economic dependence of women and the availability of new methods of birth control. **Expressive sexuality** sees sexuality as basic to the humanness of *both* women and men; there is no one-sided sense of ownership. Orgasm is important for women as well as for men. Sex is not only, or even primarily, for reproduction, but is an important means of enhancing human intimacy. Hence, all forms of sexual activity between consenting adults are acceptable. Because of the emphasis on couple intimacy in recent decades, women's sexual expression has been more encouraged than it has been earlier, and this is especially true for the white middle class (D'Emilio and Freedman 1988).

| Back Seat Dodge '38, by Edward Keinholz.

THE 1960s: SEXUAL REVOLUTION   Although the view of sex as intimacy continues to predominate, in the 1920s an alternative message began to emerge wherein sex was seen as a legitimate means to individual pleasure, whether or not it played a role in serious couple relationship. Probably as a result, the generation of women born in the first decade of the twentieth century showed twice the incidence of nonmarital intercourse than those born earlier (D'Emilio and Freedman 1988; Seidman 1991). This liberalization of attitudes and behaviors characterized the sexual revolution of the 1960s.

What was so revolutionary about the sixties? For one thing, at least for heterosexuals, laws regarding sexuality became more liberal. When the 1960s began, for instance, it was illegal in some states for married couples to use contraception. But the U.S. Supreme Court decision in *Griswold v. Connecticut* (1965) stated a concept of "marital privacy," the idea that sexual and reproductive decision making belonged to the couple, not to the state. This concept of "privacy" was extended to single individuals and to minors by subsequent decisions (*Eisenstadt v. Baird* 1972; *Carey v.*

Population Services 1977). We have rapidly come to take it for granted that sexual and reproductive choices are ours to make, although the Supreme Court has declined to extend this protection to gay male or lesbian relationships (*Bowers v. Hardwick* 1986).

People's attitudes and behavior regarding sex changed during the 1960s as well, becoming radically more permissive. For instance, in 1959 only about one-fifth of Americans surveyed said they approved of sex outside marriage. By 1970, over half were saying it was okay, a figure that has remained steady to the present (Hunt 1974; "Gallup Poll Shows" 1988). Attitudes (and information) also changed regarding female orgasm as it became clear that there is no such thing as a purely vaginal orgasm. Rather, the clitoris is involved in virtually all female orgasms; "Think clitoris" was how feminist writers of the 1960s redefined female—hence, heterosexual couples'—sexuality (Segal 1994, p. 36).

Not only did attitudes become more liberal, but behaviors (particularly women's behaviors) changed as well. The rate of nonmarital sex and the number of partners rose, while age at first intercourse dropped.

By the mid-1970s, surveys reported that the majority of unmarried adults had experienced intercourse (Darling, Kallen, and VanDusen 1984; DeBuono et al. 1990). The trend toward higher rates of nonmarital sex has continued, especially among women (men already had higher rates) (Robinson et al. 1991).

Today, sexual activity often begins in the teen years. Surveys show that teenagers in the 1970s were twice as likely to have had sex as were teens in the early 1960s. Currently more than 50 percent of unmarried teen girls report having had intercourse (Besharov 1993).[1] The proportion of sexually experienced 15-year-old females rose from less than 5 percent in 1970 to almost 27 percent in 1988. In 1990, 32 percent of girls ages 14 and 15 reported having had intercourse, as did 49 percent of the boys the same age. Teens are also having sex with more partners: By grade 12, 17 percent of girls and 38 percent of boys reported having four or more sexual partners (Besharov 1993). According to one researcher, "This isn't a rebellion. . . . They're finishing what was started in the 60s and moving toward a wider, legitimate sphere of sexual choice" (Ira Reiss, in Williams 1989, p. B11). (However, sexually active teens may be choosing by default, inasmuch as peer pressure may be the chief reason for their decision. The related issue of unmarried childbirth is addressed in Chapter 11.)

But perhaps the most significant change in sexuality since the 1960s, among heterosexuals at least, has been in marital sex. Today spouses are engaging in sexual activity more frequently than in the 1950s and in more varied and experimental ways. More married women are experiencing orgasm, and they're doing so more often (Hunt 1974; Jasso 1985; Laumann et al. 1994). Today's marrieds report taking baths or show-

ers together (30 percent), buying erotic underwear (20 percent), swimming together nude (19 percent) (Schmidt 1990); and videotaping their sexual escapades for later viewing (Toufexis 1990). In a recent study of 3,432 Americans by the University of Chicago's National Opinion Research Center (NORC), 88 percent of married partners said they enjoy great sexual pleasure (Laumann et al. 1994). If the sexual revolution of the 1960s focused on freer attitudes and behaviors among heterosexuals, more recent decades have expanded that liberalism to encompass lesbian and gay male sexuality.

THE 1980s AND 1990s: CHALLENGES TO HETEROSEXISM In 1995, *Newsweek* ran a story titled "Bisexuality." In large print, the author proclaimed that being "bi" was becoming noticeable "in pop culture, in cyberspace and on campus" (Leland 1995). The story quoted a man who said he had identified himself as gay since he was 16, even though he'd slept with both men and women. Now he is "proud to say I'm a recovering bi-phobe" (in Leland 1995, p. 49). "Bi-phobe" is a take-off on "homophobe," slang for a person who demonstrates **homophobia** (fear, dread, aversion to, and often hatred of homosexuals).

Until about two decades ago, most people thought about sexuality almost exclusively as between men and women. In other words, our thinking was characterized by **heterosexism**—the taken-for-granted system of beliefs, values, and customs that places superior value on heterosexual behavior (as opposed to homosexual) and denies or stigmatizes nonheterosexual relations (Neisen 1990). However, since "Stonewall" (a 1969 police raid on a U.S. gay bar) galvanized the gay community into advocacy, gay males and lesbians have not only become increasingly visible but also have challenged the notion that heterosexuality is the one proper form of sexual expression.

Deciding who is to be categorized as gay or lesbian, along with possible concealment by survey respondents, precludes any accurate calculation of how many gays and lesbians there are in our society. Until recently, it was generally stated that about 10 percent of adult individuals are gay or lesbian. However, recent evidence suggests that the incidence of exclusively same-sex behavior is probably lower than this and may be about 4 percent (Laumann et al. 1994). Much gay or lesbian activity is episodic, rather than ongoing or consistent, and the majority of gay men also report male–female sexual contact (Rogers and Turner 1991).

---

1. There are four principal sources of information about the sexual practices of teenagers: the National Survey of Family Growth (NSFG), a national in-person survey of females ages 15–44 conducted in 1982 and again in 1988; the National Survey of Adolescent Males (NSAM), a longitudinal survey of males ages 15–19 conducted in 1988 and 1991; the National Survey of Young Men (NSYM), a 1979 survey of 17- to 19-year-olds; and the Youth Risk Behavior Survey (YRBS), a 1990 questionnaire-based survey of 11,631 males and females in grades 9–12 conducted by the Centers for Disease Control. With minor variations caused by differences in methodology, each survey documents a sharp increase in the sexual activity of American teens. All these surveys, however, are based on the self-reports of young people and must be interpreted with care. For example, one should be wary of young males' reports of their sexual activity because they are known to exaggerate. In addition, the social acceptability of being a virgin may have decreased so much that this, more than any change in behavior, has led to the higher reported rates of sexual experience (Besharov 1993).

TABLE 5.1

How the Public Views Gay Issues

Total adults		Those who say homosexuality...	
		is a choice	cannot be changed
Jobs and Rights			
78%	Say homosexuals should have equal rights in terms of job opportunities	69%	90%
42	Say it is necessary to pass laws to make sure homosexuals have equal rights	30	58
11	Object to having an airline pilot who is homosexual	18	4
49	Object to having a doctor who is homosexual	64	34
55	Object to having a homosexual as a child's elementary school teacher	71	39
Personal Judgments			
46	Say homosexual relations between consenting adults should be legal	32	62
36	Say homosexuality should be considered an acceptable alternative life style	18	57
55	Say homosexual relations between adults are morally wrong	78	30
43	Favor permitting homosexuals to serve in the military	32	54
34	Would permit their child to play at the home of a friend who lives with a homosexual parent	21	50
36	Would permit their child to watch a prime-time television situation comedy with homosexual characters in it	27	46
Familiarity			
22	Have a close friend or family member who is gay or lesbian	16	29

NOTE: Based on telephone interviews with 1,154 adults nationwide conducted February 9–11, 1993. Those interviewed were asked, "Do you think being homosexual is something people choose to be, or do you think it is something they cannot change?" Forty-four percent said they thought it was a choice, 43 percent said they thought it wasn't possible to change, and 13 percent said they didn't know.

Source: Schmalz 1993, New York Times/CBS News Poll *The New York Times Themes of the Year,* Fall.

Conceptual problems—that is, deciding who is to be classified as homosexual (How much experience? How exclusively homosexual?)—preclude any really accurate count.

Although the percentage of gay men and lesbians in our society is relatively small, political activism has resulted in their greater visibility (Sullivan 1995). Furthermore, gay men and lesbians have won legal victories, new tolerance by some religious denominations, greater understanding on the part of some heterosexuals, and sometimes positive action by government.

Some states (California, Minnesota, Wisconsin, Vermont, Massachusetts, Connecticut, New Jersey, and Washington, DC) and communities have passed gay rights laws, extending protection against various forms of discrimination to include sexual orientation. As we go to press, the U.S. Supreme Court has declared unconstitutional a Colorado state amendment ("Colorado 2") that challenged local laws designed to protect gays from discrimination. Many states have rescinded traditional sodomy laws, which criminalized gay male and lesbian relations (and often heterosexual

For many lesbians and gays, identity assumption involves developing both a self-identity and a presented identity as a lesbian or gay man.

oral and anal sex), although twenty-two states retain them ("Gay Rights: What's the Law?" 1994). The National Education Association has endorsed a proposal to designate October as Lesbian, Gay, and Bisexual History month (Wingert and Waldman 1995).

A 1993 national poll of 1,154 American adults showed that 78 percent of us favor "equal rights in terms of job opportunities" for gay males and lesbians, up from 56 percent in 1977 and 71 percent in 1989 (Barron 1989; Schmalz 1993). As you can see in Table 5.1, however, support drops for certain occupations, such as elementary school teachers. Americans are divided over whether gay men and lesbians choose their sexual orientation, a split that shapes attitudes. As Table 5.1 shows, people who see being gay as a choice are less sympathetic to lesbians or gay men.

We are currently experiencing a challenge to heterosexism, but the goals of this challenge—equal treatment and acceptance for lesbians and gay males—

have hardly been accomplished. While 46 percent of Americans say adult gay male and lesbian relations should be legal, 55 percent find them morally wrong. Only one-third of us would let our child play at the home of a friend who lives with a gay male or lesbian parent (Table 5.1). Furthermore, the most frequent victims of hate violence today may be lesbians and gay men, with attacks ranging from being chased or spat on to murder (Herek and Berrill 1992). But federal government concern for such hate crimes has also increased (Rosenthal 1990, p A12; Salholz 1990). The controversy over gays in the military is an example of the increased visibility and activism of gay men and lesbians and also of ongoing heterosexism (Sullivan 1995).

CONSTRUCTING GAY MALE AND LESBIAN IDENTITIES AMID HOMOPHOBIA   One result of homophobia is that gay males and lesbians often negotiate, or construct their sexual identities amid hostility and maintain them within the context of a deviant subculture. According to one model (Troiden 1988, pp. 42–58), this process occurs in four stages: *sensitization, identity confusion, identity assumption* (acceptance and "coming out"), and *commitment* to homosexuality as a way of life.

Sensitization occurs before puberty. At this time most children assume they are heterosexual (if they think about their sexuality at all). But future homosexuals, feeling sexually marginal, have experiences that sensitize them to subsequent definitions of themselves as lesbian or gay: "I wasn't interested in boys (girls)"; "I didn't express myself the way other girls (or boys) would."

Identity confusion occurs in adolescence as lesbians and gays begin to see that their feelings, fantasies, or behaviors could be considered homosexual. Because the idea that they could be homosexual goes against previously held self-images, identity confusion (inner turmoil and uncertainty) results. Potentially homosexual adolescents typically engage in stigma-management strategies, such as denial or attempts to eradicate their homosexuality through therapy. Some escape through drug abuse. Others assume antihomosexual postures or establish heterosexual involvements to eliminate their "inappropriate" sexual interests: "I thought my homosexual feelings would go away if I dated a lot and had sex with as many women as possible," or "I thought my attraction to women was a passing phase and would go away once I started having intercourse with my boyfriend" (Troiden 1988, p. 48). Adolescents may define homosexual incidents as isolated cases. Another response is acceptance; in such cases behaviors, feelings, or fantasies are acknowledged as possibly homosexual, and additional information is sought.

Identity assumption, which occurs during or after late adolescence, involves developing both a self-identity and a presented identity—presented, that is, among other homosexuals at least—as homosexual. In one writer's description of this stage,

> you are quite sure you are a homosexual and you accept this fairly happily. You are prepared to tell a few people about being a homosexual but you carefully select whom you will tell. You adopt an attitude of fitting in where you live and work. You can't see any point in confronting people with your

homosexuality if it's going to embarrass all concerned. (Cass 1984, cited in Troiden 1988, p. 53)

The final stage, commitment, involves the decision to accept homosexuality as a way of life. Here the costs are less and the rewards greater for remaining a homosexual than for trying to function as a heterosexual. Having "come out," even to nonhomosexuals, "you are happy about the way you are but feel that being a homosexual is not the most important part of you. You mix socially with homosexuals and heterosexuals [with whom] you are open about your homosexuality" (Cass 1984, cited in Troiden 1988, p. 57).

The historical perspective makes it clear that sexuality is influenced by context and by social conditions. We turn to a specific look at the interactionist perspective on sexuality, which leads us to focus on negotiating cultural messages today.

An Interactionist Perspective: Negotiating Cultural Messages

"*That* we are sexual is determined by a biological imperative toward reproduction, but *how* we are sexual —where, when, how often, with whom, and why— has to do with cultural learning, with meanings transmitted in a cultural setting" (Fracher and Kimmel 1992). While it has been significantly challenged, the patriarchal sexual script persists to some extent in our society and corresponds with traditional gender expectations (see Chapter 3). If masculinity is a quality that must be achieved or proven, one arena for doing so is sexual accomplishment or conquest. A "real" man performs. A 1992 national survey, by the National Opinion Research Center (NORC) at the University of Chicago and based on a representative sample of 3,432 Americans ages 18–59, found that men were considerably more likely than women to perform, or "do" sex. For example, more than three times as many men as women report masturbating at least once a week. Three-quarters of the men reported always reaching orgasm in intercourse, while the fraction for women was nearer to one-quarter. Men are also much more likely to think about sex (54 percent of men and 19 percent of women said they think about it at least once a day) and to have multiple partners and more of them. Men are also more excited by the prospect of group sex. These differences are less pronounced among the youngest cohort, a trend that holds out the possibility of more convergence between men and women's sexual experiences (Laumann et al. 1994).

In his research on male athletes' friendships, sociologist Michael Messner (1992) found that a "Big Man on Campus" was expected to be considerably (hetero)sexually active. Other males were envious of the ease with which he could "get women." Even among high schoolers, being a "real" guy meant "you get 'em into bed" (p. 96). Messner laments that this cultural conditioning encourages men to separate sex from intimacy: Male friendship is defined as intimate but not sexual, while relationships with women are defined as sexual but not intimate. Among women, on the other hand, sexual expression more often symbolizes connection with a partner and communicates intimacy. This point becomes particularly apparent when we compare lesbian and gay male sexual behaviors.

COMPARING GAY MALE AND LESBIAN SEXUAL BEHAVIORS

In a large national sample of 12,000 volunteers from the Seattle, San Francisco, and Washington, DC areas, sociologists Philip Blumstein and Pepper Schwartz (1983) compared four types of couples: heterosexual marrieds, cohabiting heterosexuals, gay male, and lesbian couples. Among other things, they found that gay male and lesbian relationships differ as a result of socialization. The fact that lesbian couples consist of two *women* and gay couples of two *men* is significant. Gay men tend to have considerably more transitory, or casual, sex than do lesbians (or heterosexual men, whose union with a woman may require compromise on this issue), while casual sex among lesbians is relatively rare. That is, gay male sexuality is more often "body-centered" (Ruefli, Yu, and Barton 1992) and lesbian sexuality, person-centered. Blumstein and Schwartz (1983), in fact, described lesbian relationships as the "least sexualized" of the four kinds of couples compared. (Box 5.1, "A Closer Look at Family Diversity: Lesbian 'Sex'," addresses this finding.) Interestingly, meanwhile, lesbians report greater sexual satisfaction than do heterosexual women, including more frequent sexual expression and orgasms. "Their greater tenderness, patience, and knowledge of the female body are said to be the reasons" (Konner 1990, p. 26). Many—though assuredly not all—women and men today have internalized divergent sexual messages.

NEGOTIATING (HETERO)SEXUAL EXPRESSION

Today heterosexuals negotiate sexual relationships in a context in which new expectations of equality and similarity coexist with a heritage of difference. Studies in the 1970s found that, compared to before the sexual revolution, men are more interested in communicating intimately through sex while women show more interest in the physical pleasure involved in sex (Pietropinto and Simenauer 1977). Nevertheless, women may feel lonely and emotionally separated from task-oriented, emotionally reserved male partners, whereas men feel that their female partners ask too much emotionally (Tannen 1990; Gray 1995).

Moreover, if a man continues to equate sex with performance in a culture that now expects women to reach orgasm, he may feel undue pressure to sexually satisfy his partner. "So the knowledgeable man . . . instead of doing something *to* his [partner] sexually, was prepared to do something *for* her sexually. . . . Unfortunately, in the role of doing *for* rather than just doing *to* he had to assume even more sexual responsibility" (Masters and Johnson 1976, p. 6). And in a society that allows both males and female multiple partners, he may fear his mate's comparing him with other lovers (Boyer 1981).

Masters and Johnson argue nonetheless that more equal gender expectations facilitate better sex. "The most effective sex is not something a man does to or for a woman but something a man and a woman do together *as equals*" (Masters and Johnson 1976, p. 88). From this point of view the female not only is free to initiate sex but also is equally responsible for her own arousal and orgasm. The male is not required to "deliver pleasure on demand," but can openly express his spontaneous feelings.

This discussion points again to the fact that cultural messages, both about gender (Chapter 3) and about sexual expression, are negotiated. Recognizing this, sociologist Ira Reiss has developed a fourfold classification of societal standards for nonmarital sex. We should note that these are standards, or cultural prescriptions, *not* research reports of what people actually do.

Four Standards of Nonmarital Sex

Sociologist Ira Reiss' (1976) four standards—abstinence, permissiveness with affection, permissiveness without affection, and the double standard—were

Lesbian "Sex"

This essay is by lesbian theorist Marilyn Frye. It has as much to say about scientific research methods as about sexual expression.

[T]he word "sex" is in quotation marks in my title... [because] the term "sex" is an inappropriate term for what lesbians do....

Recent discussions of lesbian "sex" frequently cite the findings of a study on couples by Blumstein and Schwartz [1983, and cited by Lamanna and Riedmann in this chapter] which is perceived by most of those who discuss it as having been done well, with a good sample of couples—lesbian, male homosexual, heterosexual non-married and heterosexual married couples. These people apparently found that lesbian couples "have sex" far less frequently than any other type of couple, that lesbians are less "sexual" as couples and as individuals than anyone else. In their sample,... 47 percent of lesbians in long term relation-ships "had sex" once a month or less, while among heterosexual married couples only 15 percent had sex once a month or less. And they report that lesbians seem to be more limited in the range of their "sexual" techniques than are other couples....

The suspicion arises that what 85 percent of heterosexual married couples are doing more than once a month and what 47 percent of lesbian couples are doing less than once a month is not the same thing. And if they are not doing the same thing, how was this research done that would line these different things up against each other to compare how many times they were done?

I remember that one of my first delicious tastes of old gay lesbian culture occurred in a bar where I was chatting with some other lesbians.... One was talking about being busted out of the Marines for being gay. She had been put under suspicion

originally developed to apply to *pre*marital sex among heterosexual couples. However, they have since been more generally applied to nonmarital sexual activities of divorced and separated heterosexuals, as well as to those who do not plan to marry.

ABSTINENCE   The standard of **abstinence** maintains that regardless of the circumstances, nonmarital intercourse is wrong for both women and men. Some women have withdrawn from nonmarital sexual relationships entirely, advocating celibacy, or "the right to say no" (Johnson 1990). Many contemporary religious groups, especially the more conservative or fundamentalist Christian and Islam congregations, encourage abstinence. A 1987 Gallup poll found nearly half (46 percent) of Americans disapproving of nonmarital sex, and that figure was 7 percent higher than it had been two years earlier, in 1985. The most common reason given for opposing sex outside marriage was religious or moral beliefs (83 percent), followed by the risk of sexually transmitted diseases or STDs (20 percent) and of pregnancy (13 percent) and the belief that women should be virgins until marriage (9 percent) ("Gallup Poll Shows" 1988). With tongue only partly in cheek, one writer suggests that cybersex (sexual talk on the Internet, with or without masturbation) is a preferable "new abstinence": You can't get diseases from computers (Gerhard 1994).

Abstinence, or celibacy, receives support from counselors and feminists as a positive choice. These experts are concerned about pressures on young people to establish sexual activity before they're ready, or about pressures for men and women to engage in sexual relationships when they don't want to. They maintain that celibacy, much devalued in the context of sexual liberation, ought to remain a valid and respected option.

somehow, and was sent off to the base psychiatrist to be questioned.... To this end [the psychiatrist] asked her, "How many times have you had sex with a woman?" At this, we all laughed and giggled: what an ignorant fool he was! What does he think he means by "times"? What will we count? What's to count? ...

If what heterosexual married couples do that the individuals report under the rubric "sex" or "have sex" is something that in most instances can easily be individuated into countable instances, this is more evidence that it is not what long-term lesbian couples do ... or, for that matter, what short-term lesbian couples do....

How did the lesbians figure out how to answer the questions "How frequently?" or "How many times?" My guess is, for starters, that different individuals figured it out differently, to some degree. Some might have counted a two- or three-cycle evening as one "time" they "had sex"; some might have counted that as two or three "times." Some may have counted as "times" only the times both partners had orgasms; some may have counted as "times" occasions on which at least one had an orgasm; some may not have orgasms or have them rarely and may not have figured orgasms into the calculations; perhaps they counted as a "time" every episode in which both touched the other's vulva more than fleetingly and not for something like a health examination....

We have no idea how individual lesbians individuated their so-called "sexual acts" or encounters; we have no idea what it means when they said they did it less than once a month....

So, do lesbian couples really "have sex" any less frequently than heterosexual couples? My own view is that lesbian couples "have sex" a great deal less frequently than heterosexual couples: I think, in fact, we don't "have sex" at all. By the criteria that I'm betting most of the heterosexual people used in reporting the frequency with which they have sex, lesbians don't have sex. There is no male partner whose orgasm and ejaculation can be the criterion for counting "times." (I'm willing to draw the conclusion that heterosexual women don't have sex either, that what they report is the frequency with which their partners had sex.)

What does "have sex" mean to you? Is the term, as Frye suggests, defined principally according to the male's experience? What does this say about heterosexual women's sexual expression? What does this say about the research findings of Blumstein and Schwartz? How might Blumstein and Schwartz have asked about what they wanted to know in a better, more effective way?

Source: Frye 1992.

PERMISSIVENESS WITH AFFECTION   The standard **permissiveness with affection** permits nonmarital intercourse for both men and women equally, provided they have a fairly stable, affectionate relationship. Reiss concludes that this standard is probably the most widespread sexual norm among young college and postcollege unmarrieds today. Other social scientists, after reviewing thirty-five studies of college students' sexual behavior and attitudes conducted between 1903 and 1980, agreed (Darling, Kallen and VanDusen 1984). The NORC survey, mentioned above (and described in detail below), concluded that we have sex mainly with people we know and care about. Seventy-one percent of Americans have only one sexual partner in the course of a year (Laumann et al. 1994).

PERMISSIVENESS WITHOUT AFFECTION   Sometimes called recreational sex, **permissiveness without affection** allows intercourse for women and men regardless of how much stability of affection there is in their relationship. Casual sex—intercourse between partners only briefly acquainted—is permitted. This standard reminds us of historians D'Emilio and Freedman's (1988) argument that, from about the 1920s, American sex began increasingly to be seen as an

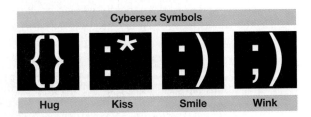

Cybersex Symbols

Hug	Kiss	Smile	Wink
{}	:*	:)	;)

Cybersex. Is it sex—cyberstyle—or is it abstinence? From an interactionist perspective, we might say that society is still constructing the answer. But we've already constructed the beginnings of a cybersex language.

"How about a kiss?"

avenue for individual pleasure. But removed from its relationship context, sex could and did become a "commercial product." That is, in the extreme, the interests and preferences of a partner could be discounted (D'Emilio and Freedman 1988). (The authors of this text do not consider this attitude to be an example of acting with sexual responsibility, a topic addressed in detail later in this chapter.) In recent class discussions on this topic, author Agnes Riedmann's marriage and family students worried that too much permissiveness without affection led people to think of sex as "no big deal." And if sex is no big deal, what is there to define a committed or marital relationship as special? Fear of HIV/AIDS and other sexually transmitted diseases (see Appendix F), meanwhile, may cause some decline in this standard.

THE DOUBLE STANDARD   According to the **double standard,** women's sexual behavior must be more conservative than men's. In its original form, the double standard meant that women should not have sex before or outside of marriage, whereas men could. More recently, the double standard has required that women be in love to have sex, or at least have fewer partners than men have.

Throughout the 1980s, researchers found the dou-

ble standard to be declining and reported expectations to be similar for men and women (Sprecher, McKinney, and Orbuch 1987; Sprecher et al. 1988; Sprecher 1989; Williams 1989). Masters and Johnson's (1966) work on human sexual response (see Appendix B) probably contributed to the sense that men and women are equally sexual creatures, similar in their sexual capacity and needs. Nevertheless, there is also new evidence that, at least among some young people, the double standard remains alive and well. Peggy Orenstein (1994) conducted naturalistic observation (see Chapter 2) in a California middle school of mostly white, middle-class students. She describes an episode in an eighth-grade sex education class on STDs:

"We'll use a woman," [Ms. Webster, the teacher] says, drawing the Greek symbol for woman on the blackboard. "Let's say she is infected, but she hasn't really noticed yet, so she has sex with three men."

Ms. Webster draws three symbols for man on the board, and as she does, a heavyset boy in a Chicago Bulls cap stage-whispers, "What a slut," and the class titters.

"Okay," says Ms. Webster, who doesn't hear the comment. "Now the first guy has three sexual encounters in six months." She turns to draw three

Intimacy and sexuality require communication—physical as well as verbal—from both partners. Sexuality has become more expressive and less patriarchal in the United States, and each generation finds itself reevaluating sexual assumptions, behaviors, and standards.

more women's signs, her back to the class, and several of the boys point at themselves proudly, striking exaggerated macho poses.

"The second guy was very active, he had intercourse with five women." As she turns to the diagram again, two boys stand and take bows.

"Now the third guy was smart—he didn't sleep with *anyone.*" She draws a happy face and the boys point at each other derisively, mouthing, "You! You!"

During the entire diagramming process, the girls in the class remain silent. (p. 61)

The more restrained sexual standards of women are usually attributed to a legacy of differential socialization, such as this example illustrates, and perhaps to greater religiosity as well (Tanfer and Cubbins 1992). One observer argues that women are more likely to have less control over a sexual encounter; hence gender differences in permissiveness may reflect differences in social power and vulnerability (Howard 1988). Meanwhile, two social scientists (Robinson and Jedlicka) have proposed an emergent new double standard, according to which both sexes feel more permissive about their own sex but expect more conservative behavior from the opposite sex. According to these researchers, men expect stricter morality of women and women expect stricter morality of men. We turn now to a brief discussion of how race influences sexual expression.

African Americans and Sexual Expression

At least one sex researcher (Belcastro 1985) has argued that African Americans and non-Hispanic whites are more similar than dissimilar in their sexual behavior and recommends that "research dedicated to classifying people on a permissiveness construct be given a low priority in favor of research describing the sexual development and behavior patterns, across the life cycle, for diversified cultural groups in diversified geographic regions" (p. 56).

Nevertheless, Robert Staples (1972), an influential scholar of black dating and family patterns, describes African Americans as less puritanical than whites about sex. Staples attributes black women's ease with their sexuality to the African heritage and, ironically, slavery and its aftermath. Unlike white women, African American women were not placed on a pedestal that required sexual innocence of them.

Not everyone agrees with Staples' perception of racial differences in sexuality. For one thing, it overlooks variations among blacks associated with social class and with religiosity. Actively religious black women may be very conservative sexually.

A study exploring the question of racial differences in sexual expression using unpublished 1970s Kinsey Institute data found that African American women and men were likely to become sexually active earlier than whites, and with greater frequency, and were more likely to engage in extramarital sex (Weinberg and Williams 1988). Black men had more premarital and extramarital partners than white men had; black and white women did not differ in this respect. This study concluded that there is a distinct black sexual culture: "Blacks were more liberal and accepting of sex, pursued it more, were more open about it, and reported fewer problems with it" (p. 197).

Interestingly, black men and women appear no closer to each other in sexual attitudes and behaviors than white men and women; in both groups, women are more conservative. Being "respectable" is important to both black adult and teen women (Weinberg and Williams 1988). But factors other than sexual attitudes shape relationships, and one observer thinks black women are currently vulnerable to sexual exploitation (Fullilove et al. 1990). As black men have lost economic standing and are significantly fewer in number than black women, they "have increased their power in relationships with women . . . and [become more] oriented to erotic fulfillment" (p. 62). Relatively unable, as a group, to be employed or to make sufficient income to support families, they are less likely to form a permanent romantic or marital attachment and more likely to press women for "casual" sex (Staples 1994).

We turn now to a consideration of sex in marriage.

Sexuality Throughout Marriage

It might surprise you that various aspects of nonmarital sex are more likely to be studied than are those within marriage (Call, Sprecher, and Schwartz 1995). Back in the early 1980s—and this may still be true—researcher Cathy Greenblatt (1983) commented that marital sex is more often the subject of jokes than of serious research. One aspect of marital sex that has consistently been researched is sexual frequency: How often do married couples have sex and what factors affect this frequency? Before we get to the answers, we need to say something about how the information is gathered.

How Do We Know What We Do? A Look at Sex Surveys

Chapter 2 points out that a *Playboy* magazine survey about sex would yield different results from one in *Family Circle.* Because the readers of these two magazines differ from each other as categories, their answers would hardly be the same. Neither survey could be interpreted to represent the entire American population. In serious social science, however, researchers strive for **representative samples**—survey samples that reflect, or represent, all the people about whom they want to know something.

The pioneer research surveys on sex in the United States were the Kinsey reports on male and female sexuality (Kinsey, Pomeroy, and Martin 1948, 1953) in the 1940s and 1950s. Kinsey believed that a statistically representative survey of sexual behavior was impossible, because many of the randomly selected respondents would refuse to answer or lie. This issue remains a problem for researchers today. For instance, footnote 1 in this chapter points out that sex surveys among teenagers needs to be interpreted with caution because, among other things, young men are likely to exaggerate their sexual activities. In the Kinsey era, however, the worry was that people would minimize their sexual behaviors. Hence the Kinsey reports were conducted with volunteers. But people who volunteer information about their sex lives very probably do not represent all Americans. They may be more sexually permissive and/or active, for example, or have a particular axe to grind.

Since the Kinsey research, other scientific studies on sexual behavior have used random samples. One was the study by Cathy Greenblatt (1983), whose sample included thirty men and fifty women, identified by random digit dialing and then sent a questionnaire. While Greenblatt's subjects were diverse in ethnicity, religion, education, income, occupation, and parental status, the sample was small (probably due to lack of major funding).

More recently, in 1992, the National Opinion Research Center (NORC) at the University of Chicago conducted interviews with a random and representative sample of 3,432 Americans, ages 18 to 59 (Laumann et al. 1994). These respondents were selected using sophisticated sampling techniques developed through decades of political and consumer polling. NORC can be reasonably certain the respondents did not lie (at least not anymore than subjects lie on surveys about other topics, always a concern for social scientists) because they were questioned in ninety-minute face-to-face interviews. Furthermore, 80 percent of the randomly selected persons actually agreed to be interviewed—an impressively high and statistically viable response rate for social science research. NORC's findings can be generalized to the U.S. population under age 60 with a high degree of confidence. Indeed, the results have been welcomed as the first-ever truly scientific nation-wide survey of sex in the United States. One unfortunate omission in the NORC data, however, is that the sample included only people under 60. As a result, the findings really cannot be generalized to the nation as a whole, nor do they tell us anything about the sexual activities of older Americans.

One study of sex among marrieds (Call, Sprecher, and Schwartz 1995) used in this section seeks to remedy the NORC study's deficiencies by using another national data set, the National Survey of Families and Households (NSFH). In 1987–88 the NSFH staff, affiliated with the University of Wisconsin, personally interviewed a representative national sample of 13,000 respondents ages 18 and over (Sweet, Bumpass, and Call 1988). The NSFH survey asked far fewer questions concerning sexual activity specifically, but some information on sexual behavior is included. Considered very reliable, the NSFH data are often used as a basis for analysis regarding many topics discussed in this text.

Conclusions based on survey research on sensitive matters such as sexuality must always be qualified by an awareness of their limitations—the possibility that respondents have minimized or exaggerated their sexuality or that people willing to answer a survey on sex are not representative of the public. Nevertheless, with data from national samples such as the NSFH and the NORC study, we have far more reliable information than ever before. Appreciating both the strengths and weaknesses of the data, let us examine what we know about sexuality in American marriages.

How Often?

Social scientists are interested in sexual frequency because they like to examine trends over time and to relate these to other aspects of intimate relationships, such as communication styles or feelings of satisfaction. For the rest of us, "How often?" is typically a question motivated by curiosity about our own sexual behavior compared to others. Either way, what do we know?

Forty-five years ago, the Kinsey reports found that the median frequency of sexual intercourse among marrieds under age 35 was 2–2.5 times a week. This means that half as many people had sex less often than that and half as many had it more often. And some couples are sexually inactive. In the 1992 NORC survey, described above, the average frequency of sex for sexually active, married respondents under age 60 was 7 times a month, or 1.6 times a week. About 40 percent of marrieds said they had intercourse at least twice a week (Laumann et al. 1994). If we take into consideration that none of these findings include older Americans, we might conclude—along with various experts in the field—that American couples average having sex about once a week. Of course, this is an *average*: "People don't have sex every week; they have good weeks and bad weeks" (Pepper Schwartz in Adler 1993). So does the ratio of good to bad weeks change over the course of a marriage? Yes: You have fewer good weeks (sorry).

FEWER GOOD WEEKS To examine sexual frequency throughout marriage, Call, Sprecher, and Schwartz (1995) looked at 6,785 marrieds with a spouse in the household (and also at 678 respondents who were cohabiting, or living together) in the NSFH data set described above. Like researchers before them (for example, Blumstein and Schwartz [1983]), they found that sexual activity is highest among young marrieds. About 96 percent of spouses under age 25 reported having had sex at least once during the previous month. The proportion of sexually active spouses gradually diminished until about age 50 when sharp declines were evident. Among 50- to 54-year-olds, 83 percent said they had sex within the previous month; for those between 65 and 69, the figure was 57 percent; 27 percent of respondents over age 74 reported having had sex within the previous month.

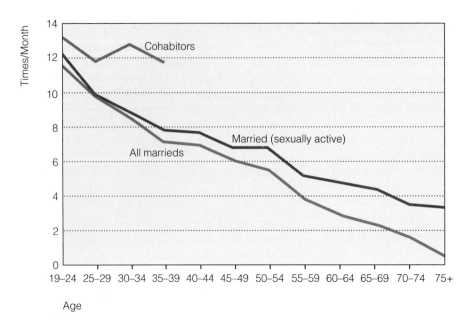

FIGURE **5.1**

Frequency of sex last month by age and marital status. (Source: Call, Sprecher, and Schwartz, 1995, p. 646)

Figure 5.1 shows the mean average frequency of sexual intercourse during the month prior to the interview by age (and marital status). When examining Figure 5.1, note that the researchers report separate mean averages for all the marrieds in the sample and also for the sexually active spouses only. (The figure also shows frequency rates for cohabitors, and we will address these findings below.) The average number of times that married persons under age 25 had sex is about twelve times a month. That number drops to about eight times a month at ages 30–34, then to about six times monthly at about age 50. After that, frequency of intercourse drops more sharply; spouses over age 74 average having sex less than once each month.[2]

It used to be that describing sexuality over the course of a marriage would be nearly the same issue as discussing sex as people grow older. Today this is not

the case. Many couples are remarried so that at age 45, or even 70, a person may be newly married. Nonetheless, we can logically assume that young spouses are in the early years of marriage.

Young Spouses

Why do young spouses have sexual intercourse more frequently than do older mates? The most important answer is simply age. On average, as people get older they have sex less often (Call, Sprecher, and Schwartz 1995). But aging does not explain the decline entirely. Young married partners, as a rule, have fewer distractions and worries, such as mortgage payments or small children. The high frequency of intercourse in this age group may also reflect a self-fulfilling prophecy: These couples may have sex more often partly because society expects them to.

After the first year, couples can expect sexual frequency to decline (Jasso 1985). Why so? It seems that a frequency pattern is set the first year. And "from then on almost everything—children, jobs, commuting, housework, financial worries—that happens to a

2. These researchers found no significant differences in this pattern that were due to gender, race, or region of the country. While higher education levels had a positive effect on sexual frequency, very low and very high educational levels had a significant negative impact on sexual frequency.

couple conspires to reduce the degree of sexual interaction while almost nothing leads to increasing it" (Greenblatt 1983, p. 294).

Spouses in Middle Age

Despite declining coital frequency, respondents in Greenblatt's study emphasized the importance of sexuality. They pointed to the total marital relationship rather than just intercourse, however—to such aspects as "closeness, tenderness, love, companionship and affection" (p. 298)—as well as other forms of physical closeness such as cuddling or lying in bed together. In other words, with time, sex may become more broadly based in the couple's relationship. During this period, sexual relating may also become more sophisticated, as the partners become more experienced and secure. The NSFH study (whose results are illustrated in Figure 5.1) found that, after age, marital satisfaction was the second largest predictor of sexual frequency. Unhappy marriages were associated with a lower sexual frequency (Call, Sprecher, and Schwartz 1995). We should be clear that it is impossible to determine which factor is cause and which is effect here: Does a lower level of marital satisfaction cause a couple to have sex less often, or is it the other way around? Another point: Middle-aged couples are apt to be more sexually active now than in the past. Over the past several decades, each age cohort has been somewhat more active than its predecessor. What appears to be declining frequency of sex with age is in part simply a generation-by-generation change toward heightened sexual activity (Jasso 1985).

Older Partners

In our society, images of sex tend to be associated with youth, beauty, and romance; to many young people, sex seems out of place in the lives of older adults. About 25 years ago, public opinion was virtually uniform in seeing sex as unlikely—even inappropriate—for older people. With Masters and Johnson's work in the 1970s indicating that many old people are sexually active, public opinion swung the other way. Then in the 1980s, researchers began to caution against the romanticized notion that biological aging could be abolished—that "old people are (or should be) healthy, sexually active, engaged, productive, and self-reliant" (Cole 1983, pp. 35, 39). Of course, biological aging cannot be eradicated, and physical changes associated

with aging do affect sexuality (Marsiglio and Donnelly 1991).

Health concerns that may particularly affect sexual activity include prostate problems, diabetes and vascular illnesses, the need to take pain-killing drugs, and—contrary to the reassuring statements we sometimes read—hysterectomies. Sexual functioning may also be impaired by the body's withdrawal of energy from the sexual system in order to address any other serious illness. Some older partners shift from intercourse to petting as a preferred sexual activity.

On the other hand, sexual intercourse does not necessarily cease with age. You can see in Figure 5.1 that for older respondents, among whom the proportion of sexually inactive couples is large, the mean average frequency seriously understates what is happening in sexually active marriages. For example, sexually active spouses over age 74 have sex about four times a month. Indeed, retirement "creates the possibility for more erotic spontaneity, because leisure time increases" (Allgeier 1983, p. 146).

When health problems do not interfere, both women's and men's emotional and psychological outlooks are as important as age in determining sexual functioning. Factors such as monotony, lack of an understanding partner, mental or physical fatigue, and overindulgence in food or alcohol can all have a profound negative effect on a person's capacity for sexual expression. Another important factor is regular sexual activity (Masters and Johnson 1966)—as in use it or lose it.

What About Boredom?

Jokes about sex in marriage are often about boredom. And among social scientists, an explanation for the decline in marital sexual frequency (after aging) is **habituation**—the decreased interest in sex that results from the increased accessibility of a sexual partner and the predictability in sexual behavior with that partner over time. There is some evidence that habituation occurs in that sexual frequency declines sharply after about the first year of marriage, no matter how old (or young) the partners are. The reason for "this rather quick loss of intensity of interest and performance" appears to have two components:

> a reduction in the novelty of the physical pleasure provided by sex with a particular partner and a reduction in the perceived need to maintain high

*"Looks like the Wilsons are putting a little excitement
back in their marriage."*

levels of sexual behavior. From this perspective, the legitimization or institutionalization of sex through marriage may affect early diminution of interest, whereas the reduction in novelty is more subtle. (Call, Sprecher, and Schwartz 1995, p. 649)

To study this theory, researchers examined the effects on coital frequency of remarriage and of cohabiting—living together in a (hetero)sexual relationship without being legally married.

FIRST-MARRIEDS COMPARED WITH REMARRIEDS AND COHABITORS

Remarried respondents reported somewhat higher rates of sex frequency, compared with people in first marriages who were the same age, and this was particularly true for those under age 40. Because people who remarry do renew the novelty of marital sex with a new partner, this finding is evidence for the habituation hypothesis, although it could also be true for other reasons (Call, Sprecher, and Schwartz 1995).

Unfortunately, the NSFH sample did not have enough cohabiting respondents over age 35 to allow for statistical analysis of older partners. Among those ages 35 and younger, however, it is clear that cohabiting respondents had considerably higher intercourse rates than did legal spouses of the same age (see Figure 5.1). This finding supports the idea that legal marriage lessens a person's interest in sex by legitimating it, or making it "perfectly okay," even expected. However, this difference could also be at least partly explained by cohabitors' generally more permissive sexual attitudes and values (Call, Sprecher, and Schwartz 1995).

All this discussion of coital frequency can tempt us to forget that committed partners' sexuality is essentially about intimacy and self-disclosure. In other words, sex between partners—heterosexual partners and gay and lesbian partners as well—both gives pleasure and reinforces their relationship.

Sex as a Pleasure Bond

Masters, Johnson, and Kolodny (1994) view sex as a **pleasure bond** by which partners commit themselves to expressing their sexual feelings with each other.

In sharing sexual pleasure, partners realize that sex is something partners do *with* each other, not to or for each other. Each partner participates actively, as an equal in the sexual union. Further, each partner assumes **sexual responsibility,** that is, responsibility for his or her own sexual response. When this happens, the stage is set for conscious mutual cooperation. Partners feel freer to express themselves sexually. Such expression may not be as easy as it seems, for it requires a high degree of self-esteem, the willingness to transcend gendered expectations, and the ability to create and maintain an atmosphere of mutual cooperation. We'll look at each of these elements in turn.

Sexual Pleasure and Self-Esteem

High self-esteem is important to pleasurable sex in several ways. First, self-esteem allows a person the freedom to receive pleasure. People who become uncomfortable when offered a favor, a present, or praise typically have low self-esteem and have trouble believing that others think well of them. This problem is heightened when the gift is sexual pleasure. People with low self-esteem may turn off their erotic feelings because unconsciously they feel they don't deserve them.

Self-esteem also allows individuals to acknowledge and accept their own tastes and preferences. This is vital in sexual relationships because there is a great deal of individuality in sexual expression. An important part of sexual pleasure lies in doing what one wants to—not necessarily doing things the ways others do them.

Third, self-esteem provides us the freedom to search for new pleasures. As we've seen, some individuals let their sexual relationship grow stale because they do not accept or appreciate their own sexuality and their needs to experiment and explore with their partner.

Fourth, high self-esteem lets each of us ask our partner to help satisfy our preferences. In contrast, low self-esteem can lead a person to be defensive about her or his sexuality and reluctant to express valid human needs. In fact, people with low self-esteem may actively discourage their partners from stimulating them effectively (Burchell 1975).

Finally, high self-esteem allows us to engage in **pleasuring:** spontaneously doing what feels good at the moment and letting orgasm happen (or not), rather than working to produce it.

Masters and Johnson (1976), who initiated contemporary sex therapy, point out that trying too hard can cause sexual problems. They use the term **spectatoring** to describe the practice of emotionally removing oneself from a sexual encounter in order to watch and judge one's productivity, and they state that this practice can be self-inhibiting.

Sexual Pleasure and Gender

A second important element in making sex a pleasure bond is the ability to transcend gender stereotypes. For instance, a man may reject tender sexual advances and activities because he believes he has to be emotionally unfeeling or "bad" or be the initiator of sexual activity.

Likewise, women may have trouble in receiving or asking for pleasure because they feel uncomfortable or guilty about being assertive. Many women have been culturally conditioned to put their partners' needs first, so heterosexual women may proceed to coitus before they are sufficiently aroused to reach climax. Such women are less likely to enjoy sex, which can detract from the experience for both partners.

Sex can be more effective as a pleasure bond when the relationship transcends restrictive gender expectations. To do this, partners must be equal and must communicate in bonding rather than alienating ways.

Communication and Cooperation

A third element in sharing sex as a pleasure bond is communication and cooperation. Partners can use conjugal sex as an arena for power struggles, or they can cooperate to enrich their sexual relationship and to nurture each other's sexual self-concept. To create a cooperative sexual atmosphere, partners must be willing to clearly communicate their own sexual needs and to hear and respond to their partners' needs and preferences as well.

When conflicts arise—and they do in any honest sexual relationship—they need to be constructively negotiated. For example, one partner may desire to have sex more frequently than the other does, which could cause the other to feel pressured. The couple might agree that the partner who wants sex more often stops pressing, and the other partner promises to consider what external pressures such as workload might be lessening his or her sexual feelings.

Other couples might have conflicts over whether to engage in oral-genital sex. It is important to communicate about such strong differences. Sometimes a couple can work out a compromise; in other cases, compromise may be difficult or impossible. Therapists generally agree that no one should be urged to do something he or she finds abhorrent. (And labeling partners "perverted" or "prudish" obviously does not contribute to a cooperative atmosphere.)

Open communication is important not only in resolving conflicts but also in sharing anxieties or doubts. Sex is a topic that is especially difficult for many people to talk about, yet misunderstandings about sex can cause genuine stress in a relationship.

Some Principles for Sexual Sharing

Some important principles can be distilled from the previous discussion to serve as guidelines for establishing and maintaining a nurturing, cooperative atmosphere even when a couple can't find a compromise that suits them both.

Partners should avoid passing judgment on each other's sexual fantasies, needs, desires, or requests. Labeling a partner or communicating nonverbally that something is disgusting or wrong can lower a person's sexual self-esteem and destroy the trust in a relationship. Nor should partners presume to know what the other is thinking or feeling, or what would be good for the other sexually. In a cooperative relationship, each partner accepts the other as the final authority on his or her own feelings, tastes, and preferences.

Another principle is what Masters and Johnson (1976) call the "principle of mutuality." Mutuality implies that "all sexual messages between two people, whether conveyed by words or actions, by tone of voice or touch of fingertips, be exchanged in the spirit of having a common cause." Mutuality means "two people united in an effort to discover what is best for both" (p. 53).

An attitude of mutuality is important because it fills each partner's need to feel secure, to know that any sexual difficulties, failures, or misgivings will not be used against him or her. "Together they succeed or together they fail in the sexual encounter, sharing the responsibility for failure, whether it is reflected in his performance or hers" (pp. 57, 89).

A final principle of sexual sharing is to maintain a **holistic view of sex**—that is, to see sex as an extension of the whole relationship rather than as a purely physical exchange, a single aspect of marriage. One woman described it this way:

> I don't quite understand these references to the sex side of life. It *is* life. My husband and I are first of all a man and a woman—sexual creatures all through. That's where we get our real and central

life satisfactions. If that's not right, nothing is. (in Cuber and Harroff 1965, p. 136)

Researchers John Cuber and Peggy Harroff found that among the couples they interviewed, those who saw sex holistically were "remarkably free of the well-known sexual disabilities" (1965, p. 136).

Recognizing its holistic emotional value is one way to keep marital sex pleasurable. Another is a commitment to discovering a partner's continually changing fantasies and needs. For, as one writer put it, besides a partner's fairly predictable habits and character,

> there is also a core of surprises hidden in us all that can become an inexhaustible source of freshness in life and love.
>
> The Greek philosopher Heraclitus once said that it is impossible to step into the same river twice. By the same token, it is actually impossible to make love to the same person twice—if we let ourselves know it. . . .
>
> Often it is enough simply to go to bed in a different, more attentive state of mind—giving and listening. (Gottlieb 1979, p. 196)

Deepening the commitment to sharing and cooperation can make sex a growing and continuing pleasure bond. This does not mean that a couple regularly has great sex by some external standard. It means that each partner chooses to try—and keep on trying—to say and to hear more. Box 5.2, "As We Make Choices: Communicating About Sex," gives tips on how to do that.

Making Time for Intimacy

Just as it is important for families to arrange their schedules so that they can spend time together, it's also important for couples to plan time to be alone and intimate (Masters, Johnson, and Kolodny 1994).

One element that often contributes to the erosion of sexual communicating, for example, is carried over from the work ethic: the principle that being productive is always more important than being pleasured. In nineteenth-century America, people thought intercourse drained vital energy and productive power from men (Barker-Benfield 1975). Few people consciously hold that belief today, yet many continue to give low priority to sex and intimacy. As two therapists put it, sex for marrieds is not always honored as an

Communicating About Sex

Negotiating our own sexuality and sexual expression in face-to-face relationships can be tricky. (Maybe that's why cybersex is so popular!) Sexual expression carries the promise of "transcending the limits of the individual and relating intimately... to other persons" (Francoeur 1987, p. 531). At the same time, sharing our sexuality makes us tremendously vulnerable. Both women and men need clear and positive feedback to know what brings their partners the greatest joy and fulfillment. One popular seminar leader on relationship communication and sex recommends that couples take a half hour every so often, particularly when they are not feeling negative about sex, and talk about it.

Here are some questions to stimulate an informative conversation:

- What do you like about having sex with me?
- How did you feel when I did _____?
- Would you like more sex?
- About how much sex would you like each week (or month, or day)?
- Would you like for us to spend more (or less) time when we have sex?
- Is there something specific you would like me to do in the next month during sex?
- Is there a new way you would like me to touch you? If yes, would you show me?

- Is there anything new you would like me to try?
- Is there anything you would like to try that we've never done?
- Is there anything I used to do that you would like me to do more of or again?

Note that a conversation like this needs to be mutual and reciprocal: *Both* partners ask and answer these questions—and they accept each other's answers without judgment either of themselves or of their partner.

What makes it difficult to talk about our sexual desires is that we don't want to hurt our partner's feelings and, conversely, we would hate to hear that we are disappointing our partner in any way at all. Meanwhile, we don't want to feel pressured to do anything that would not feel comfortable to us. So when answering these questions, it is important that you make it clear that you are not demanding more. The dialogue works best if its only purpose is to share information.

Then, too, you should not do things that do not feel right to you. But if your partner wants something that doesn't seem important to you or seems unpleasant, you might consider keeping an open mind. And when your partner does not seem open to things you favor, try to accept that at least for now. "A secret of great sex is to build on the strengths you have and not focus on the problems or what you are missing" (Gray 1995, pp. 53–55).

activity that is planned for, or counted on, for which time is set aside, or that even takes place at times when both partners are relaxed and rested. Instead, it seems for many to be relegated to a time when there is nothing else to do. This attitude explains why some couples rediscover sex when they are off alone together on vacation: They aren't distracted by business, household, or child-care responsibilities.

Planning time for intimacy involves making conscious choices. Partners can decide to move the TV from their bedroom, for example. They can choose to set aside at least one night a week for themselves alone, without work, movies, television, the VCR, the computer, another couple's company, or the children. They do not have to have intercourse during these times: They should do only what they feel like doing.

But scheduling time alone together does mean mutually agreeing to exclude other preoccupations and devote full attention to each other. Two sex therapists advise couples to reserve at least twenty-five minutes each night for a quiet talk, "with clothes off and defenses down" (Koch and Koch 1976, p. 35). This last suggestion may be easier for parents with young children who are put to bed fairly early. A common complaint from parents of older children is that the children stay up later, and that by the time they are teenagers, the parents no longer have any private evening time together even with their clothes *on*. One woman, after attending an education course for parents, found a solution. She explains:

> Our house shuts down at 9:30 now. That doesn't mean we say "It's your bedtime, kids. You're tired and you need your sleep." It means we say, "Your dad (or your mom) and I need some time alone." The children go to their rooms at 9:30. Help with homework, lunch money, decisions about what they'll wear tomorrow—all those things get taken care of by 9:30 or they don't get taken care of. We allow interruptions only in emergencies—and I'm redefining what an "emergency" is too. (personal interview)

Moreover, good sex is not something that just happens when two people are in love. Forming a good sexual relationship is a process that necessarily involves partners in open verbal and nonverbal communication. To have satisfying sex, both partners need to tell each other what pleases them.

Maintaining open communication about sex is important for couples of all ages, but it may be particularly important for young couples who are experiencing the transition from premarital sex to conjugal sex. As one young wife explained, "Married sex is different. Sometimes you get disappointed because it isn't as exciting as before marriage. Part of the thrill of premarital sex was that it was supposed to be wrong" (O'Brien 1980, p. 54). This woman found that as she and her husband began to talk about each other's needs both in and out of bed, sex got better than ever. Many young husbands agree. "Sex has changed as I've become more involved in my partner's needs," said one. "This makes the experience more enjoyable for both."

Boredom with sex after many years in a marriage may seem at least partly the consequence of a decision by default. Therapists suggest that couples can avoid this situation, perhaps by creating romantic settings— a candlelit dinner or a night away from the family at a motel—by opening themselves to new experiences, such as describing their sexual fantasies to each other, reading sex manuals together, or even renting an erotic movie. (The important thing, they stress, is that partners don't lose touch with either their sexuality or their ability to share it with each other.) Unfortunately, the myth that great sex follows naturally when a couple is "really in love" leads partners who are having problems to question whether they are well suited to each other. Therapists also point out that just because spouses find they don't desire sex together much anymore does not mean they no longer love each other.

Maintaining a satisfying sex life in marriage, then, requires mutual commitment to that end, and also communication. In some respects, young couples are expected to know all there is to know about sex, especially since the sexual revolution. An unfortunate side effect of this expectation is that partners may be reluctant to discuss their uncertainties with each other because they believe they ought to know everything.

We have been talking about human sexual expression as a pleasure bond. It is terribly unfortunate that today sexuality can also be associated with disease and death. Indeed, the fact that it is so difficult to make this transition here points to the multifaceted, sometimes even contradictory, nature of human sexual expression in the 1990s.

Sexual Expression and HIV/AIDS

HIV/AIDS was identified as such in 1981, and it is an understatement to say that our lives have not been the same since. **HIV/AIDS** is a viral disease that destroys the immune system—hence the name acquired immune deficiency syndrome. The virus that causes HIV/AIDS—called HIV, or human immunodeficiency virus—is transmitted through the exchange of infected body fluids. With a lowered resistance to disease, an HIV-infected person becomes vulnerable to infections and diseases that noninfected people easily fight off; the immediate cause of death from AIDS is often a rare form of pneumonia or cancer. Not everyone who has been infected with HIV will develop full-blown AIDS. Part of the public health problem is that a carrier of the virus can be unaware of its presence for perhaps as long as nine or more years and can infect others during that time if precautions are not taken.

Who Has HIV/AIDS?

Currently between 800,000 and 1 million Americans have HIV but are not yet suffering symptoms of AIDS (Centers for Disease Control 1994a; Laumann et al. 1994). Meanwhile, HIV/AIDS has spread quickly in the United States since 1981, when 261 cases were reported. By 1995, some 500,000 cases had been reported to the Centers for Disease Control (CDC) in Atlanta. In 1993, AIDS overtook accidents as the number one killer of men ages 25 to 44 years old, but the Centers for Disease Control also reported in that year that the rate at which HIV/AIDS is spreading has leveled off and that the number of new cases reported each year is declining ("AIDS Front" 1995).

PRIMARY RISK GROUPS   Primary risk groups in the United States are gay and bisexual men who acquire HIV/AIDS through sexual contact (about 47 percent of reported cases) and intravenous drug users who get HIV/AIDS by sharing needles with infected people (about 28 percent). An additional 5 percent of cases are attributed to men who have sex with men and also inject drugs. About 9 percent of cases are attributed to heterosexual contact, and 2 percent to receipt of infected blood from transfusions (although since 1985 donated blood is rigorously screened for HIV/AIDS). About 9 percent of cases occur among persons whose risk has not been identified (Centers for Disease Control 1994a).

GENDER AND AGE   Women comprise 17 percent of HIV/AIDS cases; that proportion is rising and will continue to rise (Centers for Disease Control 1994a). One indicator is the fact that the percent of female adolescents with HIV/AIDS has more than doubled, from 14 percent in 1987 to 32 percent in 1994 (Centers for Disease Control 1994b). About half of all HIV/AIDS-infected women got the disease from needle sharing, 34 percent from sex with infected men, and the remainder from blood transfusions or undetermined causes (U.S. Centers for Disease Control 1992, Table 5). Although men can contract HIV/AIDS from coitus with infected women, transmission from men to women is higher because of anatomical differences. Still, the risk is only about 1 in 500 for a woman to become infected from any single

sexual contact with an infected man (Laumann et al. 1994, citing the Centers for Disease Control). Repeated encounters result in about a third of the female sexual partners of infected men becoming infected. Because sexually transmitted diseases, or STDs (see Appendix F), such as syphilis, herpes, or chancroid cause open sores on the genitals, HIV/AIDS is more easily transmitted to STD-infected women.

The NORC survey concluded that "sporadic aborted breakouts of the AIDS virus into the general public are the exception, not the rule . . . we are convinced that there is not and very unlikely ever will be a heterosexual AIDS epidemic in this country" (quoted in Freeman 1994). But William Freeman, Executive Director of the National Association of People with AIDS, has responded that the NORC researchers' "implication that there is very little risk in having unprotected sex is appalling" (Freeman 1994). It is difficult to strike a balance between alerting Americans to the AIDS risk and generating unwarranted anxiety. (See Box 5.3, "HIV/AIDS: Some Precautions.")

With regard to age differentials, the highest percentage of reported AIDS cases (45 percent) is among people in their 30s. The next largest proportion (26 percent) is among individuals in their 40s. About 17 percent of reported AIDS cases are among 13- to 29-year-olds (U.S. Bureau of the Census 1995, Table 213). The number and proportion of HIV/AIDS cases among adolescents is relatively small (from 1981 through June 1994, a cumulative total of 1,768 cases). But HIV is spreading among teenagers (Centers for Disease Control 1994b).

About 6,000 cases of AIDS have been reported to date for children under age 12, and three-quarters of these are for youngsters under 5. Children under 12 comprise about 21 percent of all reported cases (U.S. Bureau of the Census 1995, Table 213). About one-fifth of children with HIV/AIDS were infected by blood transfusions. Among children under 5, the vast majority of victims are infants who contracted HIV/AIDS during birth or through breast milk. A cause for concern regarding the growing rate of AIDS among teenage females, coupled with their increasing rates of sexual activity, is the resulting heightened risk for HIV-infected infants born to teen mothers.

RACE/ETHNICITY   Racial and ethnic minority populations have been disproportionately affected by

HIV/AIDS: Some Precautions

AIDS has changed sex, relationships, and life choices. Readers will want to keep updating their information; precautions that at present seem reasonable are:

1. Sexually active individuals who are not in long-term, securely monogamous relationships predating the identification of AIDS in 1981 should use latex condoms (perhaps with spermicide) when having sex. Explicit information about protective practices is available in the *Surgeon General's Report on AIDS* and from other sources.

   The Surgeon General's report states: "Couples who maintain mutually faithful monogamous relationships (only one continuing sexual partner) are protected from AIDS through sexual transmission. If you have been faithful for at least five [to seven] years and your partner has been faithful, too, [very probably] neither of you is at risk. If you have not been faithful, then you and your partner are at risk. If your partner has not been faithful, then your partner is at risk, which also puts you at risk. This is true for both heterosexual and homosexual couples. Unless it is possible to know with *absolute certainty* that neither you nor your sexual partner is carrying the virus of AIDS, you must use protective behavior. *Absolute certainty* means not only that you and your partner have maintained a mutually faithful monogamous sexual relationship, but it means that neither you nor your partner has used illegal intravenous drugs" (Koop, n.d., p. 16).

2. Inquiry about a potential sex partner's health, HIV status, and previous partners is useful and may produce information on which to base a decision about having sex. It is entirely possible, however, that a prospective partner will not be honest. Consequently, experts argue that use of latex condoms with any partner is the most protective approach.

HIV/AIDS since the beginning of the epidemic in the United States. In 1991, AIDS was the sixth leading cause of death among African Americans and the seventh among Hispanics, while it ranked tenth among non-Hispanic whites. Figure 5.2 shows the rate of AIDS cases reported among adults and adolescents by race/ethnicity in 1992. As you can see, the rate is highest among non-Hispanic blacks at 66.6 cases for every 100,000 of them in the population. The rate for non-Hispanic blacks is more than four times that for non-Hispanic whites. The rate for Hispanics falls between those for non-Hispanic blacks and whites. At 8 cases per 100,000 of their population, Native Americans have a lower rate; Asians and Pacific Islanders have the lowest rate (Centers for Disease Control 1993a).

Minority *women* have been particularly hard hit. Of reported cases among adult and adolescent females, 74 percent are among African Americans and Hispanics. Among children under age 13, 79 percent of cases are African Americans and Hispanics. Figure 5.3 gives AIDS rates for children under age 13 by race/ethnicity. You can see from comparing Figures 5.2, and 5.3, the race/ethnicity distributions for adults are highly correlated with (although not identical to) those for children (Centers for Disease Control 1993a). This is largely because a high proportion of HIV/AIDS children, as we have seen, are infected from their mothers at birth or while nursing.

How HIV/AIDS Affects Relationships, Marriages, and Families

A theme of this text is that sociocultural conditions affect people's choices. Nowhere is this more dramatically evident today than in examining how HIV/AIDS, a societal phenomenon, has changed and will continue to alter attitudes, options, the consequences of decisions, and thereby personal decision making or choices.

Moreover, antibodies to HIV do not develop for up to six months or longer after infection with the virus, so an infected person who may appear virus-free in tests will report the possibly erroneous results in good faith.

Communication about sex and disease is going to be necessary in a way that it never was before. Americans used to risk pregnancy without discussion; the stakes are higher now.

3. It would be prudent to confine sexual activity and relationships to those worth the risk. This can mean decisions about individuals or it can mean categorical decisions about multiple partners or sex with members of high-risk groups such as homosexual and bisexual men, individuals who have multiple partners, intravenous drug users, or people known to have AIDS or HIV.

4. Decisions to take risks may involve others: your current or future sex partners, your children, and your family. A responsible sexually active individual will be voluntarily tested, and, if the test is positive, will either refrain from sex or inform the partner beforehand and use latex condoms during sex. Informing partners about one's AIDS-risk status must include discussion of sexual history—that is, of past sexual activity with infected or high-risk individuals.

5. Women planning to become pregnant or not taking precautions against pregnancy should be sure that they are free of HIV by being tested and perhaps retested over a six-month period of time. Sources of infection include not only sexual contacts but also blood transfusions obtained between 1977 and 1985.

6. Health-care workers should take the precautions recommended by guidelines for their occupation.

7. Citizens should support sex education designed to prevent the spread of AIDS. Appropriate AIDS education should be encouraged for children (because even young children can be exposed to AIDS through sexual abuse), teenagers, and adults. Videotapes intended for home viewing are available from schools, libraries, public health departments, and commercial sources.

Keep yourself informed by consulting your local public health department, the American Red Cross, student services, gay activist groups, churches, and other sources, including newspapers, radio, and television. You can also write to the U.S. Centers for Disease Control, Center for Prevention Services, Division of Sexually Transmitted Diseases, Atlanta, GA 30333, or call the Public Health Service AIDS Hotlines (800-342-AIDS, or 800-342-2437). Or you can reach the CDC hotline on the Internet at http://patients.cnidr.org/welcome.html. Or send an e-mail message containing the single word "help" to ezgate@cnidr.org.

HIV/AIDS AND HETEROSEXUALS   Heterosexuals may be responding to the threat of AIDS with changed behavior (Mosher and Pratt 1993; Ku, Sonenstein, and Pleck 1995). Sexually active singles may now expect to have a longer period of acquaintance before initiating sexual contact, hoping to experience greater attraction, security, or commitment before deciding sex is worth it. Some opt for periods of celibacy, with or without masturbation.

But dating clubs whose members are required to be regularly tested for HIV/AIDS have developed in some larger cities (Geist 1987). "Blind Dating Is Back," says one headline, as people turn to their friends rather than bars, clubs, or personals columns to meet people of the opposite sex (Foderaro 1988).

Safer sex, for heterosexuals as well as in the gay community, refers to the use of latex (they *must* be latex) condoms ("rubbers") and to the limiting of partners in number and selectivity. Latex condoms— contraceptive sheaths that fit over the penis—reduce the likelihood that HIV will be transmitted to a sexual partner during genital or anal intercourse. Latex condoms can also be used for protection during oral sex.

"Safe sex" is a misnomer, however, because condoms can be broken, although rarely, and HIV can still be transmitted. Nevertheless, latex condoms do substantially reduce the risk (Centers for Disease Control 1993b).

A side issue of safer sex has become the placement of sexual responsibility, once again, on women (Mosher and Pratt 1993). Forty percent of condom purchases are now made by women, who are encouraged to prevail upon their partners to use them. Requiring women to be sexually assertive is not always easy, especially because women have been encouraged to follow men's sexual lead in the past. Men may resist the use of condoms because folk wisdom holds that condoms cut down on

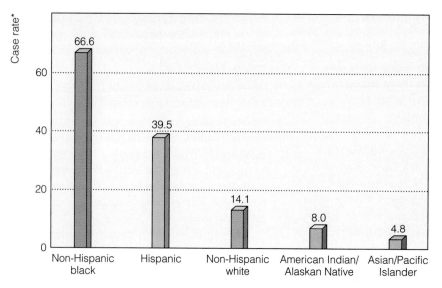

Case rate*

66.6

39.5

14.1

8.0

4.8

Non-Hispanic black | Hispanic | Non-Hispanic white | American Indian/ Alaskan Native | Asian/Pacific Islander

*Number of cases per 100,000 population in respective racial/ethnic group

FIGURE **5.2**

Rate of AIDS cases reported among adults and adolescents, by race/ethnicity, 1992. (Source: Centers for Disease Control 1993a)

sensation.[3] "We have to train women how to negotiate safe sex, . . . but we also have to focus on black and Hispanic heterosexual men who tend to resist condoms and other risk-reduction practices. It's not going to be easy" (Ronald Johnson, in Halpern 1989a, p. 85). But "if you say 'safe' and 'no' often enough, it will start to sink in," said a participant in a women's AIDS workshop (Dunning 1986).

To what extent various groups are using latex condoms, asking for their partners' sexual histories, and being selective of sexual partners is uncertain. Although some college students have evidenced more careful behavior in the form of safer sex and monogamous relationships, health educators are concerned that the reality of HIV/AIDS is not clear enough for the average college student because it usually strikes slightly older adults.

HIV/AIDS AND GAY MEN Many gay men and their families may find the coming out process more difficult, as knowledge of a gay family member's sexual orientation is also knowledge of his risk of illness and premature death (Robinson, Skeen, and Walters 1987). A new wave of homophobia has resulted in increased hostility toward gays. Gay men (indeed, many single men in high-risk geographic areas or occupations) are losing jobs, housing, and insurance.

Many gay men modified their sexual behavior in the 1980s (Kantrowitz 1987). Multiple, frequent, and anonymous sexual contacts were common elements of the lifestyle and sexual ideology for many gays (Blumstein and Schwartz 1983). But attitudes and behavior have changed somewhat, at least among men in their 30s and over. Still, recent research indicates that many gays continue to engage in risky sex (Rogers 1994). Ironically, AIDS education programs may have backfired among gays in their teens and 20s. According to one San Francisco area gay activist and psychologist with a largely gay practice, "A lot of younger men have grown up from the time they were eight years old hearing on television about gay men and HIV. They

3. In fact, some women's shelters have reported violent reactions from men when their partners ask them to use condoms. "With the coming of AIDS, the age-old battle of the sexes is literally becoming a life and death struggle for women" (Maria Maggenti, in Halpern 1989a, p. 85).

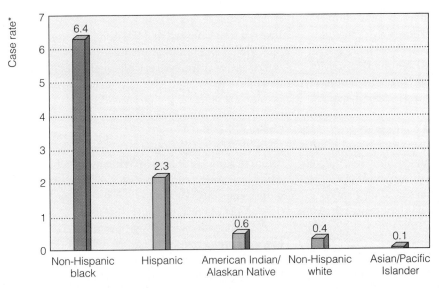

*Number of cases per 100,000 population in respective racial/ethnic group

FIGURE **5.3**

Rate of AIDS cases reported among children less than 13 years of age, by race/ethnicity, 1992. (Source: Centers for Disease Control, 1993a)

have seemlessly integrated being gay with having HIV. When you ask them if it's important not to get infected they shrug their shoulders and say, 'I guess so'" (Walt Odets, in Hagar 1995, p. 10). Then too, so much attention has gone to "HIV positives" (people who are infected) within gay communities such as San Francisco's Castro district that young gay men, "especially adolescents, may feel "left out" when they test negative for the disease (Odets, in Hagar 1995).

Moreover, many in the gay community are experiencing the frequent loss of friends and intimate partners. It is difficult for the majority of us to imagine losing as many as five—or twenty—dear friends and acquaintances to death from disease over the course of just a few years. This situation can lead a gay man to wonder whether he himself wants to go on living (Hagar 1995). Indeed, there are a substantial number of AIDS-related suicides: the bereaved, along with the ill, and sometimes those reacting to positive results from having been tested for HIV/AIDS ("Suicides Tied" 1990).

Finally, gays and their friends and families are living not only with loss, but also with the burden of community and personal care for friends, lovers, or family members with AIDS. Especially in geographically distinct gay/lesbian communities, this situation applies to lesbians as well as to gay men. Lesbians in the San Francisco area, for example, often have close ties with gay men and may consider them members of their families. In some cases gay men, who years later test positive for HIV, have donated sperm for lesbian mothers.

WIVES OF GAY MEN   Perhaps 20 percent of male homosexuals marry at least once (Buxton 1991). Recent evidence suggests that some gay soldiers have actually married for camouflage (Schmitt 1993). Consequently, some heterosexual women may be regularly exposed to the virus themselves. Although some knowingly married gays, 85 percent of a sample of wives in a support group were not aware of their husbands' homosexuality at the time of marriage.

Most of these women denied concern about infection, claiming that their husbands practiced safe sex or had only one male partner. Counselors, however, are alarmed that an unknown but possibly large number of women are knowingly or unknowingly exposed to HIV through marriage to men who have homosexual contacts. Revelation of husbands' homosexual activity through development of AIDS or through his protective revelation of risk is likely to change the dynamics of these marital relationships—and possibly end them (Dullea 1987d). One writer has pointed out that straight spouses of gays (or lesbians) have been left out of discussions and research about gay lifestyles. When a partner's coming out begins a divorce process, "straight wives and husbands experience the pain of any marital rupture, plus the anguish of being rejected as a woman or a man" (Buxton 1992).

HIV/AIDS MEANS NEW FAMILY CRISES   Some families will face unprecedented crises because of AIDS (see Macklin 1988). One woman contracted HIV from a transfusion during a difficult birth and conceived another child before the disease was diagnosed; that child was infected. Mother and child died, leaving behind a young husband and 3-year-old daughter. We will begin to see more orphaned children as parents die of AIDS (Gorman 1993).

Although monogamy has definitely increased in appeal since the advent of HIV/AIDS, some married heterosexuals have lost partners to the disease or are helping infected partners fight health battles. AIDS contracted from a blood transfusion means a prolonged medical battle for the partner, and perhaps children who may be infected. But it is a tragedy that can be shared. AIDS resulting from drug use frequently occurs in transient relationships, but in marital settings it often indicates a family that has many problems. In cases in which a partner (more often, the husband) develops AIDS from sex outside the relationship, the marriage may break down, giving rise to complicated legal issues. Women are often advised to maintain the marriage for financial reasons. But if a divorce takes place, "fault" concepts, thought to be a thing of the past, enter the picture, and the "innocent" party is likely to be awarded the couple's financial assets. Liability insurance claims and damage suits have also been filed by victims, their partners, or other relatives (Dullea 1987a; "Judge Backs Marine" 1990).

HIV/AIDS MEANS FAMILY FINANCIAL BURDENS   The burdens of AIDS will not all be emotional or involve physical care of victims; some

will be financial. For example, we may see AIDS patients or those infected by HIV dismissed from their jobs and refused health insurance. Attorneys agree that under the 1992 Americans with Disabilities Act, it is illegal for most companies to fire an employee solely because she or he is HIV positive. But many job dismissals do not have to be explained, and the burden of proof of disability discrimination is on the complainant, who may be reluctant to pursue the case for fear of making his or her condition more public. And many insurance companies have already begun to require a negative HIV test to provide life insurance, a development upheld by courts (Lambert 1991).

AIDS care is very expensive! Insurance companies have sometimes refused to pay for drugs or other treatment. Overall federal, state, and local responsibility for the health care of AIDS patients has not been determined, but most assuredly families with any financial resources will be expected to use them, leaving surviving family members hard-pressed financially and heavily in debt.

We've seen that more women are now developing AIDS, many of them single women with children. Poor and minority women, and those who use drugs, have few personal or financial resources, nor do they have the educational and organizational skills to advocate for resources as effectively as the gay community has done. People who work with AIDS patients have argued for assistance targeted to women—for example, clinics that would treat women and their children, as well as provide some help in dealing with housing and financial problems. AIDS and pregnancy prevention for those at risk of AIDS is also a major concern (Kolata 1989c). Children with pediatric AIDS have a unique array of treatment requirements and are often in the hospital (Klass 1989). Some have been absolutely and literally abandoned by their parents to hospital care.

The Politics of HIV/AIDS

HIV/AIDS and other sexually transmitted diseases are more than a medical or a family problem—they are conditions imbued with social meanings and consequences. For some, the disease is primarily a pragmatic problem requiring pragmatic solutions. For others, it is evidence of God's judgment against "sinful" lifestyles. Political and religious conflict over sexuality has characterized the 1980s and 1990s. Research and treatment of HIV/AIDS have been drawn into that

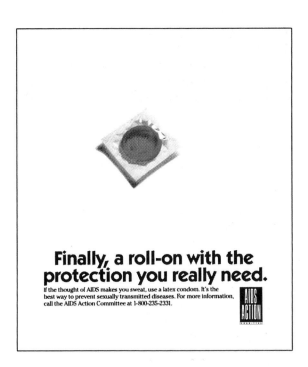

Finally, a roll-on with the protection you really need.

If the thought of AIDS makes you sweat, use a latex condom. It's the best way to prevent sexually transmitted diseases. For more information, call the AIDS Action Committee at 1-800-235-2331.

AIDS ACTION

conflict. Some, such as fundamentalist religious spokesman Patrick Buchanan (1983), who for religious reasons oppose nonmarital sexual relationships or homosexuality, may cast the AIDS epidemic into this framework. They may also rule out certain preventive measures such as condoms and explicit sex education. Other religious leaders stress compassion and work with AIDS victims.

AIDS victims have expressed anger that the government was not committed to help them because so many victims are gay. AIDS is also prevalent among minority poor, a group with little political leverage. Strong gay activism, as well as the fact that AIDS has spread to the adult heterosexual population and to children, has increased societal commitment of resources to combat AIDS, although gay activists remain convinced that not enough is being done by the government because of such discrimination, and a government advisory panel concurs (Hilts 1990b). But others argue that, given other serious health threats, the U.S. government is spending enough on HIV/AIDS—more per death, for example, than on heart disease or cancer (Stout 1992). And there is evidence that Americans generally have tired of hearing about AIDS (Krieger 1995).

How will all this influence people's personal decisions? Some campus health officials worry that a possi-ble "overreaction . . . could lead to serious emotional problems [for students]. 'Right now, the emphasis is a fear-arousing emphasis . . . , sex-negative messages. What I'm concerned about is that the young people on college campuses simply not be left with fear about sexuality'" (Kantrowitz 1987, pp. 17–18). Increasingly, however, people may choose to forgo new sexual opportunities and/or negotiate safer sex practices with partners. As we will see in the next section, politics influences individuals' decisions about sex.

Platform and Pulpit

Our entire bodies respond to sexual stimulation. Blood pressure rises and pulse rate increases, breathing becomes deeper and faster. This is true for men and women, gays and straights. More people today than thirty years ago are aware of the escalating stages of human sexual response (see Appendix B), and more people today than thirty years ago are also aware that these stages characterize both women's and men's sexualities. We owe this beneficial, heightened awareness largely to the detailed physiological information that resulted from the pioneering research of an internationally renowned research team, William H. Masters and Virginia E. Johnson. (We've quoted them quite a bit in some sections of this chapter.) In 1966, Masters and Johnson published *Human Sexual Response,* a landmark study based on laboratory research. They were the first to study sexual behavior in a laboratory. Through carefully controlled observation over a period of eleven years, they recorded in detail the bodily changes that take place as a consequence of sexual arousal. Volunteers, both married and single (382 women and 312 men), engaged in sexual activities, primarily intercourse and masturbation, in a laboratory setting. These volunteers' physiological responses were monitored by various recording devices, sometimes including tiny cameras that filmed internal changes in the women's vaginas. The researchers and their assistants also watched the sexual activities of their volunteers through one-way mirrors. When their book came out, Masters and Johnson were bombarded with hate mail; the "drop dead" letters surpassed more favorable ones at a rate of 9 to 1 (Schrof 1994). But their work did not become the political (or radio talk show!) controversy that we might expect if it first hit the bookstores today. What's different?

In our society, sexuality and reproduction have become increasingly politicized. Nowhere is this more apparent than in the intensely heated pro-life/pro-choice debate over abortion, one of the most polarizing issues in America today.

The New Christian Right

One of the most striking changes in the political climate over the past two decades has been the emergence of the **New Christian Right** (or Moral Majority or Religious Right), a loose coalition of religious fundamentalists and political conservatives who believe that American government and social institutions must be made to operate according to what they see as Christian principles.

The New Christian Right may not be "Christian" as many people define the term. The main interpretations of being Christian for the New Christian Right are specific "family" values; heterosexuality, sex only in marriage, monogamy, anti-abortion, and traditional female gender roles, both in and outside the (heterosexual) nuclear family. For example, arguing in the 1991 congressional budget hearings that the Senate should quit funding social services for gay youth, Sen-

ator Jesse Helms proclaimed that, "this money should not—and shall not, if I have anything to do with it—be used to encourage homosexuality" (quoted in National Gay and Lesbian Task Force, n.d.). We might imagine, however, that being Christian could mean having empathy for the poor or finding ways to better care for children and others at risk for serious illnesses, such as HIV/AIDS. Hence former Surgeon General Joycelyn Elders talks about the "un-Christian religious right" ("Goodbye" 1994).

Whether Christian or not, the New Christian Right is assuredly political. Members view it as a person's religious obligation to be politically active in making the United States a Christian nation, as they define it. As Pat Robertson, leader in the religiopolitical organization, *Christian Coalition,* once put it, "Not voting is a sin against God. . . . Perverts, radicals, leftists, Communists, liberals, and humanists have taken over the country because Christians didn't want to dirty their hands in politics" (in Bollier 1982, p. 70).

The New Christian Right uses political processes—lobbying, campaign contributions, getting out the vote—to influence public policy on sex matters. Leaders broaden the movement's appeal by packaging it as a "pro-family" policy group and downplaying its religious affiliations and agenda (Judis 1994). While the Moral Majority is hardly representative of a majority of Americans, the movement has had successes. For instance, in 1995 Ross Perot's political organization, *United We Stand America,* sponsored a public conference on national issues and invited contenders in the upcoming 1996 presidential campaign. Of all the speakers, it was then-Republican candidate Patrick Buchanan who wowed the crowd ("Perot Backers" 1995). Regarding HIV/AIDS, Buchanan had formerly pronounced that because "homosexuals . . . have declared war upon nature, . . . now nature is exacting an awful retribution" (quoted in National Gay and Lesbian Task Force, n.d.). This political climate has influenced both research and education about sexuality in the United States.

POLITICS AND RESEARCH   The need for more comprehensive and current data, particularly in light of HIV/AIDS, has twice led unsuccessfully to efforts to mount federally funded national sample surveys to be conducted by teams of well-respected social scientists. The NORC survey, described above, was originally meant to sample 20,000 adults, not 3,400. But Congress first insisted that questions on masturbation be omitted, then canceled the pilot study altogether on the grounds that a sex survey would be too controversial. NORC conducted the much smaller survey without federal funds. A survey of 24,000 teens in grades 7 to 11 was scrapped for the same reason ("U.S. Scraps" 1991).

POLITICS AND SEX EDUCATION   In the 1980s, then-Surgeon General C. Everett Koop, himself a fundamentalist Christian and political conservative, was moved by public health concerns to propose explicit education about AIDS to children as young as 9, including the topics of homosexuality, as well as genital intercourse and condoms. The majority of experts and some school districts embraced this plan (Putka 1990). However, the federally funded ten-session education course currently designed for junior high students, called "Sex Respect," not only appropriately stresses abstinence but also ignores controversial necessary subjects such as contraceptives and homosexuality (Nazario 1992).

In 1994, when then-Surgeon General Elders suggested that schools might teach masturbation as safe sex, President Clinton fired her, largely due to outcries from the New Christian Right. The *Traditional Values Coalition,* a religio-political association, had mailed petitions to 30,000 churches urging members to lobby for her dismissal (Fineman 1994). Labeled by talk radio's Rush Limbaugh as the "Condom Queen," Elders had also made it known that she favored giving condoms to public schoolchildren (not, she quipped, that she would put them on their lunch trays) ("Goodbye" 1994). These decisions represent our society's tendency to deny sexuality at the same time that we encourage it—mixed messages that, among other negative consequences, probably help account for the unusually high rates of teen pregnancy and abortion in the United States (Woodman 1995). Chapter 11 explores the issues of unmarried and teen pregnancy more fully.

People today are making decisions about sex in a climate characterized by political conflict over sexual issues. Premarital and other nonmarital sex, homosexuality, abortion, contraception, and reproductive technology represent political issues as well as personal choices. Furthermore, legislation or court action can render certain choices more or less openly available. Gay and lesbian couples, in particular, must struggle to create satisfying sexual and emotional relationships in situations of community disapproval or legal risk. For instance, as of this writing, committed gay male and lesbian couples are unable to marry legally in the United States (see Chapter 8).

In conclusion, both public and private communication must rise to a new level, as potential sexual partners talk about sex and disease, precautions and risk, sexual history and sexual practices. If such communication in fact develops, it would be the realization of one of the few hopeful comments about AIDS—that "out of the peril of the plague could rise a strong new American ethic of sexual responsibility" (Rosenthal 1987, p. A-23).

Sexual Responsibility

Our changing society offers divergent sexual standards, from casual sex to abstinence. In such a climate, making knowledgeable choices is a must. Because there are various standards today concerning sex outside of marriage (as well as sexual orientation itself), each individual must determine what sexual standard he or she values, which is not always easy. Moreover, today's adults may be exposed to several different standards throughout the course of their lives because different groups and individuals adhere to various standards. Even when people feel they know which standard they value, decisions in particular situations can be difficult. People who believe in the standard of sexual permissiveness with affection, for example, must determine when a particular relationship is affectionate enough.

Making these choices and feeling comfortable with them requires recognizing and respecting your own values, instead of just being influenced by others.

Whether it be due to having—or fear of contracting—a sexually transmitted disease or due to living in a culture characterized by conflicting and mixed messages, anxiety can accompany the choice to develop a sexual relationship, and there is considerable potential for misunderstandings between partners. This section addresses some principles of sexual responsibility that can serve as guidelines for sexual decision making.

It's easy to see that even though sexuality is a natural part of ourselves, it is more complex than just being a part of our bodily functions. The AIDS epidemic has brought the importance of sexual responsibility to our attention in a dramatic way. We will suggest a few guidelines that may help make sexual choices clearer.

Nelson Foote (1954), a social scientist, once observed that because sex was becoming increasingly dissociated from procreation, it was also becoming more recreational. Alex Comfort, in his book *The Joy of Sex,* called sexual expression "the most important form of adult play" (Comfort 1972, p. 85).

If we use this terminology to describe sexual expression, we should stress that it is adult play, not child play. People must take responsibility for the consequences of their behavior. Certain rules and responsibilities are important.

One obvious responsibility concerns the possibility of pregnancy. Both partners should responsibly plan whether and when they will conceive children and use effective birth control methods accordingly. (Chapter 11 and Appendixes C and D provide more information on this subject.)

A second responsibility has to do with the possibility of contracting sexually transmitted diseases (STDs) or transmitting them to someone else. Individuals should be aware of the threat and the facts concerning HIV and other serious STDs. They need to assume responsibility for protecting themselves and their partners. They need to know how to recognize the symptoms of an STD and what to do if they get one (see Appendix F). Guidelines pertaining to AIDS are presented in the preceding section of this chapter.

A third responsibility concerns communicating with partners or potential sexual partners. In one study, 35 percent of men reported lying to a partner in order to have sex with her, while 60 percent of women thought they had been lied to. Such lies involved overstating love and caring, denying other simultaneous relationships, and reporting fewer sex partners than was true (Goleman 1988c).

People should be honest with partners about their motives for wanting to have sexual relations with them. As we've seen in this chapter, sex can mean many different things to different people. A sexual encounter can mean love and intimacy to one partner and be a source of achievement or relaxation to the other. Honesty lessens the potential for misunderstanding and hurt between partners. People should treat each other as people rather than things—as people with needs and feelings. Sex should never consciously be used for exploitation or degradation.

A fourth responsibility is to oneself. In expressing sexuality today, each of us must make decisions according to our own values. A person may choose to follow values held as a result of religious training or put forth by ethicists or by psychologists or counselors. People's values change over the course of their lives, and what's right at one time may not be satisfying later. Despite the confusion caused both by internal changes as our personalities develop and by the social changes going on around us, it is important for individuals to make their own decisions about relating sexually.

In Sum

Social attitudes and values play an important role in the forms of sexual expression people find comfortable and enjoyable. This point applies to many aspects of sexuality, including even our sexual orientation—whether we prefer a partner of the same or opposite sex. Despite decades of conjecture and research, it is still unclear just how sexual orientation develops, or even whether it is genetic or socially conditioned. The 1980s and 1990s have witnessed political and other challenges by gay activists (and a few others) to heterosexism and one of its consequences, homophobia.

Whatever one's sexual orientation, sexual expression is negotiated with cultural messages about what is sexually permissible, even desirable. In the United States, these cultural messages have moved from one that encouraged patriarchal sex, based on male dominance and reproduction as its principal purpose, to a message that encourages sexual expressiveness in myriad ways for both genders equally.

African Americans may be more sexually expressive and less inhibited than other Americans. Four standards of nonmarital sex are abstinence, permissiveness with affection, permissiveness without affection, and the double standard—diminished somewhat since the 1960s, but still alive and well.

Marital sex changes throughout life. Young spouses tend to place greater emphasis on sex than do older mates. But, while the frequency of sexual intercourse declines over time and the length of a marriage, some 27 percent of married persons over age 74 are sexually active and having sex about four times a month. Making sex this kind of pleasure bond, whether married or not, involves cooperation in a nurturing, caring relationship. To fully cooperate sexually, partners need to develop high self-esteem, to break free from restrictive gendered stereotypes, and to communicate openly.

We also discussed HIV/AIDS in this chapter, with some focus on how the disease affects relationships, marriages, and families. That general discussion led us to consider how the social institutions of religion and politics today have merged somewhat to legislate for more restrictive values regarding sex. The New Christian Right advocates "family" values that include celibacy outside marriage, traditional roles for women, and heterosexism. Whether or not one agrees with the major agenda of the New Christian Right, there are certain guidelines for personal sexual responsibility that we all should heed.

In addition to the contents of this chapter, Appendixes A through G give the following information on sex-related topics: Appendix A—Human Sexual Anatomy; Appendix B—Human Sexual Response; Appendix C—Conception, Pregnancy, and Childbirth; Appendix D—Contraceptive Techniques; Appendix E—High-Tech Fertility; Appendix F—Sexually Transmitted Diseases; Appendix G—Sexual Dysfunctions and Therapy.

Key Terms

abstinence	patriarchal sexuality
double standard	permissiveness with
evolutionary psychology	affection
expressive sexuality	permissiveness without
habituation	affection
heterosexism	pleasure bond
heterosexual	pleasuring
HIV/AIDS	representative sample
holistic view of sex	safer sex
homophobia	sexual orientation
homosexual	sexual responsibility
interactionist perspective	spectatoring
on human sexuality	survival of the fittest
New Christian Right	

Study Questions

1. Give some examples to illustrate changes in sexual behavior and social attitudes about sex. What do you think future attitudes and behavior will be? Why?

2. How has the sexual revolution affected marital sex? Why?

3. Do you think that sex is changing from "his and hers" to "theirs"? What do you see as some difficulties in making this transition?

4. How do you account for the fact that younger spouses engage in coitus more frequently than older partners?

5. Discuss the relationship between sexual pleasure and (a) self-esteem, (b) gendered expectations, and (c) cooperation and communication.

6. Discuss the principles for sexual sharing that are suggested in this chapter. Are there any you would add?

7. What do we know about how sexual orientation develops? What kinds of relationships and families do gays and lesbians create in today's society?

8. Describe the process by which a lesbian or gay man constructs that self-identity. How is this process a result of homophobia in the surrounding culture?

9. How do you think AIDS will change sex and relationships? Do you anticipate that it will have a major impact on American family life or not?

10. This book stresses that people must take responsibility for the consequences of their sexual behavior. What responsibilities does the book list? Do you agree with the list? What would you add or subtract?

Suggested Readings

Beach, Frank A., ed. 1977. *Human Sexuality in Four Perspectives*. Baltimore: Johns Hopkins University Press. Excellent source on the basic biology and physiology of sexuality.

Boston Women's Health Book Collective. 1984. *The New Our Bodies, Our Selves*. New York: Simon & Schuster. Manual of women's health and sexuality that has enormous respect from the user public. Similar books have appeared for parents, gays, heterosexual men, teens, and children.

Brandt, Allan M. 1985. *No Magic Bullet: A Social History of Venereal Disease in the U.S. Since 1880*. New York: Oxford University Press. Interesting history that not only presents the facts of sexually transmitted disease in America but also captures the attitudes and fears of the public in a way that illuminates HIV/AIDS-related discrimination.

Comfort, Alex. 1987. *The Joy of Sex*. Rev. ed. New York: Crown. A sex manual that tries to be "unmanual-like." It stresses the importance of spontaneity and play in sexual expression and at the same time describes and illustrates various (heterosexual) techniques and positions. Some readers find the numerous drawings beautiful; others don't.

D'Emilio, John and Estelle B. Freedman. 1988. *Intimate Matters: A History of Sexuality in America*. New York: Harpers. Interesting history of sex in America.

Duh, Samuel V. 1991. *Blacks and AIDS: Causes and Origins*. Newbury Park, CA: Sage. Addresses major issues regarding AIDS and blacks, including AIDS in Africa and means of improving overall health status and health education in the United States.

Eidson, Ted. 1988. *The AIDS Caregiver's Handbook*. New York: St. Martin's. Advice on coping with AIDS in everyday life.

Fee, Elizabeth and Daniel M. Fox, eds. 1992. *AIDS: The Making of a Chronic Disease*. Berkeley: University of California Press. A collection of essays and articles on various aspects of the HIV/AIDS epidemic, including U.S. law, ethics, and public policy as well as international perspectives.

Francoeur, Robert. T., ed. *Taking Sides: Clashing Views on Controversial Issues in Human Sexuality*, 4th ed. New York: Guilford, CN: Dushkin, 1994. Presents both sides of controversies in various areas of human sexuality. Lots of attention to values, attitudes and norms and to the outcomes and impact of sexual behavior.

Graubard, Stephan. 1990. *Living with AIDS*. Cambridge, MA: MIT Press. Help for people who are coping with AIDS.

Haas, Kurt and Haas, Adelaide. *Understanding Sexuality*, 3rd ed. St. Louis: Mosby, 1993. Comprehensive text on all aspects of human sexuality. Authors' academic fields are psychology and communications. Textbook format with many headings and boxes makes it easy to search out particular topics.

Journal of Sex Research. Academic periodical with interesting articles on new developments.

Laumann, Edward O., John H. Gagnon, Robert T. Michael, and Stuart Michaels. 1994. *The Social Organization of Sexuality: Sexual Practices in the United States*. Chicago: The University of Chicago Press. Latest comprehensive survey of sex attitudes and behaviors among adults ages 18 to 59 in the United States. The first-ever from a national representative sample.

Masters, William H., Virginia E. Johnson, and Robert C. Kolodny. 1994. *Heterosexuality*. New York: HarperCollins. Thorough, up-to-date discussion of sexual expression and activities among heterosexuals. A nice, readable book.

Michael, Robert T., John H. Gagnon, Edward O. Laumann, and Gina Kolata. 1994. *Sex in America: A Definitive Survey*. Boston: Little, Brown. A more succinct version of the Laumann et al. (1994) book, above, but written for a popular audience and with lots of good examples added by Kolata, a science reporter at *The New York Times*.

Roscoe, Will, ed. 1988. *Living the Spirit: A Gay American Indian Anthology*. New York: St. Martin's. A collection of essays and other writings by gay Native Americans.

Rossi, Alice S., ed. *Sexuality Across the Life Course*. Chicago: University of Chicago Press, 1994. Edited volume of academic articles on various topics regarding sexuality. Special attention to biosocial and health issues. Also, attitudes. Diversity in life cycle stage; sexual topics; and some racial-ethnic diversity represented in this book.

Rothblum, Esther D. and Kathleen A. Brehony, eds. 1993. *Boston Marriages: Romantic but Asexual Relationships Among Contemporary Lesbians*. Women tell about romantic same-sex relationships that do not involve explicit sexual expression.

Silverstein, Charles. 1992. *The New Joy of Gay Sex*. New York: HarperCollins. Sexual technique and manual similar in style to Comfort's *The Joy of Sex*, but for gay men.

Sisley, Emily L. and Bertha Harris. 1978. *The Joy of Lesbian Sex*. St. Louis: Fireside. Sexual technique and relationship manual similar in style to Comfort's *The Joy of Sex*.

Sprecher, Susan and Kathleen McKinney. *Sexuality.* Newbury Park, CA: Sage, 1993. Comprehensive book on sexuality written by sociologists. Situates sexuality in social relationships. This book is one in a series on "close relationships."

Stalcup, Brenda, ed. *Human Sexuality: Opposing Viewpoints.* San Diego, Greenhaven, 1995. Covers sexual and reproductive controversies. Much coverage of legal issues. Offers debates on the influence of biology on various aspects of sexuality.

Sullivan, Andrew. 1995. *Virtually Normal: An Argument About Homosexuality.* New York: Knopf. Sullivan is a gay male with a Ph.D. in political science. His middle-line book explores "how we as a society deal with that small minority of us which is homosexual." The book examines traditional Judeo-Christian teachings on homosexuality as a major obstacle to his proposed public politics against all *public* (as opposed to private as well) discrimination against gay males and lesbians.

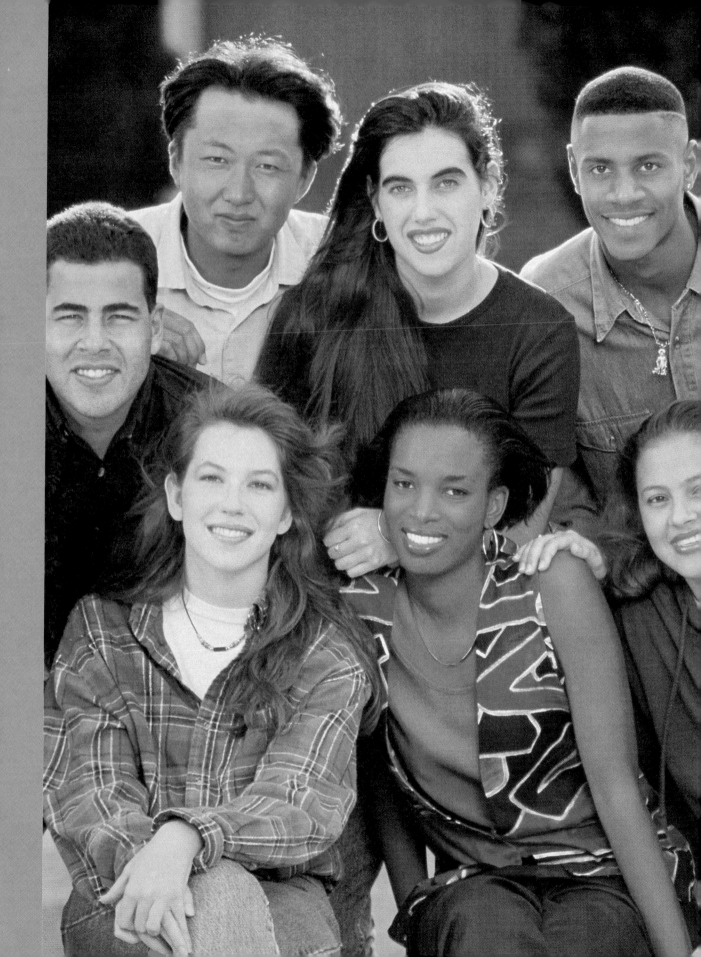

Being Single: Alone and with Others

*Today marriage and parenthood are
rarely viewed as necessary, and people
who do not choose these roles are no
longer considered as deviants.*

ELIZABETH DOUVAN, 1979

*Possibly, you thought that in this mod-
ern, enlightened age, women no longer
mope about not getting married.*

ROSE DEWOLF, 1982

| "Dawn" by Will Barnet.

Being single is one more common way of living adult life in our society. This represents a pronounced change. Throughout the first half of the twentieth century, the trend was for more and more people to marry and at increasingly younger ages. In the 1960s that trend reversed, and since that time the trend has been for more and more American adults to be categorized by the U.S. Bureau of the Census as **single:** divorced, widowed, or never-married.

In this chapter we will examine what social scientists know about singles. We'll look at reasons more people are single today and discuss changing cultural attitudes about singlehood. We will also look at the variety of singles and of single lifestyles. Many individuals who are not legally married live with partners. Some singles are parents. The distinction between marriage and singlehood has become blurred in recent years.

To begin, we examine some statistics on the increasing number of singles today.

Singles: Their Increasing Numbers

The number of singles in the United States has risen strikingly over the past thirty years. Before about 1960, family sociologists described a standard pattern of marriage at about age 20 for women and 22 for men (Aldous 1978). About 80 percent of these unions lasted until the children left home (Scanzoni 1972).

Today, in contrast, we see an increasing number of never-married young adults and of formerly married singles. The number of singles over age 17 has jumped from about 25 million at the turn of the twentieth century to nearly 75 million at present (U.S. Bureau of the Census 1995, Table 58). Singles have increased in absolute numbers partly because the population as

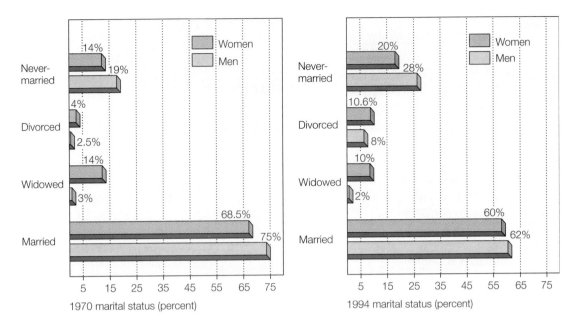

FIGURE **6.1**

❙ Marital status of the U.S. population, age 18 and over, 1970 and 1994. (Source: U.S. Bureau of the Census 1995, Table 58)

a whole has grown. There are a larger number of young people today who are in the age brackets less likely to be married.

But singles have also increased as a relative proportion of the population—from 32 percent of the total population in 1960 to 40 percent in 1994 (U.S. Bureau of the Census 1995, Table 58). Figure 6.1 depicts the marital status of the American population age 18 and over in 1994 and indicates the decrease from 1970 to the early nineties in the proportion of population that is married.

Figure 6.2 compares marital status proportions for non-Hispanic white, Hispanic, and African American women and men. As you can see in Figure 6.2, African Americans are more likely to be never-married and less likely to be married than whites. The percentage for Hispanics falls somewhere in between. African Americans are slightly more likely than non-Hispanic whites to be divorced; the proportion divorced is lowest among Hispanics, an indication of their strong cultural commitment to lifelong marriage, among other possible reasons (Oropesa, Lichter, and Anderson 1994). The fact that Hispanics are less likely to be widowed reflects the fact that, as a category, they are

younger, on average, than either non-Hispanic whites or African Americans.

As you can see from Figures 6.1 and 6.2, there are three demographic categories of singles: never-married, divorced, and widowed. Sometimes when we discuss reasons for singlehood and how singles live, we will consider only those who have never married; other times we will consider factors applying to all singles, whether divorced, widowed, or never-married. Next we explore each of these categories of singles, beginning with the never-married.

The Never-Married

There is a growing tendency for young adults to postpone marriage until they are older. By 1994 the median age at first marriage for both men and women had risen, to 24.5 for women and 26.5 for men (Saluter 1994, Table B), as high as any figures ever recorded.

As a consequence, the number of singles in their 20s has risen dramatically. In 1970, 36 percent of women ages 20–24 were single; by 1994 that figure had risen to 66 percent (see Figure 6.3). Even though traditionally, larger numbers of men than women have

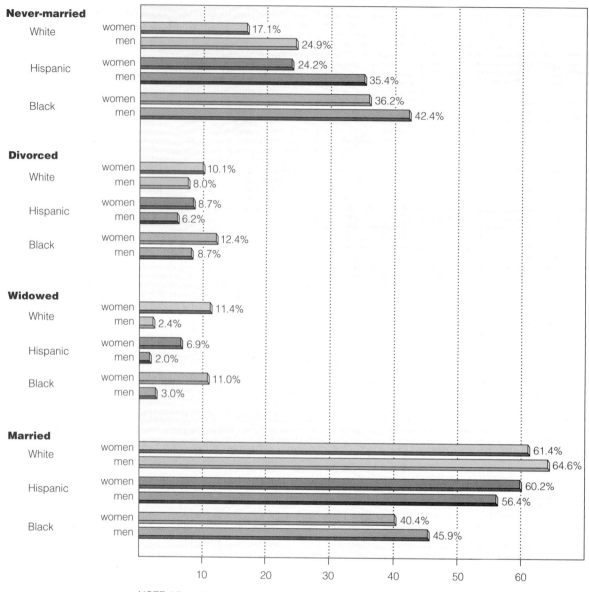

Never-married

White — women 17.1%, men 24.9%
Hispanic — women 24.2%, men 35.4%
Black — women 36.2%, men 42.4%

Divorced

White — women 10.1%, men 8.0%
Hispanic — women 8.7%, men 6.2%
Black — women 12.4%, men 8.7%

Widowed

White — women 11.4%, men 2.4%
Hispanic — women 6.9%, men 2.0%
Black — women 11.0%, men 3.0%

Married

White — women 61.4%, men 64.6%
Hispanic — women 60.2%, men 56.4%
Black — women 40.4%, men 45.9%

NOTE: Hispanics can be of any race.

FIGURE **6.2**

I Marital status of U.S. population, age 18 and over, 1993, by race/ethnicity. (Source: U.S. Bureau of the Census 1995, Table 58)

remained single in their 20s, the ranks of single men ages 20–24 have nevertheless increased from 55 percent in 1970 to 81 percent in 1994 (U.S. Bureau of the Census 1995, Table 59). This rate of singlehood is striking when compared with the 1950s, but not so unusual in a broader time frame—that is, compared with the turn of the twentieth century.

A theme of this text is that cultural values influence individual choices. Historical and demographic factors influence cultural values as well. The availability of education and jobs and the obligation to do military service have shaped twentieth-century marriage rates in complex ways (Cooney and Hogan 1991). A major historical event that affected people's decisions about

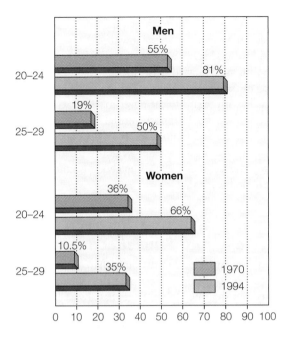

FIGURE 6.3

Change in the percentage of men and women ages 20–29 remaining single, 1970 and 1994. (Source: U.S. Bureau of the Census 1995, Table 59)

marriage and families during the 1950s was the Great Depression of the 1930s.

With its extensive unemployment and poverty, the Depression wreaked havoc on the traditional family pattern—breadwinner husband, homemaker wife, and moderate-to-large family size. One legacy of the Depression was a high cultural valuation of that threatened family form. Children born during the Depression, who came to maturity in the 1950s, appear to have acted on this high commitment to traditional family values: They married unusually early and began families soon after (Elder 1974). Expanding economic opportunity characteristic of the 1950s facilitated such choices (Easterlin 1987) because young men could get jobs easily and be relatively confident that real income would increase substantially. They were able to support stay-at-home wives and more children than their parents had been able to support and could afford better housing.

We are accustomed to thinking of the fifties as typical of American marriage patterns, partly because of the images of family life presented by popular television shows during this period and partly because

many adults can remember this era but not previous decades. The fifties were *not* typical, however, and the recent trend toward later marriage simply brings us back toward the pattern of the earlier part of this century. The percentage of never-married men and women ages 20–24 today is comparable to the proportion of young adults never married at the turn of the twentieth century.[1]

Still, the increase in young singles over the past two decades does reverse a downward trend lasting from 1900 to 1960, and we need to ask why. At least four social factors may encourage young people today to postpone marriage or not to marry at all. First, changes in the economy may make early marriage less attractive. (This reason is further explored in the next section.) Second, improved contraception may contribute to the decision to delay getting married. With effective contraception, fewer couples may find that they "have to" get married as a result of pregnancy (Cherlin 1981).[2]

A third reason for the growing proportion of singles is demographic—that is, is related to population numbers.

The **sex ratio** is the ratio of men to women in a given society or subgroup of society.[3] Historically, the United States had more men than women, mainly because more men than women migrated to this country, and, to a lesser extent, because a considerable number of young women died in childbirth. Today this situation is reversed due to changes in immigration patterns and greater improvement in women's than in men's health. Since World War II, there have been more women than men (see Figure 6.4). In 1994, for example, there were 95 men for every 100 women, whereas in 1910 there were nearly 106. The sex ratio was 100, or "even," in about 1948. In many social categories, there simply aren't enough eligible men for the number of eligible women, so in the musical chairs of marital pairing, some are left out. A fourth reason is changing attitudes toward marriage

1. For men, median age at first marriage is similar to that in 1890 when it was 26.1. For both women and men, median ages at first marriage fell from 1890 until 1960 when they began to rise again (Saluter 1994, Table B).

2. One can also argue, however, that contraception could facilitate early marriage by offering the possibility of marriage without pregnancy.

3. The sex ratio is expressed in one number: the number of males for every 100 females. Thus a sex ratio of 105 means there are 105 men for every 100 women in a given population. More specialized sex ratios can be calculated: the sex ratio for specific racial/ethnic categories or at various ages, for example, or of unmarried people only.

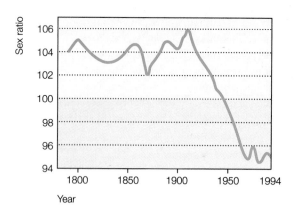

FIGURE 6.4

Sex ratios in the United States from 1790 to 1989. Because 100 represents a balanced sex ratio—an equal number of men and women—parts of the curve above the 100 line mean there are more men than women; portions below that line mean that there are more women than men. (Source: U.S. Bureau of the Census 1995, Table 15).

and singlehood, a development that has affected both the remarriage rates of divorced and widowed people and the ratio and timing of first marriage.

The Divorced and Widowed

The growing divorce rate has contributed to the increased number of singles. In 1994, 8 percent of men and 10.6 percent of women age 18 and over were divorced, a sharp increase from 1970 (Figure 6.1). Although the divorce rate is no longer increasing, it is stable at a high level, and the divorced will continue to be a substantial component of the single population. (Divorce is discussed at length in Chapter 15.)

Unlike the other singles categories, the proportion of widowed women and men has declined slightly since the 1970s. Nevertheless, the overall percentage widowed is well within range of its historical tendency. Death rates have declined throughout the twentieth century, reducing the likelihood of widowhood for the young and middle-aged. However, the proportion of older people in the population has increased, and they are at risk of losing a spouse.

Women are much more likely to be widowed than men. As you can see in Figure 6.2, this is true across racial/ethnic categories. This difference is due to women's greater longevity and their lower likelihood

of remarrying after the death of a spouse. (Chapter 14 discusses the experience of widowhood.)

Widowhood is not a status of choice, although those who are widowed may experience some positive aspects of independent living. For the divorced, and especially with regard to the never-married, we can speculate about what impact cultural change may have had on choices and feelings about singlehood. As American culture gives greater weight to personal autonomy (Bellah et al. 1985), singlehood becomes more desirable for many.

Changing Attitudes Toward Marriage and Singlehood

Social scientists used to discount changing attitudes as a serious reason for increased singlehood. They pointed out that the United States has always been a highly married society, with about 90 percent of each cohort marrying at least once. This contrasts with Europe, where considerably higher proportions of adults never marry (Bernard 1972). Sociologists also argued that although people marry at older ages now, this situation is similar to a century ago when presumably attitudes toward marriage were only favorable (Cherlin 1981). In other words, social scientists generally believed that Americans had not soured on marriage; they were just postponing it.

Sociologist Peter Stein (1976) was one of the first to argue otherwise. As one of several reasons for the increase in singles since 1960, Stein included "a shift in attitudes about the desirability of marriage . . . : people are moving away from marriage and family norms as these norms conflict with the potentials for individual development and personal growth" (pp. 5, 6). Recently, sociologist Scott South (1993) analyzed data from 2,073 never-married men and women ages 19 to 35 who were not cohabiting. As part of the National Survey of Families and Households (NSFH) (see Chapter 5), these singles were asked to respond to the statement, "I would like to get married someday." They could answer with any one of the following: "strongly agree," "agree," "neither agree nor disagree," "disagree," or "strongly disagree." Notice that this question does not measure whether a person *expects* to marry, only whether she or he *would like to* marry. Table 6.1 gives the percentages of respondents who did not agree with the statement. They either disagreed or answered that they could neither agree nor

TABLE 6.1

The Desire to Marry Among Never-Married, Non-cohabiting Individuals, by Age, Race/Ethnicity, and Sex

	White Males	Black Males	Hispanic Males	White Females	Black Females	Hispanic Females	Total
Ages 19 to 35							
Percentage not desiring marriage	15.40	23.50	8.70	17.10	21.80	25.30	17.20
Number of all respondents	566	190	79	721	389	128	2073
Ages 19 to 25							
Percentage not desiring marriage	12.60	22.80	6.80	11.20	12.70	13.10	12.60
Number of all respondents	291	97	50	288	149	51	926

Source: National Survey of Families and Households data, analyzed in South 1993, p. 362.

disagree. South cautions that this sample probably *over*estimates people's negative attitudes about marriage, because only never-marrieds are included. Presumably, the marrieds have generally positive feelings about marriage. (This overestimation is especially apparent for Hispanic females, a large majority of whom are married.) While most people *do* want to marry, 17 percent of these respondents do not. Why not?

First, compared with married people, singles hold more individualistic than familistic values. In one study, interviewers asked singles and marrieds what was important to their happiness. Perhaps not surprisingly, marrieds tended to place a higher value on marriage, children, and love, whereas singles valued friends and personal growth more (Cargan and Melko 1982, pp. 166–70). As Table 6.1 shows, singles may intensify their individualistic attitudes the longer they remain unmarried. Of those under age 26, 12.6 percent did not want to marry, but when singles up to age 35 are included, the proportion rises to 17.2 percent. Furthermore, the high divorce rates have left many gun-shy. As one 28-year-old single woman explained, "If marriage is so good, why are so many people divorced?" (in Kantrowitz 1992, p. 52).

Sociologists have applied the exchange theoretical perspective (see Chapter 2) to this question of less fa-

vorable attitudes about marriage. They argue that singles weigh the costs against the benefits of marrying. At least some people see the benefits of marriage as decreasing while the costs of being single have simultaneously declined.

One reason for the declining perceived advantages of marriage is that today society views being single as an optional rather than a deviant lifestyle. During the 1950s, both social scientists and people in general tended to characterize singles as neurotic or unattractive (Kuhn 1955, cited in Stein 1976, p. 521). That view changed so that by the late 1970s, three-quarters of those interviewed in one national poll considered it "normal" to be unmarried (Yankelovich 1981, p. 95). Virtually socially accepted alternatives to permanent marriage—being divorced, cohabiting, and permanent singlehood—have emerged.

It is also true that getting married is no longer the principal way to gain adult status. Data from the National Survey of Families and Households shows that, increasingly, twentysomethings leave their parental home for reasons other than marriage (Goldscheider and Goldscheider 1994). Figure 6.5 shows men's and women's reasons for leaving their parental home from before 1930 to the present. As you can see from Figure 6.5a and 6.5b, before 1930, 40 percent of men and

Men's reasons for leaving home, 1920s to 1980s

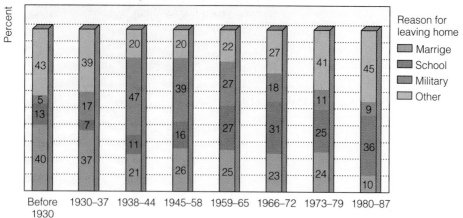

(a) Nest-leaving cohort (year reached age 18)

Women's reasons for leaving home, 1920s to 1980s

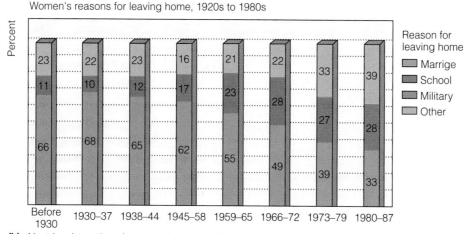

(b) Nest-leaving cohort (year reached age 18)

Other reasons for leaving home for men and women combined, 1920s to 1980s

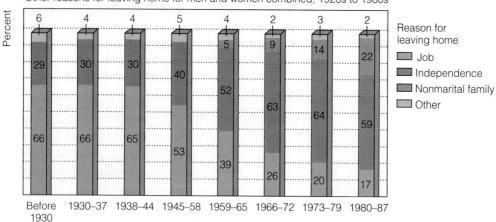

(c) Nest-leaving cohort (year reached age 18)

FIGURE **6.5**

Reasons for leaving home. (Source: Goldscheider and Goldscheider 1994, pp. 15, 17, 20)

SCHOCHET

"I don't want to live happily ever after. I want a career."

about two-thirds of women who left their parental home did so in order to marry. By the 1980s, only 10 percent of men and one-third of women left home for marriage. Moreover, the proportion who left home for reasons other than marriage, going away to school, or to join the military (Figure 6.5c) has increased steadily since about 1950. Today, 45 percent of young men and 39 percent of young women say they first left home for other reasons. Figure 6.5 breaks down these other reasons. Before 1930, the majority who left for other reasons did so for a job (66 percent), while today the majority (59 percent) do so to gain independence. The proportion leaving their parental home to establish a nonmarital family (cohabitation or single parenting) has also risen. As more and more young people choose to claim independence simply by moving out of their parental homes, marriage loses its effectiveness as the way to claim adulthood.

Accordingly, there is less parental pressure to marry than in the past. Although one researcher states that "many American parents remain vexed when their offspring approach a thirtieth birthday without at least one marriage in the record" (Shostak 1987, p. 355), one important survey found that only half of never-marrieds in their late 30s perceived parental pressure to marry (Bumpass, Sweet, and Cherlin 1991).

Feminist social scientist Barbara Ehrenreich (1983) traced the historical development of what she called the "breadwinner revolt" among men from the 1950s through the 1980s. She argued that through the 1940s men demonstrated their masculinity through family breadwinning—that is, by establishing and providing for a family. Beginning in the 1950s, divergent cultural factors, from the *Playboy* mystique to the counterculture, have legitimated singlehood for men. More recently, sociologist Kathleen Gerson (1993) has argued that our constricting economy has made family breadwinning more difficult, hence less attractive, to a growing number of men. Moreover, women have increasingly challenged men's privileges in marriage. As a category, men differ in their responses to these changes. Many will marry, but others will eschew commitment, either permanently or at least "for now."

Simultaneously, expanded educational and career options for women and, consequently, their growing commitment to paid work has given women increased economic independence (Waite, Goldscheider, and Witsberger 1986). Sociologist Frances Goldscheider has called *women's* growing lack of interest in marriage the real revolution of the last twenty years. In particular, middle-aged, divorced women with careers tend to look on marriage skeptically, viewing it as a bad bargain once they have gained financial and sexual independence. A 49-year-old divorced female executive, for example, says she loves eating popcorn for dinner rather than cooking like she used to when she was married. Once divorced, she said, "I could do anything I wanted for the first time in my life" (in Gross 1992). Box 6.1, "A Letter from a Single Student," expresses the views of one woman who chooses to be single.

In sum, it appears that much of the increase in singlehood (1) represents a return to long-term patterns of late marriage from which the 1950s deviated; (2) results from the low sex ratio and economic disadvantage, which prevent a portion of the population from marrying; and (3) results from changing attitudes toward marriage and singlehood. Experts are divided on the likely futures of singlehood and marriage. Some research suggests that today marriage has regained some ground as a favored status (Lee, Seccombe, and Shehan 1991). On the other hand, researchers point out that long periods of nonfamily living that now characterize young adult life may result in a less familistic orientation and a permanent shift toward a preference for the independence and autonomy of singlehood.

In our society, whether or not people marry—and the timing of marriage for those who do—will probably remain unsettled for a while. Says Robert Willis,

A Letter from a Single Student

Social scientist Arthur Shostak sees prejudice against singles as persistent but lessening. One thing that can help, he suggests, is for "proud singles" to "speak out and defend themselves" (1987, p. 366). A student who used our third edition wrote us to do just that. What follows is an excerpt from her letter.

Dear Drs. Lamanna and Riedmann:

I am a student at College of DuPage in Glen Ellyn, Illinois. . . . I am 37 years old and became serious about my education two years ago. . . . I am currently enrolled in Sociology 220: Sex, Marriage, and the Family. . . . The text we are using is your book, *Marriages & Families: Making Choices & Facing Change*, Third Edition.

The majority of your book discusses relationships leading to marriage and families. I was particularly anxious to study Chapter 5, which deals with singles, because I am single and occa-

sionally enjoy topics other than marriage and families. It was difficult to finish the chapter, however, because I found myself aggravated and very disappointed in the contents. Included in my course-assigned journal entry for that week is an editorial from *Chicago Life*, a singles magazine. The editor discusses the fact that society really isn't quite ready to accept singles as a group but is getting better. The editor also confirmed it was OK to be single and not want marriage. I felt the editorial had a more positive view of singles than your book.

From my reading, the chapter described being single as a holding tank for marriage. I felt the text should have presented "singlehood" more positively as a lifestyle rather than just a state of waiting. I do not feel that I've ever been openly discriminated against because I am single, but co-workers, family, and friends always expect a single person to be part of a couple. Co-workers especially are constantly waiting for me to get engaged. For that reason, I seldom discuss what I do outside the office. . . .

---

economics director of the National Opinion Research Center at the University of Chicago:

> A lot of people are trying different ways of sequencing—marriage first, then babies, then career, or getting established in a career first, then marriage or babies. . . . It seems to me until people have found a right way to travel through this, it'll be very difficult to predict how these statistics are going to change. (in Barringer 1992c, p. A-19)

African American Singles

Sociologist Robert Staples, who has specialized in research on African American families, addresses singlehood among blacks. Staples (1991) notes the high value placed on marriage in the black community, despite popular perceptions to the contrary. At the same

time, African Americans have shared in the recent trend toward greater singlehood. In fact, the proportion of married African Americans has declined sharply, from 64 percent in 1970 to 43 percent in 1994 (U.S. Bureau of the Census 1995, Table 58). (See also Figure 6.2.)

Staples explored demographic reasons for African American singlehood. For one thing, there are more black women than men in this country because young black men have relatively high mortality rates and are more likely to be imprisoned or to join the military.[4] In addition, the rate of homosexuality of black men exceeds that of black women, and more African

---

4. Much could be said about the reasons for this, but ultimately, it points to the impact of discrimination. See Staples (1991) for a more detailed discussion.

There is no denying that this is a couples' society. Even in our Sociology class I was laughed at the first night for stating I enjoyed being single and had no desire to marry. Many of the students laughed at me. I knew they were thinking: "She probably just broke up with a boyfriend"; or, "She probably just got divorced"; or, worse, "When she meets the right man, she'll know right away and she'll change her mind quickly." ...

Not everyone feels as strongly as I do. I know I represent a very small percentage of the population, but being single is not a crime. I prefer it and am very happy being single. I even like myself! I do what I want, when I want, things are my way. Perhaps this is a selfish attitude, but I've earned the right to be somewhat selfish. I work very hard at a high pressure job, am going to school, have two homes and money in the bank, I'm saving/planning for retirement, have a few good friends, am always going to seminars, meetings, or classes at the local park district—so I keep myself busy. I couldn't imagine sharing all this with another. A lot of people would not have the time or energy to keep up. Many people would not even want to! I would not want my life any other way. ...

I am not totally self-centered. I am an eternal volunteer and did volunteer work with handicapped/retarded kids and adults for 6½ years and loved every minute of it. It was very difficult and emotionally draining, but I'm glad I had the opportunity to share with these people. If I had been married, I would never have been able to give the amount of time I did to that organization. I did finally suffer burn-out with this group. However,

I'm again looking for some type of volunteer work to become involved with and share part of me with others. ...

I wanted to list my concerns about the chapter on singles and to show there really are people who enjoy being single. Marriage is not the only way of life. The book is designed to discuss marriage and families, and being single is one step in this process. Sometimes, however, it's the last step. If you should revise your book, perhaps you could try to express a little more compassion for singles. According to my recent survey, 100% of us have feelings.

Thank you very much for your time and consideration.

Sincerely,

JUDITH J. HARAZIN

Would you classify Ms. Harazin as voluntarily or involuntarily single? As a temporary single or a stable single? What evidence does the writer offer regarding prejudice and discrimination against singles?

We, the authors, took to heart Ms. Harazin's injunction that we express more compassion for singles in revising this chapter for this edition. Do you think this chapter expresses compassion and/or appreciation for singles? What might be some issues involved in expressing compassion or appreciation for singles in a pro-marriage society?

American men than women have married partners of other races. When we recognize that many lower-class, never-married black males are poorly educated and unemployed and therefore unable to support a family,[5] it becomes apparent that choices are limited for black women wanting black men as marriage partners (Staples 1991; Fossett and Kiecolt 1993).

College-educated black women have difficulty finding black mates of similar educational background (Staples 1991; see also South and Lloyd 1992a; Lichter, LeClere, and McLaughlin 1991). Further-

more, according to Staples, the sex ratio imbalance (see Table 7.1) means that "middle-class black men are able to screen out certain types of women"—perhaps those who are assertive—whereas women have fewer, if any, choices. The uneven sex ratio enables men to escape pressure to respond to changes that women are making in their lives and in their expectations of men, so that "women who want a sensitive, supportive, and affectionate mate find that men are socialized into emphasizing the values of success, leadership, and sexual performance" (Staples 1981a, p. 46).

Various conclusions could be drawn, depending on one's values and whether one agrees with Staples' characterization of black middle-class men. In writing about black women's attitudes toward marriage, Staples has been critical of black female singlehood, suggesting that middle-class black females "may be setting

---

5. The male provider role is part of many African Americans' expectations of marriage in its ideal form. But this ideal is often not realizable for lower-class men. Anthropologist Elliot Liebow's classic study of "corner men" in Washington, DC, depicts the frequently unsuccessful struggle of these men to find jobs and maintain marriages (Liebow 1967).

"This next one goes out to all those who have ever been in love, then become engaged, gotten married, participated in the tragic deterioration of a relationship, suffered the pains and agonies of a bitter divorce, subjected themselves to the fruitless search for a new partner, and ultimately resigned themselves to remaining single in a world full of irresponsible jerks, noncommittal weirdos, and neurotic misfits."

unrealistic standards in terms of the quantity and quality of the available pool [of black men]." He objected to middle-class black women's preference for "a troublesome singlehood . . . rather than compromise their standards for a mate" (Staples 1981a, pp. 27, 49). He gave priority to the social responsibility of black adults toward black children and the black community as a whole, which he defined as an obligation to marry and raise a family.

Staples' analysis focused on black women's diminishing desire for marriage. However, recent research suggests that black single women may be more desirous of marriage **than are** black single men (South 1993). As Table 6.1 shows, among never-married African Americans between ages 19 and 25, black *men* are far less desirous of marriage than are black women. While not as dramatic, this difference is true for the entire sample of respondents ages 19 through 35. Moreover, single black males are considerably less likely than their non-Hispanic white counterparts to desire marriage: 15.4 percent of single non-Hispanic white men do not want to marry, compared with 23.5 percent of single African American men. When con-

sidering responses to additional survey questions (besides the one about desiring to marry), the researcher who analyzed these data found that black men were more worried than non-Hispanic whites that they would not be able to hang out with their friends after marriage. Furthermore, African American men did not expect to improve their sex life with marriage, whereas other men did. Using an exchange perspective, South assumed that single people weigh the costs against the benefits of potential marriage and argued that men who think their living standards will improve with marriage are more likely to want to marry. Black men have less desire to marry than other men (or than black women) in large part because they expect marriage to have a less positive—or a more negative—impact on their personal friendships and sex life (South 1993).

This analysis reminds us that singlehood may be freely chosen, imposed by a structural lack of options, self-imposed, or a result of some combination of these. Again we see the theme that whereas the social structure influences people's decisions, people do have choices to make. An African American woman who

African Americans are considerably less likely to marry than are other racial/ethnic groups. Whether it is African American women or men who are less desirous of marriage depends on what research and authority you choose to believe. Sociologist Robert Staples puts the onus on black women, while recent research suggests otherwise.

faces the fact that there are fewer black men of marriageable age may choose to create a satisfying single life, to marry an African American man who is not as liberated or successful as she would like, to share a black man with another black woman, to marry a man of another race, or to establish a lesbian relationship. Of course, women of all racial/ethnic groups may have parallel choices to make.

The Variety of Singles

Factors such as age, sex, residence, religion, and economic status contribute to the diversity and complexity of single life. An elderly man or woman existing on Social Security payments and meager savings has a vastly different lifestyle from, for example, two single professionals living together in an urban area. Single life in small towns differs greatly from that in large cities.

Types of Singles

Robert Staples developed a typology of singles based largely on whether their status is freely chosen. He designated five singles types. Although his discussion was developed in the course of his work on African Americans, these types can be applied to other races and ethnic groups. The first type is the *free floating*, unattached single who dates randomly. The second type is a person in an *open-coupled relationship:* "This person has a relatively steady partner but the relationship is open enough to encompass other individuals in a sexual or romantic relationship." Staples warns that "sometimes it is an open-coupled relationship in a unilateral sense, with one of the partners pursuing other people; this may be a matter of deception or merely rest on the failure of the couple to define the relationship explicitly." In the third type—the *closed-couple relationship*—partners look only to each other

TABLE 6.2

Typology of Singlehood

	Voluntary	Involuntary
Temporary	Never-marrieds and former-marrieds who are postponing marriage by not currently seeking mates, but who are not opposed to the idea of marriage	Those who have been actively seeking mates for shorter or longer periods of time but have not yet found them Those who were not interested in marriage or remarriage for some period of time but are now actively seeking mates
Stable	Those choosing to be single (never-marrieds and former-marrieds) Those who for various reasons oppose the idea of marriage Members of religious orders	Never-marrieds and former-marrieds who wanted to marry or remarry, have not found a mate, and have more or less accepted singlehood as a probable life state

Source: Stein 1981, p. 11.

for their romantic and sexual needs. Fidelity is expected. The fourth type consists of *committed singles* living in the same household and either engaged or having an agreement to maintain a permanent relationship. The fifth type is an *accommodationist,* one who either temporarily or permanently lives a solitary life, "except for friendships, refusing all dates and heterosexual contacts." (Staples 1981a, pp. 44–45).

Staples' discussion here reminds us that singles' lives differ due to many factors. Staples makes a further point: Singlehood may be temporary or permanent. Peter Stein has grouped "the heterogeneous populations of singles according to whether singlehood is voluntary or involuntary and stable or temporary" (1981, pp. 10–12).

As you can see in Table 6.2, voluntary singles may be single either temporarily or permanently; the same is true for involuntary singles.

Voluntary temporary singles are younger never-marrieds and divorced people who are postponing marriage or remarriage. They are open to the possibility of marriage, but searching for a mate has a lower priority than do other activities such as career.

Voluntary stable singles are singles who are satisfied to have never married, divorced people who do not want to remarry, cohabitants who do not intend

to marry, and those whose lifestyles preclude marriage, such as priests and nuns.

Involuntary temporary singles are singles who would like, and expect, to marry. These can be either younger never-marrieds who do not want to be single and are actively seeking mates or older people who had not previously been interested in marrying but are now seeking mates.

Involuntary stable singles are older divorced, widowed, and never-married people who wanted to marry or to remarry, have not found a mate, and have come to accept being single as a probable life situation.

You may already have realized that throughout their lives people can move from one category to another. For example,

as younger never-marrieds who regarded singlehood as a temporary state become older, some marry. Others, unable to find an appropriate mate, remain single involuntarily and become increasingly concerned about the possibility that they will never find a mate. Others may enjoy their single state and begin to see it as a stable rather than a temporary condition. The same person can identify singlehood as a voluntary temporary status before

Those called to religious orders, such as nuns and priests, are voluntary stable singles. Sister Carol Anne O'Marie is not only a nun but also a mystery novelist.

marriage, then marry and divorce and become single again. This person may then be a voluntary stable, involuntary stable, or involuntary temporary single, depending on his or her experiences and preferences. (Stein 1981, pp. 11–12)

Then, too, singles may feel ambivalent about whether they would rather be married. Even involuntary singles may find much to enjoy in being unattached. In short, single individuals' experiences and feelings about being single, in addition to changing throughout their lives, may differ based on whether their status is voluntary and whether it is permanent.

Income and Residential Patterns of Singles

Satisfaction with single living depends to some extent on income, for financial hardships can impose heavy restrictions. Many single women, especially those with children, just do not make enough money. These women head single-parent families rather than engaging in the stereotypical singles lifestyle characterized by personal freedom and consumerism. Many work two relatively low-paying jobs, one full-time and one part-time, then take care of their homes and children. For them "career advancement" means hoping for a small annual raise or just hanging on to a job in the face of growing economic insecurity. Pursuing higher educational opportunities means rushing to class one evening a week after working all day and making dinner for the children. For reasons that we will examine in detail throughout this book, many women's experiences as heads of single-parent households can be aptly described, in former Barnard social scientist Sylvia Hewlett's words, as "a lesser life" (1986). (The problems of single-parent families are addressed in Chapter 15.)

TABLE **6.3**

Median Annual Income of Singles Compared with Marrieds, 1992

Married-couple families	$42,064
Wife in paid labor force	49,984
Wife not in paid labor force	30,326
Male householder, wife absent	27,821
Female householder, husband absent	17,221
Unrelated individuals (persons not living with any relative)	
Males	17,817
Females	12,949

I   *Source:* U.S. Bureau of the Census 1994a, Table 721.

There is evidence that occupation and income are related to marital status. As you can see in Table 6.3, married couples earn considerably more than singles, and this is true even if both spouses are not in the paid labor force. Singles who are not living with any other relatives earn significantly less, with single women earning the least. Married people are more likely to have white-collar jobs and higher incomes than are singles, regardless of age and education.

Among others, one cause of these income differences could be employment discrimination, particularly for men. "There's no question: if you're 30 and single, you'll have a harder time finding a job," said James Challenger, president of a job placement firm. Corporate managers perceive married men to be more stable; they may also be acting on prejudice against gays, on uncertainty about the sexual orientation of unmarried men. The single man's lack of a spouse provides him only one employment advantage; he is more apt to be chosen over married men for overseas assignments (Bradsher 1989, p. 21).

The situation may be different for women so far as job opportunities are concerned. Although sex discrimination lingers, unmarried women may be less disadvantaged than men by their single status. Fifty percent of higher-level female executives are single, compared to less than 10 percent of men. Single women are presumed to be able to concentrate on their work without the competing responsibility of a family; they tend to be perceived as more reliable than married women.

Discrimination on the basis of marital status is illegal in many states and cities, although it is not part of federal civil rights law. In any event, discrimination against singles may be lessening due to their growing numbers. But "because race, sex, ethnic origin, and religion are also [illegal] bases for discrimination, discrimination against singles is often hard to isolate. Is a woman not promoted because she is a woman? Is a man kept back because he has never married or because his boss suspects he is gay?" (Stein 1981, p. 236).

Whether or not job discrimination is a factor in singles' income, income and employment needs affect singles' residence patterns.

Urban and Suburban Singles

In the past, most of the singles population became concentrated in specific areas of larger cities. Today, singles move from small towns to medium-sized cities and from medium-sized cities to larger metropolitan areas not only for employment but for other reasons as well. Often, they move because they want to be near others who are like themselves.

Singles with money to spend on leisure activities migrate to larger cities because there is more to do there. More and more cities have singles-oriented special interest clubs (book, bird-watching, chess, opera, and so on) as well as singles-oriented restaurants, dance halls, travel agencies, and lecture/study groups. There is a growing number of black-oriented, middle-class singles bars and membership clubs in big cities. Reflecting their urban residential pattern, singles are more likely than marrieds to complain to pollsters of urban discontents—traffic congestion, lack of parks, air and noise pollution, and so on (Shostak 1987, p. 357).

Partly because of these dissatisfactions with city life, singles are now moving to the suburbs. At the same time, suburbs are changing. Employers, services, and cultural activities have relocated from cities to suburbs. Singles can feel at home because of the "citification" of suburbs while enjoying the outdoors and the lower cost of housing: "There's a lot of calmness here and birds chirping. It's not as stressful," says one single suburbanite (Nemy 1991, p. B-8).

The majority of singles live in apartments or other rental units, but more singles are buying their own homes or condominiums (Hardcastle 1995). Among the majority of singles who rent, meanwhile, many

TABLE 6.4

Young Adults Living with One or Both Parents, by Age, United States, 1960 and 1994

Sex and Year	Percentage Living with Their Parents	
	18–24	25–34
Total		
1960	43	9
1994	53	12
Men		
1960	52	11
1994	60	16
Women		
1960	35	7
1994	46	9

Sources: U.S. Bureau of the Census 1988, Table A-6; 1995, Table 63.

"are conspicuous in their leadership" of an emergent tenants' movement that developed over the past two decades (Shostak 1987, p. 362). Confronting landlord bans against renting to singles, rising rents, and conversion of rental units to condominiums, singles "assert thereby to 'tenants' rights'—a dramatic new frontier on which to struggle to clarify both their prerogatives *and* their responsibilities" (Shostak 1987, p. 362).

Domestic Arrangements of Singles

Wherever singles live, they must make choices about whom (if anyone) to live with. We see again the variety of singles' lives: Some live alone; some live with parents; some live in groups or communally; some cohabit with partners of the same or opposite sex; some women share a man in an ongoing relationship. We look now at this variety of domestic arrangements. In the process we will find that the distinction between being single and married is not that sharp, for people who are legally single tend to be embedded in families of one form or another.

Living with Parents

A growing proportion of young adults are living with one or both parents. In 1940 the proportion of adults under age 30 living with their parents was quite high. Sociologists Paul Glick and SungLing Lin suggest why:

> The economic depression of the 1930s had made it difficult for young men and women to obtain employment on a regular basis, and this must have discouraged many of them from establishing new homes. Also, the birth rate had been low for several years; this means that fewer homes were crowded with numerous young children, and that left more space for young adult sons and daughters to occupy. (Glick and Lin 1986a, p. 108)

Some of these same reasons apply to young people today. Table 6.4 lists the percentages of young adults living with their parents in 1960 and 1994; from it we see that the percentage of young adults living at home has increased moderately since 1960. By 1994 more than half of young adults ages 18–24 lived with parents.

The postponement of marriage discussed earlier in this chapter has meant a longer period of singlehood. Although the majority of young unmarried people maintain apartments, some grow tired of living on their own. Difficulty in finding adequate employment is also a factor in the proportion of young adults living with parents today. Others are trying to minimize expenses as they complete graduate or undergraduate education. Unmarried women who have babies, especially those who became parents in their teens, may be living with parents. Others may return after divorce, as may divorced men as well.

More men (59 percent of those ages 18–24 in 1993) than women (47 percent) live with their parents. This has led to speculation about male domestic dependency that is not always flattering:

> The rent is low and utilities are free. There is hot food on the table and clean socks in the drawer. Mom nags a little and dad scowls a lot [fathers are often not enthusiastic about the arrangement, particularly if it is because of the son's unemployment], but mostly they don't get in the way. And there's money left at the end of the month for a car payment. (Gross 1991c, p. 1)

Parents do "get in the way" of women who live at home. They are more likely to try to limit daughters' freedom to come and go or to be suspicious of suspected sexual activity. Men, on the other hand, may find living with parents a stigma when dating women who expect them to live independently (Gross 1991c).

Just as economic considerations or the need for emotional support or the need for help with child rearing may lead young singles to choose living with parents, similar pressures may encourage singles to fashion group or communal living arrangements.

Group or Communal Living

Groups of adults and perhaps children who live together, sharing aspects of their lives in common, are known as **communes.** In some communes, such as the Israeli kibbutz (Spiro 1956) or some nineteenth-century American groups such as the Shakers and the Oneida colony (Kephart 1971; Kern 1981), all economic resources are shared. Work is organized by the commune, and commune members are fed, housed, and clothed by the community. Other communes may have some private property; even some Israeli farming cooperatives that superficially resemble kibbutzes have private land plots, although members share a communal life (Schwartz 1954).

There is also variation in sexual arrangements among communes, ranging from celibacy to monogamous couples (the kibbutz, and some communes in this country) to the open sexual sharing found in both the Oneida colony and some modern American groups. Children may be under the control and supervision of a parent, or they may be more communally reared, with a deemphasis on biological relationships and responsibility for discipline and care vested in the entire community.

Communal living, either in single houses or in co-housing complexes that combine private areas with communal kitchens (Ravo 1993), may be one way to cope with some of the problems of aging, singleness, or single parenthood. Social scientist Arthur Shostak points to "the new American commune—the polymorphic, multigenerational, economically sensible middle-class commune of the 1980s" as one residential option for singles (1987, p. 364). In other words, people of different sexes and diverse ages may choose to reside together as a solution to economic and other practical concerns, such as in-home child care.

As does any living arrangement, communal living has positives and negatives. Single mothers may get help, but they also relinquish some parental control. One mother, for example, left a commune because she was opposed to other members' commitment to corporal punishment of children (Kornfein, Weisner, and Martin 1979). And, as one might expect, agreeing on standards for privacy, housekeeping, and noise may be a source of conflict among members. Still, communal living represents an attempt to provide singles with greater opportunities for social support, companionship, and personal growth.

Financial considerations and the need for social support may also encourage dating singles to share households (Shostak 1987, p. 358). We explore cohabitation, or living together, next.

Cohabitation

Singles engaging in **cohabitation,** or living together, gained widespread acceptance over the past thirty years. Although we may associate cohabitation with college students of the 1960s, the trend actually began in the late 1950s in less educated sectors of the population. The trend toward cohabitation spread widely in the 1960s, then took off sharply in the 1970s and has risen steadily ever since (Bumpass and Sweet, 1995) as Figure 6.6 illustrates. Although only 5 percent of the population is currently cohabiting,[6] in recent years about half of people married for the first time have cohabited at some time before marriage. Cohabitation is even more frequent among separated and divorced people (Bumpass and Sweet 1995).

Many cohabiting relationships are relatively short-term. A national survey of cohabiting women found that just over one-half of their relationships ended in marriage, 37.2 percent broke up, and 10 percent were still ongoing at the time of the survey (London 1991). A national survey of both sexes found that 39 percent of living-together relationships of never-married people lasted less than a year; 30 percent of cohabiting re-

---

6. The census definition on which calculations of unmarried couples are based is specific: An unmarried-couple household contains two adults, not related and of opposite sex, but no additional adults; and any children present are under 15. Besides cohabiting couples, this could include tenant and roomer, caregivers of the disabled or handicapped, and roommates of the opposite sex. It excludes gay and lesbian couples and heterosexual cohabiting couples who have other adults living with them (Saluter 1994, p. 7).

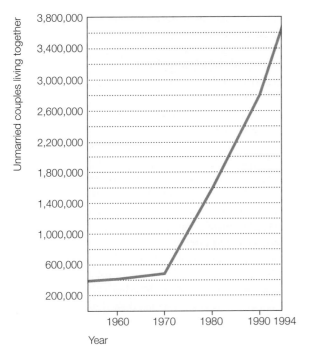

FIGURE **6.6**

Unmarried couples living together in the United States, 1960–93. (Sources: Glick and Norton 1979, courtesy of the Population Reference Bureau, Washington, DC; 1995, Table 60)

lationships of formerly married people lasted less than one year (Bumpass, Sweet, and Cherlin 1991).

Living together is a way of life that more and more people are choosing for many reasons. People from all social, educational, and age groups have at least experimented with this family form. Although approximately 80 percent of cohabitants are under age 45 (see Table 6.5), older retired couples have also found that living together without being legally married can be economically advantageous, as financial benefits contingent on not marrying are retained. Approximately 5 percent of cohabitants are over age 65 (U.S. Bureau of the Census 1995, Table 60).

Of couples currently living together, 40 percent of men and 43 percent of women were previously married (U.S. Bureau of the Census 1991b, Table 53). For this group, living together may provide a respite from the singles scene and a return to the domestic lifestyle

to which they are accustomed; or it may be, as for many never-marrieds, a way station on the road to marriage.

About three out of ten cohabiting households contain children under age 15. One-third were born to a never-married parent, one-half to a parent who was previously married, and one-sixth to the cohabiting couple. While cohabiting couples are significantly less likely to stay together than marrieds, having children in the household stabilizes unmarried couples somewhat (Wu 1995).

Unmarried couples are more likely than married couples to be interracial. Nine percent of unmarried couples and 1.5 percent of married couples are interracial: "A reasonable speculation is that interracial couples violate strongly held social norms; and, therefore, some of them may be reluctant to formalize their relationship by marriage" (Glick and Spanier 1980, p. 26). They may also receive more social support for their relationship if it is not formalized.

Some cohabitants may consider their lifestyle a means of courtship, as discussed in Chapter 7. For other cohabitants, living together can be a long-term alternative to legal wedlock. People's reasons for living together outside of legal marriage include the wish not to make the strong commitment that marriage requires and the belief that legal marriage can stifle communication and equality between partners. Some express fear of falling into traditional husband–wife roles. (Research evidence concerning the validity of these assumptions about differences between cohabiting and marital relationships is mixed and does not clearly support them; Newcomb 1979; Blumstein and Schwartz 1983; Macklin 1983; Watson and DeMeo 1987.)

Twenty percent of cohabiting relationships in the National Survey of Families and Households had lasted for five years or more. Whether an early beginning to marriage or an alternative lifestyle, "cohabitation is very much a family status, but one in which the levels of certainty about commitment are less than in marriage" (Bumpass, Sweet, and Cherlin 1991, p. 913). This uncertainty about commitment may be one reason that, compared to marrieds, cohabitants say they are less satisfied with their relationships (Nock 1995). Cohabitants express "little concern with cohabitation being a moral issue or with the disapproval of parents or friends" (Bumpass, Sweet, and

TABLE **6.5**

Selected Characteristics of Unmarried-Couple Households, 1994

Cohabitants with no children under age 15	65%
Cohabitants with one or more children under age 15	35%
Age of cohabitants	
Under 25	21%
25–44	59%
45–64	15%
65+	5%

Source: U.S. Bureau of the Census 1995, Table 60.

Cherlin 1991, p. 921). Indeed, they are more likely than others to be nontraditional in many ways, including attitudes about gender roles, and to have parents with nontraditional attitudes (Axinn and Thornton 1993; Booth and Amato 1994).

Although in responding to surveys people are articulate about their reasons for cohabitation, accounts of how cohabitation begins suggest that cohabiting does not always result from a well-considered choice (Macklin 1983). Like some of the decisions previously discussed, it can also happen by default. As one college student explained, "It just got to be too much bother to go home every morning."

Counselors and others stress the importance of clarifying motivations and goals for cohabiting before taking the step. For individuals who are not prepared, living together can lead to "misunderstanding, frustration, and resentment" (Ridley, Peterman, and Avery 1978, p. 129). Counselors also stress the importance of individuals being relatively independent before they cohabit, having clear goals and expectations, and being sensitive to the needs of their partners. (Chapter 7 discusses the useful advice of these researchers in more detail.)

As people choose cohabitation, the legal aspects of living together outside of marriage are simultaneously coming into focus and changing. In 1984, for example, Berkeley, California, became the nation's first city to extend health and welfare benefits to live-in partners of its unmarried employees. Box 6.2, "The Legal Side of Living Together," outlines some important considerations for cohabitants.

Gay and Lesbian Partners

Because census definitions of cohabitation are indirect and define cohabitation as living with someone of the *opposite sex* (see footnote 6), census statistics do not tell us anything about the domestic arrangements of gays and lesbians.

By law (U.S. Supreme Court decision in *Singer v. Hara* 1974), homosexual marriage is not legal in the United States. Attempts to legalize such marriages continue, but for now, lesbians and gay male couples are legally single. (The issue of gay and lesbian legal marriage is addressed in Chapter 8.)

Nevertheless, many gay and lesbian couples live together and share sexual and emotional commitment. They may consider themselves married as a result of personal ceremonies in which they exchange vows or rings or both, or they may form more public unions, religiously recognized by the Metropolitan Community Church (the national gay church). Some newspapers have begun to announce gay and lesbian partnerships on the society page.

Some gays and lesbians are establishing families with children, becoming parents through adoption, foster care, planned sexual intercourse, or artificial insemination. Courts vary in their receptiveness to such families; some permit lesbians to adopt their partner's children (Sullivan 1992), whereas others refuse visiting rights to a lesbian coparent when the couple breaks up (Sack 1991). With regard to the couple themselves, some communities have passed domestic partner laws (see Box 6.2), which extend some of the legal benefits of marriage, such as health insurance and bereavement leave, to unmarried same- or opposite-sex partners.

In some respects other than legal, gay and lesbian relationships are similar to heterosexual ones. One study directed by Letitia Peplau (1981) based on questionnaires administered to 128 gay men, 127 lesbians, and 130 unmarried heterosexual women and men found that both heterosexuals and gay people struggle to balance "the value placed on having an emotionally close and secure relationship" with that of "having major interests of . . . [one's] own outside the relationship [and] a supportive group of friends as well as . . . [one's] romantic sexual partner" (Peplau 1981, p. 33). Regarding gay people's similarities with—or differences from—heterosexuals, Peplau concluded:

We found little evidence for a distinctive homosexual "ethos" or orientation toward love relationships. There are many commonalities in the values most people bring to intimate relationships. Individual differences in values are more closely linked to gender and to background characteristics than to sexual orientation. (Peplau 1981, p. 33)

Blumstein and Schwartz's study (1983) of married heterosexual couples, unmarried heterosexual cohabitants, and gay male and lesbian couples reached a similar conclusion: Gender is a more important determinant of the nature of couple relationships than sexual orientation is (see also Chapter 5).

For instance, lesbians are more likely to live together (about three-fourths of coupled women compared to slightly over half of coupled men) and place a higher value on emotional expressiveness, such as sharing feelings and laughing together (Peplau and Gordon 1983). According to two major studies (Bell, Weinberg, and Hammersmith 1981, Harry 1983), about 40 to 50 percent of gays are partners in stable relationships at any one time, as compared to about 75 percent for lesbians. Likewise, lesbians find a new partner more quickly after a breakup, whereas—at least before HIV/AIDS—gays have tended to go through long transition periods of nonexclusive sexual activity (Harry 1983). Lesbian couples are likely to have met through friendship networks, with less reliance on the gay bar. And "among lesbians a sexual relationship usually arises out of a developing affectional relationship while among gay men affection may develop out of a sexual relationship" (Harry 1983, p. 227).

Psychologist Lawrence Kurdek's study of a nonrandom, convenience sample of seventy-four gay male and forty-five lesbian couples found lesbian couples to have "enhanced relationship quality" (Kurdek 1989b, p. 55). Like the other researchers, he attributes the lesbian couples'

greater relationship satisfaction, stronger liking of their partners, greater trust, and more frequent shared decision making . . . to *both* partners having been socialized to define themselves in terms of relationships with others, to regulate interactions with others on the basis of care and nurturance, to be sensitive to the needs and feelings of others and to suppress aggressive and competitive urges which may result in social isolation. (Kurdek 1989b, p. 55)

But even though there may be no distinct homosexual "ethos," gay male and lesbian relationships do differ from others in two important ways. First, gays and lesbians are less likely to adopt traditional masculine and feminine roles in their relationships. Instead, couples assume a "best friends" or "roommates" pattern (Peplau 1981; Harry 1983, p. 217). One reason many gay relationships are relatively egalitarian is that pairings of two men or two women generally provide members of the couple with similar incomes, whereas a heterosexual couple tends to be characterized by higher income, and therefore more power, for males (Harry 1983). (Chapter 10 addresses the association of income equality with egalitarian decision making.)

Same-sex partners, like singles in other situations, search for a community of friends and neighbors. Some find community in urban areas with concentrations of other gays and lesbians and strong activist organizations. But like some heterosexual singles, gay people are moving to the suburbs. A candidate for Congress in a San Francisco Bay Area suburb comments: "There are still people who clearly want to live in the gay ghettoes. . . . But a whole lot more . . . want to lead their lives like other people lead their lives" (Tom Nolan in Gross 1991b, p. E-16).

The quiet of the suburbs is particularly appealing to gay people entering middle age. And they, like other singles, contribute to the urbanization of the suburbs. Typically early activists, they have brought gay institution building and activism to suburbia. On the whole, many feel safer and more accepted in suburbia than in urban areas, where violence against gays is a constant threat (Gross 1991b).

Our discussion of heterosexual cohabiting couples and gay and lesbian relationships presumes partners who identify themselves as a couple. Even should they have a nonexclusive sexual relationship, their primary commitment is to each other. But if we look closely at intimate relationships in our society, we find other situations in which the parties in a relationship know that one of them has another primary partner. We look now at women sharing men.

Women Sharing Men

One response to women's declining chances to marry is represented by the emerging pattern of single women's sexual relationships with married men. The "new other woman" sees the married man as her only option. For the woman who would like to marry but

The Legal Side of Living Together—Some Advice

When unmarried lovers decide to move in together, they can encounter regulations, customs, and laws that cause them problems, especially if they're not prepared for them (Rowland 1994). There are no firm legal guidelines to follow. State laws vary, and new court decisions can effect changes. But a look at some of the potential trouble spots can help.

Sex Laws

Many state or local laws still forbid—under threat of fine or imprisonment—sex between people not married to each other. Even though such laws are seldom enforced these days, occasionally they are, usually on the complaint of a disgruntled partner or neighbor.

Twenty-two states have laws against "sodomy" (oral or anal intercourse) or "deviant sexual intercourse," thereby forbidding homosexual relationships. Sometimes these laws affect heterosexual as well as homosexual couples, including married couples, for such laws can prohibit *any* nongenital sexual activity.[*]

Residence

When renting an apartment or house, renters usually must sign a lease. This is a legal contract, and failure to abide by it can mean eviction. Many leases specify how many people will live in the rental unit. For a single person who later decides to take in a friend, an objecting landlord can prove troublesome.

When two or more unmarried people are looking for a place, landlords may ask each of them to sign the lease so that everyone involved is held individually responsible for its terms, including the total rent. This may also be true for utilities. Conversely, if an unmarried partner's name is not on the lease, she or he may not be entitled to continue living in the rental unit if something happens to the nominal renter. A New York state decision gave a gay partner the right to continue living in a rent-controlled apartment after his partner died (Gutis 1989c), but this decision does not necessarily apply in other states.

People who are legally unrelated also need to check zoning laws before renting or buying a house. The zoning laws of many cities require occupants in single-family residence areas to be of a certain number and relationship. These laws are being challenged with increasing frequency, but it would be wise to check out your local situation.

Bank Accounts

There are no legal restrictions against an unmarried couple's opening a joint bank account. It's important to realize, though, that one of the couple may then withdraw some or all of the money without the other's approval. Separate personal accounts could well be more sensible.

Credit Cards and Charge Accounts

Unmarried couples may find it difficult to open joint charge accounts. Creditors may not want to take a risk on the stability of the relationship, for fear that bills may not be paid if the relationship ends.

If a store or utility company does issue a two-name account, both partners are legally responsible for all charges made by either of them, even if the relationship has ended. And creditors generally will not remove one person's name from an account until it is paid in full.

Taxes

It is illegal for unmarrieds to file a joint income-tax return.

Property

When unmarrieds purchase a house or other property such as home furnishings together, it is a good idea to have a written agreement about what happens to the purchase should one partner die or the couple break up. If the property is held in "joint tenancy with the right of survivorship" and one dies, the other would take ownership without probate. If partners don't want one partner's share of the property to go to the other person, it can be held as "tenants in common," which means that it would go into the estate to be distributed according to a will. In either case, if a partner decides to sell, a buyer will want both names on the deed. (The same would hold true for a jointly owned car; any buyer would want both names on the title.)

1991

DOMESTIC PARTNERS

ON ____ SEPTEMBER 11 199.1 ____, DONALD W. DICKINSON, CLERK OF THE CITY AND COUNTY OF SAN FRANCISCO, CERTIFIES THAT

SLOANE "CHIP" BARKER III ____ and

CHRIS THORMAN ____

BECAME DOMESTIC PARTNERS BY FILING A DECLARATION OF DOMESTIC PARTNERSHIP IN THE OFFICE OF THE CLERK

DONALD W. DICKINSON
COUNTY CLERK

BY *Maura Ramirez*
DEPUTY COUNTY CLERK

CITY AND COUNTY OF SAN FRANCISCO

More and more local governments are providing the option for couples to register as domestic partners, whether they are heterosexual or gay. A domestic partner certificate usually indicates joint residence and finances, as well as including a statement of loyalty and commitment.

Surviving partners who do not have legal title to the couple's possessions or property must establish a legal right to ownership in order to keep the property. On the other hand, keep in mind that under joint title ownership creditors could take one partner's property if the other partner gets into problems with debts.

Insurance

Anyone can buy life insurance and name anyone as the beneficiary. However, insurance companies sometimes require an "insurable interest," generally interpreted to mean a conventional family tie. Moreover, the routine extension of auto and home insurance policies to "residents of the household" cannot be presumed to include nonrelatives; one should check with the company about terms of the policy.

Wills

Nearly everyone has some property. Therefore, it's good planning for each partner to have a properly drawn up will. Telling relatives or friends what to do in case of death often doesn't work out. Because handwritten wills are not recognized in some states, it's good to obtain a lawyer's advice.

Health Care Decision Making

Any individual has the right to refuse treatment, but anyone too ill to be legally competent must have an agent to act for him or her in medical decision making.

Most cohabitants want their partners to play this role. The catch is that to hospitals and doctors, "family" may mean spouses, parents, adult children, or siblings but not unmarried partners or close friends. In other cases, medical personnel are more than willing to accept the significant other preferred by the patient.

Laws in most states now offer a good solution to this uncertainty. A person can designate a decision maker through use of a "durable power of attorney for health care." Check local laws for the specifics of appointing a person of your choice to make health care decisions on your behalf.

Children

Children born to unmarried couples bring new legal issues of rights and responsibilities. Most states provide that children become legitimized if the natural parents marry or if the father

(continued)

(continued)

accepts the child into his home or acknowledges the child as his, even though he doesn't marry the mother. Even though the question of legitimacy no longer has the significance it once did, cohabiting parents would be wise to formally acknowledge paternity in conformity with the laws of the state. It is crucial in obtaining child support from an unmarried father. In fact, more and more states require that paternity be established for all newborns for precisely this reason.

Recognition of an unmarried father's interest in custody or visitation of children was established in 1972 in the case of *Stanley v. Illinois* (405 U.S. 645), so that the unwed mother is no longer entitled to sole disposition of the child in many states. Although the courts have placement discretion, couples should stipulate in writing that custody is to go to the father (if so desired) should the mother die.

Recently, some courts have permitted a lesbian partner to adopt the biological child of the other partner, ensuring legal parenthood to both members of the couple raising a child. Other courts have permitted lesbian coparents to sue for visitation rights if a couple breaks up after parenting one partner's biological child. Although the legal situation is likely to evolve in this direction, by and large, in parenthood, biology and marriage are everything. The legal rights of "biological strangers" to continue associating with children they have socially parented are very tenuous (Sack 1991).

Breaking Up

One advantage people sometimes see in cohabiting is avoiding legal hassles in the event of a breakup. But if couples do not take care to stipulate in writing—and preferably with an attorney's assistance—paternity, property, and other agreements, legal hassles *can* result. The parties may find they have legal obligations typically associated with marriage (*Marvin v. Marvin* 1976).

The Marvin case established that nonmarital partners may claim property and support if their explicit or implicit contract (the verbal or written understanding underlying their union) established these obligations. It is important to note that child

hasn't, or for the woman who values independence and career over conventional domesticity, a partial (but often lengthy) sexual and companionate relationship seems preferable to autonomous singlehood or a series of temporary relationships.

Sociologist Laurel Richardson (1985) explored this increasingly common phenomenon in interviews with fifty-five women from varied social backgrounds: high school graduates and women with postgraduate degrees; skilled workers and professional and managerial women; traditional women and feminists. For the most part, the woman did not intend to have the affair; it developed inadvertently out of sustained work or social contact. Once begun, it seemed a rational solution to the woman's dilemma: how to meet her emotional, sexual, and companionship needs in the absence of marriage. Women whose work demanded a great deal of time and energy found the married man an appealing alternative. Furthermore, married men were perceived as more stable, attractive, and willing to express feelings than were unmarried men.

Women entered the relationship with a sense of personal efficacy. Because of the distance and seeming independence built into the relationship, the affair with the married man seemed one way of resolving the tension between identity and intimacy. But over time the structure of the relationship gave most of the power to the man, and women felt this dependency and lack of control keenly. Meetings and other contacts had to depend on the man's initiative. The classic limitation of an unsanctioned relationship, holidays spent alone, was more depressing than expected.

support can never be determined by contract; it is considered an unqualified entitlement of the child.

Another complication for couples intending to remain unmarried is **common law marriage,** a legal doctrine under which couples who live together for a certain period of time (which varies by state) and "hold themselves out" to be husband and wife are legally married. Most states have abolished common law marriage, and lawyers consider it obsolete. However, like rarely enforced sex laws, common law marriage could be invoked and become an important element in a dispute between parties. Again, check your state laws, including states where you might spend extensive vacation or work time.

Domestic Partners

Although some unmarried couples want to avoid the obligations of marriage, other unmarrieds would like to have some of the privileges of marriage, benefits supportive of a commitment to a partner. Here we have some good news.

Couples who are living together in what they hope will be a permanent relationship have not previously had access to joint health care, auto and home insurance, family or bereavement leave, or other benefits supportive of a commitment to a partner. In recent years, some cities and corporations have established the concept of "domestic partner." Unmarried couples may register their partnership and then be treated as a couple so far as benefits and entitlements are concerned. The definition of **domestic partner** usually includes criteria of joint residence and finances, as well as a statement of loyalty and commitment.

Domestic partner laws or policies are a particular boon to gay and lesbian couples, who are not permitted the choice of marriage. But they meet the needs of heterosexual unmarried couples as well. In fact, 70 to 90 percent of people registering domestic partnerships have been heterosexual couples.

Registering as domestic partners has an emotional significance for some. In cities that have established domestic partner registration but have not accorded partners any privileges, couples come in to register anyway as a way of expressing their commitment to a nonmarital partner (Ames 1992).

To sum up the legal side of living together, individuals must decide how they want to live, based on their needs, values, and goals. In so doing, they will do well to be aware of the laws of their state. At the present time there are discrepancies between how people live (social reality) and the law. People may make choices that seem reasonable to them and that conform to their values—in insurance or property rights while cohabiting, for example—only to find they are not protected as they thought they were. The burden is on individuals to keep informed. At the same time, laws and policies seem to be in the process of changing to meet the needs of today's variety of families. (See also Elkin 1994.)

---

*In 1994, seventeen states and the District of Columbia had hetereosexual and homosexual sodomy laws, six states had homosexual sodomy laws, and the remaining twenty-eight states had no such laws. One state, Minnesota, had both an anti-sodomy law for heterosexuals and homosexuals and a law prohibiting discrimination against gay males and lesbians ("Gay Rights" 1994).

Although some women were able to integrate their relationship with the rest of their lives, those who maintained secrecy felt cut off from family, friends, and work colleagues. Richardson believes that in the long run the new other woman phenomenon supports male privileges while being a disadvantage to both lover and wife.

In a similar vein, sociologist Joseph Scott (1991) has studied what he terms "polygamy" among African Americans. In the African American polygamous family,[7] two (or more) women maintain separate house-

holds and are independently "pair-bonded to a man whom they share and who moves between . . . households as a husband to both women" (Scott 1980a, p. 43). In view of the particularly unbalanced sex ratio among blacks, Scott suggests "the only way many black women may have men permanently in their lives would be to share them with women who already have them as husbands or friends" (1980a, p. 48).

---

7. The term *polygamy* means more than one spouse at a time for a partner of either sex. *Polygyny* refers to multiple wives for a man, and *polyandry* to multiple husbands for one wife. *Group marriage* in this framework indicates a marriage of more than one woman to more than one man.

   Polygyny has been very common throughout history as a permissible

pattern in many cultures. Where permitted, though, it is not always that frequent, for many men cannot afford multiple wives. Polyandry is rare and tends to exist in very poor societies in which female infanticide is practiced to reduce the number of mouths to feed, with the result that there are not enough women to go around for marriage in adulthood. It is also practiced when brothers share a wife, again, often because of economic constraints. Group marriage is rarer still because of the obvious difficulties in balancing rights and duties and in working out conflicts and loyalties in this system (Stephens 1963).

Scott interviewed twenty-two African American women who either were or had been in a polygamous relationship. Half the women were single (consensual wives), and half were married (legal wives). Their relationships varied in length from two to twelve years. These women tended to be young, of low socioeconomic status, and to have experienced premarital pregnancies for which they were ill prepared. They believed men to be polygamous by nature and resigned themselves to sharing. Consensual wives saw the single black men available to them as unreliable compared to those already married. "Married men, the women say, were generally more emotionally stable, more knowledgeable about family life, and very importantly, more financially able to assist them" (Scott 1980a, p. 53). Most of the legal wives became involved in polygamy because their husbands had other close female companions before marriage and kept them. Generally, they accepted their situations as inevitable, "given the scarcity of what they call 'good men.' . . . They all shared the view that if their husbands maintained their economic priorities to their legal families, this was evidence of their giving priority to the legal marriages themselves" (Scott 1980a, p. 60).

Critics question whether polygamy among African Americans is freely chosen or imposed on the women by circumstance and men's behavior (Staples 1985). Scott replies to the critics that, in fact, the polygamy is consensual (that is, agreed to by the women) and that these unions are socially recognized as legitimate family constellations in the black community. Even though polygamy may be a logical solution to the lack of adult black males in urban settings, "many are concerned about the vulnerability of the mother/child units in 'man sharing' and see the pattern as one more instance of the exploitation of poor women" (Peters and McAdoo 1983, pp. 302–3).

Looking at singles who live with parents, in communal groups, with domestic partners, or in polygamous arrangements, we realize that the distinction between married and single is no longer very clear. Most singles are embedded in families. There are also married couples who live apart in commuter marriages (see Chapter 13), postdivorce families in which no longer married individuals still function as a couple, and married couples who are emotionally or sexually estranged.

Still, statistics classify people according to their legal marital status. Keeping that in mind, we look now at the life satisfaction of the legally single.

Aloneness, Loneliness, and Life Satisfaction

The number of one-person households has increased dramatically over the past twenty-five years. Individuals living alone now make up nearly a quarter (24 percent) of U.S. households—up from just 8 percent in 1940 (U.S. Bureau of the Census 1989a, Table 61; 1995, Table 66). As you can see in Figure 6.7, the proportion of people living alone has increased for both men and women in virtually every age category since 1970. It is also clear from Figure 6.7 that people are more likely to live by themselves as they get older. This is particularly true for women, who are most likely widowed.

Older Single Adults: The Double Standard of Aging

Because of differences in death rates, there are three women for every man over age 60. The elderly's living alone results to a significant extent from values of independence (Kramarow 1995). Meanwhile, some gerontologists have suggested that people create group marriages (Kassel 1966). Other observers have suggested that women of any age who are left without partners due to the imbalanced sex ratio might begin to explore lesbian relationships (Doudna 1981). But a study of change in preferred sexual activity from young adulthood to old age found few older adults actually shifting to homosexuality (Turner and Adams 1988). In a society that condemns homosexuality and encourages exclusive sexual relationships between one man and one woman, many older women are left without partners.

Moreover, as they grow older, women are adversely affected by the **double standard of aging** (Sontag 1976); that is, men aren't considered old or sexually ineligible as soon as women are. Being physically attractive is far more important for women than for men. Beauty, "identified, as it is for women, with youthfulness, does not stand up well to age" (Sontag 1976, p. 352). So in our society women become sexually ineligible much earlier than men do. An attractive man can remain eligible well into old age and is considered an acceptable mate for a younger woman. For

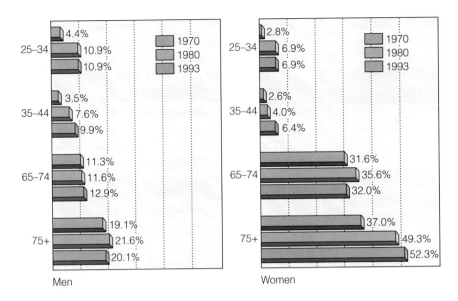

FIGURE 6.7

| Percentage living alone, by age and sex, 1970, 1980, and 1993. (Source: Saluter 1994, p. x)

older single women this situation can exacerbate more general feelings of loneliness.

Aloneness and Loneliness Are Different

Although living alone can be lonesome, we need to remember that aloneness, being by oneself, and loneliness, a subjective sensation of distress, are different (Shahan 1981). One can feel lonely in the presence of others, even a spouse, and happy alone. Moreover, living alone does not necessarily imply a lack of social integration or meaningful connections with others. In one national survey, 3,692 respondents were asked how many neighbors they knew by name, the number of relatives and friends living close by, and how often they saw them. They were also asked the number of "good friends" that could be counted on in "any sort of trouble" and the number of confidants—people with whom they could discuss just about anything— they had. Those living alone had fewer relatives living close by, and were less familiar with neighbors, than the married respondents; but they generally had more contact with friends and confidants. The researchers concluded that

contrary to conventional wisdom, we find that many persons who live alone are not socially isolated relative to others. Indeed, under most (although not all) life circumstances, they seem to show signs of an active "compensation" phenomenon, so that they visibly exceed persons living with others in their magnitude of contact with persons outside the household. (Alwin, Converse, and Martin 1985, p. 327)

Despite the "compensation phenomenon," however, singles have tended to report feeling lonely more often than do marrieds (Shostak 1987). Poor and older singles were especially likely to be lonely, perhaps because the low incomes and ill health that tend to accompany old age make socializing very difficult.

Besides age and income, being single as a result of divorce apparently affects loneliness. The newly divorced tend to suffer from depression (Menaghan and Lieberman 1986). One study found the divorced to be depressed by being alone about twice as often as other people. After expressing some surprise at these findings, the researchers explained that, "The never-married may not be lonely because they have not

TABLE 6.6

Percentages of Respondents Ages 25–39 Who Said They Were "Very Happy," by Sex and Marital Status, 1972 and 1989

	1972	1989	Change in Percentage, 1972–89
Males			
Married	32.4	38.4	+6.0
Never-Married	11.2	23.2	+12.0
Females			
Married	37.6	39.2	+1.6
Never-Married	26.2	35.2	+9.0

Source: Adapted from Lee, Seccombe, and Shehan 1991, p. 841, Table 1.

experienced marriage, and therefore have not experienced the implications of loneliness. Two-thirds of them still live with parents or with roommates of either sex" (Cargan and Melko 1982, p. 128). Moreover, as a group the never-married are younger, and, among singles, being older is related to loneliness.

Looking more broadly at life satisfaction among American adults, surveys through the 1970s generally found that singles of both sexes were less likely to say they were happy with their lives. But nationwide survey data show a steady decline from the early seventies through 1986 in the positive relationship between being married and reported life satisfaction or happiness. In 1972 only about 15 percent of singles said they were "very happy," while nearly 40 percent of the marrieds did. By 1986, the percentage of singles who said they were "very happy" had nearly doubled to almost 30 percent. Moreover, the proportion of marrieds who said they were "very happy" declined to just over 30 percent (Glenn and Weaver 1988).

More recent research (summarized in Table 6.6) indicates a return to more divergence between the married and single states. To sum up the research on the differences in happiness between married and unmarried individuals, the never-married are still less likely

to say they are "very happy," but the difference has declined in the past twenty-five years.[8]

Social scientists Norval Glenn and Charles Weaver, who conducted one of these studies, explain these changes as follows:

> In many respects, differences in the circumstances of married and unmarried persons have lessened. For instance, for at least a substantial proportion of unmarried persons, regular sexual relations without stigma have become available, and now that a divorce can rather easily be obtained by any spouse who wants it, marriage no longer provides the security, financial or otherwise, that it once did. (Glenn and Weaver 1988, pp. 322–23)

Still, married living has some clear mental health benefits compared to singlehood (Waite 1995). Speaking largely about marriage, sociologist Walter Gove and colleagues, who have examined the relationship between sex, marital status, and mental health over a period of two decades, conclude:

> The type of interaction that occurs in the family is particularly well suited to the development, maintenance, and enhancement of the self. . . . [F]amily interaction is based on a very direct and deep concern with the individual. . . . [F]amily members have a considerable investment in giving the individual support and an accurate appraisal of his or her strengths and weaknesses. . . . [T]he family [also] serves as a "back region" where the individual can expound, ponder, and complain about the characteristics of his or her [nonfamily] . . . roles. (Gove, Style, and Hughes 1990, p. 18)

Marriage seems to have physical health benefits as well (Waite 1995). Cohabitants are midway between singles and marrieds in mental and physical well-being (Kurdek 1991b).

It's possible that the relative mental health advantage of marriage may be changing, or perhaps we are

---

8. Exact comparisons of these studies are impossible because whereas Glenn and Weaver combined all unmarried individuals, Lee, Seccombe, and Shehan (1991) looked only at never-married singles. Age categorization also differs.

Because the more recent study (Lee and colleagues) emphasizes the never-married category of singles, it is important to note that the formerly married are typically less well-off than the never-married (Kurdek 1991b).

Cartoon by Chris Suddick, from *Men Are From Detroit, Women Are From Paris*, p. 94. Crossing Press. Reprinted by permission of Chris Suddick.

now more aware of negative as well as positive aspects of marriage and family life. In this vein, Gove, Style, and Hughes (1990) call attention to family violence (which we will discuss extensively in Chapter 10). Gove and colleagues also point out that marriage includes a set of obligations and the responsibility of coping with both the burdens of other family members and the disappointments and tragedies that come with family life. Increasingly, individualistic Americans may find these obligations more emotionally stressful than in the familistic past (Gove, Style, and Hughes 1990). Hughes and Gove (1989) note that social scientists have tended to focus on the benefits of social integration and ignore its costs and that there are some areas in which those living alone are consistently better off than the married. They point to less irritation;

widowed people living alone have higher self-esteem, and the never-married report a greater sense of control over their lives.

A somewhat different response to the fact that marrieds have consistently been found healthier than nonmarrieds has been offered by an Israeli sociologist, Ofra Anson. This analysis (Anson 1989) suggests that it is important to bear in mind that not all singles live alone. Living with another adult (as in group living or cohabitation) could serve as a functional alternative to marriage; that is, living with another adult, whether spouse or not, may provide the social ties and social control required for mental and physical well-being.

In this line of thinking, sociologist Catherine Ross (1995) suggests that we need to reconceptualize marital status as a **continuum of social attachment.** You can think of a continuum as a line between two opposites; the property on a continuum is not simply black or white but has many shades of gray. When we think of singlehood this way, we realize that not all singles are socially unattached, disconnected, or isolated. In Ross' research with a nationally representative sample of 2,031 adults who were interviewed by telephone in 1990, people in close relationships—whether married or not and whether living alone or not—were significantly less depressed than those with no intimate partner at all. Moreover (and this is important!), the relationship between being involved and not being depressed only held for those in happy, or supportive, arrangements. *Being alone and without a partner is less depressing than being in an unhappy relationship.*

Choice and Singlehood

Whether a person is single by choice affects satisfaction with singlehood. In 1983 and 1984 social scientist Barbara Levy Simon interviewed fifty never-married New York City women born between 1884 and 1918. When she categorized them according to Staples' classification scheme presented earlier in this chapter, she found thirty-six voluntarily stable singles. All said they were happy and satisfied. Thirty-four expressed antimarriage sentiments. As one woman put it:

> Men? Men have been important to me all my life. I have had friendship and love and sex with men since I was a young thing in Detroit. . . .
>
> You see, dear, it's *marriage* I avoid, not men. Why would I ever want to be a wife? . . . A wife is someone's servant. A woman is someone's friend. (in Simon 1987, pp. 31–32)

Predictably, the twelve women who were single involuntarily gave a less rosy picture of never-married life. Of these, seven said conflicts between caring for elderly parents and the demands of finances ended their marriage plans. These women remained somewhat bitter, resenting siblings' failure to help out and/or the inflexibility of their suitors (Simon 1987, p. 51).

Put another way, being single voluntarily must be qualitatively different from being single involuntarily. And moving from temporary to stable or permanent singlehood represents a major shift in self-identity. As one self-reported unhappy 39-year-old woman (with a master's degree in music and self-employed as a free-lance musician) put it, "I have always been brought up to believe that I would be taken care of by a husband. I just assumed that it would happen and it is not happening" (in Holmes 1983, p. 115). A voluntary temporary singlehood might gradually become either a voluntary or an involuntary stable one. You can imagine either possibility for Rich, in Case Study 6.1, "Being Single: One Man's Story."

Single Women Compared with Single Men

As Table 6.6 shows, the reported happiness of both never-married men and women has increased in the past twenty years. Nevertheless, the data in Table 6.6 support the continued truth of the conclusion that "single women of all ages are happier and more satisfied with their lives than single men" (Campbell 1975, p. 38). Thirty-five percent of never-married women report being "very happy," compared to 23 percent of men (Lee, Seccombe, and Shehan 1991). Research indicates that men are more affected than women by social isolation (study by House, Landis, and Urberson, reported in Goleman 1988a). Unmarried men may not cultivate intimate relationships, feeling they shouldn't bother others when they feel low (McGill 1985). At least among whites (see Box 6.3, "A Closer Look at Singles' Diversity: Black Men's Friendships"), "The thinness of men's friendships with each other [compared to women's] and the ways that they seem to be constantly undermined through competition and jealousy are distinctive features of modern society" (Seidler 1992, p. 17). Such isolation increases feelings of unhappiness, depression, and anxiety. "It's the 10 to 20 percent of people who say they have nobody with whom they can share their private feelings or who have close contact with others less than once a week who are at most risk" (James House, quoted in Goleman 1988a, p. 21).

Maintaining Supportive Social Networks

People's self-concepts depend greatly on other people's responses to them. People who have only secondary relationships know no one who understands them as a whole, unified person. Consequently, they may begin to feel fragmented. Moreover, human beings need others who can help confirm and clarify the meaning of situations and events. In marriage and

Being Single: One Man's Story

Social scientific research on unmarrieds is scanty, especially on working-class singles. The following is an edited interview with a 23-year-old, never-married man named Rich. He works as an auto mechanic and trade school instructor. Throughout the interview, look for themes discussed in this chapter.

RICH: I've been here in the city about two or three years now. I moved in from my home town—it was real little. I live out on about 80th Street . . . in a basement. This older couple, they were looking for somebody to just be around and kind of keep an eye on the place and do odd jobs around the house. And when I moved to the big city, that was what I found. It was reasonable, provided you did odd jobs.

INTERVIEWER: What kinds of things do you like to do?

RICH: Some buddies and I have a car we race. Weekends we take it to different towns around the circuit. We'll go to Kansas City around September, if we qualify.

INTERVIEWER: Are you dating or anything like that?

RICH: Well, it's just one of those things. Racing takes up most of the weekends. Friday night is getting ready, drive all day Saturday, figure out what isn't right with the car, work on it all Saturday night, and get in the elimination Sunday morning. Then put the car back up on the trailer and get back in time to go to work on Monday morning. . . .

I had a girlfriend. We used to make the circuit together. This was back in high school and probably a year beyond. We were planning to get married. But then one time I cracked up pretty bad, and she was there. She wanted me to quit after that. She kept saying, "You are going to get yourself killed."

INTERVIEWER: Did you want to quit?

RICH: Well, I considered it. But freak accidents happen, it's just one of those things. And what's the chances of walking out on the street and getting hit by a city bus? Actually, it's safer inside a racing car than out driving the streets during rush hour.

INTERVIEWER: So you and your girlfriend broke up because you didn't want to quit racing?

RICH: She couldn't take me for what I did or what I was. She had her things she wanted to do. I wanted to run my life the way I wanted. I suppose, looking back, I could have been a little more flexible.

INTERVIEWER: How much would you be willing to compromise now?

RICH: There are times that I think I should have quit. But racing is the way it is, that's just the way it is. After the accident, my folks thought I was crazy to keep on driving. They recommended that I get psychiatric help.

INTERVIEWER: How do you get along with your parents now?

RICH: Fine. I don't worry about their life, and they don't worry about mine. Now that's a wrong answer, because of course they worry, but they never say anything about it.

When I was dating, the heat was on to get married, "Why don't you get married?" and this and that. I wasn't ready for it. Oh, we talked about it, but it was still a couple of years down the road. There were things I had to do. There were things she had to do. But they kept asking, "When are you going to get married?" Sometimes they asked, "When are we going to expect grandchildren?" I did find a way to shut that last one off.

INTERVIEWER: How was that?

RICH: Just say, "In a few months now." They backed up and got off me about grandchildren after that.

INTERVIEWER: How do you like being single?

RICH: Lots of things you do in the single life, it's just doing it and taking it for granted. If I don't like you and you don't like me, we go our separate ways. And it's rare to find any real companionship in a single life. Oh, Wednesday nights you

(continued)

(continued)

maybe go out with the guys. You go to bars and drink, trade a couple or three rounds or something like that. But that only goes so far, and you've got several little gaps in your life. And they got to be filled....

I think there could be a whole lot deeper kind of relationship, a lot more. I don't know what it's like being married—I can only assume and go from there—but what I feel is you get an occasional comforting hand or something like that.

INTERVIEWER: Do you plan to get married then?

RICH: With the right gal, maybe. But any time you take two people, nobody has the same idea. And racing can get expensive. You can put $12,000 or $14,000 into a car plus expenses to buy your other equipment.

What does Rich share in common with other singles? How is he different from other unmarrieds? Where do his thinking and behavior reflect the traditional—and the changing—male sex role?

other primary relationships, people discuss and agree on daily events and interpretations of those events. This process may be taken for granted, but it is an important source of stability for individuals (Berger and Kellner 1970, p. 5).

Social networks involve more than simply a series of dating partners or casual acquaintances to do things with. Social scientists stress that even singles who are in love and involved in an exclusive dating relationship need a supportive network of intimate friends.

Perhaps the greatest challenge to unmarried individuals of both sexes is the development of strong social networks. Maintaining close relationships with parents, brothers and sisters, and friends is associated with positive adjustment and satisfaction among singles.

A crucial part of one's support network might be valued same-sex friendships. Other sources of support for singles involve opposite-sex friendships (either sexual or nonsexual), group-living situations, and volunteer work.

Singles may also reach out to their families of origin. Barbara Simon studied fifty never-marrieds old enough to be retired and found that they received a great deal of support from their families, especially in middle and old age. Families helped in crises, whether

of health and disability or economic loss. Black women are especially likely to be embedded in extended families, regardless of class.

In later life, single women were likely to set up joint living arrangements with siblings, something they would have been reluctant to do earlier. Twenty-three of fifty women Simon studied were living with a brother or sister in retirement. One of the women explains:

"I lost my three most intimate buddies in the space of five years. Those three were the people whom I had shared everything with since I was first employed. . . . Then I took stock of my situation. . . . The next-best thing to those friends were my sisters. They were each widows who were delighted with the prospect [of sharing an apartment]. We lived together as a trio for twelve years [until death and disability separated them]. The three of us did pretty well." (Simon 1987, p. 71)

But ties outside the family remained important. Simon (1987, pp. 53–54) found among the fifty elderly women she interviewed that "perhaps the most common thread of identity" these women shared was "their view of themselves as members of a group *larger*

For singles, it's important to develop and maintain supportive social networks of friends and family. Single people place high value on friendships, and they are also major contributors to community services and volunteer work.

than their own families." Asked what had given their lives meaning, forty-five emphasized religious, political, or humanitarian volunteer work. One woman who had been a Big Sister to sixteen Puerto Rican children over the past twenty-three years explained proudly that "all Puerto Rican kids are *my* family. . . . Of those sixteen children I have been a buddy to over the years, not one of them has gotten into trouble" (in Simon 1987, p. 54). Contributing an average of eighteen hours weekly, these women

> have acted throughout their lives as people with twentieth-century versions of a "calling," a term Robert Bellah has recently reframed in *Habits of the Heart.* These single women think of themselves as members of an integrated moral world in which

their commitments to their work, their family, their friends, their neighborhood, and their society flow from one passion—the desire to be a responsible and responsive actor in the world. (Simon 1987, p. 56)

Although individuals may be single for many reasons, they cannot remain happy for long without support from people they are close to and who care about them. This support is necessary for feeling positive about and generally satisfied with being single. Feeling satisfied with being single, of course, probably also depends on whether one is voluntarily single. At the same time, though, people who find themselves involuntarily single may feel better about their status as they choose to develop supportive networks.

The literature on men's same-sex friendships consistently concludes that male friendships center around activities and specific tasks while lacking mutual disclosure or intimacy. Do black men's friendships fit this description? African American sociologist Clyde Franklin II (1992) conducted unstructured interviews with thirty black working-class and professional men, ages 18 to 63, and asked them about their friendships. He also recorded about eighteen casual conversations with other African American men. He found that friendships among working-class African American men are different from those among black male professionals.

Working-Class Black Men's Friendships

Intense, emotional, and extremely close friendships characterize working-class black men, according to Franklin. And such relationships serve as buffers against a society that is hostile to black males. A "consciousness of kind" develops among black working-class men. For example, the greeting and response—"Yo, bro," "Hey, home"—tell a friend, "I share your daily experiences of prejudice and discrimination; . . . I, too, experience daily condescension; but I, too, feel enormous pride in being a black man. . . ."

Working-class black men told Franklin that a man's friend is "part of my life, man," or that "he would do anything for me and I will do anything for him." Others said they spent most of their time together, helping each other out: "We just chill out together, man. After work, we just hang out and be real cool to-

gether. . . . What do we talk about? . . . I guess we talk about anything I want and they won't tell a soul. I feel so close to them, man. Don't you have dudes like that? Dudes you can trust!" Expectations of loyalty, altruism, closeness, and emotional intensity are essential features of these friendships.

Upwardly Mobile Black Men and Friendship

Upwardly mobile black men, in contrast, don't say much about friendship. Instead (like non-Hispanic white men), they talk about "being the best" or "being competent." When they did talk to Franklin about their friends, they told about time spent with them watching sports or planning various business ventures. Many professionals told Franklin that they "really do not have time to cultivate deep relationships" because they spend most of their time "trying to get ahead." As one black business executive put it, "Upwardly mobile black men seem to be seekers of friends rather than actual participants in friendship relationships."

Conclusion

Working-class black men's same-sex friendships are warm and intimate; upwardly mobile black men's friendships are cool and nonintimate. Black men's friendships appear to change as they move up the social ladder. Empathy, compassion, trust, and openness give way to competitiveness, stoicism, rational thinking, and emotional independence. Franklin concludes that upwardly mobile—but not working-class—black men are similar to males in general regarding same-sex friendships.

In Sum

Since the 1960s the number of singles has risen. Much of this increase is due simply to rising numbers of young adults, who are typically single. Although there is a growing tendency for these young adults to postpone marriage until they are older, this is not a new trend but rather a return to a pattern that was typical early in the twentieth century.

One reason people are postponing marriage today is that increased job and lifestyle opportunities may make marriage less attractive. Also, the low sex ratio has caused some women to postpone marriage or put it off entirely. And attitudes toward marriage and singlehood are changing so that now being single is viewed not so much as deviant but as a legitimate choice.

Singles can be classified according to whether they freely choose this option (voluntary singles) or would prefer to marry but are single nevertheless (involuntary singles). Singles can also be classified according to whether they plan to remain single (stable singles) or marry someday (temporary singles).

Singles used to live in cities, often because employment and leisure opportunities drew them there, but recently more singles have moved to the suburbs. More and more singles are living in their parents' homes; this is usually at least partly a result of economic constraints. Some singles have chosen to live in communal or group homes. A substantial number of unmarrieds (about 5 percent) are cohabiting or living together. Some are heterosexual couples, and some are gay and lesbian couples; we compare the interpersonal patterns of these relationships. Finally, some women share a man in a polygamous arrangement.

Although research consistently finds married physically and psychologically healthier and happier than singles, this situation is slowly changing as singlehood allows the sexual expression once reserved for marriage and as marriage fails to guarantee the security that it once did. The limits that marriage puts on individuality seem more constraining to today's Americans.

However one chooses to live the single life, maintaining supportive social networks is important.

Key Terms

cohabitation	involuntary stable singles
common law marriage	involuntary temporary
communes	singles
continuum of social	sex ratio
attachment	single
domestic partner	voluntary stable singles
double standard of aging	voluntary temporary singles

Study Questions

1. How do economics and income affect a single person's life?

2. Individual choices take place within a broader social spectrum—that is, within society. How do social factors influence an individual's decision about whether to marry or remain single?

3. What are the particular circumstances constraining African American women who are single and would like to marry?

4. What do you see as the differences—advantages or disadvantages—of cohabitation compared to marriage?

5. Should the law treat married and unmarried partners alike?

6. Do you consider a relationship with a married man to be a satisfactory alternative to marriage for the single woman? Why or why not?

7. What problem do older single adults face that are not faced by younger singles? How are these problems different for men and women?

8. How is the relationship between marital status and general happiness changing? Why?

9. What are some components of a satisfying single life?

Suggested Readings

Cargan, Leonard and Matthew Melko. 1982. *Singles: Myths and Realities.* Beverly Hills, CA: Sage. This is a report of a study of singles in Dayton, Ohio. The book explores singles' relations with parents, work, leisure, sexuality, friendship patterns, health, and happiness. Although the book is now fifteen years old and the data older than that, it is unique as an extensive study of singlehood.

Elkin, Larry M. 1994. *First Comes Love, Then Comes Money.* New York: Doubleday. Written by a CPA and certified financial planner, this book explains how unmarried couples can use investments, tax planning, insurance, and wills to gain the kind of financial protection that is denied by the formal law.

Goldscheider, Frances K. and Calvin Goldscheider. 1993. *Leaving Home Before Marriage: Ethnicity, Familism, and Generational Relationships.* Madison: University of Wisconsin Press. The Goldscheiders are both professors at Brown University. They explore the many issues concerning young adults' leaving their parental homes and compare experiences across U.S. racial/ethnic subcultures.

Harry, Joseph. 1984. *Gay Couples.* New York: Praeger. Sociological study of gay relationships, with comparisons to straight relationships.

Moustakas, Clark E. 1961. *Loneliness.* Englewood Cliffs, NJ: Prentice-Hall. This important book explores loneliness as essential (or existential) to the human condition. *Everyone* experiences periods of loneliness. According to Moustakas, loneliness causes not only pangs of terror, desolation, and despair but also deeper insights and renewed awareness of the world and the self.

Nardi, Peter M., ed. 1992. *Men's Friendships: Research on Men and Masculinities.* Newbury Park, CA: Sage. Essays and research reports by social scientists on this topic.

Simon, Barbara Levy. 1987. *Never Married Women.* Philadelphia: Temple University Press. A qualitative study of fifty elderly, never-married women; includes honest talk about how being single feels to them now and how it has felt over the years; addresses their experiences as never-married women: their work, families of origin, sexual intimacy, and aging.

PART Three

Defining Your Marriage and Family

Fairy tales typically end with the hero's choosing a spouse. As children, we probably assumed there wasn't much to tell after that. But people now recognize that even after the wedding ceremony partners continue to make choices about their marriage and family.

Defining one's own marriage and family involves choosing what form the family will take. Will it be (for example) dual career, communal, extended, or nuclear? Besides creating their own family form or structure, partners choose, either consciously or by default, what kind of marital relationship they will have.

Just as deciding on a satisfying family form requires making choices, so does creating a mutually satisfying marital relationship. Chapter 7 explores knowledgeable choices about one's marriage relationship. In Chapter 8 you'll see that one creative way to meet this challenge is for couples to take time to discuss and then write a personal marriage agreement. We talk about how to begin such a process and what kinds of issues to consider.

A recurring theme in many of these chapters is communication. Chapter 9 discusses communicating and notes that partners can ignore conflicts, fight in alienating, hurtful ways, or try to fight fairly. Chapter 10 explores the role of power in marriage.

CHAPTER **7**

Choosing Each Other

Marriage: When and Why

"The Pasture"

I'm going out to clean the
pasture spring;
I'll only stop to rake the leaves away
(And wait to watch the water clear,
I may):
I sha'n't be gone long—You come too.
I'm going out to fetch the little calf
That's standing by the mother.
It's so young.
It totters when she licks it
with her tongue.
I sha'n't be gone long—You come too.

ROBERT FROST

People want to love and be loved. In our society this often means selecting someone, usually of the opposite sex, with whom to become both emotionally and sexually intimate. Although people can maintain close relationships with several others, Americans value having one special relationship with a person they love best. In our culture this relationship is supposed to have a romantic quality. It may lead to marriage.

How well do our ideals of romance describe the kind of relationship people share in marriage? As we'll see, only recently in history have people even begun to equate the two concepts of love and marriage. Although love is usually an important ingredient, successful marriages are also based on such qualities as the partners' common goals and needs, their maturity, and the soundness of their reasons for marrying.

You'll recall that Chapter 4 examines love—loving oneself and discovering a loving relationship with an intimate partner. In the following pages we'll examine some social variables that may influence choice of partners and marital stability. We'll also look at patterns by which individuals in our society develop commitments to each other.

Love—and Marriage?

That marriages should involve romance and lead to personal satisfaction is a uniquely modern idea. According to an old song, love and marriage "go together like a horse and carriage." How did our notion of romantic love come about, and why is it assumed to be the basis of marriage in our society?

Courtly love (or romantic love) flourished during the Middle Ages. Most marriages in the visible upper levels of society during this period were based on pragmatic considerations involving property and family alliances (Stone, L., 1980). Tender emotions were expressed in nonmarital relationships in which a knight worshiped his lady, and ladies had their favorites. These relationships involved a great deal of idealization, were not necessarily sexually consummated, and certainly did not require the parties to live together. In time the ideology of romantic love was adapted to a situation for which it was probably much less suitable—marriage.

As urban economies developed and young people increasingly worked away from home, arranged marriages gave way to marriages in which individuals selected their own mates. Sentiment rather than property became the basis for unions (Shorter 1975). The strong emotional and personal qualities of romantic love were in keeping with the individualism and introspection characteristic of the evolving Protestant capitalistic society of Western Europe (Stone 1980).

In the absence of arranged marriages, love provided motivation for choosing mates and forming families and thereby served important social functions. This continues to be true. The connection between love and marriage serves to harness unpredictable feelings, to the service of society (Goode 1959; Greenfield 1969).

Intense romantic feelings may serve to get a married couple through bad times (Udry 1974). However, the idealization and unrealistic expectations implicit in the ideology of romantic love can cause problems. Many Americans expect romance to continue not only through courtship but in marriage, too. Combining the practical and economic elements of marriage with developing intimacy and love is, historically speaking, a new goal.

Falling in love, then, when we expect that it could lead to marriage, may be a more practical and rational process than we think. Social scientists use the analogy of the marketplace to describe how Americans choose marriage partners.

The Marriage Market

Imagine a large marketplace in which people come with goods to exchange for other items. In nonindustrialized societies a person may go to market with a few chickens to trade for some vegetables. In modern societies, people attend hockey equipment swaps, for example, trading outgrown skates for larger ones. Americans choose marriage partners in much the same way: They enter the **marriage market** armed with resources—their personal and social characteristics—and then bargain for the best buy they can get.

In many other cultures, parents arrange their children's marriages through a bargaining process not unlike what takes place at a traditional village market. They make rationally calculated choices after determining the social status or position, health, temperament, and, sometimes, physical attractiveness of their prospective son- or daughter-in-law. In such societies the bargaining is obvious. Professional matchmakers often serve as investigators and go-betweens, just as we might engage an attorney or a stockbroker in an important business deal. The exchange is accompanied by the **dowry,** a sum of money or property brought to the marriage by the female. A woman with a large dowry can expect to marry into a higher-ranking family than can a woman with a small dowry, and dowries are often increased to make up for qualities considered undesirable (Kaplan 1985, pp. 1–13). Parents in eighteenth-century England, for instance, increased the dowries of daughters who were pockmarked.

The difference between arranged marriages and modern freely selected marriages seems so great that we are inclined to overlook an important similarity: Both involve bargaining. What has changed in modern society is that the individuals, not the family, do the bargaining.

Exchange Theory

The ideas of bargaining, market, and resources used to describe relationships such as marriage come to us from **exchange theory,** described in Chapter 2. Recall that the basic idea of exchange theory is that whether

Although the arranged marriage of this Indian couple may be a world apart from the freely chosen marriage of this Japanese American couple in Hawaii, bargaining has occurred in both of these unions. In arranged marriages, families and community do the bargaining, based on assets such as status, possessions, and dowry. In freely chosen marriages, the individuals perform a more subtle form of bargaining, weighing the costs and benefits of personal characteristics, economic status, and education.

or not relationships form or continue depends on the rewards and costs they provide to the partners. Individuals, it is presumed, want to maximize their rewards and avoid costs, so when there are choices, they will pick the relationship that is most rewarding or least costly. The analogy is to economics, but in romantic and marital relationships individuals are thought to have other sorts of resources to bargain besides money: physical attractiveness, personality, family status, skills, emotional supportiveness, cooperativeness, intellect, originality, and so on. Individuals also have costly attributes: irritability, demandingness, ineptitude, low social status, geographic inaccessibility (a major consideration in modern society), and so on. If each individual adopts a strategy of maximizing outcomes, then stable relationships will tend to exist between people who have like amounts of resources because they will strike a fair balance or bargain.

Theoretically, these bargains will last only until a better deal comes along. A weakness of exchange theory is that people who are married or with each other for a time find that considerable attachment and common identity develop, so that leaving becomes very difficult. Shared activities, a shared life, give the relationship a normative structure; that is, the patterns of everyday life become fixed in a way that deeply involves the partners. And community and interpersonal ties provide social support for the relationship and sanctions against breaking it off, even for unmarried couples.

There is also the question of rationality. Are people actually so calculating about rewards and costs, even at an unconscious level? Further, in an exchange analysis, what is rewarding and what is costly? It varies with the individual.

Descriptions of exchange behavior are reminiscent of traditional arranged marriages. Nevertheless, we present an exchange perspective as a useful tool. In fact, the basic structures of American life channel men and women into roles that have certain consequences for marital partnerships. We need to be aware of an underlying exchange structure of intimate relationships and marital ties. Let us look now at an exchange version of marital choice.

The Traditional Exchange

Individuals may bargain such characteristics as social class, age, physical attractiveness, and education, but the basic marital exchange is traditionally related to gender roles. Historically, women have traded their ability to bear and rear children and perform domestic duties, along with sexual accessibility and physical attractiveness, for masculine protection, status, and economic support (see Figure 7.1) (Sprecher, Sullivan, and Hatfield 1994). Even though more and more women are gaining status by means of their own careers or professions, many of them continue to expect greater success from their future husbands than vice versa (Ganong and Coleman 1992; Sprecher, Sullivan, and Hatfield 1994) and to attain a higher status by marrying than they would have as single individuals supporting themselves. Being a doctor's wife may offer a higher status, for example, than being a teacher or nurse.[1]

Women have had some disadvantages in the traditional exchange. For one thing, men can bargain on their potential, taking their time to shop for a partner, whereas women lose an important advantage with time. A woman's traditional bargaining assets of physical attractiveness and childbearing capacity are given less value as she ages. Men, meanwhile, can exchange promises of anticipated occupational success and power for marital security.

Sociologists point to a related bargaining advantage for men. In the traditional exchange, young men and women bargain for spouses with different degrees of available information. The man can more easily assess

1. This brief comment does not do justice to the issue of women's social status and the wife's contribution to family income. For a discussion of this issue, see Max Haller's "Marriage, Women, and Social Stratification: A Theoretical Critique," *American Journal of Sociology* 86 (1981): 766–95.

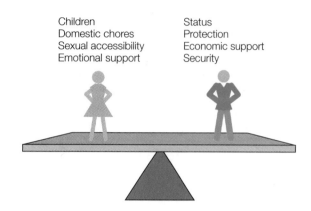

Children   Status
Domestic chores   Protection
Sexual accessibility   Economic support
Emotional support   Security

FIGURE **7.1**

I   The traditional exchange.

the woman, observing whether she offers emotional support and empathy as well as responding to her physical attractiveness. But the occupational abilities of many young men have generally not yet been tested. In this traditional exchange, then, a woman must therefore base her appraisal on such cues as a young man's grades in school and his father's occupation. Older men and women may more easily evaluate prospective partners because personal and economic characteristics are more established.

Bargaining in a Changing Society: The Optimistic View

As society changes, so does the basic exchange in marriage. Social scientists predict that, should true androgyny emerge (as was discussed in Chapter 3), exchange between partners will no longer depend greatly on practical or economic resources but instead will include "expressive, affective, sexual, and companionship resources" for both partners. If women gain occupational and economic equality with men, the exchange should become more symmetrical, with women and men increasingly looking for similar characteristics in each other (see Figure 7.2a).

Marriage based on both partners' contributing roughly equal economic and status resources should be more egalitarian. Current changes in men's roles toward greater emotional expressiveness may improve marriage. Furthermore, as gender roles change, even those exchanges that are complementary (that is,

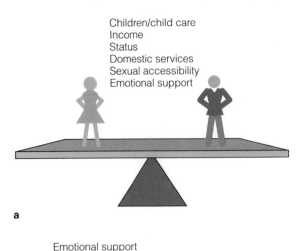

Children/child care
Income
Status
Domestic services
Sexual accessibility
Emotional support

a

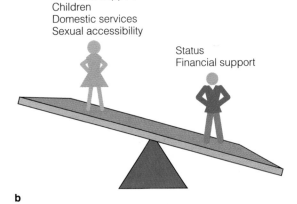

Emotional support
Children
Domestic services
Sexual accessibility

Status
Financial support

b

FIGURE **7.2**

Two interpretations of the changing traditional exchange: (a) the optimistic, egalitarian view; and (b) the more pessimistic view.

based on unlike resources) can be individualized. For example, an ambitious career woman might be comfortable bargaining for a nurturant, domestic husband, and vice versa. This is the bargaining we would anticipate in an **egalitarian** and androgynous society.

Bargaining in a Changing Society: The Pessimistic View

Another view pays more attention to the substantial inequalities that remain, particularly women's marginal position in the economic system. Although women have entered the labor market in large numbers, their jobs and incomes are inferior, on the average, to those of men.

Even though social expectations may be moving toward androgyny, women are still at a disadvantage in the marriage market. While this is changing, men continue to hold the advantage regarding access to financial security and status. Meanwhile, many of the bargaining chips women have traditionally brought to the basic exchange—children, domestic services, sexual accessibility—have been devalued because they are now available to unmarried men as well as husbands (see Figure 7.2b).

It may have been true for our grandparents, for example, that "the way to a man's heart is through his stomach." Today, however, cooking and other domestic skills and services are no longer monopolized by potential wives, and many men have discovered that they like to cook. The increased acceptance of nonmarital sex (however satisfying this may be to many women as well as men!) has weakened women's traditional bargaining position, decreasing the possibility of trading sexual accessibility for marriage. And finally, even the ability to have children may be less vital than in the past, as some people are questioning the value of having children. Moreover, with today's high divorce rate and custody of children usually going to mothers, marriage no longer assures men ongoing access to their children (Gerson 1993). In this analysis the wife's primary remaining resource in the marital exchange consists of the promise of lifelong affection and encouragement (Scanzoni 1970).

Besides cultural expectations, certain demographic features of the current marriage market negatively affect the bargaining position of women. Of particular importance is how an unequal sex ratio limits choice of marriage partners.

Sex Ratios and Bargaining for a Spouse

Chapter 6 explored the impact of *sex ratios* on singlehood. The obvious flip side of the discussion is that sex ratios affect bargaining for a spouse.

As you can see from Table 7.1, sex ratios differ somewhat for various racial/ethnic categories. Beginning with middle age, however, there are increasingly fewer men than women in every race/ethnic category. If each of these people wished to marry, many women would be left out, increasingly so in the older age groups. Among Hispanics the total sex ratio is slightly above 100, but there are still more Latinas than Latinos over 45 years old. Among blacks the discrepancy

TABLE 7.1

Ratio of Men to Women by Age and Race/Ethnicity, 1992

Age (years)	Asian/ Pacific Islander	American Indian/ Eskimo/Aleut	African American	Hispanic	Non–Hispanic White	All Races/ Ethnic Groups
14–17	105.8	103.0	103.4	105.4	106.4	105.8
18–24	104.1	104.7	97.2	116.8	103.3	104.0
25–34	95.3	97.5	88.8	113.1	100.6	100.2
35–44	89.3	92.9	86.2	103.1	100.0	98.2
45–54	91.2	92.3	82.6	94.9	97.6	95.6
55–64	81.2	88.1	77.1	87.9	92.1	90.0
65–84	79.8	76.7	65.2	72.8	72.3	71.8
85 and older	68.4	57.1	41.9	53.5	37.6	38.7

Source: Based on data from the U.S. Bureau of the Census, 1994a, Tables 20 and 21.

exists at younger ages; there are only 89 men per 100 women ages 25–34. Higher death rates of black males and the significant number of black men in prison contribute to the low sex ratio. Moreover, not only the relative number of eligible men is salient; their employment prospects also affect women's chances for marrying (South and Lloyd 1992b)—and, if they do marry, their chances of getting a good bargain (Lichter, Anderson, and Hayward 1995). You may want to look at Table 7.1 and locate yourself by age and ethnicity. If you are male, how does the sex ratio in your particular category influence your chances of finding a (heterosexual) partner? What about your chances if you are female?

Calculation of the availability of mates is complicated by both the tendency of men to be two to three years older than the women they marry and by the fact that individuals seek mates within *local* markets; thus the nationally aggregated figures presented here may not apply to certain specific regions (Lichter, LeClere, and McLaughlin 1991).

The Marriage Gradient

The **marriage gradient** is the traditional tendency for women to marry "up" with regard to age, education, occupation, and even height. But even this process takes place within culturally accepted limits. Regarding age, for example, most marriages are characterized by an age difference of no more than three or four years (Surra 1990, p. 847).

The same is true for education. College-educated women and men tend to marry each other; so do people without college educations. One analysis (Mare 1991) of census data has found evidence of increased educational homogamy (explained in the next section) from the 1930s through the 1980s. The author suggests that this situation may be due to "increasing competition in the marriage market for wives with good prospects in the labor market" (p. 15)—hence a weakening of the traditional marriage gradient with respect to education. Nevertheless, where the marriage gradient does exist, the practice sets the stage for greater marital power for husbands than wives, because husbands have greater educational and/or financial resources. Issues of marital power are explored in Chapter 10.

Besides the marriage gradient and sex ratios, an additional factor shaping marital choice is the tendency of people to marry others with whom they share certain social characteristics. Social scientists term this phenomenon *homogamy.*

Homogamy: Narrowing the Pool of Eligibles

Not everyone who enters the marriage market is equally available to everyone else. Americans, like many other peoples, tend to make marital choices in socially patterned ways, viewing only certain others as potentially suitable mates. The market analogy would be to choose only certain stores at which to shop. For each shopper there is a socially defined **pool of eligibles:** a group of individuals who, by virtue of background or birth, are considered most likely to make compatible marriage partners.

Americans tend to choose partners who are like themselves in many ways. This situation is called **homogamy:** People tend to marry people of similar race, age, education, religious background, and social class. Traditionally, the Protestant, Catholic, and Jewish religions, for example, have all encouraged **endogamy:** marrying within one's own social group. (The opposite of endogamy is **exogamy,** marrying outside one's group, or **heterogamy,** marrying someone dissimilar in race, age, education, religion, or social class.) Age and educational heterogamy are more pronounced among blacks than among whites, partly because an "undersupply" of marriageable black men prompts black women to marry down educationally and to marry considerably older or younger men (Surra 1990, p. 848).

The argument might be made that marriage has become less homogamous among whites in recent years. Although same-faith marriages are still a significant majority (McCutcheon 1988), the incidence of interfaith marriage is now considerable. And more children of blue-collar parents are marrying children from professional homes. By 1980 marriage across Euro-ethnic (for example, Irish, Italian, Polish) lines had become so common that only one in four American-born non-Hispanic whites was married to someone with an identical ethnic heritage ("Sense of Identity" 1985).

In spite of the trend toward less religious and Euro-ethnic homogamy than in the past, homogamy is still a strong force (Kalmijn 1991).

With regard to racial/ethnic intermarriage, 99 percent of non-Hispanic whites marry other non-Hispanic whites and 94 percent of African American couples are racially homogeneous (U.S. Bureau of the Census 1995, Table 61). More than 72 percent of Asian Americans and 71 percent of Hispanics marry within their group. (However, nearly 54 percent of Native Americans marry outside their race, mostly to whites ["Sense of Identity" 1985]. Native American women are the most likely of all gender/racial categories to intermarry, in fact.)

Social scientists point out that although today people are marrying across small class distinctions, they still are not doing so across large ones. For instance, individuals of established wealth seldom marry the poor. All in all, an individual is most likely to marry someone who is similar in basic social characteristics. Let's look at a hypothetical case to see why this may be so.

Reasons for Homogamy

Laura is attracted to Jeremy (and vice versa) who is a college student (like herself), two years older, and single. Laura's parents are upper middle class. They live in the expensive section of her hometown, have a housekeeper, drink wine with their meals, and frequently have parties by their pool with a live band. Catholic, they go to Mass every Sunday. Jeremy's parents are upper lower class. They are separated. His mother lives in an apartment and works as a checker in the supermarket. The family drinks iced tea at mealtime, then watches TV. They believe in "being good people" but do not belong to any organized religion.

How likely is it that Laura and Jeremy will marry? If they do marry, what sources of conflict might occur? We can help to answer these questions by exploring four related elements that influence both initial attraction and long-term happiness. These elements—propinquity, social pressure, feeling at home, and a fair exchange—are important reasons many people are homogamous.

PROPINQUITY   In our society, propinquity (or geographic closeness) has typically been a basic reason people tend to meet others much like themselves. Geographic segregation, which can result from discrimination and segregation or from community ties, contributes to homogamous marriages. Intermarriage patterns within the American Jewish community are an example. Until the 1880s, the small size of this group and its geographic dispersal limited the availability of Jewish marriage partners and led to frequent marriages with non-Jews. The large Jewish migration from eastern Europe that began in the late nineteenth century changed this. Immigrants tended to settle together in Jewish neighborhoods; for this reason, among others, intermarriage rates dropped (William Petschek National Jewish Family Center 1986b). Only about 6 percent of Jews married non-Jews in the late 1950s. Now that the barriers that used to exclude Jews from certain residential areas and colleges are gone, about half marry gentiles (Gittelson 1984).

Propinquity also helps account for social class homogamy. Middle-class people live in neighborhoods with other middle-class people. They socialize together and send their children to the same schools; upper- and lower-class people do the same. Unless they had met in a large, public university, it is unlikely that Jeremy and Laura would have become acquainted at all. The effort expended in associating with, or even meeting, people who are outside one's physical circle of friends encourages homogamy.

However, as psychologist Bernard Murstein (1986) points out, more and more people today choose spouses from open fields. An *open field* encounter refers to a situation in which the man and woman do not yet know each other well. A *closed field* encounter, on the other hand, is one in which partners are forced to interact by reason of their environment—in a small town, for example. Sociologist Judith Ericksen, in a survey of Philadelphia couples, found that 22 percent met their spouses at a "chance encounter"—not work, family, or usual social activities. She calls these meetings "marital pickups." But rather than occurring in singles bars, they happened on buses, in stores, on the street, or in elevators. Ericksen found, interestingly, that pairs who "grew up with their spouses"—that is, knew them from school and the neighborhood—had lower divorce rates than "chance encounter" couples had ("Cupid's Arrows" 1981, p. 2-B).

A more recent form of "marital pickup" can occur in cyberspace as people become acquainted on the Internet in various "chat rooms." Still, though, their mutual ability to access the Internet assures *some* degree of educational and/or financial homogamy. Box 7.1, "The New Matchmakers," addresses cyberspace courtship.

SOCIAL PRESSURE   A second reason for homogamy is social pressure. Our cultural values encourage people to marry others socially similar to themselves and discourage marrying anyone too different: Laura's parents, friends, and siblings are not likely to approve of Jeremy because he doesn't exhibit the social skills and behavior of their social class. Meanwhile, Jeremy's mother and friends may say to him, "Laura thinks she is too good for us. Find a girl more like our own kind." Sometimes, social pressure results from a group's concern for preserving its ethnic or cultural identity. When young Jews, particularly college students, began to intermarry more often in the 1960s, for example, Jewish leaders became concerned.

> From pulpits across the country rabbis lamented the disintegrating Jewish family and the "demographic holocaust" that threatened the community. Federations rushed to form task forces on intermarriage; Hillel directors sought new ways to cement the group loyalties of Jewish college students so they would not intermarry; and a spate of "how-to-stop-an-intermarriage" books appeared. (William Petschek National Jewish Family Center 1986b, p. 2)

Recent immigrants, such as Asians or various Hispanic groups, may pressure their children to marry within their own ethnic group in order to preserve their unique culture (Kitano and Daniels 1995).

FEELING AT HOME   People often feel more at home with others from similar backgrounds; couples from different social groups may struggle to communicate and may feel uncomfortable. With regard to social-class differences, Jeremy is likely to have different attitudes, mannerisms, and vocabulary from Laura. Laura won't know how to dress or behave in Jeremy's hangouts and among his friends. Each may feel uncomfortable and out of place in the surroundings the other considers natural.

STRIKING A FAIR EXCHANGE   As previously noted, exchange theory suggests that people tend to marry others whose social currency—social class, education, physical attractiveness, and even self-esteem—is similar to their own (Murstein 1986).

The questions people ask on first meeting point up their concern with each other's exchange value. They want to know whether their prospective dates are married or single, where they are employed, whether they attend college and where, what they plan to do upon graduation, where they live, and perhaps what kind of

The New Matchmakers: Formal Intermediaries in the Marriage Market

Historically, matchmaking has long claimed a commercial niche in the marriage market. Today, a modem, phone line, or advertisement might bring potential partners together.... Social scientists have long seen mate selection as occurring in a marriage market where both economic and interpersonal assets are exchanged. Commercial dating services, by making these assets explicit, further emphasize the relevance of market theory....

A market intermediary is any organization, event, or institution that takes on work that otherwise would be performed by the provider or consumer of the good. For example, help wanted advertising is an intermediary in the labor market that takes on some of the [searching and matching] tasks needed for employee selection. Searching is the process of gaining information essential for exchange (e.g., what products are available, where they can be found, how much they cost). Matching is the process of bringing together compatible exchange partners....

[In] the marriage market ... searching consists of information acquisition (learning about who is available as a potential mate). Matching is the process of using information on eligible others to determine which relationships will be pursued....

The vast majority of singles meet their dating partners through social networks, such as mutual friends or parties that function as informal marriage market intermediaries (MMIs). [But today's singles also use *formal* MMIs such as singles ads, video dating, and electronic/phone communication networks.]

Singles Ads

From 1978 to 1991, the number of social introduction services listed in the Chicago area *Yellow Pages* increased from 5 to 21. Over the same period, singles ads, once the exclusive domain of off-beat publications, have become an established feature in most major newspapers and many magazines, such as the *New York Review of Books*....

Video Dating

Of America's approximately 2,000 marriage intermediaries, about 600 use video technology.... Video dating firms usually operate by having a client first read the written descriptions of other members. At many video dating services these written descriptions contain a photograph of the single, which is often

car they drive. If prospective partners are single, they will be asked whether they are divorced or still single and, especially in the case of women, whether they have children and how many.

If meeting in the marriage market can sound like a job interview, maybe that's because in at least one important way it is: The goal is to strike a fair exchange. And even without the benefit of interviews, people learn to discern the social class of others through mannerisms, language, dress, and a score of other cues.

We've discussed some reasons for homogamy, but, as we pointed out earlier in this chapter, not all marriages are homogamous. *Heterogamy* refers to marriage between those who are different in race, age, education, religious background, or social class.

Examples of Heterogamy

How does marrying someone from a different religion, social class, or race/ethnicity affect a person's chances for a happy union? In answer to this, we first examine interfaith marriages, then look at interclass partnerships, and finally at interracial unions.

INTERRELIGIOUS MARRIAGES   Sociologist Norval Glenn (1982) estimates that about 80 to 85 percent of today's marriages are religiously homogamous. This leaves about 15 to 20 percent that are between spouses with different religious preferences, such as Protestant, Catholic, and Jewish. (Glenn considers unions between partners of different Protestant denominations as homogamous.)

placed on the back of the information sheet.... [But] clients tend to read the forms from the back to the front, doing an initial screening on the basis of the photo. If the client finds someone whose form piques his or her interest, then he or she is shown that person's video. Usually the video lasts 2 to 10 minutes and shows the client responding to a series of questions. If the suitor is still interested after seeing the video, the prospective date is contacted and then views the suitor's video. If the parties agree to meet, they are given each other's phone numbers and the rest is up to them....

Electronic/Phone Communication Networks

One of the most recent innovations in MMIs is the development of electronic communication networks, such as computer-based networks.... Unfortunately, these networks have not been well researched, so little is known about them. [There are] two kinds of computer-based networks: computer-accessed singles ads in which singles leave messages on computer bulletin boards, and on-line networks....

Like computer networks, telephone-based MMIs can also be divided into those that allow for interactive contact (i.e., party lines) and bulletin board systems (i.e., telephone singles ads) which allow singles to leave short taped messages or hear the messages left by others. As with computer networks, these systems may facilitate relationship formation by allowing people to get to know each other before screening on the basis of physical appearance takes place, but they may also cause singles to waste time and energy on a relationship that they later reject due to lack of physical attraction....

Unfortunately, work which explores... formal MMIs has often been viewed as trivial or excessively narrow.

This situation leads to two broad recommendations for future research. First, if formal MMIs are to be used as a source of data about singles in general (e.g., when personal ads are used to study what singles value in a mate), then more research is needed to determine in what ways users of formal MMIs differ from nonusers, and what differences exist between users of various formal intermediaries. Second, formal MMIs and their users should be recognized as a legitimate topic of study in and of themselves, without the need to generalize to the population of all singles....

Furthermore, longitudinal research on formal MMIs could help illuminate the impact of these services on personal and relational well-being, both for clients who form relationships and for those who don't.

How do MMIs support the view in social science that mate selection takes place in a marriage market? How do formal MMIs differ from informal ones? According to one estimate, a 5–20 percent marriage rate is typical for formal MMIs. In what ways do you think MMIs might decrease marital homogamy? In what ways do you think they might continue or even increase marital homogamy?

Source: Ahuvia and Adelman 1992, pp. 452–63; citations deleted.

Marriages are less homogamous if we consider partners' religion before the wedding. Put another way, many partners who originally differed in religion switched for the purpose of making their religions the same as their partners' (Glenn 1982).

Some of this switching no doubt took place because partners agreed with the widely held belief that interreligious marriages tend not to be as successful as homogamous ones, a belief supported by research (see, for example, Ortega, Whitt, and Williams 1988; Heaton and Pratt 1990). Exploring this question by means of statistical analysis, Glenn found that religious heterogeneity lessened marital satisfaction for husbands but not for wives. Glenn speculated that a husband's marital happiness might be more affected

because religious heterogeneity is more likely to result in religious differences between father and children; "persons who are not disturbed by their spouse's being of a different religion frequently may be disturbed by their children's being of a different religion, if only because it may relegate them to a kind of outsider or minority status in their own family" (Glenn 1982, p. 564). If the spouses are traditional, the husband may see a wife's refusal to change her religion to match his as a threat to his authority, whereas the wife does not expect this concession. The sex role ideologies of the spouses were not investigated in this study, however.

One probable reason that religious homogeneity improves chances for marital success involves value

Jeff Stahler reprinted by permission of Newspaper Enterprise Association, Inc.

consensus. "Religious orientations may come into play when deciding about leisure activity, child rearing, spending money, and many other facets of marital interaction." Then too, religious homogamy may "create a more integrated social network of relatives, friends, and religious advisors" (Heaton and Pratt 1990, p. 192).

Sociologist Tim Heaton (1984) has attributed differences in marital happiness associated with religious homogeneity almost entirely to the positive effect of church attendance: Homogamously married partners go to church more often and at similar rates. Analysis of data from the 1988 National Survey of Families and Households found that denominationally homogamous couples who attended church at similar rates had higher marital satisfaction and stability (Heaton and Pratt 1990). Another study (Sheehan, Bock, and Lee 1990) used General Social Survey data from the National Opinion Research Center to compare Catholics in heterogeneous marriages with those in homogamous ones. Heterogamously married Catholics went to Mass less frequently, but this did *not* reduce marital satisfaction. These researchers concluded that

the effect of church attendance on marital satisfaction in the general population may be due more to the integrative properties of couple-centered organizational participation than to the religious nature of the activity. Heterogamous couples in which one spouse is Catholic are unlikely to attend church together, and may compensate for this by engaging in other couple-centered activities that are equally effective in promoting marital solidarity. (Sheehan, Bock, and Lee 1990, p. 78)

Glenn concluded that

as interreligious marriages become more frequent and socially accepted, any negative effects that they have on marital quality are likely to diminish. For instance, some of the problems encountered by interreligious couples evidently have grown out of the disapproval of family and friends; and as such disapproval diminishes, so should some of the disruptive influences on interreligious marriages. (Glenn 1982, p. 564)

This is also true, as is pointed out elsewhere in this chapter, for interracial marriages.

INTERCLASS MARRIAGES   What about the marital satisfaction of partners in interclass unions?

An often-cited study of marriage in urban Chicago (Pearlin 1975) found that partners experienced more

As this Jewish and Christian couple illustrate, when elements of propinquity and cultural pressure are overcome, heterogamous marriages do take place. While some research shows that homogamous unions tend to be more stable than interreligious or interacial ones, a heterogamous couple may hold common values that will transcend differences in background, religion, or race.

stress in these class-heterogamous unions. Moreover, the spouse who had married down was more stressed than the one who had married up.

It is important to note, however, that the relationship between stress and marrying down existed only when status striving was important to the individual. Those people for whom status was important and who had married down perceived their marriages more negatively—as less reciprocal, less affectionate, less emotionally supportive, and with less value consensus—than those who had married up. In an exchange framework, the partner who had married down had not struck a good bargain in an important matter. For people for whom status was *not* important, neither marrying up nor down produced any difference in their evaluation of their marriages.

Although this research supports earlier work suggesting marital difficulty among couples with status differences, it also contradicts the often-repeated assumption that differences in cultural background necessarily produce stress and conflict. Although people marrying up and people marrying down encounter the same cultural differences, only those who married

down and who regarded social status as important experienced related stress. Status inequality and heterogeneity in and of themselves do not create problems.

We need to add a note of caution here. Chapter 4, you may recall, addresses the need to know and accept oneself as a prerequisite to forming an intimate relationship with another person. Status striving might sound like a negative characteristic and so be denied as a part of oneself, without much introspection. "Not me; I'm not into keeping up with the Joneses," one easily says. This may not be so true.

INTERRACIAL MARRIAGES   **Interracial marriages** include unions between partners of the white, African American, Asian, or Native American races with a spouse outside their own race. In June 1967 (*Loving v. Virginia*) the U.S. Supreme Court declared that interracial marriages must be considered legally valid in all states. At about the same time, it became impossible to gather accurate statistics on interracial marriages. Many states no longer require race information on marriage registration forms, so these data are incomplete at best.

TABLE 7.2

Interracial and Interethnic Married Couples, 1994

All interracial couples	3,047,000
Black–white couples	296,000
Husband black, wife white	196,000
Wife black, husband white	100,000
White and spouse of race other than white or black*	909,000
Black and spouse of race other than white or black*	78,000
Other interracial couples*	1,764,000
All Hispanic couples†	5,038,000
Hispanic–Hispanic couples	3,755,000
Hispanic and spouse of non-Hispanic origin	1,283,000

*Neither white nor black, but Asian, Native American, Aleut, Pacific Islander.

†Persons of Hispanic origin may be of any race.

Source: U.S. Bureau of the Census 1995, Table 61.

Available statistics show that the proportion of interracial to all marriages is small, about 5 percent.

Of all interracial marriages in 1994, only 10 percent were black–white. This figure represented 296,000 black–white married couples in the United States (U.S. Bureau of the Census 1995, Table 61). The remainder, for the most part, were various combinations of whites or blacks with Asians, Native Americans, and others (see Table 7.2). Three-quarters of black–white marriages involved black men married to white women (U.S. Bureau of the Census 1994a, Table 62).

Much attention has been devoted to why people marry interracially. Psychoanalytic and psychological hypotheses suggest that whites marry interracially because of rebellion and hostility, guilt, or low self-esteem. Such explanations smack of racism and are not supported by research. Another theory is **hypergamy:** marrying up socioeconomically on the part of a white woman who, in effect, trades her socially defined superior racial status for the economically superior status

of a middle- or upper middle-class black partner. It is important to note that empirical research provides no support for the theory of hypergamy.

One study of forty black–white interracially married couples found, simply, that "with few exceptions, this group's motives for marriage do not appear to be any different from those of individuals marrying . . . within their own race" (Porterfield 1982, p. 23). As you can see from Table 7.3, many respondents in this study said that they married for love and compatibility.

Heterogamy and Marital Stability

Marital success can be measured in terms of stability—whether or how long the union lasts—and the happiness of the partners. Marital stability is not synonymous with marital happiness because in some instances unhappy spouses remain married, whereas less unhappy partners may choose to separate. In general, social scientists find that marriages that are homogamous in age, education, religion, and race are most stable (Heaton, Albrecht, and Martin 1985).

Just as information on interracial/interethnic marriages is incomplete, so is information on their divorces. Only about half the states and the District of Columbia report race/ethnicity on divorce records (U.S. National Center for Health Statistics 1991a). What evidence we have is conflicting on whether interracial/interethnic marriages are more or less stable than intraracial/intraethnic unions. And while we have some census data on Hispanics who marry outside their ethnic group (Table 7.2), we do not know the divorce rates for these interethnic couples. The National Center for Health Statistics reported in 1991 that for divorces in 1988 "the distributions by race were very similar to recent marriage data" (U.S. National Center for Health Statistics 1991a, p. 5).

We can offer at least three explanations for any differences in marital stability. First, significant differences in values and interests between partners can create a lack of mutual understanding, resulting in emotional gaps. One in-depth study of seventeen married interracial couples concluded that from the view of outsiders, the union is interracial. From the inside, however, the spouses themselves may see their union as "cross-cultural" (Johnson and Warren 1994).

Second, such marriages are likely to create conflict between the partners and other groups, such as parents, relatives, and friends. Then too, continual dis-

TABLE **7.3**

Number of Interracially Married Respondents Who Mentioned Specific Motives for Their Marriage

Motives	Race and Sex of Spouse			
	Black Male	Black Female	White Male	White Female
Nonrace related motives				
Love	28	12	6	25
Capability	28	12	6	25
Pregnancy		1		1
Race-related motives				
Other race more appealing, interesting				2
Rebellion against society	3			
White female a status symbol	2			
White female less domineering	4			
Black female more independent, self-sufficient			1	
Marginality				
Desire for a husband of comparable educational/occupational status		2		
Ostracized from one's own racial group				1

NOTE: Although these categories are mutually exclusive, motives for marriage are not. Some of the respondents indicated a combination of reasons for marriage.

Source: Porterfield 1982, p. 23.

criminatory pressure from the broader society may create undue marital stress. As one interracially married husband put it, "People's heads almost whiplashed off their necks when we walked down the street together" (in Romano and Trescott 1992, p. 91). Deprived of a social network of support, partners may find it more difficult to maintain their union in times of crisis.

Finally, the higher divorce rate among heterogamous marriages may reflect the fact that these partners are likely to be less conventional in their values and behavior. Such unconventional people may divorce more readily if things begin to go wrong, rather than remain unhappily, though stably, married.

Generally, social scientists in the past have concluded that the greater the differences between partners, the more likely it is that their marriage will be conflicted (Udry 1974). These conclusions have found their way into textbooks as subtle advocacy of ho-

mogamy. Recent data have begun to challenge these conclusions, perhaps because in the past twenty years society has become more diverse. One recent study found that age heterogamy (including husbands older than their wives by eleven years and wives older by four years) has no effect on marital satisfaction (Berardo 1985).

Heterogamy and Human Values

In any case, it is important to note the difference between scientific information and values. Social science can tell us that the stability of heterogamous marriages may be lower than that of homogamous marriages, but many people do not want to limit their social contacts—including potential marriage partners—to socially similar people. Although many people may retain a warm attachment to their racial or ethnic community, social and political change has been in the direction of breaking down those barriers. People committed to an open society find intermarriage to be an important symbol, whether or not it is a personal choice, and do not wish to discourage this option.

From this perspective we can think of the negative data on heterogamy and marital stability not as a discouragement to marriage, but in terms of its utility in helping couples be aware of possible problems. Intermarrying couples such as Jeremy and Laura may anticipate and talk through the differences in their lifestyles, especially if they take a social science course!

The data on heterogamy may also be interpreted to mean that it is common values and lifestyles that contribute to stability. A heterogamous pair may have common values that transcend their differences in background. Furthermore, some problems of interracial (or other heterogamous) marriages have to do with social disapproval and lack of social support from either race.

But individuals can choose to work to change the society into one in which heterogamous marriage will be more accepted and hence will pose fewer problems. Then, too, to the degree that racially, religious, or economically heterogamous marriages increase in number, they are less likely to be troubled by the reactions of society. Again we see that private troubles—or choices—are intertwined with public issues; social changes are needed to make heterogamous marriages work. Finally, "it seems untenable . . . to assume that heterogamy can lead only to negative consequences for the marital relationship" (Haller 1981, p. 786). If people are able to cross racial, class, or religious boundaries and at the same time share important values, they may open doors to a varied and exciting relationship.

Developing the Relationship

We have seen how homogamy and a fair exchange help to narrow the field from which people choose marriage partners. This doesn't explain the whole process, however. If it did, people would inevitably marry according to very predictable patterns.

At First Meeting: Physical Attractiveness and Rapport

Why is that we can be so drawn to one person and so indifferent to another? One reason that popular opinion assumes (but social scientists find difficult to accept) is physical attractiveness. Deny it as we might, our evaluations of others, even in our early years, are

influenced by their appearance. Research shows that cute children are perceived by adults as less naughty, as well as brighter, more popular, and more likely to attend college, than are less attractive children. The same assumptions influence adults' perceptions of one another. Both women and men tend to see more socially desirable personality traits in people who are physically attractive than in those who are less attractive (Saks and Krupat 1988).

To find out how important physical appearance is in a situation in which two people meet for the first time, several social psychologists planned an experimental college dance. Students who came to the dance were under the impression that a computer had selected their dates on the basis of similar interests. In fact, they were assigned at random. During the dance intermission the students were asked how they liked their dates. For both females and males, physical attractiveness seemed to determine whether these students were satisfied with their dates (Walster et al. 1966). Physical attractiveness is especially important in the early stages of a relationship.

Physical appeal and other readily apparent characteristics serve to attract people to each other, to spark an interest in getting acquainted, which leads to an initial contact. But whether this initial interest develops into a prolonged attachment depends on whether they can develop rapport: Do they feel at ease with each other? Are they free to talk spontaneously? Do they feel they can understand each other? When two people experience rapport, they may be ready to develop a loving relationship. Common values appear to play an important role here, though whether in early stages only or throughout courtship is not clear. Couples also seem to be matched on sex drive and interest in sex (Murstein 1980), as well as on what sex means to them (Lally and Maddock 1994), suggesting that this is an important sorting factor.

As the Relationship Progresses

The wheel theory of love (Reiss and Lee 1988), described in Chapter 4, explains the role of rapport in the development of a relationship. It suggests that mutual disclosure, along with feelings of trust and understanding, are necessary first steps. As you may recall, this theory also posits that an important step in developing love is "mutual need satisfaction." Along this line, social scientist Robert Winch (1958) once proposed the **theory of complementary needs,** whereby

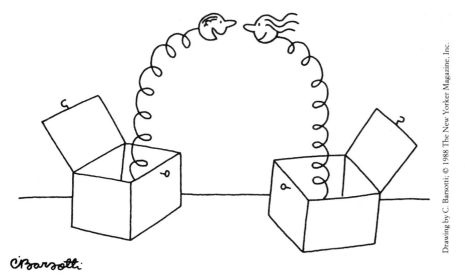

"What a nice surprise!"

we are attracted to partners whose needs complement our own. Sometimes this is taken to mean that, psychologically at least, opposites attract. Much of the popular literature on dysfunctional families is based on this premise: Alcoholics are attracted to those who would tolerate them (and vice versa); manipulators attract martyrs (and vice versa). While this idea makes intuitive sense to many of us, and while many Americans in various self-help programs would agree with it, social science researchers have found only *some* hard evidence to support complementary needs theory (McLeod 1995). Some "needs theorists" argue, on a more positive note, that we are attracted to others whose strengths are harmonious with our own so that we are more effective as a couple than either of us would be alone (Epstein, Evans, and Evans 1994).

Whatever the case regarding complementary needs, individuals gradually filter out those among their pool of eligibles who, they think, would not make the best spouse that they could find. Studies have consistently shown, for example, that people are willing to date a wider range of individuals than they would become engaged to or live with, and they are willing to live with a wider range of persons than they would marry (Schoen and Weinick 1993). Some social scientists have proposed that people go through a three-stage filtering sequence, called **SVR**—for **stimulus-values-roles** (Murstein 1986).

In the *stimulus stage,* interaction depends upon physical attraction. In the *values stage,* partners compare their individual values and determine whether these are appropriately matched. For instance, they might explore their views on marital fidelity, abortion, racism, the value of a college education, the environment, personal ambition, and work. If a couple finds a satisfactory degree of values consensus, they proceed to the final stage, that of exploring *role compatibility.* Here the prospective spouses test and negotiate how they will play their respective marital roles. It is possible, of course, that the courtship process will break down at either the values or the role compatibility stage; if it does not, however, the assumption is that the filtering process ends with one chosen marriage partner.

Relationships leading to marriage do not always show these rational characteristics, however. Nor, in fact, do they always exhibit mutual self-disclosure. For example, a person may choose a partner because of status or other benefits expected from the union. Or a person may continue a relationship with someone who discloses very little about her- or himself because it presents a challenge.

Some social scientists have wondered to what extent our courtship process has encouraged such patterns and has failed to facilitate the development of deeper, more intimate relationships. In the next section we discuss American **courtship,** the process through which the couple develops a mutual commitment to marry. We contrast getting together and cohabitation.

Courtship: Getting to Know Someone and Gaining Commitment

As romantic love has come to be associated with marriage and as parents no longer arrange their children's unions, responsibility for finding marital partners has fallen to individuals themselves. The courtship pattern that has evolved has two apparent purposes: (1) for romantic partners to try to get to know one another better, and (2) to gain each other's progressive commitment to marriage.

These two purposes can be at odds. On the one hand, courtship is supposed to lead to self-disclosure and intimacy. On the other hand, gaining a partner's commitment to marriage often involves marketing oneself in the best possible package. Before we look at the most frequent patterns of courtship in our society today, we will examine this potential contradiction.

Imaging Versus Intimacy

Whom shall I marry? That question seems obviously to be about my choice, about one of the most important controls I shall establish over my life. The more researchers probe that question, however, the more they find a secret question, more destructive, more insistent, that is asked as well: Am I the kind of person worthy of loving? This secret question is really about a person's dignity in the eyes of others, but it involves self-doubt of a peculiar kind. (Sennett and Cobb 1974, p. 63)

Many people fear not being worthy of love—a fear that is often associated with low self-esteem, as we saw in Chapter 4. As courtship progresses, these people may feel anxious, and in response they may avoid self-disclosure rather than develop it. They may "put their best foot forward," in a process called **imaging**—projecting and maintaining a façade as a way of holding the other person's interest. It is likely that everybody practices imaging to some degree.

Psychotherapists George Bach and Ronald Deutsch (1970, pp. 43–44) illustrate how this happens, using conversations between "Susan" and "Paul" (who have dated twice in the last two weeks). Here they are making themselves comfortable in Susan's apartment:

PAUL: Say, this is neat. (He looks around.) And I like that Van Gogh print. He's one of my favorite artists.

SUSAN: Is he? Mine too. I don't know just what it is. The color and vitality, I guess.

PAUL: Yes, that's what it is.

SUSAN: (She starts to put [CDs] on her [CD player].) What do you like, Paul? . . .

PAUL: Either one. (The music begins.) Why, I have that *same* [CD]! (He beams with the discovery.)

SUSAN: Really? . . . [I really like them.]

PAUL: Right. [Me too.]

Clearly, Paul and Susan are intent on selling themselves. Both are imaging, but neither seems to realize that the other is doing so. "To both, the constantly appearing bits of likeness seem amazing" (Bach and Deutsch 1970, p. 44). When they do discover important differences, the way they handle them is worth noting:

SUSAN: Well, I don't think I want any [children]. You know, with all the overpopulation and that. . . .

PAUL: No children at all? . . .

SUSAN: Well, I saw what it did to my mother, how dependent it made her, how—helpless, and I don't know. . . .

PAUL: The right man wouldn't let that happen. . . .

SUSAN: I guess it really is an experience any woman would want to have. . . . (Bach and Deutsch 1970, pp. 46–47)

Here a basic difference in values and a source of potential marital conflict have been smoothed over in a matter of minutes, in order that courting (which is intended to uncover such differences!) can proceed.

People are tempted to image in any courtship process. Social scientists agree, however, that some courting practices encourage more imaging than others. In the next few sections we will contrast three styles of courtship that are familiar to Americans today: the traditional ritual of dating, getting together, and cohabitation.

Dating

The dating system emerged in our society at the beginning of the twentieth century, prevailed through the 1950s and early 1960s, became less popular in the

The dating tradition that began in the early twentieth century has been ritualized in the high school prom. Some dating norms are changing, however, as more women share the expense of dating, as more couples and friends get together in groups, and as more individuals have less conventional ideas about appropriate partners. All these changes help create alternatives to more traditional dating practices.

late 1960s and early 1970s, and appears to have become popular again in the 1980s and 1990s.

Dating consists of an exclusive relationship developed between two people through a formal series of appointed meetings. Of course, dating can be just for fun! Dating relationships develop into marriage through a carefully orchestrated series of stages: going steady, informal engagement, formal engagement. Evolving, progressive commitment is expected, along with greater emphasis on sexual exclusivity.

A complex set of rules defines traditional dating. Males telephone females to ask them out. The male picks up the female and drives to a movie, party, drive-in, or out to dinner, where he pays. He might make sexual advances, which the female may rebuff.

A study of college students at Appalachian State University in Boone, North Carolina, found that this pattern still exists. Even though we caution against overgeneralizing from this study, the results were interesting. Responses by 130 male and female freshmen and sophomores indicated that male and female college students follow traditional gender roles in dating—the male is often expected to "pay," which he does with some resentment, and the female to "put out." These perceptions were reported for first dates, which seemed to be characterized by mistrust and uncertainty, with males and females falling back on traditional "games" but not very happy about it. The authors speculate that role playing, exploitation, and mistrust fade as dating (with the same person) continues. They also think some degree of traditional strategizing remains (Milano and Hall 1986).

There is evidence of change as well. Fifty-five percent of a sample of 400 college women reported paying for dates at least sometimes (Korman and Leslie 1982). A study comparing feminist and nonfeminist women found that the former, as expected, were more likely to share expenses on dates in both high school and college. But both sets of women perceived male sexual expectations in the same way: Sixty-two percent of the feminists and 58 percent of the nonfeminists believed that men expected women "to engage in more sexual activity on dates than the women really desired when the men paid" (Korman 1983, p. 579). Only 23 percent of each group reported fulfilling these expectations.

Although dating continues, the feminist movement and other social changes have transformed dating into a somewhat more egalitarian experience with some features of the pattern we call *getting together*. But these studies indicate considerable confusion, mistrust, and uncertainty about what rules and preferences are operating in a given situation. Such misunderstanding may contribute to date rape (Gibbs 1991a)—although we do not want to understate the force and coercion very often characterizing date rape. Colleges and other groups are beginning to deal with this serious hazard of dating (see Box 7.2, "Date Rape").

In 1949 anthropologist Margaret Mead severely criticized the dating pattern that had emerged in the United States. She perceived the process not as one in which two people genuinely try to get to know each

Date Rape

At its 1985 national convention, members of the national Pi Kappa Phi Fraternity unanimously adopted the following resolution:

Statement of Position on Sexual Abuse

WHEREAS we, the members of Pi Kappa Phi Fraternity, believe that the attitudes and behavior exhibited by members of the collegiate population have direct bearing on the quality of their present and future lives, and

WHEREAS there is an increased consciousness of sexual exploitation and violence and incidences thereof not just on the nation's college campuses but in society, and

WHEREAS the Greek community has stated its responsibility in leadership, scholarship, community service, human dignity and respect, and

WHEREAS Pi Kappa Phi is committed to excellence in the Greek community, and this requires us to identify and solve serious problems that prevent the growth and development of our brothers, and

WHEREAS Pi Kappa Phi strives to foster an atmosphere of healthy and proper attitudes and behavior towards sex and the sex roles, and wishes that the incidences of sexual abuse (mental and physical abuse—coercion, manipulation, harassment) between the men and women of the collegiate community be halted,

THEREFORE

BE IT RESOLVED that Pi Kappa Phi Fraternity will not tolerate or condone any form of sexually-abusive behavior (either physically, mentally or emotionally) on the part of any of its members, and

BE IT FURTHER RESOLVED that the Pi Kappa Phi Fraternity encourages educational programming involving social and

communication skills, interpersonal relationships, social problem awareness, etiquette and sex-role expectations; and will develop a reward system to recognize chapters and individuals that lead in fostering a healthy attitude towards the opposite sex.

In addition to the resolution, the fraternity produced the poster pictured here and distributed it to chapters across the country. The illustration is a detail from the print "The Rape of the Sabine Women." Beneath the large message a smaller one reads: "Just a reminder from Pi Kappa Phi. Against her will is against the law."

"We focused on the problem of date rape because it is a problem which needs a greater awareness among both males and females," the fraternity's executive director explained.

Date rape or acquaintance rape—being involved in a coercive sexual encounter with a date or other acquaintance—has emerged as an issue on college campuses over the past several years. Psychologists who deal with the problem agree that although we are only now acknowledging it, the phenomenon has plagued the dating scene for a long time—probably decades (Friedman, Boumil, and Taylor 1992).

In one extensive survey of 3,187 women and 2,972 men on thirty-two college campuses across the United States, 44 percent of women reported unwanted sexual contact subsequent to coercion, and 2 percent said this included intercourse. Nineteen percent of the men said they had obtained sexual contact through coercion, with 1 percent admitting to having obtained oral or anal penetration by force (Koss, Gidycz, and Wisniewski 1987, p. 166).

Most victims know their rapists (Warshaw 1995). In a recent study, half of the freshmen and sophomore college women interviewed reported unwanted attempts at intercourse by males of their acquaintance. These were usually (83 percent) men they knew at least moderately well. One-third of these attempts were accompanied by "strong" physical force, and another third by "mild" physical force. The women seemed constrained by traditional roles in their responses, which were largely passive and accepting; 37 percent did nothing. Only a minority gave a strong verbal (26 percent) or a physical response (14 percent). Half of the attacks succeeded; the stronger the victim's response, the less likely it was that the attempted rape was completed. None of the women reported the attack to the authorities and half talked to no one about it; the remainder told friends. Only 11 percent ended the relationship, whereas almost three-quarters either accepted or ignored the attack. Fifty percent continued to be friends (25 percent) or dating or sex partners (25 percent). Most blamed themselves at least partially (Murnen, Perot, and Byrne 1989).

According to Barry Burkhart, psychology professor at Auburn University, the phenomenon is prevalent because of "hidden norms" in society that condone sexual violence; that is, people have internalized gender roles according to which male aggression is acceptable. Even though rapists have typically been thought of as psychotics or criminals, Burkhart argues that research on acquaintance rape demonstrates that "normal" men are capable of rape. Other experts agree that date rape, rather than exposing aberrant behavior, exposes male students' commonly held attitudes toward women.

What does the idea that "normal" men are capable of date rape or acquaintance rape say about gender scripts today and whether, how, and how much they are changing? What does the response of the Pi Kappa Phi Fraternity to the date rape issue say about men? About gender roles today? How do men and women define rape? How can women learn to distinguish positive nurturing feelings toward others from a harmful sense of obligation to meet men's sexual needs regardless of their own feelings? Can social support encourage women to feel justified in saying no and being angry when faced with date rape?

Sources: Pi Kappa Phi Fraternity; Meyer 1984, pp. 1, 12; Koss, Gidycz, and Wisniewski 1987).

Relationships that are based on more casual forms of getting together, as opposed to formal dating, often allow for greater spontaneity and openness. If intimacy does develop, then individuals may already have a strong sense of each other's personality, moods, and values.

other but rather as a competitive game in which Americans, preoccupied with success, try to be the most popular and have the most dates. Sociologist Willard Waller (1937) had earlier termed this "rating and dating."

Mead saw at least two major problems in dating. First, it encourages men and women to define heterosexual relationships as situational, rather than ongoing. "You 'have a date,' you 'go out with a date,' you groan because 'there isn't a decent date in town'" (Mead 1949, p. 276). Because dating is formalized, women and men—even as they approach marriage—see each other only at appointed times and places. Partners look and behave their best during a date; they seldom share their "backstage behavior."[2]

2. From the interactionist theoretical perspective (Chapter 2), sociologist Erving Goffman (1959), in *The Presentation of Self in Everyday Life*, differentiates between people's front- and backstage behavior. Frontstage behavior is what we show the public; backstage behavior is more private. We can think of individuals' grooming rituals—shaving, doing their hair, applying makeup—as taking place backstage. They are preliminary preparations for meeting one's audience. Developing intimacy involves gradually allowing another person to see more of one's backstage behavior.

Second, sex becomes depersonalized and genitally oriented, rather than oriented to the whole person; that is, the salient question becomes whether a couple "went all the way"—or whether the male is "getting any"—rather than how much emotional and sensual rapport partners share.

The ways in which men and women get to know each other have changed somewhat in the forty years since Mead's insightful criticism. For one thing, an alternative script, which Libby (1976) termed "getting together," has emerged.

Getting Together

Getting together is a courtship process in which, unlike dating, groups of women and men congregate at a party or share an activity. Getting together deemphasizes relating solely to one member of the opposite sex. From childhood individuals are encouraged to develop their own interests and to share them with others, regardless of gender. These expectations continue through junior high, high school, and after. Multiple relationships are encouraged without emphasizing the dichotomy between sexual and nonsexual relationships. In getting together, females play a similar role to males, initiating relationships and suggesting activities. Women may pay; they may also either meet men at a mutually convenient spot or pick them up. Meetings are often less formal than in the traditional date.

These changes in how women and men relate to each other are associated with changing attitudes toward marriage itself. In one significant change the pattern of getting together is not as closely oriented to marriage as dating was in the 1950s. Remaining single, at least for a good part of one's 20s, is a more attractive alternative for many people today. As a result, people who are freed somewhat from the pressure to date and to marry can be more casual and spontaneous with each other. They are less likely to focus so intensely on their physical appearance and more likely to see each other in a variety of settings and moods (Murstein 1986, p. 67).

As a courtship process, getting together places less emphasis on the end result, marriage, than dating does. Ironically, this deemphasis may be effective, as it allows people to choose partners whom they really know and could be happily married to.

Next we consider a third process, *cohabitation:* living together without being married.

Power in Courtship

Chapter 10 examines power in marriages and families. Here we will look at power issues in courtship. Sociologists Naomi McCormick and Clinton Jessor (1983) comment in an essay that

> students sometimes balk when we suggest that nice people, not just sadists, use power during courtship. . . . [But] not everyone sees dating and mating in the same romantic light as these . . . students. . . . Power, the potential to influence another person's attitudes or behavior, may be an essential component of any romantic attraction or sexual relationship. (1983, pp. 66–67)

McCormick and Jessor point out that "the development of skills and knowledge, being perceived as attractive and likeable, and even acting helpless or 'needed' can all be used to influence someone else" (1983, p. 67).

These same sociologists researched, among other things, whether the rules for courtship have changed in recent years. What they conclude has some bearing on discussions of both sex roles and power. According to their research findings, the courtship game has changed in three ways:

> First, thanks to the weakening of the double standard and encouragement from feminists, women are freer to make the first move in a flirtation and to have premarital sex than in the past. Second, men seem to be encouraging women to be more assertive in initiating sexual relationships. Third, given the opportunity, men would reject sex and women would try to have sex with the same strategies that are characteristically used by the other gender.

> Despite these changes, the courtship game continues to follow gender-role stereotypes. Men ask women out more than vice versa. Men are more likely to influence a date to have sex; women are more likely to refuse sex. . . .

> As the women's liberation movement gains increasing acceptance, the courtship game will probably become less rigid. For instance, although women prefer masculine over feminine men, male college students are *not* more attracted to feminine women than they are to masculine women. . . . Even more indicative of social change, recent research contradicts earlier reports . . . that men are turned off by profeminist women. . . . As attitudes toward feminist women become more liberal, people may try out more egalitarian ways of dealing with courtship. However, such experimentation is likely to be minimal at first because out-of-role behavior is especially risky within sexual encounters where people already feel emotionally vulnerable. (1983, p. 85)

Although experimenting with egalitarian courtship practices is likely to be "minimal at first," some research on how flirtations and sexual encounters proceed suggests that, at least with beginning flirtations, "both genders have equal power"; that is, "each person takes a turn at influencing the partner and at signaling that the other's influence attempts are welcome" (McCormick and Jessor 1983, p. 76).

But such "equal power" may not last as courtship continues. In fact, popular magazines for young people, along with college counselors and others, have just begun to recognize that power struggles play a part in dating and courtship—sometimes to the point of violence.

COURTSHIP VIOLENCE   Box 7.2 discusses date rape. According to one sociologist, "it appears that violence is a common, albeit neglected, aspect of premarital heterosexual interaction" (Makepeace 1981, pp. 100–101). Due to stereotypes, courtship violence is especially neglected in research among certain ethnic groups, such as Asians for example, although it does exist (Yoshihama, Parekh, and Boyington 1991). In one study (Murphy 1988) of 485 male and female students at a university in the upper Midwest (these students were primarily from middle-income families in small towns and rural areas; hence, the study sample is not necessarily representative of all youth or even of all college youth), 40 percent reported personal involvement in an incident of courtship violence. Some of these incidents were pushes, shoves, and threats; 7 percent said they had been kicked, bitten, or hit with clenched fists. Dating violence tends to occur over jealousy, with refusal of sex, after excessive drinking of alcohol, or upon disagreement over drinking behavior (Makepeace 1981, 1988). Women are apt to report pushing or slapping a man, whereas men are more likely to beat up a date or threaten her with a weapon.

Researchers have found it surprising that about half of relationships continue after the violence rather than being broken off (Levy 1991). Given that the economic and social constraints of marriage are not usually applicable to courtship relationships, researchers have wondered about the reasons these relationships continue. Some evidence suggests that having been physically abused by one's father is highly related to men's—but not women's—verbal and physical abuse of dating partners (Alexander, Moore, and Alexander 1991). Then too, the study described next suggests some answers.

TRADITIONAL AND EGALITARIAN COURTSHIP PATTERNS   Two models of power coexist in dating relationships: the traditional and the egalitarian. Psychologist Letitia Peplau surveyed 231 college-age dating couples about their courtship behavior and attitudes. These students, a random sample from four colleges and universities in the Boston area, were asked about power in general and with respect to specific matters, such as recreation, conversation, sexual activity, amount of time spent together, and activities with other people (Peplau and Campbell 1989). Most of those questioned had *attitudes* favoring equality. Ninety-five percent of the women and 87 percent of the men agreed that "dating partners should have 'exactly equal say.'" But only 49 percent of the women and 42 percent of the men reported actually having equal power. (Forty-five percent of the men and 35 percent of the women reported more male power; 13 percent of the men and 17 percent of the women reported more female power.)

Many students pointed out that a simple "his power or her power" dichotomy did not reflect their real decision-making process. Instead, they told of taking turns, being sensitive to a partner's moods, or making concessions freely. A sense of trust of the partner was significant to feeling one had equal power in the relationship (Grauerholz 1987). Still, fewer than half the couples in the Peplau studies believed they had an equal power relationship, although most wanted to. Reasons for that, according to Peplau, appear to be, first, that many still had traditional sex role attitudes. Moreover, the "principle of least interest" (discussed in Chapter 10) was operating. Those who said they had alternatives to the relationship had greater power than their partners. Finally—and perhaps at the root of these students' failure to establish egalitarian relationships, according to Peplau—is the fact that few of them had egalitarian parental relationships as models.

Some of these dynamics probably operate to influence battered partners to maintain violent courtship relationships. Moreover, it may be true that at least some individuals in our society believe it is better to have a mate or a relationship, even a violent or unhappy one, than to have none at all.

Cohabitation and Marriage

We saw earlier that Margaret Mead was critical of the American tradition of dating. Mead proposed instead what she called the **two-stage marriage.** As the name suggests, it consists of two sequential types of marriage, each with a different license, different ceremonies, and different responsibilities. The first stage, called *individual marriage,* involves "serious commitment . . . in which each partner would have a deep and continuing concern for the happiness of the other" (Mead 1966, p. 50). It limits responsibilities, however, for the couple agrees not to have children during this time. The second stage, *parental marriage,* would follow only if a couple decides that they want to continue their relationship and to share the responsibility of children.

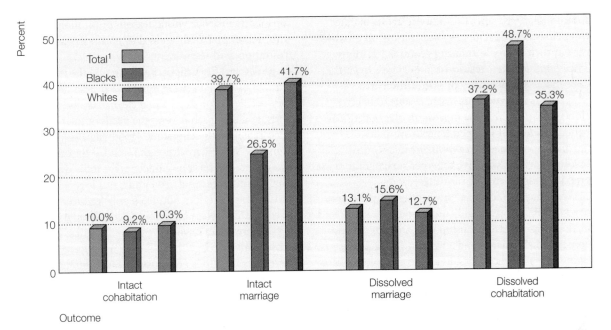

FIGURE 7.3

Percent distribution of first cohabitations of U.S. women ages 15–44, by outcome of cohabitation and race, 1988. (Source: London 1991, p. 3)

Mead's proposal is much like today's couples' decisions to live together to test and further develop their relationship before marrying. Cohabitation is also discussed in Chapter 6; this section addresses cohabitation as courtship.

A representative, nationwide survey by the National Center for Health Statistics sampled 8,450 non-institutionalized women ages 15–44 in 1988. Among other things, respondents were asked about their marital histories and about cohabiting unions relevant to those marriages (London 1991, pp. 1, 7). Overall results showed that slightly more than half (52.8 percent) of first cohabiting unions resulted in marriages (see Figure 7.3). Cohabitation was less likely to progress to marriage among black women than among white women. More than half (54.4 percent) of white women's first cohabiting unions resulted in marriage, compared with 42.1 percent of black women's; nearly half (48.7 percent) of black women's first cohabitations had dissolved, compared with a little over one-third (35.3 percent) of white women's (London 1991, pp. 2–3). Of those who said they had married

their first cohabiting partner, 13.1 percent said they had divorced—15.6 percent of blacks and 12.7 percent of whites.

Although cohabitation helps partners avoid the superficiality of dating, it does not alleviate all the problems of courting. When twenty years ago two researchers asked students their reasons for living together, they found that men's and women's goals differed. Men were more likely to indicate sexual gratification as their reason, whereas women more often hoped cohabitation would lead to marriage (Arafat and Yorburg 1973). We know of no more recent studies on this particular question, but we include these findings because we suspect that, despite gender role changes over the past two decades, this situation may not have changed significantly. In any case, when partners move in together without understanding each other's motives first, the result can be misunderstood intentions, recriminations and guilt that Margaret Mead hoped could be avoided.

In their review of research and clinical observation, Carl Ridley and his colleagues detected four common patterns of cohabitation: "Linus blanket," emancipation, convenience, and testing, all of which vary in their utility as training grounds for marriage. The "*Linus blanket*" relationship develops from the dependence or insecurity of one of the partners, who prefers a relationship with *anyone* to being alone. *Emancipation* as a motive involves using cohabitation to gain independence from parental values and influence. The *convenience* relationship involves partners who live together sexually more for practical reasons than for intimacy.

According to Ridley and his colleagues, the convenience relationship is highly traditional, involving a man who is in the relationship primarily for sex and domestic maintenance and a woman who supplies loving care and who hopes, but dares not ask, for matrimony. This form of cohabitation can be easily dismissed as exploitive and hence destructive, but Ridley and his associates claimed it can be a useful experience. Their argument was that the male gains in terms of preparation for marriage, not only learning something about domestic living but also increasing his interpersonal skills.

What does the female gain? She learns "the idea of reciprocity—mutual giving and getting in a relationship. She can learn that unconditional giving can have limited long-range payoff and that assessments of what one is giving and getting are important in certain contexts" (Ridley, Peterman, and Avery 1978, p. 132).

The *testing* mode of cohabitation is what we think of when we view cohabitation as trial marriage. Two partners, relatively mature and with a clear commitment to test their already satisfying intimate relationship in a situation more closely resembling the marriage they tentatively anticipate, move in together. If all goes well, they get married. If all doesn't go well, they separate, but because of their preparation and skills in communication they are able to understand and assimilate the experience. It becomes a beneficial experience in their development and contributes to the success of any marriage they ultimately formalize.

Ridley and his colleagues stressed the importance of clarifying motivations and goals for cohabiting before the step is taken. They stated that cohabitation could be an experience of some value as a preparation to marriage, either to test a particular relationship or to help the individual mature and become better able to sustain intimate relationships. On the other hand,

cohabitation for the wrong reasons or for individuals who are not prepared can lead to misunderstanding, frustration, and resentment.

Ridley and his associates also stressed the importance of individuals being fairly independent before they decide to cohabit, having clear goals and expectations, and being sensitive to the needs of their partners. Table 7.4 presents some "good" signs and some "concern" signs that, although intended for use by counselors, can serve as guidelines for individual decision making regarding cohabiting.

Cohabitation does not necessarily lead to marriage, of course. But when a cohabiting couple does decide to wed, how does having lived together affect their subsequent marriage? First, it does not appear that a couple's courtship is prolonged because the future marrieds choose to live together before their wedding. Studies comparing couples who cohabited with those who did not indicate that about the same length of time elapsed between their first date and marriage.

On the whole, research has found little difference in the marriages of those who cohabited before marrying and those who did not on marital adjustment generally or in terms of emotional closeness, satisfaction, conflict, equality, self-disclosure, commitment, marital satisfaction, and intimacy (Macklin 1983; Watson and DeMeo 1987).

But evidence accumulated in recent years suggests that, contrary to Margaret Mead's hopes, "trial marriage" may have a negative effect on marital success. A panel study (Booth and Johnson 1988) based on a national sample of married people interviewed in 1980 and again in 1983 found that cohabitation was negatively related to supportive marital interaction and was associated with marital disagreement and increased probability of divorce. No sex difference or effect of length of marriage was found in this study.

The researchers thought that what some cohabitants bring to marriage might explain the negative relationship between cohabitation and successful marriage. Drug, money, legal, and unemployment problems; risk taking; parental disapproval; and lesser commitment to marriage were more characteristic of cohabitants than noncohabitants. Still, much remained unexplained by the data, suggesting that further research might find the cohabitation process itself to be contributing to marital weakness.

In a subsequent study, Thomson and Colella (1992) used 1988 National Survey of Families and Households (NSFH) data to analyze the relationship

TABLE 7.4

Should You Live Together?

Counselor's Questions	"Good" Signs and "Concern" Signs
1. Could you talk a little about how each of you came to the decision to live together?	*Good signs:* Each partner has given considerable thought to the decision, including the advantages and disadvantages of living together. *Concern signs:* One or both partners have given little thought to the advantages and disadvantages of living together.
2. Perhaps each of you could discuss for a minute what you think you will get out of living together?	*Good signs:* Each individual is concerned about learning more about self and partner through intimate daily living. Both wish to obtain further information about each other's commitment to the relationship. *Concern signs:* One or both partners desire to live together for convenience only or to show independence from parents or peers.
3. Could each of you discuss what you see as your role and your partner's role in the relationship (e.g., responsibilities, expectations)?	*Good signs:* Each individual's expectations of self and partner are compatible with those of partner. *Concern signs:* One or both individuals have given little thought to the roles or expectations of self and/or partner. Individuals disagree in terms of their expectations.
4. Could each of you identify your partner's primary physical and emotional needs and the degree to which you believe that you are able to fulfill them?	*Good signs:* Each individual has a clear understanding of partner's needs and is motivated and able to meet most of them. *Concern signs:* One or both individuals are not fully aware of partner's needs. Individuals are not motivated or able to meet needs of partner.
5. Would each of you identify your primary physical and emotional needs in your relationship with your partner? To what degree have these needs been met in the past? To what extent are these needs likely to be met if the two of you were to live together?	*Good signs:* Each partner clearly understands his or her needs. Most of these needs are presently being met and are likely to continue to be met in a cohabiting relationship. *Concern signs:* One or both partners are not fully aware of their needs. Needs are not being met in the present relationship and are not likely to be met if the individuals live together.
6. Could each of you discuss what makes this relationship important to you? What are your feelings toward your partner?	*Good signs:* Partners care deeply for each other and view the relationship as a highly significant one. *Concern signs:* One or both individuals do not care deeply for their partner or do not view the relationship as a highly significant one. Partners have an emotional imbalance, with one partner more involved in the relationship than the other.
7. Could each of you explore briefly your previous dating experiences and what you have learned from them?	*Good signs:* Both individuals have had a rich dating history. Individuals have positive perceptions of self and opposite sex and are aware of what they learned from previous relationships.

(continued)

TABLE **7.4**

(continued)

Counselor's Questions	"Good" Signs and "Concern" Signs
	Concern signs: One or both partners have had minimal dating experience. Individuals have negative perceptions of self and/or of the opposite sex and do not seem aware of having learned from their previous relationships.
8. Perhaps each of you could talk for a minute about how your family and friends might react to the two of you living together?	*Good signs:* Each individual is aware of the potential repercussions from family and friends should they learn of the cohabiting relationship. Family and friends are supportive of the cohabiting relationship, or couple has considered how they will deal with opposition.
	Concern signs: One or both individuals are not fully aware of family and friends' possible reactions to their living together. Family and friends are not supportive of the cohabiting relationship.
9. Could each of you discuss your ability to openly and honestly share your feelings with your partner?	*Good signs:* Each individual is usually able to express feelings to partner without difficulty.
	Concern signs: One or both individuals have difficulty expressing feelings to partner or do not believe expressing feelings is important.
10. Could each of you discuss your partner's strengths and weaknesses? To what extent would you like to change your partner, relative to his or her strengths or weaknesses?	*Good signs:* Each individual is usually able to accept feelings of partner. Individuals are able to accept partner's strengths and weaknesses.
	Concern signs: One or both individuals are unable to understand and accept partner or have difficulty accepting partner's strengths and weaknesses.
11. How does each of you handle relationship problems when they occur? Can you give some examples of difficult problems you have had and how you have dealt with them?	*Good signs:* Both individuals express feelings openly and are able to understand and accept partner's point of view. Individuals are able to solve problems mutually.
	Concern signs: One or both partners have difficulty expressing feelings openly or accepting partner's point of view. Couple frequently avoids problems or fails to solve them mutually.

Source: Ridley, Peterman, and Avery, 1978, pp. 129–136.

between prior cohabiting and the likelihood of divorce among 714 couples in first marriages. Researchers classified these couples according to whether and for how long they had lived together before marrying. Respondents were also asked the following question: "It is always difficult to predict what will happen in a marriage, but realistically, what do you think the chances are that you and your husband/wife will eventually separate or divorce?" Response options included "very low," "low," "about even," "high," or "very high." Table 7.5 shows the results. Whereas 61.2 percent of those who had not cohabited said the likelihood of divorce was "very low," only 38.8 percent of those who had cohabited for two years or more said

TABLE 7.5

Perceived Likelihood of Divorce by Cohabitation Experience

	Did Not Cohabit	Months Cohabited			
		1–5	6–11	12–23	24+
Likelihood of divorce					
Very low	61.2%	50.9%	49.6%	36.0%	38.8%
Low	27.0	28.2	29.6	48.2	39.2
Even or higher	11.8	20.9	20.8	15.8	22.0
Valid cases	714	90	57	68	75
Percent of couples	71.8	9.0	5.7	6.4	7.0

NOTE: Respondents were couples in their first union and marriage, married less than ten years.

Source: Thomson and Colella 1992, pp. 135–36.

so. Generally, those who had cohabited were less satisfied with their marriages and less committed to the institution of marriage; their dissatisfaction increased with the length of time they had lived together before marrying. Wives who had cohabited had more individualistic views (as opposed to family-oriented views; see prologue) than those who had not.

Thomson and Colella (1992) did not disagree with explanations offered by Booth and Johnson (1988) but rather added to them. It is possible, they suggested, that the experience of cohabiting adversely affects subsequent marital quality and stability inasmuch as the experience actually weakens commitment because " 'successful' cohabitation (ending in marriage) demonstrates that reasonable alternatives to marriage exist" (Thomson and Colella 1992, p. 266). Put another way, experiencing cohabitation may lead to more individualistic attitudes and values (see also Axinn and Thornton 1992).

Another analysis of data from the same NSFH survey (Schoen 1992) compared first marriages of women who had married their first cohabiting partner with first marriages of women who had never cohabited. This analysis looked at effects for separate age categories and found that among the younger cohorts (women born between 1948 and 1957) the differential risk of divorce associated with cohabitation was insignificant. Because people born between 1948 and

1957 had the highest incidence of cohabitation, these results "suggest that the less select nature of those cohabiting played a significant role in lessening the risks of marital disruption" (Schoen 1992, p. 284). A still more recent analysis of the 1987–88 NSFH data found that only *serial cohabitation* (cohabiting) with a series of partners) was associated with higher subsequent divorce rates, at least during the first ten years of marriage (DeMaris and MacDonald 1993).

We have looked at dating, getting together, and cohabiting as courtship processes. Of course, in many real-life instances there are likely to be elements of each. A couple's courtship experience and their reasons for marrying make a difference in their long-term happiness. For this reason, when couples find themselves drifting into a default decision to marry—through dating, cohabiting or any other form of formal or informal courtship—they are wise if they stop, reconsider, and weigh the pros and cons. Box 7.3, "How to Tell When He or She Is Wrong for You," provides some guidelines. Making a conscious decision about marrying increases the chances of *not* marrying, however, and of letting go of a special relationship, and that can be a painful experience. Before evaluating qualities that make for a happy marriage, we'll look briefly at the experience of ending courtship by breaking up.

How to Tell When He or She Is Wrong for You

Sometimes, people pretend to themselves that sexual desire, the fear of being alone, or the hope for financial and other social advantages can be turned into feelings of love. Before marrying, people need to ask themselves the following questions about their partners:

1. *Does he or she have several close friends?* A person who has learned to enjoy and foster intimate friendships can put this talent to work in a marriage relationship. But if no one likes him or her well enough to be a close friend, shouldn't you wonder why *you* like him or her?

2. *Do you keep putting off introducing him or her to your friends and relatives? Does he or she put off introducing you to his or her friends and relatives?* Why? A hesitancy to show off a partner to those people who are most important to you may be a sign of uncertainty: Will the family and friends think it is a mistake?

3. *If the love relationship folded, would you still want to keep each other as friends?* For some lovers, this seems impossible. But the question here is whether the two people share enough respect and interests to want to be together even if no longer sexually intimate. Marriage involves companionship as well as sexual attraction.

4. *Do you spend most of your time trying to stay out of his or her bed?* A yes to this question can point up one of two problems: Either the chemistry just isn't there for you—and probably never will be—or your partner's emphasis on the importance of sex in an intimate relationship is considerably different from yours.

5. *Are you happy with the way he or she treats other people?* If he or she is condescending or rude—even physically violent—to others, you'll get the same treatment eventually. Watch how he or she deals with employees, food servers, maids, salesclerks, parking lot attendants, telephone operators, and so forth. Also, study his or her behavior with family members and close friends. If he or she doesn't treat them the way you want him or her to treat you, he or she is wrong for you. You may be an exception now, during courtship, but you won't be later.

6. *Do you know what he or she is like sexually?*

7. *Was your life stimulating and satisfying before you met her or him?* "Never, *never* bind yourself to someone because you need him or her to transform your unsatisfactory life into a super one."

8. *Do you often feel apprehensive about your future happiness together?* Little panics are normal, but they should be few and far between. If you are apprehensive more often than optimistic, this should serve as a warning signal.

9. *Are there taboo topics that you cannot discuss with each other?* Good relationships are built on trust, respect, spontaneity, and lack of stress. People are free to talk about almost anything. Even though they may hold very different views, not many subjects are taboo. Topics that have a bearing on the relationship are *never* taboo.

Source: "J" (Joan Garrity) 1977.

Breaking Up

Although breaking up is hard to do any time, break-ups before marriage are generally less stressful than divorce.

Research has suggested that men tend to fall in love more readily than women, and women tend to fall out of love more readily than men (Bradsher 1990). Social scientists suggest two explanations for women initiating breakups more often. First, a married woman's income and status are far more dependent on her husband than his are on her. Consequently, women must be especially discriminating, whereas men can afford to be more romantic. Second, women are more sensitive than men to the quality of interpersonal relationships. Hence, their standards for developing love may be higher than men's. A woman may experience lack of rapport or self-revelation in a relationship, for example, while the man does not. As a result, women may evaluate and reevaluate their relationships more carefully.

Whatever the end results may be, the act of breaking up can be an ordeal. Sociologists and counselors offers the following guidelines for ending a relationship:

1. Decide that terminating the relationship is what you really want to do.

2. Assuming you have definitely determined to break up, prepare yourself for wavering—but don't change your mind.

3. Plan the breakup discussion with your partner in person, but at a location from which you can readily withdraw.

4. Explain your reasons for breaking up in terms of your own values, rather than pointing out what you think is wrong with the other person.

5. Make the break final. Although some couples can break up and remain friends, hoping for a "let's be friends" relationship can be problematic. It's too easy for one or both partners to expect too much from the continued contact.

6. Seek out new relationships.

Items 5 and 6 are also good advice for those who've recently been broken up with.

Until this point in the chapter we have been discussing some of the events and motivations that bring a couple to marry or not to marry. The final section of this chapter will lay some of the groundwork for later chapters of this text by looking ahead to see how these factors influence the long-term happiness and stability of marriage.

Marriage: When and Why

A couple's happiness in marriage frequently depends on when and why they married. In this section we will look first at the relationship between marital stability and age at marriage. Then we will discuss some reasons for marrying that are less likely than others to lead to happiness.

The Stability of Early and Late Marriages

Statistics show that marriages are more likely to be stable when partners are in their 20s or older. Marriages that occur when women are over 30 may be slightly more stable than those that take place in their 20s, but the most significant distinction is between teenage and all other marriages (Norton and Moorman 1987).

Teenage marriages are twice as likely to end in divorce as marriages of those in their 20s (Norton and Moorman 1987). Social scientists generally maintain that people who marry young are less apt to be emotionally or psychologically prepared to select a mate or to perform marital roles. Low socioeconomic origins, coupled with school failure or lack of interest in school, are associated with early marriages. (Box 7.4, "A Closer Look at Family Diversity: Early Marriage in a Hmong Cohort," suggests that this overall conclusion may not apply to all ethnic groups, however.) Higher fertility and economic deprivation are also associated with early marriages (Hayes 1987; Teti and Lamb 1989).

Age itself is probably not the key variable in determining the likelihood of a marriage succeeding. Rather, it seems likely that one's age at marriage is associated with other elements contributing to marital instability, such as parental dissatisfactions accompanying precocious marriage, social and economic handicaps, lowered educational achievement (South 1995), premarital pregnancy or the female's attitude toward the pregnancy condition, courtship histories including length of acquaintance and engagement, personality characteristics, and the rapid onset of parental responsibilities. According to one study (Booth and Edwards 1985), however, marrying before age 20 remained a significant predictor of divorce even after all of these factors were statistically controlled for or taken into consideration. Respondents often mentioned sexual infidelity as a reason for conflict. As a result, the researchers hypothesized that teens who marry do so when their unmarried peers are experimenting sexually with more than one partner and before they are emotionally ready or willing to relinquish this behavior. In the words of the researchers:

It is striking that the role performance variable which best accounts for instability has to do with a lack of sexual exclusiveness. What is intriguing is

Early Marriage in a Hmong Cohort

The Hmong people make up a small minority of Asian Americans. Originally from Laos in southeast Asia, Hmongs entered the United States as political refugees after 1975; the majority have arrived since 1980. Hmongs have a high poverty rate: 64 percent live below the poverty line. About 44 percent of males and about 19 percent of females over age 25 have graduated from high school. This compares with the national average of about 75 percent for both males and females. About 7 percent of Hmong males and about 3 percent of females have a bachelor's degree or higher (compared with the national average of 23 percent for males and 18 percent for females). Hmong culture values early marriage for females and high fertility. The average family size among Hmongs is 6.6, compared with 3.2 overall ("We the American Asians" 1993; U.S. Bureau of the Census 1995, Table 73).

Noting that much of the research on Hmong families in the United States has uncritically applied a white, middle-class model and hence negatively stereotyped the Hmong, sociologists Ray Hutchison and Miles McNall (1994) set out to examine the relationship between early marriage and educational expectations and attainment among Hmong immigrants. The sociologists' data were collected as part of a longitudinal study of high school students in St. Paul, Minnesota, during the 1980s. Adolescents were surveyed annually from the ninth through the twelfth grades; bilingual questionnaires were given to the Hmong students.

Findings showed that Hmong adolescents are likely to marry and begin childbearing during high school. In the sample of 129 Hmong high school freshmen, three (6 percent of the Hmong females) were married. This proportion rose steadily. By the their senior year, 53 percent (26 of the 49 female Hmong respondents) said they were married. Eleven of these young women had two or more children; another ten had one child at home.

The researchers point out that this pattern of early marriage mirrors the cultural patterns of Hmong in Laos, where girls marry young, often to husbands several years older, and bear several children early. At the same time, the researchers emphasize that nearly half (47 percent) of the Hmong adolescent females in their sample were *not* married during high school— indicating a "substantial shift" from the stereotype that all Hmong girls" are married by age 15 or 16.

When the researchers measured the respondents on scholastic ability, educational expectations, or career aspirations, they found no noticeable differences between the married and single Hmong girls. Nor did they find significant differences on measures of psychological adjustment or stress. Furthermore, "the overwhelming majority of the married girls did not drop out of school" (p. 588).

While this is an admittedly small sample and looks only at Hmong adolescents in one geographic area, the researchers suggest that early marriage and motherhood among Hmong girls does not necessarily have a significant negative impact on school performance or even on expectations for continued education after high school graduation. Hutchison and McNall conclude that the general view that early marriage is an impediment to completing high school or further education may not apply to Hmong girls. Hmongs have a "remarkable network of family resources" to draw upon: Although they are wives and mothers, their primary child-care responsibilities are often borne by parents or siblings. "The cultural structures of Hmong society have been utilized to ensure that the younger generation will pursue the educational opportunities available to them in American society, thus enabling them to become economically independent in the second generation" (p. 588). Hutchison and McNall suggest that social scientists reevaluate the relationship between early marriage and educational achievement for other ethnic groups, such as Mexican Americans, who also continue to emphasize early marriage and large family size.

These Hmong immigrants in St. Paul, Minnesota, are celebrating the Hmong New Year, which also serves as a courting ritual. As in Laos, teenagers line up—boys on one side, girls on the other—and play catch with desirable potential mates. Catching the ball begins conversation. Tossing the ball gives girls a chance to meet boys under conditions approved by their parents. In Minnesota, however, the traditional Laotian black cloth ball is often replaced with a fluorescent tennis ball (Hopfesperger 1990, p. 1B). Because virtually all participants are Hmong, the ritual helps to ensure racial/ethnic homogamy.

that this perceived problem coincides with the peak in sexual interest, especially among males, hinting at the possibility that part of the instability experienced may have a biosocial origin. Perhaps individuals constrained to limit their sexual activity to a single individual at a time in their life when variety is important develop a pattern of acting out their impulses throughout much of their life. (Booth and Edwards 1985, p. 73)

Although one's age at marriage is only one factor contributing to marital stability or instability, age still

is associated with maturity, and sociologist David Knox (1975) has isolated four elements of maturity that he considers to be critical: emotional, economic, relationship, and value maturity.

EMOTIONAL MATURITY   The emotionally mature person has high self-esteem, which permits a greater degree of intimacy and interdependence in a relationship, as we saw in Chapter 4. Emotional maturity allows people to respond appropriately to situations. When conflict arises, emotionally mature people aim to resolve it, rather than becoming defensive or threatening to end the relationship.

ECONOMIC MATURITY   Economic maturity implies the ability to support oneself and a partner if necessary. Especially for teenagers who have had little formal training or other job preparation, economic problems can put heavy strains on a marriage. Without a decent wage, people's physical and emotional energy can be drained as they try to scrape together enough to live on. Developing a loving relationship under these conditions is extremely difficult.

RELATIONSHIP MATURITY   Relationship maturity involves the skill of communicating with a partner. People with this kind of maturity are able to (1) understand their partner's point of view, (2) make decisions about changing behavior a partner doesn't like, (3) explain their own points of view to their partner, and (4) ask for changes in their partner's behavior when they believe this is appropriate. Without the willingness and skills to understand each other and to make themselves understood, it is difficult or impossible for a couple to maintain intimacy.

VALUE MATURITY   Value maturity allows people to recognize and feel confident about their own personal values. By their mid-20s most people have developed a sense of their own values. A high school senior or a first-year college student, however, may still have a number of years of testing and experiencing before he or she reaches value maturity.

Age, then, is an important variable in determining a relationship's potential for success. We can measure age objectively; that is, statistics can tell us how age relates to marital stability. Other factors are more subjective; they have to do with the explanations people give for marrying a certain person. But these reasons are also associated with the success of a relationship.

Reasons for Marrying

The reasons people give for marrying are far more complex than "because we are in love." A combination of many complicated situations and needs motivates people to marry. As an extreme example, the media have pointed to rare instances in which resident aliens pay American citizens to marry them in order to obtain a Green card signifying permanent resident alien status. We'll look at several common reasons why people marry—first, those that are less likely to lead to a stable marriage, and second, more positive reasons—and see how each relates to the probability of a marriage's success.

PREMARITAL PREGNANCY   One problematic reason for marriage is premarital pregnancy. Recent research suggests that nomarital pregnancy today may actually thwart a woman's opportunities for later marriage, because some potential husbands may view her as stigmatized or as requiring undue financial support, and also because the time and activities of child care limit her activities in the marriage market (Bennett, Bloom, and Miller 1995). When premarital pregnancy does lead to marriage, research has identified a consistent relationship between premarital pregnancy and unhappiness in marriage (see, for example, Norton and Moorman 1987). Ironically, teenage women who marry to avoid single parenthood often become single parents after all. Research indicates that for both black and white couples, a premarital birth is most associated with subsequent marital breakup, followed by premarital pregnancy and no premarital pregnancy, in that order (Teachman 1983; Teti and Lamb 1989).

There are several reasons for this. First, the marriage is forced to occur at a time not planned. Often, pregnant brides are teenagers. At least two-thirds of all first births to teenagers were conceived out of wedlock. About 25 percent of teenage first births are legitimated by marriage before birth. Second, babies are expensive, and a couple not financially prepared for the costs can be overwhelmed. Third, "babies shatter goals": Teenage parents, for example, are less likely to attain educational goals, whether high school or beyond; and they are more likely to have lower incomes and occupational status, to have more children than they would like, and to go on welfare (Hayes 1987). Fourth, the in-law relationship may be marred if parents resent their child's marital partner for having brought about a marriage they did not want or viewed as too early. Finally, the couple may not ever have decided they were compatible enough for marriage and may resent each other, either overtly or subconsciously, during the marriage. Of course, other options—abortion, giving the child up for adoption, or raising the child as a single parent—also have an impact on the lives of the young couple. And values and emotions may vary widely among individuals. Still, it is difficult to be encouraging about young, pregnancy-inspired marriages, and some churches have been reluctant to sanction them unless the young couple is unusually mature.

There are certainly exceptions to this scenario. Factors that reduce the damaging effects of pregnancy-inspired or teenage marriage include having a supportive family (giving both financial and emotional support), being able to remain in school and then become steadily employed, controlling further fertility, and being older when the pregnancy occurs (Furstenberg, Brooks-Gunn, and Morgan 1987).

REBOUND People may tend to fall in love more easily when they're on the rebound. Sociologist Elaine Walster (1965) experimented with the concept of love on the rebound. She concluded that when people have low self-esteem, due to having been broken up with, they may be less discriminating in choosing love partners and may fall in love more easily.

Marriage on the rebound occurs when a person marries very shortly after breaking up in another relationship. To marry on the rebound is undesirable because the wedding occurs as a reaction to a previous partner, rather than being based on real love for the new partner.

REBELLION Marriage for the sake of rebellion occurs when young people marry primarily because their parents disapprove. Social-psychological theory and research show that parental interference can increase feelings of romantic attraction between partners (Katz and Liu 1988); this has been called the *Romeo and Juliet effect*. As with marriage on the rebound, the wedding is a response to someone else (one's parents) rather than to one's partner.

ESCAPE Some people marry to escape an unhappy home situation. The working-class male who hasn't gone to college, for instance, may reason that getting married is the one way he can keep for himself any money he makes instead of handing it over to his parents. Or, denied the opportunity to go away to college, working-class youths may use marriage as an escape from parental authority (Katz and Liu 1988).

PHYSICAL APPEARANCE Marrying solely because of the physical attractiveness of one's partner seldom leads to lifelong happiness. For one thing, beauty is in the eye of the beholder, and if the beholder finds he or she really doesn't like the partner, that beauty is certain

to diminish. Second, the physical beauty of youth changes as partners age. The person who married for beauty often feels she or he has been cheated. After a time, there is less to be attracted to.

LONELINESS Sometimes people, especially older adults, marry because they don't want to grow old alone. Marrying is not always the solution, for people can be lonely within marriage if the relationship isn't a strong one. In other words, it is the relationship rather than the institution that banishes loneliness.

PITY AND OBLIGATION Some partners marry because one of them feels guilty about terminating a relationship: A sense of pity or obligation substitutes for love. Sometimes this pity or obligation takes the form of marrying in order to help or to change a partner, as when a woman marries a man because she believes that her loyal devotion and encouragement will help him quit drinking and live up to his potential. Such marriages don't often work: The helper finds that his or her partner won't change so easily, and the pitied partner comes to resent being the object of a crusade.

SOCIAL PRESSURE Parents, peers, and society in general all put pressure on singles to marry. Approval (or disapproval) of a partner from family and friends may be important in a relationship's stability, especially for women (Cate and Lloyd 1992, pp. 79–80). The expectations built up during courtship exert a great deal of social pressure to go through with the marriage. As engagements are announced or as people become increasingly identified as a couple by friends and family, it becomes more difficult to back out. Still, breaking an engagement or a less formal commitment is probably less stressful than divorcing later or living together unhappily. About 100,000 couples decide to break their engagements each year (Bradsher 1990).

ECONOMIC ADVANCEMENT Marrying for economic advancement occurs in all social classes. Young divorced mothers may consider remarriage primarily because they are exhausted from the struggle of supporting and caring for their small children; and working single women often associate marrying with the freedom to stay home at least part of the time. Men, too, can marry for reasons of economic advancement. This can be especially true in some professions in which social connections provide important business ties.

Is marrying for economic advancement the right reason? The answer depends on the individuals. Certainly the thrust of this book is to encourage intimacy and a strong emotional relationship as a basis for marriage. A person going into a marriage mainly for economic reasons should be very honest with her or his partner, so that both know what the marriage means to the other.

MORE POSITIVE REASONS FOR MARRYING

We have seen that rebounding, rebellion, escape, physical appearance, loneliness, obligation, and social pressure are all unlikely bases for a happy marriage. What are some positive reasons? Knox (1975) lists three: companionship, emotional security, and a desire to parent and raise children.

Marriage is a socially approved union for developing closeness with another human being. In this environment, legitimate needs for companionship—to love and to be loved by someone else—can be satisfied. Marrying for emotional security implies that a person seeks the stable structure of marrying to help ensure the maintenance of a close interpersonal relationship over time. Although most people do not marry only to have children (and this alone may *not* be a positive reason for marrying), many regard children as a valuable part of married life. "The benefits of love, sex, companionship, emotional security, and children can be enjoyed without marriage. But marriage provides the social approval and structure for experiencing these phenomena with the same person over time" (Knox 1975, p. 143).

In Chapter 8 we will examine marriage as an institution and an intimate relationship. It seems to us that the most positive motivation toward marriage involves the goal of making permanent the relationships of love and intimacy, discussed in Chapter 4. This theme will be repeated throughout this section of the text.

In Sum

The American association of love and marriage is unique to our modern culture. Historically, marriages were often arranged in the marriage market, as business deals. Many elements of the basic exchange (a man's providing financial support in exchange for the woman's childbearing and child-rearing capabilities, domestic services, and sexual availability) remain.

What attracts people to each other? Two important factors are homogamy and physical attractiveness. Some elements of homogamy are propinquity, social pressure, feeling at home with each other, and the fair exchange. Three patterns of courtship familiar in our society are dating, getting together, and cohabitation.

Besides homogamy and the degree of intimacy developed during courtship, two other factors related to the success of a marriage are a couple's age at marriage and their reasons for marrying. People who marry too young are less likely to stay married; and there are several negative reasons for marrying that can lead to unhappiness or divorce.

If potential marriage unhappiness can be anticipated, breaking up before marriage is by far the best course of action, however difficult it seems at the time. A certain number of courting relationships will end in this fashion.

But many couples will go on to marry. Chapter 8 describes what form the marriage is likely to take and some choices the couple makes in setting up their marriage.

Key Terms

cohabitation	hypergamy
convenience	imaging
courtly love	interracial marriage
courtship	"Linus blanket"
date rape (acquaintance rape)	marriage gradient
	marriage market
dating	pool of eligibles
dowry	sex ratio
emancipation	stimulus-values-roles (SVR)
endogamy	theory of courtship
exchange theory	testing
exogamy	theory of complementary
getting together	needs
heterogamy	two-stage marriage
homogamy	

Study Questions

1. Explain how our notion of romantic love came about.

2. How are modern marriages similar to arranged marriages? How are they different?

3. Compare the way we choose marriage partners with the process at a marketplace. Why do women in the marriage market tend to be more serious shoppers than men?

4. Give four reasons people are likely to be homogamous. What difficulties are people in heterogamous relationships likely to face?

5. Why do you think homogamous marriages are more stable than heterogamous marriages? Does this necessarily mean that homogamous marriages are more successful? Why or why not?

6. Explain why the two aspects of courtship—getting to know each other better and gaining commitment to marriage—are potentially contradictory.

7. Why did Margaret Mead criticize the dating pattern that had emerged in the United States? What problems did she think were caused by dating? Do we still face these problems today?

8. Differentiate the four common patterns of cohabitation put forth by Ridley and his associates. Which are more useful as training grounds for marriage? Why?

9. If possible, talk to a few married couples you know who lived together before marrying, and ask them how their cohabiting experience influenced their transition to marriage. How do their answers compare with survey findings presented in this chapter?

10. Do you agree that age itself is not the key variable in determining a marriage's likelihood of succeeding? Describe what *you* think to be critical.

11. Discuss some problematic and more positive reasons for marrying, and point out how each relates to the probability of a marriage's success.

Suggested Readings

Bach, George R. and Ronald M. Deutsch. 1970. *Pairing.* New York: Avon. Psychoanalyst Bach founded the Institute of Group Psychotherapy in Beverly Hills, California. This interesting and readable (if old) paperback tells something of his work there and explores intimate pairing as an alternative to traditional courtship in which partners always put their best foot forward in an effort to sell themselves.

Beal, Edward and Gloria Hochman. 1991. *Adult Children of Divorce: Breaking the Cycle and Finding Fulfillment in Love, Marriage, and Family.* New York: Delacorte. Beal is a psychiatrist whose practice has brought him enough adult children of divorced parents who are afraid to marry that he's written a book about getting over the fear and tendency to isolate.

Buss, David M. 1994. *The Evolution of Desire.* New York: Basic Books. Analysis of human courtship and mating from an evolutionary psychology, or biosociological, perspective.

Cate, Rodney M. and Sally A. Lloyd. 1992. *Courtship.* Newbury Park, CA: Sage. This little and readable book, one in the Sage series on close relationships, reviews the social scientific literature on courtship.

Crohn, John. 1995. *Mixed Matches: How to Create Successful Interracial, Interethnic, and Interfaith Relationships.* New York: Fawcett Columbine. Crohn is a psychologist and therapist who gives good advice based on his clinical experience.

Cromwell, Al. 1995. *I'd Rather Be Married: Finding Your Future Spouse.* Oakland, CA: New Harbinger. Today's latest how-to book on this subject by a counseling psychologist. His overall advice: "What you see is what you get."

Cutter, Rebecca. 1994. *When Opposites Attract: Right Brain/Left Brain Relationships and How to Make Them Work.* New York: Dutton. A new version of complementary needs theory—but mainly a how-to book on getting along with someone psychologically different from you. There is a chapter on "Selecting a Life Partner" from this context.

Kaplan, Marion A., ed. 1985. *The Marriage Bargain: Women and Dowries in European History.* New York: Harrington Park. This collection of five essays by social and family historians explores the relationships between dowries and economic conditions, and between dowries and women's rights and position in society. It offers significant insights into the historical functions of marriage.

Luker, Kristin. 1975. *Taking Chances: Abortion and the Decision Not to Contracept.* Berkeley and Los Angeles: University of California Press. This book examines reasons for women's willingness to take contraceptive risks. Among other things, it explores the perceived benefits of pregnancy and sees women's willing to chance pregnancy outside marriage as one result of their diminishing currency in today's marriage market.

Murstein, Bernard. 1986. *Paths to Marriage.* Beverly Hills, CA: Sage. Well-written social science text on dating, courtship development, love, and theories of marital choice.

Rothman, Ellen K. 1984. *Hands and Hearths: A History of Courtship in America.* New York: Basic. A historical account of American courtship.

Stoop, David and Jan Stoop. 1993. *The Intimacy Factor.* Nashville: Thomas Nelson. Proceeding from the complementary needs school, these authors apply principles from the Meyers-Briggs personality test (extrovert/introvert; sensing/intuitive, thinking/feeling, and so forth) to the process of choosing a lifelong mate.

Marriage: A Unique Relationship

> *Marriage is not an answer, but a search, a process, a search for life, just as dialogue is a search for truth.*
>
> SIDNEY JOURARD

Being married is a unique relationship. Despite wide variations, all marriages have an important element in common: the commitment that partners have made—and made publicly—to each other. This is the subject of this chapter. In the discussions that follow we'll look especially at how getting married announces a personal life-course decision. In keeping with one of the themes of this text, we'll discuss how marrying is a decision-making trade-off. Brides and grooms discard prior alternatives and are no longer available in the marriage market. This is one expectation inherent in the *marriage premise,* described in this chapter. We'll look at how relationships throughout marriage reflect spouses' personal choices, either made actively or by default, for the kind of relationship a couple shares is a product of the values, expectations, and efforts partners invest in it. One way to design and retain the kind of relationship spouses want, as we will see, is to write a personal marriage agreement. To begin, we explore the relationship between marriage and kinship.

Marriage and Kinship

Who are your kin? Anthropologists have defined *kinship* as the social organization of the entire family, including blood, or **consanguineous** relatives, and **conjugal** relationships acquired through marriage. (The word *consanguineous* comes from the Latin prefix *com,* which means "joint," and the Latin word *sanguineus,* which means "of blood." The word *conjugal* comes from the Latin word *conjugere,* which means "to join together.") Parents and grandparents are consanguineous relations; spouses and in-laws are conjugal relatives; aunts and uncles may be either. Certain rights and obligations accompany one's kinship status. For example, you may expect your grown sister or brother to attend your wedding or graduation.

As you may have already noticed, the concept of kinship is closely related to that of *extended family.* An extended family includes parents and children, along with other relatives, such as in-laws, grandparents, aunts and uncles, and cousins. Some groups, such as African Americans, Hispanics, and gay male and lesbian families, also have "fictive" or "virtual" kin (a *compadrazo* in Hispanic families)—friends who are so close that they are hardly distinguished from actual relatives.

Meanwhile, it may be the case that every society has a **dominant dyad**—a centrally important twosome, or dyad, that symbolizes the culture's basic values and kinship obligations (Hsu 1971). At least among white, middle-class Americans, the husband–wife dyad is expected to take precedence over any others. According to sociologist Talcott Parsons (1943), the American kinship system is generally not based on vital extended family ties. Parsons writes that kinship in the United States is instead comprised of "interlocking conjugal families" in which married people are common members of their **family of orientation** (the family they grew up in) and of their **family of procreation** (the one formed by marrying and having children). Parsons sees the husband–wife bond and the resulting family of procreation as the most meaningful "inner circle" of Americans' kin relations, surrounded by decreasingly important outer circles. Parsons points out, however, that his model characterizes mainly the American middle class; recent immigrants and lower socioeconomic classes still rely on meaningful ties to their extended kin.

In the majority of non-European countries, the extended family (as opposed to the married couple or nuclear family) is the basic family unit (Murdock 1949). Although extended families have declined in significance for white middle-class Americans, they have continued to be especially important for various European ethnic families, such as Italians, and for African American, Hispanic, Asian American, and Native American families as well.

In the 1960s, for instance, sociologist Herbert Gans studied blue-collar Italian Americans in Boston. He concluded that their marital relationships were "qualitatively different" from the middle-class norm, with less emotion or communication. A "sexual barrier" existed: When one partner had troubles, that person was most likely to seek support from his or her same-sex friends in the extended family or from sib-

| "The Dinner Quilt," Faith Ringgold, 1986.

lings, not from the other partner (Gans 1982 [1962]). The dominant dyad was seldom the married couple.

Similarly, Hispanics today do not necessarily subscribe to the middle-class norm of the primary conjugal bond; **la familia** ("the family") means the extended as well as the nuclear family. Like the Italians Gans studied forty years ago, many Mexican Americans and other Hispanics live in comparatively large, reciprocally supportive kinship networks. For example, most Puerto Rican families live in "ethnically specific enclaves" (largely in New York City) and rely more on extended, consanguineous kin than on conjugal ties (Wilkinson 1993). Puerto Ricans are similar to working-class African Americans in this way.

Recent Asian immigrants are also likely to emphasize extended kin ties over the marriage relationship.

Among Chinese and Japanese Americans, *hsiao* defines the dominant dyad—that of adult child (especially son) and aging parent (especially father). *Hsiao* requires that an adult child provide aid and affection to parents even when this might conflict with marital obligations (Lin and Liu 1993).

In the following, a Vietnamese refugee describes his reaction to U.S. housing patterns, which reflect nuclear family norms and husband–wife as the dominant dyad:

> Before I left Vietnam, three generations lived together in the same group. My mom, my family including wife and seven children, my elder brother, his wife and three children, my little brother and two sisters—we live in a big house. So when we

came here we are thinking of being united in one place. But there is no way. However, we try to live as close as possible. (in Gold 1993, p. 303)

American housing architecture is similarly discouraging to many Muslim families—from India, Pakistan, or Bangladesh, for example—who would prefer to live in extended households (Nanji 1993).

Kinship Obligations and Marriage Relationships

Thirty years ago, family sociologist Jesse Bernard noted what she called a **parallel relationship pattern** among spouses in the working class; she distinguished this pattern from the **interactional pattern** of middle-class marriages. In a parallel relationship, the husband was expected mainly to be a hard-working provider and the wife, a good housekeeper and cook. "Companionship in the sense of exchange of ideas or opinions or the enhancement of personality by verbal play or conversation is not considered a basic component in this pattern" (Bernard 1964, p. 687). In the interactional pattern, partners expect companionship and intimacy as well as more practical benefits.

Depending on individual preferences and socioeconomic backgrounds, American spouses choose to emphasize either practical or emotional benefits of marriage. The more practical style has been called **utilitarian marriage;** the more emotional, **intrinsic marriage** (Cuber and Harroff 1965). In intrinsic unions, the husband–wife relationship is "interactional" and an end in itself. Intrinsic marriages rest on intimacy and mutual affection between partners. Mates strive to fulfill as much as possible of each other's personal needs for sexual expression and companionship. In utilitarian marriages, the emphasis is not necessarily on the husband–wife dyad as centrally important; probably the relationship is "parallel." Marriage may be a means to other ends, such as satisfying one's parents' wishes according to *hsiao* or uniting two distinct kinship networks of similar ethnicity. As they become upwardly mobile and/or remain for two or three generations in the United States, however, recent immigrants to some extent gradually assume (that is, become acculturated to) non-Hispanic white middle-class norms that more highly value intrinsic marriages (Chilman 1993; Wilkinson 1993). We should not forget, meanwhile, that even for fifth- or sixth-generation non-Hispanic white Americans, utilitarian marriage

may provide material luxuries or career advancement or simply basic economic security.

Moreover, in real life very few marriages are either completely utilitarian or completely intrinsic. Rather, these two types represent polar opposites of an imaginary line or continuum. Real-life marriages fall somewhere on the continuum, combining elements of pragmatism and emotional sharing in various degrees. Between the two extremes is an almost limitless variety of types of marriage relationships. With an appreciation for both utilitarian and intrinsic unions, we turn now to some characteristic relationship patterns in American marriages.

Five Marriage Relationships

We can't look at every possible kind of relationship, but we can look at a few representative types based on the research of Cuber and Harroff, who conducted extensive interviews with upper middle-class married couples. From their interviews with 107 men and 104 women, Cuber and Harroff (1965) classified five kinds of marital relationships that act as useful prototypes. Two are closer to the utilitarian pole on the continuum; two are intrinsic; and one (conflict-habituated) is difficult to define as either utilitarian or intrinsic.

Although this classic study was done in the 1960s, the interactional qualities of marriage, which are its focus, are not so timebound. This research continues to be useful in the study of marriage and other long-term relationships.

In identifying five types of marriage—conflict-habituated, devitalized, passive-congenial, vital, and total—Cuber and Harroff emphasize that their findings represent "different kinds of adjustment and conceptions of marriage, rather than degrees of marital happiness" (1965, p. 61). In other words, couples living in any of these relationships might or might not be satisfied with them. The relationships differ according to how spouses feel about their marriages. Some researchers also have found these relationship types among cohabitants (Clatworthy 1975, p. 77ff).

Conflict–Habituated Marriages

Couples with a **conflict-habituated marriage** experience considerable tension and unresolved conflict. Spouses habitually quarrel, nag, and bring up the past.

As a rule, both spouses acknowledge their incompatibility and recognize the atmosphere of tension as normal.

Conflict-habituated relationships differ from those in which conflicts arise over specific issues. In conflict-habituated relationships, the subject of the argument hardly seems important, and partners generally do not resolve, or expect to resolve, their differences. "Of course we don't settle any of the issues," said a veteran of twenty-five years of this type of marriage. "It's sort of a matter of principle *not* to. Because somebody would have to give in then and lose face for the next encounter" (Cuber and Harroff 1965, p. 45).

These relationships do not necessarily end with divorce. In fact, it has been suggested by some psychiatrists that for some partners this kind of marriage fulfills a need for conflict. In this sense, conflict-habituated relationships are intrinsic relationships, for they fulfill partners' emotional needs. Such relationships cannot be called intimate, however, for they are not based on mutual acceptance and support or on honest self-disclosure.

Devitalized Marriages

Partners in a **devitalized marriage** have typically been married several years, and over the course of time the relationship has lost its original zest, intimacy, and meaning. Once deeply in love, they recall spending a great deal of time together, enjoying sex, and having a close emotional relationship. Their present situation is in sharp contrast: They spend little time together, enjoy sex less, and no longer share many interests and activities. Most of their time together is "duty time" spent entertaining, planning and sharing activities with their children, and participating in community responsibilities and functions. Once an intrinsic union, the marriage has become utilitarian.

Cuber and Harroff found devitalized marriages to be exceedingly common among their respondents. They also found several reactions among devitalized partners. Some were accepting and tried to be "mature about it." As one wife explained, "There's a cycle to life. There are things you do in high school. And different things you do in college. Then you're a young adult. And then you're middle-aged. That's where we are now" (in Cuber and Harroff 1965, pp. 47–48). Others were resentful and bitter; and still others were ambivalent. From his study of a group of fifty marriages, psychoanalyst George Bach provides this description:

> Essentially they had resigned themselves to . . . ritualized routines. . . . They were mutual protective associations who looked on their marriage as pretty much of a lost cause but felt it would be disloyal and ill-mannered to complain about it, especially since the loneliness of being unmarried would probably be worse. (Bach and Wyden 1970, pp. 47–48)

Emotional emptiness does not necessarily threaten the stability of a marriage. Many people, like the accepting wife just quoted, believe that the devitalized mode is appropriate for spouses who have been married several years. Devitalized partners frequently compare their relationship with others who have similar relationships, concluding that "marriage is like this— except for a few oddballs or pretenders who claim otherwise" (Cuber and Harroff 1965, p. 50).

Passive-Congenial Marriages

Partners in a **passive-congenial marriage** (utilitarian), like devitalized partners, emphasize qualities other than emotional closeness. These qualities may be different for different groups. Upper middle-class couples tend to emphasize civic and professional responsibilities and the importance of property, children, and reputation; working-class people focus on their need for economic security, the benefits of the basic exchange, and their hopes for their children.

Unlike the devitalized marriage, passive-congenial partners never expected marriage to encompass emotional intensity. Instead, they stress the sensibility of their decision to marry. There is little conflict, but that does not mean there are no unspoken frustrations. And while there is little intimacy, these unions fulfill partners' needs for more casual companionships.

Passive-congenial marriages are less likely to end in divorce than are unions in which partners have unrealistic expectations of emotional intensity. Partners may decide to terminate passive-congenial unions, however, if they feel the marriage is not adequately fulfilling their more practical needs such as for economic support or professional advancement. Or one partner may discover that he or she wants greater intimacy from a relationship or may inadvertently fall in love with someone else.

Vital Marriages

In a **vital marriage,** a type of intrinsic marriage, being together and sharing are intensely enjoyable and important. As one husband said:

> The things we do together aren't fun intrinsically—the ecstasy comes from being *together in the doing.* Take her out of the picture and I wouldn't give a damn for the boat, the lake, or any of the fun that goes on out there. (Cuber and Harroff 1965, p. 55)

This statement should not lead you to believe that vital partners lose their separate identities, or that conflict does not occur in vital marriages. There is conflict, but it is more apt to center on real issues rather than the "who said what first and when" and "I can't forget when you . . ." that characterize conflict-habituated marriages. And vital partners try to settle disagreements quickly so they can resume the relationship that means so much to both.

Typically, vital partners consider sex important and pleasurable. "It seems to be getting *better* all the time," said one spouse, age 55. Instead of being dutifully performed as a ritual, sexuality pervaded vital partners' whole lives. A grandmother spoke about it this way:

> You can't draw the line between being in bed together and just being alive together. You can touch tenderly when you pass; you wait for the intimate touch in the morning. Even the scents you make in bed together are cherished. (in Cuber and Harroff 1965, pp. 135–36)

Enduring vital marriages may be a minority. Those in Cuber and Harroff's research often reported feeling that their lifestyles were neither experienced nor understood by their associates. Even rarer than vital marriages are total marriages.

Total Marriages

Total marriages are also intrinsic. They are like vital marriages, only more multifaceted. "The points of vital meshing are more numerous—in some cases all of the important life foci are vitally shared" (Cuber and Harroff 1965, p. 58).

Spouses may share work life (similar jobs, same employer, or projects such as writing a book, making a film, or running a family business) and friends and leisure activities, as well as home life. They may organize their lives to make it possible to be alone together

for long periods. Both vital and total marriages are emotionally intense, but the total marriage is more all-encompassing, whereas the vital marriage leaves areas of individual activity to each partner. Total marriages are rare, but they do exist and can endure.

On the negative side, total marriages are vulnerable to rapid disintegration if marital quality changes. They foster a mutual dependency that makes it difficult for the remaining partner to adjust in case of death or divorce. On the positive side, they provide a wide range of fulfillment focused on the couple as a unit.

This discussion points up three important facts. First, marriage is different from other sexual relationships, even cohabiting. Second, partners expect marriage to offer certain advantages, such as companionship, psychic and sexual intimacy, and emotional support. Third, marriage involves more responsibilities than do other sexual relationships.

The Marriage Premise

By getting married, partners accept the responsibility to keep each other primary in their lives and to work hard to ensure that their relationship continues. Essentially, this is the **marriage premise.** We can look more closely at the two important elements of this definition: permanence and primariness.

Expectations of Permanence

A wedding is a community event marking a bride and groom's passage into adult family roles. Whether spouses love each other has only recently been considered important, as we saw in Chapter 7. And with few exceptions, marriages were always thought of as permanent, lifelong commitments.

Although it is statistically less permanent now than it has ever been in our society, marriage, more than any other relationship, holds the hope for permanence. The marriage contract remains a legal contract between two people that cannot be broken without permission of society or the state. Most religions urge permanence in marriage. In our society, couples usually vow publicly to stay together "until death do us part" or "so long as we both shall live." People enter marriage expecting—hoping—that mutual affection will be lasting.

Expectations for permanence derive from the fact that, historically, marriage has been a practical social institution. Economic security and responsible child

In total marriages, couples often work together, as does this couple, and share all important areas of life.

rearing required marriages to be permanent, and even at the turn of the twentieth century parents only occasionally outlived the departure of their last child from home.

Today, as we've seen in earlier chapters, marriage is somewhat less important for economic security. However, another function has become more important for most people. They expect marriage to provide emotional support and love. It might be said that we are coming closer to realizing a prediction, made by social theorist Herbert Spencer in 1876, that some day "union by law" would no longer be the essential part of marriage. Rather, "union by affection will be held of primary moment" (Spencer, H., 1876, cited in Burgess and Locke 1953, p. 29).

Today, marriages of intimacy are usually held together by more than mutual affection, and that is why marriages remain more permanent than do other intimate relationships. If there were no legal contract, marriage would be no different from living together.

Today's marriages continue to be bolstered by *mores* (that is, by strongly held social norms), public opinion, and the law, although less so than in the past. Tradition, religious beliefs, and social pressure from family and friends all encourage couples to spend considerable time and energy working to improve their relationship before they contemplate divorce. As a result, marriages, more than other intimate relationships today, last for life.

As far back as 1949, Margaret Mead pointed out that "all the phraseology, the expectation of marriage that would last, 'until death do us part,' has survived long after most states have adopted laws permitting cheap and quick divorces" (1949, p. 334). The result is "great contradictoriness" in American culture. People "are still encouraged to marry as if they could count on marriage's being for life, and at the same time they are absorbing a knowledge of the great frequency of divorce" (Mead 1949, p. 335). The recognition of this cultural contradiction is a step toward self-understanding and a help in making personal choices about permanence in marriage. Today "marriage *may* be for life, *can* be for life, but also may not be" (Mead 1949, p. 338).

Marital relationships can be permanently satisfying, counselors advise, only if spouses give up the romantic myth of the effortless, conflict-free, and happy relationship. Loving, but not romanticizing, can grow with time, knowledge, intimacy, and shared experience. Spouses who want to stay together must learn to care for the "unvarnished" other, not a "splendid image" (Van den Haag 1974, p. 142). In this regard, psychoanalyst Sidney Jourard addressed an audience about his own marriage of twenty-six years. Here is part of what he said:

> Marriage is not for happiness, I have concluded after 26½ years. It's a many splendored thing, a place to learn how to live with human beings who differ from oneself in age, sex, values, and perspectives. It's a place to learn how to hate and to control hate. It's a place to learn laughter and love and dialogue. (Jourard 1979, p. 234)

Getting married, then, can encourage partners to continue to commit themselves to learning how to live with each other, to developing, through the course of their individual life cycles, an ongoing love relationship.

Expectations of Primariness

Marriage involves the expectations of **primariness:** the commitment of both partners to keeping each other the most important people in their lives. Usually couples agree that primariness will include expectations of **sexual exclusivity,** in which partners promise to have sexual relations only with each other. "Many people still feel that the self-disclosure involved in sexuality symbolizes the love relationship and therefore sexuality should not be shared with extramarital partners" (Reiss 1986a, pp. 56–57).

Sexual exclusivity emerged as a cultural value in traditional society to maintain the patriarchal line of descent; the wedding ring placed on the bride's finger by the groom symbolized this expectation of sexual exclusivity on her part. Although historically polygamy was common (remember the Old Testament patriarchs), the Judeo-Christian tradition extended expectations of sexual exclusivity to include both husbands and wives.

For example, the Book of Common Prayer asks both partners to "forsake all others."

Polygamy has been illegal in the United States since 1878, when the U.S. Supreme Court ruled that freedom to practice the Mormon religion did not extend to having multiple wives (*Reynolds v. United States* 1878).

Although the Mormon church no longer permits polygamy, there are dissident Mormons (not recognized as Mormons by the mainstream church) who follow the traditional teachings and take multiple wives. They tend to live in remote areas for fear of prosecution, but occasionally members of these families have talked to the media to explain and advocate their lifestyle. Some multiple wives have argued that polygamy is a feminist arrangement because the sharing of domestic responsibilities benefits working women (Johnson 1991; Joseph 1991).

Polygamy has not received the public acceptance that heterosexual and even homosexual cohabitation have. Bigamy is vigorously prosecuted. Yet, civil libertarians argue that the Supreme Court should rescind its *Reynolds* decision on the grounds that the right to privacy permits this choice of domestic lifestyle as much as any other. But nonetheless, the majority of husbands and wives today feel that extramarital sex for either partner is wrong (Macklin 1987).

How closely do the actions of married couples support these beliefs? Although monogamy remains the cultural ideal, it is not as widespread in practice as many people think.

Extramarital Sex

Taboos against extramarital sex are widespread among the world's cultures, but the proscription against extramarital sex is stronger in the United States than in many other parts of the world. This is probably due to the emphasis, discussed earlier, on the marital dyad as the seminal relationship in American families. Currently, about 78 percent of Americans believe that extramarital sex is "always wrong," an increase of 7 percent since 1974 ("Changes in Sexual Permissiveness" 1992).

Although a large majority of Americans publicly disapprove of extramarital sex, in practice the picture is somewhat different. Statistics are probably less than totally accurate because they are based on what people report: Some spouses hesitate to admit an affair; others boast about affairs that didn't really happen. Nevertheless, in the recent national survey on sexuality in the United States conducted by the National Opinion Research Center (see Chapter 5), one-quarter of all husbands and 15 percent of wives ages 18 through 59 reported having had at least one affair. Among men in their 50s, the figure was 37 percent (Laumann et al. 1994). According to sexologists Masters, Johnson, and Kolodny, "Extramarital sex hasn't disappeared in the 1990s. In fact, there is little evidence that participation in extramarital sex has even slowed down a bit in the age of AIDS" (1994, p. 483). Some people may regard extramarital sex as an openly acknowledged freedom for one or both spouses. Box 8.1, "Sexually Nonexclusive Marriages," addresses consensual extramarital sex.

Gender differences are evident in any discussion of extramarital sex. Wives are more likely to have an affair because they feel emotionally distanced by their husbands; men are far more likely to have an affair for the sexual excitement and variety they hope to find (Masters, Johnson, and Kolodny 1994, pp. 494, 499). Although gender roles are changing, there is evidence that the sexual double standard persists. In one survey of midwestern college students whose mean age was 19 years, males were generally more tolerant of a man's infidelity than a woman's. Also, males were more tolerant of men's affairs than females were of women's (Margolin 1989).

REASONS FOR EXTRAMARITAL AFFAIRS   After reviewing the literature on extramarital sex, Macklin (1987) concludes that having an affair depends on the following factors:

- Opportunity—whether there is sufficient privacy and a potential partner is available.

- Willingness to take advantage of the opportunity. Generally, spouses who are less satisfied with or dependent on their marriages and have more liberal values regarding monogamy are more likely to have affairs.

- Expectations for satisfaction—influenced by role models and past experiences, as well as attraction to the potential partner.

- Expectations about negative consequences—based on the perceived likelihood of being found out or rejected by one's spouse or friends if found out.

Some extramarital affairs are short term; others are more enduring. Many *short-term affairs* are "situation

specific": a fling while attending an out-of-town convention or business meeting, a sexual encounter at an office holiday party, a surprise telephone call from a former boy- or girlfriend that leads to sex but ends as abruptly as it began, a cruise-ship romance that occurs when spouses take separate vacations (Masters, Johnson, and Kolodny 1994, pp. 483–86).

Other short-term affairs are motivated by the need for conquest (a new "notch on his or her gun barrel") or to get revenge for a spouse's real or imagined injustices. Husbands are more likely to engage in conquest affairs while wives are more likely to have anger/revenge affairs. Another type of short-term affair precedes a possibly anticipated divorce; predivorce affairs are "like test flights—transient forays into the world of sex outside marriage as a prelude to making the final decision to terminate a relationship that is already on a shaky foundation" (Masters, Johnson, and Kolodny 1994, p. 489).

A final type of short-term affair is the male bisexual affair. While a wife's bisexual affair is more likely to be long term, a husband's bisexual affair(s) is/are much more likely to be quick and anonymous (Masters, Johnson, and Kolodny 1994, p. 489). Male bisexual affairs can occur among husbands who are predominantly heterosexual but occasionally are drawn to the danger, variety, or intrigue of same-sex relations as a means of experiencing a different form of sexual excitement. On the other hand, there are married men who might appear to be heterosexual but who are really closeted gay men using marriage to hide their sexual orientation (Masters, Johnson, and Kolodny 1994, p. 489).

Long-term affairs tend to be more complex than short-term liaisons. Among others, Masters, Johnson, and Kolodny list the following types, depending on their purpose: hedonistic, marriage maintenance, intimacy reduction, and reactive affairs.

Hedonistic affairs rarely lead to emotional entanglements:

> The affair is an indulgence, a creative act of playfulness, an oasis of sensual energy in a world fogged over by trivial details of everyday life. . . . The participants often have happy and sexually fulfilling marriages of their own. (Masters, Johnson, and Kolodny 1994, p. 492)

Marriage maintenance affairs are convenient arrangements that provide something that is missing

Sexually Nonexclusive Marriages

Marriages in which partners agree that primariness will not necessarily involve sexual exclusivity are of various types. Chapter 6 discussed "the new other woman," or "man sharing." Besides this relationship type, sexually nonexclusive unions include swinging and sexually open marriages.

Advocacy of sexually nonexclusive marriage, as well as research on such unions, is largely a product of the pre-AIDS era. We present this lifestyle not only in the spirit of discussing alternatives available in our culture but also to highlight our theme of conscious versus unconscious decision making. The choice of having a sexually nonexclusive marriage, which many people reject in the forms presented here, is a choice nevertheless being made by many American spouses with little deliberation and without consensus. Yet all forms of sexual nonexclusivity carry similar emotional and health/mortality risks.

Swinging is a marriage arrangement in which couples exchange partners in order to engage in purely recreational sex. In the 1970s (our most recent data), it was estimated that about 2 percent of adults in the United States had participated in swinging at least once. Swinging gained media and research attention in the late sixties and seventies. Since then, little research has been done on swinging.

What we do know is that active swingers emphasized its positive effects—variety, for example. Former swingers who had given up the lifestyle pointed to problems with jealousy, guilt, competing emotional attachments, and fear of being discovered by other family members, friends, or neighbors (Macklin 1987).

Like swinging, sexually open marriage received considerable publicity in the late sixties and seventies. In a **sexually open marriage,** spouses agree that each may have openly acknowledged sexual relationships with others while keeping the marriage relationship primary. Unlike in swinging, one or both spouses go out separately, and outside relationships can be emotional as well as sexual. Partners in sexually open unions tend to be highly independent, willing to take risks, noncon-

forming, and committed to nonpossessiveness and personal growth. Couples usually establish limits on the degree of sexual and/or emotional involvement of the outside relationship, along with ground rules concerning honesty and what details to tell one another (Macklin 1987, p. 335).

Spouses in sexually open marriages often say that the arrangement has many personal and relationship benefits. They also report complications: jealousy, guilt, difficulty in apportioning time and attention, pressure from the extramarital partner, need for continuous negotiation and accommodation, and loneliness when the spouse is not home (Masters, Johnson, and Kolodny 1994).

Research suggests that sexually open marriages endure and are satisfying to both spouses only when the marital relationship is highly affectionate and mutually understanding and respectful. Perhaps it goes without saying that both partners need to be in equal agreement about this lifestyle, or it will not work. At least one researcher (Buunk 1982) has argued that sexually open marriages may be more satisfying overall for husbands than for wives. Moreover, both spouses need to have the personality characteristics and communication skills necessary to deal effectively with complex and potentially stressful relationships. This is a tall order and may be why many sexually open marriages have unanticipated difficulties (Masters, Johnson, and Kolodny 1994) or apparently become monogamous after a period of time. Some spouses alternate between sexually exclusive and nonexclusive marriage states (Macklin 1987).

Finally, all of us realize that the advent of AIDS means that sexual adventurousness can have potentially lethal consequences, for children as well as for consenting adults. The willingness to consider a consensual agreement to have a sexually nonexclusive marriage now must take into account more than personal values and relationship management challenges, for it involves the risk of disease and even death. (See Chapter 5 for a more detailed discussion of AIDS and marriage.)

There are many types of extra-marital affairs, just as there are different reasons for having an affair. While an affair may have positive consequences for the individuals involved and even for the marriage, the negative effects are likely to outweigh any positive ones.

from the marriage, such as kinky sexual experimentation. By supplying this element, the affair actually stabilizes the marriage and makes it less likely that a marital breakup will occur. "Although common wisdom has it that affairs often lead to marital dissolution, we have encountered hundreds of marriages that were held together and solidified by affairs" (Masters, Johnson, and Kolodny 1994, p. 491).

Intimacy reduction affairs are undertaken by spouses who are ambivalent about the intimacy demanded by their husband or wife. Creating a safety zone of emotional distance within the marriage, the affair serves as a buffer against too much closeness:

> When tension and anxiety over too much intimacy mount, it can be defused by more involvement with the affair. In contrast, when enough emotional space has developed in the marriage for it to feel comfortable, rather than smothering, the affair

may be ignored for awhile, since there is at least temporarily safe harbor at home. (Masters, Johnson, and Kolodny 1994, p. 492)

Reactive affairs are motivated by a spouse's desire to be reassured amid changing life circumstances. For instance, a middle-aged husband who is feeling over the hill may seek to prove his youthfulness by having an affair with a younger woman. Or a wife may rediscover her sexuality once her children have left home and opt for the excitement of extramarital sex. A wife's bisexual affair often fits this category (Masters, Johnson, and Kolodny 1994, p. 493).

EFFECTS OF EXTRAMARITAL AFFAIRS   Extramarital sex *can* have positive effects. As we have seen, an affair can sometimes help keep a marriage together by providing some missing need or reducing intimacy anxiety. Because an affair does not necessarily offer better sex, a spouse might find renewed appreciation for the marriage. But recognizing that affairs can have a positive side is hardly the whole story: "We are firmly convinced that the downside of extramarital sex usually looms larger than any potential benefits" (Masters, Johnson, and Kolodny 1994, p. 502).

First, an affair that is nonconsensual involves deceit and can be considered a form of theft: "What is stolen is the bond of trust and its attendant consent to mutual vulnerability between spouses. Such vulnerability is based largely on the assumption that neither partner is out to hurt the other" (Masters, Johnson, and Kolodny 1994, p. 503). Breaking that trust can result in unanticipated—and unwanted—complications. For one thing, the discovery of a spouse's affair (which happens in a large number of cases) is likely to seriously undermine marital trust and intimacy. The reaction is often shock and outrage, setting off a series of negative consequences that reverberate over time through the course of the marriage. Anger over a spouse's affair may linger and undermine the relationship for years. To some partners, extramarital sex is such a profound violation of moral and religious principles that it shatters a fundamental marital pillar that can never be fully repaired. In other unions the problems have nothing to do with moral or religious beliefs but are grounded in what the affair takes from the marital relationship. Sometimes an affair becomes known to the entire family, including the children. Clearly, this situation is not good (Masters, Johnson, and Kolodny 1994, p. 502).

Moreover, affairs victimize the uninvolved partner without giving her or him prior warning or any way to avoid being injured. Not only has trust been eroded and feelings been hurt, but the uninvolved spouse may have been exposed to various sexually transmitted diseases—not a rare occurrence. For many spouses, concern about AIDS heightens anger and turmoil over affairs. The uninvolved spouse is often exploited financially as well, because what were thought to be joint funds have been unilaterally spent on dinners, gifts, hotel rooms, or weekends away.

JEALOUSY **Jealousy** is emotional pain, anger, and uncertainty arising when a valued relationship is threatened or perceived to be threatened. Sociologist Ira Reiss sees a spouse's jealousy as "a boundary-setting mechanism." When boundaries are violated (in this case, those circumscribing the married couple as pri-

mary), "jealousy occurs and indexes the anger and hurt that are expected to be activated by a violation of an important norm" (1986a, p. 47).

Research has found differences in how men and women experience and react to jealousy. Men are more reactive to the sexual threat, whereas women are more anxious about losing a primary relationship. Women, being more likely to monitor relationships, are more apt to try to change to please their partner so as to avoid the threat of another relationship. Men, meanwhile, are more likely to seek solace or retribution in alternative relationships (White and Mullen 1989). Several research projects suggest that those who feel insecure already or have poor self-images are more inclined toward jealousy (Macklin 1987; Buunk 1991).

We tend to equate feeling jealous with sexual threats to a relationship, but other threats also evoke jealousy. Friends, colleagues, work, education or leisure commitments, and even a couple's children can be perceived as "trespassers" (McDonald and Osmond 1980, p. 4). Box 8.2 addresses marriage and opposite-sex friends.

Even though marital jealousy can probably never be completely eliminated, mutually supportive encouragement may lessen feelings of insecurity and jealousy. Nurturing one's self-esteem can allay what may indeed be unwarranted jealousy. In sum, jealousy, like pain, needs to be viewed as a warning signal. Jealousy may mean that one or both partners' interests in outside activities need to be counterbalanced with activities within the relationship.

RECOVERING FROM AN EXTRAMARITAL AFFAIR

Given that affairs do occur, a good number of people will perhaps rethink the issue when they discover that their spouse has had (or is having) one. The uninvolved mate will need to consider how important the affair is relative to the marital relationship as a whole. Can she or he regain trust? In some cases, the answer is "no"; trust never gets reestablished and the heightened suspicion gets incorporated into other problems the couple might have.

Whether trust can be reestablished depends on several factors. One is how much trust there is in the first place. One researcher (Hansen 1985) has suggested that for this reason new marriages may be especially vulnerable to breaking up after an affair. Many marriages can and do recover from affairs, however. Therapists suggest that doing so requires the following:

In a society publicly dedicated to monogamy, spouses may be wary of opposite-sex friendships because of their potential for sexual involvement. Moreover, the absence of clear role expectations for an opposite-sex friend of a married person makes it difficult to know what to say and do in such a situation.

Partners need to negotiate carefully and discuss openly their feelings about the nature of the activities and degree of emotional and physical intimacy they find acceptable in their spouse's friendships. Couple boundaries should be clarified. What confidential information can be shared? Can the couple's sex life be discussed with others?

Same-sex friendships can present the same emotional threat to gay and lesbian couples that cross-sex friendships do to heterosexuals. Yet they are different in that same-sex friendships are culturally acceptable, even encouraged. There may be more social support and less self-consciousness in same-sex friendships among gays and lesbians, but there is also an ambiguity that can give rise to insecurity. Lesbian and gay couples are frequently embedded in a larger gay or lesbian community, occasioning frequent contact with same-sex friends who are potential partners. As with heterosexual couples, open discussion, agreement, and clarification of boundaries between a couple and their friends may prevent misunderstandings and conflict.

The following ten statements can help clarify what you believe is appropriate in opposite-sex friendships for heterosexual partners and in same-sex friendships for homosexual partners. Answer "always," "sometimes," "occasionally," or "never" to the following statements as they apply to someone of the sex from which you also choose sexual partners:

A wife/husband/female partner/male partner should be able to:

1. go to lunch with a work colleague to discuss business
2. go to lunch with a work colleague just because they are friends
3. go to lunch with a friend, not a work colleague
4. go to dinner with a work colleague
5. go to dinner with a friend, not a work colleague
6. go to a movie with a friend
7. go dancing with a friend
8. spend a full day with a friend
9. vacation with a friend
10. have sex with a friend

To evaluate your responses, look first at where you draw the line. Although you may have said "always" to all ten statements, it is more likely that you did so with those listed earlier than later. Examine if and how your responses differ according to whether a partner is married, is heterosexual or gay, or female or male. If you are in a relationship now, you might ask your partner to respond to these statements and then compare answers. Doing so will let you know where opinions differ and might stimulate communication and negotiation on these topics.

- The offending spouse needs to apologize sincerely and without defending her or his behavior.

- The offending spouse needs to allow and hear the verbally vented anger and rage of the offended partner (but never permit physical abuse).

- The offending spouse needs to allow for trust to rebuild gradually and to realize that this may take a long time—up to two years or more.

- The offending spouse needs to do things to help the offended partner to regain trust—keep agreements, for example, and calling if he or she is running late.

- The offended spouse needs to decide whether she or he is committed to the marriage and, if so, needs to be willing to let go of resentments as this becomes possible.

- The couple should consider marriage counseling (described in Appendix H).

On the one hand, data on extramarital affairs suggest that lifelong fidelity is far from universal; on the other hand, specific acts of infidelity may not be routine or continual. The contrasting data on extramarital sex suggest that today, as in the past, many spouses are torn between lifelong commitment to a sexually exclusive marriage and desire for outside sexual relationships. For the majority of mates, however, the decision to marry involves the promise to forgo other sexual partners. If we interpret fidelity to mean a primary commitment to one's partner and the relationship, then maintaining emotional intimacy becomes essential to being faithful. Without continued intimacy and self-disclosure, partners may not remain in love or keep a central place in each other's lives. Emotional commitment, in this view, is the essential element of primariness, as much or more than sexual fidelity per se. This ideal of primariness can make marriage relationships more satisfying.

To sum up, marriage entails making ongoing choices about permanence and primariness. In the next section, we'll examine whether same-sex couples should marry legally.

Should Same-Sex Couples Marry?

Many gay male and lesbian couples live together in long-term, committed relationships (Allen and Demo 1995). Partners may exchange vows or rings or both. Couples may publicly declare their commitment in ceremonies among friends or in some congregations and churches, such as the Unitarian Universalist Association or the Metropolitan Community Church, the latter expressly dedicated to serving the gay community. Catholics have access to a union ceremony designed by Dignity, a support association for gay and lesbian Catholics. In a 1993 gay rights march in Washington, DC, 1,500 same-sex couples participated in a mass wedding with ministers and rice (Salholz 1993). None of these couples is legally married, however. By law, dating to a U.S. Supreme Court decision in 1974, marriage is defined as a union between one man and one woman; hence same-sex marriage is not legally recognized in the United States.

Meanwhile, some cities and counties have passed *domestic partner* laws (defined in Box 6.2). According to this concept, unmarried couples may register their partnership and then receive some of the legal benefits of marriage, such as joint health or auto insurance or bereavement leave, for example. Besides practical benefits, registering as domestic partners has emotional significance for some same-sex couples, who do so as a way of publicly expressing their commitment (Ames 1992). But registering as domestic partners lacks the deep symbolism of marriage. "Gays and lesbians were raised in the same culture [as] everyone else," notes gay historian Eric Marcus. "When they settle down they want gold bands [and] legal documents" (in Salholz 1993).

Attitudes toward gay rights have generally become more liberal. Nevertheless, a 1993 poll by the *Washington Post* found that 53 percent of Americans oppose homosexual relationships in general and 70 percent are against same-sex marriage (Salholz 1993). For religious fundamentalists and other conservative groups, the case against gay marriage is unambiguous: Marriage is intrinsically straight. From this point of view, the move to legalize marriage is an "attempt to deconstruct traditional morality." As the posters proclaim, "God made Adam and Eve, not Adam and Steve" (Salholz 1993; Gross 1994).

Legal Marriage and Hawaii

While several western European nations are beginning to consider it, the only countries that legally recognize gay male and lesbian marriages are Denmark and Norway (Singer and Deschamps 1994; Wyman 1994). This situation may be about to change, however. U.S.

| Should same-sex couples marry? The question is controversial, even among gay men and lesbians.

gay activist Craig Dean and his lover, Patrick Gill, formed the Equal Rights Marriage Fund after they were denied a marriage license in 1991. At about the same time, after being denied a marriage license in Hawaii, Patrick Lagon, a graphic artist in his 30s, and Joseph Mellio, a chef in his late 40s, sued to gain Hawaii's legal recognition of their sixteen-year relationship. Two lesbian couples joined the suit. Lagon explained that both he and Mellio were raised to expect to fall in love and marry "like anybody else" (De Lama 1994). The men talk proudly of their "simple" life together, characterized by TV, pet dogs, a flower garden, and weekend drives (Gross 1994).

In response to the suit, in 1993 the Hawaii Supreme Court voted 3-2 that refusal to recognize same-sex marriages may violate sex discrimination laws. Among other things, the court said that defining

marriage as necessarily between one man and one woman was "circular and unpersuasive" reasoning. The court sent the case back to a lower court where the state was required to prove that it had a "compelling interest" in denying same-sex couples the right to marry.

As a result, in their 1994 legislative session, Hawaii's lawmakers sought to clarify the state's marriage statute. They considered, on the one hand, a state constitutional amendment that would limit marriage to heterosexuals and, on the other, a broad domestic partnership act (see Box 6.2) that would appease same-sex couples without giving them the right to marry legally. Legislators finally agreed on a bill restating that marriage is meant for "one man and one woman" but also creating a commission to propose remedies for any discrimination against same-sex couples that might result from denying them the legal right to marry (Gross 1994). The outcome will be either legalization of same-sex marriage by Hawaii's Supreme Court or a broad and far-reaching domestic partnership act—and perhaps both (Gross 1994). At the time of this writing, many observers think that Hawaii is about to legalize same-sex marriage.

By itself, any court ruling in Hawaii would not affect other states. But many same-sex couples from throughout the United States could marry legally in Hawaii, then seek recognition of their marital status in their home states. States usually recognize one another's legal decisions, according to the *principle of reciprocity.* The only exception is when one state claims a violation of local public policy, which is expected to be the explanation for not recognizing gay marriage in the twenty-three states where sodomy remains a criminal act (see Chapter 5). Within the next several years the issue is likely to find its way to the U.S. Supreme Court (Gross 1994).[1]

Views of Gay Rights Activists on Same-Sex Marriage

Lesbians and gay men who favor legalized same-sex marriage argue that legal marriage yields economic and other practical advantages. Marrieds can lower their taxes by filing a joint return; if one spouse dies or is disabled, the other is entitled to Social Security benefits; legal partners can inherit from one another without a will; spouses are immune from subpoenas requiring testimony against each other. Some companies offer health insurance to an employee's legal spouse and dependents. And when one or both partners suffer from AIDS or other serious illness, the protections of marriage could prove highly beneficial (Stoddard 1989; Wyman 1994).

Moreover, unlike heterosexual couples, homosexual couples receive little social support for continuing long-term relationships. On average, relationships for both gay males and lesbians last two years to three years, and a pattern of serial monogamy exists (Harry

1983). As Mary Mendola, lesbian author of the *Mendola Report* (1980), explains:

> The major difference separating us as heterosexual and lesbian couples: our lesbian marriage had none of the support systems Mr. and Mrs. Next Door enjoyed. I had not had a bridal shower or a bachelor party, depending on how one looks at it. . . . Aunts and uncles did not come to visit and admire our home. We never received anniversary cards. As trivial as these things may seem, they represent something vitally important: heterosexual couples are encouraged to stay together. . . . Lesbian and homosexual couples have no such support systems. Rather than being encouraged to stay together, we are conditioned to believe there is no future for us as couples. (p. 4)

Being able to marry legally could help to change this situation.

Furthermore, denying lesbians and gay men the right to marry implies that they are not as good or valuable as heterosexuals. The right to marry legally could remedy this. Some lesbians and gay men believe that legal marriage would be their "most important civil rights victory" yet; related successes would follow. For instance, laws banning same-sex sodomy (see Chapter 5) would very likely be repealed or declared unconstitutional. "And social custom—already in flux—would undoubtedly change" (Wyman 1994).

Gay and lesbian opponents of legal same-sex marriage, such as gay rights attorney Paula Ettelbrick, object to mimicking a traditionally patriarchal institution based on property rights and institutionalized husband–wife roles. As one lesbian explained,

> Within my home I feel married to Frances, but I don't consider us "married." The marriage part is still very heterosexual to me. One of the reasons I don't like to associate with marriage is because heterosexual marriage seems to be in trouble. It's like booking passage on the *Titanic.* Frances is my life partner; that's how I'm accustomed to thinking of her. (in Sherman 1992, pp. 189–90)

More generally, opponents object to giving the state power to regulate primary adult relationships. They stress that legalizing same-sex unions would further stigmatize any sex outside marriage, with unmarried lesbians and gay men facing heightened discrimination (Ettelbrick 1989).

---

1. As we go to press, Congress is considering the Defense of Marriage Act, a proposed federal statute declaring marriage to be a "legal union of one man and one woman," denying gay couples many of the civil advantages of marriage, and also relieving states of the obligation to grant "full faith and credit" to marriages performed in another state. How the U.S. Supreme Court's ruling of May 20, 1996, which struck down a Colorado challenge to local gay rights ordinances, will impact the proposed Defense of Marriage Act is "anybody's guess" (Vicki Haddock, "Activists Hail Shift by U.S. Justices," San Francisco Examiner, May 21, 1996, pg. A-10).

"We are gathered here to join together this man and this woman in matrimony—a very serious step, with far-reaching and unpredictable consequences."

Whether one's commitment is same-sex or heterosexual, being—and staying—married involves choices. The following section examines some of these.

A Matter of Choices Throughout Life

Our theme of making choices throughout life surely applies both to early years of marriage and to couples preparing for marriage. Surprisingly, there is less social science research on these topics than on divorce! Along with the available social science data, we will incorporate our own ideas, those of other family sociologists, and information available in the media.

Preparation for Marriage

Given today's high divorce rate, clergy, teachers, parents, policymakers, and others have grown increasingly concerned that individuals be better prepared for a marital relationship. Conventional dating in which the focus is on the enjoyment of social events does not prepare couples for dealing with practicalities, decisions, and problems. Yet, as we also saw in Chapter 7, living together does not seem to prepare couples well either; at least, the marital success rates of couples who cohabited before marriage are no higher, and are often lower, than those of couples who did not cohabit premaritally.

Family life education courses, which take place in high school and college classrooms, are designed to prepare individuals for marriage. Premarital counseling, which generally takes place at churches or with private counselors, is specifically oriented to couples who plan to be married. Many Catholic dioceses, for example, require premarital counseling before a couple may be married by a priest. Catholic dioceses have also organized weekend engagement encounters during which couples who plan to marry learn about and discuss various aspects of married life. Such programs are designed and generally conducted by professionally trained people. Other churches have adopted these programs, and some couples may seek premarital counseling on their own initiative.

Premarital counseling has two goals: first, to evaluate the relationship with the possibility of deciding against marriage; and second, to sensitize partners to potential problems and to teach positive ways of communicating about and resolving conflicts. Even though common sense suggests that these kinds of programs help, they have not been in existence long enough for us to scientifically evaluate their impact on subsequent marriages (Darling 1987). Nevertheless, family experts see these programs as important, especially for adult children of troubled, dysfunctional, or divorced families (Spanier 1989).

Premarital education or counseling can help make the first years of marriage go more smoothly.

The First Years of Marriage

In the 1950s marriage and family texts characteristically referred to the first months and years of marriage as a period of adjustment, after which, presumably, spouses had learned to play traditional marital roles. Today we view early marriage more as a time of role *making* than of role *taking*.

Role making involves issues explored fully in other chapters of this text. Newlyweds negotiate expectations for sex and intimacy (Chapter 5), establish communication (Chapter 9) and decision-making patterns (Chapter 10), balance expectations about marital and job responsibilities (Chapter 13), and come to some agreement about childbearing (Chapter 11) and how they will handle and budget their money. When children are present, role making involves negotiation about parenting roles (Chapter 12). Role-making issues peculiar to remarriages are addressed in Chapter 16. Moreover, the time of role making is not a clearly demarcated period but rather continues throughout marriage.

Even though the early stages of marriage are not a distinct period, sociologists continue to speak and write about them as such. One thing we know is that this period tends to be the happiest, with gradual declines in marital satisfaction afterward. (Perhaps this is where we get the popular expression, "The honeymoon's over.") Why this is is not clear. One explanation points to life-cycle stresses as children arrive and economic pressures intensify; others simply assume that courtship and new marriage are periods of emotional intensity from which there is an inevitable decline (Whyte 1990, pp. 190–95).

We do know something about the structural advantages of early marriage, and it is likely that these contribute to high levels of satisfaction. For one thing, partners' roles are relatively similar or unsegregated in early marriage. Spouses tend to share household tasks and, because of similar experiences, are better able to empathize with each other. Early marriage may also be less vulnerable to stress arising from dissatisfaction with finances. Typically, both partners are employed, and though their salaries may be relatively low, so are their expenses.

But early marriage is not characterized only by happiness. Couples must also accomplish certain tasks during this period. In general, "the solidarity of the new couple relation must be established and competing interpersonal ties modified" (Aldous 1978, p. 141). The couple constructs relationships and interprets events in a way that reinforces their sense of themselves as a couple (Wallerstein and Blakeslee 1995).

Changing friendships is one indication that getting married changes—sometimes subtly, sometimes surprisingly—what people expect from themselves and their partners. Marriage changes a couple's relationship. To some extent this is true for all newlyweds, even those who lived together before marrying. One reason for this change is that weddings usher in a host of conscious and unconscious beliefs about wives' and husbands' roles. One husband, for example, said that before his wedding he and his partner ate out regularly. He enjoyed this and had expected it to continue. His wife, however, felt that married people ought to eat at home. Some spouses (both husbands and wives) said they became more possessive. They felt they should now go out as a couple rather than separately, for example. One new husband surprised both himself and his wife by getting very angry when his wife refused to vote his way in a presidential election (Arond and Pauker 1987).

Because being married is a different relationship from dating or living together, the wedding may begin a new cycle of adjustment and change even for couples who have lived together for some time. Newlyweds may feel as if they're falling in love all over again. Subsequently, they may find themselves disappointed about both old and new issues. The increased security of marriage can encourage partners to let their hair down. Sometimes this comes as a shock to a spouse and raises new problems. Said one young husband, "I never knew my wife could be so sloppy." Getting through this stage requires making requests for change and negotiating resolutions, along with renewed acceptance of one another.

For spouses who previously lived together, an unexpected problem may arise. One may define the wedding and honeymoon as ho-hum events; to her or him nothing has changed. The mate, meanwhile, may see the wedding as important, marking the advent of a different relationship. One wife described her honeymoon as frustrating for this reason. The couple went separate ways because he wanted to shop and she wanted to see the sights. This would have been acceptable to her, she said, had they not been on their honeymoon. Although they had traveled together before, she had expected increased togetherness this time; but he behaved as if it were just another trip. Similarly, a groom reported being upset on his honeymoon when his wife, with whom he had lived for three years, "jumped out of bed at 7:00 A.M." each day to jog. "I wanted her to lounge around with me in bed," he explained. "I wanted to play the romantic honeymoon image up more" (Arond and Pauker 1987, p. 219).

Newlyweds typically reported heightened feelings of responsibility, along with increased expectations for responsible behavior from their mates. A woman explained that before marriage she was reluctant to challenge her partner's habit of taking off alone on motorcycle trips. As his wife, however, she felt "entitled to a more predictable and stable life" (Arond and Pauker 1987, p. 13). As one wife put it, "I don't feel like such a kid anymore" (p. 26). One man (who said he felt a lot closer to his wife than before the wedding) said being married meant feeling "rooted in a community" and "a lot more settled and future-directed" (p. 12).

Choosing a Utilitarian or Intrinsic Marriage

It is important to keep in mind that utilitarian and intrinsic marriage types are poles on a continuum. A couple's relationship lies somewhere along that continuum (between these poles) and combines utilitarian with intrinsic characteristics. In Chapter 4 we saw that sociologist Francesca Cancian (1985) views love as combining both instrumental and affectionate qualities. Similarly, a marriage relationship fulfills both practical and intimacy needs. Where partners place

The early stages of marriage tend to be the happiest, as couples have time to enjoy each other and to pursue common interests. During this time, couples also need to establish a sense of solidarity and to modify competing interpersonal ties.

their relationship on the utilitarian–intrinsic continuum depends on choices they make throughout their lives—and on societal conditions that limit or expand their options as well.

Some couples choose to emphasize the utilitarian functions of their marriage. For example, a passive-congenial union can be a deliberate arrangement for spouses who want to direct their creative energies outside of their relationship—into obligations to parents or other kin, raising children, or careers.

Working-class marital relationships are likely to be more utilitarian because spouses are less free to develop according to their personalities and lifestyle choices. They seem channeled by restricted childhoods and current economic pressures into a preoccupation with day-to-day survival and with attaining a minimum of order and stability, and they seem handicapped by the spiral of alienation likely to develop out of their early struggles. Many working-class couples have the desire but not the skills, models, or freedom from worry to sustain a more intrinsic relationship (Rubin 1976).

Among middle-class partners, however, particularly non-Hispanic whites, more utilitarian unions can be chosen mainly by default. Couples allow themselves to drift, ignoring their initial commitment to keep each other primary in their lives. Increased emphasis on other matters, such as career advancement or children, results in slow emotional erosion (Sternberg 1988b, p. 273). Among Cuber and Harroff's devitalized couples, husbands who were promoted grew busier, came home later, and brought with them "bulging briefcases." The nonworking wives of "organization men" busied themselves with children and community organizations. Today both partners may be concentrating on careers rather than building intimacy.

Although many utilitarian marriages are chosen by default, successful intrinsic couples often shape their relationship "deliberately, consciously, sometimes even ruthlessly" (Cuber and Harroff 1965, p. 134). At different times throughout their marriage, a couple's relationship may be more (or less) couple-centered, parenting-centered, or relatively loosely connected as one or both partners pursue careers or other personal interests (Marks 1989). We have seen that the first years of marriage are often more couple-centered than are later years. Spouses who desire enduring, more intrinsic relationships must keep the primariness of their mate and marriage as a high priority. For example, a spouse or partner may choose to turn down career opportunities that could threaten the relationship. In one instance reported by Cuber and Harroff, a husband married for twenty-two years had passed up two promotions because one would have required some traveling and the other would have taken evening and weekend time away from his wife.

Keeping one's marriage vital requires that partners consciously and continuously strive to maintain primariness and intimacy, which can entail personal sacrifice in career goals or other areas of life. As a result, intrinsic partners see their relationship as the product of an enormous investment: in time, effort, and priorities. In other words, an emotionally meaningful relationship does not often develop "by drift or default" (Cuber and Harroff 1965, pp. 142–45).

Clearly, the nature and quality of a marital relationship has a great deal to do with choices partners make. They risk disillusionment less when they make their choices consciously and knowledgeably. Partners are freer today to design their own unions, both the form their marriages and families will take and the kind of relationship they will develop. They therefore need to be as honest as they can be with each other (and with themselves) about their expectations and goals, both before and throughout a marriage. We saw in Chapter 1 that as people change throughout life, the kind of relationship they need also changes. Thus, marriages need to be flexible if they are to continue to be satisfying to changing partners. In the next section we'll see what this flexibility entails.

Static Versus Flexible Marriages

For years, writes Sidney Jourard,

> spouses go to sleep night after night, with their relationship patterned one way, a way that perhaps satisfies neither—too close, too distant, boring or suffocating—and on awakening the next morning, they reinvent their relationship *in the same way.* (1976, p. 231)

Spouses who choose to behave in this way are shaping **static (closed) marriages** (O'Neill and O'Neill 1972). Partners in static marriages rely on their formal, legal bond to enforce permanence, strict monogamy, and rigid husband–wife role behavior. Spouses who rely on formal rules to maintain their feelings of intimacy, however, have unrealistic expectations about marriage. They may expect, for example, that their loving relationship will (in spite of the rising divorce rate) last forever. They may falsely believe that marriage guarantees their own and their partner's total commitment. Often they begin to think that their mate belongs to them.

People do change, however, and promises will not prevent change. Partners who vowed never to change may try to hide their personal growth from each other, with the result, of course, that intimacy diminishes.

They may be left with a permanent, but stale, relationship.

An option is to actively pursue a **flexible marriage,** one that allows and encourages partners to grow and change. In a flexible marriage, spouses' roles may be renegotiated as the needs of each change (Scarf 1995).

Flexible marriages are more likely to be intrinsic, as partners are freer to reveal their changing selves and the parts of themselves that no longer fit into their established pattern. They can continue to be in touch at a deep emotional level while they alter the outer framework of their lives. Utilitarian marriages are also likely to benefit from flexibility, as utilitarian needs are likely to change over time.

How can marriages remain or become flexible? Some people may have an intuitive knack for achieving this kind of marriage. Others are able to act and react creatively on realizing the need for flexibility, without following any particular pattern. Some partners may object in principle to the idea of consciously and formally considering their marriage arrangements in order to make the marriage more flexible. But again, we want to stress the way in which decisions, in the absence of conscious reflection, often are made by default, and we suggest negotiating a personal marriage agreement. Even for those who do not wish to exercise this option, the following section outlines some of the issues likely to arise in a flexible marriage.

Contracting for Flexibility— A Contradiction?

How do people design flexible marriages? It may sound contradictory, but one way is purposefully to contract for ongoing flexibility. Partners begin by consciously negotiating personal marriage agreements.

When partners marry, they agree to two contracts: a legal–social one between the couple and the state or society and a personal one between the partners themselves. "The legally married nuclear family has access to this ready-made blueprint for building a family: familiar models, designated roles, and socially approved patterns of behavior" (Kornfein, Weisner, and Martin 1979, pp. 284–85). But at the same time, the legal–social contract is hidden, or generally outside people's conscious awareness; it is seldom verbalized or evaluated. "It is in many ways an unconscious contract, agreed to by default in the sense that it is unwitting" (O'Neill and O'Neill 1972, p. 51). But that does not make the contract any less binding on how partners think, feel, and behave toward each other. Indeed, the power of the hidden legal–social contract to constrict spouses' relationships is tremendous.

One way for couples to develop flexible marriages is to write **personal marriage agreements,** which involve articulating, negotiating, and coming to some agreement on expectations about how you and your partner will behave. Moreover, to develop flexible marriages, partners need to renegotiate their agreements often, to keep the relationship pliable enough to accommodate the changes in two people over time. Static marriage contracts too often lead to disillusionment (Crosby 1985). When the rules begin to seem too rigid or it becomes clear that unrealistic expectations will not be met, closed contracts leave only two alternatives: divorce or a devitalized relationship.

In drawing up a personal marriage agreement, people begin by separately writing down their expectations about their relationship. Even if they are single and not seriously involved in a relationship, thinking about and writing down their expectations about intimate relationships is important for self-understanding. Later they can compare and negotiate the differences in their expectations. The result is a working marital agreement.

The goal of personal marriage agreements is to help partners actively define their relationship. Because unmarried partners can also benefit from this process, such personal agreements could also be termed **relationship agreements.** (Research indicates that the need to resolve such issues as sexual exclusivity, division of labor, and power and decision making is not much different in gay or lesbian pairings than in heterosexual pairings or marriage [Peplau 1981; Harry 1983].)

Writing an agreement may seem rational and cold. Or it may seem to define a commitment that is limited, like a business contract, not one that is open-ended and at least intended to last forever, like marriage. But there are important reasons for defining a marriage with some degree of consciousness and awareness.

Reasons for a Marriage Agreement

One important reason for writing a marriage agreement is that it helps partners to be aware of and avoid choosing closed marriage by default. There are other reasons a contract can be helpful.

First, partners need to articulate the primary focus of their marriage. Will their union be—or is it—mainly utilitarian or intrinsic? Partners who want intrinsic unions need to articulate that, or they may unwittingly allow occupational, household, educational, or other responsibilities to drain their energies.

Second, writing a marriage agreement can allow partners to understand each other's role expectations. Differentiated roles can be satisfying if both partners accept the arrangement. Spouses who wish to develop egalitarian relationships, however, must battle remnants of patriarchal tradition, both in society and in their own previously internalized attitudes. Women who are willing to share the provider role may still expect a husband to do the household repairs and plan the long-term finances. Men who in many ways truly appreciate their wife's achievements may wish they didn't have to worry about the wife's schedule or her need to travel, or they may wish she were more available to play hostess, or run to the cleaners, or deal with the children.

Third, partners may have different ideas of what marriage means and different expectations about how they will behave in marriage. For example, one partner may expect to continue going out alone with his or her separate friends; the other may expect to build a joint social network. Or one partner may assume they share a desire to buy a house and to sacrifice small pleasures to build up a nest egg, whereas the other may place more emphasis on the day-to-day sharing of movies, dinners out, small gifts, and travel experiences.

Discussing and negotiating these differences is important. Writing an agreement before marrying can point up differences, many of which can be worked out. If partners uncover basic value differences and cannot work them out—for example, about whether or not to have children—it would probably be better to end their relationship before marriage than to commit themselves to a union that cannot satisfy either of them. Writing an agreement after having been married for some time can point up previously unrecognized disagreements. Facing these openly and honestly is important in maintaining a mutually supportive relationship.

Fourth, love before and during the early years of marriage is often blind. Courting partners often romanticize each other, creating images and refusing to see anything that contradicts those images. Negotiat-

ing personal agreements helps cut through the tendency to overromanticize.

Finally, in areas such as property ownership and distribution, a personal agreement clarifies the intent of couples, regardless of marital status. For married couples, this can be particularly important in separate property states (states in which by law property belongs to the spouse who has title or who earned the money that purchased the property); community property states consider all property acquired during the marriage to be jointly owned (see Weitzman 1981).

A premarital agreement can also determine division of property between a current spouse and children of an earlier marriage. Prospective remarriage partners are more likely to see a need for a premarital contract to clarify financial obligations because the potential for future conflict is obvious. There is more ambiguity in attempting to plan in advance the financial outcomes of a divorce, and such premarital contracts are not always well-received by the courts. To adapt to the reality of divorce, the law is beginning to move in this direction, however. In some states, spousal support in the event of divorce can be waived in a prenuptial contract. Child support can never be waived, as the state reserves the right to protect the economic interests of children (Dullea 1988b).

Sometimes prenuptial contracts about property—usually between an older, wealthy man and a not-so-wealthy woman—are signed in a coercive atmosphere of "sign or stay single" (Dullea 1988b). Such potential for abuse should remind us that negotiation of a premarital contract or relationship agreement should take place in an atmosphere of equality, respect, and full disclosure, with each party free to seek professional legal advice on the matter. Nonfinancial aspects of a marital agreement are not legally enforceable; parts relating to property may or may not be. At the very least, though, they are likely to carry some weight by manifesting the partner's intentions at the time of marriage. The following paragraphs provide an overview of some of the most crucial aspects of a marital agreement.

Some Questions to Ask

An initial marriage agreement can be short or long, general or detailed. Some agreements include such fine points as exactly how spouses will allocate money and household chores. Others are more general, emphasizing marital goals, values, and attitudes. (A general agreement may become more specific in the course of renegotiations or further development.) However it evolves, the agreement should consider the following basic questions, along with any others that are important to particular couples (Crosby 1991, pp. 321–22).

1. *Preliminary statement of marital goals.* Will you aim primarily at developing intimacy? Or do you want to emphasize more practical advantages from marriage, such as economic security or someone to share a household with, for example? What kind of relationship do you and your partner realistically anticipate: conflict-habituated? passive-congenial? vital? total? How do you feel about devitalized relationships? Should partners learn to accept these gracefully, work toward changing them, or divorce?

2. *Provisions for revision and renewal of contract.* Because spouses change, marriage agreements need to be rewritten often. Couples can plan to discuss and revise their agreement periodically—every six months for example. Ideally, they agree that their contract is a living document to be renegotiated and changed any time one partner feels the need. Is there any part of this contract that you think now you would never, under any circumstances, consider changing?

3. *Provisions for dissolution of the legal marriage.* When should a marriage be dissolved and under what conditions? How long and in what ways would you work on an unsatisfactory relationship before dissolving it?

4. *Decision making and division of labor.* Will decisions be made equally? Will there be a principal breadwinner? Or will partners equally share responsibility for earning money? How will funds be allocated? Will there be his, her, and our money? Or will all money be pooled? Who is owner of family property, such as family business(es), farms, or other partnerships? Will there be a principal homemaker? Or will domestic chores be shared?

5. *Religious beliefs and practices and educational goals.* What are your religious values? Do you expect your partner to share them? Will you attend church services together? How often? If you are of a different religion from your mate, whose church will you attend on special religious holidays? What about the children's religion?

   What are your educational goals? What educational goals do you expect your partner to have? Under what circumstances could you or your partner put aside wage-earning or housekeeping responsibilities to pursue advanced education?

6. *Relationships with other relatives.* How will you relate to your own and to your spouse's relatives? How do you expect your partner to relate to them? Will you expect to share many or most activities with relatives? Or do you prefer more couple togetherness, discouraging activities with relatives?

7. *Children.* Whose responsibility is birth control, and what kind of contraception will you use? What is your attitude toward unwanted pregnancy: abortion? adoption? keeping and rearing the child? Do you want children? If so, how many and how would you like to space them? How will you allocate child-rearing responsibilities and tasks? Will either spouse be primarily responsible for discipline? What are some of your values about rearing children?

8. *Expectations for sexual relations.* Can you discuss your sexual needs and desires openly with your partner? Are there sexual activities that you consider distasteful and would prefer not to engage in? How would you expect to deal with either your own or your partner's sexual dysfunction if that occurred?

9. *Extramarital friendships and sexual relations.* How much time and how much intimate information will you share with friends other than your partner? What is your attitude toward friendships with people of the opposite sex? How about Internet (cyberspace) friends? Would you ever consider having sex with someone other than your mate? If so, under what circumstances? How would you react if your partner were to have sex with another person?

10. *Privacy expectations.* How much time alone do you need? How much are you willing to allow your partner? Will you buy a larger house or rent a bigger apartment so that each partner can have private space?

11. *Communication expectations.* Will you purposely set aside time to talk with each other? What topics do you like to talk about? What topics do you dislike? Are you willing to try to become more comfortable about discussing these? If communication becomes difficult, will you go to a marriage counselor? If so, what percentage of your income would you be willing to pay for marriage counseling?

12. *Vacations.* What kinds of vacations will you take? Will you take couple-only vacations? Will you take separate vacations? If so, how often and what kind?

13. *Definition of terms.* What are your own and your partner's personal definitions of *primariness, intimacy, commitment,* and *responsibility*?

Addressing these questions and others like them is important to keeping a marriage relationship vital. In negotiating with each other, people must try to keep an open mind and use their creativity. Creatively negotiating not only an agreement but a working relationship is difficult, if only because two people, two imaginations, and two sets of needs are involved. Differences *will* arise because no two individuals have exactly the same points of view. (Some methods for reconciling these differences, and other conflicts, are discussed in Chapter 9.)

In Sum

Although marriage is less permanent and more flexible than it has ever been, it is still unique and set apart from other human relationships. Although legal marriage is not possible for same-sex couples in the United States, this situation may soon change. Meanwhile, the marriage premise includes expectations of permanence and primariness. As both of these expectations come to depend less on legal definitions and social conventions, partners need to invest more effort in sustaining a marriage.

Because few data are available on the period just before marriage, there is much concern about preparation for marriage—is it adequate in today's society? Premarital counseling and family life education in the schools are two approaches that have been developed, but we need more research data on their effectiveness.

We need to learn more about the transition into marriage, although it appears that the early years of marriage tend to be a happy time.

Two opposite poles on a continuum of marriage are the utilitarian marriage and the intrinsic marriage. Most marriages fall somewhere in between. Some frequently occurring marital types are the conflict-habituated, the devitalized, the passive-congenial, the vital, and the total marriage.

Partners change over the course of a marriage, so a relationship needs to be flexible if it is to continue to be intrinsically satisfying. Static marriages are usually devitalized. Marriage or domestic partner agreements, which can be renegotiated as the need arises, are one useful way of coming to mutual agreement. Working on a marriage agreement together can help partners develop a couple identity—one of the tasks of early marriage.

Key Terms

conflict-habituated
   marriage
conjugal relationship
consanguineous
   relationship
devitalized marriage
dominant dyad
expectations of permanence
extramarital sex
family of orientation
family of procreation
flexible marriages
interactional relationship
   pattern
intrinsic marriage
jealousy

la familia
marriage premise
parallel relationship pattern
passive-congenial marriage
personal marriage agree-
   ments
primariness
relationship agreements
role making
sexual exclusivity
sexually open marriage
static (closed) marriages
swinging
total marriage
utilitarian marriage
vital marriage

Study Questions

1. Discuss the expectation of permanence in a marital relationship. How does it affect a relationship?

2. Partners' expectations of primariness in marriage may be interpreted sexually or emotionally. How do these two interpretations differ? Which do you agree with? Why?

3. Give some reasons why people enter utilitarian marriages. How do these differ from reasons for entering intrinsic marriages?

4. Describe the five kinds of marital relationships classified by Cuber and Harroff. Which kinds are utilitarian and which are intrinsic? Explain your reasoning.

5. Why do you think that total marriages are rare?

6. Do you think that legalizing same-sex marriage is a good idea? Give evidence to support your opinion.

7. What are some intentional choices people make that lead to utilitarian relationships? How do these differ from choices by default?

8. Do you think it is easier to develop a utilitarian relationship or an intrinsic relationship? Why?

9. Differentiate static and flexible marriages, and discuss the reasons a marriage agreement can be helpful.

10. This chapter lists thirteen questions that the authors think are important to consider in any marriage agreement. Which do you think are the most important? Which do you think are the least important? Why?

11. Design your own marriage agreement. What does it tell you about your ideas on marriage?

12. What do you think couples should do before marriage to improve their chances of having a happy and stable marriage?

13. What do you think are likely to be the most important issues in the early years of marriage?

Suggested Readings

Cuber, John and Peggy Harroff. 1965. *The Significant Americans*. New York: Appleton-Century-Crofts. (Published also as *Sex and the Significant American*. Baltimore: Penguin, 1965). Classic study of five types of marriage.

Milardo, Robert M., ed. 1988. *Families and Social Networks*. Newbury Park, CA: Sage. Scholarly articles about family connections to others at various points in the family life course.

O'Neill, Nena. 1977. *The Marriage Premise*. New York: M. Evans. A reconsideration of the values of traditional marriage.

Salovey, Peter, ed. 1991. *The Psychology of Jealousy and Envy*. New York: Guilford. Reports the important research programs in this area of psychology.

Scanzoni, John, Karen Polonko, Jay Teachman, and Linda Thompson. 1989. *The Sexual Bond: Rethinking Families and Close Relationships*. Newbury Park, CA: Sage. A thorough consideration of issues of primariness and permanence in marriage that is anything but conventional. The authors propose reconceptualizing "family," given the pervasive departure from the cultural ideals of permanent marriage and the nuclear family.

Sherman, Suzanne, ed. 1992. *Lesbian and Gay Marriage: Private Commitments, Public Ceremonies*. Philadelphia: Temple University Press. A collection of first-person accounts by lesbian and gay male couples regarding their views on marriage and their various experiences with wedding/marriage ceremonies.

Wallerstein, Judith S. and Sandra Blakeslee. 1995. *The Good Marriage: How and Why Love Lasts*. New York: Houghton Mifflin. These two marriage counselors and researchers offer their own typology of marriage relationships and list tasks that every couple needs to accomplish in a long-term, successful union.

CHAPTER 9

Communication and Conflict Resolution in Marriages and Families

I know you believe you understand what you think I said, but I am not sure you realize that what you heard is not what I meant.

ANONYMOUS, FROM A POSTER

When two people always agree, there's no need for one of them anyway.

BEN BROWN, 54, MARRIED THIRTY-ONE YEARS

One important effect of loving is that it provides individuals with a sense of being personally and specially cared for in an impersonal society. Providing emotional security and feelings of belonging is an important function of the family today.

Families are powerful environments. Nowhere else in our society is there such power to support, hurt, comfort, denigrate, reassure, ridicule, love, and hate. For most Americans, belonging to a family probably yields both positive and negative feelings. At one moment a person feels supported and reassured; at another, unappreciated and misunderstood. For virtually all members, family living involves striking a delicate balance between belonging and feeling constrained. Striking such a balance involves effective communication, among other things.

Today, countless books and articles offer advice to help spouses and partners learn how to relate to and behave toward each other. Family-life courses in high schools and colleges teach future spouses how to communicate. Marriage counseling is a thriving profession. Marriage enrichment programs assist couples who have no serious conflict but who want to enhance their relationship. And social scientists and counselors are finding that improving and enriching an intimate relationship often centers around learning to communicate effectively.

Research on couple communication has found that unhappily married couples are distinguished by their failure to manage conflict. Distressed couples tend toward negative exchanges that put these couples' marriages on a downward spiral. It is very important, then, to learn to fight—that is, to communicate about conflict. The emotional tone of everyday communication is also important (Noller and Fitzpatrick 1991, pp. 45–46).

Communication between mates is addressed throughout this text. Chapter 4, for example, emphasized the need for self-esteem as a prerequisite to self-disclosure. Chapter 8 discussed negotiating flexible marriage agreements. In this chapter we'll focus on couple communication generally, especially communicating about conflicts. We'll see that the sulking that characterizes unhappiness and boredom in marriage often results from spouses' attempts to deny or ignore conflict, and we'll examine several other outcomes of refusing to deal openly with conflict. We will also explore some alienating practices that couples should avoid when fighting. Finally, we will discuss some healthy attitudes and propose some guidelines for communicating constructively, and we'll look at some differences among marriages in their preferred communication styles. To begin, we will address the idea of families as powerful environments.

Families as Powerful Environments

Using the interactionist theoretical perspective (Chapter 2), Charles Horton Cooley (1902) used the term **looking-glass self** to describe the process by which people adopt as their own, and gradually come to accept, the evaluation, definitions, and judgments of themselves they see reflected in the faces, words, and gestures of those around them.

Cooley's description has broad applicability in social environments, but perhaps nowhere is it more evident than in the family, for two reasons. First, people reveal more of themselves in families than they do in superficial relationships; thus, they are more vulnerable to evaluation, encouragement, and criticism. And second, the opinions of family members have a significant influence on the way people perceive themselves. Family members are **significant others:** people whose opinions about each other are important to each individual's self-esteem.

Attributions

The family has a tremendous influence on self-concept and behavior. So powerful is the family in shaping people's self-images that psychoanalyst R. D. Laing compares the process to hypnotism. In hypnosis one person gets another not only to *do* but also to *be* what the first person wants. Family members exercise hypnotic power over one another by implying that a member *is* a kind of person—selfish, kind, competent, lazy, and so on. This is accomplished through the process of **attribution:** ascribing certain character traits to people. For example, a person is told she or he *is* intelligent, not directly instructed to *act* intelligently. In fact, attributions can be considered part of a family's archives. Anecdotes reinforce a member's identity as reckless, helpful, or unlucky (Weigert and Hastings 1977).

Attributions are often more powerful in a primary group of more than two: One person says or implies something about a second person to a third person in front of the second. For example, a mother might tell her children in front of their father that he is a skilled handyman. With repetition, the father may soon have a family reputation, which even he believes, of being able to fix anything—even if he can't.

As the process of attribution occurs again and again, family members unconsciously begin to behave according to their reputations. In a sense, family interaction can be viewed as a play in which the same theme is repeated in scenario after scenario, such as being late for work "again" or becoming flustered before guests arrive. Each member, without thinking, plays out the part attributed to him or her (Laing 1971, pp. 78–80).

Consensual Validation

Family members' definitions of the world around them are also influenced by their families. People see things differently from one another, and as a result, they define reality in many different ways. They also depend on others, especially significant others, to help them affirm their definitions. This process is called **consensual validation,** and it is an important dimension of family functioning. Through conversation, members can help one another feel comfortable about how each perceives the world. Husbands and wives may feel stronger in their political convictions—or in their views on social issues such as the energy crisis or the death penalty—if their spouse agrees; or a teenager may feel more comfortable when she refuses to drink with her friends if she has a sibling who thinks she's doing the right thing.

Consensual validation also takes place in a much broader sense. In Berger and Kellner's (1970) words, "the reality of the world is sustained through conversation with significant others." Family interaction

> validates over and over again the fundamental definitions of reality once entered into, not, of course, so much by explicit articulation, but precisely by taking the definitions silently for granted and conversing about all conceivable matters on this taken-for-granted basis. (pp. 52–53)

One of the alarming aspects of marital breakup is that the taken-for-granted world is no longer validated by the partner (Vaughan 1986).

Negative Family Power

Even in an ongoing family, interaction does not always validate members' perceptions of situations and events. Significant others in a family have power, if they choose to use it, to make one member feel out of step. One way families do this is by a process Bach and Wyden (1970) call *gaslighting.* The term comes from the old movie *Gaslight,* in which a husband attempts to drive his wife insane so she can be put in an asylum. He gradually turns down their gaslight over a period of time but tells his wife she's wrong when she perceives that their home is becoming dimmer.

In **gaslighting,** one partner chips away at the other's perception of him- or herself and at the other's definitions of reality. Typically, this occurs through destructive snipes at a partner, sometimes faintly camouflaged as humor. Often the gaslighter uses mixed messages or sarcasm. When a spouse protests or questions, the gaslighter denies everything and insists that the spouse must be crazy. In the movie the husband *wanted* to drive his wife insane; in real families, members can accomplish the same results without intending to.

A related destructive family behavior is **scapegoating:** consistently blaming one particular member for virtually everything that goes wrong in the family (Vogel and Bell 1960).

The power of the family over its members is due partly to the modern definition of the family as a sanctuary and retreat. Ironically, the family's tendency to create boundaries not only gives members a place to belong but also increases family members' power to undermine one another's self-esteem (Pruchino, Burant, and Peters 1994). Family systems theory, described in Chapter 2, represents one way of explaining how this occurs. The whole family, as a system of interacting parts, works together to create complementary roles, or family personalities, for each of its members. Some of these roles, such as family rescuer or scapegoat, may be antithetical to building self-esteem.

In calling attention to the negative power of the family to shape self-concept and behavior, we do not mean to leave the impression that powerful family forces are invariably malevolent. Rather, we want to balance the "folk concept of the family" (Edwards 1991), which conveys an image of warmth, understanding, support, and happy holiday rituals, with a description of the darker side of family life (Worchel 1984). Some readers may be dealing with this dark side.

Antidotes to Destructive Behavior

An antidote to destructive behavior such as gaslighting or scapegoating is for a victim to begin thinking more independently and challenging negative attributions. People do this more readily when they have sources for building self-esteem outside their families. Among adolescents, for example, the peer group is a common resource. Other sources are grandparents and other relatives, teachers, work colleagues, or therapists. With such help from the outside, people can more easily resist the temptation to let destructive significant others define their identities.

A second antidote is to begin to foster a family atmosphere of encouragement rather than criticism. This does *not* mean that families are successful only if they are unvarying centers of solidarity, consensus, and harmony—and never of conflict, anger, or fighting. But according to an idealized image, as we will see in the next section, this is what many people expect the family to be.

Our society holds out a **conflict taboo** that considers conflict and anger morally wrong, and it discourages these emotions even (or especially) within the family. The assumption that conflict and anger don't belong in healthy relationships is based partly on the idea that love is the polar opposite of hate (Crosby 1991). But emotional intimacy necessarily involves feelings of both wanting to be close and needing to be separate, of agreeing and disagreeing.

Conflict and Love

Marital anger and conflict are necessary forces and a challenge to be met rather than avoided. This is especially true in the early years of marriage, when individuals are often still engaged in the process of getting to know each other. (At this point, a complete lack of conflict might even be cause for concern!) It is also true at points throughout a good relationship. As we saw in Chapter 1, partners do not necessarily undergo the same changes at the same time. In one observer's words, "for a marriage to be reasonably productive, a couple needs to arrive at a balance of love *and* hate, honor *and* dishonor, obedience *and* disobedience," with partners acting not only as each other's companions and lovers but also as each other's critics (Charny 1974, p. 55).

We don't want to give the impression that communication and loving are *all* or *only* conflict resolution. (Box 9.1, "As We Make Choices: Some Rules for a Successful Relationship," gives ideas about communicating more generally.) Crosby points out that people can misinterpret the idea of working at marriages and other relationships. "Instead of working *at* marriage we may, with all good intentions, end up making work *of* marriage" (Crosby 1991, p. 287). Along with resolving conflicts, being married involves play—humor, spontaneity, fun. Being able to feel playful and behave playfully involves feeling safe in the presence of a partner. But how conflicts, which do necessarily arise, are addressed and resolved has much to do with how secure mates feel in their relationship.

Denying Conflict: Some Results

Many married couples are reluctant to fight. This reluctance can have destructive effects on the partners as individuals and on their relationship. For example, recent research examined the communication approaches of 133 African American and 149 white newly married couples. Couples who believed in avoiding marital conflict were less happy than others two years later (Crohan 1992). Other longitudinal research found that disagreement and anger exchange were related to current marital unhappiness but led to measurable improvement in marital satisfaction at the three-year follow-up (Gottman and Krokoff 1989).

The implication of this research is that couples may have gone through a painful period in order to deal with their dissatisfactions, but engaging in conflict enhanced the long-run health of their marriages.

In a later section we'll review some research on marital communication that suggests that the dynamics of conflict or conflict avoidance differs depending on the nature of the marriage. Here we'll look at several potential negative side effects of conflict avoidance: first those that relate to the individual, then those that affect the marriage relationship.

ANGER "INSTEADS"   Most people know when they are angry, but they feel uncomfortable about expressing anger directly. Through years of being socialized in the conflict taboo, people learn not to raise their voices and not to make an issue of things. One result is that many people resort to anger substitutes, or **anger "insteads"** (Kassorla 1973), rather than dealing directly with their emotions.

"Insteads" such as overeating, boredom, depression, physical illness, and gossip are probably familiar to most readers. Even though these "insteads" may be more socially acceptable than the direct expression of anger, they can be self-destructive. Because of this, they are costly, both to the individual who uses them and to any intimate relationship of the individuals. For example, sex therapists report that one of the most common complaints they hear from married couples these days is that "we don't feel much like having sex anymore" (Masters, Johnson, and Kolodny 1994). Repressing one's anger can contribute to this sexual boredom.

PASSIVE-AGGRESSION   The anger "insteads" can be subtle forms of what psychologists call **passive-aggression.** When a person expresses anger at someone but does so indirectly rather than directly, that behavior is passive-aggression. People use passive-aggression for the same reason they use anger substitutes—they are afraid of direct conflict.

Chronic criticism, nagging, nitpicking, and sarcasm are all forms of passive-aggression; like overeating, these actions momentarily relieve anxiety. In intimate relationships, however, they create unnecessary distance and pain. Most people use sarcasm unthinkingly, and they often aren't aware of its effect on a partner. But being the target of a sarcastic remark can be painful; it also can result in partners feeling alienated from each other.

Some Rules for a Successful Relationship

Psychologists Nathaniel Branden and Robert Sternberg, both of whom are mentioned in our discussions about love in Chapter 5, have developed some rules for nourishing a romantically loving relationship. Here are ten:

1. *Express your love verbally.* Say "I love you" or some equivalent (in contrast to the attitude, "What do you mean, do I love you? I married you, didn't I?").

2. *Be physically affectionate.* This includes making love sexually as well as hand-holding, kissing, cuddling, and comforting—with a cup of tea, a pillow, or a woolly blanket.

3. *Express your appreciation and even admiration.* Talk together about what you like, enjoy, and cherish in each other.

4. *Share more about yourself with your partner than you do with any other person.* In other words, keep each other primary (see Chapter 8).

5. *Offer each other an emotional support system.* Be there for each other in times of illness, difficulty, and crisis; be generally helpful and nurturing—devoted to each other's well-being.

6. *Express your love materially.* Send cards or give presents, big and small, on more than just routine occasions. Lighten the burden of your partner's life once in a while by doing more than your agreed-upon share of the chores.

7. *Accept your partner's demands and put up with your partner's shortcomings.* We are not talking here about putting up with physical or verbal abuse. But demands and shortcomings are part of every happy relationship; and so is the grace with which we respond to them. Love your partner, not an unattainable idealization of him or her.

8. *Make time to be alone together.* This time should be exclusively devoted to the two of you as a couple. Understand that love requires attention and leisure.

9. *Do not take your relationship for granted.* Make your relationship your first priority and actively seek to meet each other's needs.

10. *Do unto each other as you would have the other do unto you.* Unconsciously we sometimes want to give less than we get, or to be treated in special ways that we seldom offer our mate. Try to see things from your lover's viewpoint so that you can develop the empathy that underlies every lasting close relationship.

Source: Branden 1988, pp. 225–28; Sternberg 1988b, pp. 272–77).

Sex becomes an arena for ongoing conflict when mates habitually withhold it or use it as passive-aggressive behavior. A partner makes a disparaging comment in front of company. The hurt spouse says nothing at the time but rejects the other's sexual advances later that night because "I'm just too tired." It is much better to express anger at the time an incident occurs. Otherwise, the anger festers and contaminates other areas of the relationship.

Other forms of passive-aggression are sabotage and displacement. In **sabotage** one partner attempts to spoil or undermine some activity the other has planned. The husband who is angry because his wife invited friends over when he wanted to relax may sabotage her evening, for example, by acting bored. In **displacement** a person directs anger at people or things the other cherishes. A wife who is angry with her husband for spending too much time and energy on his career may hate his expensive car, or a husband who feels angry and threatened because his wife returned to school may express disgust for her books and "clutter." Often, child abuse can be related to displaced aggression felt by a parent. Child abuse is discussed further in Chapter 10.

DEVITALIZED MARRIAGES   Another possible consequence of **suppression of anger,** the devitalized marriage, was discussed in Chapter 8. The result of suppressing anger over a long period of time can be indifference, the opposite of both love and hate (Crosby 1991). Because partners refuse to recognize or express any anger toward each other, their penalty may be the emotional divorce of a devitalized marriage.

This kind of gradual erosion does not take place only within marital relationships. Unmarried lovers who feel progressively less enthusiastic about their partners may be refusing to accept and voice anger, too. Little irritations build up until finally one decides, "I'm tired of his (or her) always doing _____ ." If the irritations had been brought into the open, the offending lover could have chosen to change, and the relationship might have continued.

Although the suppression of anger can be a source of boredom and devitalization, partners can go too far in the opposite direction and habitually or violently hurt each other in angry outbursts. Chapter 10 addresses spouse abuse at the extreme of destructive conflict. Here we'll look at some emotionally alienating behaviors that may come up during conflicts.

Some Alienating Practices

Alienating fight tactics are those that tend to create distance between intimates. They don't resolve tension or conflict; they increase it. Many of the tactics discussed next are prevalent among conflict-habituated couples (although almost everyone has probably had firsthand experience with some of them).

Fight Evading

Chronic fight evaders react to their partner's attempts to raise disputed or tension-producing issues by refusing to engage with the partner's initiatives. They fear conflict and hesitate to accept their own and others' hostile or angry emotions.

Fight evaders use several tactics to avoid fighting, such as

1. Leaving the house or the scene when a fight threatens

2. Turning sullen and refusing to argue or talk

3. Derailing potential arguments by saying, "I can't take it when you yell at me"

4. Flatly stating, "I can't take you seriously when you act this way"

5. Using the "hit and run" tactic of filing a complaint, then leaving no time for an answer or for a resolution (Bach and Wyden 1970)

Research indicates that a common pattern in distressed marriages is a repeated cycle of negative verbal expression by a wife and withdrawal by the husband (Kurdek 1995); which came first is hard to say. One researcher speculates that wives, being more attuned to the emotional quality of a marriage and having less power, attempt to bring conflict out into the open by initiatives that have an attention-getting negative tone. Husbands try to minimize conflict by conciliatory gestures. Either a healthy problem-solving dialogue may ensue, or, more likely, the husband's minimization of conflict may seem to the wife to be a lack of recognition of her emotional needs and her concern about the marriage (Noller and Fitzpatrick 1991). In any event, withdrawal from interaction about conflict is related to the deterioration of a marriage over time (Kurdek 1995).

Fight evaders often argue that they avoid conflicts because they don't want to hurt their partners. Often, however, the one they are really trying to protect is

Evading and suppressing conflict can devitalize both a marriage and an individual's self-worth. Chronic fight evaders hesitate to acknowledge their own and their partner's angry emotions and may fear rejection or retaliation.

themselves. (Case Study 9.1 illustrates the anxiety that can accompany sharing a grievance.) "A great deal of dishonesty that ostensibly occurs in an effort to prevent pain actually occurs as we try to protect and shield ourselves from the agony of feeling our own pain, fear, fright, shame, or embarrassment" (Crosby 1991, pp. 159–60).

Psychologist John Gottman has argued that wives and husbands have different goals when they disagree. "The wife wants to resolve the disagreement so that she feels closer to the husband and respected by him. The husband, though, just wants to avoid a blowup. The husband doesn't see the disagreement as an opportunity for closeness, but for trouble" (in Goleman 1986). "I just don't know what she wants." In one husband's words, "When she comes after me like that, yapping like that, she might as well be hitting me with a bat" (Rubin 1976, p. 113). And as another explained, "I just got mad and I'd take off—go out with the guys and have a few beers or something. When I'd get back, things would be even worse" (p. 77). From his wife's perspective, "The more I screamed, the more he'd withdraw, until finally I'd go kind of crazy. Then he'd leave and not come back until two or three in the morning sometimes" (p. 79). But evading fights can make partners who want and need to fight feel worse, not better.

Gunnysacking and Kitchen-Sink Fights

Fight evading encourages **gunnysacking:** keeping one's grievances secret while tossing them into an imaginary gunnysack that grows heavier and heavier over time. Martyring (see Chapter 4) is typically accompanied by gunnysacking. When marital complaints are toted and nursed along quietly in a gunnysack for any length of time, they "make a dreadful mess when they burst" out (Bach and Wyden 1970, p. 19).

A typical result is a destructive and ineffective kind of fighting that Bach and Wyden call the **kitchen-sink fight,** in which partners don't focus on resolving specific issues. Instead, "the kitchen plumbing is about all that isn't thrown in such a battle." The following is an illustration of a kitchen-sink fight:

HE: I need more socks.

SHE: So get some.

HE: I thought you could pick them up since you'll be out this afternoon anyway.

SHE: You're just like your father—always ordering somebody around.

HE: Oh yeah? At least my mother could cook!

SHE: Well, she's a mess herself! She hasn't had a decent haircut in years.

Sharing a Grievance

Janet, 28, is married to Joe, who has two sons from a previous marriage. Here Janet talks about her need to share a grievance with Joe and the anxiety that taking such a risk can cause.

Joe's sons live with their mother in Colorado. They come to stay with us Christmas, spring break, and summers. Last summer when they were here, I was a wreck. I was jealous of Joe's relationship with the boys. He did so much with them, it seemed, and I was feeling left out. I was so upset, I was feeling sick.

I went to a physician, trying to figure out whether I really was physically ill or whether it was my head. He said, "Just take some time out and think about where you stand. And as time goes by and you feel the time is right, certainly discuss these feelings with your husband." Which is exactly what I did!

It was really kind of tricky though. Joe wanted so desperately for everybody to be happy, for us to have the one great big happy family. It was tricky because I had to find the right moment, the right atmosphere to speak what was on my mind. But I did. And it worked!

I remember so vividly reaching the point where I knew that I had to say something real soon or else it was gonna be bad for everybody. I knew Joe and I had to be away from the boys when we talked. And Joe had to be away from the phone.

So one Saturday afternoon he had to go to his office to pick something up, and I volunteered to go with him. But the atmosphere wasn't right in the car. I didn't say anything. I thought, "Well, as soon as—when we get to the office—I'll just sit down and kind of open up then."

Well, the janitor that opened up the outside door also had to open the door to Joe's office. He just kept standing there, so I couldn't open up. I can remember the anxiety. My hands were sweating.

Finally, back in the car, before we got home, I said, "We have to stop for a drink because there's something on my mind and I really have to talk to you about it." We did. I think I opened the conversation by saying, "You know, I finally realize what it feels like to be a father when there is a newborn in the house." Because there was this bonding between him and the boys, and you hear all the time about the father's feeling left out with a newborn. And that was just exactly how I felt.

He said, "Well, let's see. How shall we approach this?" He was very open. I laugh now at all my anxiety because there was a tremendous effort on his part to really understand what my feelings were.

Why do you think Janet felt anxiety about sharing her feelings with Joe? What risk did she take in being honest? What did she gain? What did Joe gain?

Suppose Joe had been hostile rather than supportive. In your opinion, would the risk still have been worth it for Janet? What do you believe might have happened in the long run had Janet not risked initiating a fight?

HE: You ought to do something with *your* hair. It's ridiculous.

SHE: Not that there's room in the bathroom with your stuff scattered all over. . . .

These fighters reached into their gunnysacks to drag totally irrelevant and past occurrences into their argument. The immediate issue—who will buy the socks—is not resolved. In fact, it seems to have been temporarily forgotten (or itself gunnysacked to be used as future ammunition!). Obviously, this fight did not make the partners feel better about themselves as individuals or about their relationship. Gunnysacking couples might do well to try to focus their arguments and keep up to date on their grievances as a way of discouraging this habit.

Drawing by Weber; © 1991 The New Yorker Magazine, Inc.

Mixed, or Double, Messages

A third alienating tactic is the use of **mixed, or double, messages:** simultaneous messages that contradict each other.[1] Contradictory messages can be verbal, or one can be verbal and one nonverbal. For example, a spouse agrees to go out to eat with a partner but at the same time yawns and says that he or she is tired and had a hard day at work. Or a partner insists "Of course I love you" while picking an invisible speck from her or his sleeve in a gesture of indifference.

Senders of mixed messages may not be aware of what they are doing, and mixed messages can be very subtle. They usually result from simultaneously wanting to recognize and to deny conflict or tension. Mixed messages allow senders to let other people know they are angry at them and at the same time to deny that they are. A classic example is the *silent treatment.* A spouse becomes aware that she or he has said

or done something and asks what's wrong. "Oh, nothing," the partner replies, without much feeling, but everything about the partner's face, body, attitude, and posture suggests that something is indeed wrong.

Besides the silent treatment, other ways to indicate that something is wrong while denying it include making a partner the butt of jokes, using subtle innuendos rather than direct communication, and being sarcastic (also defended as "just a joke" by mixed-message senders).

Sarcasm and other mixed messages create distance and cause pain and confusion, for they prevent honest communication from taking place. Expressing anger is a far better tactic, for it opens the way for solutions. The next section presents guidelines for fighting more constructively.

Bonding Fights

The tactics described previously are likely to alienate partners, but some goals and strategies can help make fighting productive rather than destructive. This kind of fighting, which brings people closer rather than pushing them apart, is called **bonding fighting.** The key to creating a bonding fight is for partners to

1. Communication scholars and counselors point out that there are two major aspects of any communication: *what* is said (the verbal message) and *how* it is said (the nonverbal "metamessage"). The metamessage involves tone of voice, inflection, and body language. In a mixed message, the verbal message does not correspond with the nonverbal metamessage.

try to build up, not tear down, each other's self-esteem while fighting.

Social groups within the United States vary considerably in the endorsement their cultures give to the open expression of emotion, which may make bonding fighting more or less difficult for people within or across cultures. Deborah Tannen's best-selling book *You Just Don't Understand,* which drew wide attention for its comparison of men's and women's communication styles, also points out communication differences among, for example, New York Jews, Californians, New Englanders, and Midwesterners, and among Scandinavians, Canada's native peoples, and Greeks.

And culturally based communication patterns may change over time (Tannen 1990, pp. 201–10). Some black writers (see, for example, Franklin 1986; Chapman 1988) have emphasized the importance of open communication in their culture, one historically supportive of expressiveness. Sociologist Clyde Franklin II argues that empathy between black men and women has deteriorated in recent years: "As black males have attempted to become 'men' in America, they have shed some of the important qualities of humanity," becoming increasingly nonexpressive and nonempathetic in their male–female relationships. And, according to Franklin, black women "who have embraced the feminist perspective also have discarded altruism," by which he appears to mean concern for the other (Franklin 1986, pp. 112–13). Data from a national sample of 2,059 married individuals shows African Americans reporting somewhat lower levels of marital harmony or satisfaction than whites (Broman 1993). Nevertheless, there are countless supportive marriages within the black community (Cose 1995).

Some Guidelines

Now we turn to several specific guidelines for constructive fighting.[2] Each of them has as its goals the development of a mutuality in communication and conflict resolution.

2. These guidelines are largely taken from Bach and Wyden's *The Intimate Enemy* (1970) and Crosby's *Illusion and Disillusion: The Self in Love and Marriage* (1991). These books were written for a popular audience and represent the authors' opinions and experience rather than a synthesis of psychological theory or research. However, academic communications research has supported (or at least failed to contradict) this advice. See Noller and Fitzpatrick (1991) for a review of academic research on marital communication.

LISTEN   Listening is so basic we often assume we're doing it effectively. But in couple conflict, listening may be forgotten as both partners strive to put forward their points of view. Accuracy in hearing what the speaker intended to communicate is noticeably lower in less well-functioning marriages (Noller and Fitzpatrick 1991).

Communication involves both a sender and a receiver. Just as the sender gives both a verbal message and a nonverbal metamessage, so also does a receiver give nonverbal cues about how seriously she or he is taking the message. Listening while continuing to do chores, for example, sends the nonverbal message that what is being heard is not very important. Thinking about what listening does may help a person listen as well as talk.

According to sociologist/counselor Carlfred Broderick, good listening has the following important results:

1. The attitude of listening itself shows love, concern, and respect. . . . Any act that expresses a positive attitude is likely to trigger a sequence of positive responses back and forth. . . .

2. The avoidance of interrupting and criticism prevents the sending of negative messages such as "I don't care how you feel or what you think." "You're not worth listening to." . . .

3. You discover how things actually look from your spouse's or partner's point of view. There's a risk, because what you hear may be surprising and even unsettling. But it is nearly always worth it. In fact, it's hard to imagine how any couple can become close without achieving insight into each other's feelings.

4. You lose your status as chief expert on what your spouse really thinks, wants, fears, and feels. Instead, your spouse takes over as the final authority on his or her own feelings. . . . [Furthermore] if you listen sympathetically to your spouse, he or she is able to develop greater clarity in areas that may have been confused and confusing. . . .

5. You set an example for your spouse to follow in listening to your . . . feelings (Broderick 1979a, pp. 40–41).

LEVEL WITH EACH OTHER   Partners need to be as candid as possible; counselors call this **leveling**—being transparent, authentic, and explicit about how one feels, "especially concerning the more conflictive or hurtful aspects" of an intimate relationship (Bach and Wyden 1970, p. 368). Leveling is self-disclosure in action.

Various studies indicate that because of boredom (an anger "instead"), indifference, or the mistaken impression that the partner already knows how the other feels, spouses often overestimate how accurately their partner understands them and then fail to understand their partner. Underlying conflicts often go unresolved because partners fail to voice their feelings, irritations, and preferences—and neither is aware that the other is holding back. The solution to this problem is to air grievances—to candidly explain where one stands and how one feels about a specific situation.

Being candid does not mean the same thing as being tactless. Leveling, in fact, is never intentionally hurtful and is always consistent with the next guideline.

USE I-STATEMENTS TO AVOID ATTACKS

Attacks are assaults on a partner's character or self-esteem. Not surprisingly, research has found "contemptuous remarks or insults" to be destructive fighting tactics (John Gottman, research reported in Goleman 1989, p. 1-B).

A rule in avoiding attack is to use *I* rather than *you* or *why.* The receiver usually perceives I-statements as an attempt to recognize and communicate feelings, but you- and why-statements are more likely to be perceived as attacks—whether or not they are intended as such. For example, instead of asking "Why are *you* late?" a statement such as "*I* was worried because you hadn't arrived" may allow more communication.

I-statements are most effective if they are communicated in a positive way. A partner should express his or her anger directly, but it will seem less threatening if he or she conveys positive feelings at the same time negative emotions are voiced. The message comes across, but it's not as bitter as when only angry feelings are expressed.

GIVE FEEDBACK AND CHECK OUT YOUR INTERPRETATION

Partners must also ensure that complaints and other messages are properly understood. Communication consists of a sender, a receiver, and a message. A message is sent not only through the content and emotional tone of words but through facial expressions, gestures, body language, and voice quality as well. In fact, the words of an emotional message have far less effect than does the speaker's facial expression or tone of voice. The receiver must evaluate all the components of the message and then make some judgment about what the message means. Obviously, there is much room for error in interpreting a message because words and nonverbal cues alike can have different meanings in different contexts.

Partners can ensure that they receive messages accurately by giving **feedback:** They repeat in their own words what the other has said or revealed. For example, a husband says, "I don't like it when you talk on the phone for a long time in the evenings, when we have so little time together." To give feedback, his wife would respond with something like, "I realize it irritates you when I'm on the phone in the evenings instead of visiting with you." Giving feedback requires self-esteem because it involves facing and dealing with an issue rather than becoming defensive or avoiding it.

Because studies consistently show that partners in distressed marriages seldom understand each other as well as they think they do (Noller and Fitzpatrick 1991), a good habit is to ask for feedback by a process of **checking-it-out:** asking the other person whether your perception of her or his feelings or of the present situation is accurate. Checking-it-out often helps avoid unnecessary hurt feelings. As the following example shows, the procedure can also help partners avoid imagining trouble that may not be there:

SHE: I sense you're angry about something. (*checking-it-out*) Is it because it's my class night and I haven't made dinner?

HE: No, that's not why, but I am angry. I'm angry because I was tied up in traffic an extra half hour on my way home.

Bach and Wyden propose a fight plan in which partners take turns airing grievances, checking things out, and giving feedback. First, the complainant states a grievance. Without interrupting or becoming defensive, the recipient listens, then asks questions for clarification if necessary and gives feedback. When both are satisfied that they understand each other, the recipient begins to level, responding to the complainant's grievance. Now the complainant listens quietly, then offers feedback. Partners take turns leveling, listening, checking things out, and giving feedback until both positions are clearly understood.

Open communication involves listening—without judgment, without formulating a response while the other talks, and without interrupting. The goal isn't agreement, but acknowledgment, insight, and understanding.

CHOOSE THE TIME AND PLACE CAREFULLY

Fights can be nonconstructive if the complainant raises grievances at the wrong time. One partner may be ready to fight when the other is almost asleep or is working on an important assignment, for instance. At such times, the person who picked the fight may get more than he or she bargained for.

Partners should try to negotiate *gripe hours* by pinning down a time and place for a fight. Fighting by appointment may sound silly and may be difficult to arrange, but it has two important advantages. First, complainants can organize their thoughts and feelings more calmly and deliberately, increasing the likelihood that their arguments will be persuasive. Second, recipients of complaints have time before the fight to prepare themselves for some criticism.

FOCUS ANGER ONLY ON SPECIFIC ISSUES

Constructive fighting aims at resolving specific problems that are happening *now*—not at gunnysacking. To maintain their self-esteem, recipients of complaints

need to feel that they can do something specific to help resolve the problem raised. This will be more difficult if they feel overwhelmed.

KNOW WHAT THE FIGHT IS ABOUT

Fighting over petty annoyances, such as who should have put gas in the car, is healthy and can even be fun as an essentially harmless way to release tension. But partners sometimes unconsciously allow trivial issues to become decoys, so that they evade the real area of conflict and leave it unresolved. An irate husband, for example, who complains about how his wife treats their children may really be fighting about his feelings of rejection because he feels that his wife isn't giving him enough attention. Before partners bring up a grievance, they need to ask themselves what they are really fighting about.

ASK FOR A SPECIFIC CHANGE, BUT BE OPEN TO COMPROMISE

Initially, complainants should be ready to propose at least one solution to the problem. Recipients, too, need to be willing to come up with possible solutions. If they are careful to keep proposed solutions pertinent to the issue at hand, partners can then negotiate alternatives.

Resolving specific issues involves bargaining and negotiation. Partners need to recognize that there are probably several ways to solve a particular problem, and backing each other into corners with ultimatums and counterultimatums is not negotiation but attack. John Gottman found that happily married couples reach agreement rather quickly. Either one partner gives in to the other without resentment or the two compromise. Unhappily married couples tend to continue in a cycle of conflict. Stubbornness is associated with deterioration of the marriage over time (Gottman and Krotkoff 1989).

BE WILLING TO CHANGE YOURSELF   Communication, of course, needs to be accompanied by action. The romantic belief that couples should accept each other completely as they are is often merged with the individualistic view that people should be exactly what they choose to be. The result is an erroneous assumption that if a partner loves you, he or she will accept you just as you are and not ask for even minor changes. On the contrary, partners need to be willing to change themselves, to be changed by others, and to be influenced by their partner's feelings and rational arguments. Defensiveness and refusing to change are dysfunctional responses that contribute to marital deterioration (Gottman and Krotkoff 1989). Every intimate relationship involves negotiation and mutual compromise; partners who refuse to change, or who insist they cannot, are in effect refusing to engage in an intimate relationship.

DON'T TRY TO WIN   Finally, partners must not compete in fights. American society encourages people to see almost everything they do in terms of winning or losing. Yet research clearly indicates that the tactics associated with winning in a particular conflict are also those associated with greater marital unhappiness (Noller and Fitzpatrick 1991).

Bonding fights, like dancing, can't involve a winner and a loser. If one partner must win, then the other obviously must lose. But losing lessens a person's self-esteem and increases resentment and strain on the relationship. This is why in intimate fighting there can never be one winner and one loser—only two losers.

Both partners lose if they engage in destructive conflict. Both win if they become closer and settle, or at least understand, their differences.

If a couple cannot designate a winner and a loser, they can be at a loss to know how to end a fight. Ideally, a fight ends when there has been a mutually satisfactory airing of each partner's views. Bach suggests that partners question each other to make sure that they've said all they need to say.

Sometimes, when partners are too hurt or frightened to continue, they need to stop fighting before they reach a resolution. Women often cry as a signal that they've been hit below the "beltline" (the beltline being Bach's image for a person's tolerance limit) or that they feel too frustrated or hurt to go on fighting. Men experience the same feelings, but they have learned from childhood not to cry. Hence, they may hide their emotions, or they may erupt angrily. In either case it is necessary to bargain about whether the fight should continue. The partner who is not feeling so hurt or frightened might ask, "Do you want to stop now or to go on with this?" If the answer is "I want to stop," the fight should be either terminated or interrupted for a time.

Changing Fighting Habits

Social scientist Suzanne Steinmetz traced the patterns of how families resolve conflict in fifty-seven intact urban and suburban families. Her research shows not only that individual families assume consistent patterns or habits for facing conflict but also that these patterns are passed from one generation to the next (see also VanLear 1992). Parents who resort to physical abuse teach their children, in effect, that such abuse is an acceptable outlet for tension. In this way the "cycle of violence" is perpetuated (Steinmetz 1977). Irene Kassorla (1973) points out that a similar process takes place as people learn anger "insteads" from their parents.

Generational Change

Even though this generalization about the transmission of marital communication patterns from parents to children is correct, it is important to note variations from this pattern. VanLear (1992) found that young married men tend to rebel against their parents' marital conflict style. (Women are more apt to follow their parents' lead.) VanLear's comparison (in fifty-eight

Learning to express anger and to deal with conflict early in a relationship or marriage are challenges to be met rather than avoided. Acknowledging and resolving conflict is painful, but it often strengthens marriage in the long run.

families) of married couples in their late 20s to parental couples in their 50s found that the younger married men tended to choose different conflict styles from their parents. Most often, that meant they evidenced less conflict avoidance. These men also married women very different in conflict style from their own mothers. Moreover, the younger couples also reported more sharing and disclosure than was characteristic of the parental generation (VanLear 1992).

Couple Change

The change in couple communication and conflict management just described is the unplanned outcome of social change and family dynamics. But couples also may consciously act to change marital interaction patterns. Some training programs in communication, conducted by psychologists, have proven quite effective in helping people to change these behaviors. One such program, which is based on psychologist John Gottman's research, is a program for engaged couples at the University of Denver. Thirty-five couples who received training in expression of feelings, constructive conflict, and communication skills were reported to be happier one year later than couples who received no training (Brandt 1982). The following steps

are given to help people work for change on their own.

The first step in changing destructive fighting habits is to accept the reality of conflict. Feeling that conflict is wrong or that it should be kept out of the children's sight, for example, can cause dangerous buildups of tension.

A second, related step is to begin to use the guidelines for bonding fighting we have described. But changing can be a confusing and frightening process. For example, partners who have been hardened fight evaders may suddenly need to argue over just about *everything*. Afraid that they won't like themselves this way, that their family and friends "won't even know me," or that their partners won't understand or cooperate, these spouses will be tempted to return to their old and familiar ways. They need to realize two important things.

First, the feeling that one wants to fight over everything is largely a short-term result of releasing long-harbored resentments. The best way to counteract this tendency is to concentrate on keeping arguments focused only on current and specific issues. Recognizing old resentments is an indication of growth and well-being, but fighting about old resentments isn't necessary.

Second, as partners grow more accustomed to voicing grievances regularly, their fights may hardly seem like fights at all: Partners will gradually learn to incorporate many irritations and requests into their normal conversations. Partners adept at constructive fighting often argue in normal tones of voice and even with humor. In a very real sense their disagreements are essential to their intimacy.

When partners are just learning to fight, however, they are often frightened and insecure. One way to begin is by writing letters or using a tape recorder. In this way the complainant is not inhibited or stopped by a mate's interruption or hostile nonverbal cues. Then the mate can read or listen to the other's complaints in privacy when she or he is ready to listen. Letters and tape recorders are no substitute for face-to-face communication, but they don't intimidate either. Once partners become more comfortable with their differences, they can begin to fight face to face.

Another option is to record a fight and play it back later. This exercise can help spouses look at themselves objectively. They can ask themselves whether and where they really listened, offered feedback, and stuck to specifics or resorted to hurtful, alienating tactics.

Although all the suggestions made in this text may help, learning to fight fair is not easy. A good number of fights are even about fighting itself. Sometimes one or both partners feel that they need outside help with their fighting, and they may decide to have a marriage counselor serve as referee. See Appendix H, "Marriage and Close Relationship Counseling," for a discussion of this alternative.

Gender Differences

Deborah Tannen's book *You Just Don't Understand* (1990) argues that men typically engage in **report talk,** conversation aimed mainly at conveying information. Women, on the other hand, are likely to engage in **rapport talk,** speaking to gain or reinforce rapport or intimacy. The resulting "men and women talking at cross-purposes" (p. 287) Tannen identifies is played out in marital or other relationship communication. "Gender is a category that will not go away," says Tannen (p. 287).

To those who ask, "Is change possible?" Tannen argues that it is. "But those who ask this question rarely want to change their own styles. Usually, what they have in mind is sending their partners for repair"

(p. 297). Instead, "a more realistic approach is to learn how to interpret each other's messages and explain your own in a way your partner can understand and accept" (p. 197). "Opening lines of communication" may involve understanding as well as change.

The Myth of Conflict-Free Conflict

By now, so much attention has been devoted to the bonding capacity of intimate fighting that it may seem as if conflict itself can be free of conflict. It can't. Even the fairest fighters hit below the belt once in a while, and it's probably safe to say that all fighting involves some degree of frustration and hurt feelings. After all, anger is anger and hostility is hostility—even between partners who are very close to each other.

Moreover, some spouses are married to mates who don't want to learn to fight more positively. As we saw in Chapter 8, partners can indeed be satisfied with passive-congenial, devitalized, or even conflict-habituated unions. In marriages in which one partner wants to change and the other doesn't, sometimes much can be gained if just one partner begins to communicate more positively. Other times, however, positive changes in one spouse do not spur growth in the other. Situations like this often end in divorce.

Even when both partners develop constructive habits, all of their problems will not necessarily be resolved, and their marriage might not last. It is possible that good communication is a *characteristic* of satisfying relationships and marriages, rather than a *cause* of marital satisfaction (Stephen 1985).

Even though a complainant may feel that he or she is being fair in bringing up a grievance and discussing it openly and calmly, the recipient may view the complaint as critical and punitive, and it may be a blow to that partner's self-esteem. The recipient may not feel that the time is right for fighting and may not want to bargain about the issue. Finally, sharing anger and hostilities may violate what the other partner expects of the relationship.

A study comparing mutually satisfied couples with those experiencing marital difficulties found that when couples are having trouble getting along or are stressed, they tend to interpret each other's messages and behavior more negatively. Satisfied partners did not differ from distressed ones in how they intended their behavior to be received by mates. But distressed partners interpreted their spouse's words and behavior as being more harsh and hurtful than was intended (Gottman 1979; Noller and Fitzpatrick 1991).

Then, too, not all negative facts and feelings need to be communicated. Some psychologists suggest that one characteristic of happier couples is the restraint each uses with regard to expressing negative thoughts and feelings (Lauer and Lauer 1985). Before offering negative information, it is important to ask yourself why you want to tell it (to win?) and whether the other person really needs to know the information. This is particularly true with regard to revealing a past extramarital affair. And there is some evidence that a wife's informing her husband about a pregnancy and planned abortion does not enhance—and in some cases, may damage—the marital relationship (Ryan and Plutzer 1989).

Moreover, not every conflict can be resolved, even between the fairest and most mature fighters. If an unresolved conflict is not crucial to either partner, then they have reached a stalemate. The two may simply have to accept their inability to resolve that particular issue.

Finally, George Bach, who founded and directs an institute to teach constructive aggression, warns that "fight training does not always end 'happily.'" In one instance a couple attending Bach's fight training sessions realized through their better communication that the husband felt he should not have married at all. He believed he was "not the marrying type." His former passive-aggressive behavior had been a signal that he wanted to leave his wife. When this was exposed in fight training, the couple decided on a trial separation and eventually got a divorce (Bach and Wyden 1970). Like love, fair fighting doesn't conquer all. But it can certainly help partners who are reasonably well matched and who want to stay together. Success in marriage has much to do with a couple's skills in relating to one another—perhaps much more than the social similarity, financial stress, and age at marriage often emphasized by social scientists in earlier studies of marital adjustment (Gottman 1979; Noller and Fitzpatrick 1991).

We turn now to look at what some recent communication research has to say about comparing successful and unsuccessful marriages.

Communication, Happiness, and Marital Type

In his great novel *War and Peace* (1911), Leo characterized happy marriages as being all al while suggesting that unhappy marriages exhibit infinite variety. Scholars of marital communication find that happy marriages do share some common qualities, especially the expression of positive emotions and affection, but unhappy marriages also have some common features.

In distressed marriages, couples are less accurate in decoding, or understanding, their partner's communications, and they also had a misplaced confidence in their accuracy. Moreover, "the emotional climate of distressed marriages includes less positive affect, more negative affect, and more reciprocity of negative, but not positive, affect" (Noller and Fitzpatrick 1991, p. 47). The body of research reviewed by Noller and Fitzpatrick was sufficient to answer one age-old question: Which came first, poor communication or poor marriage? "[L]ongitudinal investigations . . . show that poor communication skills precede the onset of distress" (p. 47).

It also appears from research on marital communication that Tolstoy was wrong in presuming that happy marriages are all alike. One of the most prominent researchers of marital communication, Mary Anne Fitzpatrick, has developed a research-based typology of marriage based on the Relational Dimension Instrument, a questionnaire measuring ideology, interdependence, and conflict avoidance/expressivity in marriage (Fitzpatrick 1988).

Fitzpatrick found variation among married couples in their ideology of marriage, and thus in their expectations of closeness or distance and in their attitudes toward conflict. She identified four marital types: *traditionals, independents, separates,* and *mixed couples.* The more the partners' communication matched their ideology of marriage, the more satisfied the couple was with their marriage. Let's examine each of these marital types next. (Case 9.2, "Making a Marriage Encounter Weekend," describes how one couple improved their relationship through Marriage Encounter.)

Traditional Couples

Traditionals are conventional in their beliefs about marriage and are relatively conservative regarding sex

Making a Marriage Encounter Weekend

According to its promotional literature, Marriage Encounter is a weekend for couples "who have a good thing going for them, and who want to make it better." Worldwide Marriage Encounter is a movement devoted to the renewal of the sacrament of matrimony in and for the Catholic church. However, at least ten other faiths offer similar weekends. The movement began in the New York area in 1968. Since then, more than 1 million couples throughout the United States have made the weekend, and the movement has spread into fifty-five foreign countries. Here, a woman tells about her Marriage Encounter weekend.

We were going along fine. But I think we both had begun to feel we were maybe taking each other pretty much for granted. So we signed up for a Marriage Encounter weekend. Believe me, it was a big step: finding a babysitter and everything just to go *talk*!

Marriage Encounter is an extensive forty-four hours of uninterrupted togetherness. You do nothing else but eat, sleep, and share with each other. There's virtually no socializing with the other couples. It's just the two of you.

You write letters to one another. You write them alone, privately, and then you exchange them. You aren't using your body language while you're writing this, of course, or being interrupted or interrupting the other person. You have a train of thought. With just trying to talk sometimes, you know, the other person jumps in or tells you what you're saying is positive or negative, and you've gotten off the track. So we would write for ten minutes, picking a topic like, "How do I feel about sex?" Or, "How do I feel about anything really?" And both write.

Then you exchange notebooks and you each read it through twice—once with the heart and once with the head, as they say. And then you what they call dialogue on it for ten minutes. It really leads into neat areas....

I was amazed at how little I had been into feelings, and how much I was into thoughts. And the same for Jerry. It was just really weird how we both thought we were expressing our feelings before the encounter. But we weren't. Not completely anyway. We thought we'd been letting it all hang out, but we weren't. We were expressing our thoughts and our grudges and our hang-ups. But not our true—not all our true—feelings....

I prayed too. I don't want to leave that out. I don't want to minimize that. I want to have the courage to emphasize that I prayed; I prayed hard that weekend, partly because when I went into the room where I was going to write down my feelings about something, I sometimes found myself surprisingly afraid.

And then when we'd start writing, I'd say to myself, "Do I feel like *this*?" I would be surprised at how I felt, and I'd be surprised at how he felt too. It's such a revelation—almost as exciting for you alone as it is for your relationship, because you find out a lot of stuff about yourself that you had no idea about. Anyway I did. And Jerry did too.

When we left that weekend, they gave us a list of ninety topics, suggestions for continuing the dialogue. Things like sex, kids, finances. You can think up anything you want to dialogue about, but we've been more or less going by this list. We recognized, though, that for a while we were skipping around on the list. And then we realized that we were avoiding subjects that we were afraid to talk about. So we've decided—and this took courage, believe me—to go just straight down the list and take the topics as they come....

It's really been a revelation. But maybe the very best thing about it is just that we're making the time to do it.

How might weekend encounters such as this one enhance family commitment? What might be some drawbacks or problems with a Marriage Encounter weekend? In what ways does intimacy (described in Chapter 4) affect family cohesion, do you think? Why?

roles. They are very interdependent; that is, they like to do a lot of things together, rather than leading independent lives or being concerned about autonomy in decisions and choices. As part of their interdependence, they favor sharing and disclosure. Perhaps as part of their conventionality they avoid conflict, *except* on important issues, which they do not hesitate to address. These couples are unhappy with conflict avoidance of major problems and readily address serious issues.

Separate Couples

Separates, as the label indicates, are not very interdependent. They avoid both conflict and more positive modes of engagement with each other. Fitzpatrick describes them as emotionally divorced (Fitzpatrick 1988) and "ambivalent about their family values" (Noller and Fitzpatrick 1991, p. 48).

In Fitzpatrick's typology, separates are described as having conservative sex roles and beliefs. Perhaps involved in activities that differ by gender, these couples have no counterbalancing ideology of interdependency to bring them together for sharing and self-disclosure. It's possible to imagine a more modern version of separateness in which each member of a couple is involved in his or her own individual career and social world to the exclusion of any real intimacy with the partner (Marks 1989).

Independent Couples

Independents are the liberals of marital communication. Independent couples have less conventional values and value autonomy and change. They favor nontraditional sex roles and are open to engaging in conflict; certainly they are assertive with each other. Independence does not mean separateness, however, and these couples can be described as moderately interdependent.

Mixed Couples

In any typology of couples there will be some partners whose ideologies are not the same. Over time the number of traditional couples has declined and that of independents has increased, but perhaps the most important factor affecting marriage today is the likelihood that members of a couple will *differ* in their orientation (Fitzpatrick 1988; Noller and Fitzpatrick 1991; VanLear 1992). Differences in marital expecta-

tions probably pose a greater challenge to contemporary marriage than do changes in roles per se.

Marital Type and Marital Communication

Traditionals scored highest on the Dyadic Adjustment Scale (Spanier 1976; Fitzpatrick 1988, pp. 102–3, 107). Fitzpatrick asks "Who is happiest of them all?" and answers "pure type traditionals" (p. 107). However,

> it is important to note that couples in all of [the pure] types score above average in marital happiness or satisfaction. Each of these types may represent functional or workable marriages in that each has achieved a level of marital adjustment that appears to operate without debilitating impasses or blockages. The . . . stressors that could potentially affect each of the various marital types will differ. (Fitzpatrick 1988, p. 107)

When looking at marital communication we find that an important element is the match between the couples' ideology and their communication behavior. The marital happiness of the more interdependent couples depends on their level of sharing and disclosure, whereas separates are more satisfied if they can avoid jarring conflicts. Sociologist Stephen Marks (1989) posits that the pivotal task for all marrieds is to balance each partner's need for autonomy with the simultaneous need for intimacy and togetherness. The happiest couples are those who manage to do this—by negotiating personal and couple boundaries through supportive communication (Scarf 1995).

Family Cohesion

A clear implication from the body of research on marital communication is that marriages are held together not only by the successful management of conflict but by positive communication and expression of feelings (Erickson 1993; Langhinrichsen-Rohling, Smutzler, and Vivian 1994).

Yet it is often difficult to know what it is, exactly, that husbands, wives, and other members of the family should do to create **family cohesion**—the emotional bonding of family members. How do people create family cohesion?

In an effort to find out what makes families strong and cohesive, social scientist Nick Stinnett researched

130 "strong families" in rural and urban areas throughout Oklahoma (Stinnett 1979, 1983). Obviously, this limited sample, selected with help from home economics extension agents, has no claim to representativeness. The concept of "strong family" is equally subjective. Various individuals or groups have their own ideas about family strengths that are likely to vary by religion, ethnicity, and political persuasion. Nevertheless, Stinnett's interest in strong families was very influential in stimulating research on the specifications and correlates of family solidarity and satisfaction (see, for example, Meredith et al. 1989 on holiday celebrations and other family rituals).

Stinnett's suggestions for building positive family bonds have a simplicity that is deceptive because the common sense that suggests that positive family-building efforts are important often takes second place to the drama of dealing with conflict and crisis. Yet, in the long run, research makes it clear that the existence or nonexistence of positive feelings is the most important determinant of marital and family happiness.

When Stinnett made his observations of family strengths, six qualities stood out. First and most important, members often communicated *appreciation for one another.* They "built each other up psychologically" (p. 25). In Stinnett's assessment, "each of us likes to be with people who make us feel good about ourselves" (p. 25); thus, the family liked being together.

Stinnett suggests an exercise to help family members express appreciation for each other. Family members sit in a circle. Taking turns, each tells something he or she likes about the person to her or his left. When they have gone around the circle once or twice, each member then shares something he or she likes about himself or herself. After that, the direction is reversed, and members say something they like about the person to their right. The exercise needn't take long, and even when all members are not present, it helps build a sense of family togetherness and appreciation for one another. Studies show that families maintain more supportive interaction patterns for a time after the experience. To create togetherness, then, families might use this technique periodically.

Second, Stinnett found that members of strong families *arranged their personal schedules* so that they could do things, or simply be, together. The members of some families agreed to save Sundays for one another. Or members could agree to reserve every other weekend strictly for family activities. What families do together at home doesn't have to be routine, habitual, or boring: They might have a winter picnic in front of the fireplace, for example.

A third characteristic Stinnett found was *positive communication* patterns. Family members took time to talk with and listen to each other, conveying respect and interest. They also argued, but they did so openly, sharing their feelings and talking over alternative solutions to their problems.

Fourth, members of strong families had a *high degree of commitment* to promoting one another's happiness and welfare and to the family group as a whole. And they "put their money where their mouths were," investing time and energy in the family group. When life got so hectic that members didn't have enough time for their families, they listed the activities they were involved in, found those that weren't worth their time, and scratched them off their lists, leaving more free time for their families. Stinnett comments on the element of knowledgeable decision making in that action:

> This sounds very simple, but how many of us do it? We too often get involved and it's not always because we want to be. We act so often as if we cannot change the situation [but] we do have a choice. . . . [T]here is a great deal that families can do to make life more enjoyable. (Stinnett 1979, p. 28)

A fifth characteristic of the families Stinnett studied was *spiritual orientation.* Findings from a large national sample show that being religious does not necessarily make for happier marriages (Booth et al. 1995). The small study described in Box 9.2, "A Closer Look at Family Diversity: Religion's Role in Rural African American Marriages," offers some evidence to the contrary, however. Although Stinnett's "strong families" were not necessarily members of an organized religion, they did have a sense of some power and purpose greater than themselves.

And sixth, Stinnett found that strong families were able to *deal positively with crises.* Members were able to see something good in a bad situation, even if only gratitude that they had each other and were able to face the crisis together, supporting one another. (Chapter 14 discusses dealing creatively with crises.)

In general, Stinnett's families took the initiative in structuring their lifestyles to enhance family relationships. Instead of drifting into a family relationship by

Religion's Role in Rural African American Marriages

Much research shows that the church plays a prominent role in African Americans' lives. The church gives moral guidance and political leadership, and it is the center of community life. Religion also offers African Americans a salient coping mechanism for negotiating life's stresses. Knowing all this, a group of social scientists at the University of Georgia (Brody et al. 1994) hypothesized that religion might also play a role in marital satisfaction among black Americans, particularly those in the rural South.

Ninety African American families with married parents were recruited from areas with populations of fewer than 2,500 in Georgia and South Carolina. To do this, a black researcher contacted community members, such as pastors and teachers, to explain the research project and request their help. The community leaders then spoke with potential participant families and asked whether they would volunteer. Names of those who said "yes" were given to the researcher, who subsequently contacted the families. The final sample included an economic cross-section of rural married African American families: Family annual incomes ranged from $2,500 to $57,500.

Because most questionnaires and other tools used to evaluate family processes have been developed with white, middle-class families in mind, the scientists were concerned about their validity for this project. Consequently, they formed *focus groups*, or discussion groups, comprised of rural black community members. The focus groups decided which instruments were best to use.

Ultimately, the researchers decided to videotape interaction sessions in the families' homes. Couples were videotaped playing the board game "Trouble" for fifteen minutes, and also discussing and issue that had been presented to them by the researchers. Later the videotapes were played at the University of Georgia and coded for whether interaction between the couples was warm and supportive. Undergraduate and graduate African American university students served as home visitors to collect data from the families. Prior to data collection, the visi-

Providing moral guidance, encouragement, political leadership and community, religion is important to African American families.

tors received one month of training in administering the self-report instruments and observational techniques.

The scientists measured religiousness, or *religiosity*, with two questions. The first asked how often each respondent had attended church services during the previous year; answers ranged across a 7-point scale, from 0—"never" to 7—"more than

(continued)

(continued)

once a week." The second question asked the respondents to rate the importance of church attendance; answers ranged along a 5-point scale from 1—"not very important" to 5—"very important."

This study found that for both husbands and wives, religiosity was associated positively with the quality of marital communication. That is, spouses who attended church more often and found it important to do so were more warm and harmonious when communicating with each other. The researchers con-

cluded that religiosity is an important predictor of marital communication, at least among this sample of rural, black families. This was true even after family financial resources were taken into account.

Researchers concluded that the church in African American communities contributes to norms that promote supportive family relations. These, in turn, help the family to cope with the economic and social stressors that accompany life in the rural South.

---

default, they played an active part in carrying out their family commitments.

In Sum

Families are powerful sources of support for individuals, and they reinforce members' sense of identity. Because the family is powerful, however, it can cause individuals to feel constrained. Tactics such as gaslighting, scapegoating, or negative use of the looking glass can all be stressful or denigrating to an individual.

Even though such family interaction tactics can reach the point of pathology, family conflict itself is an inevitable part of normal family life. Americans are socialized to respect a conflict taboo, but both sociologists and counselors are recognizing that to deny conflict can be destructive to both individuals and relationships.

Although fighting is a normal part of the most loving relationships, there are right and wrong ways of fighting. Alienating practices should be avoided; they hurt a relationship because they lower partners' self-es-

teem. Bonding fights, in contrast, can often resolve issues and also bring partners closer together by improving communication. There are three important bonding techniques: leveling, using I-statements, and giving feedback. In bonding fights both partners win.

Research on marital communication indicates the importance to marriage of both positive communication and the avoidance of a spiral of negativity. Stinnett's research on "strong families" suggests that what makes families cohesive are expressing appreciation for each other, doing things together, having positive communication patterns, being committed to the group, having some spiritual orientation, and being able to deal creatively with crises.

Partners need to be attentive to typical gender differences in communication and to other cultural differences as well. There is some variation in the patterns of marital communication that work for couples, depending on their model of marriage. Traditionals, independents, and separates have different ideologies of marriage, prefer different levels of interdependence, and are comfortable with different levels of overt conflict.

Key Terms

alienating fight tactics	kitchen-sink fight
anger "insteads"	leveling
attribution	looking-glass self
bonding fighting	mixed, or double, messages
checking-it-out	passive-aggression
conflict taboo	rapport talk
consensual validation	report talk
displacement	sabotage
family cohesion	scapegoating
feedback	separates
gaslighting	significant others
gunnysacking	suppression of anger
independents	traditionals

Study Questions

1. Explain why families are powerful environments. What are the advantages and disadvantages of such power in family interaction?

2. Do you agree that marital anger and conflict are necessary to a vital, intimate relationship? Why or why not?

3. Discuss the possible negative side effects of denying conflict in a marriage relationship. Include the effects on the individuals and on their relationship.

4. Define passive-aggression and give examples of behavior that could be passive-aggressive.

5. Describe some alienating fight tactics. If someone you care for used such methods in a disagreement with you, how would you feel? What might you say in response?

6. Discuss your reactions to each of the guidelines proposed in this chapter for bonding fights. What would you add or subtract?

7. Why might it be difficult for people to change their fighting habits? What do you think is necessary for change to occur?

8. In your opinion, when should couples visit a marriage counselor?

9. Do you agree with psychologist John Gottman's statement that husbands and wives have different goals in marital conflict? Why or why not? Give examples to support your response.

10. Do Stinnett's suggestions for "strong families" work? Put another way, do they seem realistic; would they really have an impact on family life?

11. Describe traditional, independent, and separate marital types. What kind of communication and conflict management does each seem to want?

Suggested Readings

Fitzpatrick, Mary Anne. 1988. *Between Husbands and Wives: Communication in Marriage.* Newbury Park, CA: Sage. Reports academic research on marital relationships, including communication; difficult for the nonprofessional reader, but one of the most important books in this field.

Imber-Black, Evan, Janine Roberts, and Richard A. Whiting, eds. 1988. *Rituals in Families and Family Therapy.* New York: Norton. Series of articles with creative ideas for rituals in families at various stages of the life cycle; includes the use of family rituals to create a new family identity in remarriage and to intervene in families having problems around adoption and alcoholism. The book's perspective and analysis is directed toward therapeutic use but should prove interesting to social scientists as well.

Kantor, David and William Lehr. 1975. *Inside the Family: Toward a Theory of Family Process.* San Francisco: Jossey-Bass. Theoretical treatment of family interaction based on observation of families. Good for those who would like a thorough academic treatment of family interaction, heavy going for others.

Laing, Ronald D. 1971. *The Politics of the Family.* New York: Random House. Psychiatrist analyzes family interaction, particularly the power of the family to shape identity and definitions of the situation.

Lerner, Harriet Goldhor. 1985. *The Dance of Anger.* New York: Harper & Row. A well-written book on anger, its causes, and its consequences for individuals and relationships. This book is hopeful because it gives practical advice on how to deal constructively with old and current anger.

Scarf, Maggie. 1995: *Intimate Worlds: Life Inside the Family.* New York: Random House. A counseling psychologist, Scarf explores the issues of intimacy, power, and conflict as these affect marital/family communication and levels of health in families. A good read.

Tannen, Deborah. 1990. *You Just Don't Understand: Women and Men in Conversation.* New York: Basic Books. On the best-seller list for years, this very readable book is based on research, yet directed toward a popular audience. It presents information on common gender differences in communications practices in a way that would be useful to people in any relationship involving any combination of the sexes.

Power and Violence in Marriages and Families

To the extent that power is the prevailing force in a relationship—whether between husband and wife or parent and child, between friends or between colleagues—to that extent love is diminished.

RONALD V. SAMPSON,
THE PSYCHOLOGY OF POWER

A letter to an advice column told this story: A wife wrote that her husband "hates to get out of bed in the morning" and depends on her to get him up. "Yesterday it took nearly an hour and I was worn out." When her husband finally does get up, she wrote, he "throws things around, spills coffee on the floor, tears down curtains and screams at me because I made him late for work." The woman was very upset, yet the problem persisted.

This chapter examines power in relationships, particularly marriage. We will discuss some classic studies of decision making in marriage. We'll look at what contemporary social scientists say about conjugal power. We will discuss why playing power politics is harmful to intimacy and explore an alternative. Finally, we will explore one tragic result of the abuse of power in families—family violence, including wife abuse, husband abuse, child abuse, and elderly abuse. We begin by defining power.

What Is Power?

Power can be defined as the ability to exercise one's will. There are many kinds of power. Power exercised over oneself is **personal power,** or autonomy. Having a comfortable degree of personal power is important to self-development. **Social power** is the ability of people to exercise their wills over the wills of others. Social power can be exerted in different realms, even within the family. Parental power, for instance, operates between parents and children. In this chapter our discussion of power in families will focus on power between married partners, or **conjugal power.** What are the sources of conjugal power? We'll seek to answer that question in this chapter.

Power Bases

Two social scientists (French and Raven 1959) have suggested six bases, or sources, of power: coercive, reward, expert, informational, referent, and legitimate. **Coercive power** is based on the dominant person's ability and willingness to punish the partner either with psychological-emotional or physical violence or, more subtly, by withholding favors or affection. Slapping a mate or spanking a child are examples; so is refusing to talk to the other person—giving the silent treatment. **Reward power** is based on an individual's ability to give material or nonmaterial gifts and favors, ranging from financial support (food, clothing, and shelter) to college tuition to a night out at a concert or the movies.

Expert power stems from the dominant person's superior judgment, knowledge, or ability. Although this is certainly changing, our society has traditionally attributed expertise in important matters, such as finances or career decisions, to men. Consequently, wives have been encouraged to assign expert power to husbands more often than the reverse. **Informational power** is based on the persuasive content of what the dominant person tells another individual. A husband may be persuaded to stop smoking by his wife's giving him information on the health dangers of smoking.

Referent power is based on the less dominant person's emotional identification with the more dominant individual. In feeling part of a couple or group, such as a family, whose members share a common identity, an individual gets emotional satisfaction from thinking as the more dominant person does or behaving as the "referent" individual wishes. A husband who attends a social function when he'd rather not "because my wife wanted to go and so I wanted to go too" has been swayed by referent power. In happy relationships, referent power increases as partners grow older together (Raven, Centers, and Rodrigues 1975). Finally, **legitimate power** stems from the more dominant individual's ability to claim authority, or the right to request compliance. Legitimate power in traditional marriages involves both partners' acceptance of the husband's role as head of the family.

Throughout the rest of this chapter, we will see these various power bases at work. The consistent research finding, for instance, that the economic dependence of one partner on the other results in the dependent partner's being less powerful can be explained by understanding the interplay of both *reward power* and *coercive power:* If I can reward you with financial support—or threaten to take it away—I am more likely than otherwise to exert power over you. (Box 10.1, "A Closer Look at Family Diversity: The Power of Speaking English in Immigrant Families," applies some of these concepts.) We turn now to look more specifically at conjugal power and decision making.

Conjugal Power and Decision Making

Research on marital power began well before the second wave of the feminist movement (see Chapter 3), the development of family therapy, and the identification of family violence as a social problem. In the 1950s, social scientists Robert Blood and Donald Wolfe had an academic interest in conjugal decision making. Their book *Husbands and Wives: The Dynamics of Married Living* (1960) was based on interviews with wives only. Nevertheless, it was a significant piece of research and shaped thinking on marital power for many years.

Egalitarian Power and the Resource Hypothesis

THE RESOURCE HYPOTHESIS   Blood and Wolfe began with the assumption that although the American family's forebears were patriarchal, "the predominance of the male has been so thoroughly undermined that we no longer live in a patriarchal system" (Blood and Wolfe 1960, pp. 18–19). On this basis they developed their major hypothesis about what had replaced the system of patriarchy.

They reasoned, in their **resource hypothesis,** that the relative power between wives and husbands results from their relative resources as individuals: The spouse with more resources has more power in marriage. These resources, according to Blood and Wolfe, include education and occupational training. Within marriage a spouse's most valuable resource would be the ability to provide money to the union. Another

The Power of Speaking English in Immigrant Families

As the United States becomes an increasingly diverse society, social scientists have realized that some—perhaps many—of our time-worn concepts and assumptions are no longer applicable. One example is how sociologists have always thought about power among family members, or **family power.** They have assumed that parents exercised power over their children. Because children were less experienced and had fewer resources, their parents exerted over them *expert power* (as well as other forms, such as *coercive* and/or *reward power*). Because this was simply assumed to be so, family power research focused almost exclusively on the dynamics behind conjugal power.

Current immigration patterns may change this situation, however. Increasingly, children who immigrate with their parents from non-English-speaking countries may be assuming power-wielding roles in their families. Many adult immigrants observe that their children seem to absorb English almost effortlessly. Rather than risk embarrassment or costly misunderstandings, the parents begin to rely on their children for tasks large and small.

At first, even young children may be told to memorize the family's address and phone number, for example, so that they can relate it to a taxi driver for another family member. As the children's vocabulary grows, so does their family's dependence on them. Before long they may be going through the mail, sorting bills from ads, or negotiating rent with the apartment manager. Immigrant children translate for their parents during doctor appointments and job interviews. Some assist English-speaking customers in the family business. They serve as their families' language and cultural interpreters, translating not only English but also unfamiliar American folkways (Alvarez 1995).

The children's expertise turns the traditional parent–child power relationship upside down: The family's adults are dependent while the youngsters take charge. There can be comic moments: "I once told my mother that the F on my report card stood for *fabuloso*," confesses Sandino Sanchez from the Do-

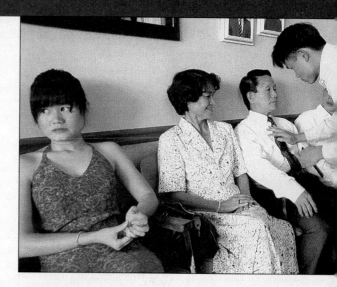

As the Ly family waited to have a studio portrait taken, Justin straightened his father's tie—his mother Bik Yong Ly and older sister Pam sat nearby. The simple act of straightening his father's tie illustrates immigrant children's role of translator of United States culture to their parents—a role that can significantly affect family power dynamics. This is an area for future research.

minican Republic, now an adult. However, humor aside, mediating between the family and the outside world gives a person tremendous control—control that was once mostly reserved for husbands. As 16-year-old Sara from Bangladesh who lives in Astoria, New York, puts it, "I have the most power in my family. I speak English and they don't."

Some mental health professionals warn that this power reversal can damage a formerly functional family structure: Parents feel impotent and children, overly anxious. This development in American family power has yet to be systematically researched, however. If you were to scientifically explore this subject, how might you design your research project?

Source: Alvarez 1995.

Who wields more power in this relationship, do you think? According to the resource hypothesis, it's the partner with more resources—higher education or better-paying job, for example. While things are changing, this is still more often the man than the woman.

resource would be good judgment, probably enhanced by education and experience. (Recall the discussion of exchange theory in Chapters 2 and 7 and note that the resource hypothesis is a variation on exchange theory.)

To test their resource hypothesis, the researchers interviewed about 900 wives in greater Detroit and asked who made the final decision in eight areas, such as what job the husband should take, what car to buy, whether the wife should go to work (or quit work), and how much money the family could afford to spend per week on food. From their interviews Blood and Wolfe drew the conclusion that most families (72 percent) had a "relatively equalitarian" decision-making structure, although there were more families in which the husband made the most decisions (25 percent) than there were wife-dominated families (3 percent).

The resource hypothesis was supported by the finding that the relative resources of wives and husbands were important in determining which partner made more decisions. Older spouses and those with more education made more decisions. Blood and Wolfe also found the relative power of a wife to be greater after she no longer had young children or when she worked outside the home and thereby gained the wage-earning resource for herself. They reported little on black families, except to say that black husbands, when compared with white husbands, had unusually low decision-making power in comparison to their wives.

CRITICISM OF THE RESOURCE HYPOTHESIS   In all, Blood and Wolfe's study had the important effect of encouraging people to see conjugal power as shared rather than patriarchal. The power of individual partners was seen as resting on their own attributes or resources rather than on social roles or expectations, a perspective that changed the thinking of social scientists. But Blood and Wolfe's study has also been strongly criticized.

One criticism concerns Blood and Wolfe's criteria for conjugal power. Critics stated that power between spouses involves far more than which partner makes the most final decisions; it also implies the relative autonomy of wives and husbands, along with questions about division of labor in marriages (Safilios-Rothschild 1970). Moreover, Blood and Wolfe have come under heavy fire for their assumption that the patriarchal power structure has been replaced by egalitarian marriages.

Resources and Gender

Feminist Dair Gillespie (1971) pointed out that power-giving resources tend to be unevenly distributed between the sexes. Husbands usually earn more money even when wives work, and so husbands control more economic resources. Husbands are often older and better educated than their wives, so husbands are more likely to have more status, and they may be more knowledgeable, or seem so. Even their greater physical strength may be a powerful resource, although a destructive one, as we will see later in this chapter.

Moreover, women are likely to have few alternatives to the marriage if they cannot support themselves or are responsible for the care of young children. Consequently, according to Gillespie, the resource hypothesis, which presents resources as neutral and power as gender-free, is simply "rationalizing the preponderance of the male sex." Furthermore, women are socialized differently from men so that by the time they marry, women have already been "systematically trained to accept second best," and marriage is hardly a "free contract between equals" (p. 449).

Research tends to support Gillespie's insight that American marriages continue to be inegalitarian even though they are no longer traditional. True, resources make a difference, and an important factor in marital decision making is whether or not a wife is working. Wage-earning wives have more to say in important decisions (Blumberg and Coleman 1989). As one wife put it: "I think we share in the decision making. I make decisions at the office from nine to five and I think it would be a little strange if I came home and was treated like a pussycat" (Blumstein and Schwartz 1983, p. 141).

But resources are gender influenced, and one way in which women come to have fewer resources is through their reproductive roles and resulting economic dependence (Blumberg and Coleman 1989). Wives who do not earn wages, especially mothers of small children, have considerably less power than do women who earn an income. Relationships tend to become less egalitarian with the first pregnancy and birth (Coltrane and Ishii-Kuntz 1992). Just after marriage the relationship is apt to be relatively egalitarian, with the husband only moderately more powerful than the wife—if at all. Often at this point the wife has considerable economic power in relation to her husband; that is, she is working for wages and may even have established herself in a high-paying career.

But during the childbearing years of the marriage, the practical need to marry and remain married is felt especially strongly. Divorce is likely to mean that the woman must parent and support small children alone. In some cases, expectant mothers lose not only economic power but also "sexual attractiveness" power (Blumberg and Coleman 1989). Moreover, they have fewer alternatives and less energy or physical strength to resist dominance attempts, and they may be vulnerable to the manipulative offer or withdrawal of the husband's help with child care. (On the other hand, a mother may exert power over her husband by threatening to leave and take the infant with her.)

Working, then, contributes to marital power. But even though the working wife is less obliged to defer to her husband and has greater authority in making family decisions, she still does not necessarily participate equally in decision making and is still unequally burdened with housekeeping, child rearing, and caring for her partner (Greenstein 1995). In one study, social scientists Dana Hiller and William Philliber (1986) found that the husband's expectations determine how and whether housekeeping roles will be shared. Even if she is employed, a wife does more housework if her husband believes she should than if her husband thinks he too should be involved (Brines 1994). Also, if her husband believes her wages are needed, a wife has more power (Blumberg and Coleman 1989). Hence, working for wages does not necessarily give a wife full status as an equal partner. There are a variety of explanations for this continuation of male dominance in a society that, on the whole, articulates an ideology of marital equality.

While one generation may hold more traditional conjugal power roles, the next generation may renegotiate and consciously change those roles, especially as women assume more autonomy and make gains in the workplace. Here several generations display both traditional and changing roles.

Resources in Cultural Context

The family ecology theoretical perspective (Chapter 2) stresses that family interaction needs to be examined within the context of the society and culture in which it exists. Accordingly, the cultural context may determine whether or not a resource theory explains marital power. Studies comparing traditional societies, such as in Greece, with more modern ones, such as in France, suggest that in a traditional society norms of patriarchal authority may be so strong that they override personal resources and give considerable power to all husbands (Safilios-Rothschild 1967; Blumberg and Coleman 1989). Put another way, *legitimate power* predominates.

Even in the United States, we must recognize the continuing salience of tradition and the assumption that it is legitimate for husbands to wield authority in the family. A study of Japanese American couples in Honolulu, for instance, found that even though wives actively participated in making countless decisions, they did so by virtue of delegation of power from their husbands (Johnson 1975). The continued importance of traditional legitimations of husbands' authority is apparent in some religious groups; the new Religious Right (Chapter 5) is committed to the restoration of traditional family norms and roles.

IMMIGRANTS AND CONJUGAL POWER   Male dominance is also strong in ethnic groups that have recently immigrated to the United States. A study of Puerto Rican families (Cooney et al. 1982), for example, found that the norms of the parent generation born in Puerto Rico were patriarchal. "[These] norms emphasize the generally superior authority of the man within the family" (Cooney et al. 1982, p. 622). The specific socioeconomic and personal resources of the husband are irrelevant: He has the power. However, the generation born in the United States has moved to a "transitional egalitarian society" typical of the rest of the United States; that is, "husband-wife relationships are more flexible and negotiated . . . [and] socioeconomic achievements become the basis for negotiation within the family" (1982, p. 622).

Because new immigrants are increasing as a proportion of our population, we may expect a temporary increase in traditional patriarchal families. Subsequent generations, however, can be expected to adopt the more common American pattern, in which husbands wield greater power by virtue of their greater access to resources.

We must note another way in which cultural context conditions the resource theory; it explains marital power only in the absence of an overriding egalitarian norm as well as a traditional one. Put another way, if traditional norms of male authority are strong, husbands will almost inevitably dominate, regardless of personal resources. Similarly, if an egalitarian norm of marriage were completely accepted, then a husband's superior economic achievements would be irrelevant to his decision-making power because both spouses would have equal power (Cooney et al. 1982). It is only in the present situation, in which neither patriarchal nor egalitarian norms are firmly entrenched, that marital power is negotiated by individual couples and the power of husbands and wives may be a consequence of their resources. Because husbands tend to have more resources, they are more likely to dominate.

We now look at the influence of social class and racial/ethnic diversity on conjugal power.

Social Class, Racial/Ethnic Diversity, and Conjugal Power

The majority of Americans, regardless of class, do not see power inequities in their own marriages. And although some sociological studies do find middle-class husbands to have more power than working-class husbands, differences among classes are slight (Peplau and Campbell 1989). Relative economic power of husband and wife *is* important, however (Blumberg and Coleman 1989).

THE MYTH OF BLACK MATRIARCHY   Patterns of power among blacks and Mexican Americans are similar to those among whites. During the 1950s and 1960s, social scientists believed that African American marriages were characterized by a matriarchal power

structure in which wives and mothers were dominant; however, recent research suggests otherwise. Black marriages tend to be more egalitarian than whites' (John, Shelton, and Luschen 1995). Talk about "black matriarchy" was an exaggeration. One reason black marriages are more egalitarian than whites' is that proportionally more black than white wives are wage earners (Peplau and Campbell 1989).

MEXICAN AMERICANS AND ROLE MAKING

Just as black matriarchy is a myth, so may be the belief that Mexican Americans behave only according to patriarchal standards (Hondagneu-Sotelo 1992). Indeed, as part of the ideology of la familia (Chapter 8), Mexican American families do generally uphold the principle of the male as household head (Wilkinson 1993, p. 35). Nevertheless, one study (Williams 1990) based on in-depth interviews and participant observation (see Chapter 2) among Mexican Americans ages 25–50 in Austin and Corpus Christi, Texas, focused on husbands' and wives' role making—negotiating and changing their respective roles. Respondents saw themselves as different from their parents in terms of conjugal power: "Years back it used to be the man. He made all the decisions, the women had no say so"; "My mother se quedaba en la casa [stayed at home]. She had no voice in family affairs" (p. 85). Today, despite some resistance from husbands, power relationships are changing as personal identity and autonomy emerge for wives. As one woman explained,

> I am 44, I've got money, a good car and my kids are behind me. Now, if I am not home in time to make dinner, he can make a sandwich. We've been together a long time and he knows how to make a sandwich. I have had to stand my ground. Two years ago I wanted to go to school to study for a radiologist. He didn't let me. He said I didn't have to. I got my GED. Recently . . . I made a stand that I'm going out and I do. I go out during the day with friends to go out to eat. Not at night. . . . I always felt the obligation that I had to come home to cook. (Williams 1990, p. 94)

Love, Need, and Power

Some have argued that a primarily economic analysis does not do justice to the complexities of marital power. Perhaps a wife has considerable power through her husband's love for her:

The relative degree to which the one spouse loves and needs the other may be the most crucial variable in explaining total power structure. The spouse who has relatively less feeling for the other may be the one in the best position to control and manipulate all the "resources" that he [sic] has in his command in order to effectively influence the outcome of decisions, if not also to dominate the decision making. Thus, a "relative love and need" theory may be . . . basic in explaining power structure. (Safilios-Rothschild 1970, pp. 548–49)

This theory is congruent with what sociologist Willard Waller termed the **principle of least interest.** The partner with the least interest in the relationship is the one who is more apt to exploit the other. The spouse who is more willing to break up the marriage or to shatter rapport and refuse to be the first to make up can maintain dominance (Waller 1951, pp. 190–92). Dependence on a relationship can be practical and economic as well as emotional. For example, older women have less probability of remarriage after divorce or of significant employment (if they have not already established a career), so they may be reluctant to leave a marriage (Blumberg and Coleman 1989).

Like resource theory, the **relative love and need theory** is a variation of exchange theory. (See Chapter 2.) Each partner brings resources to the marriage and receives rewards from the other partner. These may not balance precisely, and one partner may be gaining more from the marriage than the other partner, emotionally or otherwise. This partner is more dependent on the marriage and thus is more likely to comply with the other's preferences.

The relative love and need theory does not predict whether husbands or wives will generally be more powerful. In other words, it assumes that women are as likely to have power as men are: "The man who desires or values the woman as a mate more than she desires or values him will be in the position of wanting to please her. Her enchantment in his eyes may be physical attractiveness, pleasing personality, his perception of her as a 'perfect' wife and mother" (Hallenbeck 1966, p. 201).

Generally, however, the wife holds the less powerful position. How does the relative love and need theory account for this? One explanation offered is that "love has been a feminine specialty" (Cancian 1985, p. 253). Women are more socialized to love and need

their husbands than the reverse. They also tend to be more relationship oriented than men are (Tannen 1990). According to the principle of least interest, this puts women at a power disadvantage.

Moreover, Americans *believe* that women are more dependent on the marital relationship than men are. "If most people believe that women need heterosexual love more than men, then women will be at a power disadvantage" (Cancian 1985, p. 257). In our society, women are encouraged to express their feelings, men to repress them. Men are less likely, therefore, to articulate their feelings for their partners, and "men's dependence on close relationships remains covert and repressed, whereas women's dependence is overt and exaggerated" (p. 258). Overt dependency affects power: "A woman gains power over her husband if he clearly places a high value on her company or if he expresses a high demand or need for what she supplies. . . . If his need for her and high evaluation of her remain covert and unexpressed, her power will be low" (p. 258).

Micropolitics: The Private Sphere

In sum, according to Francesca Cancian, "men dominate women in close relationships [and] . . . husbands tend to have more power in making decisions, a situation that has not changed in recent decades" (1985, p. 259). But, argue several social scientists, men are far less powerful in the private, intimate sphere than they are in the public world. Therapists and mass media, and to an increasing degree the public, support women's desire for more expression of feeling, reducing men's ability to minimize the emotional sphere in which women are deemed more competent. Men's ability to get what they want may also be limited by the very "avoidance of dependence" built into the male role. "He may not even know what he wants or needs from her, and therefore, may be unable to try to get it" (Cancian 1985, p. 259). Men are as likely to feel controlled by women as women believe they are controlled by men:

Insofar as love is defined as the woman's "turf," an area where she sets the rules and expectations, a man is likely to feel threatened and controlled when she seeks more intimacy. Talking about the relationship, like she wants, feels like taking a test that she made up and he will fail. The husband is likely to react with withdrawal and passive aggression. He is blocked from straightforward counter-attack insofar as he believes that intimacy is good. (Cancian 1985, p. 260)

Cancian's view of this dilemma is presented with much empathy for men. Psychotherapist and social scientist Lillian Rubin's work on men and women in intimate relationships (1976, 1983, 1994) looks at the differences between men's and women's styles of expressing love neutrally in terms of the problems they produce for couples. A more acid view is presented by sociologist Jean Lipmen-Blumen (1984) as she discusses women's greater relationship skills from the perspective of conflict theory. (Conflict theory is presented in Chapter 2.) In her view, women are oppressed by men and respond with "micromanipulation"—that is, manipulation of interpersonal power in the micro, or private, sphere.

Lipmen-Blumen argues that men dominate the public sphere of work and political leadership, whereas women occupy the private sphere. If men dominate public policy, then women,

> as well as other powerless groups, become well versed in interpreting the unspoken intentions, even the body language, of the powerful. . . . By the various interpersonal strategies of micromanipulation, women have learned to sway and change, circumvent, and subvert the decisions of the powerful to which they seem to have agreed. . . . Women have also mastered how to "obey without obeying" those rules they find overly repressive. When necessary, they cooperate with men to maintain the mirage of male control. True, a growing minority of women have also consciously rejected the tools of micromanipulation in favor of a more direct assault [by becoming involved in the public arena of work and politics]. The majority, however, continue to operate primarily at the interpersonal level. (Lipmen-Blumen 1984, pp. 30–31)

With Lipmen-Blumen's analysis, we have come back to the question of men's and women's power in the larger society. Sociologists see marital power as affected by position and resources in the larger society, the public world of work and politics (Huber and Spitze 1983; Blumberg and Coleman 1989). If women have relied on micromanipulation, it is primarily because, as we noted in Chapter 3, they are not yet integrated into the higher levels of the occupational structure or political leadership.

Lipmen-Blumen's description reinforces our initial premise: Power disparities discourage an intimacy based on honesty, sharing, and mutual respect. Realization of the American ideal of equality in marriage, however, would seem to support the development of intimacy in relationships.

The Future of Conjugal Power

Equalization of the marital power of men and women can occur in a number of ways. First, women can attain equal status in the public world and develop resources that are truly similar to men's. Despite the persistence of occupational segregation, the trend of the times is in this direction. Women's increase in income should "lead to an increased sense of self, sense of control over [their] lives, and expectation of achieving greater bargaining power within the relationship" (Blumberg and Coleman 1989, p. 239).

Second, society can come to value more highly women's resources of caring and emotional expression. Even though women's care of children, the elderly, and other dependents, their creation of a warm, comfortable home, and their emotional support of family and other social bonds have not been much rewarded with either money or status, feminists forcefully argue the social importance of these contributions. One impact of feminism on American culture may be the way women's values are increasingly incorporated into the culture (Lenz and Myerhoff 1985); men's liberation movements now articulate expressive values as well (see Chapter 3). Although there may be more rhetoric than reality to this cultural change, women's traditional assets could increase in worth in such a cultural climate.

And finally, norms of equality can come to be so strong that men and women will have equal power in marriage regardless of resources. Equality is an important value in American culture. Our society could come to legitimate norms of equality in marriage as strongly as it endorsed patriarchal authority in the past.

Some American Couples

Having looked at theories and research on conjugal power, we might look more closely now at a study of American couples to observe the workings of power in the everyday lives of some contemporary couples.

As you may recall from Chapter 5, sociologists Philip Blumstein and Pepper Schwartz (1983) undertook a comparison of four types of couples: heterosexual married couples, cohabiting heterosexual couples, gay couples, and lesbian couples. Twenty-two thousand questionnaires were sent to couples who responded to media advertisements for participants or to the researchers' solicitation for participants at various events and meetings. Over 12,000 questionnaires were returned and some 300 couples were interviewed; data were used only if both parties participated.

The results confirm some of our ideas about power. Gender was by far the most significant determinant of the pattern of power. Composed, by definition, of a male and a female, marital and cohabiting couples tended to be the least egalitarian. As the resource theory suggests, money was a major determinant of power, and two men or two women were far more likely to have similar incomes. But money strongly affected power even in a same-sex couple. High-earning individuals (or the employed partner with an unemployed companion) tended to be excused from tiresome household chores and got to pick leisure activities. The way in which one partner's success in the wider world serves as a basis for claiming relationship benefits is illustrated by the reaction of a gay partner:

> Our biggest arguments are about what I haven't done lately. . . . I am willing to help out when I can, but my career is not just nine to five and usually I either don't have the time or I'm so tired when I come home that the last thing I'm going to do is clean the kitchen floor. . . . We had this one discussion where he suggested I get up earlier in the morning to help clean up if I'm too tired at night. I blew up and told him that I was the one with the career here and it was my prospects and my salary that gave us his vacations and you just can't be a housewife and a success all at the same time. (Blumstein and Schwartz 1983, pp. 152–53)

The impact of gender was apparent in a comparison of gay and lesbian couples. Gay men tended to be extremely competitive, very aware of each other's earning power and other signs of status in the public world. Lesbians, on the other hand, with feminine values of cooperation and pleasing others, worked very hard at their relationship and often deferred to each other. In fact, nonassertiveness sometimes

It is still true that more wives than husbands perform the bulk of household chores. However, some married couples find that power inequalities diminish over time and that relative commitment, energy, and rewards balance out in the long run.

became a problem—neither partner liked to initiate sex.

Lesbian couples illustrate another principle of conjugal power: the importance of norms and the cultural context. Committed to egalitarianism and cooperative decision making, these couples were the only ones to transcend the principle that economic resources determine decision-making power. Norms of equality were simply so strong that lesbians made strenuous efforts not to let differential earning power or unemployment affect control over the relationship.

Finally, this study tells us that commitment influences power. In marriages, representing the highest level of commitment, low-resource partners (usually the women) felt much freer to spend money earned by the partner than did individuals in a cohabiting relationship. They were likely to view resources as joint ones. The formal commitment represented a barrier to separation that limited the principle of least interest. Because both partners would have found it difficult to leave the relationship, both felt secure in the relationship. They were less apt to believe that money or other resources were divided into "yours" and "mine." As couples in the other relationship types intensified their commitment over time, the partner with fewer resources also gained in power (Blumstein and Schwartz 1983).

Perhaps because of their greater sense of separate identity and need to be economically independent if necessary, maintaining equality was very important to individuals in gay, lesbian, and cohabiting couples. "Only married couples do not rely on equality to hold them together" (Blumstein and Schwartz 1983, p. 317). Resources, rather than a commitment bond, ensured their security and power within the relationship. Married couples took a longer view of the rewards obtained from marriage, which perhaps explains why another study (Schafer and Keith 1981) found that the power inequity perceived by wives diminished over time. Regardless of actual power in the marriage, it seems that as the years pass, spouses develop an increased sense of identity with each other and so are more apt to view marital decisions as joint ones. They also believe, despite periods of inequality, that relative commitment, energy, and marital rewards even out over time.

Blumstein and Schwartz emerged from their project impressed by the advantages of marriage. The stability of a formal commitment enabled couples to survive difficulties and required a sharing and negotiation of conflict that in the long run strengthened the relationship. Still, they noted that, at least into the 1980s, the institution of marriage had been organized around inequality; attempts at change had led to spouses' frustration and unhappiness (1983, p. 324). Here's one husband's experience:

> We have a very traditional labor pattern. Very traditional. She does all of the household chores—at least up to a short time ago. Then all I did was take out the garbage. . . . Now she's at work and I have to help out more, but I resist doing things I'm not supposed to do. I'll do the outside work, but the

inside work has always been her territory and I don't think I should have to learn things that she has spent twenty years perfecting. . . . So we have got me in this mode of helper—which I don't like very much—but it is more necessary right now if the house is going to be well kept. . . . I am willing to make her life a little easier but I can't say I enjoy any of it. (Blumstein and Schwartz 1983, p. 146)

According to Blumstein and Schwartz, "Even couples who willingly try to change traditional male and female behavior have difficulty doing so. They must not only go against everything they have learned and develop new skills, but they have to resist the negative reaction of society" (1983, p. 324).

We turn now to a discussion of the process of changing power relationships in marriage.

Power Politics Versus No-Power Relationships

Through the early 1980s, researchers found "very little consistent evidence that egalitarian couples are more satisfied with their marriages than husband dominant couples" (Gray-Little 1982, p. 634). Since that time few researchers have investigated this question, and attention has turned to conjugal power as evidenced in division of labor in household tasks, discussed in Chapter 13. But evidence indicates that equitable relationships are generally more apt to be stable and satisfying (Lennon and Rosenfield 1994; Greenstein 1995), as we argue throughout this chapter.

With a goal for couples of developing relationships best suited to partners' individual needs and assets, marriage counselors today are virtually unanimous in asserting that intimacy takes place only insofar as partners are equal. Social scientist Peter Blau terms this situation *no-power*. **No-power** does not mean that one partner exerts little or no power; it means that both partners wield about equal power. Each has the ability to mutually and reciprocally influence and be influenced by the other. As we use the term, *no-power* also implies partners' unconcern about exercising their relative power over each other. No-power partners seek to negotiate and compromise, not to win (see Chapter 9). They are able to avoid **power politics,** a term that relates broadly to the power bases and tactics discussed in this chapter (see Box 10.2, "As We Make Choices: Disengaging from Power Struggles").

Disengaging from Power Struggles

Carlfred Broderick, sociologist and marriage counselor, offers the following exercise to help people disengage from power struggles. He begins by pointing out that almost nothing is more frustrating and resentment-inducing than an elaborate, unilaterally developed plan for yourself, your spouse, your children, and your friends. It is frustrating because when you try to live by such a script, or set of rules, you are condemned to seeing yourself as a failure (because neither you nor your children ever measure up) and to feeling rejected (because your spouse and friends never come through). Moreover, you are likely to imagine that you are surrounded by lazy, selfish, unfeeling, stubborn, underachieving, low-quality people.

The object of this exercise is to get you out of the business of monitoring everyone else's behavior and so free you from the unrewarding power struggles resulting from that assignment. Here is the exercise:

1. Think of as many things as you can that your spouse or children *should do, ought to do,* and *would do if they really cared,* but *don't do* (or do only grudgingly because you are always after them). Write them down in a list.

2. From your list choose three or four items that are especially troublesome right now. Write each one at the head of a sheet of blank paper. These are the issues that you, considerably more than your spouse, want to resolve (even though he or she, by rights, should be the one to see the need for resolution). Right now you are locked in a power struggle over each one, leading to more resentment and less satisfaction all around.

3. In this step you'll consider, one by one, optional ways of dealing with these issues without provoking a power struggle. Place an A, B, C, and D on each sheet of paper at appropriate intervals to represent the four options listed below. Depending on the nature of the issue, some of these options will work better than others, but for a start, write a sentence or paragraph indicating how each one might be applied in your case. Even if you feel like rejecting a partic-

ular approach out of hand, be sure to write something as positive as possible about it.

Option A: Resign the Crown

Swallow your pride and cut your losses by delegating to the other person full control and responsibility for his or her own life in this area. Let your partner reap his or her own harvest, whatever it is. In many cases your partner will rise to the occasion, but if this doesn't happen, resign yourself to suffering the consequences.

Option B: Do It Yourself

There's an old saying, "If you want something done right, do it yourself." Accordingly, if you want something done, and if the person you feel should do it doesn't want to, it makes sense to do it yourself the way you'd like to have it done. After all, who ever said someone should do something he or she doesn't want to do just because you want him or her to do it?

Option C: Make an Offer Your Partner Can't Refuse

Too many interpret this, at first, as including threats of what will happen if the partner doesn't shape up. The real point, however, if you select this approach, is to find out what your partner would really like and then offer it in exchange for what you want him or her to do. After all, it's your want, not your spouse's, that is involved. Why shouldn't you take the responsibility for making it worth your spouse's while?

Option D: Join with Joy

Often the most resisted task can become pleasant if one's partner shares in it, especially if an atmosphere of play or warmth can be established. This calls for imagination and good will, but it can also be effective in putting an end to established power struggles.

Source: Broderick 1979a, pp. 117–23.

Power Politics in Marriage

Unequal relationships discourage closeness between partners: Exchange of confidences between unequals may be difficult, especially when self-disclosure is seen to indicate weakness and men have been socialized not to reveal their emotions (Henley and Freeman 1989). Wives, feeling less powerful and more vulnerable, may resort to pretense and withholding of sexual and emotional response (Blumberg and Coleman 1989).

Unequal power relationships in marriage are said to encourage power politics: partners' struggles to gain or keep power over each other. This may also be true of relatively egalitarian unions. As gender norms move from traditional to egalitarian, all family members' interests and preferences gain legitimacy, not only or primarily those of the husband or husband-father. The man's occupation, for example, is no longer the sole determining factor in where the family will live or how the wife will spend her time. Thus, decisions formerly made automatically, or by spontaneous consensus, must now be consciously negotiated. A possible outcome of such conscious negotiating, of course, is greater intimacy; another is locking into power politics and conflict. If the essential source of conjugal power is (as the relative love and need theory suggests) "the greater power to go away," then "politics in marriage has to do with suggesting the use of that power to leave the marriage." A spouse plays power politics in marriage by saying, in effect, "This is how it would be if I were not here" (Blumberg and Coleman 1989; Chafetz 1989).

Both equal and unequal partners may engage in a cycle of devitalizing power politics. Partners come to know where their own power lies, along with the particular weaknesses of the other. They may alternate in acting sulky, sloppy, critical, or distant. The sulking partner carries on this behavior until she or he fears the mate will "stop dancing" if it goes on much longer; then it's the other partner's turn. This kind of seesawing may continue indefinitely, with partners taking turns manipulating each other. The cumulative effect of such power politics, however, is to create distance and loneliness for both spouses.

Few couples knowingly choose power politics, but this is an aspect of marriage in which choosing by default may occur. Our discussion of power in marriage is designed to help partners become sensitive to these issues so that they can avoid such a power spiral, or reverse one if it has already started.

Alternatives to Power Politics

There are alternatives to this kind of power struggle. Robert Blood and Donald Wolfe (1960) proposed one in which partners grow increasingly separate in their decision making; that is, they take charge of separate domains, one buying the car, perhaps the other in charge of disciplining their children. This alternative is a poor one for partners who seek intimacy, however, for it enforces the separateness associated with devitalized marriage.

THE NEUTRALIZATION OF POWER

A second, more viable alternative to perpetuating an endless cycle of power politics is for the subordinate spouse to neutralize the dominant partner's power. In **neutralizing power,** a subordinate weakens the powerful person's control by refusing to cooperate in that power. Neutralizing power is different from playing power politics. Mates involved in power politics seek to wield power over each other, usually by combining coercive and reward tactics. A subordinate partner who seeks only to neutralize power does not work to gain dominion or control but instead tries to move toward an equal position.

One way to neutralize conjugal power is for the less powerful spouse to obtain needed services from some other source. For example, the wife who is compliant and deferent because she needs her husband's financial support can neutralize this situation by becoming employed—a process that enhances her relative power in the marriage, as we have seen. A second neutralizing technique is to accept the services from the powerful one but—realizing there is an element of voluntarism in power—to refuse to feel reciprocally obligated, deferent, or compliant. For example, a husband who needs his wife's cooperation in entertaining business associates can accept her services but refuse to feel consequently obliged to paint the house. The less powerful partner can also resign him- or herself to doing without the service or can perhaps find a substitute for it. If one spouse wields power because the other finds it difficult to handle social situations and depends on the spouse to arrange social activities, the socially dependent spouse may work toward improving his or her social skills and self-esteem to the point that he or she can create some rewarding social contacts on his or her own.

Partners in a no-power relationship work at dong things on equal terms and seek to negotiate and compromise, thus avoiding deadly power games. By not competing with each other, both partners win.

THE IMPORTANCE OF COMMUNICATION Techniques of power neutralizing often bring the risk of devitalizing a relationship, depending on how spouses undertake them. Subordinate mates who go about neutralizing their situation without explaining what they are doing and why risk estrangement. The reason is that dominant partners may mistake a subordinate mate's acts of deference and compliance as evidence of love rather than fear. If signs of deference are withdrawn, a dominant partner can conclude that "she (or he) doesn't love me anymore." The harder the subordinate partner works to neutralize power, the more effort will the dominant one invest in regaining control. The result is estrangement. Case Study 10.1, "An Ice-Skating Homemaker in a Me-or-Him Bind," illustrates this pattern.

One wife reported that she had lived ten years in a husband-dominated marriage before she returned to school, and when she did return, their power relationship changed. Receiving substitute nourishment for her self-esteem in good grades and new friends, she no longer relied totally on her husband's signs of affection. In subtle ways she showed decreasing deference and felt herself growing toward equality in the marriage. But the couple did not discuss these changes. Meanwhile, her husband had begun to drink heavily.

The wife remained convinced he would "come along." Four years later, the couple were hardly speaking to each other and began marriage counseling. "I thought she didn't love me anymore," her husband said. "After she went back to school, she stopped doing things for me."

This couple might have avoided estrangement through mutual self-disclosure. The husband could have shared his anxiety. The wife could have explained that, from her point of view, signs of deference—doing things for him—did not show love. Perhaps the couple could have negotiated an agreement whereby he continued to feel loved while she proceeded to gain equality. As couples assert their interests and bargain in marriage, communication is especially important in establishing and maintaining trust.

THE CHANGE TO A NO-POWER RELATIONSHIP Even when couples discuss power changes, living through them can be difficult (Blaisure and Allen 1995). Changing conjugal power patterns can be difficult, even for couples who talk about it, because these patterns usually have been established from the earliest days of the relationship. Although partners may not have discussed them directly, they set up unconscious agreements by sending countless verbal and nonverbal

An Ice-Skating Homemaker in a Me-Or-Him Bind

Joan is an attractive, full-time homemaker who has been married nineteen years. Recently she began taking courses at a local university and also became involved in learning to figure-skate.

INTERVIEWER: When did you begin ice-skating?

JOAN: Well, I took one year when I was a kid, but I had to ride the bus and the streetcar and all that.... And then I didn't skate again until last year. I've been skating for two years now and I just love it. I'm getting better. I can do three turns real well and three of the very basic dances....

I try to skate twice a week. But it's created a problem with Chuck. Last year he was working days during the skating season. But now he works midnight to eight and he just hates for me to go there during the day.... I don't know whether it's because I like it real well or what. There's nothing there to be jealous of because I skate with the housewives.... I don't know whether it's that I enjoy something he can't do well or what it is. But he doesn't like it. When he works midnight to eight, he knows every time I go. When he was working days and I went, as long as my work was done and I had dinner on the table, there was no problem when he came home from work.

Now he knows every time I leave this house. Every place I go he knows. Every time the garage door opens, it wakes him up. It's almost like being in prison without the doors being locked....

One time I stayed too late. I got home at five thirty. My brother was there—I got home at five thirty and no dinner or nothing. He told me, he said, "If you ever do this again—there's no dinner—if you ever do this again, I'm going to cut your skates." Oh boy! So I try to avoid doing that. I come home about four o'clock, so I can get dinner on okay. But the trouble [with the ice-skating] is I can go there and it's almost like on that ice nothing—I just get totally absorbed in it and I forget I'm a mother, forget I'm a wife, I forget everything, I'm just there. I felt that way about golf and water-skiing too, but those things didn't bother Chuck because I was doing them with him I think. Skating excludes him....

At first I thought it was jealousy. No, it's not jealousy. It's possessiveness. He wants to control what I do:... "This is a possession now; I own this person; I can control her mind and her body."

My daughter's starting to want to skate now too. The rink is open for the public this summer and I'm not going to be skating very much because I'd have to go in the evenings and that's just not going to work out with our schedule. But I'll try to go. Like this weekend Chuck will be out of town, so I'm going to go then. Anyway, my daughter's going to go with me when she's out of school. She's getting so she can skate pretty well and she's starting to like it. So she wants a pair of skates. Now he won't buy her the skates. He says, "No, we're not going to spend the money on something like that." Now I think that's terrible....

I could just go buy them because I definitely bought my own skates. I just went out and bought them. And then at first I lied to him—this is awful—I told him, "Oh, these are just my sister-in-law's skates...." Then finally once I told him they were my own. He said, "Oh! You can afford those skates and I can't afford a jacket." I said, "You can afford a jacket. Go buy one if you want...."

A lot of times he'll say, "What are you going to do today?" And I say, "Well, it's Tuesday and I skate on Tuesday." He's known that all year, but every time he wants to take me to lunch or go somewhere, it is always on Tuesday.... One time I said, "Why can't we do it on Monday or Wednesday?" He said, "Oh, I never thought about that...."

He gets mad about everything I really like. Like when I started bowling and I really liked that, he gave me a hard time. It's really not just the skating. If he took the ice-skating away, and I replaced it with something else I liked equally well, that would be the thing he'd be against.

Whether a spouse is free to spend time in self-actualizing pursuits, how much time is allowed, and which pursuits are allowed depend largely on conjugal power. Chuck is exercising coercive power in threatening to cut Joan's skates. Such a threat is psychological abuse, or psychological violence.

cues. As partners experience recurring subtle messages, they build up predictable behavior patterns. From the interactionist perspective, certain behaviors not only come to be expected but also to have symbolic meaning. For many couples, initiating or responding to sexual overtones, spending holidays in a certain way, or buying favorite foods or other treats symbolizes not just who has how much power but love itself. Fifteen years ago, sociologist William Goode had an insight that is still relevant. He wrote that the most important change in men's position, as they themselves see it, is a "loss of centrality," a decline in the extent to which they are the center of women's attention. According to Goode:

> Men have always taken for granted that what they were doing was more important than what the other sex was doing, that where they were, was where the action was. Their women accepted that definition. Men occupied the center of the stage, and women's attention was focused on them. . . . [But] the center of attention shifts to women more now than in the past. I believe that this shift troubles men far more, and creates more of their resistance, than the women's demand for equal opportunity and pay in employment. (Goode 1982, p. 140)

It is possible that spouses have never experienced a no-power man–woman relationship. By experience each knows either the dominant or the submissive role; vicariously, each knows how to play the other's role. When undergoing changes in power, they may be more inclined to reverse roles, moving to behavior they know vicariously, rather than creating new egalitarian roles.

Husbands may respond to wives' power challenges by abdicating interest in decision making, assuming a submissive rather than an equal stance. Their reaction is "It's up to you" or "Whatever you want. I don't have anything to say around here anymore anyway." They relinquish, along with husbandly authority, their willingness to influence the relationship. Much of this is probably passive-aggressive behavior.

The best way to work through power changes is to openly discuss power and to fight about it fairly, using techniques described in Chapter 9. The partner who feels more uncomfortable can bring up the subject, sharing his or her anger and desire for change but also stressing that he or she still loves the other. Meanwhile, partners need to remember that this is easier said than done. Attempts at communication—and open communication itself—do not solve all marital problems. Changing a power relationship is a challenge to any marriage and is painful for both partners. One option is to seek the help of a qualified marriage counselor (see also Appendix H).

The Role Marriage Counselors Can Play

Today many marriage counselors are committed to viewing couples as two human beings who need to relate to each other as equals. In other words, they are committed to helping couples develop no-power relationships. They realize that once *both* spouses admit—to themselves and to each other—that they do in fact love and need each other, the basis for power politics is gone. On this assumption, counselors help spouses learn to respect each other as people, not to fear each other's coercive withdrawal.

Couples need to be aware that, like everybody in society, marriage counselors have internalized masculine and feminine biases. Also, they have personal reactions to and feelings about women's emerging rights. Some counselors are angry about past injustices. They may be more committed to urging a wife's separate autonomy through divorce, if necessary, than to helping her reach a no-power marital relationship. In the first session (or before, if possible), clients should ask prospective counselors about their feelings on this issue. One counselor may assume, for example, that whenever marriage seriously threatens personal growth, it should be dissolved. Another may assume that partners need to learn to communicate and, only after that, to determine whether they want to stay married or to divorce.

African American social scientists and social workers have pointed to issues concerning potential racial and cultural bias on the part of therapists. It may be that black couples and families have culturally specific needs. For example, counseling for African Americans should recognize the validity of extended-family forms and propose specific therapeutic approaches that are appropriate for black clients and families (Taylor et al. 1990, pp. 1000–1001). Bias issues no doubt apply to other racial/ethnic minority groups as well.

Choosing an appropriate counselor can be difficult for reasons other than their differing assumptions. A dominant husband, fearful that "it's going to be two against one," may feel threatened by a female counselor. On the other hand, a wife may fear that a male counselor will be too traditional or unable to relate to her. In this situation, counselors sometimes work as a team, woman and man.

It is important that both partners feel comfortable with a counselor from the beginning, for two reasons. First, clients will hesitate to be frank with counselors they mistrust or don't like. Without honesty and openness, marriage counseling is only an expensive hour away from work or home. Second, as counselors gradually probe deeper into the couple's problems, one or both spouses may begin to feel threatened. Problems they had never recognized emerge into consciousness. As one counselor put it, "Things get worse before they get better." A common reaction is to dismiss the counselor's insight as all wrong. If couples quit at this point, they may find themselves more estranged than before they began. Understanding and accepting a counselor's assumptions from the beginning help to avoid failure. Nevertheless, marriage counseling does not keep all marriages intact.

No marriage, indeed no relationship of any kind, is entirely free of power politics. But together partners can choose to emphasize no-power over the politics of power. One spouse can't do it alone, however. No-power involves both spouses' conscious refusal to be exchange oriented or to engage in psychological bookkeeping. No-power involves honest self-disclosure and negotiation rather than passive-aggressive maneuvering. As Chapter 9 points out, the politics of love requires fighting so that both partners win. When power politics triumphs over no-power, one result can be family violence—psychological (emotional) and/or physical.

Family Violence

The use of physical violence to gain or demonstrate power in a family relationship has occurred throughout history, but only recently has family violence been labeled a social problem. The discovery of child abuse in the 1960s was followed in the 1970s by attention to spouse abuse as a widespread problem with roots in assumptions about marital power. With the 1980s came concern about elderly abuse.

The National Crime Survey by the Justice Department estimates that about 623,000 cases of **violence between intimates**—murders, rapes, robberies, or assaults committed by spouses, ex-spouses, boyfriends,

or girlfriends—occur each year.[1] This number represents more than 13 percent of all these violent crimes committed annually. Among the victims of intimate violence, about 50 percent of the victims are attacked by boyfriends or girlfriends; 35 percent, by spouses; and 15 percent, by ex-spouses (U.S. Department of Justice 1994). Although we focus primarily on spouse abuse in this chapter, it is worth noting here that the violence rate between cohabitors is higher, in fact, than that between spouses; as a category, cohabitors are younger than marrieds, less integrated into family and community, and more likely to have psychobehavioral problems such as depression and alcohol abuse (Stets 1991).

As Table 10.1 shows, 92 percent of the victims of nonfatal intimate violence are women. Compared with women who have lower educational achievement, women who graduated from college are the least likely to be victimized by intimates. Similarly, women with annual family incomes under $9,999 are most likely to be victims of intimate violence, while those with family incomes over $30,000 are the least likely to be victimized. When education and income are taken into account, non-Hispanic white, African American, and Hispanic women have equivalent rates of intimate violence committed against them. Of any age group, women ages 20–34 have the highest rates of violent victimization attributable to intimates. All else being equal, women living in central cities, suburbs, and rural areas are equally likely to be victims of intimate violence (U.S. Department of Justice 1994).

In 1992, an estimated 2,167 murder victims were killed by intimates. Of these, 70 percent were women. Actually, murder rates between intimates have declined over the last twenty years. However, the rate for female victims of intimate homicide dropped only slightly, while the rate for male victims was cut in half. Among African Americans, more husbands were killed by wives than vice versa in 1977; but by 1992, the opposite was true. Among whites, wives have consistently outnumbered husbands as victims of intimate murder (U.S. Department of Justice 1994).

The Department of Justice notes that violence between intimates is difficult to measure; it often occurs

1. This estimate may be low. A different estimate by the Senate Judiciary Committee is that at least 1.1 million assaults, aggravated assaults, murders, and rapes against women are committed in the home and reported to police annually, nationwide (Committee on the Judiciary, United States Senate 1992, p. iii). Also see Table 10.2.

TABLE 10.1

Violence Between Intimates, Reported to Department of Justice, National Crime Victim Survey (NCVS)

		Sex of Victim	
		Female	Male
Average annual number of nonfatal violent victimizations, 1987–1991	621,015	572,032	48,983
Estimated number of intimate homicides, 1992	2,167	1,510	657

	Percentage of All Victims Who Are Female
Nonfatal violent crimes	92
By spouse	93
By boyfriend/girlfriend	91
By ex-spouse	89
Homicides	70

Source: U.S. Department of Justice 1994.

in private, and victims may be reluctant to report incidents because of shame or fear of reprisal. As a result, these statistics may not be accurate. Probably the best estimate of the extent of violence within family households is still provided by two National Family Violence surveys, the first in 1975 and the second in 1985. Together the surveys gathered data from national probability samples of 8,145 husbands, wives, and cohabiting individuals (Straus, Gelles, and Steinmetz 1980; Straus and Gelles 1986, 1988). The authors defined *violence* as "an act carried out with the intention, or perceived intention, of causing physical pain or injury to another person." This definition is synonymous with the legal concept of assault. Respondents were specifically asked about the following acts: threw something at the other; pushed, grabbed, or shoved; slapped or spanked; kicked, bit, or hit with a fist; hit or tried to hit with something; beat up the other; burned or scalded (for children) or choked (for spouses); threatened with knife or gun; and used a knife or gun (Straus and Gelles 1988, p. 15). *Severe violence* was defined as acts that have a relatively high probability of causing an injury. The acts constituting severe violence are kicking, biting, punching, hitting with an object, choking, beating, threatening with a knife or gun, using a knife or gun, and, for violence by parents against children, burning or scalding the child (Straus and Gelles 1988, p. 16).

Together, the 1975 and 1985 surveys found that in 16 percent of the couples (161 cases per 1,000 couples; see Table 10.2, Section A), at least one of the partners had engaged in a violent act against the other during the previous year. In other words, each year in about one out of every six couples in the United States, an individual commits at least one violent act against his or her partner. And if the period considered is the entire length of the marriage rather than just the previous year, a violent act is committed in 28 percent, or between one out of four and one out of three couples. In short, if you are married, the chances are almost one out of three that your husband or wife will hit you.

National Family Violence survey data also yield information on violence directed toward children by parents and by siblings. For example, 7 percent of children (70 cases per 1,000 children) between ages 15 and 17 undergo severe violence from their parents in any given year (see Table 10.2, Section C). Sections D

TABLE 10.2

Annual Incidence Rates for Family Violence and Estimated Number of Cases Based on These Rates

Type of Intrafamily Violence*	Rate per 1,000 Couples or Children	Number Assaulted, 1994[†]
A. Violence between husband and wife		
any violence during the year (slap, push, etc.)	161	8,700,000
severe violence (kick, punch, stab, etc.)	63	3,400,000
any violence by the husband	116	6,250,000
severe violence by the husband ("wife beating")	34	1,800,000
any violence by the wife	124	6,800,000
severe violence by the wife	48	2,600,000
B. Violence by parents—child ages 0–17		
any hitting of child during the year	Near 100% for young child[†]	
very severe violence (Child Abuse-1)	23	1,541,000
severe violence (Child Abuse-2)	110	7,370,000
C. Violence by parents—child ages 15–17		
any violence against 15–17-year-olds	340	3,536,000
severe violence against 15–17-year-olds	70	728,000
very severe violence against 15–17-year-olds	21	218,400
D. Violence by children ages 3–17 (1975–76 sample)		
any violence against a brother or sister	800	53,600,000
severe violence against a brother or sister	530	35,510,000
any violence against a parent	180	12,060,000
severe violence against a parent	90	6,030,000
E. Violence by children ages 15–17 (1975–76 sample)		
any violence against a brother or sister	640	6,656,000
severe violence against a brother or sister	360	3,744,000
any violence against a parent	100	1,040,000
severe violence against a parent	35	364,000

Source: Adapted from Straus and Gelles 1988, pp. 18–19; data are from the 1975 and 1985 National Family Violence surveys.

*Section A rates are based on a nationally representative sample of 6,002 currently married or cohabiting couples interviewed in 1985. Section B rates are based on the 1985 sample of 3,232 households with a child age 17 and under.

[†]The column giving the "Number Assaulted" was computed by multiplying the rates in this table by the 1994 population figures as given in the 1995 Statistical Abstract of the United States. The population figures (rounded to millions) are 54 million couples (Table 62) and 67 million children ages 0–17. The number of children 15–17 was estimated as 10.4 million. This was done by taking .75 of the number ages 14–17, as given in Statistical Abstract Table 16.

[†]The rate for 3-year-old children in the 1975 survey was 97%.

and E of Table 10.2 give information on violence by children. Clearly, violence among siblings is the most common form of family violence. More than half (53 percent) of children between ages 3 and 17 annually engage in at least one act of severe violence against a brother or sister. For those between ages 15 and 17 the incidence of violence is 36 percent, and 35 percent have engaged in at least one severely violent act against a parent.

Family violence exists in all social classes. There is statistical evidence to suggest that it occurs more often in blue-collar and lower-class families, a finding partly attributable to the fact that middle-class families have greater privacy than lower-class families and hence are better able to conceal family violence (Fineman and Mykitiuk 1994). Also, middle-class individuals have recourse to friends and professional counselors to help deal with their violence; consequently, their altercations are less likely to become matters for the police (Buzawa and Buzawa 1990).

We will examine issues of spouse abuse, child abuse, and elderly abuse in some detail in the sections that follow. Often two or more of these forms of domestic violence occur together in the same family.

Wife Abuse

Wife abuse is a serious and significant problem whose most visible manifestation occurs in wife beating. Other forms of violence, such as kicking, pushing down stairs, and hitting with objects, are also considered abuse. It is difficult to define wife abuse so that it includes all possible acts of serious physical violence while excluding verbal abuse because verbal aggression (threats, foul name calling, and other vocal assaults) virtually always occurs with physical aggression (Stets 1990). Although verbal aggression is of great psychological importance, we concern ourselves here primarily with repetitive, physically injurious acts. Many of the alienating fight practices discussed in Chapter 9 can also be verbal, or emotional abuse.

Physical wife abuse results in serious injuries. In the physical damage typical of battered women surveyed in the early 1980s in California shelters,

> most injuries were to the head and neck and, in addition to bruises, strangle marks, black eyes, and split lips, resulted in eye damage, fractured jaws, broken noses, and permanent hearing loss. Assaults to the trunk of the body were almost as common and produced a broken collarbone, bruised and broken ribs, a fractured tailbone, internal hemorrhaging, and a lacerated liver. (Jaffe, in National Institute of Mental Health 1985, p. 1)

MARITAL RAPE   Wife abuse may take the form of sexual abuse and rape, and sexual abuse is often combined with other physical violence. Ten percent of wives in a Boston study reported that their husbands used physical force or threats to compel sex (Finkelhor and Yllo 1985). Another study found that 35 to 50 percent of marital rapes involved beating, hitting, or

kicking (Russell 1982). Incredibly sadistic attacks are reported by many wives.

Under traditional law **marital rape** was not considered rape at all because the wife was considered her husband's property and he was entitled to unlimited sexual access. The laws against rape in general have improved during the past two decades, however, mainly as a result of the women's movement. Although husbands can now be accused of marital rape in slightly over half the states, this still means that forcing sex upon a wife is legal in nearly half the states, at least while spouses are living together (Basow 1992, pp. 320–21).

THE THREE-PHASE CYCLE OF DOMESTIC VIOLENCE   Experts are finding that the initial violent episode usually comes as a shock to a wife, who treats her mate's violence as an exceptional, isolated outburst. He promises it will not happen again; she believes him. She also tries to figure out what she did to cause his reaction so there will be no reason for it to happen again. The likelihood is that it will, however, because of what counselors call the **three-phase cycle of violence.** First, tension resulting from some minor altercations builds over a period of time. Second, the situation escalates, eventually exploding in another violent episode. Third, the husband becomes genuinely contrite, treating his wife lovingly. She wants to believe that this change in him will be permanent. This cycle repeats itself, with the violence worsening, if nothing is done to change things (Sonkin, Martin, and Walker 1985).

Two questions arise: Why do husbands beat their wives? And why do wives tolerate it? Both can be answered by the relative love and need theory.

WHY DO HUSBANDS DO IT?   Studies indicate that husbands who beat their wives are attempting to compensate for general feelings of powerlessness or inadequacy—in their jobs, in their marriages, or both. As Chapter 3 indicated, our cultural images and the socialization process encourage men to be strong and self-sufficient. Husbands may use physical expressions of male supremacy to compensate for their lack of occupational success, prestige, or satisfaction. Husbands' feelings of powerlessness may also stem from an inability to earn a salary that keeps up with inflation and the family's standard of living or from the stress of a high-pressure occupation. Put another way, absent the

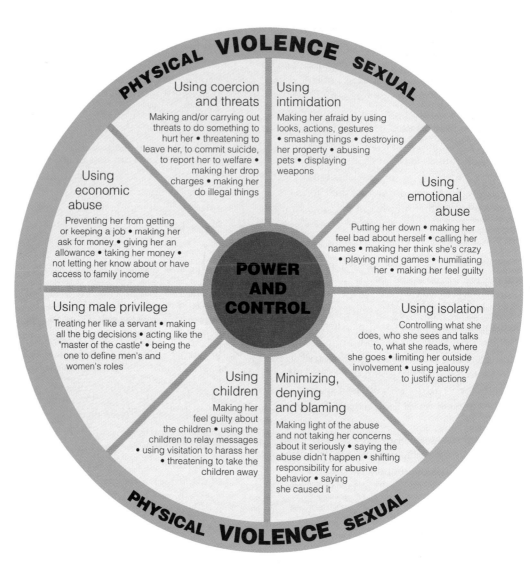

PHYSICAL VIOLENCE SEXUAL

Using coercion and threats

Making and/or carrying out threats to do something to hurt her • threatening to leave her, to commit suicide, to report her to welfare • making her drop charges • making her do illegal things

Using intimidation

Making her afraid by using looks, actions, gestures • smashing things • destroying her property • abusing pets • displaying weapons

Using economic abuse

Preventing her from getting or keeping a job • making her ask for money • giving her an allowance • taking her money • not letting her know about or have access to family income

Using emotional abuse

Putting her down • making her feel bad about herself • calling her names • making her think she's crazy • playing mind games • humiliating her • making her feel guilty

POWER AND CONTROL

Using male privilege

Treating her like a servant • making all the big decisions • acting like the "master of the castle" • being the one to define men's and women's roles

Using isolation

Controlling what she does, who she sees and talks to, what she reads, where she goes • limiting her outside involvement • using jealousy to justify actions

Using children

Making her feel guilty about the children • using the children to relay messages • using visitation to harass her • threatening to take the children away

Minimizing, denying and blaming

Making light of the abuse and not taking her concerns about it seriously • saying the abuse didn't happen • shifting responsibility for abusive behavior • saying she caused it

PHYSICAL VIOLENCE SEXUAL

FIGURE *10.1*

I Coercive power some male partners resort to for power and control.

reward power base, some husbands resort to *coercive power*. Figure 10.1, from a program for male batterers in Duluth, Minnesota, illustrates how a male partner's need for power and control can result in both psycho-emotional and physical violence.

Men may also use violence to attempt to maintain control over wives or partners trying to become entirely independent of the relationship (Dutton and Browning 1988). Richard Gelles (1994) lists among "risk factors" for men who abuse women those who

are between 18 and 30 years old, unemployed, users of illicit drugs or abusers of alcohol, and high school dropouts. With few legitimate avenues toward personal or social power, such men may resort to coercive power over the women in their lives in an attempt to gain control.

WHY DO WIVES PUT UP WITH IT? Women do not like to get beaten up. They do not cooperate in their own beatings, and they often try to get away.

Fifty thousand women participated in this rally at the National Mall in Washington, DC. The women gathered to protest all types of violence against women.

They may, however, stay married to husbands who beat them repeatedly. For the most part, battered wives seek divorce only after a long history of severe violence and repeated conciliation. There are several reasons for this, and they all point to these women's lack of personal resources with which to take control of their own lives.

Fear Battered wives' lack of personal power begins with fear. "First of all," reports social scientist Richard Gelles, "the wife figures if she calls police or files for divorce, her husband will kill her—literally" (Gelles, in C. Booth 1977, p. 7). This fear is not unfounded. An estimated 75 percent of murders of women by their male partners occurred in response to the woman's attempt to leave (de Santis 1990). Husbands or ex-husbands have shown enormous persistence in pursuing and beating or killing women

who try to leave an abusive situation (Johann 1994).

Cultural Norms Furthermore, our cultural tradition historically has encouraged women to put up with abuse. English common law, the basis of the American legal structure, asserted that a husband had the right to physically chastise an errant wife, provided the stick was not bigger than his thumb (a legal norm still prevalent in the nineteenth century) (Straus and Gelles 1986). A cross-cultural analysis of domestic violence in ninety different societies found family violence to be virtually absent in sixteen. These sixteen societies were characterized by economic and decision-making equality between the sexes, norms encouraging nonviolence generally, and regular intervention by neighbors and kin in domestic disputes (Levinson 1989). Although the legal right to physically abuse women has long since disappeared in our society, our cultural heritage continues to influence our attitudes.

Love, Economic Dependence, and Hopes for Reform   Wives may tolerate abuse, not because they enjoy being battered, but because they love their husbands, depend on their economic resources, and hope they will reform. Battered wives who stay with their husbands fear the economic hardship or uncertainty that will result if they leave. They hesitate to summon police or to press charges because of the loss of income or damage to a husband's professional reputation that could result from his incarceration. This economic hardship is heightened when children are involved. For mothers, leaving requires either being financially able to take her children—who may also be in danger—with her, or leaving them behind. Consequently, even though a dramatic rise in a wife's earnings may prove threatening to a low-earning husband, in the long run mutual awareness of the wife's economic independence may ultimately deter wife abuse (Blumberg and Coleman 1989).

Childhood Experience   Another frequent factor in women's tolerance of abuse is childhood experience. Research suggests that people who experience violence in their parents' home while growing up may have an increased tolerance for violence and regard beatings as part of married life (Straus, Gelles, and Steinmetz 1980). One study of 204 recently divorced Iowa women concluded that girls subjected to abusive parenting tend to develop a hostile, rebellious personality and are likely to affiliate later with men who are like them. These men, in turn, tend to be violent toward their dates and, later, their wives (Simons et al. 1993).

Gendered Socialization   Another factor that helps perpetuate wife abuse is that women accept the cultural mandate that it is primarily their responsibility to keep their marriage from failing. Believing this, wives are often convinced that their emotional support can lead husbands to reform. Thus, wives often return to violent mates after leaving them (Herbert, Silver, and Ellard 1991).

Low Self-Esteem   Finally, unusually low self-esteem interacts with fear, depression, confusion, anxiety, feelings of self-blame (Andrews and Brewin 1990) and general helplessness to create the **battered woman syndrome,** in which a wife feels incapable of making any change (Johann 1994).

A WAY OUT   A woman in such a position needs to redefine her situation before she can deal with her problem, and she needs to forge some links with the outside world to alter her relationship.

Although there are not enough of them (Jones 1994), a network of shelters for battered women provides a woman and her children temporary housing, food, and clothing to alleviate the problems of economic dependency and physical safety. These organizations also provide counseling to encourage a stronger self-concept so that the woman can view herself as worthy of better treatment and capable of making her way in the outside world if need be. Finally, shelters provide guidance in obtaining employment, legal assistance, family counseling, or whatever practical assistance is required for a more permanent solution.

This last service provided by shelters—obtaining help toward more long-range solutions—is important, research shows. Two face-to-face interviews with the same 155 wife-battery victims (a "two-wave panel study") were conducted within eighteen months during 1982 and 1983 in Santa Barbara, California. Each of the women interviewed had sought refuge in a shelter. Findings showed that victims who were also taking other measures (for example, calling the police, trying to get a restraining order, seeking personal counseling or legal help) were more likely to benefit from their shelter experience. "Otherwise, shelters may have no impact or perhaps even trigger retaliation (from husbands) for disobedience" (Berk, Newton, and Berk 1986, p. 488). The researchers conclude:

The possibility of perverse shelter effects for certain kinds of women poses a troubling policy dilemma. On the one hand, it is difficult to be enthusiastic about an intervention that places battered victims at further risk. On the other hand, a shelter stay may for many women be one important step in a lengthy process toward freedom, even though there may also be genuine short-run dangers. Perhaps solutions can be found in strategies that link together several different kinds of interventions. (1986, p. 488)

Husband Abuse

Both wives and husbands sometimes resort to violence. One study based on a national sample of 960 men and 1,183 women (not married to each other) found that nearly one husband in eight and about the same number of wives (11.6 percent of husbands, 12.1 percent of wives) had committed at least one violent act during the year in which the research was conducted. In 49 percent of the violent situations, both partners engaged in some form of physical abuse. Although men may be more likely to deny using violence (DeMaris, Pugh, and Harman 1992), surveys show a pattern of mutual violence between spouses (Stacey, Hazlewood, and Shupe 1994). In the remaining situations, about an equal number of husbands and wives were violent while their partner was not (Straus, Gelles, and Steinmetz 1980). More recent data confirm this basic pattern (Flynn 1990). Straus and Gelles found it "distressing" that, "in marked contrast to the behavior of women outside the family, women are about as violent within the family as men" (1986, p. 470). They noted that criticism from feminist scholars (see for example, Yllo and Bograd 1988) has led most researchers to avoid publishing articles on battered husbands (Gelles and Conte 1990, p. 1046). Furthermore, battered husbands have fewer community resources, such as shelters, than do battered wives (Cose 1994a). Responding to the controversy, Straus and Gelles have stated their position:

> Violence by women is a critically important issue for the safety and well-being *of women*. Let us assume that most of the assaults by women are of the "slap the cad" genre and are not intended to and do not physically injure the husband. The danger to women of such behavior is that it sets the stage for the husband to assault her. Sometimes this is immediate and severe retaliation. But regardless of whether that occurs, the fact that she slapped him provides the precedent and justification for him to hit her when *she* is being obstinate, "bitchy," or "not listening to reason" as he sees it. Unless women also forsake violence in their relationships with male partners and children, they cannot expect to be free of assault. Women must insist as much on non-violence by their sisters as they rightfully insist on it by men. (Straus and Gelles 1988, pp. 25–26)

Indeed, it may be that there are two forms of heterosexual violence against women—"patriarchal terrorism" and "common couple violence" (Johnson 1995). Meanwhile, "it would be a great mistake if our new awareness of husband abuse were to deflect attention from wives *as victims*" (see Yllo and Bograd 1988). This is so for the following reasons, all based on research:

1. Husbands have higher rates of inflicting the most dangerous and injurious forms of violence, such as severe beatings (Straus, Gelles, and Steinmetz 1980).

2. Violence by husbands does more damage, even if it is an exchange of slaps or punches, because of a man's generally greater physical strength; the woman, therefore, is more likely to be seriously injured (Brush 1990).

3. Violent acts by the husband tend to be repeated over time, whereas those by wives do not. (Statistics showing about equal incidences of husband and wife violence are in response to a research question about "at least one" incident; Straus, Gelles, and Steinmetz 1980.)

4. Much of wives' violence appears to be in self-defense. "One of the most fundamental reasons why some women are violent within the family, but not outside the family, is that the risk of assault for a typical American woman is greatest in her own home" (Straus and Gelles 1988, p. 19).

5. Husbands are more apt to leave an abusive relationship within a short time. Having more resources, men rarely face the choice women do of choosing between poverty (for their children as well as themselves) and violence (Straus, Gelles, and Steinmetz 1980).

For all these reasons, women are rightly concerned about losing public attention and resources to what is perceived as a less drastic need (Dobash et al. 1992). Yet there may be a need for some programmatic support for male victims of spouse abuse. Both needs must be met.

Abuse Among Lesbian and Gay Male Couples

Domestic violence has long been one of the lesbian and gay communities' "nastiest secrets" (Island and Letellier 1991, p. 36; Obejas 1994, p. 53). Lesbians may have denied the issue because they believe in the inherent goodness of lesbian relationships or are afraid of giving fuel to homophobia. As for violence in the relationships of gay men, there may be even more silence and denial:

> After two and one-half years of researching and learning about gay men's domestic violence, we have reached the conclusion that as a community we are responding to domestic violence generally the same way a victim responds to domestic violence when it first happens to him. Our community is minimizing the problem. (Island and Letellier 1991, pp. 36–37)

Research on lesbian and gay male relationship violence is very scanty, but the few studies that have been done suggest that violence between same-sex partners occurs at about the same rate as it does in heterosexual relationships. As is also true for straights, domestic violence can be found in all racial/ethnic categories, social classes, education levels, and age groups. Among lesbians, neither "butch/femme" roles nor the women's physical size has been found to figure into violence (Obejas 1994).

Some of the relationship dynamics in same-sex abusive partnerships are similar to those in abusive straight relationships (Kurdek 1994). Alcohol and/or drug abuse are often involved, for instance. And the abusive partner is typically jealously possessive, using violence or threats of violence to keep the partner from leaving him or her. Furthermore, the couple is likely to deny or minimize the violence, along with believing that the violence is at least partly the victim's fault (Island and Letellier 1991; Renzetti 1992).

Meanwhile, there are other relationship dynamics that are specific to same-sex domestic abuse. For one thing, both men and women in same-sex relationships are freer to be either dominant or submissive, so it is more difficult to allocate violence to culturally influenced gender roles. Lesbians as well as gay men may fight back more often than do heterosexual women, a situation that leads to confusion about who is the battered and who is the batterer. Furthermore, some lesbians and gay men who are battered in one relationship may become batterers in another relationship (Island and Letellier 1991; Obejas 1994).

Lesbian violence is different in still other respects. For one thing, according to psychologist Vallerie Coleman, who works with battered lesbians, heterosexual

men tend to feel they have a right to abuse their mates, while lesbians do not (Obejas 1994). And a greater proportion of lesbian batterers seek help than do heterosexual male abusers. Lesbians are more likely to go into treatment on their own, according to Coleman. "Heterosexual men go in because they're court-mandated" (in Obejas 1994).

A special problem for lesbian and gay male domestic violence victims is that few resources exist to serve their needs, although some services are beginning to be developed, particularly in San Francisco and New York City (Island and Letellier 1991; Obejas 1994). This is an issure that is just beginning to be discussed and researched; studies should enhance what we know about all the various forms of domestic violence.

Stopping Relationship Violence

According to one estimate, 32 percent of victimized women will be revictimized within a relatively short time in the absence of intervention (Langan and Innes 1986). Moreover, intimate abuse tends to escalate, growing more severe over time.

The interests of men and women in curbing spousal violence converge when we consider the issue of male victims of spousal homicides. Even though men kill wives and girlfriends at nearly twice the rate at which women kill husbands and boyfriends, some victimized women do murder their male partners. Although few would endorse murder as the solution of choice, lawyers have recently been successful in using the battered woman syndrome as a defense and convincing the courts that there are no alternatives (Walker 1988; Johann 1994).

Police protection and arrest of wife abusers constitute an alternative that may be in the long-term interests of both abuser and abused. If a wife has an effective means of coping with violence, she will not be forced into self-protective acts. There had been little legal protection for battered women until fairly recently. In the past, police typically avoided making arrests for assault that would be automatic if the man and woman involved were not married. The laws themselves contributed to police reluctance: Statutes might require a police officer to witness the act before making an arrest at the scene, or more severe injury might be required for prosecution for battery. In some cases restraining orders required additional court action before they could be enforced. Over the past twenty years, however, laws have been changed to make arrests for domestic violence more feasible. At least thirteen states have enacted policies that mandate (require) arrest in certain situations involving family violence (Buzawa and Buzawa 1990, p. 96).

Some studies (Sherman and Berk 1994; Berk and Newton 1995) have found that mandatory arrests deter new assaults. Furthermore, mandatory arrest policies, when publicized, may deter some other men —but not those who see wife abuse as legitimate— from physically assaulting their wives because they fear public humiliation (Williams 1992; Bureau of Justice Assistance 1993). On the other hand, some battered mates insist that arresting an abusive partner or pressing charges will only aggravate the situation and result in escalating violence later. Research continues as to the circumstances in which a mandatory arrest policy can be effective (Sherman, L. W. 1992).

Moreover, the courts now permit civil suits against police failure to protect. Although not all research has supported the original study (Prial 1988), "perhaps presumptory arrest should be the operational policy until strong evidence is presented showing that other forms of police intervention are more effective" (Berk and Newton 1985, p. 262). There is also some evidence that negative social sanctions from either partner's relatives or friends can help stop wife abuse (Lackey and Williams 1995).

More direct assistance to receptive wife abusers is also offered. Counseling and group therapy, thought to be ineffective with male abusers, have now been tried with some success. Here we mean not conventional marital therapy but group therapy for abusive men who wish to stop. Many abusing husbands have difficulty controlling their response to anger and frustration, dealing with problems, or handling intimacy. Even though many, perhaps most, abusers are not reachable (Hueschler 1989), others have a sincere desire to stop. Group therapy reduces stigma and provides a setting in which abusive husbands can learn more constructive ways of both coping with anger and balancing autonomy and intimacy, often another area of difficulty. Some men's therapy groups have emerged in which former batterers help lead male abusers toward recovery (Allen and Kivel 1994).

We turn now to another family violence issue in which the more powerful abuse the less powerful— that of child abuse.

Child Abuse and Neglect

Perceptions of what constitutes child abuse or neglect have differed throughout history and in various cultures. Practices that we now consider abusive were accepted in the past as the normal exercise of parental rights or as appropriate discipline. Until the twentieth century, children were mainly considered the property of parents. Aristotle wrote that a son or a slave is property and there is no such thing as injustice to one's property. In ancient Rome a man could sell, abandon, or kill his child if he pleased. Infanticide has been practiced from earliest antiquity to limit family size, to ensure crop growth through religious sacrifice, or to relieve financial burdens (DeMause 1975, p. 87). In colonial Massachusetts and Connecticut, filial disobedience was legally punishable by death.

Today, standards of acceptable child care vary according to culture and social class. What some groups consider mild abuse, others consider right and proper discipline. In 1974, however, Congress provided a definition in the Child Abuse Prevention and Treatment Act. The act defines child abuse and neglect as the physical or mental injury, sexual abuse, or negligent treatment of a child under the age of 18 by a person who is responsible for the child's welfare under circumstances that indicate that the child's health or welfare is harmed or threatened (U.S. Department of Health, Education, and Welfare 1975, p. 3).

People use the term **child abuse** to refer to overt acts of aggression, such as excessive verbal derogation, beating, or inflicting physical injury. Section B of Table 10.2 contains National Family Violence survey data on child abuse; note that it includes two alternative operational definitions of child abuse—very severe violence (Child Abuse-1) and severe violence (Child Abuse-2). Both definitions include extreme acts, such as kicking, biting, punching, beating, burning or scalding, threatening with a knife or gun, and using a knife or gun. Spanking or hitting a child with an object such as a stick, hair brush, or belt is not "abuse" according to either the legal or informal norms of American society, although the latter is in Sweden and in several other countries (Haeuser 1985, in Straus and Gelles 1988, p. 16). Therefore, to take such normative factors into consideration, the researchers omitted hitting or trying to hit a child with an object from Child Abuse-1. Child Abuse-2 includes this factor "on the grounds that hitting with an object carries a relatively greater risk of injury than spanking or slapping with the hand. For this reason we think that the more inclusive measure (Child Abuse-2) is the better indicator of the extent of physical abuse of children" (Straus and Gelles 1988, p. 26).

Another form of child abuse is **sexual abuse:** forced, tricked, or coerced sexual behavior—exposure, unwanted kissing, fondling of sexual organs, intercourse, rape, and incest—between a young person and an older person (Gelles and Conte 1990). **Incest** involves sexual relations between related individuals, including steprelatives. The definition of child sexual abuse excludes mutually desired sex play between or among siblings close in age; coerced sex by strong and/or older brothers *is* sexual abuse and may be more widespread than parent–child incest (Canavan, Meyer, and Higgs 1992).

Incest is the most emotionally charged form of sexual abuse; it is also the most difficult to detect. The most common forms are father–daughter incest and incest involving a stepfather or older brother. The latter is very probably the most common (Canavan, Meyer, and Higgs 1992). Incest appears to be related to a variety of sexual, emotional, and physical problems among adults who were abused as children (Wyatt and Powell 1988; Simons and Whitbeck 1991; Gilgun 1995). One study of 137 battered wives found that those who had been incest victims were significantly more likely to have suffered marital rape (Shields and Hanneke 1988). Sexual abuse by paid caregivers is also a problem being addressed by policymakers and child-care professionals; sexual exploitation of homeless children is yet another (Shamim and Chowdhury 1993).

Child neglect includes acts of omission—failing to provide adequate physical or emotional care. Physically neglected children often show signs of malnutrition, need immunizations, lack proper clothing, attend school irregularly, and need medical attention for such conditions as poor eyesight or bad teeth (U.S. Department of Health, Education, and Welfare 1975, p. 8). Often these conditions are due to economic stress (Baumrind 1994; Kruttschnitt, McLeod, and Dornfeld 1994), but child neglect can also be willful neglect. **Emotional child abuse or neglect** involves a parent's often being overly harsh and critical, failing to provide guidance, or being uninterested in a child's needs. Emotional child abuse might also include allowing children to witness violence between partners (Carlson 1990; Rubiner 1994). Although emotional

abuse can occur without physical abuse, physical abuse always results in emotional abuse as well.

HOW EXTENSIVE IS CHILD ABUSE?

Interpreting rising rates of reported child abuse to mean that child abuse is rapidly increasing may be inaccurate.[2] For one thing, all states now have compulsory child-abuse reporting laws, so a growing proportion of previously unreported cases now comes to the attention of child-welfare authorities. For another, new standards are evolving in respect to how much violence parents can use in child rearing. The definition of child abuse is being gradually enlarged to include acts that were not previously thought of as child abuse. In fact, some evidence suggests that among whites, although not blacks (Hampton 1991), the incidence of child abuse has actually declined over the past ten to twenty years (Straus and Gelles 1986; Gelles and Straus 1988, p. 111). Whatever the precise figures, maltreatment is a significant problem. Section B of Table 10.2 shows that nearly 100 percent of young American children are spanked, hit, or slapped by parents. Eleven percent of children under age 17 experience severe violence at the hands of a parent. About 2.3 percent have experienced very severe violence in the previous year. Moreover, "as in the case of beaten wives, the actual rate and the actual number is almost certainly greater because not all parents were willing to tell us about instances in which they kicked or punched a child" (Straus and Gelles 1988, p. 26).

When these researchers include violence toward children from siblings as well as parents, they conclude that among children ages 3–17, nearly one child in twenty-two is a victim of physical abuse from another family member (Gelles and Straus 1988, p. 103). A child's statistical chances of being kidnapped or molested by a stranger are minuscule compared to that same child's statistical chances of being physically or sexually abused by a family member or a friend of the family.

There are no inclusive national statistics on the extent of sexual abuse of children; estimates come from various small samples and range from 6 to 62 percent

---

2. There is speculation that some child-abuse charges are false. In some cases, according to family-law experts and health professionals, fathers and stepfathers involved in divorce cases are being wrongly accused, usually as part of "the already emotionally charged atmosphere of court contests over child custody" (Dullea 1987c, p. 5; Wexler 1991; Gardner 1993; Heves; 1994).

of all female children and from 3 to 31 percent of all male children. This wide variation in estimates results from different definitions of sexual abuse used in various studies and from methodological factors, such as the conditions under which people are interviewed (Gelles and Conte 1990).

Although boys are also victimized, it appears that girls are between two and five times more likely to be sexually abused. Research has found that a very small proportion of abusers are female—at most 4 percent when the victims are girls and 20 percent when the victims are boys (Glaser and Frosh 1988, p. 13). "Especially since contacts with female children occur with at least twice or three times the frequency as male children, the presumption that sexual abusers are primarily men seems clearly supported" (Finkelhor 1984, in Glaser and Frosh 1988, p. 13). One study (Finkelhor and Baron 1986) found that girls who have been separated from or have poor relationships with their mothers are more likely to be victimized. "The data point to the importance of mothers in protecting children from sexually aggressive men" (Gelles and Conte 1990, p. 1052).

Abused children live in families of all socioeconomic levels, races, nationalities, and religious groups, although child abuse is reported more frequently among poor and nonwhite families than among middle- and upper-class whites. This situation may be due to some of the same reasons, discussed earlier, that spouse abuse is reported to police more frequently by the poorer classes. Another reason may be unconscious discrimination on the part of physicians and others who report abuse and neglect (Hampton and Newberger 1988).

ABUSE VERSUS "NORMAL" CHILD REARING

One writer has observed that "a culturally defined concept of children as the 'property' of caregivers and of caregivers as legitimate users of physical force appears to be an essential component of child abuse" (Garbarino 1977, p. 725). It is all too easy for parents to go beyond reasonable limits when angry or distraught or to have limits that include as discipline what most observers would define as abuse (Baumrind 1994). Hence, child abuse must be seen as a potential behavior in many families—even those we think of as "normal" (Gelles and Straus 1988).

Bearing this in mind, consider the following societywide beliefs and conditions that, when exaggerated,

can encourage even well-intentioned parents to mistreat their children:

- A belief in physical punishment is a contributing (but not sufficient) factor in child abuse. Abusive parents have learned—probably in their own childhood—to view children as requiring physical punishment.

- Parents may have unrealistic expectations about what the child is capable of; often they lack awareness and knowledge of the child's physical and emotional needs and abilities. For example, slapping a bawling toddler to stop her or his crying is completely unrealistic.

- Parents who abuse their children were usually abused or neglected themselves as children. Violent parents are likely to have experienced and thereby learned violence as children. Currently, researchers are focusing on specifically how this tendency toward family violence is transmitted. One factor is an inappropriate conception of justice—of deserved consequences for "right" and "wrong" (Blackman 1989).

  This does *not* mean that abused children are predetermined to be abusive parents (Gelles and Straus 1988, pp. 48–49). A good estimate is that 25 to 35 percent of abused children grow up to abuse their own children. Although this is not a majority, it is considerably higher than the child abuser rate of 2 to 4 percent in the general population (Gelles and Conte 1990).

- Parental stress and feelings of helplessness play a significant part in child abuse. Financial problems can cause stress, especially for single mothers; overload, so often related to financial problems, also creates stress that can lead to child abuse. "Economic adversity and worries about money pervade the typical violent home" (Gelles and Straus 1988, p. 85). Other causes of parental stress are children's misbehavior, changing lifestyles and standards of living, and the fact that a parent feels under pressure to do a good job but is often perplexed about how to do it.

- Families have become more private and less dependent on kinship and neighborhood relationships. Hence, parents and children are alone together, shut off at home from the "watchful eyes and sharp tongues that regulate parent-child relations in other cultures" (Skolnick 1978, p. 82). In neighborhoods that have support systems and tight social networks of community-related friends—where other adults are somewhat involved in the activities of the family—child abuse and neglect are much more likely to be noticed and stopped (Gelles and Straus 1988).

COMBATING CHILD ABUSE   Circumstances that are statistically related to child maltreatment include parental youth and inexperience, marital discord and divorce, and unusually demanding or otherwise difficult children (Baumrind 1994). Two approaches to combating child abuse and willful neglect exist: the punitive approach, which views abuse and neglect as crimes for which parents should be punished, and the therapeutic approach, which views abuse as a family problem requiring treatment.

Those who favor the punitive approach believe that one or both parents should be held clearly (that is, legally) responsible for abusing a child. A complicated issue emerging with regard to this approach involves holding battered women criminally responsible either for abusing their children or for failing to act to prevent such abuse at the hands of their male partners. A 1981 national survey by the U.S. Department of Health and Human Services found the mother or mother substitute to be the only involved parent in 24 percent of child-abuse cases. Fathers or father substitutes were the perpetrators in 22 percent of cases, and both parents participated in 41 percent of the cases (in Coontz and Martin 1988, p. 77). One small study of sixty-six Wisconsin parents (two-thirds of whom were women) who had been investigated by protective service agencies and subsequently agreed to be interviewed found that the men and women were similar in the extent of their involvement in abuse, in the methods used to mistreat children, in the extent to which they injured children (Kadushin and Martin 1981), and in their attitudes toward the abuse and their role in it (Coontz and Martin 1988). However, as we have seen, women are more likely to report their own violence than are men; and the men who were willing to participate in this study may have been less violent than other male child abusers.

Moreover, research is beginning to uncover a "complexity of connections between spouse [abuse] and child abuse" (Erickson 1991, p. 198). Although some studies have found low correlations between a man's

battering his wife and battering his children, other research has found that in about half the cases in which a woman was being battered, the batterer was also abusing the children in the household. Some critics have begun to question whether the law should hold a battered woman responsible for failing to prevent harm to her children when, as a battered woman, she cannot defend even herself. A battered woman is likely to believe that the man of the house has a right to discipline his wife and children and hence is unlikely to challenge him; moreover, by failing to act, the battered woman may be making a reasonable decision under the circumstances:

> Perhaps it appears very reasonable to the battered woman to allow the batterer to assault the child as long as she can see that the battering is not "going too far." The alternatives might appear to be worse: interfering and thereby causing an acceleration of the battery so that it does "go too far," or fleeing with the children to an unimaginable life, with no money, no shelter, and no protection from the batterer, should he follow them. . . . When the law punishes a battered woman for failing to protect her child against a batterer, it may be punishing her for failing to do something she was incapable of doing. . . . She is then being punished for the crime of the person who has victimized her. (Erickson 1991, pp. 208–9)

Child abuse has by no means been decriminalized (all states have criminal laws against child abuse). Meanwhile, the approach to child protection has gradually shifted from punitive to therapeutic (Stewart 1984). Not all who work with abused children are happy with this shift, however. These critics prefer to hold one or both parents clearly responsible. They reject the family system approach to therapy because it implies distribution of responsibility for change to all family members, including the child. Nevertheless, social workers and clinicians—rather than the police and the court system—increasingly investigate and treat abusive or neglectful parents.

The therapeutic approach involves two interrelated strategies: (1) increasing parents' self-esteem and their knowledge about children and (2) involving the community in child rearing (Goldstein, Keller, and Erne 1985). One voluntary program, Parents Anonymous (PA), holds regular meetings to enhance self-esteem and educate abusive parents. Another voluntary association, CALM, attempts to reach stressed parents before they hurt their children. Among other things, CALM advocates obligatory high school classes on family life, child development, and parenting, and it operates a 24-hour hotline for parents in stress.

Involving the community involves getting people other than parents to help with child rearing. One form of relief for abused or neglected children is to remove them from their parents' homes and place them in foster care. More and more, however, people are choosing alternatives, such as *supplemental mothers,* who are available to babysit regularly with potentially abused children. Another community resource is the *crisis nursery* where parents can take their children when they need to get away for a few hours. Ideally, crisis nurseries are open 24 hours a day and accept children at any hour without prearrangement.

We turn now to another topic that is gaining more and more attention: elder abuse and neglect.

Elder Abuse and Neglect

Granny bashing, or elder abuse, first caught the public's attention in Great Britain in 1975. This research area is new, and many questions—even including how to define elder abuse and neglect—remain unanswered. As with child abuse and neglect, **elder abuse** involves overt acts of aggression, whereas **elder neglect** involves acts of omission or failure to give adequate care. Elder abuse by family members can include physical assault, emotional humiliation, purposeful social isolation (for example, forbidding use of the telephone), or material exploitation and theft (Johnson 1986). Although various studies have concluded that between 1 and 10 percent of individuals over age 60 are abused or neglected, as yet there are no accurate national statistics (Hudson 1986). The emerging profile of the abused or neglected elderly person is of a female, 70 years or older, who has physical, mental, and/or emotional impairments and is dependent on the abuser–caregiver for both companionship and help with daily living activities. Studies have found that the neglected elderly are older and have more physical and mental difficulties (and hence are more burdensome to care for) than are elder abuse victims (Pillemer 1986).

There are many parallels between elder abuse and other forms of family violence. In fact, "there is reason to believe that a certain proportion of elder abuse is actually spouse abuse grown old" (Phillips 1986,

p. 212). In some cases, marital violence among the elderly involves abuse of a caregiving partner by a spouse who has become ill with Alzheimer's disease (Pillemer 1986). Other causal factors include stress from outside sources, such as financial problems or a caregiver's job conflicts, and social isolation (lack of connectedness with friends and community). As we have seen, these factors are also associated with child abuse. We do not as yet know whether children who neglect or abuse their parents are more likely to have experienced neglectful or violent upbringings (Pillemer 1986, p. 243).

Elder abuse victims, in contrast to the neglected, are relatively healthy and able to meet their daily needs. The common denominators in cases of physical elder abuse are shared living arrangements, the abuser's poor emotional health (often including alcohol or drug problems), and a pathological relationship between victim and abuser. Indeed, data from one study of 300 cases of elderly abuse in the Northeast found that abusers (frequently an adult son) were likely to be financially dependent on the elderly victim. As a result, abusive acts may be "carried out by abusers to compensate for their perceived lack or loss of power" (Pillemer 1986, p. 244). "In many instances, both the victim and the perpetrator were caught in a web of interdependency and disability, which made it difficult for them to seek or accept outside help or to consider separation" (Wolf 1986, p. 221).

Researching and combating elder abuse generally proceeds from either of two models: the caregiver model and the domestic violence model. The caregiver model views abusive or neglectful caregivers as individuals who are simply overwhelmed with the requirements of caring for their elderly family members. This model tends to focus on differences in mental and physical health and neediness between elderly people who are abused and/or neglected and those who are not. The domestic violence model, in contrast, views elder abuse and neglect as one form of family violence and focuses both on characteristics of abusers and on situations that put potential victims at increased risk (see Finkelhor and Pillemer 1988; Whittaker 1995). In discussing it here, we have placed elder abuse within the context of family violence. And in addressing family violence in a chapter on power in marriage and families, we have assumed that "all forms of abuse have at their center the exploitation of a power differential" (Glaser and Frosh 1988, p. 6; Whittaker

1995). We close this chapter with a reminder of everyone's basic rights in any relationship.

In Sum

Power, the ability to exercise one's will, may rest on cultural authority, economic and personal resources that are gender based, love and emotional dependence, interpersonal manipulation, or physical violence.

The relative power of a husband and wife in a marriage varies by national background and race, religion, and class. It varies by whether or not the wife works and with the presence and age of children. American marriages experience a tension between male dominance and egalitarianism. Studies of married couples, cohabiting couples, and gay and lesbian couples illustrate the significance of economic-based power, as well as the possibility for couples to consciously work toward more egalitarian relationships.

Physical violence is most commonly used in the absence of other resources. Although men and women are equally likely to abuse their spouses, the circumstances and outcomes of marital violence indicate that wife abuse is a more crucial social problem. It has received the most programmatic attention. Recently, programs have been developed for male abusers, but less attention has been paid to male victims. Studies indicating that arrest is an important deterrent to further wife abuse illustrate the importance of public policies that meet family needs.

Economic hardships and concerns (among parents of *all* social classes and races) can lead to physical and/or emotional child abuse—a serious problem in our society and probably far more common than statistics indicate. One difficulty is drawing a clear distinction between "normal" child rearing and abuse.

Elder abuse and neglect is a new area of research, but initial data suggest that abused elderly are often financially independent and abused most frequently by dependent adult children or by elderly spouses. Although some scholars view elder abuse and neglect as primarily a caregiving issue, this chapter presents elder abuse as a form of family violence and as a family power issue.

Key Terms

battered woman syndrome
child abuse
child neglect
coercive power
conjugal power
elder abuse
elder neglect
emotional child abuse or neglect
expert power
family power
incest
informational power
legitimate power
marital rape
neutralizing power
no-power
personal power
power
power politics
principle of least interest
referent power
relative love and need theory
resource hypothesis
reward power
sexual abuse
social power
three-phase cycle of violence
violence between intimates

Study Questions

1. What is Blood and Wolfe's resource hypothesis of conjugal power? Do you agree with this hypothesis? What are your specific agreements and criticisms?

2. How does newer research on conjugal power either confirm or refute the ideas and findings of Blood and Wolfe? Include specific examples of more recent research.

3. What is the principle of least interest? How does it fit into the relative love and need theory? Do you agree with this theory more or less than you agree with the resource hypothesis? Why?

4. How is gender related to power in marriage? How do you think recent social change will affect power in marriage?

5. What does Blumstein and Schwartz's study of four types of couples tell us about power in relationships?

6. How does the relative love and need theory explain wife beating and wives' reactions to the beatings? Do you think that battered women's shelters provide an adequate way out for these women? Why or why not? What about arresting the abuser?

7. How likely do you think it is for a couple to develop a no-power relationship? What are some of the difficulties that you see in trying to develop such a relationship?

8. Should husband abuse receive more attention in the form of social programs? Why or why not?

9. Describe the battered woman syndrome. Do you think it should be used as a defense in homicide cases? In failure to act to protect a child from abuse cases?

10. Do you think there is a battered husband syndrome? Why or why not?

11. Explain some similarities and differences between same-sex couple violence and heterosexual couple violence.

12. Differentiate between child abuse and child neglect. Do you think that physical abuse is more or less, or equally damaging to children than emotional abuse?

13. It is not always easy to draw a line between "normal" child rearing and abuse. In your own opinion, at what point can this line be drawn? Where would you draw the line regarding emotional abuse, and why? Be specific; give examples.

14. Discuss some societywide beliefs and conditions that, when exaggerated, can encourage even well-intentioned parents to mistreat their children.

15. What can we as individuals and as a society do to combat child neglect that is really due to poverty?

16. Explain the therapeutic approach to combating family violence. Why do you think this approach is gradually overtaking the punitive approach to combating child abuse? What about the punitive model and wife beating?

17. How is elder abuse and neglect similar to other forms of family violence? How may it be different?

Suggested Readings

Bart, Pauline B., Patricia Y. Miller, Eileen Moran, and Elizabeth Anne Stanko, eds. 1989. *Violence Against Women*. Special issue of *Gender and Society* 3 (December). Articles on wife abuse, rape, and other violence against women from a feminist perspective.

Blumberg, Rae Lesser and Marion Tolberg Coleman. 1989. "A Theoretical Look at the Gender Balance of Power in the American Couple." *Journal of Family Issues* 10:225–50. Academic article that summarizes just about every idea sociologists have had about marital power.

Blumstein, Philip and Pepper Schwartz. 1983. *American Couples: Money, Work, and Sex*. New York: Morrow. "Power" should have been in the title because the book is about the dynamics of power in four types of relationships: married heterosexual couples, cohabiting heterosexual couples, gay couples, and lesbian couples. Money, work, and sex, of course, have a lot to do with power. Only lesbian couples succeed in having no-power relationships, but their strategies could be used by any couple.

Dobash, R. Emerson and Russell P. Dobash. 1992. *Women, Violence, and Social Change*. New York: Routledge. Sociological perspective on family violence, including need for and efforts toward social change.

Fineman, Martha A. and Roxanne Mykitiuk, eds. 1994. *The Public Nature of Private Violence*. New York: Routledge. This collection covers all forms of domestic violence with essays and research articles by well-known researchers and writers in the field.

Finkelhor, David and Kersti Yllo. 1985. *License to Rape: Sexual Abuse of Wives*. New York: Holt. Research on marital rape.

Flynn, Clifton P. 1990. "Relationship Violence by Women: Issues and Implications." *Family Relations* 39(2) (April): 194–98. This article presents the issues involved in researching and analyzing women's acts of family violence.

Hampton, Robert L., ed. 1991. *Black Family Violence*. Lexington, MA: D. C. Heath/Lexington Books. A readable and interesting collection of research articles and essays by experts on violence in African American families.

Hotaling, Gerald T., David Finkelhor, John T. Kirkpatrick, and Murray A. Straus, eds. 1988. *Family Abuse and Its Consequences: New Directions in Research*. Newbury Park, CA: Sage. A collection of research reports on various aspects of family violence, including child abuse, wife abuse, elder abuse, and courtship violence.

Island, David and Patrick Letellier. 1991. *Men Who Beat the Men Who Love Them: Battered Gay Men and Domestic Violence*. New York: Haworth. Theory, research, and practical advice on abuse and violence in gay men's relationships.

Johann, Sara Lee. 1994. *Domestic Abusers: Terrorists in Our Homes*. Springfield, IL: Charles C. Thomas. Johann is an attorney who examines police and judicial policy concerning domestic abuse.

Komter, Aafke. 1989. "Hidden Power in Marriage." *Gender and Society* 3(2): 187–216. Well-written, fascinating theoretical analysis of the hidden impact of gender roles on conjugal power.

Renzetti, Claire M. 1992. *Violent Betrayal: Partner Abuse in Lesbian Relationships*. Newbury Park, CA: Sage. Theory, research, and practical advice on lesbian battering.

Schneider, Elizabeth. 1986. "Describing and Changing: Women's Self-Defense Work and the Problem of Expert Testimony on Battering." *Women's Rights Law Reporter* 9 (3, 4): 195–222. Excellent resource on homicide provoked by wife abuse as a legal, ethical, and feminist issue.

Schuler, Margaret, ed. 1992. *Freedom from Violence*. OEF International. A global analysis of violence against women around the world and strategies they use to combat it.

Straus, Murray Arnold and Richard J. Gelles. 1990. *Physical Violence in American Families*. New Brunswick, NJ: Transaction.

Walker, Lenore E. 1984. *The Battered Woman Syndrome.* New York: Springer. Definition, description, and causes of the battered woman syndrome, which is being used by some attorneys as a defense in marital homicide cases.

White, Evelyn C. 1985. *Chain, Chain, Change.* Seattle: Seal Press. Self-help book by and for female African American victims of family violence.

Yllo, Kersti and Michele Bograd, eds. 1988. *Feminist Perspectives on Wife Abuse.* Beverly Hills, CA: Sage. Argues against the gender-neutral concept of "spouse abuse" and suggests reasons why wife abuse is the real and pressing problem.

PART Four

Experiencing Family Commitment

One of the most vital issues that couples face is the decision to have children. In Chapter 11 we will look at the value of children to parents, along with some of the costs they can impose on the marital relationship. We will discuss various choices couples are making about having children.

Chapter 12 discusses parenting. Here the emphasis is not so much on the dos and don'ts of good parenting as on building a supportive relationship between parents and children. We examine how this relationship changes over the life course.

For many people today, decisions about having children and the process of rearing them are intertwined with work lives. Chapter 13 explores how work—both inside and outside the home—affects marital relationships.

The theme that family relationships change over the years runs through all of these chapters. Furthermore, partners' attitudes about themselves and their marriages change. For instance, a couple may arrange their work roles inside and outside the home one way during one phase of their marriage and prefer to do things differently at another time. Making these changes requires that each partner be willing to change his or her attitude about what to expect in the marriage and how things should be done.

To Parent or Not to Parent

And now I want a child.
And I want that child to carry me in his head forever, and to love me forever. . . .

Is that a sentence of life imprisonment?
A lifetime of love?
The way the world is structured?

JOYCE CAROL OATES, Do with Me What You Will

A remarkable change has taken place in American childbearing patterns over the past thirty-five years. The U.S. total fertility rate—the number of births a typical woman will have over her lifetime—dropped sharply from a high of 3.6 in 1957 to the lowest level ever recorded (1.738) in 1976.[1] It remained below 1.9 from 1972 through 1987, rising to 2.08 in 1990. Now birthrates are declining again. The total fertility rate was 2.05 in 1994 (Ventura et al. 1995, Table 4).

1. Explanation is in order here about birth and fertility rates. The *birthrate*, or **crude birthrate,** as it is technically termed, is the number of births per thousand population. The total number of births and the birthrate depend not only on how many births each woman has but also on how many women of childbearing age there are in the population. For example, as the large baby boom cohort (babies born during the post–World War II era, 1946–1964) reached childbearing age, the crude birthrate rose. Now many baby boomers have passed the peak fertility ages, and the crude birthrate has been declining since (Ventura et al. 1995).

A more specific measure is the *general fertility rate*, which is the number of births per thousand women in their childbearing years (ages 15–44). However, even this statistic does not tell us much about family size.

The **total fertility rate** for a given year—the rate to which we will refer most often in this text—is the number of births that women would have over their reproductive lifetimes if all women at each age had ba-

Beginning in the late 1950s, cultural attitudes about ideal family size changed, and couples began to favor two-child families (Blake 1974). On the average, American women are now having just two children each, slightly less than the population replacement level of 2.1. Moreover, attitudes about whether to have children at all have changed in recent years, so that choosing not to be a parent is more acceptable today.

At the same time that overall fertility[2] levels have dropped, childbearing has increasingly shifted to later ages. Changes in fertility are related to the fact that married women are waiting longer to have their first babies, are allowing more time between births, and are choosing to have smaller families. In 1992, about 30 percent of women age 30 were still childless, compared to 15 percent in 1970. A substantial number of these women, however, will eventually have children. Women in their 30s, who had postponed parenthood, were now having first, second, and in some cases, third children.

Throughout this chapter we'll be looking at the choices individuals and couples have about whether or not to have children and how many. Among other things we'll see that modern scientific and technological advances have both increased people's options and added new wrinkles to their decision making. We'll see, too, that technological progress does not mean that people can or do exercise complete control over their fertility. To begin, we'll review some fertility trends in the United States. Then we'll examine the decision whether to parent.

bies at the rate for each age group that year. In reality, this is an artificial figure: the family size that the average woman would have at average childbearing rates provided those rates continued.

A final measure is *completed fertility*—that is, the number of babies per woman of a given cohort (women born in a given year, such as 1946). However, to actually ascertain completed family size, we need to wait until the cohort is age 45 or 50 to know what these women's completed fertility is. So although completed fertility rates might be more accurate, they are long in coming; and the figures most often used to grasp family size is the total fertility rate.

2. The term **fertility** is used by demographers to refer to actual births. Even though everyday language uses fertility to mean ability to reproduce, the technical term for reproductive capacity is **fecundity**. Confusingly, *infertility* and *infecundity* are both used to denote inability to reproduce. *Subfecundity* designates reduced reproductive ability, as for those, for example, who have demonstrated their fecundity (by having had children) but who have difficulty bearing more children when they wish to.

Fertility Trends in the United States

Declining U.S. fertility appears to be a sudden change when we compare current birthrates to those of the 1950s. But, as Figure 11.1 indicates, the trend is actually a continuation of a long-term pattern dating back to about 1800. Causes of this long-term trend include the fact that alternatives to the motherhood role began to open up with the Industrial Revolution and the resulting creation of a "labor force" (see Chapter 13). Women could combine productive work and motherhood in a preindustrial economy, but when that work moved from home to factory, the roles of worker and mother were not so compatible. Consequently, as women's employment increased, fertility declined. Since the 1940s women have entered the labor force in growing proportions, partly in response to an expanding postindustrial economy that required more clerical and service workers. A second cause was declining infant mortality, a result of improved health and living conditions. Gradually, it became unnecessary to bear many children to ensure the survival of a few. Changes in values accompanying these transformations in U.S. society (and in other industrialized societies) made large numbers of children more costly economically and less satisfying to parents (Easterlin and Crimmins 1985).

In the face of this long-term decline, the rapid surge in fertility in the late 1940s and 1950s requires explanation. It appears that parents who had grown up during the Great Depression, when family goals were limited by economic factors, found themselves as adults in an affluent economy—one that promised support for their own children. Thus, they were able to fulfill childhood dreams of happy and abundant family life to compensate for deprivations previously suffered as children (Easterlin 1987). The family was so highly valued during this period that marriage and motherhood became primary cultural goals for American women. Men also concentrated their attention on family life.

Family Size

Women born during the Depression years of the 1930s favored larger families. Compared with previous cohorts, many more had three or four or five or six children. Before this time the trend had been toward smaller families. The percentage of families with

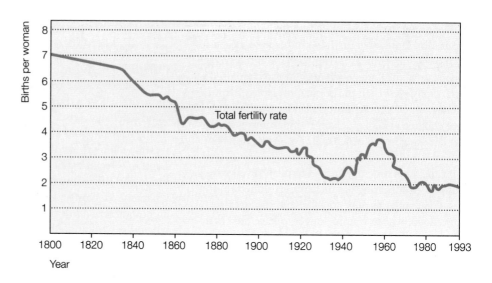

FIGURE **11.1**

Annual total fertility rate of white women, 1800–1993. The rate has been declining for over 100 years, except for the baby boom period from the late 1940s until the early 1960s. (Sources: Thornton and Freedman 1983, p. 13; U.S. Bureau of the Census 1986a; 1995, Tables 91 and 92; U.S. National Center for Health Statistics 1985, 1987, 1989; Population Reference Bureau 1992)

five or more children dropped from about 35 percent in the cohort of women born in 1870 to 12 percent for the women born in 1910. The percentage of one- to three-child families increased from 35 to 60 percent (Thornton and Freedman 1983). Today, there is a strong preference for two-child families by married women of all ages. In fact, this is the principal reason for the current decline in our fertility rate.

Differential Fertility Rates

Fertility rates and decisions whether or not to parent have social, cultural, and economic origins. In general, the more highly educated and well-off families have fewer children. Although they have more money, their children are also more costly; these parents expect to send their children to college and to provide them with expensive possessions and experiences. Moreover, people with high education or income have other options besides parenting. For example, they may be involved in demanding careers or enjoy travel, activities they weigh against the greater investment required in parenthood. This tendency for the more highly educated and wealthier to have fewer offspring is characteristic of all racial groups in the United States.

FERTILITY RATES AMONG AFRICAN AMERICANS At the end of the eighteenth century in this country, white and black women appear to have borne children at approximately the same rates. At about this time the white population began to reduce its fertility, so that by the end of the nineteenth century childbearing among whites had declined significantly. The birthrate among blacks, meanwhile, did not decline significantly until after 1880, when it began to drop rapidly; by the 1930s it was close to that of whites (see Figure 11.2). Since then, black fertility has generally paralleled white childbearing patterns, increasing during the post–World War II baby boom and decreasing after the late 1950s (Wells 1985). Generally, however, blacks have had higher fertility rates than whites. Although the fertility rate of blacks has risen and declined along with that of the white majority, a considerable difference between the two groups remains. The total fertility rate for African Americans in 1993 was 2.47, as compared to a total fertility rate of 1.97 for whites.

We suggested earlier that since about 1800 expanding opportunities and changing values have made large numbers of children less economically rewarding and less personally satisfying to individuals. Put an-

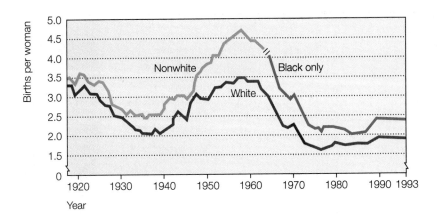

FIGURE **11.2**

Total fertility rates for nonwhite and white women, 1917–1993. (The Census Bureau used the category "nonwhite" until 1964, so that black births were combined with Asian, Native American, and other nonwhite births. Blacks, however, made up 90 percent of this category, so nonwhite rates can be used as a fairly accurate indicator of black rates. Beginning in 1964, black births were counted separately.) (Sources: Reid 1982; U.S. National Center for Health Statistics 1985, pp. 17–18; 1989, pp. 18–19; 1990c; U.S. Bureau of the Census 1995, Table 92)

other way, although economic and social pressures on families can impel them to limit fertility, as the sharp drop in fertility during the Depression indicates, long-term decline in birthrates is mostly a consequence of increased opportunity for economic advancement, more easily taken advantage of by small families (Weeks 1992). The differing decline in fertility rates between white and black populations in this country suggests that although education and other opportunities opened up for whites with the Industrial Revolution earlier in the nineteenth century, they did not do so for blacks until well after the Civil War. "Beginning about 1880 black Americans began to be better educated, more highly urbanized, and began to earn better wages, all phenomena that have been associated with the control of fertility among white Americans" (Wells 1985, pp. 47–50). In other words, it is only when individuals have options that are satisfying, other than parenthood, that they choose to limit their childbearing.

With regard to the current fertility rate among black Americans, the same explanation holds true (Moore, Simms, and Betsey 1986). There is some evidence that blacks hold more sexually permissive attitudes than do whites (see Chapter 5). Due to advances

in birth control techniques, of course, sexual permissiveness need not result in high fertility rates, but individuals must be motivated to use available contraception. What causes high fertility is lack of opportunity —the absence of viable and desirable alternatives to childbearing.

FERTILITY RATES AMONG HISPANICS Hispanics have the highest fertility rate of any racial/ethnic minority. Their total fertility rate of 2.9 in 1994 (U.S. Bureau of the Census 1995, Table 92) was 50 percent above that of non-Hispanic white women. Rates for Mexican American women are highest among Hispanics. Puerto Ricans and Cubans have relatively lower rates (Vega 1990, p. 1016). It has been predicted that, due to high fertility rates coupled with continuing immigration, Hispanics could become our nation's largest minority group (surpassing African Americans) within about fifteen years (U.S. Bureau of the Census 1995, Table 19).

Reasons for their relatively high birthrate include the fact that Hispanics immigrate from nations with high birthrates and still-powerful Catholic and rural traditions that value large families. Then, too, among poor minorities large families may serve important

Native American women who live on reservations have significantly higher fertility than those who do not. Differential birthrates reflect the fact that people in various cultures have different beliefs and values about having children.

functions. One study (Sharff 1981) of Hispanics on New York's Lower East Side indicated that children provide valuable services for parents. A child might be an insurance policy against a parent's old age, for example. Once educated and earning a steady income, children were expected to contribute to the support of their parents. (This is a phenomenon common to less industrialized societies without bureaucratized welfare systems such as Social Security.) Even while growing up, these children's wages were an important part of the family income; children worked, for example, as translators and representatives to bureaucratic agencies. This phenomenon is also true for other immigrant populations, such as Filipino families in which 30 percent have three or more workers, including children ("We the American Asians" 1993).

Moreover, the lifetime fertility of Latinas varies strongly with their educational attainment. "If lower education tends generally to make for higher fertility, and if the Spanish origin women are relatively more concentrated in the lower educational categories, it is not difficult to see why higher fertility persists within many of the groups" (Bean and Tienda, 1987, cited in Vega 1990, p. 1017).

ASIAN AMERICAN AND NATIVE AMERICAN FERTILITY These two groups represent a small proportion of total U.S. births, less than 2 percent each. Younger Asian American and Pacific Islander women have lower fertility rates than do non-Hispanic whites, although there is considerable variation by country of origin. Chinese, Filipino, and Japanese American rates are low, whereas Hawaiian and "other Asian" (a category that includes Pacific Islanders) rates are relatively higher. Foreign-born Asian and Pacific Islander women have relatively high birthrates, but as immigrant groups assimilate, their birthrates converge with those of whites.

Fertility rates of Native Americans (including Eskimo and Aleut) are 40 percent higher than those of non-Hispanic whites (U.S. Bureau of the Census 1995, Table 92). Native American births have increased dramatically since 1970, unlike birthrates for the total population, but part of the apparent increase may be a tendency for mixed-race individuals (or even non–Native Americans) to report their race as Indian. Native American women who live on reservations have significantly higher fertility than those who do not (Taffel 1987), probably for the reasons of limited educational and economic opportunity noted earlier.

All in all, differential birthrates reflect the fact that people in various cultures hold different beliefs and values about having children.

The Decision to Parent or Not to Parent

The variations in birthrates just described reflect values and attitudes about having children. But in traditional society couples didn't decide to have children. Children just came, and preferring not to have any at all was unthinkable. Earlier in the twentieth century, family planning efforts still focused on the timing of children and family size rather than on whether to have them. Now choices include "if" (that is, whether to have children) as well as "when" and "how many."

Not all choices can be realized, whether they reflect a desire to have children or to avoid having children. The extent to which people today consciously choose (or reject) parenthood or experience it as something that simply happens to them is uncertain. Certainly educated and affluent people have more control over their lives generally, so they may be more apt to approach parenthood as a conscious choice. Among others—teenagers, for example, parenthood is usually less thought-out. Some people may be philosophically disinclined to plan their lives generally (Luker 1984). Nevertheless, more so than in the past, our society presents the possibility of choice and decision making about parenthood. The decision to parent or not to parent (or when to parent) is one in which many couples and individuals invest a great deal of thought and emotion.

Although social change and technology provide more choices, they also present dilemmas. It is not al-

ways easy to choose whether to have children, how many to have, and when to have them. In the following pages we'll look more closely at some of the factors involved in deciding about children: first, the social pressures; then, the personal pros and cons.

Social Pressures

We saw in Chapter 6 that single people in our society often feel strong pressures to conform by marrying. The same kinds of pressures exist for married people who don't want to have children.

Although these pressures are less than in the past, our society may still have a **pronatalist bias:** Having children is taken for granted, whereas not having children must be justified. Some of the strongest pressures can come from the couple's parents, who often have difficulty accepting and respecting their children's choices about whether to have children, let alone when and how many. Hopeful prospective grandparents are not always subtle: One gave his wife a grandmother photo album for Christmas, even though his childless daughter and her husband were not aware of any grandchildren on the way!

Although socialization still inclines people toward parenthood, the expectation for married couples to have children is becoming less pronounced. For example, the term *child-free* is often used now instead of the more negative-sounding term *childless*.[3] Some observers, in fact, argue that U.S. society has become antinatalist—that is, against having children.

IS AMERICAN SOCIETY BECOMING ANTINA-
TALIST?   Some observers have suggested that the social pressures *not* to have children are becoming too strong, especially for some young, highly educated women. In the late 1970s Betty Friedan expressed concern that the element of choice was being taken away and saw little hope for improvement if some women "have to give up motherhood to keep on in jobs or professions as my generation gave up jobs or professions to make a career of motherhood" (Friedan 1978, pp. 196, 208).

In the years since Friedan voiced these worries, other observers have raised similar—and broader—concerns. They warn that American society is charac-

---

3. Each term conveys an inherent bias. For that reason and because there are no easy-to-use substitute terms, we make equal use of both *childless* and *child-free* in this text.

terized by **structural antinatalism** inasmuch as "our values, laws, employment policies and culture are inimical to children and disastrous for committed parents" (Leach 1994). Sociologists Janet Hunt and Larry Hunt warn that corporate institutions are "committed to 'masculine' values." Their "primary goals are power and profit . . . over family well-being" (1986, p. 281). Should American values not change, the Hunts predict

> not only a sense of exclusion on the part of those with families, but a widening gap in the standard of living between parents and non-parents. . . . Those with children will tend to have lower incomes, in addition to absorbing the expenses of children, and will fall further and further behind their child-free counterparts. (Hunt and Hunt 1986, p. 283)

How, exactly, might those with children come to be excluded? One example comes from the National Advocacy Project on Family Day Care. According to officials, controversy over home-based day-care facilities is dividing communities throughout the country. Neighbors' opposition to child-care homes center on noise and added commotion, increased traffic as parents deliver and pick up their children, and the fact that a residentially zoned house is being used as a business (Brooks 1984a, p. C8).

A second example involves work leave for parents, an issue touched on in Chapter 13. Seventy-one percent of all U.S. women who become pregnant are employed at the time. Half of these women work into their third trimester. Under the 1978 federal Pregnancy Discrimination Act, an employer may not fire or harass a worker because she is pregnant. But many women (for example, those working in companies with fewer than fifteen workers) are not covered by this law; some women are fired, demoted, or denied benefits simply because they are pregnant. More likely, a pregnant employee may experience subtle discrimination, such as being given less important work assignments (Hughes 1991).

Observers point out that as a nation we have continued high levels of military spending and given tax breaks to the wealthy while cutting health, nutrition, social service, financial aid, and education programs directly affecting the welfare of our children. Partly as a result of such cuts—and of the absence of national support for a family wage—children under 6 years of age have the highest rate of poverty (24 percent in 1992) of any age group (U.S. Bureau of the Census 1994a, Table 729), and children in the United States are more likely to die before age 1 year than those in most other industrialized nations (Population Reference Bureau 1992).

Antinatalism may have other, potentially detrimental consequences as well. Psychoanalyst Erik Erikson, who has focused on adult development in his work, expresses concern that as the birthrate continues to fall (in response to perceived antinatalism or for other reasons) some couples will become self-absorbed and will neglect other outlets that exist to help them take some responsibility for the generations to come. To avoid what he calls stagnation, Erikson advises people who choose not to have children to channel their procreativity "in active pursuits which universally improve the condition of every child chosen to be born" (Erikson 1979). Similarly, sociologist Christopher Lasch argues that Americans' pervasive feelings of "emptiness and insignificance" arise essentially from "the erosion of any strong concern for posterity. . . . We are fast losing the sense of historical continuity, the sense of belonging to a succession of generations originating in the past and stretching into the future" (Lasch 1980, p. 5).

There are social pressures in our society to have children and not to have children. Either kind of pressure can influence a person's choice and act as a source of guilt or self-doubt. The decisions people make should reflect not only external social pressures but also their own needs, values, and attitudes about becoming parents. In the next three sections we will look at some of the advantages and disadvantages associated with parenthood.

Children's Value to Parents

Traditionally, children were viewed as economic assets; in a farm economy, more hands added to the work that could be produced in the fields and kitchens. The shift from agricultural to industrial society and the development of compulsory education transformed children from economic assets to economic liabilities. But as their economic value declined, children's emotional significance to parents increased (Zelizer 1985), probably because declining infant mortality rates made it safe to become attached to children, to invest in them emotionally (Rundblad 1990). Parents' quest was for "a child to love" (Zelizer 1985, p. 190).

Children can bring vitality and a sense of purpose into a household. Having a child also broad-ens a parent's role in the world; mothers and fathers become nurturers, advocates, authority figures, counselors, caregivers, and playmates.

In a study of 100 white working-class and middle-class biological and adoptive parent couples in the Midwest, coauthor Mary Ann Lamanna found par-ents reporting a variety of emotional satisfactions from having children. Many stated that children gave their lives meaning and purpose, a sense of destiny. Chil-dren provide a sense of continuity of self—as one par-ent put it, "the advantage of seeing something of yourself passed on to your children" (although this can be a mixed blessing: "Sometimes you see some bad things in yourself that have been passed on to your children").

Others reported that their satisfaction is in belong-ing to a close family unit, which they associate with having children, not just with being married. One parent commented: "It gives us one more thing to talk about all the time together. . . . Since the daughter, we all do things together more than we used to—the zoo, picnics, and things like that."

Many parents in the study enjoyed the satisfaction of nurturing the emotional and physical growth of their children. The challenge involved in child rearing can help fill needs for creativity, achievement, and competence: "It's interesting to see our children react-ing to us and our ideas and their response to the way

we treat them; it's a challenge" (personal interview, Lamanna 1977). Others spoke of the joy of loving and being loved by their children.

In her study of men's feelings about family and work, sociologist Kathleen Gerson (1993) found that some men value close relationships with their children because they missed having a close emotional bond with their own fathers. Others seemed to find that children offered them an emotional base that work or romantic relationships could not promise (p. 177).

Children provide several other satisfactions: They may represent (though to an ever-lessening degree) a potential means of support and security for parents once parents are no longer able to provide for them-selves. This was and is still extremely important in less-modern societies in which government programs of old-age support do not exist. (However, the parents in the American study thought this was *not* a good rea-son for having children.) Also, even more than com-pleting school, taking a first job, or getting married, having children is tangible evidence that one has reached adulthood.

Parents may perceive rewards in having children that they may not get in jobs they do not find mean-ingful. Children give parents a chance to influence the course of others' lives and to feel looked up to; they can be symbols of accomplishment, prestige, or wealth; and they attest to the parents' sexuality. Family life also offers an opportunity to exercise a kind of authority and influence that one may not have at a job.

Children can add considerable liveliness to a household, and they have fresh and novel responses to the joys and vexations of life. Focusing on her enjoyment of her grandchildren, one grandmother, whose two grandchildren and daughter have lived with her for about five years, said, "They're fun kids. . . . I'm less apt to sit down and watch TV; I'm more apt to be sitting down playing Legos" (Jendrick 1993, p. 616). Children provide a sense that something new and different is happening, which may help to relieve the tedium of everyday life. Playing with them can give parents the feeling of reliving their own childhoods.

Costs of Having Children

Although the values of having children can be immeasurable, the experience can also be costly. On a purely financial basis, children decrease a couple's level of living considerably. One estimate of the cost of raising a child in the United States through age 17—and before sending him or her to college—in 1993 was $132,660 for a middle-income family (Kalish 1994).

Added to the direct costs of parenting are **opportunity costs:** the economic opportunities for wage earning and investments that parents forgo when rearing children. These costs are felt most by mothers. A woman's career advancement may suffer as a consequence of becoming a mother in a society that does not provide adequate day care. One observer suggests that many women, as a result of having difficulty finding good jobs in today's sluggish economy, are starting families instead. But they face lost pension and Social Security benefits later ("Poverty Looms" 1991). Conversely, the loss of free time is one important cost of trying to lead two lives, as a family person and as a career person, when domestic responsibilities fall largely on the mother (Hochschild 1989).

Parents in Lamanna's study identified some other costs of having children: They add tension to the household and restrict parents' activities outside the home. Children require a more efficiently organized residence and a daily routine that can limit the parents' spontaneity. Children also make for substantial additional work—not only physical care but also the work of parenting: teaching cultural norms, guiding the child's social and emotional growth, and dealing

with anxiety about such pitfalls to healthy development as school difficulties and drug use. Adding to these emotional costs is the parent's recognition that once assumed, the parent status is one that a person cannot easily escape.

But as the burdens of parental responsibility for children are augmented by managing two jobs or two careers within the family or a single parent's need to sustain both job and family, children may again become useful. Their instrumental value as household helpers may again become important. Children may come to have utilitarian as well as emotional value to parents, offsetting some of the practical costs of parenting. Still, children are typically a challenge to their parents' couple relationship.

How Children Affect Marital Happiness

A common cost of having children is marital strain. Evidence shows that children—especially in the first year after their birth, or when there are many, or they are of preschool age—stabilize marriages (White 1990, pp. 906–7); that is, such couples are less likely to divorce.[4]

But a stable marriage is not necessarily the same as a happy and satisfying one. Many couples report that the happiest time in marriage was before the arrival of the first child and after the departure of the last. Spouses' reported marital satisfaction tends to decline over time whether they have children or not. But serious conflicts over work, identity, and domestic responsibilities can erupt with the arrival of children (Glenn 1990).

When they have children, spouses may find that they begin responding to each other more in terms of more traditional role obligations. Now more than before—and particularly among couples who have children earlier in their marriages (Coltrane 1990)—the husband is responsible for breadwinning and the wife for house and child care. Spouses who as parents are not only busier but also sexually segregated begin to do fewer things together and to share decision making less (White, Booth, and Edwards 1986). Dissatisfac-

---

4. At least one analysis, in fact, concludes that most of the difference in marital happiness between parents and nonparents is not due to a decline in marital happiness after children are born but to the fact that unhappy parents tend to stay married, whereas unhappy nonparents more often divorce. Therefore, when parents and nonparents are compared in cross-sample studies, still-married nonparents appear happier as a group (White and Booth 1985a).

tion after the arrival of the first child seems more pronounced and longer lasting for wives than for husbands (Glenn 1990, p. 825).

Studies comparing marriages with and without children consistently report marital happiness to be higher in child-free unions (Glenn and McLanahan 1982; Houseknect 1987; Somers 1993). Even though the addition of a child necessarily influences a household, the arrival of a child is less disruptive when the parents get along well and have a strong commitment to parenting. (Chapter 12 looks more closely at the relationships between parents and their children.) Of course, choices about parenthood involve several options, as we will see in the next section.

Three Emerging Options

We have been discussing several factors that influence the decision to have or not to have children. In this section we will look at three emerging options: choosing to remain childless, postponing parenthood, and having a one-child family.

Remaining Child-Free

According to a 1990 U.S. Census Bureau survey, 22 percent of women ages 18–34 who do not have children plan to remain child-free ("And Two" 1991). Despite some people's beliefs, individuals who choose to remain childless are usually neither frustrated nor unhappy. Voluntarily childless couples typically have vital relationships. Often they believe that adding a third member to their family would change the character of their intense personal relationship.

Child-free women tend to be attached to a satisfying career. Childless couples value their relative freedom to change jobs or careers, move around the country, and pursue any endeavor they might find interesting. A study comparing 74 voluntarily child-free women and men with 127 fathers and mothers found that the child-free couples felt negatively stereotyped by society but meanwhile were more satisfied than the parents were with their relationship as a couple (Somers 1993).

REMAINING CHILDLESS: DIFFERENCES IN COMMITMENT   Not all couples are equally committed to remaining child-free. In one study (done twenty years ago) of thirty voluntarily childless couples, it was apparent that the degree of resolve varied among the respondents (Nason and Poloma 1976).

Some—those who had been sterilized—were irrevocably committed. Others were "strongly committed." They used effective contraception, discussed sterilization, and agreed that if contraception failed, the wife would have an abortion. As a couple they showed no signs of ambivalence about not having children, but they did not entirely rule out the possibility of parenting if their current relationship ended.

The "reasonably committed" couples expressed some minor doubts about the permanence of their decision, but they used contraceptives regularly and effectively and said the wife would probably have an abortion if contraception failed. Some couples were "committed with reservations." They used contraceptives effectively but expressed some doubt about their decision and stated flatly that they would not consider an abortion in the event of unwanted pregnancy.

It is easier to feel strongly committed to remaining childless when in one's 20s or even early 30s. Because people may change their minds, sterilization may not be wise before about age 35.

WOMEN'S AND MEN'S REASONS   From a number of studies over the past two decades of couples who decided not to have children, we can draw the following conclusions:

1. It is more often the woman who first takes the child-free position (Seccombe 1991). Often, she is an achievement-oriented only child or a firstborn child who had to help raise her younger brothers or sisters.

2. Men who want to remain childless tend to be more confident about their decision than women, who express more ambivalence. Men like to have freedom and privacy to enjoy life, and they like to have more money. Men in egalitarian marriages may prefer having fewer or no children over sharing the work of raising them (Gerson 1993).

3. When a couple disagrees about having children, substantial conflict may occur before this issue is resolved. It is, of course, possible that it cannot be resolved. A difference in the desire for children is such a serious matter that it ought to be talked out before marriage.

THE DECISION NOT TO PARENT: DELIBERATE OR BY DEFAULT   In the past, most child-free couples decided after they married not to have children. In some cases the couple did not consider the subject before marriage, but shortly after marrying they began to discuss the possibility of staying child-free. The greatest number of these couples probably made their decision after several years of marriage, as a result of continued postponement. Initially, they took parenthood for granted but put it off until the idea faded in favor of childlessness (Nason and Poloma 1976). This kind of decision by default often results when the wife wants to devote the first several years of marriage to her career. By her 30s or early 40s, she may be so involved in her career that she does not want to take time for children, or she may worry about her own health and the effect of her age on the fetus. Also, fertility tends to diminish after about the mid-20s, so it may be more difficult for older couples to conceive. Moreover, the couple may be concerned about adapting their relationship to the presence of children. Sometimes, partners use prolonged postponement to avoid confronting the issue of permanent childlessness, although they eventually realize that they will not have children.

The main disadvantage of not having children earlier in marriage is that for some couples, a time may arrive when they want but cannot have children. Many couples, of course, change their minds and have children before it is too late.

Postponing Parenthood

The proportion of first births to women in their 30s has increased fivefold since 1970. Delayed parenthood is a second option available to couples.

Later age at marriage, together with the desire of many women to complete their education and become established in a career, appear to be important factors in the high levels of postponed childbearing. Older first-time mothers are better educated and better-off financially than younger mothers (Kobren 1988; Ventura 1988). With the availability of more reliable contraception and legalized abortion, and with the trend toward having only two children, couples can now plan their parenthood earlier or later in their adult lives.

Because fertility declines gradually with age, older couples tend to take longer to conceive. Older women also have higher rates of miscarriage and run a greater risk of conceiving children with certain genetic defects (Winslow 1990). Still, the medical bottom line is that "it is relatively safe for women to postpone childbearing" into their late 30s (Winslow 1990).

POSTPONING PARENTHOOD: BY DEFAULT OR PROGRAMMATIC   Prospective parents are also interested in the social and psychological effects of the timing of parenthood. To assess the arguments for and against late first-time parenthood, Pamela Daniels and Kathy Weingarten (1980) interviewed seventy-two couples, half of whom had their first child in their late 20s or later.

They found that more than half of the thirty-six couples who postponed parenthood did not deliberately choose late parenthood but spent years "suspended and waiting." For them, the value of their early child-free time became apparent only in retrospect. "If I had been more thoughtful about that time I was fooling around not having a baby," said a 40-year-old mother of three young children, "I might have done something more constructive with that time."

The remaining number of late first-time parents were **programmatic postponers:** They arrived at late first-time parenthood by a deliberate, self-conscious process of mutual intention, negotiation, and planning. They knew beforehand exactly why they were putting off having their first baby: the need for psychological readiness and a time to be free to explore and experiment; the desire to find the right partner and create a strong marriage first; and the desire to prolong a career (especially for wives).

A TWO-SIDED COIN   But programmatic late parenthood, the researchers concluded, is a two-sided coin. Whereas early first-time parents in the study "couldn't remember a time when they weren't parents," programmatic postponers reported a sharp sense of before and after. Although they were not sorry they had children, they missed the child-free lifestyle. Moreover, the mothers in this study found that combining established careers with parenting created unforeseen problems. Those who set their careers temporarily aside to be full-time mothers met with criticism from their work colleagues and peers. Those who continued to work, even though they reduced their hours, felt they missed important time with their children or were generally overloaded. One special difficulty with choosing late parenthood is that career commitments may ripen just at the peak of parental responsibilities.

Furthermore, late parents reported their impatience for the empty nest stage of life, when they could return to the personal privacy and freedom from responsibility they enjoyed before their children were born. Early first-time parents, on the other hand, felt they reaped definite pluses later on. One woman, who had been a first-time mother in her early 20s and was now older and a systems analyst, said,

> I like the fact that my children are as old as they are, and that I'm as young as I am and my career is so open ahead of me. I'd hate to be in my career position, wanting to have children and not knowing when to make the break. (Daniels and Weingarten 1980, p. 60)

In this respect, postponing parenthood means making a trade-off: More free time before having children means less time after the children are grown. An awareness of this fact—plus an understanding that having children, although rewarding, can cause logistic and emotional complications—is important for people who consciously decide to postpone becoming parents.

Being born to older parents affects children's lives as well. They usually benefit from financial and emotional stability that older parents can provide and the attention given by parents who have waited a long time to have children. "In many ways . . . offspring of older parents are children of economic and emotional privilege. . . . When their playmates' fathers and mothers are thinking about promotions, mortgages, and their own identities, children of older parents are more likely to take center stage" (Yarrow 1987, p. 17, citing Iris Kern). But children of older parents also report embarrassment that their parents look older than their friends' parents and anxiety about their parents' health and mortality. Parents may become frail while children are still young. Older parents need to plan for guardianship. And children of older parents may have to assume the burden of caring for their elderly parents before they have established themselves in their adult lives.

Of course, we may be overstating the uniqueness of older parents. Older first-time mothers are somewhat of a novelty, but during the baby boom era many women had babies in their late 30s or 40s.

The One-Child Family

Throughout this discussion we've been talking about the choice between having or not having children, or waiting to have children. We have used the plural, re-

flecting a common assumption in our society that when a couple become parents, they will have at least two children. Although it is essentially true that there's no such thing as halfway parents, it is also true that there are some differences in degree between having one child and having several children. We'll consider the one-child family as a third option available to couples.

Thirteen percent of women ages 18–34 surveyed in 1990 by the U.S. Census Bureau said they planned to have one child ("And Two" 1991). The proportion of one-child families in America appears to be growing because of a constricting economy, the high and rising cost of raising a child through college, and some women's increasing career opportunities and aspirations. Divorced people may end up with a one-child family because the marriage ended before more children were born.

Negative stereotypes present only children as spoiled, lonely, dependent, and selfish. To find out whether there was any basis for this image, psychologists in the 1970s produced a staggering number of studies. "The overall conclusion: There are no major differences between only children and others; *no* negative effects of being an only child can be found" (Pines 1981, p. 15). One study interviewed adults between ages 17 and 62 who had been only children and concluded that despite many people's beliefs, only children turn out to be no more selfish, lonely, or less well adjusted than those with siblings (Falbo 1976).

Another study (Hawke and Knox 1978) compared the positive and negative aspects of having an only child by interviewing 102 parents of only children and 105 only children. These only children had better verbal skills and higher IQs than children in any other size family, whether eldest, youngest, or whatever. Their grades were as good or better. They displayed more self-reliance and self-confidence than other children and were often the most popular among elementary school students. They were as likely to have successful college experiences and careers, happy marriages, and good parenting experiences as children with brothers and sisters.

ADVANTAGES   Parents with only one child found that they could enjoy parenthood without being overwhelmed and tied down. They had more free time and were better-off financially than they would have been with more children. The researchers found that

Many families choose to have only one child, a decision that can ease time, energy, and economic concerns. There may be extra pressure on only children, and they do not experience sibling relationships. But only children tend to receive more personal attention from parents, and parents may enjoy not feeling so overwhelmed as they might with more offspring to care for.

family members shared decisions more equally and could afford to do more things together (Hawke and Knox 1978). Then too, a second child presents a new set of family challenges, among them managing an entirely new family relationship—that between siblings (Kreppner 1988).

DISADVANTAGES   There were disadvantages, too, in a one-child family. For the children, these included the obvious lack of opportunity to experience sibling relationships, not only in childhood but also as adults, and the extra pressure from parents to succeed. They were sometimes under an uncomfortable amount of parental scrutiny, and as adults they had no help in caring for their aging parents. Disadvantages for parents included the constant fear that the only child might be seriously hurt or die and the feeling, in some cases, that they had only one chance to prove themselves good parents.

To this point in the chapter, we have taken for

granted a very important factor: the element of choice in having or not having children. Indeed, the extent to which we have this choice is itself remarkable. Until a few decades ago (in some cases a very few years ago), most of the technology on which this ability to choose depends simply didn't exist. The next sections explore some personal and relationship ramifications of fertility technology. We'll look first at some issues surrounding contraception. (A description of contraceptive methods and their effectiveness can be found in Appendix D.)

Preventing Pregnancy: History and Issues

During the 1830s two books describing birth control techniques were published in the United States. Twenty years later, newspaper advertisements offered various mechanical and chemical birth control devices. Of course, a majority of these were relatively ineffective and/or dangerous. Although both the condom and the diaphragm were relatively safe and effective (see Appendix D), it was not until the 1960s, with the development of the pill, that a significant technological breakthrough in preventing pregnancy was achieved. With the pill, women could be more certain of controlling fertility; they did not need male cooperation to do so; and they could engage in sexual activity for its own sake, disconnected from reproduction (Luker 1990).

Before the pill, many of women's abortive or contraceptive efforts were no doubt made privately and in secret. Nineteenth-century America was obviously a more patriarchal society (Chapter 3) than today. Indeed, men were only beginning to recognize a woman's right to have a say in how many children she bore. Women's gaining some control over fertility decisions thus was a move toward sexual equality.

Although heavy reliance in the twentieth century on the condom and the diaphragm implied male as well as female responsibility for contraception, since the advent of the pill in the sixties contraception has been viewed as mostly the female's responsibility. Writing in *The New Our Bodies, Ourselves,* the Boston Women's Health Book Collective (1992) takes the position that placing total responsibility for birth control on women is unfair. The authors add that women may still choose not to use birth control for many reasons, among them their hesitation to inconvenience a partner. But the fear of displeasing him attests to the inequality of the relationship.

This, of course, is the traditional double standard, which assigns women the responsibility for controlling sexual behavior. The double standard may be in effect here because men can more readily escape the consequences of an unwanted pregnancy. And the physical and opportunity costs of children tend to be higher for women than for men. Furthermore, it is women who physically experience abortion or carry and bear children and who more often relinquish careers and education for parenthood. (See Box 11.1 for a discussion of some public policy issues and pregnancy.)

Feminists note that not only do couples often assign contraceptive responsibility to the woman but also that researchers and manufacturers tend to focus on female rather than male contraceptives, even though systemic contraceptives such as the pill may entail substantial health risks for women. There has been some interest in male contraceptives; tests have begun on a male hormonal contraceptive ("Male Con-

traceptive" 1990), and there is a strong resurgence of interest in condoms because of the protection they provide against AIDS and other STDs.

Meanwhile, sterilization, especially female sterilization (Forste, Tanfer, and Tedrow 1995), is the leading birth control method among married couples in the United States. The pill, followed by condoms, are the methods of choice among never-married women (Mosher and Pratt 1990). Still, between 1.2 million and 3 million unwanted pregnancies occur in the United States each year.

We have seen that people take many factors into consideration in their decision to have or not to have children. With several methods available both for contraception and for attempting to increase fertility (discussed later in the chapter), people today are more able to make decisions about parenthood than they have ever been. However, as with other kinds of personal choices we have considered throughout this text, people don't always choose knowledgeably about having children. Default decisions are one (but not the only) reason pregnancy and parenthood can profoundly disrupt lives.

Society has a considerable stake in the next generation, and issues related to procreation tend to be central. Pregnancy outside of marriage and abortion are two reproductive issues that are social as well as personal concerns.

Pregnancy Outside of Marriage

Although childbearing in marriage has declined, birthrates have risen among unmarried women, making births outside of marriage an increasing proportion of total births. Figure 11.3 shows the percentage of children born to unmarried mothers since 1940. By 1993, 31 percent of all births were to unmarried women (Ventura et al. 1995, Table 11). This situation represents a profound change in our society over the past fifty years. In 1940, fewer than 5 percent of all births were to unmarried women.

The majority (60 percent in 1993) of births outside marriage are to white mothers (Ventura et al. 1995, Table 16). The unwed birthrate began to decline among blacks[5] toward the end of the 1970s while continuing to rise among whites. In the 1980s both rates rose, but more rapidly for whites than

5. Note that blacks make up 90 percent of the nonwhite rate; therefore, the nonwhite rates used in Figure 11.3 can be used as a fairly accurate indicator of black rates.

The Pregnant Mother, the Fetus, and the State

Reproduction is a private matter, says the U.S. Supreme Court, and most Americans agree. At the same time, society has an important stake in the next generation. And it is the state's, or government's, responsibility to protect children and other vulnerable people against abuse and neglect.

These two principles have come into conflict in recent policy debates concerning various behaviors by pregnant women (excluding abortion, discussed elsewhere in this chapter). For example, some physicians, convinced of the necessity of a cesarean birth (defined in Appendix C) in specific cases, have obtained court orders to enable them to proceed against the mother's will (Gallagher 1987; Pollitt 1990). Appendix C also discusses state regulation of midwife-attended home births. And some public health experts have proposed routinely testing pregnant mothers for HIV/AIDS unless they forcefully object (Almond and Ulanowsky 1990).

A more public set of issues involves circumstances in which an infant is physically and psychologically damaged by the mother's prenatal use of alcohol or other drugs. An estimated 50,000 babies are born each year with alcohol-related defects (Dorfman 1989), and over 375,000 babies are born to mothers who have used drugs during pregnancy (Toufexis 1991, p. 57).

Infants exhibiting fetal alcohol syndrome and cocaine babies face short-term distress and long-term disabilities, even though these conditions are preventable by maternal abstinence. Drug-exposed newborns, many born prematurely, stay in the hospital almost five times longer than normal infants, and their care is thirteen times as expensive. Often learning-disabled, impulsive, and chronically irritable, drug-exposed children are likely to be taught in special education classes at costs more than twice that of regular schooling (Toufexis 1991, pp. 57–59).

Outraged hospital personnel have contacted local prosecutors, who have either placed babies into foster care or terminated parental rights and charged mothers with various criminal offenses. These charges include fetal abuse in a felony complaint of drinking to the point of abusing an unborn child in Nebraska

(Flanery 1992a), drug trafficking to a fetus in Michigan ("Pregnant Woman Can't" 1991), and fetal murder in a California case involving a stillborn infant born to a cocaine-using mother ("Charge in Fetus's" 1992).

Most charges like these have been dismissed by judges because the laws involved were not intended to apply to a fetus. Nevertheless, incarceration has been used to punish or prevent alcohol and drug use by pregnant women (Pollitt 1990). A North Dakota woman was sentenced to nine months in prison for reckless endangerment after repeatedly inhaling paint fumes while pregnant ("Pregnant Woman Draws" 1992).

Alcohol abuse among Native American women is thirty-three times that of white women; blacks have six times the alcoholism rate of whites. But drug abuse by pregnant women seems to occur at the same rate across class and racial lines; 11 percent of pregnant women were found to have used illegal drugs during pregnancy. Legal measures have not usually been directed at white middle-class mothers, however (Brody 1988). "Cocaine abuse is common among members of the white upper and middle classes, but it is hidden better. Their babies are usually born at private hospitals that rarely ask mothers about drug use or screen them and their children for illegal chemicals" (Toufexis 1991, p. 56).

What should be done? What are the broader social implications? Native American anthropologist Michael Dorris, who has an adopted son with a host of medical and behavioral problems due to his biological mother's alcohol abuse during pregnancy, has urged incarceration of substance-abusing pregnant women and supports a reservation social worker's proposal for forced sterilization (Dorris 1989). Proponents of coercive measures cite the heartbreaking impacts on drug-exposed children and their burden to society.

Others believe we should emphasize education and treatment. Substance-abusing pregnant women are often aware of the damage inflicted on a fetus and want to do something about their problem, but addiction treatment programs often refuse to accept pregnant women. Moreover, greater prenatal damage results from the low priority our society gives to prenatal care among the poor than results from drug abuse (Pollitt 1990).

Feminists and civil libertarians are concerned about the priority given to fetal health over maternal well-being and the erosion of the common-law right to refuse treatment. They are concerned that control over pregnant women or those who may become pregnant may extend to lifestyle choices that are low risk (even trivial) and of little relation to infant health outcomes. They are concerned about what this implies for women's roles: Does pregnancy take priority? And they argue that coercive measures will only discourage drug-using women from seeking prenatal care and/or will push them to have otherwise unwanted abortions.

Difficult questions. Complex answers. Who should control prenatal well-being? Mothers? Families? The state? This matter will remain a serious issue and a social dilemma for a long time.

blacks; consequently, the racial differential has been reduced. Still, African American rates remain significantly higher.

In 1993, 69 percent of African American births occurred outside marriage (Ventura et al. 1995, Table 16). This is due to two demographic changes. First, because the overall length of time that African American women spend in marriage has shortened, the length of time during which a nonmarital pregnancy can occur has increased. Second, fertility has declined more among married than among unmarried black women, which results in an increase in the percentage of total births to unmarried black women. "Therefore, it is erroneous to interpret the increase in the percentage of births to unmarried black women as a rise in their birth rate. In reality, the birth rate of unmarried black women actually declined during the seventies and early eighties" (Taylor et al. 1990, p. 1001).

Nevertheless, for black women, who have lower marriage rates as well as higher unwed birthrates, marriage and parenthood have become separate experiences (Cherlin 1981). In fact, differential tendency to

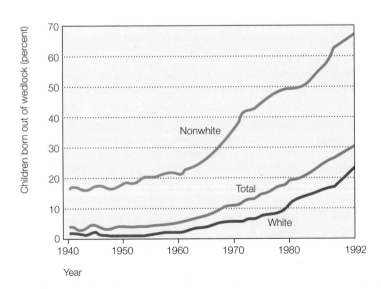

FIGURE *11.3*

Percent of children born out of wedlock, 1940–1992. (Sources: Thornton and Freedman 1983, p. 22; U.S. National Center for Health Statistics 1984, 1989; U.S. Bureau of the Census 1995, Table 94)

marry to legitimate a pregnancy accounts for a substantial part of the difference in unwed birthrates among racial/ethnic groups (Cutright and Smith 1988).

Various images come to mind when we think of pregnancy outside of marriage. Surely one of them is that of the older professional woman, unmarried, who chooses to bear and rear a child on her own. A contrasting image is that of the teenage mother. Concern about an epidemic of teen pregnancy has been a major focus of research and program development since the 1970s. Next we will address both childbearing by older single women and teen pregnancy.

Older Single Mothers

Although unwed birthrates are highest among young adult women (ages 20–24), they have increased dramatically for older women in recent years. Rates for women ages 25–34 increased by more than 50 percent between 1970 and 1992 (U.S. Bureau of the Census 1995, Table 94). The increase in childbearing among older single women is largely a white phenomenon, with an increase of 128 percent for white women in their 30s between 1970 and 1994, compared to a 20 percent increase for black women (Ventura et al. 1995, Table 15).

Although many of these pregnancies and births are unintended, others are planned. Both situations are affected by social change. As opportunities grow for women to support themselves, and as the permanence of marriage becomes less certain, the principle of legitimacy becomes less important economically; that is, there is less motivation for a woman to avoid giving birth out of wedlock because she cannot count on lifetime male support for the child even if she is married. Furthermore, stigma and discrimination against unwed mothers have somewhat lessened, and the distinction in legal terms between legitimate and illegitimate children has been virtually eliminated (Weitzman 1981). Nevertheless, the burden of responsibility for support and care of the child remains on the mother.

Case Study 11.1, "A Baby? A Single Woman Thinks About It," describes how one woman considers becoming a single mother by choice. Generally, these women seek out men perceived as good biological fathers and expect little or no continued involvement. They consciously establish social networks and carefully plan child care. Research findings are mixed regarding solo mothers' children's outcomes (Gringlas

and Weinraub 1995). Some older single mothers are partners in committed lesbian relationships, and it appears that their children are not significantly different from those raised by heterosexuals (Patterson 1992).

A special case of unwed parenthood, presenting especially challenging problems, is that of teen parenthood. The next section discusses teen pregnancy and parenthood.

Teenage Pregnancy

American females reach puberty earlier now than in the past. The average age of the onset of menstruation was almost 17 in nineteenth-century Europe. In the United States today, the average age is about 12 (Alvardo 1992). Moreover, teens have become sexually active at younger ages (Chapter 5). Because much teen sexual activity takes place without contraception, birthrates rose rapidly in the 1960s as sexual behavior liberalized. However, by the time a teen pregnancy epidemic was identified in the mid-1970s, teen birthrates had actually begun to decline.

In 1993, births to teen mothers constituted 13 percent of all births and 29 percent of births outside marriage (Ventura et al. 1995, Table 11). These figures are down from 16 percent and 41 percent in 1980 (U.S. Bureau of the Census 1991b, Tables 92 and 93). Teen births have seemed epidemic because teen birthrates have not declined as rapidly as births to older married women. Moreover, as the large baby boom cohort came of age, there were simply more teenagers giving birth, despite declining rates.

The perception of a teenage pregnancy *problem* remains accurate despite lowered rates. The United States has by far the highest teen pregnancy, abortion, and birthrates of any industrialized country. Part of the decline in teen fertility beginning in the mid-1970s was not the avoidance of pregnancy but rather the termination of 45 percent of teen pregnancies by abortion. Moreover, the age of marriage has risen, as have educational expectations, so that the gap between premature pregnancy and the assumption of adult roles is significant. Also, more young unmarried white women are keeping their babies than in the past, and teens of both races are more likely to set up housekeeping as single parents. In other words, more babies are now being cared for by their teenage mothers.

Meanwhile, teen pregnancy presents serious health hazards for the mother and child: Complications of pregnancy, miscarriage and stillbirth, prematurity, and

U.S. teenagers have higher rates of pregnancy and parenthood than do teens in other developed countries; high schools, such as this one in California, and community centers are beginning to provide effective support.

birth defects and neurological disabilities are more likely with teenage mothers than mothers in their 20s and early 30s, attributable to lack of prenatal care and poor maternal nutrition (Hale 1990; Cramer 1995; Leland et al. 1995). (Low birth weight, a significant factor in infant mortality, is associated with unplanned and unwed pregnancy, regardless of age or race [Eberstadt 1992].)

Teenage parents face a bleaker educational future, a stunted career, and a very good chance of living in poverty, compared with peers who do not become parents as teenagers (Zabin et al. 1992). Also, teen mothers are less likely than their counterparts to marry later, and teen marriages have higher divorce rates (Manning 1993). It is unclear to what extent these consequences result from teen pregnancy per se and not from various background characteristics such as having lived in poverty (Geronimus 1991). But we can logically assume that the responsibilities of moth-

erhood present further educational and occupational obstacles to already disadvantaged teens (Chase-Lansdale, Brooks-Gunn, and Palkoff 1991). Moreover, these negative consequences tend to be passed on to the next generation. Long-term prospects for children of teenage parents include lower academic achievement and a tendency to repeat the cycle of early, unmarried pregnancy (Hayes 1987; Hofferth and Hayes 1987).[6]

Social scientists and others have pointed to the ongoing need for support programs to help young parents and to help their families help them. Such

6. "However, while early childbearing increases the risk of ill effects for mother and child, it is unclear that the risk is so high as to justify the popular image of the adolescent mother as an unemployed woman living on welfare with a number of poorly cared-for children. To be sure, teenage mothers do not manage as well as women who delay childbearing, but most studies have shown that there is great variation in the effects of teenage childbearing" (Furstenberg, Brooks-Gunn, and Morgan 1987, p. 142).

A Baby? A Single Woman Thinks About It

Mary, a pharmacy student, is 29, attractive, and never married. Having always viewed herself as temporarily single, she is beginning to see a possibility that she will never marry—that she may become an "involuntary stable single" (Stein 1981, pp. 10–12). Still wanting to parent a child, she talks here about artificial insemination.

MARY: The problem with somebody of my age is that most of the men that I meet have been married before and divorced and don't want to get married for a very long time or never want to be married again. Or they are confirmed bachelors. Or they are married and they want to have a fling with somebody my age. Meeting a man who is 29 years old, single, and wants to get married is the most difficult thing I have ever encountered. It is! It is rare!

So I have begun to consider artificial insemination. I have talked to my sister at length about it because she is older than I am. She is 30. She has already been married (and divorced) and doesn't want to get married again unless she really meets somebody that is really perfect for her. And the older I get, the more I think—and she agrees with me—that artificial insemination is ideal. I mean I don't want to be deprived of the joy of motherhood just because I don't have the proper mate, because I can't find somebody. And I don't want to be pushed into a marriage either. I think a lot of people do that. They get married because they are desperate so they compromise and fall into relationships that are not good for them because they want to have kids.

INTERVIEWER: Would you want to raise a child by yourself?

MARY: Maybe it is naive, but I don't think it is. I will never consider artificial insemination until I am so financially stable that my child would not suffer economically. I am working on that now. So, I will wait until I am 35 years old. I feel, though, that I have so much love to give, because of the person that I am, that I don't think my child would ever suffer from lack of love from a father.

And I think that also what is important—what would be really helpful to my child by the time my child is old enough to understand that she or he does not have a father—is the fact that five years from now there are probably going to be more kids (born to single mothers) who were artificially inseminated. So my child will not be alone. I would feel worse by going out and dating somebody and getting pregnant and having a kid (out of wedlock). How do you explain that to your child? That would be more difficult to explain, I think, don't you? . . .

If I were to become artificially inseminated and my child came to me and said, "Mommie, why don't I have a dad?" I guess I would tell my child that I really tried to find the right person—a loving, warm, giving person—for a daddy but I wasn't able to do that. And I would say, "I wanted to have a special child like you, and I have love to give you."

How do you feel about Mary's plans for coping with being an "involuntary stable single"? You may want to reread this box when considering other subjects in this text.

programs include health centers, family-planning clinics, nutritional services, primary and preventive health care, vocational counseling, and legal and educational services and parental training courses in the public schools.

Support programs in the schools and programs targeted to teen parents have, in fact, produced good outcomes. Teen parents who remain with their families for a time and who do not marry seem to do better in the long run. A seventeen-year follow-up of teen mothers found a substantial proportion doing surprisingly well in terms of education, income, and general quality of life (Furstenberg, Brooks-Gunn, and Morgan 1987).

Many social workers and policymakers point to the need for including the teen father in programs and

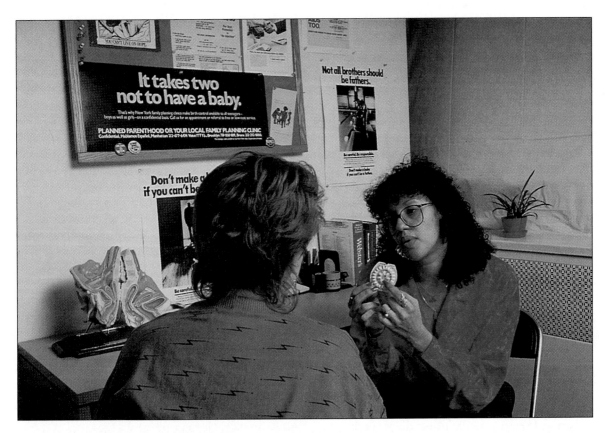

Studies have shown that unmarried partners who have a higher level of commitment and a greater sense of self-esteem and acceptance of their sexuality are more likely to use contraceptives. While males are becoming more involved in contraceptive decisions, there still remains the double standard that females are responsible for birth control.

services. In 1986, for example, New York City launched a "Young Fathers Program" for unwed fathers. The program offers birth control information, helps young men to find jobs in order to help support their offspring, and generally encourages "accepting the consequences of one's actions" (Freedman 1986). Some young fathers want very much to be part of their children's lives: "I don't want her to call anybody else daddy. . . . She's my kid" (in Martin 1990, p. A-27)—to which the high school counselor replied: "You guys might not have one dime in your pocket. . . . But what it's all about is rocking a baby or reading a 3-year-old a story" (Leonard Mednick, in Martin 1990, p. A-27).

Still, prevention remains the better option so far as teen pregnancy is concerned. One solution is to give teenagers more information about contraceptives and increase their availability. Early data indicated that school sex education programs have some effect in preventing teenage pregnancy. More recent reviews of research have indicated no effect of classroom sex education programs on deferring sexual activity, using contraception, or preventing pregnancy. Community-wide sex education programs integrating a variety of settings may be more effective ("Sex Education" 1989).

A hospital-based outreach program targeted to eighth graders was effective in postponing sexual activity and so preventing pregnancy. In this program, older teens (male–female pairs, eleventh or twelfth graders) led discussion sessions on peer pressures and how to resist them, including skill practice in how to say no. Basic information on reproduction, contraception, and STDs was also included. The comparison group who did not participate in the program were five times as likely as program participants to have become sexually active by the end of eighth grade. Differences between the program and comparison groups continued through ninth grade; 24 percent of program participants compared to 39 percent of nonparticipants became sexually active (Howard and McCabe 1990).

Reasons for Teenage Nonmarital Pregnancy

Five out of six—and in some samples as many as nine out of ten (Hardy et al. 1989)—teen pregnancies are unintended (Trussell 1988). Figures from a 1988 national sample show that of never-married women ages 15–44 who "had been exposed to the risk of unintended pregnancy" within three months before being surveyed, about 70 percent were using some form of birth control. Black women were less likely to use contraception than were white women, but those using contraception were more likely to use the most effective methods: female sterilization and the pill (Mosher and Pratt 1990). Perhaps less than 20 percent of sexually active teen women always use effective contraception (Trussell 1988).

In the 1988 National Survey of Adolescent Males ages 15–19, 57 percent reported using a condom the previous time they had sexual intercourse; 77 percent reported using an effective method of birth control (including female methods, such as the pill or diaphragm). These figures are up from 1979, when 21 percent of male teens said they used a condom during their previous intercourse and 58 percent reported having used effective birth control (Ku, Sonenstein, and Pleck 1995). Thus,

> it appears that young males may be making more responsible decisions about contraception and in the process lowering their chance of experiencing an unwanted outcome (e.g., pregnancy, STD, or HIV/AIDS). However, another interpretation of these data would suggest that large numbers of males continue to be irresponsible and place themselves and others at risk of experiencing unwanted outcomes. (Marsiglio 1992a, p. 5)

It is ironic that at a time when contraceptives are so available, the rates of premarital pregnancy are increasing. A number of researchers have tried to explain this contradiction. They found that males may often believe, according to the sexual double standard, that the "real" parent and the one responsible for contraception is the female. In the 1988 National Survey of Adolescent Males, 76 percent said they agreed "a lot" that men should know whether their partner was using contraception before having sex (Marsiglio 1993). This leaves nearly one-quarter who did not entirely agree; and if we assume that at least some who agreed said so because they thought it was the most acceptable response, the proportion of those who do not fully agree is probably higher.

Eighty-seven percent of teens in one survey gave not anticipating intercourse as their reason for not using contraception, although they wished to avoid pregnancy. Teens also believed their risk of pregnancy was smaller than it actually is (Trussell 1988). And teen women may be reluctant to antagonize a partner who is unenthusiastic about, say, condoms.

PARADOXICAL PREGNANCY   Some explanations for the female partner's failure to use contraceptives are provided by a fifteen-year-old study of 18–19-year-old undergraduate women at Indiana University (Byrne 1977). The term **paradoxical pregnancy** was used to summarize the finding that the more guilty and disapproving these women were about premarital sex, the less likely they were to use contraceptives regularly, if at all. Such paradoxical pregnancies resulted because personal disapproval of premarital sex was often not strong enough to inhibit sexual behavior, but it did inhibit the use of contraceptives.

The researchers had four explanations for why such pregnancies happen. First, the sexually negative individual avoids the expectation that intercourse will occur. Because sex becomes a spontaneous event, he or she is unprepared. A second reason is that "adolescents have an intense desire to keep their own sexual activity private and are embarrassed to discuss sexual matters with others, including their partners, friends, parents, counselors, and physicians" (Trussell 1988, p. 267). Procuring birth control devices means going to a doctor, drugstore, or clinic and asking for them. To some individuals, this is like giving public notice of their private affairs.

So also, a third element of paradoxical pregnancy is a lack of communication. Sexual partners need to talk to each other about contraception to make sure somebody has done something. Yet partners who disapprove of premarital sex are less likely to talk to each other about either sex or contraception. A fourth reason, the researchers found, has to do with using contraceptive devices. The pill requires an individual to think about sex at least once daily, and mechanical devices require some direct contact with the genitals. People who disapprove of premarital sex may shy away from this.

More recent research has explored the relationship between sexual abuse and teen pregnancy. As was pointed out in Chapter 10, a Los Angeles study found that for one woman in four, their first experience of intercourse occurred as a rape (Gail Wyatt, in Brody 1989c). Rates of prior sexual molestation among pregnant teens are higher than one in four. In one sample of 535 pregnant teen women in Washington state, 55 percent had been molested, 42 percent had been victims of attempted rape, and 44 percent had been raped as children or young adolescents (Boyer and Fine 1992). Other researchers, using small samples of pregnant mothers in prevention or support programs across the country, are reporting similar findings. Sexual abuse in childhood may lead to lowered self-esteem and less caring and assertiveness about contraception among adolescent females (Small and Kerns 1993). Then too, nonvoluntary sexual intercourse is unlikely to be accompanied by contraception and after puberty is the cause of some unplanned pregnancies (Moore, Nord, and Peterson 1989; Woodman 1995).

Other research has found relationships between teen sexual activity and other behaviors of which parents would likely disapprove (Rodgers and Rowe 1990; Billy, Brewster, and Grady 1994) and between teen pregnancy and having ever moved, run away from home, been suspended or expelled from school, been stopped by police, or used illegal drugs (Mensch and Kandel 1992; Luster and Small 1994; Small and Luster 1994; Stack 1994). A recent study of single women ages 14–21 has found that for Hispanic, black, and non-Hispanic white women alike, being in school or employed increases the likelihood that they will use birth control (Kraft and Coverdill 1994).

Contraception can provide a solution to the potential problems associated with pregnancy outside of marriage. When contraception isn't used, however, many women who don't want to remain pregnant decide to have an abortion. We will look next at this option, which is itself a very controversial social issue.

Abortion

Abortion is the expulsion of the fetus or embryo from the uterus either naturally (spontaneous abortion or miscarriage) or medically (induced abortion). This section addresses induced abortion.

About 1.5 million legal abortions were performed in the United States in 1992, down from a peak of about 1.6 million in 1990 (U.S. Bureau of the Census 1994a, Table 111).[7] There were 25.9 abortions per 1,000 women ages 15–44 (childbearing age) in 1992, down from a peak of 29.3 in 1981. There were 379 legal abortions for every 1,000 live births in 1992, down from a peak of 436 in 1983 (U.S. Bureau of the Census 1995, Table 111).

More than four-fifths (82 percent) of abortions are obtained by unmarried women (U.S. Bureau of the Census 1995, Table 112). Over half (55 percent) of abortions are obtained by women in their 20s, and 23 percent are obtained by teens (U.S. Bureau of the Census 1995, Table 112). The abortion rate per 1,000 live births is considerably higher among nonwhites (655 for nonwhite women versus 318 for white women); however, nearly two-thirds of all abortions (65 percent) are obtained by white women (U.S. Bureau of the Census 1995, Tables 111 and 112).

The Politics of Abortion

From earliest history, abortion has been a way of preventing birth. The practice was not legally prohibited in the United States until the mid-nineteenth century. At that time it was outlawed because of its mortality risk, because it was provided by female medical entrepreneurs at a time when physicians sought to establish their own professional respect and control, because many physicians believed that fetal development was a continuous process and that human life should be protected from the time of fertilization, and because of fears that decreased childbearing among resident Protestant white women would permit the overwhelming of American society by newly emigrated,

7. Latest valid data. The most accurate source of data is the Alan Guttmacher Institute's regular survey of abortion providers (e.g., Henshaw 1987; Henshaw, Forrest, and Van Vort 1987). This survey is difficult to execute and reports of data are usually two or more years behind; the 1985 data are the most recent. These data, rather than the government's own surveillance data, are cited in the definitive *Statistical Abstract of the United States.*

Various other social surveys ask questions about abortion, but an estimated 47 percent of abortions known to have taken place are not reported to interviewers (Forrest 1987).

mostly Catholic white ethnic groups with high birthrates (Mohr 1978).

Laws prohibiting abortion established during the nineteenth century stood relatively unchallenged until the 1960s, when an abortion reform movement directed toward legislative change resulted in modification of some state laws. This movement culminated in the 1973 U.S. Supreme Court decision *Roe v. Wade,* which legalized abortion throughout the United States. Contrary to what some people may think, however, *Roe v. Wade* did not legalize any and all abortions in any and all situations. *Roe v. Wade* allows abortion to be obtained without question in the first trimester of pregnancy, but abortion is subject to regulation of providers and procedure in the second trimester and can be outlawed by states after fetal viability (when the fetus is able to live outside the womb), which occurs in the third trimester.

Today the future of legalized abortion is in doubt. The Hyde Amendment in the late 1980s restricted federal funding for abortion through Medicaid. About two-thirds of the states today refuse state Medicaid funding for abortions unless the woman's life is in danger. Government funding restrictions have the result of limiting the abortion option for the poor but not for others (Henshaw 1995).

The Webster decision (*Webster v. Reproductive Health Services* 1989) signaled permission for states to place restrictions on abortion, and many have done so. In a subsequent decision (*Casey v. Planned Parenthood of Southeastern Pennsylvania* 1992), the Court reaffirmed women's overall right to legal abortions; but it also came within one vote of overturning *Roe v. Wade* and allowed states to restrict abortion further. The decision upheld provisions in a Pennsylvania abortion law that requires (1) doctors to provide information on possible alternatives to abortion, such as adoption; (2) a woman to wait 24 hours between the time she receives such information from her physician and the time an abortion is performed; (3) a woman under age 18 to have the consent of at least one parent or a judge before having an abortion; and (4) a married woman in most circumstances to notify her husband before having an abortion (Barrett 1992; Ness 1992).[8]

Many states besides Pennsylvania have restrictions like these. Some states require a married woman to get her husband's consent before an abortion; others require minors seeking an abortion to notify one or both parents first and/or to obtain their or a judge's consent. Those favoring this latter restriction argue that if a minor needs parental permission for procedures like having her ears pierced or taking an aspirin at school, then surely we should require it for an induced abortion. But those working in the health care field have argued that young women in supportive families are likely to discuss options with their parents anyway, and that when communication is forced it is more likely to be hostile and nonsupportive. Hence the regulation may more often lead to family conflict—and perhaps violence—than to harmony (Rodman 1991). The press has reported at least one case in which a pregnant teen chose suicide rather than informing her parents or a judge (Carlson 1990).

An abortion alternative that would make the procedure more private, somewhat safer, and potentially less expensive is the drug RU-486. Developed ten years ago in France, RU-486 is taken orally. When taken either monthly or soon after conception, the drug blocks implantation of a fertilized egg. RU-486 is an effective abortifacient for up to nine weeks after a missed menstrual period (Hilts 1992).

Threatened boycotts of U.S. pharmaceutical companies have discouraged their interest in RU-486, although the drug has been widely used in France. Ultimately, the Food and Drug Administration will either approve or reject the sale of RU-486 in the United States. As of this writing, the drug cannot legally be sold or used in the United States and can be brought into the country only for certain tests that have been approved by the FDA. Meanwhile, the American Association for the Advancement of Science adopted a resolution in 1991 calling for "freedom of medical research" and urging pharmaceutical companies and the FDA to make RU-486 available for "further research and use as medically indicated" ("RU-486 . . . Again" 1991). And some state legislators have introduced resolutions calling for the removal of federal obstacles to RU-486, but these would have no force of law and are intended only to show support for the drug (Tanouye 1992a; 1992b). Meanwhile, it appears that other drugs, now legally prescribed in the United States for other medical conditions, can cause abortion; some doctors may be prescribing them for medical abortions.

As virtually everyone is aware, pro-choice and pro-life activists—those who favor or oppose legal abor-

---

8. Prompted by this U.S. Supreme Court decision, Congress stepped up work on the Freedom of Choice Act, an abortion rights bill designed to limit further erosion of a woman's legal right to abortion (Eaton 1992).

TABLE 11.1

Percentage of U.S. Adults Approving of Legal First-Trimester Abortions Under Certain Circumstances

If the woman's health is seriously endangered by the pregnancy	80%
If the woman became pregnant as a result of rape	70
If the woman became pregnant as a result of incest	70
If there is a strong chance of a serious defect in the baby	59
If the woman's mental or emotional health might be damaged by the pregnancy	55
If the family has a very low income and another child would create a financial burden	29
If the pregnancy would require a teenager to drop out of school	28
If the woman has been abandoned by the father of the unborn child	27
If the pregnancy is unplanned and would interrupt a professional woman's career	19
If a couple is using abortion as a repeated means of birth control	9
If the couple wants a boy and a test has revealed that the baby will be a girl	6

Source: July 1991 Gallup Survey; reported in Kelly, 1991.

tion—have made abortion a major political issue. Legislation and other public policy have been shaped by the conflict. Compromises proposed by scholars (Glendon 1987; Tribe 1990) have made little headway, and the future of abortion in the United States is difficult to predict (Segers and Byrnes 1995). But the American public generally takes a centrist position that favors abortion under certain circumstances (Jelen and Chandler 1994).

Social Attitudes About Abortion

Polls show that a substantial majority of Americans believe that abortion should be legal; 61 percent are opposed to overturning the *Roe v. Wade* decision ("Americans Became" 1990).

Table 11.1, which lists the results of a typical national poll, shows that the particular circumstances involved influence people's attitudes about abortion. There is more support for the so-called "hard reasons" of health, genetic defect, and pregnancy resulting from rape and less support for the "soft" reasons of low income, not being married, or simply not wanting more children in the family (Kelly, J. R., 1991). Thirty-nine percent support abortion "for any reason" ("Most Americans Remained Opposed" 1984). Yet, even

some pro-choice activists are morally uncertain about repeat abortions (Luker 1984).

Attitudes vary according to the stage of pregnancy. Majority opinion favors abortion in the second and third trimesters only when the woman's life is endangered. Support for rape, incest, and serious fetal deformity as motives for abortion remains high, but below 50 percent (Kelly 1991, p. 311).

Ninety percent of all legal abortions occur within the first trimester (U.S. Bureau of the Census 1995, Table 112). Second-trimester abortions create greater moral problems for many people because they interrupt the fetus's development at a later stage. But only during this period can doctors check for diseased or defective fetuses by taking samples of amniotic fluid. Other second-trimester abortions are obtained by women who may not have realized they were pregnant or who could not make arrangements for a first-trimester abortion, often because they did not have the money (Torres and Forrest 1988). The U.S. House of Representatives has voted to ban third-trimester or "late-term" abortions, and the issue promises to stimulate hot debate in the Senate (Rogers 1995).

What all this public discussion about abortion means for individuals is that women making decisions about abortion today do so in a far more political climate than before (Henshaw 1995). But the fact remains that legal abortions are physically quite safe. The safety of abortions is our next topic of discussion.

The Safety of Abortions

In current abortion procedures, during the first trimester of pregnancy (the first twelve weeks) the fetus is removed by suction (vacuum aspiration) or in a surgical process in which the cervix is dilated and the contents of the uterus scraped out with a sharp instrument (dilatation and curettage, or D&C). In the second trimester (months four, five, and six), a small amount of amniotic fluid is replaced by a salt solution, or a woman is injected with a powerful drug called prostaglandin. Both procedures induce contractions of the uterus, causing the fetus to be expelled. A method infrequently used in the second trimester is *hysterotomy* (not to be confused with hysterectomy), in which the woman's abdomen is opened and the fetus and placenta are removed, much as in a cesarean delivery.

Before abortion was legalized in 1973, illegal and self-induced abortions were accompanied by risk of death or serious injury, including damage to reproductive capacity. But death rates for legal abortions properly performed are lower than those for normal childbirth (Henshaw 1990). Abortion is now one of the safest medical procedures when it is performed in a hospital or a clinic in the first trimester, as 90 percent of abortions are. For every 100,000 women who become pregnant, 57 die due to complications of pregnancy. There is fewer than one death per 200,000 pregnancies ended by abortion in the first trimester (Hilts 1992). Abortions in the second trimester are about ten times more dangerous than those performed in the first trimester.

According to recent and compelling evidence, abortion has no impact on the ability to become pregnant; sterility following abortion is very uncommon (Stubblefield et al. 1984). Reviews of research also conclude that there is no risk to future pregnancies from a first-trimester vacuum aspiration abortion of a first pregnancy (Hogue et al. 1983). Techniques used since 1973 have averted the risk of miscarriage initially found to be associated with multiple abortions (Kline et al. 1986). But questions of emotional and psychological distress are more often raised regarding abortion.

Emotional and Psychological Impact of Abortion

It's safe to say that for most women (and for many of their male partners) abortion is an emotionally upsetting experience (Shostak and McLouth 1984). Recently, the media have featured women who tell of "a sense of guilt and sorrow that wouldn't leave" after an abortion (Lacayo 1989a, p. 22). But journalistic stories also feature women who say they have never regretted their choice to abort a fetus: "I was 19, with one year of college. I had dreams and plans, and wanted children when I was old enough and smart enough. I didn't feel remotely able financially to take care of a child" (in Lacayo 1989a, p. 22). The decision to abort is often very difficult to make. But the emotional distress involved in making the decision and having the abortion does not typically lead to severe or long-lasting psychological problems.

Some married couples may consider aborting an unwanted pregnancy if, for example, they feel that they have already completed their family or could not afford to raise another child.[9] The couple may believe an additional baby would place too much emotional strain on their marriage or prevent them from meeting the needs of other family members.

Abortion can also be an issue for couples—particularly older ones and those in later stages of pregnancy—who through prenatal diagnosis techniques find out that a fetus has a serious defect. (Prenatal diagnosis techniques and the developmental problems they can detect are described in Appendix C.) Results of ultrasound and amniocentesis are favorable in 95 to 99 percent of cases, and not all parents choose abortion when they are not. In fact, some parents exclude the possibility of abortion in advance but want the information provided by prenatal testing to prepare themselves to care for a disabled infant. For those faced with evidence of a defective fetus, abortion is not easy, for these are usually wanted pregnancies. "Patients and genetic counselors alike report that mourning a pregnancy ended because of an abnormality is almost identical to mourning the death of a newborn infant" (Powledge 1983, p. 40).

But abortion decisions are primarily made within the context of unmarried, accidental pregnancy. And

---

9. A survey of married women who obtained abortions found that almost 90 percent had told their husbands. The researchers expressed concern, however, that only slightly more than half the couples engaged in any meaningful discussion of alternatives and feelings (Ryan and Plutzer 1989). More research is needed on this important topic.

studies have found positive educational, economic, and social outcomes for teen women who resolve pregnancies by abortion rather than giving birth. In a recent study, "those who obtained abortions did better economically and educationally and had fewer subsequent pregnancies than those who chose to bear children. Those who had abortions even fared better than those who were not pregnant at the start of the research project" (Holmes 1990). These outcomes are related to the reasons women have abortions. In a 1987 survey, three-quarters of women obtaining abortions cited educational needs or work responsibilities, two-thirds said they could not afford to have a child; half said they were in a poor relationship and/or did not want to be a single parent. Some did not want others to know they had had sex or were pregnant. Teens were likely to say they were not mature enough to be a parent (Torres and Forrest 1988).

Several major reviews of extensive research on the emotional and psychological impacts of abortion do not find significant negative consequences. Emotional stress is more pronounced for second-trimester than earlier abortions and for women who are most uncertain about their decision. Still, a panel of experts reporting in the well-respected journal *Science* concluded that "for a vast majority of those who have voluntary abortions, 'severe negative reactions are infrequent in the immediate and short-term aftermath'" ("Study Sees" 1990, p. A-16). Although longer-term studies are needed, the panel did not anticipate any change in this conclusion (Adler et al. 1990). An American Psychological Association review also concluded that "significant psychiatric sequelae to abortion are rare" (American Psychological Association 1987, p. 18). In 1989, Surgeon General C. Everett Koop, after a comprehensive review of research findings to date, concluded that the available data "do not support the premise that abortion does or does not cause or contribute to psychological problems" (in Kelly 1990, p. 543).

Regardless of predicted outcomes, women (and men) making decisions about abortions are most likely to make them in the context of their values, and a detailed review of the ethical and social issues involved is outside the scope of this text.

Involuntary Infertility

While society is currently focused on the prevention of conception or birth, a growing minority of individuals and couples face a different problem. They want to have a child, but either they cannot conceive or they cannot sustain a full-term pregnancy.

Involuntary infertility is the condition of wanting to conceive and bear a child but being physically unable to do so. It is usually defined in terms of unsuccessful efforts to conceive for at least twelve months or the inability to carry a pregnancy to full term.[10] (A parallel concept of **subfecundity,** or secondary infertility, indicates the problem of parents who have difficulty having additional children.) Those with reproductive incapacity will not be identified unless they seek treatment or define themselves as infertile.

The problem originates with females in 30–40 percent of cases, with males in 40 percent of cases, and with both partners or is unknown in the remaining cases (Benson 1983). About half of couples defined as infertile will eventually conceive and deliver, either with or without medical intervention (Collins et al. 1984).

The physiology of conception, pregnancy, and childbirth is described in Appendix C and should be referred to in conjunction with this section. Several factors can contribute to involuntary infertility. A man may have relatively few sperm in his semen (low sperm count) or they may be abnormal. A woman may fail to ovulate, may have a blockage or scarring in her fallopian tubes, or may have a condition called *endometriosis,* in which uterine tissue migrates to other parts of the abdomen. There are some cases in which each partner might be capable of conceiving with another individual. When both partners have even a minor problem, however, their chances of conceiving together are lowered. Then there are cases of idiopathic infertility—that is, infertility of no known cause.

Through advances in health and medicine, the incidence of involuntary infertility declined for over eighty years. There is a perception of a recent increase, but in fact infertility has not increased since 1965, remaining at about 10 percent of all married couples of childbearing age and at about 20 percent of married couples with no children (Mosher 1990a).[11]

---

10. Miscarriages end almost one-third of pregnancies, for reasons that are largely unknown. Most women who miscarry, however, are fertile and go on to have successful pregnancies (Beck 1988a; Kolata 1988b).

11. These figures are estimates and take into account the increasing likelihood that older partners have been surgically sterilized, which masks the existence of reproductive impairment that might have been

Neither modern contraceptive means (other than the IUD) nor abortion impair fertility (Hogue et al. 1983; Population Information Program 1984; Kline et al. 1986). There are new factors contributing to infertility today, however. One is the rising incidence of STDs, which can damage the reproductive systems of women and sometimes men; another is the fairly widespread use of the IUD in the 1970s and 1980s, which can cause uterine infection and scar the fallopian tubes. Exposure to various drugs, chemicals, and radiation can also cause infertility, and smoking and excessive exercise (which inhibits ovulation) can contribute to it. There is also concern that various environmental pollutants are lowering males' sperm counts and generally damaging human reproductive systems (Begley and Glick 1994; Connor 1994; Lemonick 1994). Furthermore, infertility appears to have increased because fecundity (the ability to conceive) declines with age while miscarriages increase with age. The present tendency to postpone childbearing until one's 30s and 40s creates a class of potential parents who are intensely hopeful and also financially able to seek treatment.

When faced with involuntary infertility, an individual or a couple experiences a loss of control over life plans and feels helpless, damaged or defective, angry, and often guilty (Becker 1990).[12] Research shows that this is especially true for women (Abbey, Andrews, and Halman 1991). The cultural "centrality of the motherhood role" tends to render involuntary infertility among women "more central to their lives and more frequently on their minds" (p. 310).

Then too, according to one counselor, the psychological burden of infertility may fall more heavily on professional, goal-oriented individuals. These are people "who have learned to focus all their energies on a particular goal. When that goal becomes a preg-

nancy that they cannot achieve, they see themselves as failures in a global sense" (Berg 1984, p. 164).

Besides having an effect on each partner's self-esteem, the situation can hurt their relationship and can create a marital crisis (Abbey, Andrews, and Halman 1992). As one wife explained, "We went through a heavy discussion, considering—not considering divorce, but we discussed it. The fact that we'd even discussed it was, to us, pretty far gone. He said to me, 'If you want to get somebody else to do it [impregnate her] . . . If you want to get somebody else' . . ." (in Becker 1990, p. 79).

For many couples, the partners slowly become aware that they confront a situation "of which they can make no sense." Moreover, this creeping awareness often arises at about the time when other couples they know are publicly planning their pregnancies without apparent difficulty. "It is under these social conditions that most couples embark on the quest for a medical solution to their problem" (Matthews and Matthews 1986, p. 643).

For many couples, the first step in treatment is to go to a doctor. After an examination the doctor often recommends that they relax for another six months or so and also try to have sexual intercourse at scheduled times, when the wife is most likely to be ovulating. This often involves keeping accurate temperature charts, for a woman's body temperature rises after ovulation. Unfortunately, scheduling sexual intercourse in this way is not likely to help a couple relax as they were advised to do.[13]

If after a period of time conception has still not occurred, further medical treatment may be sought. This might include various drugs (to induce ovulation, for example) or microscopic surgery (to repair blocked fallopian tubes in the female, for instance or to repair a blockage in the testicles in the male). Other medical procedures are becoming more available today. These include artificial insemination, in vitro fertilization, and related techniques (see Appendix E). These procedures can be successful, but they can be difficult and expensive, and they raise the question of whether and

discovered had those couples tried to have children. Raw data on infertility, which do *not* calculate the probable existence of this masked infertility, show a *reduction* in infertility from 1965 to 1982 (Mosher and Pratt 1985; Mosher 1990a).

Critics of the notion of an infertility epidemic point out that the definition of infertility has changed since the 1950s, when couples were considered infertile after five (and later, two) years of unsuccessful attempts to get pregnant. The reduction of the time period to one year confuses "The inability to conceive with difficulty in conceiving quickly. . . . The only infertility epidemic is of infertility specialists; between 1965 and 1988, membership in the American Fertility Society jumped from 2,400 to 10,300" (Raymond 1991, p. 29).

12. A national organization called Resolve provides support groups, counseling, and referral services to infertile people.

13. Scheduling sex for the main purpose of conception can feel depersonalizing (Becker 1990, p. 93) and can also add conflict to a relationship. As one wife explained, "All the things you read about—that men feel like they are just a tool. You have to have an erection and ejaculate at a certain time whether you want to or not. He has said to me in times out of genuine anger, 'I feel like all you want me for is to make a baby. You don't really want me, you just want me to do it'" (in Becker 1990, p. 94).

For couples who want to become pregnant, involuntary infertility is a painful obstacle. For some who can afford it and choose it, reproductive technology can be an answer, although there are social and there may be ethical issues to consider.

to what extent technology should be involved in human procreation.

Reproductive Technology: Social and Ethical Issues

Reproductive technologies such as those described in Appendix E enhance choices for some couples and can reward them with much desired parenthood. But reproductive technologies have tremendous social implications for the family as an institution and raise serious ethical questions as well.

COMMERCIALIZATION OF REPRODUCTION

A general criticism is that the new techniques, performed for profit, commercialize reproduction; prospective parents (or, more specifically their parts—testicles, sperm, ovaries, eggs, wombs, and so on), children, and embryos are treated as products and are thereby dehumanized (Lauritzen 1990; Merson

1995). An example is the selling of sperm to for-profit sperm banks and their deliberate overstocking of sperm by certain donor characteristics, such as better-than-average athlete or IQ. In 1995, the highly respected and renowned Center for Reproductive Health at the University of California, Irvine, was accused of several profit-driven "ethics lapses," such as stealing eggs or embryos from some of their patients to create pregnancies in others (Cowley 1995).

Concern has also been expressed about **selective reduction** (also called *selective termination*). In selective reduction, some (but not all) fetuses in multiple pregnancies resulting from ovulation-stimulating fertility drugs or the GIFT procedure (see Appendix E) are selectively aborted, usually in the first trimester (Overall 1990).

Furthermore, the freezing and saving of embryos raises the possibility not only of posthumous or post-divorce fertility but also of "leftover embryos" (Annas

FIRST ROW: BABY X. SECOND ROW: SURROGATE MOTHER, FOSTER FATHER, ADOPTIVE MOTHER, BIOLOGICAL FATHER, FOSTER MOTHER, GUARDIAN. THIRD ROW: ATTORNEYS.

1991, p. 35). Lawsuits involving the disposition of frozen embryos left by a couple killed in a plane crash and a divorced couple's dispute over the wife's desire to bear children from their frozen embryos are examples (Elson 1989b; Cooper 1992).[14]

MEDDLING WITH NATURE   A related issue involves whether unlimited technological meddling with nature is prudent. For example, one clinic's survey reported that up to two-thirds of its clients were single professional women and/or lesbians (Isaacs and Holt 1987). An outcome of fertility technology is that a single woman can conceive and bear a child without ever developing any relationship, sexual or otherwise, with a man. Although some herald this as a major breakthrough, others worry about what it means for the future of males in the family.

14. In one three-year legal battle, *Davis v. Davis,* the custody of seven frozen embryos was contested in a divorce action. The divorcing wife (but not her husband) wanted the embryos implanted so she could have a child. The judge in the original trial ruled that the embryos were children and awarded custody of them to the wife. But that decision was reversed by an appellate court, which ruled that the state cannot force a man to become a father against his will and gave the Davises equal control over the embryos' disposition. At this writing the embryos remain in frozen storage at the clinic, which was seeking to be relieved of responsibility for them. The wife has remarried and has said she may begin fertility treatments with her new husband (Cooper 1992; "The Fate of" 1992).

As a second example, we are now faced with the question of how old is too old to become a mother. Menopause, which stops ovulation, has traditionally been considered a natural barrier to pregnancy. But using a younger woman's eggs and an embryo transfer, some women past menopause—one as old as 59, who had twins in England—have carried and delivered infants (Carlson 1994). Older mothers who bear children this way are likely to be financially stable, but postmenopausal pregnancies have more health risks, although these risks are usually manageable. Observers are concerned that a mother giving birth in her 50s will be 70 when her child graduates from high school (Gormon 1991c). This has always been true for fathers, who can remain fertile into old age. Nevertheless, critics wonder whether women who postpone parenthood should be encouraged to "wait until they are on the cusp of reproductive senility to try to procreate" (Halpern 1989c, p. 148).

THE STATUS OF PARENTHOOD   More specific issues arise regarding the legal and social status of parenthood when reproductive technologies are involved. For example, who is legally responsible for the baby should it be born deformed and neither mother wants the baby?

With AID (see Appendix E), courts have had to address the question, "Who is the father?" The legal status of the procedure itself, of the child conceived, and of donor and nondonor fathers varies from state to state. By 1990, thirty states had enacted artificial insemination laws in which sperm donors, with the exception of the husband, have no parental rights (Marcus 1990). In these states children born to a married couple by AID are considered fully legitimate, provided the husband has consented to the procedure; the husband has full parental rights and responsibilities.

But these laws were enacted under the assumption that the woman would not know the sperm donor and failed to take into account situations in which a male donates sperm to a female friend. In 1989 an Oregon court granted legal rights as a father (including regular visitation) to such a donor, whose friend had promised that he could help raise the child (Marcus 1990). Hence, in many states the rights and obligations of the parties involved remain ambiguous. Even when the couple has executed a written agreement to the contrary, sperm donors known to the mother have at times received parental rights. In one situation a lesbian whose partner had conceived a child through artificial insemination was ordered to pay child support after the couple separated because she had agreed to be the father (Isaacs and Holt 1987). Indeed, some states have not changed old laws treating AID as adultery and the child as illegitimate. So it is important to obtain legal advice if this choice is made.

Moreover, surrogacy may mean that "maternity will be as much in dispute as paternity ever was" (Lauritzen 1990, p. 41). Surrogacy, along with embryo transfer technology, creates the possibility that a child could have three mothers: the genetic mother, the gestational mother, and the social (child-raising) mother. In such a situation, how do courts define the "real" mother?

The 1987 Baby M and the 1990 Anna Johnson cases raise this kind of question. In the Baby M case, Mary Beth Whitehead had contracted to bear a baby through artificial insemination for the Sterns. Stern's sperm was used to inseminate Whitehead, and she was to be paid $10,000 for her services. But later she turned down the money, preferring to keep the baby. The trial that ensued centered around whether a birth mother has a right to her baby that supersedes legal

contracts. The state court's decision was that surrogate contracts are contrary to the law and public policy.

In the case of Baby M, the surrogate was also the baby's genetic mother. This was not so in a subsequent case in which the Calverts were granted full custody of a baby born to a gestational mother, Anna Johnson. Johnson had contracted with the Calverts to receive an embryo transfer, then carry and deliver the test-tube baby, genetically the Calverts', for a fee of $10,000. Several months into her pregnancy, Johnson said she had bonded with the fetus and sued to be legally recognized as a parent. The judge denied parental rights to Anna Johnson (Stevens and Geyelin 1990; Annas 1991, p. 36).

The vast majority of surrogates fulfill the contract. But partly due to conflicts over parental status, many states drew up surrogacy laws after the Baby M case. Several states (for example, California) have passed laws to regulate paid surrogacy, thereby permitting it; other states (for example, New York) have outlawed or refused to enforce surrogacy contracts ("Surrogacy for Pay" 1992).[15]

INEQUALITY ISSUES Reproductive technologies also raise social class and other inequality issues (Rae 1994; Ragone 1994). Of the clients of the broker involved in the Baby M surrogacy, one-third had annual family incomes of over $90,000 (Langley 1988). Paid surrogates ordinarily receive between $10,000 and $15,000 if the child is born alive and healthy. For nine months of gestation, $10,000 is equivalent to $1.57 per hour. "In the Baby M trial, M stands for money," wrote one columnist. "The wealthy don't become surrogates and the poor do not buy surrogates and the hired matchmakers do not work for love" (Goodman 1987).

The issue of race inequality has also been raised. Anna Johnson, who was granted no parental rights, was African American; the Calverts were white. Some have wondered whether this situation influenced the court's decision and warn that poor, black women could become "the surrogate class" (Sanders 1992).

---

15. One surrogacy broker called the New York law "a step backward for women," arguing that if a man can sell his sperm, a woman should be allowed to sell the use of her uterus. Meanwhile, some have opposed paid surrogacy as baby selling. In a contrasting opinion by the judge in the Anna Johnson case, surrogates "are not selling a baby, they are selling . . . the pain and suffering, the discomfort, that which goes with carrying a child to term" (in Annas 1991, p. 36).

Other critics have pointed to the developing international surrogacy market in which mostly nonwhite, third world women are encouraged to carry and deliver babies for mostly white, First World mothers (Raymond 1989).

Moreover, feminists fear new forms of exploitation of women if what is now optional technology becomes expected of infertile women whose husbands, for example, desire a child (Lauritzen 1990). Critics argue that the cultural and moral celebration of women as nurturant and altruistic may be subtly coercive in family situations:

> The potential for women's exploitation is not necessarily less because no money is involved and reproductive arrangements may take place within a family setting. The family has not always been a safe place for women, and there are unique affective "inducements" in familial contexts that do not exist elsewhere. (Raymond 1990, p. 10)

Despite all this, it seems possible that, with public and private deliberation, the potential for good in reproductive technologies—reducing involuntary infertility—can be retained without at least some of the negative effects prophesied.

Reproductive Technology: Making Personal Choices

Fertility treatment can be financially, physically, and emotionally complex. Those who are successful are euphoric: "Our son is so gorgeous!" marveled one mother (McCarthy, in Halpern 1989b, p. 151). (Reports of fraud and other professional violations in fertility centers [see, for example, Chartrand 1992; Cowley 1995] make it important to understand that an individual seeking treatment is a consumer and should interview the doctor and investigate the facility.)

Choosing to use reproductive technology depends on one's values and circumstances. For example, African Americans have tended to view high-tech infertility treatments as "a white thing," recalling former slave conditions where blacks were studs and breeders for whites. Even though blacks' infertility is higher than non-Hispanic whites', infertile African Americans tend to be incorporated into and perform parenting roles in extended families rather than pursue infertility treatments (Sanders 1992).

Moreover, religious beliefs may influence decisions. The Catholic church officially views AID as adultery and argues that reproduction should be kept within the marriage bed (McCormick 1992). The Jewish tradition requires physical union for adultery and hence does not define AID as adulterous. But Judaism does view masturbation as sinful; hence a man's obtaining sperm either to sell or to artificially inseminate his wife is morally problematic (Newman 1992). Meanwhile, Protestantism notes that the Bible sees infertility as cause for sorrow and exhalts increasing human freedom beyond natural barriers (Meilander 1992).

Besides racial and religious influences, costs are a consideration. Entirely possible is a "$30,000" baby—a result of two GIFT procedures totaling $18,000, eight artificial inseminations totaling $8,000, one frozen embryo transfer at $1,000, and miscellaneous tests totaling $3,000 (Elmer-Dewitt 1991b; Brant, Springen, and Rogers 1995). Often these costs are not covered by insurance. Furthermore, the need for frequent physicians' visits can interfere with job attendance, and women undergoing infertility treatments report being discriminated against in the workplace (Shellenbarger 1992g).

Then too, infertility treatment can be physically daunting. Women report that artificial insemination can be "very uncomfortable" (Halpern 1989c, p. 149); IVF procedures are painful. Moreover, a woman undergoing IVF

> is hostage to it. She will have her blood drawn on selected days. . . . She will be injected at least once and often twice a day with powerful hormones. . . . She will be subject to regular ultrasound examinations. She will probably lose time from work, even if her physician is near her workplace. If she chooses to venture farther afield, to one of the nationally known programs . . . , she will have to live out of a suitcase, away from friends, workmates, and often, family, until the sixteenth day of the cycle, when the fertilized eggs are transferred to the uterus. (Halpern 1989c, p. 150)

Furthermore, the very availability of reproductive technology can make it difficult to know when to quit. One woman's reproductive history included three cesarean sections, two microscopic surgeries, IVF, four embryo transfers, selective termination of two of the fetuses, revelation via ultrasound that one of the remaining twins had no bladder or kidneys,

TABLE *11.2*

Among Children Born to Never-Married Women, Percentages Relinquished for Adoption, by Race and Year of Birth

	Before 1973	1973–1981	1982–1988
Black women	1.5	0.2	1.1
White women	19.3	7.6	3.2
All women	8.7	4.1	2.0

NOTE: Percentages are based on data combined from the 1982 and 1988 National Surveys of Family Growth and refer to premarital births to women ages 15–44 at either survey.

Source: Adapted from Bachrach, Stolley, and London 1992, p. 29.

spontaneous miscarriage of the abnormal twin, and intrauterine death of the remaining fetus (Overall 1990).

For some, infertility treatment can become the problem instead of the solution (Brant, Springen, and Rogers 1995). Coming to terms with infertility has been likened to the grief process in which initial denial is followed by anger, depression, and ultimate acceptance: "When I finally found out that I absolutely could not have children . . . it was a tremendous relief. I could get on with my life" (Bouton 1987, p. 92). Some people gradually choose to define themselves as permanently and comfortably child-free. A second way to get on with life is through adoption. Indeed, some couples explore adoption options even as they continue infertility treatments (Williams, L. S., 1992).

Adoption

Adequate adoption statistics do not exist because since 1975 no federal agency regularly collects data. The National Committee for Adoption estimates that about 50,000 children are legally adopted by nonrelatives (not including stepparents) each year. Legally adopted youngsters make up about 2 percent of U.S. children (Gibbs 1989); 4 percent of families

contain adopted children (Moorman and Hernandez 1989). More highly educated women, those not working full-time, and those with higher family incomes are more likely to adopt. White women are more likely to adopt through formal adoption procedures than are African American women or Latinas (Bachrach et al. 1990).[16] Adoptions increased steadily through much of this century and reached a peak in 1970 of about 175,000, but the number has steadily declined since then (Bachrach et al. 1990). Fewer infants are now available due to more effective contraception and legalized abortion.

Furthermore, unmarried mothers are increasingly likely to keep their babies. Data from the National Survey of Family Growth, which interviewed about 8,000 women in both 1982 and 1988, show that women in the 1980s were far less likely than in the early 1970s to place their children for adoption. As Table 11.2 indicates, before 1973 nearly 9 percent of infants born to never-married women were relinquished for adoption, and nearly 20 percent of babies born to white mothers were given up for adoption. By the mid-1980s only 2 percent of all women and 3 percent of whites did so. The levels of relinquishment among black women remained low throughout this period but also declined slightly overall. Relinquishment among Latinas may be virtually nonexistent (Bachrach, Stolley, and London 1992). Box 11.2, "Deciding to Relinquish a Baby," explores this option from the birth mother's perspective.

Some couples pursue international adoption; about 15 percent of U.S. adoptions (about 8,000 a year) are of children from outside the country (Bogert 1994). A second option is to adopt difficult-to-place, or "special

16. These surveys on adoption were limited to women to avoid double counting. Still, it is interesting to note that much data on childbearing and other ways of becoming parents are reported as data on women, as though men had nothing to do with becoming parents. This tends to occur for two reasons: One is the traditional mode of thinking that associates children and women. The other has to do with methodological strategies. Women are thought to be more reliable sources of data on fertility, contraception, abortion, and so on. Men, even husbands, may not be fully informed about the female partner's pregnancy history or use of contraception; women, of course, may become pregnant when not in a regular relationship with a man. In the past, women were also more easily interviewed, as they were at home more often. If it was more efficient to interview only one member of the couple, the wife was chosen.

We have recently become more sensitive to the different perspectives that husbands and wives may have about the same events. Many studies now interview both husbands and wives and compare their responses. We have also become less neglectful of the male's role in reproduction.

Deciding to Relinquish a Baby

Young women who carry a pregnancy to term can either parent the child themselves, often with help from extended kin such as the baby's grandmother, or place the baby for adoption. Teenage parenting has raised concerns, and some professionals view adoption as the better alternative (Resnick 1984). This view holds that relinquishment enables young women to pursue the developmental tasks of adolescence and early adulthood unencumbered by parenthood. However, clinicians often argue that regret associated with relinquishment can be debilitating and can damage a birth mother psychologically (Watson 1986).

A theme of this text is that making decisions knowledgeably is preferable to making them by default. Although currently there are no national data from which to draw random samples of relinquishers and parenters, two studies help shed light on this subject. A pair of studies on the consequences of relinquishment versus parenting for 190 teenage birth mothers (78 relinquishers and 112 parenters) from one agency found that, after controlling for background variables, relinquishers did better on a variety of sociodemographic outcomes than did those who kept their babies. The two groups were similar on psychological outcomes. The majority in both categories were satisfied with their pregnancy resolution decision, but relinquishers were sig-

nificantly less satisfied with their decision than were parenters (McLaughlin, Manninen, and Winges 1988a, 1988b).

In another study (Kalmuss, Namerow, and Bauer 1992), researchers analyzed the short-term effects of relinquishing versus parenting in a sample of 527 unmarried non-Hispanic white and black pregnant women (311 of whom parented and 216 of whom placed their babies) ages 21 or younger. The women were interviewed before their baby's birth and again six months after delivery. At the second interview, relinquishers were significantly more likely to be enrolled in school and were more optimistic about achieving positive educational, financial, occupational, and marital outcomes by age 30.

In terms of comfort with the pregnancy resolution choice six months after delivery, relinquishers experienced more regret than parenters and were less likely to say they would make the same decision again. But 56 percent of relinquishers reported little or no regret about their decision, and 78 percent said they would make the same decision again. The researchers concluded that "our findings regarding comfort with the pregnancy resolution decision are not consistent with those from the clinical literature indicating that relinquishers experience a great deal of dissatisfaction with their decision to place the baby for adoption" (Kalmuss, Namerow, and Bauer 1992, p. 89).

needs" children—those who are older, nonwhite, must come with siblings, and/or are disabled.

The Adoption Process

The experience of legal adoption varies widely across the country, partly because it is subject to differing state laws. Generally, adoptions can be public or private. **Public adoptions** take place through licensed agencies that place children in adoptive families. **Private adoptions** (also called *independent adoptions*) do

not involve a formal agency and are arranged directly between adoptive parent(s) and the biological (or birth) mother, usually through an attorney. Because there is no central registry of available babies, adopting couples themselves find a woman who is willing to relinquish her child by contacting doctors, lawyers, or social workers or by placing newspaper ads. Legal fees and often the birth mother's medical costs are paid by the adopting couple. Many couples pursue public and private adoption simultaneously (Williams 1992).

Licensed agencies may be more likely to offer extensive counseling and explicitly to inform both relinquishing and adopting parents of their legal rights. Agencies have been criticized, however, for cumbersome bureaucratic procedures and unrealistic standards of income, housing, and lifestyle for prospective adoptive parents. Waiting periods for a healthy, white infant can be up to seven years; older prospective adoptive parents (and the majority of them are) are likely to be screened out by an agency (Williams 1992).

Private adoptions may offer both the birth and adopting parents more personal control. But private adoptions can be expensive for an adopting couple. Attorney's fees and medical expenses typically range from $8,000 to $20,000 or more. Some pay fees up to $50,000 or even $100,000 (Waldman and Caplan 1994). Moreover, private adoptions may render both the birth and adoptive parents more vulnerable to exploitation. The birth mother may feel manipulated into relinquishing her baby when she doesn't really want to, for example. (The biological mother's consent is especially an issue in overseas adoptions because it is more difficult to be sure that the mother has willingly placed her child for adoption [Bogert 1994].) Or, even after her medical and living expenses during pregnancy have been paid by the prospective parents, the birth mother might capriciously choose a different adoptive couple (Crossen 1990).

Whether public or private, an adoption process can be closed, open, or semi-open. A **closed adoption** is one in which the adoptive and biological families have no communication and do not know one another's identities. Until fairly recently, the adoption process was virtually always closed and shrouded in secrecy.[17] An **open adoption** involves some direct contact between the biological and adoptive parents, ranging from one meeting before the child is born to lifelong friendship. In **semi-open adoptions,** biological and adoptive families exchange personal information, such as letters or photographs, but they do not have direct contact.

Even when an adoption is closed, many states now have laws permitting the adoptee access to records at a certain age or under specified conditions. The issue of opening previously closed records remains controversial, however. Advocacy groups of adoptees support open records. Meanwhile, those who encourage adoption over abortion fear that pregnant women will choose the latter unless they are guaranteed confidentiality, and some adoptive parents find open records threatening. Adoption professionals take various positions.

Another concern that has arisen in the last several years—with 1994 media attention to Baby Jessica and Baby Richard—is whether birth parents can claim rights to a biological child after the child has been adopted. However, of all domestic adoptions, fewer than 1 percent are ever contested by biological parents (Ingrassia 1995).

Adoption of Racial/Ethnic Minority Children

Although African Americans make up 12 percent of the population, 40 percent of children awaiting adoption are black. "Some advocacy groups insist that there are enough black homes for black kids" but that agencies do not feel pressed to find homes for minority children, and prospective minority parents encounter biases in the placement system (Beck 1988b). Over the past two decades, efforts to recruit minority adoptive parents have been somewhat successful. Such efforts not only require public relations programs but also must address problems that prospective minority adoptive parents often face: middle-class and racial bias, along with sometimes unreasonable eligibility standards and bureaucratic red tape (Kalmuss 1992). Meanwhile, many minority children will remain in foster care indefinitely if not adopted by whites (Simon 1993).

In 1971, agencies placed more than one-third of their black infants with white parents (Nazario 1990b). But interracial adoptions, having increased rapidly in the 1960s and early 1970s, have been much curtailed since 1972, when the National Association of Black Social Workers strongly objected. Suggesting that transracial adoptions amounted to cultural genocide, minority advocates expressed concern about identity problems and the loss of children from the black community. Native American activists have successfully asserted tribal rights and collective interest in Indian children rather than allowing them to be relinquished to adoptive parents of other races. In addition

---

17. Twenty years ago some delivery-room personnel covered the mirrors and draped towels in front of a woman relinquishing her baby, or even blindfolded her, so she could not see the baby; in the nursery the infants were marked DNS (do not show) or DNP (do not publish the mother's name) (Gibbs 1989).

to concerns about identity, they have expressed the fear that coercive pressures might exist to provide adoptable children to white parents. Similar concerns surfaced in the mid-1970s when many unaccompanied children left Vietnam as refugees, and these concerns are expressed today regarding overseas adoptions from the third world.

As a result of this controversy, adoption agencies have shied away from transracial adoptions; today thirty-five states somehow restrict them (Nazario 1990b). Consequently, only about 8 percent of current adoptions are interracial, usually adoption by white parents of mixed-race, African American, Asian, or Native American children (Bachrach et al. 1990).

The debate over transracial adoption persists, however, as growing numbers of white parents want to adopt and minority children wait in foster care. The National Committee for Adoption is lobbying to remove legal restrictions on transracial adoptions (Nazario 1990b). In 1990 a white couple denied the opportunity to adopt an African American baby filed a federal discrimination suit ("Couple Sues a County" 1990).

Some transracially adopted children, now in their 20s, relate painful racial experiences. For example, one young woman never wanted to go out to dinner alone with her father because people might think she was a prostitute (Nazario 1990b). But studies indicate that transracial adoptions are largely successful so far as outcomes to children are concerned. One longitudinal study (Simon and Alstein 1987) observed African American children in white homes from 1971 and concluded that the children developed a positive sense of black identity and a knowledge of their history and culture. Similar research (McRoy, Grotevant, and Zurcher 1988) found transracially adopted children to have high self-esteem and confidence, although they did struggle with identity issues. Some researchers have suggested that rather than causing serious problems, transracial adoptions may produce individuals with heightened skills at bridging cultures. In the summary words of one researcher, "the message of our findings is that transracial adoption should not be excluded as a permanent placement when no appropriate permanent inracial placement is available" (Simon 1990).

The problems and concerns regarding interracial adoption are indicative of societywide racial prejudice, of course. As African American sociologist Joyce Ladner has observed,

This couple formed "The Up with Down's Syndrome Foundation," and they have adopted many handicapped children.

ultimately the future of transracially adopted children, as with all children in the society, is inextricably linked to the future of the American people. Their growing-up years can be as problem free or as problematic as the majority of Americans decide. The racial attitudes and behavior—as well as their attitudes toward adoption itself—will, more than anything else, determine these children's outcomes. (Ladner 1977, p. 78).

Adoption of Older Children and of Disabled Children

Together with ethnic minorities, children who are no longer infants and disabled children make up the large majority of youngsters now handled by adoption agencies. According to estimates, at least 36,000 and perhaps up to 100,000 children are awaiting adoption in foster care (Fishman 1992). About one-third are ever adopted (Lacayo 1989b).

Special needs adoptions occur not only to couples who are infertile but also to those with altruistic motives. Gay men have adopted infants with HIV/AIDS, for example (Morrow 1992). National adoption exchanges for children with Down's syndrome and spina bifida have waiting lists of would-be parents. In some cases, lesbian and gay male couples adopt such hard-to-place children because law or adoption agency policy denies them the ability to adopt other children.

The majority of adoptions of older and disabled children work. Overall, only about 2 percent of agency adoptions end up being **disrupted adoptions** (the child is returned to the agency before the adoption is legally final) or **dissolved adoptions** (the child is returned after the adoption is final). But disruption and dissolution rates rise with the age of the child at the time of adoption—from 10 percent for children older than 2 to about 25 percent for those between 12 and 17 (Sachs 1990). When we consider that in many cases the child's departure from the adoptive home is not reported to the agency, estimates of adoption failure among older and disabled children range between 4 and 40 percent (Barth and Berry 1988).

What causes these high failure rates? For one thing, more children available for adoption are emotionally disturbed or developmentally impaired due to drug- or alcohol-addicted biological parents (see Box 11.1) or to physical abuse from biological or foster parents. Furthermore, older children have undergone more previously broken attachments, as they have been moved from one foster home to another or through disrupted adoption attempts. Many of these children gradually develop **attachment disorder,** defensively shutting off the willingness or ability to make future attachments to anyone (Barth and Berry 1988).

Moreover, some parents who had hoped to adopt a healthy infant may be manipulated by agencies into accepting an older or disabled child instead, without being properly advised of potential problems. For example, prospective parents might be told that a child is mildly hyperactive when in truth the problems are much worse, ranging from showing virtually no emotion to destructive rampages—behaviors with which parents are unprepared to cope (Fishman 1992).

According to developmental psychologist Advid Brodzinsky, who has compared adopted with non-adopted children, even normal adopted children often go through a process of emotional turmoil in middle

adolescence as they grieve not ha
raised by biological parents (ir
for adopted children with a
tional and behavioral probl
cult to deal with. One father de
his 9-year-old adopted son, Chris, a
severe child abuse:

> [At first] he started making friends with ou
> dren in the neighborhood. We read with him
> fully every day and started a tradition of back rubs
> that he much enjoyed. And while he was still extremely guarded, there were glimpses of trust. . . .
> [But as time progressed,] Chris became more and more uncontrollable . . . , and his play became both more agitated and more mechanical. . . . Discipline became less and less effective as he gradually pulled himself outside the orbit of the family. . . . Games that started innocently enough would escalate in agitation until someone got hurt "accidentally." . . . Because of the great difference in age, we began to fear for the safety of our younger children.

Throughout all this, we struggled to understand what made Chris tick and what was happening to our family. We modified our approach countless times (including home schooling for four months) in the hope of finding a strategy that worked. We searched, ultimately in vain, to find some nugget—an interest, a passion—on which to build a relationship and to foster his development. He seemed, though, to be stuck at the emotional level of a three-year-old, and we were being drawn, much against our will, into round-the-clock vigilance to protect ourselves and the rest of our family. . . . (Hotovy 1991, p. 541)

Although Chris' adoption was not dissolved, he was placed in a psychiatric hospital with hopes that eventually he could return to his adoptive home. An adopted child's genetic disorders and psychiatric problems can be extremely expensive for adoptive parents to treat. In one reported case, a family reached its $1 million limit in health insurance coverage and spent another $100,000 of its own (Woo 1992a).

Adoption professionals point out that parents are willing to adopt all kinds of children, so long as they know what they are getting into. Although adoption agency negligence is probably not widespread, some state courts have allowed parents to sue agencies for withholding information about children. Judges had traditionally disallowed suits over children with

etic defects or psychiatric problems on the grounds
.at biological parents have no such recourse when
hey have a disabled child. Some critics hold that parents who really wanted a perfect infant are just disappointed with reality. But adoptive parents argue that
agencies are expected to match children's backgrounds
with couples who know how to help them (Woo
1992a).

Any adoption entails both responsibility and risk.
Prospective adoptive parents need to think carefully
about what they can put up with. Prospective parents
who imagine themselves accepting the child as she or
he is need to know, in as much detail as possible, what
that really means in daily life. Many states have passed
medical disclosure laws, which make it easier to obtain
accurate information about a child. Agencies are attempting to gather more data, are asking biological
mothers about drug usage, heavy drinking, and blood
transfusions. Meanwhile, prospective adoptive parents
have a responsibility to question case workers thoroughly about a child, perhaps consulting genetic
counselors for testing, information, and advice.

In Sum

Today, individuals have more choice than ever
about whether, when, and how many children to have. Although parenthood has become more of an option, there is no evidence
of an embracement of childlessness. The majority of
Americans continue to value parenthood, believe that
childbearing should accompany marriage, and feel social pressure to have children. Only a very small percentage view childlessness as an advantage, regard the
decision not to have children as positive, believe that
the ideal family is one without children, or expect to
be childless by choice.

Nevertheless, it is likely that changing values concerning parenthood, the weakening of social norms
prescribing marriage and parenthood, a wider range of
alternatives for women, the desire to postpone marriage and childbearing, and the availability of modern
contraceptives and legal abortion will eventually result
in a higher proportion of Americans remaining childless. In fact, some observers have begun to worry that
American society may be drifting into a period of
structural antinatalism.

Children can add a fulfilling and highly rewarding
experience to people's lives, but they also impose complications and stresses, both financial and emotional.
Couples today are faced with options other than the
traditional family of two or more children: remaining
childless, postponing parenthood until they are ready,
and having only one child. Often, people's decisions
concerning having a family are made by default.

Birthrates are declining for married women. Especially white women are waiting longer to have their
first child and are having, on the average, two children. Although pregnancy outside of marriage has increased, many unmarried pregnant women choose
abortion.

Methods of contraception are discussed in Appendix D. The anatomy and physiology of pregnancy and
childbirth are presented in Appendix C. For couples
who have difficulty in conceiving, reproductive technologies offer both hope and anxiety (Appendix E).
These technologies include artificial insemination, in
vitro fertilization, embryo transfer, and use of a surrogate mother. There are many social and ethical issues
surrounding these procedures. Adoption is a way of
becoming a parent without conceiving; some families
have both adopted and biological children.

Key Terms

abortion
attachment disorder
closed adoption
crude birthrate
disrupted adoptions
dissolved adoptions
fecundity
fertility
involuntary infertility
open adoption
opportunity costs

paradoxical pregnancy
private adoptions
programmatic postponers
pronatalist bias
public adoptions
selective reduction
semi-open adoptions
structural antinatalism
subfecundity
total fertility rate

Study Questions

1. Discuss reasons why there aren't as many large families as there used to be.

2. How is a pronatalist bias shown in our society? What
effect does such a bias have on people who are trying
to decide whether to have children? Are there antinatalist pressures in our society?

3. Discuss the advantages and disadvantages of having
children. Which do you think are the strongest reasons for having children? Which do you think are the
strongest reasons for not having children?

4. What do you see as the major differences among couples choosing to have children soon after marriage, those choosing to remain childless, those postponing parenthood, and those choosing to have only one child?

5. What can cause involuntary infertility? What might account for the recent attention to infertility?

6. What are some important issues surrounding abortion? Do the data on who obtains abortions and on public attitudes on abortion add anything important to the debate?

7. What are some important issues surrounding adoption? How do societywide issues influence personal decisions about adoption? Give examples.

8. How might infertility technologies affect adoption decisions? How might situations regarding adoption affect decisions about whether to begin or continue infertility treatments?

Suggested Readings

American Psychological Association. Public Interest Directorate. 1987. *Research Review: The Psychological Sequelae of Abortion.* Washington, DC: American Psychological Association. Psychology addresses an important question: What are the emotional and psychological outcomes of abortion?

Bartholet, Elizabeth. 1993. *Adoption and the Politics of Parenting.* Boston: Houghton Mifflin. Interesting combination of personal experience, social policy, and academic reflection on parenthood. Bartholet, a Harvard law professor, is both a biological and adoptive parent. In her book she presents data on adoption and discusses its changing social context. She describes her personal experiences with international adoption. She discusses the possibilities that exist for adoption today and what is required of prospective adoptive parents. She proposes policy and legal changes. Finally, she reflects on the differences and similarities of adoptive and biological parenthood.

Becker, Gay, M.D. 1990. *Healing the Infertile Family: Strengthening Your Relationship in the Search for Parenthood.* New York: Bantam. Discusses relationship issues and provides help regarding infertility in couples.

Edwards, John N. 1991. "New Conceptions: Biosocial Innovations and the Family." *Journal of Marriage and the Family* 53(2):349–60. A review of infertility technologies, followed by a theoretical discussion of their profound implications for how we think about and define the family.

Havemen, Robert H. and Barbara Wolfe. 1994. *Succeeding Generations: On the Effects of Investments in Children.* New York: Russell Sage Foundation. A policy book that pleads for more economic, time, and emotional investment in U.S. children today—from parents, the community, the workplace, and the larger society.

Hayes, Cheryl D., ed. 1987. *Risking the Future: Adolescent Sexuality, Pregnancy, and Childbearing,* vol. I. Washington, DC: National Academy Press. National Academy of Science's review of virtually all research on teen pregnancy and parenting; excellent comprehensive overview of this issue.

Kral, Ron and Judith Schaffer. 1988. "Treating the Adoptive Family." Pp. 185–203 in *Variant Family Forms,* edited by Catherine S. Chilman, Elan W. Nunnally, and Fred M. Cox. Newbury Park, CA: Sage. Summary of clinical issues regarding adoptive families.

Leach, Penelope. 1994. *Children First.* New York: Knopf. A critical look at how the United States treats its children today with pleas and suggestions for policy change in what Leach sees as an essentially antinatalist society.

Luker, Kristin. 1975. *Taking Chances: Abortion and the Decision Not to Contracept.* Berkeley and Los Angeles: University of California Press. Classic research study based on interviews with women obtaining abortions, but really about how contraceptive use is influenced by relationship pressures, desire for spontaneity and romance, and attitudes toward self and sexuality.

———. 1984. *Abortion and the Politics of Motherhood.* Berkeley: University of California Press. Interesting study of pro-life and pro-choice activists and their social worlds; tells us a lot about women's roles; good history of abortion in America.

Martin, April. 1993. *The Lesbian and Gay Parenting Handbook: Creating and Raising Our Families.* New York: HarperPerennial. Just what the title says, this handbook is full of concrete advice both for lesbians and gay men, both on the options for becoming parents and also on raising children.

Ragone, Helena. 1994. *Surrogate Motherhood: Conception in the Heart.* Oxford: Westview Press. An ethnographic study of surrogate mothers that investigates the political economy of surrogate motherhood.

Rothman, Barbara Katz. 1989. *Recreating Motherhood: Ideology and Technology in a Patriarchal Society.* New York: Norton. Critical discussion of reproductive technology and its implications by a sociologist who has written on many aspects of conception, pregnancy, and birth.

Schaffer, Judith and Ron Kral. 1988. "Adoptive Families." Pp. 165–84 in *Variant Family Forms,* edited by Catherine S. Chilman, Elan W. Nunnally, and Fred M. Cox. Newbury Park, CA: Sage. Review of research.

Shostak, Arthur and Gary McLouth. 1984. *Men and Abortion: Lessons, Losses, and Love.* New York: Praeger. Essays and research reports on men's attitudes toward, feelings about, and roles in abortion.

Tribe, Lawrence H. 1990. *Abortion: The Clash of Absolutes.* New York: Norton. Excellent overview of the abortion issue: history, law, politics, social factors, and moral arguments. Tribe, a noted scholar of constitutional law, is pro-choice. The book is essential reading for *anyone* interested in the abortion issue.

Walter, Carolyn Ambler. 1986. *The Timing of Motherhood.* Lexington, MA: Heath. A readable book, based on scholarly research, investigating the effect of timing on motherhood. The book looks at women's perception of the mother role, satisfaction with parenthood, balancing the dual roles of motherhood and employment, and "energy of youth versus perspective of maturity."

Zelizer, Viviana A. 1985. *Pricing the Priceless Child: The Changing Social Value of Children.* New York: Basic Books. Interesting historical analysis of the place of children in parents' lives and in American society.

Parents and Children Over the Life Course

It's difficult to have faith in the miracle of growth when so much is at stake.

EDA LE SHAN

For most of human history, adults reared children simply by living with them. From a very early age children shared the everyday world of adults, working beside them, dressing like them, sleeping near them. The concept of childhood as different from adulthood did not emerge until about the seventeenth century (Ariès 1962). Children were increasingly assigned the role of student and were gradually drawn away or segregated from the adult world. Although children may continue to do household chores, especially in rural areas, the move to school has been a move away from participation in the everyday lives of adults. One result is that people today regard children as people who need special training, guidance, and care.

But at the same time, our society does not offer parents or stepparents much psychological or social support. Our society is often indifferent to the needs of parents, including the economic support of families with children. The rate of child poverty now exceeds that of the nation as a whole. And the mood of the country and politicians seems not to favor social programs for children (Samuelson 1995; Sandalow 1995).

In this chapter we will discuss a limited range of parenting issues. We'll begin by looking at some difficulties of parenting. Next we'll examine the roles of mothers and fathers, and then we'll see that parenting takes place in social contexts that vary with regard to parents' marital status, social class, racial/ethnic background, and sexual orientation. (The special concerns of divorced single parents are discussed in Chapter 15; those of stepfamilies are considered in Chapter 16.) We will then look at parenting over the life course. We'll discuss some common parenting issues as we go along and consider how parents can make relationships with their children more satisfying.

Parents in Modern America

Although raising children can be a joyful and fulfilling enterprise, parenting today takes place in a social context that can make child rearing an enormously difficult task.

We would not want to point out the difficulties of today's parents without first noting some advantages. Because health conditions are better, parents today are less likely to have to cope with serious childhood illnesses (Bianchi 1990). Parents now have higher levels of education and are likely to have had some exposure to formal knowledge about child development and child-rearing techniques. Many fathers are more involved than earlier in the twentieth century. Many families are smaller, and parents' tasks may be lessened in this respect.

Nevertheless, the family ecology theoretical perspective (Chapter 2) leads us to point to ways that the larger environment makes parenting especially difficult today. Here we list eight features of the social context of child rearing that can make modern parenting difficult:

1. In our society the parenting (or stepparenting) role typically conflicts with the working role, and employers place work demands first. Hours for full-time work seldom coincide with hours when children are in school, so that an employed parent often leaves home in the morning before children leave for school and returns after the children do. Taking time off from work to nurse a sick child at home or keeping in touch by phone with children after school hours may lead to negative job evaluations.

   The average employed person is now on the job an additional 163 hours a year more than she or he was two decades ago, or the equivalent of an extra month a year. Some executives and professionals find work obligations stretching to eighty or more hours weekly (Schor 1991). And more and more parents are taking second jobs to keep up with inflation. (Chapter 13 presents work–family conflicts in more detail.)

2. Parenting today requires learning attitudes and techniques that are different from those of the past. There is greater emphasis on using positive communication techniques, for example, with corporal punishment criticized. Moreover, today's parents are probably judged by higher standards than in the past. Having children's teeth straightened, for instance, or producing offspring who can succeed in college is an expectation rather than an option.

3. Today's parents rear their children in a pluralistic society, characterized by diverse and conflicting values. Parents are only one of several influences on children. Others are schools, peers, television, movies, music, rented videos, books, the Internet, travel, and, yes, drug dealers and guns. Even though parents want to pass their own values on, children may reject these for other, quite different ones. Not all parents have the same values, and many communities witness struggles over sex education programs in the schools and the availability of rap or heavy metal music in the record stores.

4. The emphasis on the malleability of children tends to make parents feel anxious and guilty about their performance. Psychologists have publicized the fact that parents influence their children's IQs and self-esteem. Parents may lose confidence, believing that they will make mistakes that have dire consequences.

5. Child-rearing experts sometimes disagree among themselves. Over the years they have shifted their emphasis from one child-rearing goal to another, from one best technique to another. Today there is some consensus among university-based child psychologists concerning disciplinary techniques. But given that values shape ideas about what children should be like as young adults, or what parents might do to get them there, psychologists and parents may have conflicting perspectives.

6. Parents today are given full responsibility for raising successful or good children, but their authority is often put to question. The state may intervene in parental decisions about schooling, discipline and punishment, and medical care (LeMasters and DeFrain 1989).

   As an example, about 600 children die annually in auto accidents due to not being properly strapped into safety seats or not wearing seat belts; some states have begun to prosecute parents for disobeying these laws (Smolowe 1991).

7. The fact that people live longer today than in the past means that parents, especially mothers, are responsible not only for working outside the home and child care but also for the needs of aging

parents. Middle-aged parents in this **sandwich generation** may find themselves especially pressed as role conflicts arise between parenting and elder care. Caring for aging parents is further addressed in Chapters 13 and 14.

8. Many families are composed of combinations other than the idealized nuclear family model. Divorced parents and stepfamilies must cope with divergent norms as children go back and forth between homes or are now living with stepsiblings raised with different rules. There is likely to be lingering emotional distress to cope with (Wallerstein and Blakeslee 1989), and perhaps continued legal struggles as well. Unwed parents face the problem of working out the level of involvement of the other parent, of explaining to children why they are not a marital family, and they may also have ongoing court battles. Gay male and lesbian parents face the questions of schools, neighbors, and their children's playmates, perhaps legal challenges, and the need to work out their position as child rearers with respect to sexuality and gender role issues. And despite growing equality and similarity of the sexes in daily life, our society combines an ambivalent commitment to equality with a continuing sense that the mother's role and father's role are different.

Mothers and Fathers: Images and Reality

If we look at men and women as parents, we find a range of ideas and practices.

Old and New Images of Parents

Our cultural tradition stipulates that mothers assume primary responsibility for child rearing. In the United States the mother is expected to be the child's primary *psychological parent,* assuming the major emotional responsibility for the safety and upbringing of her children (Tavris 1992). The "enduring image of motherhood" includes the ideas that

> a woman's identity is tenuous and trivial without motherhood. A woman enjoys and intuitively knows what to do for her child; she cares for her child without ambivalence or awkwardness. Motherhood is a constant and exclusive responsibility. A mother is all-giving and all-powerful. Within the "magic circle" of mother and child, the mother de-

votes herself to her child's needs and holds her child's fate in her hands. (Thompson and Walker 1991, p. 91)

Fathers were once thought to be mainly providers, or breadwinners, neither competent nor desirous of nurturing children on a day-to-day basis (Atkinson and Blackwelder 1993), but see Case Study 12.1 for "A Full-Time Father's Story." Moreover, racial/ethnic stereotyping continues to give us an exaggerated image of African Americans and Latinos as parents: black matriarchs; aloof or absent black fathers; macho, authoritarian Latino fathers (Gibbs 1993; Case 1994b). But

> the images are myths: black couples share childrearing no less, and perhaps more, than white couples, and black husbands are as intimately involved with their children as white husbands, although it is more difficult for them to provide for and protect their children. (Thompson and Walker 1991, p. 91)

There are also emerging images of fathers and mothers:

> "New" fathers are intimately, actively involved with their children; they are responsible and care for their children day-by-day [sharing child-care responsibilities with working mothers]. Both "new" parents can have it all, including a sense of fairness, respect, and cooperation with one another. (Thompson and Walker 1991, p. 91)

Sharing Parenthood

Diane Ehrensaft (1990) searched for these new parents to learn more about them. She interviewed and observed sharing parents in the San Francisco Bay Area and some other large urban centers or academic communities. She defined **shared parenting** not in terms of time, but rather as an identity. (Still, the time split between mothers and fathers in her sample was seldom 50–50, although it did not exceed 65–35.) Her central question was, "Were both father and mother **primary parents**—a couple "mothering together"—rather than one parent and one helper?

The sample of forty shared-parenting couples differed in their timing of parenthood. Some parents, then in their mid-30s to 40s, had had children ten or more years earlier—in the early 1970s, when they were in their 20s or early 30s. The other group of par-

The enduring image of motherhood is of mother and child as defined by extensions of one another. According to this image, women intuitively know how to nurture their children and willingly devote themselves to mothering.

ents were over 35 and had postponed having children until recently. Their children were typically under 5.

In searching for couples who share parenthood, Ehrensaft encountered many near misses—couples who had decided to share but did not succeed at it. Those couples who chose shared parenting and were able to do it were not distinguished by the wife's work expectations or previous experience with shared housekeeping. Three factors may have affected both a father's dedication to everyday parenting and the mother's commitment, too.

First, many men and their wives were strongly influenced by the feminist movement or other 1960s ideologies that kept them going during the more difficult times: adapting to breast-feeding, for example,

which imposes some difference between the parents. Second, many of the fathers were in occupations related to children, such as academic child psychology. Finally, both parents tended to have good job security, so that they could risk the displeasure of their supervisors in giving time to parenthood.

Besides the early challenge of breast-feeding—and Ehrensaft noted that shared-parenting couples did not prolong it—a major difficulty was the lack of confidence of both parties in the father's competence as a parent—not an unrealistic concern given differential socialization. And there were residual differences despite the overall characterization of shared parenting. Women *did* devote more time to parenting even in these couples, and they resented it. There was also a

A Full-Time Father's Story

Bob is 31, comes from a working-class background, and has been married seven years. At the time of this interview, he had just gotten out of the Air Force and was collecting unemployment benefits. His wife is a clerical worker. His future goal, he said, is to be a child psychologist. He has two children—Tim, 6, and a daughter, Nicole, 18 months. His father deserted his family when Bob was a baby, and he feels he suffered greatly as a result. Here he describes what it's like for him to be a temporary househusband and full-time father.

Today I took a bunch of kids from my son's school to the zoo. Kindergartners. You know I like kids (*laughs*).... It was something hectic. The kids, they all want to run six different ways. They were climbing over the fences and through the fences to the elephants....

I think fathers should be more involved with their kids when it is possible. In fact, a couple of mothers commented [at the zoo] that they were happy to see a father along for a change. I never had a father.... I'm determined that my son and I should have time together—which is tough when I'm working.... [But for a while now] I've got the two kids. I fix them lunch and all that. You fix a can of soup, you give her half, you give him half. Hers you don't have any soup in it—just noodles and stuff because she eats with her fingers and you don't want any soup in there. And while they're eating I usually get dressed to take them out. And after they're done eating, I dress them.

I change her diaper and dress her and I dress my kid for school and take him to school. And while the kid's at school, we go somewhere like the park—me and my baby.... There aren't too many dads in the park, I'll tell you. Most of them are out working. I get some strange looks once in a while, but it doesn't bother me. It really doesn't. Besides, if some fox comes by, I'll just tell her my wife and I were divorced and I got the kids, right? (*laughs*).... Then about two thirty, quarter to three, I drive over to the school from the park, and it's time for him to get out. Not the stereotype you'd think, huh?

Sometimes it's a pain, you know. I can't get anything done. And there's a lot of things I could do, but it's really a chore to keep track of the baby. She's at the age now where she wants outside all the time, and it's really a hassle. I mean I'll bring her in the back door, and while you're locking the back door, she goes out the front. She's one door to the other, and she's out again. Yesterday our son came in, yelling, "She's out walking down the sidewalk!" She was having herself a good old time....

I would do it [though] anytime I was off work because, well, I look at the economical aspects. I can't afford to pay for child care when I'm at home. It's much, much cheaper for me to go ahead and do it myself, and it's not really that much of a hassle ... unless you've gotta go to the library or something like that. You can't drag them with you because—I took them to the library to get some books and I about went nuts chasing the little one around the shelves. She's really too young to understand "Now you stay there!" And she smiles and she's gone by the time you can turn around....

I'm going to keep doing this until my twenty-six weeks is up or I find a job. I'm not saying I'm not looking for a job now, but I am saying I'll probably keep doing this all the twenty-six weeks. ... I love those kids to pieces. I can just sit there and just look at them, and just smile inwardly at what they do, you know. I don't know if it detracts from my image or anything....

In what ways might Bob and his wife's arrangement be better than day care, do you think? Not as good as day care? As a parent, what would you do differently from Bob? What do you like about Bob's style of fathering? Do you think Bob is a househusband and full-time father now only because of economics? What else might motivate him?

Fathers who share parenting represent a significant change from the past.

tendency to backslide as the husband's income outpaced the wife's. Finally, mothers, more than fathers, engaged in "worrying" (anticipating and coping with problems) and gave more attention to the "psychological management" of children. They seemed to identify more strongly with children and thus to devote more attention to the child's clothing, birthday parties, and other activities from which a father could distance himself. There was a tendency for mothers to "take over" in these areas.

Fathers tended to have more psychological separation from children, whereas mothers had a communal relationship with them, a sense of shared identification. Men were "intimate" with children, seeking *emotional closeness,* whereas women's relationships with children could be better described as "nurturing" and originating from their identification with a *child's needs.* Interestingly, before their first child was born, both fathers and mothers hoped for daughters. Women wanted daughters with whom they could identify; men wanted daughters to actualize their commitment to feminism.

Despite lingering differences between these fathers and mothers, fathers who share parenting represent a significant change from the past. Occasionally moth-

ers felt squeezed out by an involved father, but by and large mothers felt that their parenting was enhanced. The couples felt that their relationship was strengthened by sharing such an important family activity.

However, Ehrensaft noted some potential strains of shared parenting. Couples had very little time alone. Ehrensaft speculated that this could represent avoidance of the marital relationship, leaving a vacuum that would be difficult to deal with at the empty nest stage of family life.

There were also some important socialization issues. Children tended to switch attachments, and often they were more attached to one parent than to another at any one time. Parents were left with a difficult choice between respecting the child's wishes or imposing their preference for equal parenting.

Ehrensaft was also concerned about the "overparenting" of children, who might have two hovering parents. "Parenting by committee" was time-consuming and sometimes confusing to the child (see, for example, Kutner 1989b). Moreover, consultation often extended to the child, leaving little differentiation between adult and child roles. Parents were somewhat isolated; they found it difficult to engage with other, more traditional parents in their social environments (Ehrensaft 1990).

Mothers and Fathers: What Do They Do?

Unlike these pioneers of shared parenting, most mothers and fathers in our society continue to be somewhat different parents. One study of fathers reported that 74 percent thought they *should* share child rearing equally with mothers, but only 13 percent actually did so (Thomas and Wilcox 1987; see also Hyde and Texidor 1994).

Mothers typically engage in more hands-on parenting and take full responsibility for children, whereas fathers are often viewed as helping. Mothers are more constantly present and available, while fathers come and go. In fact, fathers in married-couple families are not often alone with their children, and the mother serves as a sort of mediator of the parent–child relationship (Thompson and Walker 1991).

From the child's point of view, fathers are "novel, unpredictable, physical, exciting [and] engaging." Not surprisingly, children prefer to play with fathers rather than mothers. Mothers' activities include "attentive love . . . preserving life, fostering growth, and molding an acceptable person" (Thompson and Walker 1991, pp. 91–92). In fact, Thompson and Walker sum up their review of research on parent roles by terming fathers "playmates" and mothers "caregivers and comfort givers." The intense daily contact with children is viewed ambivalently by mothers: as a source of great life satisfaction but also a source of a great deal of frustration and stress.

Fathers view their still-powerful breadwinning role as an important contribution to their children. They are very emotionally invested in their children, even when they are less active in parenting. They tend to compare themselves favorably to their own fathers (who were much less involved with children), even if they are not yet "new" fathers or 50–50 parents. As children grow older, fathers become more involved in instruction and discipline (Thompson and Walker 1991).

Good Dad–Bad Dad

There is yet another side to contemporary fatherhood. Sociologist Frank Furstenberg (1991) notes two aspects of fathers, the "good dad" and "bad dad" images of fathers in family life today. Succeeding the breadwinner father of the industrial era, we now have both the new father and the deadbeat dad, as they are termed in the media.

As women have entered the labor force in greater numbers, men have been encouraged by their family's need and by the redefinition of male roles (see Chapter 3) to want to play a larger part in the day-to-day care of the family. Although few families have realized the shared-parenting pattern described earlier, more and more men are trying. Surveys of male employees at the DuPont Corporation in 1986 and 1991 found sharp increases in the percentage of men who "expressed personal interest" in flexible work options (37 percent versus 56 percent); in leave to care for newborn children (15 percent versus 35 percent); and in leave to care for sick children (40 percent versus 64 percent) ("New Fathers" 1991, p. 13).

At the same time, more and more divorced or unwed fathers are distancing themselves from their children (Blankenhorn 1995)—are not seeing them and not supporting them:

There are two sides to male liberation. As men have escaped from the excessive burdens of the good provider role, they have been freed to participate more fully in the family. They have also been freed from family responsibilities altogether. (Furstenberg 1992, p. 347)

One result is the recent push by some interest groups, such as the National Fatherhood Initiative, for example, to reemphasize the provider role of fathers as valued and necessary (Chira 1994a).

The Parenting Alliance

Recently, researchers have turned their attention from either the separate parenting roles of men and women or the small group of role-sharing parents to the coparenting relationship more generally. Whether or not parents attempt to share parenting, whatever their division of labor is in the home, or whether or not they are even married, the relationship between a mother and father affects their parenting. Even though this relationship is different from their relationship as a couple, the dynamics of their couple relationship spill over into parenting effectiveness.

Gable, Belsky, and Crnic (1992) examined the **parenting alliance** of parents of 2-year-olds. Observing parents, they noted instances of parents' both supporting each other and undercutting the parental acts of the other, often without awareness. Parents can inadvertently or sometimes intentionally support and undermine one another. Moreover, "[M]arital and interparental relations both pre- and postdivorce, rather than the breakup of the family per se, may be more informative in accounting for children's behavior problems and adjustment differences" (p. 286).

The interaction between mothers and fathers as they coparent children should be the subject of much future research.

The Diversity of Parents

We have been looking at some differences between mothers and fathers and how they work together as parents. But parents differ by more than sex, and the social context in which they parent is variable.

Most obvious are differences in social class and racial/ethnic background (to be discussed momentarily), which are associated both with differential resources available to families and children and with cultural differences. Religious commitments of parents can affect their child rearing, especially with respect to

Many relationship dynamics affect parenting. Fathers and mothers not only relate individually to their children and to each other but also relate as coparents to their children, as do the parents of this joyful dancer.

discipline, sex education, or a sense of estrangement or inclusion in the public school milieu.

Parents vary in age. Another way of looking at early versus later parenthood is the variation in age difference between parent and child. Parents may be young and inexperienced (teenagers even), young adults building careers, or middle-aged parents of infants (in which case they will be senior citizens when their children are adolescents).

Parents vary in sexual orientation and marital status. We consider divorced single parents in Chapter 15 and remarried parents in Chapter 16. Here we look at two categories of parents not discussed elsewhere: gay parents and unwed single parents.

Gay Male and Lesbian Parents

Gay men and lesbians can become parents in several different ways. A substantial number of gays (20 percent) and lesbians (33 percent) have been married,

and some have children from those marriages (Harry 1988). About 56 percent of lesbian couples have children living with them (Harry 1983), often from previous marriages. Custody issues involving lesbians have had various resolutions. The courts have increasingly taken the position that the sexual orientation of mothers is not an issue if the children are well cared for and well adjusted. To date, gay men have seldom been awarded custody unless uncontested, and visiting rights have been an issue as well. Nevertheless, many gays are noncustodial parents.

Gays and lesbians have also sought parenthood as single adoptive parents and as birth parents, as one partner gives birth to a baby they both parent. As Urvashi Vaid, staff member of the National Gay and Lesbian Task Force, commented: "There's an enormous interest in parenting as we enter our middle years" (in Dullea 1988a, p. 26). This baby boom includes not only lesbian coparents but an array of combinations of

This gay couple is parenting these two adopted boys. Gay and lesbian parents have sought to legalize their parenting relationships with varying results.

lesbian mothers and biological fathers, surrogate mothers and gay biological fathers (less frequent), and at least one commitment to parenthood of four parties: a gay couple and a lesbian couple. In some cases, one lesbian partner is inseminated with the sperm of a male relative of the other partner, genetically linking the baby to both female parents (Baker 1989; Kolata 1989d; Salholz 1990).

Gay and lesbian parents have tried to legalize their parenting relationships with varying results. Some courts have permitted joint adoption by gay or lesbian parents, whereas others have not permitted a partner to adopt the other's adoptive or biological child. Some states have refused to place children with gay male or lesbian foster parents (Dullea 1988a; Sack 1991; Marks 1992).

Parenting concerns that arise as children grow older tend to center on the child's sexual development and

peer relations as adolescents, and on stability of the couple relationship of the parents (Gross 1991d). Courts have been divided on whether or not a nonbiological or adoptive coparent can retain visiting rights or compete for custody in the event of a breakup (Sack 1991; Marks 1992).

Research finds children of gay male and lesbian parents to be well-adjusted, with no noticeable differences from children of heterosexual parents (Goleman 1992b; Patterson 1992; Flaks et al. 1995). Nor are they more likely to be gay as adults. They may, however, encounter challenges from peers. A child's need to blend in may clash with a political parent's desire for activism on gay rights issues. But concerning parenthood for gays and lesbians, eminent pediatrician T. Berry Brazelton contends that "if the parents really love a child and think about the child's issues rather than their own, there is no reason to shy away from it"

(Gross 1991d, p. A-12). In some large cities, such as San Francisco or Minneapolis, there are now workshops for lesbians considering parenthood, for lesbians ready to get pregnant, for gays and lesbians who want to parent together, and for new gay parents. There are also discussion groups on rearing children and choosing child care, and there are play groups and organized events for children of gay parents.

Never-Married Single Mothers

Never-married single mothers are still a minority (one-third) among female single parents; the remainder are divorced, widowed, or separated. But as the proportion of single mothers who have never married increases (see Figure 12.1), there is more interest in what might set these families apart from the divorced single parents described in Chapter 15.

For one thing, mothers who are not married when they give birth are usually younger, less well-educated, and more likely to be on welfare for an extended period (Lewin 1992). They are often teen mothers, although as the large baby boom cohort has grown older, single motherhood is more frequent among women in their 20s (U.S. Bureau of the Census 1994a, Table 100).

Studies of young, never-married mothers abound, and most indicate multiple disadvantages to both mothers and children (Polakow 1993). The complexity of young, never-married parenthood is illustrated by sociologist Frank Furstenberg's follow-up study of black women who gave birth as teenagers in Baltimore in the mid-1960s. Many overcame this initial setback: Although they were still strongly disadvantaged after five years, eventually many completed high school and established themselves in terms of income and as parents, although their children seem at risk of living less productive lives than they (Furstenberg, Brooks-Gunn, and Morgan 1987).

Still, one researcher has expressed the thought that young, never-married mothers who have family support are as capable of providing attentive and nurturing parenting to their infants as are frantic working parents (Smith 1988). Many others have noted the economic and related difficulties encountered by poor black never-married mothers (Luker 1990; Geronimus 1991).

Teen pregnancy outside marriage is also discussed in Chapter 11. Considering the context of young, never-married parenthood in race and class terms

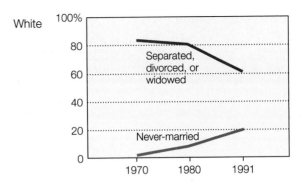

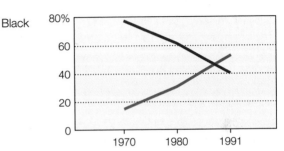

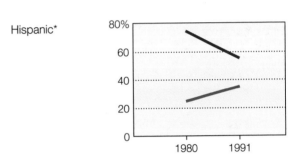

*These data were not compiled for Hispanic women before 1980.

FIGURE *12.1*

Single mothers: the changing mix. (Source: Adapted from Doniger 1992, p. A-16)

leads us into a more direct discussion of the variation in the social context of parenthood.

Social Class and Parenting

Virtually all life experiences are mediated or influenced by social class, and parenting is no exception (Garrett et al. 1994). You'll recall a theme of this text: Decisions are influenced by social conditions that limit or expand a decision maker's options. This

TABLE *12.1*

Inequality in Family Income: Gains and Losses

Income Category	Family Income			Family Income Adjusted for Family Size		
	Share of Total Income		Share of the Average Gain in Income Between '77 and '89	Share of Total Income		Share of the Average Gain in Income Between '77 and '89
	1977	1989		1977	1989	
Highest 20 percent:						
Top 1 percent	7%	12%	70 %	8%	13%	44%
Next 4 percent	11	12	25	12	13	19
Next 5 percent	10	10	10	10	10	8
Next 10 percent	16	15	11	16	15	11
Second-highest 20 percent	23	22	8	22	21	15
Middle 20 percent	16	15	2	15	15	11
Second-lowest 20 percent	12	10	−7	11	10	3
Lowest 20 percent	6	4	−11	6	4	−7
No income or negative income			−8			−4
Total	100%	100%	100 %	100%	100%	100%

I   Sources: Adapted from Nasar 1992b, p. C-5; data from Congressional Budget Office. Sums may not exactly equal 100% due to rounding error.

section examines some ways in which the conditions of social class—of work, income, and education—affect a parent's decisions and options.

The Family in a Changing Economy

Since the mid-1970s the U.S. economy has been "adrift in the doldrums" (Rudolph 1989). This has meant a declining economic foundation for the majority of American families. The rise of a service (as opposed to an industrial) economy spells lower wages for many workers. Although some service jobs—those that involve technological expertise or a fairly high level of education—pay well, many new jobs in data entry and word processing that accompany the information age are low-paying, with few benefits and limited prospects for advancement. The combination of inflation, recession, unemployment, and structural transformation of the economy (Eitzen and Baca Zinn 1992; Samuelson 1992) over the past two

decades put an end to the era of the postwar boom.

Analysis of the past decade's incomes reveals another feature affecting families' sense of well-being—increased inequality. The wealthy gained during the 1980s while the less well-off became even more so (Karoly and Burtless 1995). Seventy percent of the total rise in family income between 1977 and 1989 went to the top 1 percent of families. Even taking into account that family incomes may support different-sized families, the wealthiest fifth gained at the expense of the poorest fifth (Nasar 1992b; see Table 12.1 and item 15 in Box 1.1).

As Figure 12.2 shows, minority families are less well-off economically when family incomes are compared. They are even less relatively well-off when assets (such as a home and savings and investments) are considered. African American families have only 10 percent of the amount of assets that white families have,

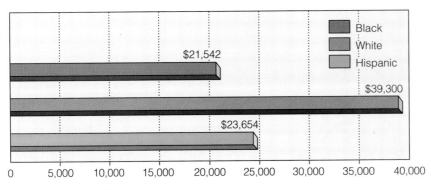

Income (1992 dollars)

FIGURE **12.2**

Median family income for black, Hispanic, and white families in 1993. Although middle-class black incomes have improved over the past twenty years, the economic situation of lower-class blacks has not. During this period the proportion of black families headed by women increased while the proportion of black men with jobs fell. (Source: U.S. Bureau of the Census 1995, Table 735)

and Latino families have a similar level of assets (Eargle 1990, p. 8). The decline in wages and the rise in unemployment have hurt minority families to an even greater extent than they have the white majority. Women—along with the disabled, some of the retired, and nonproperty owners in general (Aldous, Marsh, and Trees 1985)—have also been hit especially hard, particularly those who had recently entered the job market and fell into the "last hired, first fired" category.

After declining to 4.9 percent in March 1989 (Hershey 1989), unemployment rose again to 7.8 percent in June 1992 (Greenhouse 1992). The nineties have been characterized by a long-lasting recession, which has severely affected both white-collar and blue-collar jobs. High- and mid-level managers (those employees with technical expertise and college educations) and factory workers alike have found it difficult to find and hold jobs. One in five college graduates in the 1980s had jobs for which their degrees were not required; that is, they were underemployed (Newman 1988; Hinds 1992; Nasar 1992c).

The decline in the U.S. economy has had a strong effect on young families. Young workers have low seniority, often low educational attainment, and "many of them are competing with low-wage workers around the world for jobs" ("Behind Jobless Figures" 1991, p. A-11). Further,

job losses among young workers, often the parents of America's youngest and most vulnerable children, are creating a devastating cycle of declining earnings, declining family incomes and rising child poverty. (Marian Wright Edelman, president of the Children's Defense Fund, in "Behind Jobless Figures" 1991, p. A-11)

Poverty-Level Parents

As Figure 12.3 shows, the poverty rate was close to 25 percent in 1959 but dropped consistently during the 1970s to a low of 11 percent. It began to rise in the late 1970s and reached 15 percent in 1993. If we include those "near poor" who are just above the official poverty line, the figure rises to nearly 20 percent (U.S. Bureau of the Census 1995, Table 744; see items 15 and 16 in Box 1.1). Children make up 40 percent of the poor; their poverty rate (at 23 percent in 1993) is above that of the nation as a whole (U.S. Bureau of the Census 1995, Table 747).

Families maintained by women and minority families have even higher rates of poverty. Close to half (47 percent) of African American children and 41 percent of Hispanic children are living in poverty (U.S. Bureau of the Census 1995, Table 747). Yet even though we may think in terms of the stereotypical black female-headed family on welfare in an urban setting, even more poor children are white, Asian, or Native

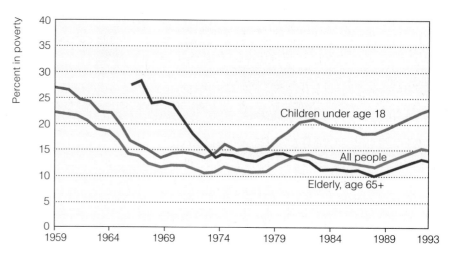

FIGURE **12.3**

| U.S. poverty rates, 1959–1993. (Source: U.S. Bureau of the Census 1995, Table 747)

American; have at least one working parent; and live outside the central city (Johnson et al. 1991). Despite higher *rates* of minority poverty, in sheer numbers white families predominate. The latter make up 65 percent of poor families, and rates of poverty are as high in predominately white rural areas as in the inner cities (Kilborn 1992b). Journalists (Whitman and Friedman 1994) and others have begun to talk about "the white underclass." But the poorest county in the United States is a South Dakota county covering most of the Pine Ridge Indian reservation (Kilborn 1992d), calling our attention to the high poverty and unemployment rates of Native Americans.

Over the past decade a shortage of affordable housing has helped create a significant and visible number of homeless families—a phenomenon that twenty years ago would have been unthinkable (see Figure 12.4). Thirty to fifty percent of the homeless today are mothers and children. It is difficult to imagine raising children while living on the street. For example, carefully monitoring children's homework becomes an impossible expectation for a parent whose concern is finding the children's next meal or keeping them warm.

Furthermore, some 5.5 million children age 12 or under are hungry or in families that experience food shortages, which lead to weight loss, fatigue, irritability, and headaches, all of which affect educational per-

formance (Chira 1991; Pear 1991). Poor children may drop out of school to try to help their families; this is one explanation for the especially high dropout rate of Latino youth (Celis 1992; Huston, McLoyd, and Coll 1994).

But the majority of poor parents have homes and are employed, if sporadically. These "working poor" have minimum- or less-than-minimum-wage jobs with irregular and unpredictable hours and no medical insurance or other benefits. They may not have completed high school and are more likely than other Americans to be functionally illiterate—unable to read or write well enough to complete a job application or fill out other forms. Many move often from city to city in search of milder weather (because it is more difficult to be poor in cold climates) or access to jobs. This makes it difficult for a parent to establish support systems and hinders children's chances for school continuity and success (Moylneux 1995).

Assuredly, raising children in poverty is qualitatively different from doing so otherwise. Items that other Americans take for granted (relatively safe neighborhoods, adequate housing, heat, air conditioning, eyeglasses, dental work, music lessons) are simply not available. Poverty-level parents (and their children) have poorer nutrition, more illnesses, schools that are less safe, and limited access to quality medical care. A parent's not feeling well must make

Blue-Collar Parents

Blue-collar, or working-class, families usually take their designation from the husband-father (or stepfather) who is employed in blue-collar work: factory production, maintenance, construction, truck driving, or appliance repair, for example. Although blue-collar parents typically have only high school educations, their incomes often exceeded those of white-collar Americans in the past. Protected by powerful trade unions, blue-collar parents enjoyed benefits such as seniority, pensions, paid vacations, and fully covered medical care.

But our economy has become increasingly a service economy over the past several decades, and manufacturing has declined, with many companies moving their work overseas. Blue-collar parents have declined both in number and in resources. Layoffs have become a major concern, along with wage rollbacks and loss of benefits. Furthermore, unlike in the past, today's blue-collar children may not have access to their parents' occupations. Many will move into white-collar jobs; others will take relatively low-paying service jobs. Hence blue-collar parents face problems associated with guiding children into occupational and educational worlds (college, for example, with the unfamiliar language of majors, minors, and credit hours) that they have neither experienced nor fully understand. LeMasters and DeFrain see blue-collar parents as an emerging minority group: "When all of the social leaders are white collar, when the mass-media idols are white collar, how does a blue-collar father or mother portray his or her way of life as something for a child to emulate or admire?" (1989, p. 108).

Blue-collar parents (and lower middle-class parents)[1] are more likely than upper middle-class parents to be strict disciplinarians, expecting conformity, obedience, neatness, and good manners from their children (Kohn 1977; Luster, Rhoades, and Haas

REMEMBER THOSE HUNGRY KIDS IN CHINA? NOW THEY'RE IN OMAHA.

Currently, an estimated 5.5 million American kids don't regularly get enough to eat. Still for most people, the idea of hungry children is a foreign one. To join our campaign to Leave No Child Behind, call 1-800-CDF-1200.
Kids can't vote. But you can. **THE CHILDREN'S DEFENSE FUND**

Poverty translates into hunger and, increasingly, homelessness. Poor childhood nutrition leads to health problems, which, in turn, affect educational potential.

raising children in poverty conditions even more difficult than it would be otherwise. Then, too, poverty-level parents do not have access to the legal services of those in higher social classes. Upper middle-class parents, for example, can hire the family attorney to represent a teenager caught for drug possession or speeding. Poverty-level parents must rely on public defenders and the workings of a legal system that they may not understand. This is partly why poorer children are not only more apt to be arrested for minor offenses but also to receive more severe punishments (LeMasters and DeFrain 1989, pp. 101–2).

1. The line between blue-collar and lower levels of white-collar employment (lower middle-class) blurred in the post–World War II era of rising blue-collar incomes. Also blue-collar men (factory workers, for example), frequently marry lower middle-class women (secretaries, for example).

Child development advice literature became widely dispersed across classes through the popular media during this period. Despite a convergence of knowledge about child rearing between white-collar and blue-collar parents, the economically insecure position of blue-collar and lower middle-class parents and their differences in education still leave a gap in parenting values and behavior between these strata and the upper middle-class.

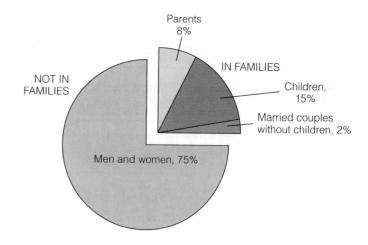

FIGURE *12.4*

U.S. homeless population by family status in the 1980s. Most of those designated "parents" are mothers. (Source: Burt 1992 in Ahlburg and De Vita 1992, p. 35)

1989). Middle-class parents, in contrast, tend to emphasize a child's happiness, creativity, ambition, achievement, and independence. They are less restrictive and more "affectionate and responsive" (Belsky 1991, p. 122). But employed wives of blue-collar husbands often have white-collar office or service jobs. Hence, they are exposed to middle-class values. This can make for what LeMasters and DeFrain call a "value stretch" between blue-collar men and their mates: Husbands may value a more restrictive parenting style than do their wives.

Lower Middle-Class Parents

The lower middle-class includes grocery checkers, retail clerks and other lower-level salespeople, small-business owners and managers, secretaries, receptionists and word processors, government clerks, post office workers, and the like. They earn moderate wages; benefits and job security depend on the employer. Today employers are cutting back, so most workers at this level have fewer benefits than in the past and little job security.

Even in two-paycheck marriages, lower middle-class parents often face time and money problems. Single mothers in this social class may be working at full-time *and* part-time jobs in order to meet everyday expenses. LeMasters and DeFrain (1989) suggest that

problems for these parents are worsened by their aspiring to an upper middle-class standard of living.

Besides being ambitious for themselves, lower middle-class parents are typically ambitious for their children, encouraging them to go to college even though they themselves usually are not college graduates. This means that their children may eventually adopt different values and behaviors—from how to spend leisure time to political or religious values—and this can create strain between a family's generations.

Upper Middle-Class Parents

Upper middle-class parents do not have considerable family wealth but earn high salaries as corporate executives and professionals: physicians or attorneys, for example. Other professionals, such as college professors, accountants, engineers, architects, and psychotherapists, may not earn quite so much money but have a comfortable income that supports an upper middle-class lifestyle.[2] These parents and stepparents have more options than do those in lower social classes. They can choose the neighborhood in which

2. The "minor professionals"—teachers, nurses and allied health professionals, social workers and human services personnel, technical specialists, and middle managers—tend to identify with the upper middle-class, although some would classify them as lower middle-class. Perhaps a middle middle-class designation would be most appropriate.

they want to live and rear their children. They can afford to send their offspring to college and sometimes to private schools. On a different level (but one that points up the myriad of daily-life advantages), upper middle-class parents can hire household help or purchase cars for their adolescent children so that sharing cars is unnecessary.

In general, upper middle-class parents have the material and educational resources to prepare their children for occupational success. Despite this, however, they cannot positively ensure that each child will eventually enjoy a position in the upper middle class. An executive's child will more than likely have to finish college and maybe even attain a higher degree in order to match the parent's occupational status. A concern of parents in the upper middle class, then, is that their adult children not have to move down in social class. Related to this is the problem of how to teach a child raised in affluence to live in less luxurious conditions should she or he be unable to afford luxury in adulthood (LeMasters and DeFrain 1989). The current generation of college graduates is facing both a job shortage and lower pay and living conditions unknown to their parents (Hinds 1992).

Upper-Class Parents

The wealthiest 20 percent of U.S. households earn about 47 percent of the nation's total income and hold 44 percent of the total wealth (U.S. Bureau of the Census 1991a, pp. 17, 20). Upper-class parents not only have disproportionately high shares of wealth and income, but often their families have been established as wealthy for several generations and possess significant power and influence. Wives of the elite class do not usually work outside the home, although they may commit large amounts of time to social activities and volunteer work.

Social scientists have little access to the upper class. Consequently, relatively little is known about them for certain. One thing we do know is that their unique challenge is motivating children who are guaranteed a comfortable living no matter what—and whose accomplishments will likely never surpass those of their forebears (Bedard 1992).

This section has described various parenting issues that are peculiar to the different social classes in the United States, regardless of ethnicity. The next section addresses racial/ethnic diversity with regard to parenting issues.

Racial/Ethnic Diversity and Parenting

There is considerable overlap among class and racial/ethnic categories. The upper class is almost entirely white. The upper middle class, still overwhelmingly white, now includes substantial numbers of people of color, particularly Asians. Other Asians, especially Southeast Asians, remain in lower social classes.

Many African Americans are now solidly middle class as a result of greater educational opportunities and the opening of employment, particularly public employment, to blacks following the civil rights movement and subsequent affirmative action policies. Middle-class black families are more likely to have two earners. The African American middle class remains vulnerable to discrimination in employment and housing as well as to economic dislocation (Blauner 1992) and has the special burden of exposure to everyday slights (McClain 1992). Health researchers have speculated that the workaday stress resulting from subtle (or not so subtle) discrimination accounts for some of the difference in heart disease rates among whites and blacks (Goleman 1990b; Leary 1991).

African Americans, Native Americans, Latinos, and some Asians are heavily represented in the working class. And there is an overrepresentation of people of color, particularly female-headed families, in the poverty ranks. Even within the broad "poverty" category, blacks are much more likely to live in neighborhoods of "extreme poverty," those where two-fifths of the population is poor (Passell 1991). Yet a majority of African Americans are now members of the working or middle class (Bray 1992).

Race and ethnicity affect parents' options and decisions. When an ethnic group is disproportionately poor (or wealthy), parenting differences due to ethnicity become confused with those due to social class. For example, because African Americans have been disproportionately poor, we have tended to think of "the black family" as one in which poverty-level mothers are always teenagers and fathers take little responsibility for their children. As more blacks move into the working class and middle class, however, it becomes increasingly apparent that to say "a black family is a black family is a black family" is inappropriate (Willie 1986, p. 231).

African American Parents and Children

Evidence suggests that African American (as well as Hispanic and Asian American) parents' attitudes, behaviors, and hopes for their children are similar to those of other parents in their social class. (Julian, McKenry, and McKelvey 1994). After reviewing the literature, sociologist John McAdoo concluded that "when economic sufficiency rises within Black families, an increase in the active participation of the Black father in the socialization of his children is observed" (1988, p. 266). Studies indicate that when we control for class, the child-rearing style of African American parents compares to that of whites (Taylor et al. 1991).

In research exploring the topic of parenthood among African Americans, 136 black men were approached and questioned in barber shops, churches, shopping centers, parks, and so on in southern California. The vast majority (77 percent) were married. Eighty-two percent of these fathers mentioned education and obtaining "the good life" as major goals for their offspring. Over 90 percent said they regularly spent positive leisure time with their youngsters. Overall, they felt they were actively involved with their families. Interestingly, however, these men tended to perceive *other* black fathers as lacking in parental responsibility. According to the author of this study,

> one possible explanation for this "I'm okay, you're not" attitude is that perhaps Black men have accepted the stereotype put forth by social scientists in the past. This is most alarming. . . . Black men must begin to understand and question such notions. (Connor 1986, p. 64)

Actually, "the affluent black family is ingenious in its inner direction as it strives to overcome, and to cultivate those personal characteristics that will demonstrate qualification to be part of the system" (Willie 1981, pp. 178–79).

Although minority parents resemble other parents in their social class, the impact of race or ethnicity remains important. Parents of color face additional challenges. For example, the status of middle-class black parents does not suffice to protect them from intermittent demeaning or suspicious behavior on the part of whites, and such incidents may occur in front of children or be directed at them (McClain 1992). Even so simple a matter as buying toys becomes problem-

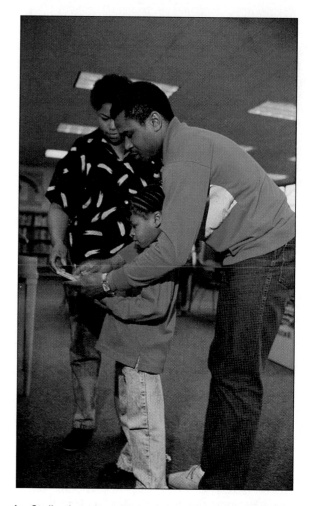

Studies that take social class into account show that black and white parenting styles are similar. However, this African American couple face concerns that white parents don't have to think about: They must balance protecting their daughter from racism against preparing her to challenge it.

atic. Black dolls only? Should the child choose? What if the choice is a white Barbie doll?

Native American Parents and Children

Native American teens have a suicide rate four times that of other teens, and many report "extreme hopelessness" or sadness and drinking problems. According to University of Minnesota researcher Michael Resnick, "Native American youths have a familiarity and intimacy with death and loss within the family comparable to few other young people in our society"

("Young Indians" 1992, p. D-24). These psychological stresses are associated with the high unemployment and poverty levels prevailing in this group, particularly on the reservation ("American Indians" 1992).

Mexican American Parents and Children

Mexican American parents cope with a generational gap that can include differential fluency and different attitudes toward speaking Spanish (Anti-Defamation League of B'nai B'rith 1981). As in other bicultural families, conflicts can extend into many matters of everyday life:

> We could never get stuff like pizza at home, just Mexican foods [complains a 15-year-old Mexican teen]. . . . My mother would give me these silly dresses to wear to school, not jeans. No jewelry. No make-up. And they'd always say, "Stick with the Mexican kids. Don't talk to the Anglos. They'll boss you." (Suro 1992, p. A-11)

Some think the lower educational levels of Hispanics (compared to blacks, for example; Valdivieso and Davis 1988, pp. 6–7) result from a comparison of Mexican American achievements and standard of living in the United States to past conditions in Mexico. Current well-being and even a partial high school education seem so superior that young Mexican Americans may not be inclined to persevere through high school and into college (Suro 1992). And yet much has changed, as attachment to and involvement with the extended family and family rituals have declined under the pressure of an urban, bureaucratic work life (Williams 1990).

Asian American Parents and Children

Asian American children have achieved above-average education levels in the United States and have done relatively well in the professions, so they are thought to have few problems. Nevertheless, like other ethnic minorities, they have suffered from discrimination. Current data indicate that college-educated Asian American men earn lower salaries than white men with comparable education (Barringer 1992b). During World War II Japanese Americans were confined to concentration camps and suffered the confiscation of their property. Anti-Asian vandalism, harassment, and intimidation persist, according to a 1986 report by the Commission on Civil Rights (in LeMasters and DeFrain 1989, p. 142).

Moreover, contemporary Asian American youths must contend with the high expectations created by the stereotyping of Asians as a "super-minority." Some fall victim to the strong pressures for educational and occupational success, experiencing distress that sometimes leads to suicide (Rigdon 1991a).

Raising Minority Children in a Racist and Discriminatory Society

For minority parents who are poor, discrimination compounds the circumstances of poverty. Poverty-level minority parents find themselves confined to reservations (in the case of Native Americans) or urban ghettos, characterized by danger, high rates of mental illness and crime, illegal drugs, and general lack of opportunity. Some inner-city schools find themselves working with zero-parent families, with perhaps more than half of enrolled children living with neither mother or father, or with a parent who is nonfunctional due to drugs, AIDS, or some other damaging condition. These children live with foster parents, with devoted but aging and exhausted grandparents, or with other relatives and friends, and they change residence frequently. A 1990 survey for the Annie E. Casey Center for the Study of Society estimates that close to 10 percent of American children live with someone other than a parent; the U.S. Census Bureau estimates that 4.5 percent are "zero-parent children" (in Gross 1992).

Whatever its social class, the minority family—whether black, Native American, Hispanic, Asian American, or multiracial—must serve as an insulating environment, shielding members as much as possible from racial slurs and injustices. Minority parents must prepare their children "for survival in an environment that is hostile, racist, and discriminatory" (Peters 1988). Rearing children in a white-dominated society creates situations that white families never encounter, regardless of their social class. For example, a parent of a child of color must decide whether to warn the youngster who is going off to school for the first time about the possibility of classmates' racial slurs (Ambert 1994). As a result, minority parents are acutely concerned about the need to develop high self-esteem in their children, along with pride in their subcultural heritage. Some minority parents decide not to discuss racism or discrimination with their children; they do not want them to become unnecessarily bitter or resentful. Instead, these parents prepare to help their offspring cope with racism when it arises (Moore 1993; White 1993). Other minority parents believe it is important to teach their youngsters the history of discrimination against them. Black author James Baldwin, for example, would teach black children about their history and the racism they will encounter. In his words,

> any Negro who is born in this country and undergoes the American educational system runs the risk of becoming schizophrenic. On the one hand he is born in the shadow of the stars and stripes. . . . He pledges allegiance to that flag which guarantees "liberty and justice for all." He is part of a country in which anyone can become president, and so forth. But on the other hand he is also assured by his country and his countrymen that he has never contributed anything to civilization—that his past is nothing more than a record of humiliations gladly endured. He is assured by the republic that he, his father, his mother, and his ancestors were happy, shiftless, watermelon eating darkies who loved Mr. Charlie and Miss Ann, that the value he has as a black man is proven by one thing only— his devotion to white people. . . . All this enters the child's consciousness much sooner than we as adults would like to think it does. (Baldwin 1988, pp. 4–5)

More and more parents of color (as well as some white parents) are asking for a multicultural curriculum. They want all children to gain a broader perspective on the pluralistic racial/ethnic heritage of our society. Nebraska, a predominantly white state, has passed legislation mandating multicultural education in the public schools.

Changing education is not always easy, and frequently there is little consensus on goals and materials. An example was the conflict centering on the five-hundredth anniversary of Columbus' arrival in the Americas—celebrated by some, protested by others who view our "discoverer" as an exploitative intruder. Columbus Day is the principal holiday of Italian Americans, but for Native Americans it marks an invasion.

Even parents within the same racial/ethnic group do not necessarily agree on the best approach to racial issues. About one-third of African American parents do not attempt any explicit racial socialization. Others differ as to whether they emphasize forewarning or take a more militant position toward the elimination of social inequality (Taylor et al. 1991). Novelist Baldwin, for example, advocates telling children early that blacks "were brought here as a source of cheap labor" and that racism has been "a deliberate policy hammered into place in order to make money from black flesh" (1988, p. 7). Baldwin would assure black youth that the "agonies by which they are surrounded" are "criminal" (p. 11). Some parents in other ethnic minorities would agree.

A dilemma faced by all minority parents is to address the balance between loyalty to one's subculture and individual advancement in the dominant society (Jackson, McCullough, and Gurin 1988). For example, in Native American families few have been willing to define the dominant culture as valuable. Insisting on the superiority of their own cultural heritage, they are reluctant to assimilate into the broader society. More commonly, in many ethnic families the dialect or language spoken at home is neither used nor respected in the larger society. Hence, in order to succeed educationally and occupationally, children must become bilingual or forsake the language of their ancestors. (For a personal view of discrimination, see Case Study 12.2, "An Asian Immigrant Talks About Discrimination.")

Valuing one's subcultural heritage while simultaneously being required to deny or "rise above" it in order to advance poses problems both for minority individuals and between minority parents and their children. Native Americans must choose between the reservation and its high poverty level (some have 90 percent unemployment; Bedard 1992, p. 99) and an urban life that is perhaps alienating but presents some economic opportunity. Latinos may see a threat to deeply cherished values of family and community in the competitive individualism of the mainstream American achievement path. Asian Americans may live out the "model minority" route to success but experience emotional estrangement from still-traditional parents, as portrayed in Amy Tan's novel, *The Joy Luck Club* (1989).

What parents of all social contexts have in common is the joy and complexity of parenting, as children progress through various stages of development. We turn next to these stages of parenting.

An Asian Immigrant Talks About Discrimination

In the following essay, written by a university student, the writer tells about discrimination he experienced as an Asian immigrant child. The essay points to the special challenges of ethnic/racial minority parents as they raise children in a dominant society that is often discriminatory.

For millions of people who face discrimination in the land in which they were born, there is only one dream—to raise their children in a land where there is not such discrimination. My family is one of those millions of families. We faced many obstacles in life being "foreign-blooded" in Taiwan where I was born. With my parents being immigrants (in Taiwan) from mainland China, my family was an ethnic minority there. With that come numerous disadvantages. Therefore, my family tried to change the situation for the better.

The country my parents decided to move to was the United States. After all, every immigrant believes that the United States is the melting pot of the world. What's more, they decided on moving to New York, the melting pot of the United States. Knowing that there are so many Chinese immigrants in New York that there is actually a part of the city called Chinatown, my parents thought there would be no more discrimination against them in their future. Now we realize just how naive we were.

I entered this country a boy of 7 knowing three words of this foreign language: *dog, cat,* and *fish.* However, I learned the Pledge of Allegiance, every word and meaning, by the end of the first week I was here. When the first day of classes came for me, I remember standing up and reciting it along with the rest of the class so proudly that I shook with excitement the whole way through. I remember looking around at my classmates and thinking to myself that I was equal to any one of them. Unfortunately, that was to change before the day was over.

After I got my lunch later that day, I sat down at the first available table where I saw a seat. The kids there told me to leave. I just looked at them because I didn't understand what they said. I knew they were talking to me because they got my attention by saying, "Hey, Ching Chong." Next thing I knew, a few of them came up to me and said they'd show me how to eat my lunch. It seemed like a friendly gesture so I accepted. Well, the kids ended up mixing everything in my plate together and then pouring the milk into it. Their explanation was that it looked more like Chinese food that way so I could feel more comfortable.

Everyone at the table was laughing so I hit the boy who was pushing my head toward the plate. A fight broke out, and we both had our parents called into school. That's where my family's waking up process began. And we saw just how equal we weren't to the Joneses.

When my parents got to school, the first question the assistant principal asked was, "What restaurant do you folks own?" For my father who had been a captain in the merchant marine for sixteen years, it was an insult bigger than life. My father very sarcastically explained what he does for a living and the assistant principal had the nerve to tell my father that, "in this country, people should have respect in offices."

Then the assistant principal suggested that I was psychologically troubled and antisocial because I made no attempt to talk to anybody. However, the assistant principal made no attempt to talk to me while I sat for an hour in his office waiting for my parents to arrive. If he did, maybe he would have realized that I spoke no English. . . .

Our original dream (of living without discrimination) no longer exists in our minds. This year marked the thirteenth year since our immigration to this "melting pot." In the thirteen years, my family and I have seen and faced one act of discrimination after another. Now we have another dream. My mother told me after that first day of school, "If you can't fly with the flock, make sure you fly ahead of them." This new dream is basically

(continued)

(continued)

that idea: that I will be able to get ahead of what level this or any other society believes that my kind of people should stand at.

How does this student's essay illustrate the particular challenges of racial/ethnic minority parents? What other examples or challenges can you think of that might apply here? How might all parents work to eradicate racism and ethnic discrimination in neighborhoods and schools?

Stages of Parenting

People are accustomed to thinking of childhood as a developmental process. The early milestones of learning to walk, talk, and think abstractly, for example, are widely recognized. The developmental perspective also applies to adults, and it clearly applies to the parenting role, which is characterized by different tasks and needs at different stages.

Transition to Parenthood

Becoming a parent can be difficult, often more difficult than the parents had anticipated. New fathers and mothers report being bothered by the baby's interruption of such activities as sleeping, going places, and sexual expression. New mothers are distressed about their personal appearance, the additional amount of work required of them, and the need to change plans for their own lives and futures.

Almost thirty years ago, in what has become a classic analysis that still applies, social scientist Alice Rossi (1968) analyzed the transition to parenthood, comparing the parent role with other adult roles, such as worker or spouse. The transition to parenthood, Rossi asserts, is more difficult than the transition to either of these other roles for several reasons:

1. Cultural pressure encourages adults to become parents even though they may not really want to. But once a baby is born, especially to married couples, there is little possibility of undoing the commitment to parenthood.

2. There is little preparation for parenting. Most mothers and fathers approach parenting with little or no previous experience in child care. Fantasy often takes the place of realistic training for future parents. Romanticizing leads to disillusionment and often to an emotionally painful cycle of anger, depression, and guilt. But preparation for parenthood that comes solely from book learning is not always the answer either. Paradoxically, today's well-educated and well-read parents may overreact to normal behavior. Some 35 percent of children brought to the doctor by concerned parents because they were crying frequently had nothing physically wrong with them. What they had were parents who were perhaps too well aware of the many potential health threats and psychological problems (Kutner 1991).

3. Unlike other adult roles, the transition to parenting is abrupt, whether one is tackling it alone or as part of a couple. New mothers suddenly are on 24-hour duty, caring for a fragile and mysterious and utterly dependent infant. This is also true for fathers, although perhaps to a lesser degree.

4. Adjusting to parenthood necessitates changes in the couple's relationship. Husbands can expect to receive less attention from their wives. Employed wives who have established fairly egalitarian relationships with their husbands may find themselves in a different role, particularly if they quit working to become full-time homemakers. The transition

can be extra difficult when a spouse's expectations for couple involvement in parenting are not fulfilled (Kalmuss, Davidson, and Cushman 1992).

5. Today's parents do not have clear guidelines about what constitutes good parenting. They can learn about the child's nutritional, medical, and other physical needs, and they can follow the general prescription that a child needs loving contact and emotional support. But what should parents do to guide their children toward adult competency? The proliferation of experts in child rearing can lead parents to believe they could not possibly be competent.

Contemporary research supports the continued validity of Rossi's analysis (Wallace and Gotlib 1990), although investigators are also finding that subgroups of parents with better or poorer adjustment can be identified (Belsky and Rovine 1990; Cowan and Cowan 1992; Levy-Shiff 1994). But a point not mentioned by Rossi (because it was not so true in the 1960s) is that more of today's parents will be caring for fragile infants. Smaller and smaller infants are surviving, thanks to modern technology. But they remain in the hospital in intensive care for some time after birth, may need special care, are likely to suffer some disability, and do not fit easily into a family routine (Rosenthal 1991).

Parents with Babies

It is often difficult for new parents to acquire a sense of perspective because their experience is so new and there is little for them to compare it to. It helps to know that babies are different from one another even at birth; the fact that one baby cries a lot does not necessarily mean that he or she is receiving the wrong kind of care. Through the interactive perspective, parents can realize that their own ideas and attitudes are affected by the infant's appearance and behavior, and vice versa.

It is clear the home environment—for example, poverty status, parents' self-esteem and intelligence, and stress levels—affect an infant's irritability (Luster, Boger, and Hannan 1993). Nevertheless, from birth infants have different "readabilities"—that is, varying clarity in the messages or cues they give to tell caregivers how they feel or what they want (Bell 1974). They also have different temperaments at birth: Some

are "easy," responding positively to new foods, people, and situations and transmitting consistent cues (such as tired cry or hungry cry). Other infants are more "difficult." They may have irregular habits of sleeping, eating, and elimination, which sometimes extend into childhood; they may adapt slowly to new situations and stimuli; and they may seem to cry endlessly, for no apparent reason. Still other babies, of course, are neither easy nor particularly difficult.

Generally, an "easy" baby contributes to the caregiver's conviction that she or he is a good parent, whereas difficult babies may cause the parents to blame themselves. A parent's relationship with a difficult child may lack many of the expected rewards of parenthood. These parents may be tempted to treat the growing baby as a problem child or to respond by oversheltering the child, two reactions that may cause the child to internalize the sense that he or she is different. Instead, they need to strike a balance between the child's needs for security and predictability and their own needs. These parents may also need more time away from their child than do parents of easier children (Thomas, Chess, and Birch 1968).

All infants have certain needs. Although not all social scientists are in complete agreement (Eyer 1992), there is considerable evidence that infants need to bond, or attach, to one consistent and dependable caregiver, usually the mother (Bowlby 1969; Ainsworth 1973; Karen 1994). Like anyone else, they need positive, affectionate, intimate relationships with others in order to develop feelings of self-esteem. They need encouragement, conversation, and variety in their environment in order to develop emotionally and intellectually. They need at least one reliable, warm caregiver to establish the basic human trust that is a prelude to normal development and a protection against psychological harm (Peterson and Rollins 1987; Werner and Smith 1992). Discipline is always inappropriate for babies.

Parents with Preschoolers

Preschool children continue to have many of the same needs they had as infants. They need opportunities to practice motor development. They also need wide exposure to language, especially when people talk directly to them. And they need to experience consistency in the standards they are trying to learn. (Many parents today worry about whether the standards they set are too harsh or too lenient, but the

Mike Luckovich/*Atlanta Constitution*. Reprinted by permission of Creators Syndicate.

content of the standards is less important than the consistency.)

During the preschool years, parents and stepparents will establish a parenting style. Haim Ginott, author of *Between Parent and Child* (1965), distinguishes between permissiveness and overpermissiveness. **Permissiveness** is "an attitude of accepting the childishness of children"; it is fundamental to intimate parent–child relationships. Permissiveness means accepting that a clean, tucked-in shirt on most children will not stay clean or tucked in for long. Permissiveness is also "the acceptance of children as persons who have a constitutional right to have all kinds of feelings and wishes." Permissive parents do not admonish or restrict their children's wishes or feelings. They do, however, set limits on unacceptable actions. For example, it is permissible for a child to be angry, but it is never permissible for a child to hit his or her parents in anger. In contrast, **overpermissiveness** allows undesirable acts as expressions of feelings.

Both parent and child need clear definitions of what behavior is unacceptable. Limits are best set as house rules and stated objectively in third-person terms. A parent can say, for example, "Chairs are for sitting in, not for jumping on." With preschoolers, limits need to be set and stated very clearly: A parent who says "Don't go too far from home" leaves "too far" to the child's interpretation. "Don't go out of the yard at all" is a wiser rule. As the child learns to distinguish what is "too much," limits can be more flexible.

Parents of School-Age Children

School-age children need to be encouraged to accomplish goals appropriate to their abilities. Schools expect parents' cooperation and support in the form of helping the child acquire good study habits, showing interest in his or her work and progress, and assisting with schoolwork—though parents should not *do* the homework or complete the projects. The extracurricular activities of school-age children also demand a significant amount of support and time from parents who are already overloaded.

Children today face demands for achievement from schools, church and recreational organizations, peers, and, of course, parents. This can produce the "hurried child" who is forced to assume too many challenges and responsibilities too soon. Hurried children may achieve in adult ways at a young age, but they also acquire the stress induced by the pressure to achieve. Or they may "drop out" and abandon goal-directed academic and extracurricular activity. Parents can help not only by checking on whether they have realistic expectations for their children but also by moderating any unreasonable outside demands on the children.

Children also need to feel they are contributing family members. Those who are not given responsibility for their share of household chores may have trouble feeling they belong, or they may become demanding because they have learned to belong as consumers rather than as productive family members.

Children need to be assigned tasks and taught how to do them. One of the pluses of single-parent families as child-rearing institutions is that children's help is usually genuinely needed if the family is to function successfully.

Children also need to learn how to get along with others. One particular area of conflict is between siblings. Parents should recognize the inevitability of sibling rivalry and make an effort *not* to overrespond either by punishing competitiveness or by creating an artificially equal environment (taking one child out for lunch, for instance, because the other has been invited to a birthday party). Children should be encouraged to work out disputes by themselves whenever possible (Felson and Russo 1988). Often, in fact, the dispute loses its momentum when no parents are there to listen.

Having said this, we need to emphasize that physical violence *should* be kept in check by parents. Research on family violence (Straus, Gelles, and Steinmetz 1980) indicates that violence between siblings, particularly boys, and some of it quite serious, is the most pervasive form of family violence. Parents should provide children with an environment in which nonviolent methods of solving conflicts are learned (Gelles and Straus 1988; see also Ambert 1994). Rather than spankings, or hitting children, timeouts, when all interaction stops for a few minutes, can help. Moreover, unlike spankings, timeouts do not show a child that hitting is an acceptable way to resolve conflict.

A further need of school-age children is for increasing recognition of their individuality and emerging autonomy. Like adults, children need some privacy. They can be assigned a spot in the house that is theirs and no one else's, or they can regularly be given some time alone when parents agree not to interrupt them. As growing individuals, children also need to practice making their own decisions. For example, the parent

may offer the child a choice between wearing alternative pants or shirts.

The question of care or supervision of school-age children of working parents is addressed in Chapter 13.

Parents of Adolescents

Adolescence is the time when children search for identity—who they are and will be as adults (Collins 1990; Steinberg 1990). While the majority of teenagers do not cause or undergo familial "storm and stress" (Larson and Ham 1993), the teen years do have the special potential for being a time of conflict between parent and child, for in this stage both are involved in periods of transition. Both are experiencing biological shifts: The child is undergoing the physical and hormonal changes accompanying puberty; the parent, those of midlife. From a social-psychological viewpoint, the child is getting ready to move into the adult world, and the parent is increasingly aware of what he or she has yet to accomplish in life. A result may be an increased level of parent–child conflict as they struggle to accommodate the very different needs of each (Hatfield 1985).

Another complicating factor is that our society offers no clear guidelines or prescriptions for relinquishing parental authority: The exact time when authority should be relinquished, or how much, is not culturally specified.[3]

Any struggle between parent (or stepparent) and adolescent over when and how parental authority should be rescinded characteristically involves values. Because our society increasingly requires an adult to display a distinctly personal set of values, adolescents may feel the need to reject their parents' values, at least intellectually and for the time being, to gain adult status.

Research shows that teens and parents conflict more often over everyday matters, such as chores, than over bigger issues like sex and drugs. One study that

---

3. When to relinquish authority is, however, linked to certain economic factors. Upper middle-class families can afford to defer their children's financial independence, paying for some or all of their college education. Working-class parents, however, are experiencing a financial crunch (called the "life cycle crunch" by Valerie Oppenheimer 1974) just at this time. Because blue-collar income typically decreases during late adulthood, children are pushed to earn their own living. It also appears that stepfamilies launch children earlier than intact married-couple families; the exit of an older adolescent may be a way of resolving family conflicts (White and Booth 1985b; Goldscheider and Goldscheider 1989).

analyzed National Survey of Families and Households (NSFH) data (see Chapter 5) found white parents reporting more conflict with their teens than did African American or Hispanic parents. The author speculates that this may be because non-Hispanic white parents expect and cultivate more autonomy in their offspring while other cultural groups are more inclined to teach obedience (Barber 1994). Meanwhile, collisions with adolescents over values are especially difficult to negotiate because they usually are not conflicts over specific behaviors. Teenagers wonder why they should change what they believe when it does not actually do the parent any harm. If parents insist on having their own way, the adolescent does the same. Parent and child thereby may engage in a power struggle in which the adolescent feels honor-bound to do just the opposite of what the parent demands. Many decisions—whether homework will get finished, what school the youngster will attend, what career choices are available, what the consequences of experimentation with alcohol, other drugs, and premarital sex will be—do of course have far-reaching implications for the young person. These may be viewed by parents as intolerable risks, but by the children as valid choices to make for themselves. The next section examines the parenting process—and parenting styles—more generally.

The Parenting Process

Considering the lack of consensus about how to raise children today, it may seem difficult to single out styles of parenting. From one point of view there are as many parenting styles as there are parents. Yet certain elements in relating to children can be broadly classified.

Five Parenting Styles

One helpful grouping is provided in E. E. LeMasters' listing of five parenting styles: the martyr, the pal, the police officer, the teacher–counselor, and the athletic coach (LeMasters and DeFrain 1989).[4] Individual parents probably combine elements of two or more of these styles in their own personal parenting styles.

MARTYR Martyring parents believe "I would do anything for my child." Some common examples of martyring are parents who habitually wait on their children or pick up after them, parents who nag children rather than letting them remember things for themselves, parents who buy virtually anything the child asks for, and parents who always do what the children want to do.

A martyring parenting style presents some problems. First, the goals martyring parents set are impossible to carry out, and so the parent must always feel guilty. Second, as Chapter 4 points out, martyring tends to generate manipulative behavior.

PAL Some parents, mainly those of older children and adolescents, feel that they should be pals to their children. They adopt a **laissez-faire discipline** policy, letting their children set their own goals, rules, and limits, with little or no guidance from parents.

Pal parenting is unrealistic. For one thing, parents in our society *are* responsible for guiding their children's development. Children deserve to benefit from the greater knowledge and experience of their parents, and at all ages they need some rules and limits, although these change as children grow older. If parents are too permissive, children do not learn to develop self-control:

> "Permissiveness alone doesn't work well because the parents, while warm and loving, do not provide enough structure or monitoring of their child's behavior." . . . When parents failed to set limits, "the child doesn't know when to stop and acts . . . out." (psychologist Dr. James Bray, in Brody 1991, p. B-8)

Much research relates laissez-faire parenting to juvenile delinquency and child behavior problems (Baumrind 1978; Maccoby and Martin 1983; Loeber and Southamer-Loeber 1986).

LeMasters and DeFrain point out that there are also relationship risks in the pal–parent model. If

---

4. There are some similarities between LeMasters' police officer, coach, and pal, and psychologist Diana Baumrind's authoritarian, laissez-faire, and authoritative parenting styles (Baumrind 1971; 1978). The reader who would like a more academic presentation of modes of parental control should pursue Baumrind's articles and related psychological research and theory.

Baumrind's typology and sociologist Murray Straus' writings on "parental support and control" (1964) stimulated much social science research and theorizing about parental effectiveness (Patterson 1982; Maccoby and Martin 1983; Parke and Slaby 1983; Peterson and Rollins 1987; Brody 1991). Psychologist Gerald Patterson (1982) at the Oregon Social Learning Center has been particularly innovative in pursuing the use of research findings in actual work with parents.

things don't go well, parents may want to retreat to a more formal, authoritarian style of parenting. But once they've established a buddy relationship, it is difficult to regain authority.

POLICE OFFICER   The police officer model is just the opposite of the pal. These parents make sure the child obeys all the rules at all times, and they punish their children for even minor offenses.

Being a police officer as a parent doesn't work very well. Spanking is ineffective and seems to lead to subsequent problems (Brody 1991). Whether or not it involves physical punishment, **autocratic discipline,** which places the entire power of determining rules and limits in the parents' hands, has been associated—like laissez-faire parenting—with juvenile delinquency and child behavior problems (Gove and Crutchfield 1982; Maccoby and Martin 1983).

There are several reasons why the police officer role doesn't work today. For one thing, adolescents are far more likely to be influenced by their parents' knowledge and expertise or a wish to identify with parents' values than by the parents' authority. Even for younger children, parental warmth leads to greater obedience (Brody 1991) and has been positively linked to children's overall social competence (Boyum and Parke 1995).

TEACHER–COUNSELOR   The parent as teacher–counselor acts in accord with the **developmental model of child rearing,** in which the child is viewed as an extremely plastic organism with virtually unlimited potential for growth and development. This model conceptualizes the parent(s) as almost omnipotent in guiding children's development. If they do the right things at the right time, their children will more than likely be happy, intelligent, and successful.

The teacher–counselor approach has many fine features, and children do benefit from environmental stimulation, as well as from parental sensitivity to their needs (Brody 1991). Yet this parenting style also poses problems. For one thing, it puts the needs of the child above those of the parent(s). Also, parents who respond as if each of their child's discoveries is wonderful may give the child the mistaken impression that he or she is the center of everyone's universe.

Moreover, this view exaggerates the power of the parent and the passivity of children. Children also have inherited intellectual capacities and needs. In-

Parents who act as teachers focus on guiding their children's development, while coaching parents encourage their children to develop and practice their own talents. Those who combine the best characteristics of the different styles, as does this father, are often effective parents.

stead, many child development experts believe a bidirectional or **interactive perspective** to be most useful, for it regards the influence between parent and child as mutual and reciprocal (Peterson and Rollins 1987, pp. 487–91). The athletic coach model proceeds from this view.

ATHLETIC COACH   Athletic coach parenting incorporates aspects of the developmental point of view. The coach (parent) is expected to have sufficient ability and knowledge of the game (life) and to be prepared and confident to lead players (children) to do their best and, it is hoped, to succeed.

This parenting style recognizes that parents, like coaches, have their own personalities and needs. They establish team rules, or *house rules* (and this can be done somewhat democratically with help from the players), and teach these rules to their children. They enforce the appropriate penalties when rules are broken, but policing is not their primary concern. Children, like team members, must be willing to accept discipline and, at least sometimes, to subordinate their own interests to the needs of the family team.

Coaching parents encourage their children to practice and to work hard to develop their own talents. But they realize that they cannot play the game for their players. One of today's most widely read child psychologists, John Rosemond, has written column after column on this point, cautioning parents, for example, not to get too involved in homework (Rosemond 1991a, 1991b). LeMasters and DeFrain (1989) recommend the athletic coach parenting style as the most realistic and effective.

One criticism may be that it does not give enough emphasis to parental warmth. The coach image suggests impersonality, and subsequent researchers have found the parent's warmth and supportiveness to be a major predictor of positive outcomes for children (Patterson 1982; Maccoby and Martin 1983; Parke and Slaby 1983). Children of parents who are sensitive to their needs are more obedient (Brody 1991).

The ideal parent is the authoritative parent who demands maturity and effectively punishes forbidden behavior after having clearly stated the rules, but who listens to the child's point of view, respects the child, and encourages the child's self-development and independence. Such parents tend to have children who are socially competent—that is, with high self-esteem and cooperative, yet independent, personalities.

It is important that parents agree, at least on their basic goals and parenting style. Children can live with minor differences (and in fact learn something about people from these variations), but truly inconsistent discipline results in ineffective socialization into the rules and values of society. Parents who are not married and living together must work especially hard at this (Kutner 1989b, p. 92).

Instead of engaging in power struggles, parents need to recognize that by the very nature of their relationship they exercise decreasing control over what their children choose to do. Parents do continue to influence their adolescents, as well as their younger children, as appropriate models and as consultants.

Parents as Models and Consultants

Values are not just taught; they are caught from those around us whom we admire. That is why it is important for parents to practice what they preach.

Besides acting as models, parents and stepparents or other adults in the household can influence adolescents by getting "hired as a consultant" (Gordon 1970, p. 263). As consultants they discuss, for example, important world issues, such as the environment and racism, with the adolescents. Consultants, like athletic coaches, are prepared with facts and information about the alternatives under consideration; they inform their clients about the potential consequences of each alternative so that the client can make knowledgeable decisions. Consultants can offer their own opinions and viewpoints if they explain that this is what they are doing. But as the coach must allow the players to play the game, the consultant must allow the clients to make their own decisions. (Box 12.1, "Communicating with Children," offers some important advice on parenting.)

Child psychologist John Rosemond (1991b) describes the "over-involved parent" who "hovers, assumes responsibility [for the child's tasks], encourages dependence, and sends negative messages" (p. 35). Such parenting undercuts the child's development of his or her own competency and suggests that the parent mistrusts the child's capabilities. In contrast, the "consulting parent . . . is available, assigns responsibility, encourages independence, and sends positive messages" about the child's competence and motivation (Rosemond 1991b, p. 35).

Setting Limits Through Democratic Discipline

As children reach adolescence, limits can be based more and more on **democratic discipline,** in which all family members involved have some say. Rules are discussed ahead of time, and both parents and children attempt to compromise whenever possible. Moreover, it appears that problem solving is more effective when a parent strives to limit negative emotional blaming and/or outbursts (Forgatch 1989).

Social scientists and counselors hold that democratic or authoritative discipline is more effective with adolescents than is either laissez-faire or autocratic discipline. They also point out that parents need not feel guilty about occasionally setting limits without

Communicating with Children—How to Talk So Kids Will Listen and Listen So Kids Will Talk

Either knowledgeably or by default, we choose how we communicate with children. There are more and less effective ways to communicate, so a knowledgeable choice would very probably involve choosing more, not less, effective ways. What are some of these methods?

Helping Children Deal with Their Feelings

Children—even adult children—need to have their feelings accepted and respected.

1. *You can listen quietly and attentively.*

2. *You can acknowledge their feelings with a word.* "Oh . . . Mmm . . . I see. . . ."

3. *You can give the feeling a name.* "That sounds frustrating!"

4. *You can give the child his wishes in fantasy.* "I wish I could make the banana ripe for you right now!" Or, "I wish I could make your boss give you the promotion you want."

5. *You can note that all feelings are accepted, but certain actions must be limited.* "I can see how angry you are at your brother. Tell him what you want with words, not fists."

Engaging a Child's Cooperation

1. *Describe what you see, or describe the problem.* "There's a wet towel on the bed."

2. *Give information.* "The towel is getting my blanket wet."

3. *Say it with a word.* "The towel!"

4. *Describe what you feel.* "I don't like sleeping in a wet bed!"

5. *Write a note.* (above towel rack)
   Please put me back so I can dry.

   Thanks!
   Your Towel

Instead of Punishment

1. *Express your feelings strongly—without attacking character.* "I'm furious that my saw was left outside to rust in the rain!"

2. *State your expectations.* "I expect my tools to be returned after they've been borrowed."

3. *Show the child how to make amends.* "What this saw needs now is a little steel wool and a lot of elbow grease."

4. *Give the child a choice.* "You can borrow my tools and return them or you can give up the privilege of using them. You decide."

5. *Take action.* Child: "Why is the tool box locked?" Father: "You tell me why."

6. *Problem solve.* "What can we work out so that you can use my tools when you need them, and so that I'll be sure they're here when I need them?"

Encouraging Autonomy

1. *Let children make choices.* "Are you in the mood for your gray pants today or your red pants?"

2. *Show respect for a child's struggle.* "A jar can be hard to open. Sometimes it helps if you tap the side of the lid with a spoon." "A tax return can be hard to fill out. Sometimes it helps to tear it into a million pieces and drop them from an airplane."

3. *Don't ask too many questions.* "Glad to see you. Welcome home."

4. *Don't rush to answer questions.* "That's an interesting question. What do you think?"

5. *Encourage children to use sources outside the home.* "Maybe the pet shop owner would have a suggestion."

6. *Don't take away hope.* "So you're thinking of trying out for the play! That should be an experience."

Praise and Self-Esteem

Instead of evaluating, describe.

1. *Describe what you see.* "I see a clean floor, a smooth bed, and books neatly lined up on the shelf." Or, "I see your car parked exactly where we agreed it would be." **(continued)**

(continued)

2. *Describe what you feel.* "It's a pleasure to walk into this room!" Or, "I've enjoyed learning a little about the music you like."

3. *Sum up the child's praiseworthy behavior with a word.* "You sorted out your pencils, crayons, and pens, and put them in separate boxes. That's what I call *organization!*"

Freeing Children from Playing Roles

1. *Look for opportunities to show the child a new picture of himself or herself.* "You've had that toy since you were 3 and it looks almost like new!"

2. *Put children in situations in which they can see themselves differently.* "Sara, would you take the screwdriver and tighten the pulls on these drawers?"

3. *Let children overhear you say something positive about them.* "He held his arm steady even though the shot hurt." Or, "She's back home for a while, and I do enjoy her company."

4. *Model the behavior you'd like to see.* "It's hard to lose, but I'll try to be a sport about it. Congratulations!"

5. *Be a storehouse for your child's special moments.* "I remember the time you..."

6. *When the child acts according to the old label, state your feelings and/or your expectations.* "I don't like that. Despite your strong feelings, I expect sportsmanship from you."

Source: Adapted from Faber and Mazlish 1982.

---

providing rational explanations. Parents should recognize that it is not always possible to give children watertight explanations for every limit they set. In some situations, a parent has a vague feeling that he or she should not permit something but cannot give a rational explanation why, or the explanation does not stand up to teenagers' rebuttal. In these cases, parents should recognize that it is all right simply to ask the child to accept and respect their feelings of anxiety. I-statements from parent to teenager, as discussed in Chapter 9, can help. For example, "I get angry when you leave the car without gas in it, and I go out to go to work and find the tank empty." Or, "I was worried because I hadn't heard from you."

The next section explores some ways that all parents can improve their relationships with their children by building self-confidence and sharing the child-rearing role with others.

Toward Better Parent–Child Relationships

Studies generally show that good parenting involves at least three factors: (1) adequate economic resources, (2) being involved in a child's life and school, and (3) using supportive, rather than negative, communication in the family (DuBois, Eitel, and Felner 1994; Klebanov, Brooks-Gunn, and Duncan 1994; Simons, Johnson, and Conger 1994). At the same time, American parents today need to work to build their confidence and self-esteem. Important to this is a parent's learning to accept his or her limitations and mistakes as inevitable and human. Instead of feeling guilty about how poorly he handled a fight between his children, for example, a father can praise himself for the patience he showed in helping his child with a homework assignment.

As parents gain self-acceptance, they will also become more accepting of their children. When parents make positive attributions to their children, such as "She [or he] wants to do the right thing," children internalize them. This sense of respect contributes to better parent–child relationships *and* improved self-images for both parents and children.

In the past twenty-five years, national organizations have emerged to help parents with the parent–child relationship. One program is Thomas Gordon's Parent Effectiveness Training (PET), which applies the guidelines for no-win intimacy to the parent–child relationship. Another is Systematic Training for Effective Parenting (STEP). Both STEP and PET combine instruction on effective communication techniques with emotional support for parents. These programs are offered in many communities, and related books are also available.

An evaluation of a parent training program developed by child psychologist Carolyn Webber-Stratton at the University of Washington Parenting Clinic reported that parents can become more effective through self-help techniques presented on videotape—and without professional support. Parents in the program came from all economic, marital, and educational backgrounds (Webber-Stratton 1992).

In addition to gaining self-acceptance, self-esteem, and knowledge of good parenting techniques, parents can involve other members of their communities in child rearing. The first step, perhaps, is to insist that society recognize the positive achievements of conscientious parents, and not just criticize parents and children for inadequacies.

Then, too, parents can encourage more cooperation among friends and neighbors. For instance, parents might exchange homework help—"I'll help Johnny with math on Tuesday evenings if you'll help Mary with English"—and of course form car pools for children's lessons and activities. Such practical exchanges provide the occasion for children to form supportive relationships with other adults, and they are also a basis for shared parenting.

Another source of support is the community itself: teachers, school counselors and principals, police officers, adolescents' employers, the library, and the public in general. Parents and stepparents are encouraged to seek general information, answers to specific questions, and avenues for assistance from local agencies or counselors, in books, or in parent advocacy organizations. Programs such as Big Brothers/Big Sisters and Foster Grandparents involve volunteers who care

about and can help with child rearing. If children learn to relate well to other human beings and to take some personal responsibility for these relationships, they will be prepared to have good relationships as adults.

Finally, parents need to be willing to seek professional help when their efforts seem to be unsuccessful. Problems are best addressed early in the child's life, when intervention or the changes parents make in response to advice will have the most effect (research evidence supports this truism; Brody 1991). Psychotherapists are now prepared to help with problems of children still in infancy (Gelman 1990).

The Resilient Child

This textbook is written for people interested enough in parenting and family life to take a course. But no one is perfect; most parents have failings of one sort or another. Parents may encounter serious problems in their lives, problems that affect their children: poverty, discrimination, divorce, unemployment, legal and financial conflicts, scandal, sudden change of residence, crime victimization, military service, war, a death close to the child—and yes, mental illness, drug or alcohol abuse, or family violence.

The hope that research in child development offers to parents who make mistakes—or worse—is that children can be surprisingly resilient (Anthony and Cohler 1987; Furstenberg and Hughes 1995). A long-term study was conducted in Hawaii based on a sample of all the children (698) born on the island of Kauai in 1955—one-half of whom were in poverty, and one-sixth of whom were physically or intellectually handicapped. A smaller group of 225 children was identified as being at high risk of poor developmental outcomes. Researchers found that even one-third of this last group "grew into competent young adults who loved well, worked well, played well, and expected well" (Werner 1992, p. 263). The researchers spoke of a "self-righting tendency" whereby children lucky enough to have certain characteristics—a sociable personality, self-esteem derived from a particular talent, a support network, and above all, a good relationship with at least one caring adult—emerged into adulthood in good shape (Werner 1992; Werner and Smith 1992). Risk is risk, but hope is not unrealistic (Furstenberg and Hughes 1995). Other research indicates that "[a]dults who acknowledge and seem to have worked through difficulties of their childhood

Children have a certain degree of resilience, which is enhanced by strong familial bonds. Although these children experience economic hardship, the self-esteem they gain through their family and friends can help them in becoming self-confident adults.

are apparently protected against inflicting them on their children" (Belsky 1991, p. 124).

Graham Spanier, a recent president of the National Council on Family Relations, took as his theme the resilience of children and families in unpromising circumstances. "Many children who experience abuse, family disruption, or poverty nevertheless reach adulthood, against great odds, with a strong commitment to family life" (1989, p. 3). "Some widely held beliefs about the negative impact on children of divorce or parental absence may be overstated, if not wrong" (p. 10). Spanier also suggested that individuals can move on from difficult childhoods to create a satisfying family life:

> Hope stems from the evidence that much childhood trauma has the prospect of being left behind; that some of the setbacks of childhood are short-lived; that we should be skeptical about any intellectual conviction about the superiority of the intact nuclear family; that growing insight into protective factors may help us help children to negotiate risk situations better; that individual and family resilience, or hardiness, exists and can be learned or enhanced; that . . . [responsible individuals in extended family networks] can forge continuity out of chaos; and that the intergenerational dynamics of the future may dictate stronger familial bonds. (Spanier 1989, p. 10)

This growth and resiliency will not happen automatically but rather only through effective social intervention and society's commitment to the family.

When Parents and Children Grow Older

We tend to assume that once individuals reach a certain age (21, for example), marry, or graduate, they will have the resources to function comfortably in our society. But parenting doesn't end when a child reaches 18, 21, or even 25. This is complicated by the fact that ideas differ on just what a parent owes an adult child. Our culture offers few guidelines about when parental responsibility ends or how to withdraw it.

Parents with Young Adult Children

More and more young adult children either do not leave the family home or return to it—after college, after divorce, or upon finding first jobs unsatisfactory. Unemployment and underemployment, along with a decline in affordable housing, make launching oneself into independent adulthood especially difficult today (Goldscheider and Goldscheider 1994).

Parents who anticipated increased intimacy or personal freedom may be disappointed when the nest doesn't empty. Said one widower who is of retirement age but is keeping the large house and his university position so his youngest son can live at home and finish college tuition-free:

He seems to think I enjoy working now, that I'd go on forever. I can't find the gratitude in him for the fact that I'm keeping this job just so he can finish school. He's a fifth-year senior. He tells me there're scores of these folks around now. What happened to graduation after four years? He says I'm being unrealistic. He works, and I should be glad about that, I guess. So over Christmas he's driving to Minnesota to visit his girlfriend. He's twenty-two. I guess he doesn't need to ask my permission, but it gripes me anyway. It's a case of his money is his and my money is ours. I buy the food, pay the mortgage and utilities. He puts gas in his car and drives across the country to see his girlfriend! Besides, I thought he'd be around over Christmas. . . . (Interview with Agnes Riedmann, 1989)

Whether or not parents continue to share their homes with adult children, the relationship will probably be enhanced if they relinquish parental authority, recognizing that their children's attitudes and values may differ from their own (Miller and Glass 1989). Relinquishing parental authority, however, does not mean allowing absolutely any behavior to go on in the family home. For instance, parents who believe that alcohol and other drug usage is immoral have the right to disallow it under their roof. In general, parents should feel comfortable in setting reasonable house-

hold expectations. One way to do this is to negotiate a parent–adult child residence-sharing agreement.

PARENT–ADULT CHILD RESIDENCE-SHARING AGREEMENTS   Some issues to address and negotiate are the following:

1. How much money will the adult child be expected to contribute to the household? When is it to be paid? Will there be penalties for late payment?

2. What benefits will the child receive? For example, will the family's laundry soap or anything in the refrigerator be at the child's disposal?

3. Who will have authority over utility usage? Who will decide, for instance, when the weather warrants turning on an air conditioner or where to set the thermostat?

4. What about the telephone? At what time is it too late to call? At what point is a conversation too long? Will telephone rules apply to everyone in the household? (One solution to telephone problems is for the adult child to pay for a second line.)

5. What are the standards for cleanliness and orderliness? For instance, what precisely is the definition of "leaving the bathroom (or kitchen) in a mess"?

6. Who is responsible for cleaning what and when? What about yardwork?

7. Who is responsible for cooking what and when? Will meals be at specified times? Will the adult child provide his or her own food?

8. How will laundry tasks be divided?

9. If the adult child owns a car, where will it be parked? Who will pay the property taxes and insurance?

10. What about noise levels? How loud may music be played and when? If the noise associated with an adult child's coming home late at night disturbs sleeping parents, how will this problem be solved?

11. What about guests? When are they welcome, with how much notice, and in what rooms of the house? Will the home be used for parties? (First, though, what *is* a party? Three guests? Six? Five hundred?)

12. What arrangement will be made for informing other household members if one will be unexpectedly late? (As an adult, the child should have a right to come and go as he or she pleases. But courtesy requires informing others in the household of the general time when one can be expected home. This avoids unnecessary phone calls to every hospital emergency room in the region.)

13. What about using the personal possessions of other members in the household? Can a mother borrow her adult daughter's clothes without asking, for example?

14. If the adult child has returned home with children, who is responsible for their care? How often, when, and with how much notice will the grandparent babysit? Who in the household may discipline the children? How, when, and for what?

Although a residence-sharing agreement can help temporarily, the goal of the majority of parents is for their adult children to move on. But even when the adult child does not reside at home, children benefit from parents' emotional support and encouragement through their 20s and after.

THE NEED FOR ONGOING EMOTIONAL SUPPORT

Parents typically have had more experience with solving problems and handling complex emotions than their adult children have had. The young adult is facing difficult life tasks—for example, finding meaningful full-time employment or grieving a loss—for the first time and hence without experience. At trying times, the adult child who has moved out may seek a renewed sense of security by touching home base—either by calling often by telephone or in person.

Just listening and using other positive communication skills (described in Chapter 9) can help. (It also helps not to jump to conclusions at a time like this. A child who calls in desperation because his or her marriage is breaking up may *not* be asking to move home.) Parents can help to replenish self-confidence in frustrated children by reminding them of their past successes and commenting on the strength and skills they demonstrated (Haines and Neely 1987). A parent might say, for example, "I remember your persistence as you worked toward first-chair in band." A grieving

Many grandfathers, including those who may have been pressured for time when their own children were young, welcome the chance to be involved with their grandchildren. Some grandparents perform a role more like parenting, providing daily care and support for their grandchildren.

young adult will need to be informed that what she or he is feeling is normal.

At times parents may choose to confront their adult children: "It sounds as if drinking is beginning to cause problems for you," or "It sounds as if you feel stuck in a job you don't like." Serious problems, such as dealing with an adult child's chronic depression or chemical addiction, require counseling and/or support groups designed for this purpose.

In later life, parents may continue to play a vital role in the lives of their adult children and grandchildren, as the next section discusses.

Grandparenting

Surveys show that parents over age 65 prefer to live in their own homes but near their adult children and grandchildren. Middle-aged adult children sometimes return to the geographical area they grew up in to be near their aging parents.

Among whites, grandparents usually assume that role sometime in their 40s or 50s. Because blacks tend to be younger than whites when they have their first child, they are likely to become grandparents at a younger age. Because people are living longer, a grandparent spends more time in that role than in the past. Also, grandparenting is less likely to overlap with the parenting role than in previous decades. Besides these demographic changes, this century has seen increased emphasis on love, affection, and companionship with grandparents. Hence many grandparents find the role deeply meaningful: Grandchildren give personal pleasure and a sense of immortality. Some grandfathers see the role as an opportunity to be involved with babies and very young children, an activity discouraged when their own children were young (Cunningham-Burley 1987). Overall, the grandparent role is mediated by the parent; not getting along with the parent dampens the grandparent's contact, hence relationship with his or her grandchildren (Whitbeck, Hoyt, and Huck 1993).

Sociologists Andrew Cherlin and Frank Furstenberg surveyed 510 grandparents across the country by telephone. (Just over one-fifth were under 60 years old. Forty-three percent were between 65 and 74, and 17 percent were over 75.) About 20 percent of their sample said they saw their grandchildren less often than every two or three months. One-fourth saw a grandchild once a week or more, although not daily. Another 12 percent saw a grandchild daily; often they lived together (Cherlin and Furstenberg 1986, p. 72).

Grandparents continue to provide practical help (King and Elder 1995). They may serve as valuable "family watchdogs," ready to provide assistance when needed (Troll 1985). In low-income and minority families, parents and children readily rely on grandparents and other kin. Even among white middle-class families, it is not unusual for grandparents to contribute to the cost of a grandchild's tuition, wedding, or first house. As you can see from Figure 12.5, 82 percent of those interviewed by Cherlin and Furstenberg had given grandchildren money in the previous year. (Grandparents try not to meddle, however. Only

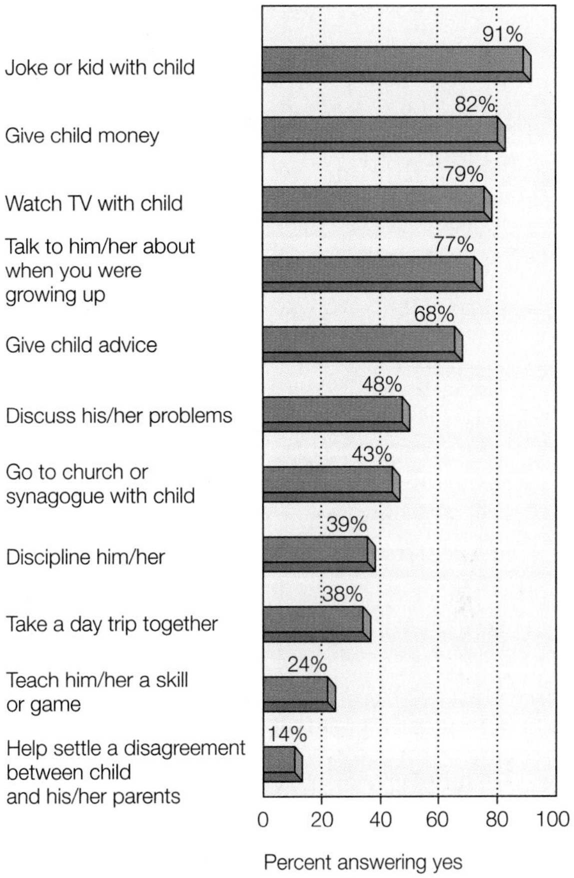

FIGURE 12.5

Percentage of grandparents engaging in various activities with their gandchildren during the previous twelve months. (Source: Cherlin and Furstenberg 1986, p. 74)

14 percent said they had helped settle a disagreement between grandchildren and parents.) As we saw in the previous section, divorced parents may turn to their parents for practical and emotional aid. Then, too, when the adult child of a divorced parent divorces, assistance may be more readily available from a grandparent.

Grandparenting styles vary and are shaped by health, employment status, and personality (Troll 1985). Some grandmothers, particularly older and more traditional ones who live close by, are heavily involved with their grandchildren. For others grandmotherhood is a significant but primarily symbolic role; these tend to be younger grandmothers who are

"You're real good with kids, Grandma...you ought to have some of your own."

involved in work, leisure, or social activities of their own.

Also, race affects grandparenting. In low-income African American families, grandmothers often provide crucial assistance to unmarried daughters with children. Black grandparents were much more likely than whites to assume a parentlike role with their grandchildren. For one thing, the grandchild's parents are more likely to be adolescents. In the Cherlin and Furstenberg study, 87 percent of black grandparents felt free to correct a grandchild's behavior, compared to 43 percent of white grandparents. As one black grandmother said of her 14-year-old grandson, "He can get around his mother, but he can't get around me so well" (1986, p. 128).

Moreover, in the past five years the rapid spread of AIDS and crack, which has affected women more than previous drugs and has resulted in increased incarceration of women, has left many more grandmothers to assume the responsibility of child rearing

(Jendrek 1993; Holloway 1994). Estimates suggest that 3–4 million grandmothers and greatgrandmothers are primary caregivers. And such relationships are not limited to inner-city settings: "Trust me, it's everywhere," comments Sylvie de Toledo, a social worker who founded Grandparents as Parents. "Everyday her phone rings with frantic older couples, who had been contemplating peaceful retirement, now being physically, financially, and emotionally drained by their newfound, albeit familiar, child-rearing responsibilities" (Malcolm 1991, p. B-6).

Research on primary parenting by grandparents is just beginning (Jendrek 1993), but we can return to Cherlin and Furstenberg's study (1986) to identify three general styles of grandparenting among more conventionally situated grandparents: remote, companionate, and involved. About 30 percent of Cherlin and Furstenberg's respondents had "remote" relationships with their grandchildren. Usually this was because they lived far away. About 55 percent of the relationships were "companionate." These grandparents did things with their grandchildren but exercised little authority over the child and allowed the parent to control access to the youth. Another 15 percent of the respondents were more "involved." Some lived with their grandchildren. (In most of these cases, the grandparent is household head with an adult child and his or her children having moved in, rather than vice versa.) Involved grandparents who do not reside with their grandchildren tend to frequently initiate interaction with a grandchild.

Of course, a grandparent may have different relationship styles with different grandchildren. In general, grandparents are most actively involved with preadolescents, particularly preschoolers (although 8 percent of Cherlin and Furstenberg's sample preferred teenagers). Preschoolers are more available and respond most enthusiastically to a grandparent's attention. There is evidence that, after a typically disinterested adolescence, adults renew relationships with grandparents (Cherlin and Furstenberg 1986, p. 89).

How does an adult child's divorce affect the grandparent relationship? Evidence suggests that the news hits hard, and grandparents worry over whether to intervene on behalf of the grandchildren. As might be expected, effects of the divorce are different for the **custodial grandparent** (parent of the custodial parent) than for the **noncustodial grandparent** (parent of the noncustodial parent) (Spitze et al. 1994). According to Cherlin and Furstenberg's study, during the marital breakup only 6 percent of custodial grandparents saw their grandchildren less often, compared with 41 percent of noncustodial grandparents (1986, pp. 142–48). This pattern persists. Several years after the divorce, 37 percent of custodial and 58 percent of noncustodial grandparents were seeing their grandchildren less often than before the breakup. But because of pressure from noncustodial grandparents, all fifty states have now passed laws giving grandparents the right to seek legalized visitation rights.

With the current trends in child custody, the most common situation is for maternal grandparent relationships to be maintained or enhanced while paternal ones diminish. Because divorce appears to strengthen maternal intergenerational ties, children of divorced parents may develop stronger bonds with *some* of their grandparents than do other youngsters (Cherlin and Furstenberg 1986, p. 164).

Then, too, remarriages create stepgrandparents. There is very little information on stepgrandparents, but what data there are suggest that younger stepgrandchildren and those who live with the grandparent's adult child are more likely to develop ties with the stepgrandparent. Meanwhile, stepgrandparents typically distinguish their "real" grandchildren from those of remarriages (Cherlin and Furstenberg 1986, pp. 156–57).

Adult Children and Their Parents

In discussing child rearing and grandparenting, we take the perspective of the older generation looking at the younger. But as parents, children, and their children grow older, things change.

Marriage, and then parenthood, redefine the relationship between parents and children. For many mothers and daughters, motherhood for the daughter creates a closer bond than had existed in adolescence—perhaps ever—as these women who may have very different views and styles may now have something very important in common (Kutner 1990d).

Whether or not parent–child relationships become more egalitarian is debatable. L. R. Fisher's (1986) study of mothers and daughters found that generally mothers never completely relinquish a parental role and that the adult daughter–mother relationship had both peer and parental elements. But other psychologists have found the contrary. One interviewee described reminding her visiting parents to keep their feet off the couch, a spontaneous remark that symbolized that she was "on the same level as my parents" in their relationship (Kutner 1990d, p. C-8). Even when parents need their adult children for care, however, role reversal is not very likely to occur (Kutner 1988a).

Some parents and children never successfully negotiate the transition to an adult–adult relationship and become estranged. Such a parent–child rupture seems even more difficult to talk about than divorce. In some families, the reality of past abuse, a conflict-filled divorce, or simply fundamental differences in values or lifestyles make it seem unlikely parents and children will spend time together (Kutner 1990a). Money matters can also cause tension (Kutner 1990c). Sometimes adult children must try to intervene in parental problems such as marital conflict (Kutner 1990b).

Parents may become more controlling as they grow older, a common reaction to loss of bodily and social power with aging and retirement. Parent–child relations enter a new stage when parents become disabled, frail, or suffer cognitive impairment that affects their ability to care for themselves or to live alone. Children are often uncertain about when to step in. A study of 242 adult children and 66 elderly parents found that children tended to let parents live independently (usually the parents' preference) as long as it seemed safe (Hansson et al. 1990). Elderly people are often uncomfortable receiving help, as it represents a threat to self-esteem and autonomy, especially if the caregiver is controlling or if earlier conflicts reemerge (Brubaker, Gorman, and Hiestand 1990; Cicerelli 1990).

One recent small study of 387 elderly parents in Florida found that they expected help from their adult children in proportion to the aid they had once given their adult children (Lee, Netzer, and Coward 1994). Except among some Asian American families, the adult child involved in a parent's care is far more likely to be a daughter, or even a daughter-in-law, than a son, raising issues of gender equity similar to those of parenting and domestic work discussed in Arlie Hochschild's *The Second Shift* (1989) (see Chapter 13). In fact, women today spend more years in elder care than in caring for young children (Abel 1991). A fairly large (905) New York sample of individuals over age 40 found that being divorced does *not* decrease a daughter's help to parents, as some observers had predicted it would (Spitze et al. 1994).

The parent–child relationship ends in a visible way with the parent's death (or, less frequently, with the child's). The relationship is such a meaningful one that it often continues emotionally after death. Sibling relations may be much affected as well. Adult children need to be aware of the issues likely to follow a parent's death (see Suggested Readings for one source of information).

In Sum

Raising children is both exciting and frustrating. The family ecology theoretical perspective reminds us that societywide conditions influence the relationship, and these factors can place extraordinary emotional and financial strains on parents. Formerly, children were expected to become more help and less trouble as they grew older; today, it is different. Most of what children need costs more as they grow: clothes, transportation, leisure activities, and schooling.

This chapter began by presenting some reasons why parenting and stepparenting can be difficult today. Some things noted are that work and parent roles often conflict. Middle-aged parents, especially mothers, may be sandwiched between dependent children on one hand and increasingly dependent, aging parents on the other.

Not only mothers' but fathers' roles can be difficult, especially in a society like ours in which attitudes have changed so rapidly and in which there is no consensus about how to raise children and how mothers and fathers should parent. In addition, both children's and parents' needs change over the course of life. One thing that does not change as children mature is their need for supportive communication from a parent. Grandparents, particularly companionate and involved ones, can be helpful.

The need for supportive—and socially supported—parenting transcends social class and race or ethnicity. At the same time, we have seen that parenting differs in some important ways, according to economic resources, social class, and whether parent and child suffer discrimination due to minority status or sexual orientation of the parents.

To have better relationships with their children, parents need to recognize their own needs and to avoid feeling unnecessary guilt; to accept help from others (friends and the community at large as well as professional caregivers); and finally, to try to build and maintain flexible, intimate relationships using techniques suggested in this chapter, along with those suggested in Chapter 9.

Key Terms

autocratic discipline	noncustodial grandparent
custodial grandparent	overpermissiveness
democratic discipline	parenting alliance
developmental model of child rearing	permissiveness
	primary parents
interactive perspective	sandwich generation
laissez-faire discipline	shared parenting

Study Questions

1. Describe reasons why parenting can be difficult today. Can you think of others besides those presented in this chapter?

2. How does parenting differ according to social class? Race or ethnicity? How would *you* prepare a minority child to face racial discrimination?

3. Compare the advantages and disadvantages of the five parenting styles discussed by Le Masters and Defrain.

4. Why might the transition to parenthood be more difficult than the transition to other adult roles, such as worker or spouse?

5. Compare the parenting challenges of parents with babies, preschoolers, school-age children, teenagers, and young adult children. Which stage do you think would be the most challenging? The least? Why?

6. What have you observed about the relationships of adult children and their parents? What do they enjoy? What are areas of conflict?

Suggested Readings

Blankenhorn, David. 1995. *Fatherless America: Confronting Our Most Urgent Social Problem.* New York: Basic Books. Blankenhorn is founder and president of the conservative organization, Institute for American Values, and this book presents his arguments and position.

Brazelton, T. Berry. 1983. *Infants and Mothers: Differences in Development.* 2d ed. New York: Delacorte. Child development made clear to parents, by a respected pediatrician.

Child Development. 1994. Vol. 65 (2). Entire issue devoted to children and poverty. An excellent special issue.

Cowan, C. P. and P. A. Cowan. 1992. *When Partners Become Parents.* New York: Basic Books. Engaging and very readable book on the transition to parenthood. Supportive with good advice.

Dinkmeyer, Don and Gary D. McKay. 1990. Series of books on parenting, using the STEP (Systematic Training for Effective Parenting) method. Circle Pines, MN: American Guidance Service. Worthwhile workbooks for all parents. Useful and encouraging.

Eyre, Linda and Richard Eyre. 1994. *Teaching Your Children Joy. Teaching Your Children Responsibility. Teaching Your Children Values.* New York: Simon & Schuster, Fireside Books. Three good books written with wit and humor and including specific tasks and ideas.

Faber, Adele and Elaine Mazlish. 1982. *How to Talk So Kids Will Listen and Listen So Kids Will Talk.* New York: Avon. As the title says, how to communicate with children.

Ginott, Haim G. 1965. *Between Parent and Child.* New York: Macmillan. A classic. Techniques of parent–child communication, with sections on rebellion, anger, teenage sex, driving, drinking, and drugs, among other topics.

Gottfried, A. E. and Allen W. Gottfried. 1994. *Redefining Families: Implications for Children's Development.* New York: Plenum. An important collection of articles on the diversity of parenting experiences in the United States, from father in intact families to lesbian and gay male parents.

Griswold, Robert L. 1993. *Fatherhood in America: A History.* New York: Basic Books. Well-written and interesting account of fathering in the United States from 1800 to now, by a history professor at the University of Oklahoma.

Hansen, Shirley M. H., and Frederick W. Bozett, eds. 1985. *Dimensions of Fatherhood.* Beverly Hills: Sage. Discusses fathers of children at every stage of the family life cycle: dual-earner fathers, househusband fathers, stepfathers, custodial single-parent fathers, gay fathers, widowed fathers, noncustodial fathers, adolescent fathers, and fathers in the military.

Hutchinson, Earl Ofari. 1992. *Black Fatherhood: The Guide to Male Parenting.* Los Angeles: Middle Passage Press. A good little book that addresses issues faced by black fathers and their children.

Lowinsky, Naomi Ruth. 1992. *Stories from the Motherline: Reclaiming the Mother–Daughter Bond, Finding Our Feminine Souls.* Los Angeles: Jeremy P. Teacher, Inc. This book by a Jungian psychologist combines issues of contemporary feminism with those of traditional motherhood. The premise is that in order for a woman to be fulfilled, she needs to recognize the deep influence upon her of other women, whose purposes both expanded and limited her own.

Martin, April. 1993. *Lesbian and Gay Parenting Handbook: Creating and Raising Our Families.* New York: HarperCollins. Based on the author's psychotherapy practice; workshops; consultation with experts in law, psychology, adoption, and reproductive medicine, as well as interviews with fifty-seven families across the country. Contains information and advice in many areas of gay/lesbian family formation and parenting. Enlivened by stories about individual families.

Polakow, Valerie. 1993. *Lives on the Edge: Single Mothers and Their Children in the Other America.* Chicago: University of Chicago Press. An important and readable study of the economically desperate conditions faced by many single-mother families.

Poussaint, Alvin F. and James P. Comer. 1992. *Raising Black Children: Questions and Answers for Parents and Teachers.* New York: NAL/Dutton. Advice to parents from African American psychiatrists at Harvard and Yale.

Qvortrup, M. B., G. Sgritta, and H. Wintersberger, eds. 1994. *Childhood Matters: Social Theory, Practice, and Politics.* Vienna: European Centre for Social Welfare Policy and Research. Cross-cultural examination of childhood around the globe that uses a classic sociological, or family ecology, theoretical perspective.

Register, Cheri. 1991. *"Are Those Kids Yours?": American Families with Children Adopted from Other Countries.* New York: Free Press. If you are raising an adopted child from another country, or thinking about it, this is a good book. Realistic in the challenges to be faced, yet optimistic and at times even inspirational.

Rosenberg, Elinor B. 1992. *The Adoption Life Cycle: The Children and Their Families Through the Years.* New York: Free

Press. Not a how-to book, this is a scholarly review of the literature on adopted children and their families over the life cycle.

Spock, Benjamin and Michael B. Rothenberg. 1985. *Raising Children in a Difficult Time.* New York: Pocket. Dr. Spock's classic updated; forty years' worth of advice to parents.

Stockman, Larry and Cynthia S. Graves. 1990. *Adult Children Who Won't Grow Up.* Rocklin, CA: Prima. Authors point to social and economic conditions that make this problem more common, and they offer good advice to parents.

Willie, Charles Vert. 1991. *A New Look at Black Families,* 4th ed. Dix Hills, NY: General Hall. Detailed discussion of class differences in African American families provides a context for understanding the diverse child-rearing circumstances of black families.

Zahn-Waxler, Carolyn, ed. 1995. *Sexual Orientation and Human Development.* Special issue of *Developmental Psychology.* Vol. 31. Academic journal articles including biosocial perspectives and social construction of homosexuality; articles on pressures and problems of gay youth; gay and lesbian relationships and parenthood, as well as some integrative essays. Of general interest regarding sexual orientation. Useful for gay males and lesbians contemplating parenthood; useful for parents of gay male or lesbian children.

Work and Family

> *I look out across the yard. I see Sofia dragging a ladder and then lean it up against the house. . . . Got her head tied up in a headrag. She clam up the ladder to the roof, begin to hammer in nails. Sound echo cross the yard like shots.*
>
> ALICE WALKER, THE COLOR PURPLE

"Where do you work?" is a new question in human history. Spouses have always worked. Until recently, cooperative labor for survival was the primary purpose of marriage (Tilly and Scott 1978). Providing and caring for all family members, including dependent children and the old, is integral to our definition of families and to the functional theoretical perspective on families as well (see Chapter 2).

Only since the Industrial Revolution has working been considered separate from family living, and only since then have the concepts "employed" and "unemployed" emerged. For those who stayed home (usually wives and mothers), working in the public sphere was beyond their experience and took on an aura of mystery. For those who labored outside the home and earned money (mainly husbands), partners who stayed at home seemed unproductive; they were not "employed." "His" work has been in the public sphere, for money; "her" work has taken place in the private sphere of the household, for free. Spouses' work roles have been sexually segregated (see Chapter 3).

Today, although occupational segregation and sex discrimination in employment persist, the trend is away from distinguishing work based on sex. A majority of married women work outside the home.

Many chapters in this book discuss social changes and how they affect people's attitudes and family life. This chapter looks at one aspect of modern living that is profoundly affecting marriages and families. We'll explore traditional employment patterns that have characterized our society until recently, then look at newer patterns and the interrelationship of work and family roles for both women and men. We'll see that the trend toward women's working outside the home offers new options for families. And with new options come new responsibilities for making knowledgeable decisions. To begin, we will examine the concept of "labor force" as a social invention.

The Labor Force—A Social Invention

Although human beings have always worked, it was not until the industrialization of the workplace in the nineteenth century that people characteristically became wage earners, hiring out their labor to someone else and joining a so-called **labor force.** The labor force, then, is a social invention.

Social scientists from Karl Marx to Sigmund Freud have agreed that working is inherently self-actualizing. To be so, however, working must be meaningful to the worker; that is, it must be characterized by some appreciable degree of creativity and self-direction. But with industrialization, capitalists (those who owned factories and/or other means of production) organized the labor force in order to enhance profit margins. "Individuals became organized in accordance with the requirements of technological production" (Berger, Berger, and Kellner 1973, p. 32). Efficiency—not worker creativity, self-direction, or workers' other individual or family needs—became an overriding principle.

The Labor Force in Postindustrial Society

Gradually throughout the twentieth century, our society has moved from an industrial one that manufactured products to a postindustrial one that transmits information and offers other services. In such a service society, production is transferred to the office. Synchronized clerical workers, administrators, and other bureaucrats, working according to a job description, perform one step in a sequence of activity. Individuals employed in traditional women's jobs—in the clerical field, for example—may find themselves with little control over tasks and conditions of work. These are situations that tend to give rise to stress and depression (Barnett and Baruch 1987). Similar to an industrialized society, postindustrial workers' individual and family needs are secondary to owners' goals of efficiency and profit.

Meanwhile, a great proportion of the new jobs created by the service economy pay less than did industrial work. Many are part-time and offer no employee benefits such as contributions to retirement or health care (Grubb and Wilson 1989). Although other factors are involved, American workers' average hourly wages, when adjusted for inflation, have declined since 1973. As we will see, one way families have adapted to this decline is for both wives and husbands to be employed.

At the same time, however, employees are increasingly aware that they are replaceable, if not dispensable. Partly in order to compete with foreign markets and partly to maintain corporate profits in a generally constricting economy, employers have dedicated themselves to reducing costs by streamlining operations. One way they have done that has been to let some workers go. By the 1990s, job security has become an issue even for highly educated managerial employees (Ehrbar 1993).

Many Americans find their positions or careers challenging and satisfying. But a significant (and growing) proportion of workers have difficulty finding personal satisfaction or security in their jobs (Davis and Milbank 1992). Moreover, especially in single-parent families and in unions in which both partners are employed and have dependents, the separation of the labor force from family living creates role conflict and tension. Increasingly, stressed and overloaded workers juggle what have become conflicting obligations: providing for and caring for family members.

It is in this context (which flows from the family ecology theoretical perspective described in Chapter 2) that we explore the relationship between work and family. We will look first at what has been the traditional work model in industrialized society.

The Traditional Model: Provider Husbands and Homemaking Wives

As discussed in Chapter 3, competing for success in a chosen occupation has been essential to the traditional masculine gender role. A husband has been culturally and legally expected to be his family's principal breadwinner. Meanwhile, wives have been culturally and legally bound to husband care, house care,

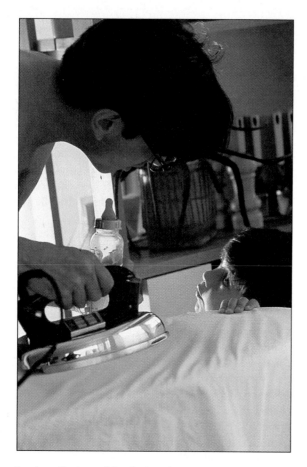

According to traditional roles, wives are the principal home-makers, and husbands the primary economic providers. Even as the majority of wives are employed outside the home today, they continue to do the lion's share of the housework.

and child care. One 1962 court ruling, for example, maintained that a wife "must perform her household and domestic duties" as well as be her husband's "helpmate, to love and care for him in such a role, to afford him her society and her person, to protect and care for him. . . ." (*Rucci v. Rucci* 1962).

Husbands and the Provider Role

What sociologist Jessie Bernard (1986) terms the **good provider role** for men emerged in this country during the 1830s. Before then a man was expected to be "a good steady worker," but "the idea that he was *the* provider would hardly ring true" (Bernard 1986, p. 126). The good provider role lasted into the late 1970s. Its end, again according to Bernard, was offi-

cially marked when the 1980 U.S. Census declared that a male was no longer automatically assumed to be head of the household. Indeed, the proportion of families in which men are the sole breadwinner has declined from 42 percent in 1960 to 15 percent. This decline "is greatest among men handicapped in the labor market by older age, low education, and minority status" (Wilkie 1991, p. 111).

Although the role of family wage earner is no longer reserved for husbands, many Americans still believe that the man should be the *principal* provider for his family. Sociologist Jane Hood (1986) identified three provider role systems in dual-worker marriages. They vary according to the couples' attitudes toward husbands' and wives' responsibilities. Some working couples see their roles in terms of a main provider/secondary provider division. For the **main/secondary provider couple,** providing is the man's responsibility; the home, the woman's. Whatever wages her employment brings in are nice, but extra. In the two other models, the wife's income is recognized as essential to the couple's finances. In a **coprovider couple,** both partners are seen as equally responsible for providing. In the **ambivalent provider couple,** the wife's providing responsibilities are not clearly acknowledged.

Recent research suggests that coprovider couples, although emerging, are a statistical minority—about 15 percent of all two-earner unions (Potuchek 1992).

Table 13.1 shows the average number of hours per week spent in the paid labor force for men and women. Whether single or married, parent or not, men worked more hours than women. We will return to this table later in this chapter. Here we need only point out that men continue to be primary breadwinners. "Even in two-earner couples, the division of labor still reflects the conventional assignment of paid employment more to the husband and unpaid work more to the wife" (Ferree 1991, pp. 154, 167).

REWARDS AND COSTS The good provider role entailed both rewards and costs for men. Rewards included social status and reinforcement of the husband's authority in the family. Moreover, the traditional exchange, described in Chapter 7, meant that the male exchanged breadwinning for the female's homemaking, child rearing, sexual availability, and more general husband care.

A serious cost was that the good provider role en-

TABLE *13.1*

Hours Per Week Men and Women Spent in the Paid Labor Force, by Marital Status and Number of Children

	Men	Women	Women as Percentage of Men
Grand mean	48 hours	39 hours	81%
Marital status			
Married	48	37	77
Never-married	47	43	91
Number of children			
None	47	40	85
One	49	39	80
Two or more	50	38	76

Source: Adapted from Shelton 1992, Table 3.4, p. 39.

couraged a man to put all of his "gender-identifying eggs into one psychic basket"; that is,

> success in the good-provider role came in time to define masculinity itself. The good provider . . . was a bread*winner*. . . . Men were judged as men by the level of living they provided. . . . The good provider became a player in the male competitive macho game. (Bernard 1986, p. 130)

Consequently, failure (or even mediocre performance) in the role meant one had failed *as a man,* indeed as a (male) *person* (Bernard 1986). What can feel like personal failure as a breadwinner, of course, is actually a result of having to cope with societal expectations that do not mesh with the reality of economic opportunities. This situation is especially applicable to blue-collar and minority, particularly African American and Native American, husbands (Gerson 1993; Grimm-Thomas and Perry-Jenkins 1994).

Meanwhile, social scientist Joseph Pleck has argued that "the most obvious and direct effect" of the male's breadwinning role is "the restricting effect of the male occupational role on men's family role" (1977, p. 420). Society has encouraged men to give primacy to their work and to let their family relationships come second. This situation has many effects. One effect is that it creates absentee fathers—fathers who are away from home for extended periods of time and who sel-

dom see their children, especially when the children are young.

Husbands who want to share household work and child care will not find it easy (Brayfield 1995). "The male feels not just conflicts, but intense pressures," said a Los Angeles lawyer and father. "Society hasn't lowered its level of job performance, but it has raised its expectations of our roles in our children's lives" (Gregg 1986, p. 48). Men in a seminar conducted by James Levine, one of the first to undertake formal study of fathering, had this to say:

> I have so little time with my child. I feel so guilty.
> The stress factor is astronomical. There is so much pressure on my time at work and at home.
> I feel like I'm stretching myself as far as I can. I fear it's going to take a toll on my health. (in Lawson 1990, p. B-1)

For this reason, partners who want to create new options for themselves need to work for changes in the public sphere, an option explored later in this chapter.

Meanwhile, "men are becoming more vocal regarding work pressures on the family," wrote columnist Anna Quindlen (1990, p. 19). She continued:

> The role fathers have carved out for themselves today is a vast improvement. . . . Those of us obliged to convert behavior into trends have probably been a little heavy-handed on the shared childbirth and

egalitarian diaper-changing. But fathers today do seem more emotional with their children, more nurturing, more open. Many say, "My father never told me he loved me," and so they tell their own children they love them all the time. . . . The men are saying: "I don't want to live this way anymore. I want to be with my kids." I think the corporate culture will have to begin to respond to that. (Quindlen 1990, p. 19)

NEW OPTIONS FOR MEN   Some husbands today are rejecting the idea that dedication to one's job or occupational achievement is the ultimate indicator of success. Some are choosing less competitive careers and are spending more time with their families. Such men place the needs of their families and their own desire for time with their families ahead of career success (Gerson 1993).

While some husbands persist in fighting to decrease the demands of the workplace in order to participate more at home, a very small minority are relinquishing breadwinning completely to become **househusbands:** husbands who stay home to care for the house and family while their wives work. About 2 percent of fathers are stay-at-home dads (Hochswender 1990).

Reasons younger househusbands give for staying home include unemployment, poor health, disillusionment or dissatisfaction with the competitive grind, the desire to spend more time with their children, and their wives' desire to pursue full-time careers. These situations are still considered role reversal rather than normal family situations. But acceptance may be growing. One househusband is a Native American whose wife is a legislative aide for the Native American Rights Fund. He believes that his wife's work is of such significance that it deserves priority, so he looks after their 5-year-old son as well as the house. He claims, "I get kidded, but I think people pretty much respect me for it."

The various options for men who would like to give more time to families are limited by the simple fact that in our society men typically earn more than women do, a point explored more fully later in this chapter. Consequently, whatever their values, many families find themselves needing to encourage the husband's dedication to his job or career in the interests of the family's overall financial well-being. Thus, our society's gender system places constraints on family

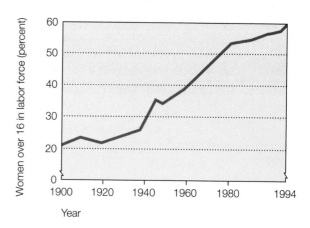

FIGURE   *13.1*

The participation of women over age 16 in the labor force, 1900–1994. (Sources: Thornton and Freedman 1983; U.S. Bureau of Labor Statistics 1986, 1990, 1995)

choices. The next section examines wives' traditional work role.

Wives as Full-Time Homemakers

Historically, the homemaker, or housewife—the married woman who remains in the home to do housework and rear children—is a relatively modern role. Before industrialization, women produced goods and income by working on the family farm or, for example, by taking in boarders. As the Industrial Revolution removed employment from the household, women could less easily "combine employment with care of the home. At the same time, the increase in real income made it feasible for most wives to devote their time solely to housekeeping. As a result, even as recently as 1940, only . . . 14 percent of married women were in the labor force" (Thornton and Freedman 1983, p. 24). By 1994, 60 percent of married women were in the labor force (U.S. Bureau of Labor Statistics 1995, Table 1). Women who head households are even more likely to work outside the home; 85 percent of divorced mothers of children between ages 6 and 17 are in the labor force (U.S. Bureau of the Census 1995, Table 638). Figure 13.1 shows the overall participation of women in the labor force during 1900–1994.

Although the trend has been for wives to join the labor force, currently more than one-third of married women are full-time homemakers. Although traditional legal assumptions minimize the value and importance of housework (Ahlander and Bahr 1995), the economic benefits of a full-time homemaker to her family have been evaluated in many ways.[1] One is to figure the value of her projected earnings were she to join the labor force. Another is to calculate the replacement cost of the homemaker's services. Economists conclude that a full-time housewife with at least one child under age 6 contributes household services equal to more than 100 percent of mean family income.

Even these calculations ignore important economic benefits to society provided by homemakers. For example, older wives often provide nursing services for their husbands or other relatives.

Homemakers are not direct producers; they serve others (husbands, children, and other kin), some of whom participate directly in the economic and political world by meeting their needs for everyday maintenance (food, clothing, and so on) and emotional support. Among some business, professional, and political families, wives are expected to help their husbands professionally by cultivating appropriate acquaintances and by being charming hostesses and companions. These wives are part of a **two-person single career;** their contributions advance their husbands' careers and benefit the spouses' employers. Wives in two-person single careers typically do volunteer work in the arts and for charitable organizations; they may also provide their husbands career support services such as typing, bookkeeping, researching, and writing.

But work that is directly productive (and paid for) is more highly valued and rewarded in most industrialized societies. It is therefore not surprising to find that the job of homemaker has ambiguous status and low economic rewards. An essential feature of the housewife role is constant availability to meet *others'* needs, and female houseworkers report a lower sense of being in control of their lives than do employed women (Bird and Ross 1993).

One study investigated the hypothesis that the homemaker role is particularly stressful in modern American society—which emphasizes individual achievement—but that this may not be so in more

traditional, familistic societies in which the homemaker role is more highly valued (Ross, Mirowsky, and Ulrich 1983). Mexican culture, for example, places more importance on the family than on work for both sexes. And "the Mexican female, as the center of the family, is accorded honor and prestige that is not available to her American counterpart" (p. 672). It is therefore not surprising that data from a survey of 330 married people conducted in El Paso, Texas, and Juarez, Mexico, suggested that women of Mexican ethnic identity were somewhat less psychologically distressed by the homemaker role than were Anglo women. In both cultures, however, employed women were found to be less distressed than homemakers, a pattern found in other studies as well (for example, Gore and Mangione 1983).

ECONOMIC COMPENSATION? Social scientist Jessie Bernard once suggested that housewives be paid for their job of keeping the work force motivated and in good condition (Bernard 1982). Presently, homemaking is not formal employment; no financial compensation is associated with this position.

In the United States today only a small minority of states have community property laws in which all of the couple's property is considered to be owned equally and in common (whatever the actual work, earnings, or contribution of homemaking activities or support services). A wife, even one who participates in her husband's career, may not be legally entitled to the resulting economic rewards either during the marriage or after divorce or the husband's death.

Steps could be taken to address these inequities. Changes in the Social Security system favorable to homemakers have been proposed but not enacted. Politically active groups have proposed that Congress pass a Homemaker Bill of Rights, granting—among other things—legal recognition of homemakers as equal economic partners in marriage, salaries for homemakers, revision of discriminatory Social Security provisions so that a homemaker can be covered in her own name in and out of marriage and the labor force, eligibility for unemployment compensation, and inclusion in the gross domestic product of the value of the goods and services homemakers produce.

Although some of these goals are probably unrealistic in today's political and economic climate, legal changes regarding inheritance and property rights are gradually effecting a move toward an economic part-

---

1. Economists have interested themselves in this issue, often as a service to insurance companies or personal injury trial attorneys who want to establish the projected value of a disabled homemaker's loss of services.

nership concept of matrimony (Weitzman 1985). For example, guidelines have changed so that full-time homemakers are permitted to set up individual retirement accounts (IRAs). In 1984 Congress passed the Retirement Equity Act (REACT), which acknowledged "marriage as an economic partnership, and the substantial contribution to that partnership of spouses who work both in and outside the home." REACT better protects the spouse of an employed participant in private retirement plans. For example, some retirement plans offer the participant a choice of whether to include lifetime survivor income, similar to life insurance benefits. REACT provides that a married participant cannot waive the survivor benefit—thus taking more in retirement income while living—without the written consent of the spouse. Moreover, REACT supports the existing power of state courts to split pension benefits between divorced and separated partners. Nevertheless, more changes are required—both legal and social—before homemakers can claim equal rights in marriage.

HOMEMAKERS BY CHOICE   In a study of 317 mothers of infants, researchers interviewed the mothers whose babies had just been born and had them complete questionnaires three months later. The majority of new mothers sampled believed that they alone could best meet their children's needs. At the first interview almost 70 percent of the mothers said they would rather stay home with the baby than seek outside employment; after three months this proportion had risen to 75 percent (Hock, Gnezda, and McBride 1984).

Clearly, a significant proportion of homemakers enjoy their work (Glass 1992). And a number of employed women would rather be full-time homemakers but, presumably because of financial constraints, do not have this option. The critical difference between full-time homemakers who enjoy their work and those who don't is *choice*. Satisfied homemakers are women who are exercising their preference to work in the home.

As full-time homemaking has become a minority pattern rather than the taken-for-granted role of adult women, reasons for this preference are more consciously thought through. One small study of housewives elicited two major themes. These mothers, college-educated women who had been employed prior to parenting, wished to stay deeply involved in

One key to finding satisfaction in a full-time homemaker role is choice. Women who choose to work as homemakers, and who do not face economic constraints, are often comfortable with and enjoy their role.

their children's daily lives and did not anticipate that this would change as the children grew older. They wanted to help out at their children's schools and to be home when the children returned from school. They had also assessed their situations and were pessimistic about the chances of getting their husbands to share housework and child care in the event they went to work. Staying at home is one response to our society's pervasive lack of support for working mothers. And although these women exhibited considerable confidence about their ability to reenter the labor market successfully, they saw paid employment as unsatisfying, preferring to pursue other activities, such as arts and crafts, as a supplement to domesticity (Milner 1990).

Some contemporary housewives consider this role a stage: "I'm learning skills and enhancing relationships

are pointing out—both from research and from personal experience—that expecting to have it all in a two-career marriage, particularly one with children, may be unrealistic.

The earlier discussion of the two-person single career suggests that a career is not an individual phenomenon. Rather, it has traditionally been a lifestyle that largely depends for its success on the active assistance of one's spouse. Partners in two-career marriages can find themselves trying to fill the expectations of a two-person single career work world. The public nature of careers—the requirement to entertain, for example, or to be available as a couple to attend career-related events—presents more time demands on both partners.

For two-career couples with children, family life is hectic and often tense as partners juggle schedules, chores, and child care. "Career and family involvement have never been combined easily in the same person" (Hunt and Hunt 1986, p. 280). Career wives, in particular, often find themselves in a catch-22 situation. The career world tends to view the person who splits time between work and family as being less than professional, yet society encourages working women to do exactly this.

Self-Employment

Whether or not both spouses have a career, in some two-earner couples at least one spouse is self-employed. Eight percent of the labor force was self-employed in 1993 (based on U.S. Bureau of the Census 1994a, Tables 628 and 630). Although more and more older retirees are starting their own businesses (Shaver 1991), married mothers are too (Alexander 1991). "The reason given most often by women is competing domestic demands, such as taking care of children or household chores" ("Both Sexes" 1990). Mothers employed at home, however, report interruptions. Like full-time homemakers, they are often asked to do chores formerly shared by many: to watch neighbors' children when bad weather closes the school, for example, or to keep an eye out for the older kids (Kutner 1988b).

Part-Time Employment

Another employment pattern among two-earner couples involves one spouse (usually the wife) being employed part-time. One-quarter of employed women are working part-time (U.S. Bureau of Labor Statistics 1994, Table 1). Many women choose part-time employment to better accommodate family needs. Indeed, recent studies find that mothers employed part-time are more traditional and similar to full-time homemakers than to full-time employed mothers in their attitudes about wife and mother roles (Glass 1992; Muller 1995). However, some women are channeled into part-time work because they cannot find full-time jobs and/or adequate child care.

Although greater family and personal time is a clear benefit of part-time employment, there are costs. As it exists now, part-time work seldom offers job security or other benefits. Employers do pay Social Security for part-timers, but the latter are generally excluded from private retirement funds or health insurance. And part-time pay is rarely proportionate to that for full-time jobs. For example, a part-time teacher or secretary usually earns well below the wage paid to regular staff.

Shift Work

In still other cases, one or both spouses engage in **shift work,** defined by the Bureau of Labor Statistics as any work schedule in which more than half an employee's hours are before 8 A.M. or after 4 P.M. It has been estimated that in one-quarter of all two-earner couples, at least one spouse does shift work. In about 15 percent of all dual-earner couples, only the husband works shift; in 6 percent only the wife does so; and in 3 percent both spouses do shift work. Blacks, younger workers, and those with working-class occupations are more likely to be involved in shift work (White and Keith 1990). Because shift work is more common among young people, nearly half of all new parents in which both are employed full-time are affected by shift work.

Some spouses work shift for higher wages or to ease child-care arrangements. A parent's doing shift work substantially increases the likelihood that fathers will participate in child care (Presser 1988). But shift work reduces the overlap of family members' leisure time. Analysis of responses of 1,668 individuals in a national survey designed to examine the relationship between labor force participation and marital quality (White and Keith 1990) found that—after controlling for education, race, age, number of children, years married, wife's employment, and family income—a partner's doing shift work reduced satisfaction with the sexual relationship and increased the probability of divorce.

Leaving the Labor Force and Reentry

Finally, a common work pattern among marrieds has been for the wife to be employed full-time until shortly before the birth of her first child and then to leave the full-time labor force either permanently or temporarily. In 1990, 47 percent of formerly employed mothers did not return to the labor force within one year of having a baby (U.S. Census Bureau data, reported in Wadman 1992). As an example, one mother left her job as a sales representative to stay home with her 15-month-old daughter after becoming frustrated with trying to balance even part-time work and day care. "I never saw myself as a stay-at-home mom, but my child's development means more to me than any job," she said (in Saltzman 1991, p. 46). "The 'homecoming' movement is a way of living life to the fullest," explained another mother. "It's a move that protects . . . marriages. It gives the couple time to strategize, time to get things done, and time for sex" (in Richardson 1988, p. C-10).

Yet it is important to realize that there are costs associated with temporarily leaving the labor force. Anne Machung, research associate for a study of graduating women at the University of California, describes her respondents:

> Almost all plan to have careers and children at the same time. . . . There is enormous confusion among young women about how they're going to do this. They have this fantasy that they can get their careers established by age 28 or 29, have children, then re-enter the labor force. They're not aware of the difficulties of staying home and then re-entering the work force. They don't realize there are consequences. (in Rimer 1988b, p. 30)

For one thing, women who leave the labor force face lost pension and Social Security benefits ("Poverty Looms" 1991). Moreover, after years of being at home, reentering women find that they have lost job or professional contacts, confidence, and, in some cases, skills. Potential employers question the career or work commitment of reentering women.

All this results in decreased earning power and, in the case of professional women, reduced potential career advancement. As one mother who had graduated in the top 10 percent of her law school class and left a high position in a law firm to stay home for ten years said, "I'll probably never achieve the pay scale or the prestige I would have had if I'd kept working those 10 years" (in Wadman 1992). Partly because college-educated women face greater loss in employment opportunities by staying at home, they are far more likely than others to return to the labor force within one year of having a baby. Sixty-eight percent of college-educated women returned to employment within one year in 1990, compared with 52 percent of high school graduates (U.S. Census Bureau data, reported in Wadman 1992). Whether college-educated or not, mothers who had earned a greater proportion of their family's income are more likely to return quickly to the labor force (Wenk and Garrett 1992).

This discussion of ways that two-earner couples organize their provider and family roles illustrates that unpaid family work is indeed a factor to be considered. The following section examines unpaid family work.

Unpaid Family Work

Unpaid family work involves the necessary tasks of attending both the emotional needs of all family members and the practical needs of dependent members (such as children or elderly parents) as well as maintaining the family domicile.

Caring for Dependent Family Members

As discussed in Chapter 12, our cultural tradition and social institutions give women principal responsibility for child rearing. This situation has not changed even as more and more mothers have entered the labor force. Even in two-career marriages, wives are far more likely to make all the child-care arrangements (Peterson and Gerson 1992).

Moreover, our culture designates women as "kin-keepers" (Hagestad 1986) whose job it is to keep in touch with—and, if necessary, care for—adult siblings and other relatives. The vast majority of informal elderly care is provided by female relatives, usually daughters and (albeit less often) daughters-in-law (Dwyer and Seccombe 1991; Keith 1995). One small, qualitative study (Aronson 1992) of women over 35 who were caring for aging mothers pointed to the power of the stereotypical link between femininity and caregiving in our society (see Chapter 3). Women, more so than men, feel obliged to care for frail parents and experience guilt when they do not. One woman said she had "yelled at" her brothers, "and they did go down and visit [her ailing mother], . . . but not one of them go down and clean or bring laundry" (p. 17). Respondents cited men's primary commitment to employment and relative inability to anticipate their mothers' needs as explanations for their husbands' and brothers' lesser sense of obligation and involvement. A woman teaching full-time, in poor health, and feeling very overextended explained her brother's lack of involvement with her failing mother this way: "But, you see, he's in a different situation. He's head of a company. He's under a lot of pressure. He's inherited some of the problems my father had physically. . . . So I'm not going to put a load of anything on him" (p. 19).

Helping an elderly parent can have a profound effect on the primary caregiver. "In fact, just when many women on the 'Mommy Track' thought they could get back to their careers, some are finding themselves on an even longer 'Daughter Track,' with their parents, or their husband's parents, growing frail" (Beck 1990b, p. 49). Typically, caregivers to disabled parents give up leisure activities and contacts with friends in order to fulfill work and family responsibilities (Burnley 1987). At home a wife can feel torn between caring for her children, her husband, and her aging parent (Kleban et al. 1989)—not to mention the housework.

Housework

Utopians and engineers alike shared a hope that advancing technology and changed social arrangements would make obsolete the need for families to cook, clean, or mind children (Hayden 1981). But instead, middle-class women spend more time on housework today than they did at the turn of the twentieth century. Collective arrangements proposed by utopians and feminists never caught on. Housewives were left alone in their free-standing suburban homes, as the postwar economy thrived on the construction of houses and automobiles (Wright 1981). Servants, who did much of the work for earlier middle-class housewives, entered factory work or took more challenging jobs, eventually benefiting from changed roles for women. Technology seems merely to have raised the standards; instead of changing clothes at infrequent intervals, we now do so daily (Cowan 1983).

Moreover, part of the pleasure and emotional significance of home life is that people can express their individual tastes in decor, food, lifestyle (casual or formal), and daily routine. Consequently, it is difficult to collectivize or commercialize housework, although employed women and their families are certainly trying. Commercial services have sprung up to do errands, arrange household repairs, and chauffeur children. Recently, increased immigration has provided a class of women to do child care and cleaning for affluent dual-career families (Rimer 1988a). But only upper middle-class women rely on paid help, so contemporary women typically spend as much time on housework as did women in earlier times.

WHO DOES HOUSEWORK? Despite changing attitudes among couples (Ferree 1991) and media portrayals of two-earner couples who share housework, women in fact continue to do the bulk of it. Table 13.2 presents data from a representative national sample of adults interviewed in the National Survey of Families and Households. Respondents were asked about amounts of time spent in the previous week on nine selected household tasks: preparing meals, washing dishes, cleaning house, outdoor chores, shopping, laundry, paying bills, car maintenance, and driving. As you can see, employed women spend 38 hours per week, compared with 22 hours spent by employed men, in these 9 household tasks. For women—but not men—marriage increases household labor hours to 42 per week. The number of children in the family is associated with both women's and men's household work. Although men with children spend more time on household labor than those without, children's impact on women's housework time is greater; the result is a larger gap between mothers and fathers than between child-free partners. Among employed parents with two or more children, women spend 27 more hours per week on household chores than men do (Shelton 1992, pp. 67–68).

Table 13.2 also shows that employed husbands do about 52 percent as much household labor as employed wives. Put another way, employed wives do nearly twice as much. Overall, employed men do 58 percent as much housework as employed women. These statistics represent some improvement in sharing of household work over previous years. In 1965, women spent six times as much time doing housework as men did (Pleck 1985). In 1975, men did 46 percent as much housework as women; in 1981, the figure was 54 percent (Shelton 1992, p. 65). But the fact remains that despite the assumption of an earning role by a majority of women, there has been little

TABLE *13.2*

Hours Per Week Spent in Household Labor by Employed Men and Women, by Marital Status and Number of Children in the Household

	Men	Women	Men as Percentage of Women
Grand mean	22 hr/wk	38 hr/wk	58%
Marital status			
Married	22	42	52
Unmarried	22	32	69
Number of children			
None	19	28	68
One	26	44	59
Two or more	24	51	47

Source: 1987 National Survey of Families and Households data, adapted from Shelton 1992, Tables 4.2 and 4.3, pp. 67–68.

change in responsibility for household work. As few as one in ten husbands (Blair and Johnson 1992)—and perhaps fewer than one in twenty (Coltrane and Ishii-Kuntz 1992)—do as much housework as their wives. "In spite of some variation in husbands' responsiveness to wives' employment, the overall pattern is one of little change in husbands' housework and childcare tasks or time. The constancy, rather than the variability, requires explanation" (Shelton 1990, p. 132).

RACE/ETHNICITY AND HOUSEWORK   Shelton's analysis did not initially examine the effect of race or ethnicity on couples' division of housework labor, but later she and a colleague proceeded to compare black, Hispanic, and non-Hispanic white men's household labor time (Shelton and John 1993). Using the same data set (the National Survey of Families and Households), the researchers found that, overall, African American men in families spend about 25 hours each week on household labor compared to about 23 hours for Hispanic men and 20 hours for non-Hispanic white men. For all these groups, the time spent on household labor is between one-third and 40 percent of what their wives spend (Shelton and John 1993, p. 138). We can conclude that the pattern of men's spending considerably less time than women in housework is similar for all three racial/ethnic categories, al-

though black and Hispanic men spend more time in unpaid family work than do white men.

Shelton and John went on to look at whether men's employment status affected their participation in household labor—and they found some interesting results. As you can see in Table 13.3, Hispanic and non-Hispanic white men who are not employed spend more time in household labor than do their employed counterparts. This is what the researchers had hypothesized. However, the pattern is different for African American men. Black men who are not employed spend less time in household labor (19.5 hours) than their full-time employed counterparts (27 hours). The researchers speculate that this counterintuitive finding may indicate that "black men who are not employed are different from nonemployed white and Hispanic men." They may be younger, for example, or "be a distinct group of black men characterized by both low time investments in paid labor and low investments in household labor" (p. 140). It is also possible that, especially for African American males, being a breadwinner is such an important aspect of proving one's manhood (see Chapter 3) that unemployed family men are reluctant to do anything else that might seem to undermine their manhood, such as labor traditionally considered women's work (see also Orbuch and Custer 1995).

TABLE *13.3*

Hours Per Week Spent in Household Labor by White, Black, and Hispanic Men in Families, by Employment Status

	White	Black	Hispanic
Employment Status	23.5	19.5	23.0
Not employed	19.1	26.6	22.7
Employed part-time (1–39 hr.)	18.2	27.0	22.3
Employed full-time			
Men's % of household labor time	34%	40%	36%
n	2,798	183	164

Note: We use 39 hours as our break between part-time and full-time in order to ensure an adequate *n* for the part-time category.

Source: 1987 National Survey of Families and Households data, adapted from Shelton and John 1993, Tables 7.1 and 7.2, pp. 139, 141.

It is important to note that when Shelton and John examined factors other than race/ethnicity that affect men's household labor input—such as their age, number of children, sex role attitudes, and their wives' sex role attitudes—race/ethnicity was no longer significantly associated with household labor time. In other words, the differences among white, black, and Hispanic men's household labor time that we observe in Table 13.3 reflect other differences among them, such as social class, education, the presence of children, or their wives' paid labor time. The authors conclude that, "We need to examine differences among black, white, and Hispanic men's perceptions of their family responsibilities if we are to understand how they balance work and family responsibilities" (Shelton and John 1993, p. 147). Doing so would likely involve in-depth, unstructured interviews with smaller samples, rather than analyzing data only from large data sets, such as the National Survey of Families and Households.

THE SECOND SHIFT   Many employed wives (but few husbands) put in a **second shift** in unpaid family work that amounts to an extra month of work each year (Hochschild 1989). In a study (Ferree 1991) of 382 mostly white, two-earner couples interviewed in 1989, husbands and wives together averaged a total of about 60 hours per week of employment and housework (excluding child care). Nearly one-quarter of the wives held part-time positions and did the lion's share of the housework, whereas another 9 percent of the

couples hired people to do much of the housework. Thirty-eight percent had what the researcher called two-housekeeper marriages (in which full-time employed wives did 60 percent or less of the housework, excluding child care). But 29 percent of the women were "drudge wives," holding full-time jobs and doing over 60 percent of the housework, excluding child care. When one adds child care to the equation, consistent research findings make it indisputably clear that today's employed wives do "more than their share" (Piotrkowski, Rapoport, and Rapoport 1987).

Furthermore, the household chores that men do (mowing the lawn, household repairs, auto maintenance) are more likely to have clear and identifiable boundaries, allow greater discretion in how and when they are accomplished, and have a leisure component (such as taking the children to the zoo). Wives, meanwhile, spend most of their time in tasks characterized by the opposite qualities. Washing dishes, cooking, and bathing and dressing children are tasks allowing little discretion as to when they must be done. And research shows that men remain reluctant to do the least desirable household tasks such as cleaning the bathroom (Berk 1988).

Moreover, "someone must do the attentive and coordinative work if people, relationships, and the family itself are to survive" (Thompson 1991, p. 190). Even when husbands share household labor, wives continue to take responsibility for it. As one employed mother put it:

The second shift is probably more enjoyable when shared by both partners.

You need to get food in the house, you spend Saturday morning doing grocery shopping. And when I don't, we end up somewhere around Sunday night saying, why isn't there any bread for breakfast tomorrow, or for lunches tomorrow? (in DeVault 1991, p. 64)

A substantial majority of women in two-earner marriages continue to see housekeeping standards as a personal reflection on them and hence to assume responsibility for enforcing them (Ferree 1991; Manke et al. 1994). When husbands take part in family labor, they are more likely to join in child care than in housework, but more mothers than fathers in Hochschild's study "kept track of doctors' appointments and arranged for playmates to come over. More mothers than fathers worried about the tail on a child's Halloween costume or a birthday present for a friend" (Hochschild 1989, p. 7).

The second shift for women means a "leisure gap" between husbands and wives, as the latter sacrifice avocational activities (Schnittger and Bird 1990), leisure—and sleep—to accomplish unpaid family work (Hochschild 1989; Shelton 1992). As a result,

women tend to talk more intently about being overtired, sick, and "emotionally drained." Many women I know could not tear away from the topic of sleep. They talked about how much they could "get by on" . . . six and a half, seven, seven and a half, less, more. They talked about who they knew who needed more or less. Some apologized for how much sleep they needed—"I'm afraid I need eight hours of sleep"—as if eight was "too much." . . . They talked about how to avoid fully waking up when a child called them at night, and how to get back to sleep. These women talked about sleep the way a hungry person talks about food." (Hochschild 1989, p. 9)

Women's revolutionary entry into the labor force ought to mean a concurrent restructuring of household labor. As we saw earlier in this chapter, husbands *are* doing somewhat more around the house than twenty years ago. But "women, rather than men, continue to adjust their time to accomplish both paid and unpaid work" (Shelton 1990, p. 132; Spitze and Ward 1995). Hochschild (1989) calls this situation a **stalled revolution.**

Ways in which individual families manage during this stalled revolution vary. Some wives scale back their paid work, and others quit entirely (Hochschild 1989). (These may seem the best choices, given the options. But the reality is that "no-fault divorce laws combined with rising divorce rates have substantially increased the risks for women [who do so]. . . . Divorced women are now expected to be able to support themselves" [Peterson 1989, p. 2].) Many busy spouses lower housework standards and food preparation time after a wife becomes employed (Shelton 1990). Some two-earner couples hire household help (Ferree 1991), although usually it is the wife who ends up coordinating things (Thompson 1991). Especially in single-parent families, older children help (Fassinger 1990), but children's contributions are limited (Berk 1988), and they actually do little to lessen the load (Piotrkowski, Rapoport, and Rapoport 1987). Several studies have found that when family size and the wife's employment hours increase to the point that she can no longer do it all, her husband begins to pitch in

(Peterson and Gerson 1992)—but often only to do some of the "nicer" tasks such as playing with the kids while she makes dinner (Berk 1985; Hochschild 1989).

Husbands who see breadwinning as their major role and themselves as the family's primary breadwinner are less likely to do housework. Wives get more help when their earnings are considered essential to family finances (Perry-Jenkins and Crouter 1990). The larger the share of family income provided by a wife, the more household work her husband does (Piotrkowski, Rapoport, and Rapoport 1987; Kelly 1990). When a wife is thought to be working mainly for self-fulfillment, household help is less likely (Hochschild 1989). Husbands who telecommute or have other kinds of flexible schedules do more (Kelly 1990). Generally, husbands follow "the path of least resistance" (Peterson and Gerson 1992, p. 532): To secure her spouse's help with the second shift a wife generally must take the initiative, in essence demanding his participation. Wives' overload or husbands' egalitarian beliefs seldom translate into actual help without this push (Hochschild 1989; Guelzow, Bird, and Koball 1991).

WHY WOMEN DO THE HOUSEWORK   Those doing housework generally find the tasks boring. But household labor is of keen interest to social scientists because struggles over who does it are central to analyses of class and gender.[4] We might apply the conflict and feminist theoretical perspectives discussed in Chapter 2 to this question. Doing so would mean arguing that women are more likely than men to do housework—and other unpaid family labor—because they have less power in their families than do men. Meanwhile, a number of related perspectives on the division of household labor have emerged.

The **ideological perspective** points to the effects of cultural expectations on household labor (Hiller and Philliber 1986; Hardesty and Bokemeier 1989). In this view, who does the housework at least partly reflects stereotypes and normative scripts about "who should do what" (Berk 1985). Some research has found that more highly educated husbands are more

willing to do housework, a finding in support of the ideological perspective inasmuch as more education results in less traditional gender attitudes (Ross 1987). As discussed in Chapter 3, girls are socialized to do housework more often than boys are. Researchers (see, for example, Blair and Lichter 1991; Shelton 1992) have found that wives do the vast majority of traditionally feminine household tasks, or "women's work." Wives do more than 96 percent of making beds, for example, and 94 percent of diapering children. Husbands, who do 86 percent of household repairs, 75 percent of the grass cutting, and 77 percent of snow shoveling, spend most of their household labor time on chores considered masculine (Berk 1985). In a study of 382 two-earner marriages in 1989 (Ferree 1991), wives were more likely to cook, clean, and do laundry, whereas husbands helped clean up after meals or shopped for groceries.

As pointed out in Chapter 3, however, adults do not simply live out cultural scripts learned in childhood. Rather, people weave the fabric of their lives with many threads, among them opportunities and chance (Gerson 1985). For example, a woman who is experiencing career success, or who is divorced, or whose husband is unemployed may gradually come to see herself as a primary breadwinner. At the same time, a wife with children and a higher-earning husband is less likely to redefine formerly learned gender boundaries (Potuchek 1992).

A second view holds simply that the partner with more time does the housework (England and Farkas 1986). Expanding on this idea, the **rational investment perspective** argues that couples attempt to maximize the family economy by trading off between time and energy investments in paid market work and unpaid household labor (Becker 1991). Spouses agree that each partner will spend more time and effort in the activities at which she or he is more efficient. Similarly, the **resource hypothesis,** proposed by Blood and Wolfe (1960) and described in Chapter 10, suggests that one spouse's household labor is a consequence of her or his resources compared to those of the other. Thus, the partner with higher resources (such as greater income or long-term earning potential) will have more power and hence spend less time on housework.

In principle, both the rational investment perspective and the resource hypothesis are gender neutral: Either husband or wife might invest more time and energy in the labor force and less in housework. But as

---

4. All the perspectives on who does the housework assume, of course, that the household division of labor is not natural but socially constructed and subject to change. Moreover, current perspectives differ from the 1950s approach, which assumed that women's roles were simply expressive and housework was not real family labor. Hence, researchers now examine what was once taken for granted.

we have already seen, husbands are likely to be receiving higher salaries than are their wives even within the professions and in sales and management positions. It follows that, as Table 13.1 indicates, men and not women invest relatively more time in the labor force. Furthermore, as was discussed in Chapter 10, resources are unevenly distributed between women and men. As a result, husbands are likely to have more status and power in marriages and can therefore resist housework demands (Coverman 1985).

Findings from the National Survey of Families and Households data support all of these perspectives to some degree. "This suggests that divisions of household labor are, in fact, the result of multiple causal forces. Who does what around the house is shaped by time availability, relative resources, and ideology" (Coltrane and Ishii-Kuntz 1992, p. 53).

A REINFORCING CYCLE   Data from large, representative national samples show that married men spend more time on a career track and/or in paid labor than do single men; the opposite is true for women. Similarly, men with more children spend more time on a career track and/or in paid labor than men with fewer or no children. Again, the opposite holds true for women (Cooney and Uhlenberg 1991; Shelton 1992; see Tables 13.1, 13.2, and 13.3).

When we consider men's and women's paid and unpaid work, it becomes apparent that a **reinforcing cycle** emerges: For a number of reasons—including cultural expectations, persistent discrimination, and the fact that employed wives interrupt their paid work and relocate to accommodate their husbands' careers far more often than the reverse (Shihadeh 1991)— men employed full-time average higher earnings than women employed full-time. Because most husbands actually or potentially earn more than their wives, couples allow her paid work role to be more vulnerable to family demands than his. This situation, in turn, has the effects of lowering the time and energy a wife spends in the labor force and of giving employers reason to pay women less than men. This lower pay, coupled with society's devaluation of family in favor of job demands, encourages husbands to see their wives' work (paid and unpaid) as less important than their own—and to conclude that they really shouldn't be asked to take responsibility for homemaking. Disproportionately burdened with household labor, wives find it difficult to invest themselves in the labor force

From the *Wall Street Journal.* Permission, Cartoon Features Syndicate

". . . And as part of your community service, you're to help your wife with the cooking and cleaning around the house."

to the same degree that husbands do. This situation remains true even for the majority of two-career couples, including, for example, physicians married to one another (Grant et al. 1990).

Given the general finding of disparity between husbands' and wives' contributions to household labor, "it is important to keep in mind that there is considerable variability in the amount of housework that men do, and that there are men who make significant contributions to the work of the household" (Blair and Johnson 1992, p. 575). In the next section we will examine how partners juggle household labor demands.

Juggling Employment and Unpaid Family Work

The concept of juggling implies a hectic and stressful situation. Virtually all research and other writings on the subject suggest that today's typical American family is a hectic one. This is particularly true when there are children in the home.

The Speed-Up of Family Life

For one thing, people are spending significantly more hours in paid employment now than twenty years ago (Schor 1991). The number of hours in a typical work

TABLE 13.4

Working Overtime: Percentage of Full-Time Employees Who Work More Than 49 Hours Per Week

All full-time employees	23.5%
Executives, administrators, managers	36.5
Men	44.3
Women	23.9
Professionals	29.4
Men	37.4
Women	19.5
Production workers	19.6
Salespeople	33.9

Source: Bureau of Labor Statistics data in the *New York Times,* June 3, 1990, p. 3E.

week has declined since the beginning of the twentieth century; still, in a constricted economy characterized by "the pervasive fear of dismissal" (O'Boyle 1990), professionals, managers, and lower-level workers as well feel compelled to demonstrate their skills and job commitment by working more hours (Michaels and Willwerth 1989). Nearly one-quarter of all employees now work more than 49 hours per week (see Table 13.4). More than 7 million Americans, or 6 percent of the labor force, held two or more paid jobs in 1994 (U.S. Bureau of the Census 1995, Table 646). Although some Americans are working two *part*-time jobs, workers with more than one job average 52 hours per week. Furthermore, paid time off (vacation, holidays, and sick leave) fell by about 15 percent during the 1980s (Schor 1991); leisure hours shrunk from 26.2 in 1973 to 16.6 in 1987 (Kilborn 1990). In short, "Americans are starved for time. Increasing numbers of people are finding themselves overworked, stressed out and heavily taxed by the joint demands of work and family life" (Schor, in Coston 1992).

Among families following the traditional model, this added workload translates into greater father absence. "The 5 o'clock Dad of the 1950s and '60s has been replaced by an 8 or 9 o'clock Dad, who very likely also attends to his job on Saturday or Sunday, if

he's not holding down an altogether different weekend job" (Schor, in Coston 1992).

Sociologist Arlie Hochschild's (1989) qualitative study of 500 two-earner couples points to the frantic pace and sense of pressure for everyone involved, but especially for mothers and for children, whose harried mothers often rush them:

> "Hurry up! It's time to go," "Finish your cereal now," "You can do that later," "Let's go!" . . . "Let's see who can take their bath the quickest!" Often a younger child will rush out, scurrying to be first in bed, while the older and wiser one stalls, resistant, sometimes resentful: "Mother is always rushing us." (Hochschild 1989, pp. 9–10)

We will look now at how children in two-earner marriages are doing more generally.

How Are Children Faring?

Before women with children entered the work force in large numbers, working mothers were considered problematic by child development experts. Now they are taken for granted (Brazelton 1989b). But the psychological development of children without a parent at home full-time is frequently a subject of concern.

Early research supported the notion that working mothers were not detrimental and might even be advantageous to their children's development (Hoffman and Nye 1974). Overall, this continues to be the prevailing view. Several studies show that, contrary to common assumptions, employed mothers spend almost as much time in direct interaction with their children and in visiting their children's schools as other mothers do, and no significant difference exists in the quality of care provided by employed mothers and full-time homemakers (Wilkie 1988; Chira 1994b; Muller 1995). Furthermore, employed mothers foster independence in their children and provide more structure to daily life, parenting qualities favored by child development experts. They are especially good role models for adolescent daughters, who have better social and personality adjustment and higher academic motivation and accomplishment than do daughters of homemakers (Wilkie 1988). Then too, the economic benefit to children of working mothers cannot be overlooked. Family income tends to be favorably associated with various child outcome measures.

Important for parents, though, is keeping their child's needs in the forefront in the face of daily pres-

Many two-earner and single-parent families find life a constant race to finish one task and get to the next, resulting in little time for conversation or leisure. Employed mothers in particular have to create ways to juggle child care, household, and kin-care concerns with their own demanding jobs.

sures. Recent studies, in fact, have found that mothers who work part-time are better at this than those who work full-time—and may indeed spend more time helping their children with homework than even full-time homemakers (Muller 1995). Suransky, while granting the importance and equity of women's current opportunities for work achievement, cautions about eclipsing children's needs in the process. Hochschild seemed astonished to hear one mother talk about her 9-month-old daughter's need for independence; clearly, a child's needs were being minimized to benefit the mother's schedule (1989, p. 198).

Eminent pediatrician T. Berry Brazelton urges parents to respect the process of attachment to a new infant. "When new parents do not have the time and freedom to face this process and live through it successfully, they may indeed escape emotionally. In running away, they may miss the opportunity to develop a secure attachment to their baby and never get to

know themselves as real parents" (Brazelton 1989b, p. 68). Discipline, sleep issues, feeding, competition with the caregiver, and perfectionism are other issues Brazelton sees as possible pitfalls for working families. But "understand that there is no perfect way to be a parent. The myth of the supermom serves no real purpose except to increase the parents' guilt" (Brazelton 1989b, p. 69). Brazelton urges the same policy measures we outline later in this chapter, and Box 13.1, "Bringing Up Baby: A Doctor's Prescription for Busy Parents," presents his guidelines for individual parents. We turn now to the question of how parents are faring as they juggle paid and unpaid work.

How Are Parents Faring?

This chapter focuses on work and family in *marriages* for two reasons. First, the vast majority of research on the interface between paid employment and family labor concerns marrieds; second, single parenting is ad-

Bringing Up Baby: A Doctor's Prescription for Busy Parents

Juggling work and family life can often seem overwhelming. Dr. Brazelton offers some practical advice for easing the strain on harried parents:

1. Learn to compartmentalize—when you work, be there, and when you are at home, be at home.

2. Prepare yourself for separating each day. Then prepare the child. Accompany him to his caregiver.

3. Allow yourself to grieve about leaving your baby—it will help you find the best substitute care, and you'll leave the child with a passionate parting.

4. Let yourself feel guilty. Guilt is a powerful force for finding solutions.

5. Find others to share your stress—peer or family resource groups.

6. Include your spouse in the work of the family.

7. Face the reality of working and caring. No supermom or superbaby fantasies.

8. Learn to save up energy in the workplace to be ready for homecoming.

9. Investigate all the options available at your workplace—on-site or nearby day care, shared-job options, flexible-time arrangements, sick leave if your child is ill.

10. Plan for children to fall apart when you arrive home after work. They've saved up their strongest feelings all day.

11. Gather the entire family when you walk in. Sit in a big rocking chair until everyone is close again. When the children squirm to get down, you can turn to chores and housework.

12. Take children along as you do chores. Teach them to help with the housework, and give them approval when they do.

13. Each parent should have a special time alone with each child every week. Even an hour will do.

14. Don't let yourself be overwhelmed by stress. Instead, enjoy the pleasures of solving problems together. You can establish a pattern of working as a team.

Source: T. Berry Brazelton, M.D., *Newsweek,* Feb. 13, 1989, pp. 68–69. Courtesy *Newsweek* Magazine.

dressed in some detail in Chapter 15. But here for a moment we want to point to the work–family conflict for the single custodial parent, and the following experience captures it well:

I am stopped dead on the San Francisco Bay Bridge. I shut off the engine; I have not moved for twenty minutes. The minutes and then the hours tick by. My heart beats faster and louder as I stare at the dashboard clock. I was due to pick up my four-year-old son at five o'clock. The hands are crawling past six o'clock, seven o'clock, eight o'clock. Finally the cars break free and I race to the baby-sitter's house. She primly informs me that she had no idea where I was and she has called the police to take my son. Choking with anxiety, I drive

to the police station. The cruel-mouthed sergeant tells me that my son has been sent to his father. The police have determined that I am clearly not a fit mother and will no longer be allowed to take care of him.

This is my recurring nightmare. It is founded in a real incident which occurred ten years ago when I was living through a painful divorce and trying to learn the difficult role of single-parent working mother. The real-life incident did not end disastrously; the baby-sitter was merely annoyed. But clearly the sharp anxiety and profound feelings of inadequacy which were evoked by that very troubled period have not yet been put to rest. (Mason 1988, pp. 11–12)

Although the rough edges of the work–family conflict may be particularly sharp for single parents, two-earner marriages assuredly have them also. When national surveys ask people what is most important in their lives, women tend to put family ties first, whereas white men rank family ties first or second, along with work. For highly successful black husbands, marital happiness may be more important to overall life satisfaction than work success (Thomas 1990). "For both sexes, love is very important" (Cancian 1987, p. 75; Barnett, Marshall, and Pleck 1992). But whether for a male or female, "career and family involvement have never been combined easily in the same person" (Hunt and Hunt 1986, p. 280).

The media have given us the new man, the baby boom era (or younger) husband who shares wage-earning and family responsibilities on an egalitarian basis and relates warmly to his children. Indeed, more fathers are taking time off work following the birth of a child, and they are more visible in parenting classes, in pediatricians' offices, and dropping off and picking up children in day-care centers (Levine, in Lawson 1990). But "the image of the new man is like the image of the supermom: it obscures the strain" (Hochschild 1989, p. 31).

The primary source of strain is apparently role conflict: Employers consciously and unconsciously fail to see unpaid family work as important and do not believe that employees, especially males, should allow family responsibilities to interfere with labor force involvement. For example, one husband related that when he'd told his office manager he had to stay home one morning to await the delivery of new furniture, his boss asked, "Why can't your wife wait?" (in Hochschild 1989, p. 102). In the mid-1980s a research organization found that of the 119 major corporations surveyed that offered paternity leave, 41 percent responded that "no time" was the appropriate amount of time for a man to take off at the birth of his child (Gregg 1986, p. 50). As a result of these workplace attitudes, some husbands report having lied to bosses or taking other evasive steps at work to hide conflicts between job and family. One man told his boss that he has "another meeting" so that he can leave the office each day at 6 P.M.; "I never say it's a meeting with my family." Another man said he parked his car in the back lot to avoid having to pass his boss at 5:30 P.M. while leaving for the child-care center. The fact that "men are not supposed to feel these sorts of conflicts," according to Families and Work Institute researcher James Levine, makes matters worse, increasing personal stress (Shellenbarger 1991a). As Hochschild concluded after her observations in two-earner homes,

> men who shared the load at home seemed just as pressed for time as their wives. . . . As long as the "woman's work" that some men do is socially devalued, as long as it is defined as woman's work, as long as it's tacked on to a "regular" work day, men who share it are likely to develop the same jagged mouth and frazzled hair as the [full-time employed] mom. (Hochschild 1989, p. 31)

Many (but not all) of the two-earner partners interviewed by Hochschild were two-career couples. In the next section we will look at some stresses peculiar to two-career marriages.

Two-Career Marriages—Juggling Even Faster

Twenty years ago, Hunt and Hunt (1977) noted that dual-career families require a support system of child-care providers and household help that depends heavily on ability to pay, which is inherently limited to a small number of families. Today's two-career union is premised on the existence of a labor pool of low-paid, but highly dependable, household help. The vast majority of such help is provided by women, many of whom have their own families to worry about (Romero 1992). (See Box 13.2, "A Closer Look at Family Diversity: Diversity and Child Care.") Even parents who can afford to pay for it find that locating such help can be difficult.

Two-career couples with children may find themselves juggling even faster than spouses in other two-earner marriages. For example, two careers requiring travel can present added problems as parents "scramble to patch things together" for overnight child care (Shellenbarger 1991b). Two-career partners need the dexterity to balance not only career and family life but also her and his careers so that both spouses prosper professionally in what they see as a fair way. In the 1980s sociologist Rosanna Hertz (1986) conducted an in-depth study of twenty-one two-career couples in the metropolitan Chicago area. "For two careers to flourish, equity in these marriages becomes crucial," she concluded (p. 55).

Diversity and Child Care

Every weekday evening in affluent homes across America, two groups of women trade places.

Mothers who follow careers come home and the women who are paid to care for their children prepare to depart or step aside....

For Ruth Sarfaty, 31 years old, a Manhattan public relations executive with a 2¹/₂-year-old daughter, the other woman is Cheryl Ryan, 40, a mother of five from Trinidad. "I'm completely dependent on her," said Ms. Sarfaty, whose husband is a real estate broker. "She's in my home more than I am. We could not earn a living without her."

Yet, fundamental as these arrangements may be, there are few working relationships that are more ambiguous, complex or ultimately fragile....

[T]he mothers who seek such help typically have few examples to rely on. Their mothers stayed home. These daughters are not even sure what the other women in their lives should be called. Families often refer to them as baby sitters. The experts prefer nanny or caregiver. There are rarely credentials or contracts. Many caregivers are illegal aliens who are grateful for work that is paid for off the books.

A Gamble, and for Low Pay

While in-home caregivers in cities like Washington and New York receive an average of about $250 a week, the national average is $94; among work categories on which statistics are kept, only people of the clergy are paid less. Sixty-two percent leave their jobs each year; the only higher turnover is among gas station attendants. "It's a gamble," Professor Zigler said. "If you get a wonderful one, it's like having a new, valued family member. If you get an awful one, you and your child are in trouble." ...

Caregivers can wield tremendous power over these women. A 35-year-old investment banker, Maureen White, described "the tyranny of the nannies." She said: "You think, if the nanny is happy, the baby is happy. If the baby's happy, you're happy. If you're happy, your husband's happy."

When problems come up, they can throw a family into chaos. Stories of caregivers' abrupt departures abound....

Even if the arrangement is running smoothly, women who employ caregivers often wonder about their influence, especially as the children start talking. "You're always re-evaluating the choice," said Rhea Paul, a professor of speech and hearing science at Courtland State University, Oregon, who has hired a grandmother from the Netherlands to care for her three children.

"Every day you come home and ask yourself: 'Am I doing the right thing?' Sometimes I think I should have stayed home."

Special Fabrics of Living

Women who strive to be model mothers learn to make accommodations as their children and caregivers develop their own rituals. Amy Samuelson's 3-year-old son, Zachary, has looked forward to weekly outings to McDonald's with his caregiver. Ms. Samuelson, who is a nutritionist, said, "It's not my favorite place, but it's important to Zachary."

Such domestic arrangements, while unfamiliar to many professional women, have long been a part of the fabric of life for the wealthiest Americans and for white Southern families.

Today, while there are no reliable estimates of the number of in-home caregivers, it seems clear that not enough skilled ones are available. Caregivers often tell of being approached in playgrounds with their charges by other mothers seeking to lure them away from their employer.

Yet for all the demand, many caregivers complain bitterly about how they are treated. "In this country baby sitting is not seen as a job," said a Jamaican caregiver who keeps a photograph of her employer's child on her dresser at home along with a photograph of her own son. "People think it's just domestic work."

(continued)

(continued)

"It's long hours," said Marie Gaston, 50, who works as a live-in caregiver in Aspen, Colo. "When you travel with them, it's 24 hours. Sometimes it's two or three weeks before you get time off. There's no overtime." ...

In some cases, such employment can lead to legal status in the United States, but even that has its drawbacks, according to one caregiver. "They seem nice at first," the caregiver said of such employers. "And then when they know they're going to sponsor you they start to treat you differently because they know if you don't work with them you're going to have to start over with someone else again. Some of them treat you like a slave. I have friends who come for baby sitting, and they make you rake the leaves."

Caroline Brownell, who directs an employment agency in San Rafael, California, said she recently decided to stop placing caregivers because of the unrealistic demands of "corporate mothers."

A few efforts are under way to introduce standards. Mary Starkey, who runs a placement agency in Denver, helped found the 300-member International Nanny Association to seek better pay and working conditions. "We're really starting at the beginning," Ms. Starkey said. ...

Five years ago, the 92d Street Y in Manhattan started what may be the first support and education course for caregivers. Roanna Shorofsky, director of the nursery school there, said, "I realized what was happening when I looked down the hallway one day and saw all the caregivers picking up the kids." Of the 150 children enrolled in the nursery school last year, a third had in-home care.

Among the caregivers in last year's course were women from Ireland, Haiti, West Germany, Brazil, China and Jamaica. With the prevalence of foreign-born caregivers, many children move in and out of distinctly different cultural worlds.

"There is no Mary Poppins," said a 39-year-old Manhattan advertising sales executive who interviewed 50 women before she found a 50-year-old woman from Trinidad, who has worked for her for seven years and now attends to a 7-year-old and a 4-year-old.

"You have to constantly make compromises," said the advertising woman, who asked not to be identified. "This is one of the most important relationships I'll have in my entire life. I work at it all the time." ...

Sometimes no effort can keep arrangements from falling apart. That was the experience of Ms. Samuelson, 38, a corporate nutritionist who lives in Riverdale with her husband, a college professor, and Zachary, their son. The first woman she hired had taken excellent care of a friend's children but became pregnant in the Samuelson's employ and left after seven months.

The second woman, whom Ms. Samuelson found through an advertisement in The Irish Echo newspaper, lasted one week. A third woman left after six months to return to her native Jamaica to care for her sick mother.

Finally, nearly two years ago, Ms. Samuelson found a 35-year-old Jamaican who was devoted to Zachary. But three months ago the woman announced that she was five months pregnant and not sure she would continue working after she had the baby.

"When child care breaks down," Ms. Samuelson said, "everything else breaks down."

Sources: Sara Rimer, © *New York Times,* Dec. 26, 1988; see also Romero 1992; Lipman 1993.

The two-career couples in Hertz's sample were not necessarily ideological egalitarians, but had fallen into a new (egalitarian) style of marriage as a consequence of pragmatic management of work and family lives pursued with a keen desire for success. Once embarked on this route, however, dual-career couples face a constant struggle not to fall back on the old rules and roles of marriage that they witnessed as children. Marital equity is not taken for granted; it is worked for. As one husband said, "I certainly don't think this is a gloriously equal marriage marching off into the sunset. I think we struggle for equality all the time. And we remind each other when we are not getting it." . . .

Couples rarely spoke directly about equality. They spoke instead of trying to strike a balance between careers and family commitments by keeping each other in check, so that neither spouse could tip the scale in favor of his or her own career. "You both make a trade-off. It's hard because we both frequently talk about it—how much trade-off each one of us should be making, whether we are both making the same amount, or whether one of us feels that the other one is spending too much time on the job. . . . If one of us thinks the other one is not doing what they should be doing overall, he has to come out and say it. She's got to tell me as well." (in Hertz 1986, pp. 55–57)

The balance between partners can be upset by career fluctuations as well as family time allocations. The contrast between one career that is going well and one that is not can be hard on the partner on the down side. But the marriage can "operate as a buffer, cushioning the negative impacts of failures or reversals in one or the other career" (Hertz 1986, p. 59). When the marriage is rewarding, compromises such as turning down opportunities that would require relocation are acceptable because of the importance given to marriage as well as career.

Hertz (1986) found that couples were realistic, though sometimes regretful, about some benefits of the traditional relationships they are giving up. Although men acknowledged that their wives provided less husband care, they appreciated the excitement, and also the status, associated with an achieving wife. Some of them had considered or made career changes that would not have been possible if wives had not been successful wage earners. Both partners claimed fulfillment from and assigned emotional meaning to an egalitarian dual-career marriage: " 'She has a sense of a full partnership and she should'" (p. 75).

Commitment was perceived as truer: "'Working . . . has decreased my dependence. . . . That makes it into much more of a voluntary relationship than an involuntary one of my staying with him because I need to stay with him. He's not staying with me because he's obligated to somehow, and it just makes it into a much clearer situation'" (p. 75). Very visible to Hertz was the way in which communication was enhanced by similar lives, making possible a higher level of mutual support than in conventional couples. "These couples . . . [had] a different level of understanding about each other's lives, a level that is intimate and empathic" (p. 77).

The couples studied by Hertz did report conflict over balancing time, commitment, and career moves. Indeed, the geography of two careers presents a significant challenge to couples.

THE GEOGRAPHY OF TWO CAREERS   Juggling two careers and a relationship can be especially difficult if each member of a couple lives in widely separated places. One couple, for example, now in their early 30s and friends since college, say they would be married now if it weren't for their careers. Instead, they maintain a long-distance relationship between the East Coast, where he is an architect, and Chicago, where she is a television producer. "Who's going to move and who's going where? It's sort of like a big game of chicken," he explains. "My heart is with him, but I'm also having the best time of my life right now at work," she explains (in Stern 1991, p. B1).

Because career advancement often requires geographic mobility—and even international transfers— juggling two careers can prove difficult for marrieds. A career move for one partner can make the other a **trailing spouse** who relocates to accommodate the other's (but not one's own) career. Increasingly, couples turn down transfers because of two-career issues. As a result, some large companies now offer career-opportunity assistance to a trailing spouse, such as hiring a job search firm, intercompany networking, attempts to locate a position for the spouse in the same institution, or career counseling (Lublin 1992).

Although wives still move for their husband's careers more often than the reverse (Shihadeh 1991), the number of trailing husbands has increased in recent years, perhaps to 10 to 30 percent of all trailing spouses. Counselors who work with trailing husbands note that these men have few role models and must confront norms and social pressures that conflict with their decisions. Financial pressures when a trailing spouse cannot find a job may intensify the strain. But more two-career marriages today are based on a conscious mutuality to which partners have become accustomed by the time a career move presents itself. Such couples are less likely to have problems with a female-led relocation than are more traditional marrieds. For many spouses, trailing is preferable to commuting, another solution to the problem of opportunities in two locations.

TO COMMUTE OR NOT TO COMMUTE? Social scientists have called marriages in which spouses live apart **commuter marriages** or **two-location families.** The vast majority of commuting couples would rather not do so but endure the separation for the sake of career or other goals. Since research began on commuter marriages in the early 1970s, social scientists have drawn different conclusions. Some studies suggest that the benefits of such marriages—greater economic and emotional equality between spouses and the potential for better communication—counter its drawbacks. Other research suggests a different view, focusing on difficulties in managing the lifestyle. One conclusion to be drawn from the research is that commuters who are able to have frequent reunions are happier with the lifestyle than those who cannot.

One study (Bunker et al. 1992) compared life satisfaction for 90 commuting and 133 single-resident, two-career couples. Almost three-fourths of the commuters saw their partner weekly. The researchers were surprised to find that commuters experienced less stress and overload than the single-residence couples. "Perhaps there is some restructuring in the commuting two-residence couple that simplifies life or perceptions of it. Perhaps short separations facilitate compartmentalization, allowing commuters to keep work life and family life in well-separated spheres, and to confront the demands of each role in alternation rather than simultaneously" (p. 405). Then too, the commuter couples had significantly fewer babies and young children than did single-resident couples; commuter marriages probably work better in the absence of dependent children (Stern 1991).

Generally, the researchers concluded that commuting has both rewards and costs. Commuters reported more satisfaction with their work life than did single-residence, two-career respondents, but commuters were significantly less satisfied with their partner relationships and family life (Bunker et al. 1992). Some commuters say they can put their respective careers first for only so long before their relationship frays:

"Career-wise it was absolutely fantastic," said one spouse who had given it up. "Personally, it was absolutely horrible. It was kind of like a teeter-totter on both extremes" (in Stern 1991, p. B3). Consider the following example:

> Michael and Judy Casper intended to celebrate their 25th wedding anniversary last month with a three-week vacation in Alaska. But their conflicting careers scotched the trip.
>
> Michael couldn't get away from his new job—as a Gulf Bank general manager in Kuwait City. He says he spent the anniversary watching a rented videotape.
>
> Judy lives 7,700 miles away in Houston, where she owns a thriving home-interior shop—her first paid job after following her husband around the world for 17 years. She celebrated the anniversary by going out for Italian food with their three college-age children. Later, she cried. (Lublin 1992, p. B1)

"Perhaps reflecting the strong influence of these relationship issues on overall quality of life," commuters were less satisfied with life as a whole than were single-residence couples (Bunker et al. 1992, p. 405).

Couples who have been married for shorter periods of time seem to have more difficulties with commuter marriages. Younger couples who are simultaneously beginning their careers and their relationship often lack not only experience but also information about managing a two-career marriage. Perhaps because of their history of shared time, more "established" couples in commuter marriages have a greater "commitment to the unit" (Gross 1980).

This general discussion of juggling paid and unpaid labor points to the fact that despite the benefits of employment to women and their families, and despite the broad societal pressures and gender role changes leading to high female employment rates, neither families nor public policy have fully adapted to this change. The next section examines policy issues regarding work and family.

Social Policy, Work, and Family

Our discussions so far show that employment and household labor conflict; that is, "time spent in one sphere means less time spent in another. If commitments to paid labor and household labor call for

Day-care centers and preschools provide care for many children during the workday. While children may not receive as much direct adult attention as they might with a single caregiver or at family day care, they benefit from greater interaction with other children and a preschool curriculum.

full-time participation in both, that time must come either at the expense of leisure or else some of the demands of paid labor and household labor must go unmet" (Shelton 1992, p. 143). Corporate policy expert Ellen Galinsky recounts the following not unusual episode from the life of a working family:

One father told of a morning when his four-year-old daughter lost her shoes. Amid all the other normal complications of getting two young children dressed, fed, and ready for daycare, he and his wife searched everywhere but the shoes were nowhere to be found. The mother began to go through the house a second time while the father made calls to find someone to take over his carpool to the metal factory where he worked. Because his factory had strict time policies (warnings were issued; after several warnings a worker could be fired), he felt frantic, as did his wife who also had a job tied to the time clock. Eventually they settled for last year's too small shoes, put them on their daughter, and they all rushed out the door. Both parents were late for work, and were censored by their supervisors. When they got home from work, they found a note from their daughter's teacher—it asked them please to buy shoes for their child that fit. (Galinsky 1986, pp. 109–10)

Although single mothers and married women in the labor force "are an established fact, the institutions of society still are geared to meet the needs of two-parent families with only one employed partner" (Vannoy-Hiller and Philliber 1989, p. 101). This section examines public or social policy issues as society responds (or fails to respond) to today's work–family conflicts. Policy issues center on two questions: "What is needed?" and "Who will provide it?"

What Is Needed to Resolve Work–Family Issues?

Researchers and other work–family experts are in general agreement that single-parent and two-earner families are in need of more adequate provisions for child and elder care, family leave, and more flexible employment scheduling. We will examine each of these separately.

CHILD CARE   Policy researchers define **child care** as the full-time care and education of children under age 6, care before and after school and during school vacations for older children, and overnight care when employed parents must travel. Child care may be paid or unpaid and provided by relatives or others. About two-thirds of children in child care are cared for in their own homes by relatives, neighbors, or paid nannies (Clinton 1990). Low-income (Otten 1990), rural (Atkinson 1994), black, and Hispanic parents are especially likely to have relatives take care of their children (Wilkie 1988).

Paid care may be provided in the child's home by a caregiver who comes to the house daily or lives in (see Box 13.3). The term **family day care** refers to care provided in a caregiver's home, often by an older woman or a mother who has chosen to remain out of the labor force to care for her own children. Family day care supplies an estimated 41 percent of child care used outside the home and is the most widely used type of care by nonrelatives for toddlers and infants (Atkinson 1992). Parents who prefer family day care seem to be seeking a familylike atmosphere, with a smaller scale, less routinized setting. Perhaps they also desire social similarity of caregiver and parent to better ensure that their children are socialized according to their own values. **Center care** provides group care for a larger number of children in day-care centers. Use of day-care centers has increased rapidly; in 1992, 4 million American children were in child-care centers (Shellenbarger 1992e), partly a result of the growing scarcity of in-home caregivers as relatives or neighbors who formerly cared for children now join the labor force themselves. Increased center-care use is also due to the perception that center day care offers greater safety and a strong preschool curriculum. About half of parents using paid care change their arrangements each year because a caregiver quit, the cost was too high, the hours or location were inconvenient, the child was unhappy, or the parent disliked the caregiver (Shellenbarger 1991d).

Research by the Families and Work Institute has found four sources of parental stress regarding child care: (1) It is difficult to find, (2) some arrangements are of lower quality than others, (3) child care is expensive, and (4) "parents are forced to put together a patchwork system of care that tends to fall apart" (Galinsky and Stein 1990, pp. 369–70).

Talk to employed parents of young children for any length of time, and the subject turns to the difficulty of finding day care. Some parents pay providers months in advance to reserve space for babies not yet born. Other couples may work at timing pregnancies to correspond with rare day-care vacancies. Day care is even more difficult to find for mildly ill youngsters too sick to go to their regular day-care facility, although a few centers have opened in various parts of the country to begin to fill this need (Galinsky and Stein 1990). Then too, family day care and day-care centers are usually open weekdays only and close by 7 P.M. Some parents, such as single mothers on shift work or those who travel, need access to 24-hour care centers.

Adding to the difficulty of finding day care is the fact that parents are looking for *quality* day care, and some facilities are of better quality than others. Some research points to concern that infants in their first year who are in nonparental care for 20 or more hours per week "are at elevated risk of being classified as insecure in their attachments to their mothers at 12 or 18 months of age and of being more disobedient and aggressive when they are from 3 to 8 years old" (Belsky 1990, p. 895). Meanwhile, studies have generally supported the conclusion that children in day care and children cared for by their own parents differ little in development and emotional stability. "The consensus at recent conferences has been that good day care— that is, day care with adequate staffing by trained responsive adults—not only has no ill effects, but can be beneficial" (Lewin, 1989b, p. 91).[5]

Nevertheless, "day care with adequate staffing by trained responsive adults" does not describe all day care in this country. For example, family day-care providers can be unusually stressed due to high demands and limited family and financial resources (Atkinson 1992). It is widely assumed among experts that high provider turnover and repeated staff

---

5. Despite a smattering of confirmed cases, concerns about abuse of children in day care have largely proved unfounded; studies indicate that children are at greater risk of abuse in their own homes (Finkelhor, Hotaling, and Sedlak 1991).

Selecting a Day-Care Facility

The demand for day-care facilities has grown steadily in the past decade as more mothers join the labor force. "In a wave of fundamental social change, day care is becoming a basic need of the American family" (Watson, 1984, p. 14). Although some employers offer child care, finding quality facilities at an affordable cost has become a serious problem for many parents.

Comprehensive and universal government-funded day care does not exist in the United States today, and it is a complicated and controversial issue. The two types of group day care that are available are home-based care, in which one or more individuals care for several children in their own homes, and center-based care, in which children take part in a nursery school program or in an after-school program with a teaching staff and a classroom. Center-based facilities can be local operations or nationally affiliated franchises.

Most child-care experts agree that when day-care facilities are clean, safe, and adequately staffed, and when children are allowed to play together comfortably with some respect for individual needs, such environments are as healthy for children as home care. Besides these general considerations, parents should look for the following criteria in a day-care facility:

- *A stable staff.* Some research indicates that children do best when cared for consistently by the same people. Ask about staff turnover. Because some turnover is inevitable, ask how the situation is handled. How are children informed of and prepared for upcoming staff changes?

- *Low staff-to-child ratio.* Although experts disagree on the proper staff-to-child ratio, federal guidelines for day-care

centers recommend ratios of 1:3 for infants, 1:4 for toddlers, and 1:8 for 3- to 6-year-olds. For home-based facilities, recommended ratios are 1:5 for children under 2 years old and 1:6 for youngsters between 2 and 6 years.

- *A well-trained staff.* Because day-care workers are poorly paid, it is difficult to find centers with highly educated staff. Nevertheless, the ideal situation is for staff to be specifically trained in such fields as child development and psychology.

- *The right kind of attention.* Babies need a responsive adult who coos and talks to them. One-year-olds need a staff member who will name things for them. Two-year-olds need someone who reads to them. Older children need adults as well but can also profit from supervised exchanges (such as games) with other children.

- *Appropriate activities.* Learning the alphabet is fine, but children don't need to be immersed only in academics. Look for a facility that also fosters play and community activities, such as trips to the zoo or fire station.

- *A role for parents.* Any day-care facility should welcome parental involvement. Be wary of centers that do not allow unannounced visits.

- *References.* Ask for names and phone numbers of other parents who have children enrolled there and talk with them about the facility. Be wary of a center that refuses to give you this information.

Source: Adapted from Watson, 1984.

departures associated with day-care centers can result in a child's emotional withdrawal and can otherwise interfere with development. Staff turnover is high and increasing, rising from 43 percent in 1990 to 69 percent in 1992 (Shellenbarger 1992e). Low salaries are a factor; average pay for center teaching assistants actually fell slightly between 1990 and 1992—to

$6.36 an hour (Shellenbarger 1992e). A related issue is that low salaries notwithstanding, keeping center day care affordable may necessitate caregiver–child ratios that are too low for quality care (Lewin 1989b; Clinton 1990). (Box 13.3, "As We Make Choices: Selecting a Day-Care Facility," discusses things parents should look for in choosing day care.)

Indeed, paid child care is expensive. For those paying for child care, costs amount to between 10 and 35 percent of the family budget. Lower-income families pay a disproportionately higher percentage of their earned income than do middle- or upper-income families—often equal to what they pay for housing (Galinsky and Stein 1990). "Some women have had such a tough time affording childcare that they have had to leave the job" (Aldous 1990, p. 361; Joesch 1991).

As they struggle to find quality, affordable child care, many parents must make more than one arrangement for each child (Folk and Yi 1994). As they patch together a series of child-care arrangements, the system becomes increasingly unpredictable and is inclined to fall apart (Galinsky and Stein 1990). This same situation very probably applies to elder care.

ELDER CARE   **Elder care** involves providing assistance with daily living activities to an elderly relative who is chronically frail, ill, or disabled (Galinsky and Stein 1990). Increasingly, employees will need forms of elder care, such as day care for the elderly (Ettner 1995). The fact that more and more people now live longer means that middle-aged children (especially daughters) are increasingly responsible not only for working outside the home and for child care but also for the needs of aging parents. Employees in this sandwich generation are especially pressed as conflicts arise among roles as employee, parent, and adult child of aging parent. The strains are showing up in the workplace. In a 1990 survey of 7,000 federal workers, nearly half said they cared for dependent adults. Of those, three-quarters had missed some work: Sixteen percent had missed 40 hours of work or more in one year, and nearly half of those had missed the equivalent of two weeks or more of work in a year (Beck 1990b, p. 51). Research in one company indicated that one of every four employees responsible for the care of elderly relatives had problems with the cost of services needed, and one employee in five worried about the relative while at work (Galinsky and Stein 1990). In recent years, about 14 percent of caregivers to the elderly have switched from full- to part-time jobs and 12 percent have left the work force, according to the American Association of Retired Persons. Another 28 percent have considered quitting their jobs. About 3 percent of U.S. companies offer employees some help with elderly dependents, such as lo-

cating outside assistance in cases of emergency (Beck 1990b).

FAMILY LEAVE   **Family leave** involves an employer's allowing workers to take an extended period of time from work, either paid or unpaid, for the purpose of caring for a newborn, for a newly adopted or seriously ill child, or for an elderly parent, with the guarantee of a job upon returning. The concept of family leave incorporates maternity, paternity, and elder-care leaves.

The 1993 Family and Medical Leave Act mandates up to twelve weeks of unpaid family leave for workers in companies with at least fifty employees. Mandated *unpaid* leave will not solve the problem for a vast majority of employees, however, as most working parents need the income. Indeed, some union officials and others worry that concern with corporate work–family policies may be shifting attention away from more central issues, such as wages. "In fact, women with low pay cannot afford to take advantage of unpaid family benefits programs like the . . . unpaid leave legislation, even when they are available. And many women receive low pay" (Aldous 1990, p. 361).

FLEXIBLE SCHEDULING   About 12 percent of full-time workers have flexible schedules. **Flexible scheduling** includes options such as **job sharing** (two people share one position), working at home or telecommuting, compressed work weeks, flextime, and personal days. Compressed work weeks allow an employee to concentrate the work week into three or four, sometimes slightly longer, days. **Flextime** involves flexible starting and ending times with required core hours, usually 10 A.M. to 3:30 P.M. A frequently requested flexible scheduling provision involves **personal days.** Employees dislike feeling they have to lie to stay home with a sick child or an elderly parent; they want these days away from work to be legitimate days off. Some companies have responded positively with employee personal days, which typically are unpaid (Christensen and Staines 1990).

Flexible scheduling, although not a panacea, can help parents share child care or be at home before and after an older child's school hours. Some types of work do not lend themselves to flexible scheduling (Christensen and Staines 1990), but the practice has begun to catch on among federal government workers and in a few large companies, partly because it offers employee-recruiting advantages and prevents turnover (Trost 1992).

Another important reason that companies are now offering some workers flexible scheduling is that it saves the company money. For example, twenty employees at one firm who work at home two days each week saved their employer thousands of dollars in leasing costs by sharing office space. As another example, it is not uncommon for employees on compressed work weeks to accomplish their entire job for reduced pay (Shellenbarger 1992d). Furthermore, job sharing and reduced work weeks allow a company to keep workers rather than laying them off permanently. According to a DuPont spokesperson, "We have widely communicated that flexible work practices are an alternative for avoiding layoffs and keeping valued employees while going through current cost-cutting activities" (in Trost 1992). Some union officials charge that moves toward flexible schedules may actually be corporate ploys to replace most costly full-time jobs with part-time work offering fewer benefits (Christensen and Staines 1990).

Companies are finding that employees on flexible schedules like them and "will do almost anything" to make them work (Shellenbarger 1992d). Yet jobs with flexible schedules often pay less, and there are proportionately fewer takers in today's weak economy (Trost 1992). Moreover, although flextime has gained currency as one solution to work–family conflicts, research shows that it is not as beneficial as often hoped. Employees who get flextime report enhanced job satisfaction—possibly due to greater feelings of autonomy—but do not find that flextime alleviates all or even most family–work conflicts (Christensen and Staines 1990).

A MOMMY TRACK? In a *New York Times* survey, 83 percent of women "say they [sometimes] feel torn between the demands of their job and wanting to spend more time with their family"; still, 55 percent agree that "a woman who takes time out of her career for child care can never make up for it on the job." Forty-nine percent "say women have had to give up too much to get ahead" (Belkin 1989, p. 26). A career mother who was nursing her baby reported having weaned it earlier than she had planned in order to make a business trip (Shellenbarger 1991b), for example. As one mother said, "I think that we have been granted an equality that is not yet livable and a set of 'choices' that are mostly irreconcilable" (in Triedman 1989, p. 61).

An effort to create formal alternatives involved the

mommy track, the instant label given to corporate career consultant Felice Schwartz's proposal in the *Harvard Business Review* (1989a). To meet the needs of women who wished to devote more time to their families during the childbearing and early child-rearing years, Schwartz proposed that business (and by extension, professions such as law) create positions that demand less time and overall involvement. "Career and family women" would elect that track while "career primary women" continued to meet the heavy demands of the "fast track." Schwartz acknowledged that "career and family women" would most likely not make it to the top ranks of management or the professions. But they could enjoy middle-management jobs while maintaining their families with less stress, a trade-off many might welcome (Schwartz 1989a).

Schwartz herself did not use the term *mommy track,* and her proposal was well received at higher levels of the corporate world. But the term conveys the negative response from many policy analysts and from corporate and professional women as well. Concern (not to mention distress) was expressed that such a policy would reinforce society's assigning to women (but not men) primary social responsibility for child care. By assuming that they would likely have problems combining work and family, the proposal also risked condemning women in general to continued second-class employment status (Kantrowitz 1989; Lewin 1989a). In reply, Schwartz argued that

the danger of charting our direction on the basis of wishful thinking is clear. Whether or not men play a greater role in child rearing, companies must reduce the family-related stresses on working women. The flexibility companies provide for women now will be a model in the very near future for men— thus women will not be forced to continue to take primary responsibility for child care. . . . What I advocate is that companies create options that allow employees to set their own pace. (Schwartz 1989b, p. 27)

In her subsequent book written with corporation managers in mind, Schwartz (1992) argued again that "as things stand now, the business world must enlarge its infrastructure to support the family needs of employees." To that end, Schwartz recommended that corporate America should:

• Accept the fact that you must be flexible and must provide family supports. The alternatives include unacceptable rates of turnover, terrible losses in productivity, and exclusion from the leadership pool of high-potential, high-performing women (and, increasingly, men) who want to be involved in their children's lives.

- Provide the full range of ongoing benefits to women on disability and maternity leave and to those who return part-time.

- Let women who have babies return when they're ready. By "ready" I mean when they feel well psychologically and physically, when they're getting enough sleep to function effectively during the day, when they feel they've bonded with their babies, and when they have located, tested, and are satisfied with whatever kind of child care they've chosen.

- Let women return from maternity leave on less than full-time or other alternative schedules—part-time, shared, telecommunicating arrangements—for as long as they want.

- Permit new fathers to take parental leaves, sequencing them with those of their wives.

- Establish a policy that permits parents to cut back to half-time (at prorated pay) and re-enter the competition for senior management levels, partnership, or tenure if they choose. (Translation: Don't put them on a "mommy track.")

- When a woman (or a man) is out on leave or working part-time for a significant period, provide the additional heads and hands that are necessary to get the work done.

- Take responsibility—in partnership with parents, communities, and governments—for making high-quality, affordable child care available for every child.

- Find every opportunity you can to enable parents (and other executives who need uninterrupted time to think and work) to work at home.

- Contract out as much work as you can.

- Finally, the *sine qua non* for all of the above: Learn how to measure productivity instead of time in the office and put systems in place to do so. (Schwartz 1992, pp. 203–4)

Controversy has surrounded the suggestion by some of a separate career path, often labeled "the mommy track," for women who want to combine family and career. This graduate will probably find the next few years to be demanding, challenging, and rewarding.

Although concerns about mommy tracks persist (Trost and Hymowitz 1990), some experts view mommy track options as more advantageous than their absence (Raabe 1990). Meanwhile, a few companies have begun to make "family-friendly" policies, such as those listed above, explicitly available to both male and female employees (Shellenbarger 1992f). In sum, "It is not that women have different roles to play but that parenting is patently impossible if both men and women take on the traditional male role" (Moen 1992, p. 128).

THE LIFE COURSE SOLUTION   As an alternative to the mommy track, one policy analyst (Moen 1992) has offered the radical suggestion that all of us—parents, educators, corporate executive officers (CEOs), middle managers, and regular employees—push for a complete redefinition of the individual life course. "Women's entrance into the workplace has highlighted the conflict between paid employment and family life. It also suggests a question which is seldom asked: Are there conflicts between men's work and the family? For many families, the answer is a resounding yes" (Quarm 1984, p. 205). According to the **life course solution,** policymakers and the rest of us would rethink "the lockstep pattern of education, employment, and retirement." Doing so

> could lead to a variety of arrangements, including a return to school at various ages and a continuation of paid work well beyond the usual age of retirement. It could also encourage both men and women to cut back on their working hours or to take extended sabbaticals while their children are young. The changes we as a society are experiencing call loudly for a thoughtful reappraisal of existing life patterns. This could lead to a reconfiguration of the life course in ways that create more options and a greater diversity for both men and women in youth, early adulthood, mid-life, and the later years. (Quarm 1984, p. 129)

The life course solution assumes that all Americans need changed social policies with regard to work–family conflicts. Who will lead in making policy changes is a matter of debate.

Who Will Provide What Is Needed to Resolve Work–Family Issues?

Federal policymakers hotly disagree over whose ultimate responsibility it is to provide what is needed regarding various work–family solutions. A principal conflict concerns whether solutions such as child care or family leave constitute employees' rights, or privileges for which a worker must negotiate.

The countries of northwestern Europe, which have a more pronatalist and social-welfare orientation, are far more likely than the United States to view family concerns as a right. There is "the pervasive belief . . . that children are a precious national resource for which society has collective responsibility . . . to help develop and thrive" (Clinton 1990, p. 25). Putting this belief into practice, most European countries are committed to *paid* maternity leave for up to a year or longer (Aldous and Dumon 1980). In 1974, the Swedish government extended its one-year paid maternity leave to men and changed the term to *parental leave,* arguing that the move was "an important sign that the father and mother share the responsibility for the care of the child" (Haas 1990, p. 402). In fact, the United States is the only major industrialized nation that does *not* provide paid maternity leave, and many less wealthy nations provide such leave (Seager and Olson 1986). Among industrialized nations, further-

more, only the United States and South Africa lack a national child-care policy. In the absence of any integrated federal family policy in the United States, the result is a hodgepodge of piecemeal legislation aimed at solving specific, critical problems.

In the wake of President Bush's 1992 veto of the Family and Medical Leave Act—and prior to President Clinton's subsequent signing of it—a consortium of 137 companies, cities, and agencies voluntarily formed the American Business Collaboration for Quality Dependent Care. The group announced that it would spend $25 million to make major improvements in the quality and supply of child-care and elder-care services in forty-four communities across the nation (Shellenbarger and Trost 1992a). But according to a spokesperson at the Children's Defense Fund, the effort (although notable) does little to address the nation's most fundamental child-care problems, such as affordability and low teacher salaries, and hence "doesn't cure the overall day-care problems for American families" (Shellenbarger and Trost 1992a).

The collaborative effort is an example of some large corporations' apparent growing interest in effecting **family-friendly workplace policies** that are supportive of employee efforts to combine family and work commitments. Such policies include on-site day-care centers, sick-child-care facilities, other organizationally or financially assisted child-care services, flexible schedules, parental or family leaves, workplace seminars and counseling programs, and support groups for employed parents (Galinsky and Stein 1990). Because such programs are voluntary, companies that initiate them tend to have CEOs (often females) who are committed to helping resolve work–family conflicts (Kingston 1990). Research on family-friendly outcomes for employers is currently inadequate (Raabe 1990), but what there is suggests that such programs help in recruitment, reduce employee stress and turnover, and enhance morale and thus increase productivity (Galinsky and Stein 1990; Shellenbarger and Trost 1992b).

But family-friendly policies are hardly available to all American workers, of course (Trczinski and Finn-Stevenson 1991; Shellenbarger 1993). Companies that help provide or subsidize child care are likely to be large and in service industries rather than in, for example, manufacturing. And employees at corporate headquarters typically have greater access to on-site day care than do workers in branch offices (Shellenbarger 1992b). Furthermore, family-friendly policies are not necessarily available to all employees even within a company that offers them. Professionals and managers are much more likely than technical and clerical workers to have access to leave policies, telecommuting, or other forms of flexible scheduling (Christensen and Staines 1990)—or to be able to afford tuition at corporate, on-site day-care centers. "Just as top managers get more stock options and nicer offices, they get better child-care deals than the working class. One company supplying child care even has a division—called ExecuTots—catering to professionals. . . . In factory work, once you punch your time card, your children don't exist" (Guyon 1992).

Moreover, companies have tended to leave the administration of work–family policies to the discretion of individual managers or supervisors. Although some aggressively promote work–family benefits, others view them less favorably (Galinsky and Stein 1990). Thus, many employees are reluctant to lobby for child-care benefits or request time off to attend to family needs (Auerbach 1990; Beck 1990b). "Some women are so nervous that they try to appear biologically correct at all times," according to one vice-president for human resources (in Guyon 1992). Equal Employment Opportunities Commission guidelines state that employers offering leave to mothers must offer the same leave to fathers. But some fathers say their managers have denied them parental leave, and other employees believe that "even discussing childcare issues at work is freighted with potential for embarrassment, censure and career-limiting exposure" (Guyon 1992; see also Schwartz 1992, pp. 39–62). Then too, supervisors approve various benefits on a case-by-case basis. Not defined as employee rights, flexible scheduling and other benefits become rewards for prized performers (Shellenbarger 1992b). "Winning benefits . . . has become a delicate stealthy art, played best by working parents in positions of power" (Guyon 1991).

A few experts point to family-friendly companies as indicative of positive future trends. But many analysts warn that voluntary programs and benefits depend on cost constraints and corporate self-interest (Kingston 1990) and are not likely to become widespread (Aldous 1990; Shellenbarger 1993). According to then Secretary of Labor William Bruck, speaking in the late 1980s, "It's just incredible that we have seen the femi-

nization of the work force with no more adaptation than we have had. . . . It is a problem of sufficient magnitude that everybody is going to have to play a role: families, individuals, businesses [and] government" (in O'Connell and Bloom 1987, p. 11). Even though some positive changes have occurred since Bruck made this point, the situation remains very much as he described it. Nevertheless,

> we cannot leave to a few progressive business leaders the problem of easing the well-documented conflicts between fulfilling job requirements and caring for family members. Governmental action requiring businesses to give these benefits and subsidies to fund them is necessary. . . . The business of America is *families,* and especially the children they nurture. If we skimp on their welfare, there will be no capable workers to carry on business. (Aldous 1990, p. 365)

"One important step in realizing a more satisfying balance between work and family lives is to recognize the full scope of the challenge" (Kingston 1990, p. 453).

We have devoted considerable attention to work–family policy because these issues so strongly influence the options and choices of individual families. Working in the political arena toward the kinds of changes we want is one aspect of creating satisfying marriages and families. Given social constraints, employed couples want to know what *they* can do to maintain happy marriages. We turn to that topic now.

The Two-Earner Marriage and the Relationship

We have been addressing problems associated with two-earner marriages. But research shows that—provided there is enough time to accomplish things—a person's having multiple roles (such as employee/spouse/parent) does not add to stress and in fact can enhance personal happiness (Small and Riley 1990; Guelzow, Bird, and Koball 1991; O'Neil and Greenberger 1994). Research also points to the heightened satisfaction, excitement, and vitality that two-earner couples can have because these partners are more likely to have common experiences and shared worldviews than are more traditional spouses, who often lead very different everyday lives (Chafetz 1989; Hughes, Galinsky, and Morris 1992). At the same time, however, conflict can arise in two-earner marriages as couples negotiate the division of household labor, or second shift (Hochschild 1989), and more generally adjust to changing roles (Orbuch and Custer 1995).

Negotiating the Second Shift

How a couple allocates paid and unpaid work represents a combination of each partner's **gender strategy**—a way of working through everyday situations that takes into account an individual's beliefs and deep feelings about gender roles, as well as her or his employment commitments (Hochschild 1989). In today's changing society, one's beliefs and deeper feelings about gender may conflict. For example, a number of men in Hochschild's study articulated egalitarian sentiments but clearly had traditional gut-level feelings.

Even when each spouse's feelings correspond with his or her beliefs, partners may disagree with one another. Tensions exhibited by many of Hochschild's respondents were a consequence of "faster-changing women and slower-changing men" (1989, p. 11). These couples often engaged in prolonged struggles, which did not always end happily. Some husbands distanced themselves to avoid pressure; some justified their reluctance to participate, claiming no time, incompetence, fatigue, clumsiness, squeamishness, and inexperience (Thompson 1991). Some wives in Hochschild's study mentioned divorce statistics, clearly seeing that to press husbands too hard was to risk divorce (Hochschild 1989). Some gave up efforts to get husbands to share, but did so with a resentment that poisoned the marriage.

Moreover, even when spouses share similar attitudes about gender, circumstances may not allow them to act accordingly. In one couple interviewed by Hochschild, for example, both partners held the traditional belief that a wife should be a full-time homemaker. Yet because the couple needed the wife's income, she was employed and they shared housework on a nearly equal basis. How couples manage their everyday lives in the face of these various contradictions reflects a consciously or unconsciously negotiated gender strategy.

One gender strategy noted by Hochschild involves a husband's consistently praising his wife's homemaking skills rather than actually sharing the tasks. As another example, Hochschild observed that some wives who earned more than their husbands "made up" for

their mate's possible loss of self-esteem by doing more of the second shift than their proportionate income would have predicted. Still another gender strategy, used by wives who would like a husband to do more but are reluctant to insist, is to compare their husbands to other men "out there" who apparently are doing even less (Hochschild 1989). Some men suggest an even more favorable comparison: They compare themselves to their father or grandfathers (Thompson 1991). Ironically perhaps, one study that used national survey data found that the more a husband relies on his wife's income, the less housework he does. The author argues that, when being principal breadwinner is in jeopardy, a husband demonstrates his manhood by not doing "women's work" (Brines 1994).

A fairly common gender strategy, according to Hochschild (1989), is to develop **family myths**—"versions of reality that obscure a core truth in order to manage a family tension" (p. 19). (Box 13.4, "Nancy and Evan Holt: A Family Myth and Ongoing Resentments," illustrates how and why a family myth might be created.) For example, when a husband shares housework in a way that contradicts his traditional beliefs and/or feelings, couples may develop a myth alleging the wife's poor health or incompetence in order to protect the man's image of himself. A common family myth defines the wife as an organized and energetic superwoman who has few needs of her own, requires little from her husband, and congratulates herself on how much she can accomplish (Thompson 1991).

Study after study show that marital satisfaction is greater when wives feel that husbands share fairly in the household work (Pina and Bengtson 1993; Barnett 1994; Perry-Jenkins and Folk 1994; Suitor and Pillemer 1994; Ahlander and Bahr 1995). For example, a study of 452 married couples in Cincinnati concluded that

> the factor which most strongly affects the marital quality experienced by both spouses appears to be the ability to give and receive support. Husbands with sensitive personalities experience higher quality marriages and produce higher quality marriages for their wives. Women also experience higher marital quality the more egalitarian she believes her husband's expectations to be. (Vannoy and Philliber 1992, p. 397)

Nancy and Evan Holt: A Family Myth and Ongoing Resentments

The following is excerpted from the story of one couple in Arlie Hochschild's qualitative study, The Second Shift *(1989).*

Nancy Holt arrives home from work, her son, Joey, in one hand and a bag of groceries in the other. As she puts down the groceries and opens the front door, she sees a spill of mail on the hall floor, Joey's half-eaten piece of cinnamon toast on the hall table, and the phone machine's winking red light: a still-life reminder of the morning's frantic rush to distribute the family to the world outside. Nancy, for seven years a social worker, is a short, lithe blond woman of thirty who talks and moves rapidly. She scoops the mail onto the hall table and heads for the kitchen, unbuttoning her coat as she goes. Joey sticks close behind her, intently explaining to her how dump trucks dump things. . . .

Having parked their red station wagon, Evan, her husband, comes in and hangs up his coat. He has picked her up at work and they've arrived home together. Apparently unready to face the kitchen commotion but not quite entitled to relax with the newspaper in the living room, he slowly studies the mail. . . .

From the beginning, Nancy describes herself as an "ardent feminist," an egalitarian (she wants a similar balance of [work and family] spheres and equal power). Nancy began her marriage hoping that she and Evan would base their identities in both their parenthood and their careers. . . . Evan felt it was fine for Nancy to have a career, if she could handle the family too.

As I observe in their home on this evening, I notice a small ripple on the surface of family waters. From the commotion of the kitchen, Nancy calls, "Eva-an, will you *please* set the table?" The word *please* is thick with irritation. . . .

In the sixth year of her marriage, when Nancy . . . intensified her pressure on Evan to commit himself to equal sharing, Evan recalled saying, "Nancy, why don't you cut back to half time, that way you can fit everything in." At first Nancy was baffled: "We've been married all this time, and you *still* don't get it. Work is important to me. I worked *hard* to get my MSW. Why *should* I give it up?" . . .

She couldn't see it Evan's way, and Evan couldn't see it hers.

In years of alternating struggle and compromise, Nancy had seen only fleeting mirages of cooperation, visions that appeared when she got sick or withdrew, and disappeared when she got better or came forward.

Women who perceive their housework load as fair also experience higher psychological well-being (Lennon and Rosenfield 1994).

Maintaining Intimacy During Role Changes

We saw earlier in this chapter that some two-earner couples see themselves as coproviders, with each spouse equally responsible for earnings. Other couples see themselves as a main/secondary provider team; still other couples have mixed or ambivalent attitudes about their provider and caregiving roles (Potuchek 1992). Coprovider couples are likely to have egalitarian beliefs about household labor, whereas main/secondary marriages are more likely to hold to traditional attitudes. But the majority of two-earner couples today are mixed, or what Hochschild (1989) terms *transitional.* Such mixed couples may be at the highest risk for conflict over work and family roles (McHale and Crouter 1992).

Some husbands carry a good share of the responsibility for family work (Ferree 1990). But getting comfortable with transitional marital roles is not a quick and easy process. One of Hochschild's points is that not only do women suffer from overload when men don't share the second shift; so do husbands, "through the resentment their wives feel toward them and through their need to steel themselves against that re-

After seven years of loving marriage, Nancy and Evan had finally come to a terrible impasse. Their emotional standard of living had drastically declined: they began to snap at each other, to criticize, to carp. Each felt taken advantage of: Evan, because his offering of a good arrangement was deemed unacceptable, and Nancy, because Evan wouldn't do what she deeply felt was "fair." ...

This struggle made its way into their sexual life. ...

The idea of a separation arose, and they became frightened. [Two divorced couples she knew were] less happy after the divorce than before, and both wives took the children and struggled desperately to survive financially. Nancy took stock. She asked herself, "Why wreck a marriage over a dirty frying pan?" Is it really worth it? ...

Not long after this crisis in the Holts' marriage, there was a dramatic lessening of tension over the issue of the second shift. It was as if the issue was closed. Evan had won. Nancy would do the second shift. Evan expressed vague guilt but beyond that he had nothing to say. Nancy had wearied of continually raising the topic, wearied of the lack of resolution. Now in the exhaustion of defeat, she wanted the struggle to be over too. ...

One day when I asked Nancy to tell me who did which tasks from a long list of household chores, she interrupted me with a broad wave of her hand and said, "I do the upstairs, Evan does the downstairs." What does that mean? I asked. Matter-of-factly, she explained that the upstairs included the living room, the dining room, the kitchen, two bedrooms, and two baths. The downstairs meant the garage, a place for storage and hobbies—Evan's hobbies. She explained this as a "sharing" arrangement,

without humor or irony—just as Evan did later. Both said they had agreed it was the best solution to their dispute. Evan would take care of the car, the garage, and Max, the family dog.... Nancy took care of the rest.... For Nancy and Evan, "upstairs and downstairs," "inside and outside," was vaguely described like "half and half," a fair division of labor based on a natural division of their house.

The Holts presented their upstairs-downstairs agreement as a perfectly equitable solution to a problem they "once had." This belief is what we might call a "family myth," even a modest delusional system. Why did they believe it? I think they believed it because they needed to believe it, because it solved a terrible problem. It allowed Nancy to continue thinking of herself as the sort of woman whose husband didn't abuse her—a self-conception that mattered a great deal to her. And it avoided the hard truth that, in his stolid, passive way, Evan had refused to share. ... This outer cover to their family life, this family myth, was jointly devised. It was an attempt to agree that there was no conflict over the second shift, no tension between their versions of manhood and womanhood, and that the powerful crisis that had arisen was temporary and minor. ...

The wish to avoid such a conflict is natural enough.... After Nancy and Evan reached their upstairs-downstairs agreement, their confrontations ended. They were nearly forgotten. Yet, as she described their daily life months after the agreement, Nancy's resentment still seemed alive and well.

What are some of the benefits to the Holts of having created their upstairs-downstairs family myth? Some of the costs? Was the Holts' family myth an avenue to intimacy, or not? Why?

sentment" (1989, p. 7). If the transition proceeds from a mutual commitment to achieve an equitable, no-power relationship, however, the result can be greater intimacy. As pointed out in Chapter 9, a first step is to address the conflict.

ACCEPT CONFLICT AS A REALITY   The idea that mates can sometimes have competing interests and needs departs from the more romanticized view that sees marriages and families as integrated units with shared desires and goals. Nevertheless, as a first step toward maintaining intimacy during role changes, partners need to recognize their possibly competing interests and to expect conflict (Paden and Buehler

1995). The following letter to "Working Woman" columnist Niki Scott from a 34-year-old husband married eight years illustrates this point:

There's a pervasive attitude among people like you who write primarily that most women are accommodating, submissive, fragile, non-assertive doormats who need encouragement to stand up for themselves in even the most basic ways.

What I want to know is: where do you find these pitiful, downtrodden women? . . .

My wife—and all the wives of all the men I know—have no trouble defending their "boundaries." They are all too eager to confront every tiny issue, all too unwilling to keep the small stuff of life small.

If I forget to take out the trash . . . the problem is not that I didn't take out the trash. It's that I've somehow proven that I don't care about my wife, our home, or the rotten examples I set for our children.

What could have been a two-paragraph exchange turns into a recitation of . . . all the ways I take my poor working wife for granted.

How do we men defend our boundaries? Not a single man in this country does enough around the house, judging by your columns. But . . . where are all these women who work all day and spend their nights and weekends raising their kids virtually alone and cooking and cleaning and waiting on their selfish, lazy, no-good husbands?

I know far more men who do *more* of the cooking, shopping, cleaning and looking after the kids than their wives do.

I work for a Fortune 500 company, furthermore, where every man I know spends a great deal of time and energy watching every word he says because if he isn't on guard every minute, he can be accused of everything from sexism to sexual harassment. (Scott 1992, p. 404.[6]

Sociologists Janet and Larry Hunt (1986), two of the first to research two-career marriages, argue that the low status of household work makes it symbolically unattractive to most husbands, who have become used to greater conjugal power. Furthermore, women have moved into and thereby altered one sphere of men's lives, the workplace. As a result, men may press for stability, resisting change in another sphere—at home. Moreover, many husbands believe their wives are employed to fulfill personal needs, such as getting out of the house (Weiss 1987). Hence they consider accommodation to their wives' wage work not as something wives are entitled to, but as a sacrifice to themselves because they have to make do with a less orderly home life (Thompson 1991). And finally, research on one small sample of 314 dual-income spouses found that women cope with overload by talking about it, while for men, talking about it only

seemed to make them feel worse (Paden and Buehler 1995). Today's domestic stage is set for conflict.

ACCEPT AMBIVALENCE   The next step in maintaining intimacy as spouses adjust to two-earner marriages is for both to recognize that each will have ambivalent feelings. The following excerpt from one young husband's essay for his English composition class is illustrative: "I'm in school six days a week. My wife works between 40 to 50 hours a week. So I do the majority of the cooking, cleaning, and laundry. To me this is not right. But am I wrong to think so? I'm lost in my own mind."

Women may also be ambivalent. Conflicting feelings become evident: They want their husbands to be happy; they want their husbands to help and support them; they feel angry about past inequalities; and they feel guilty about their declining interest in housekeeping and husband care and their decreasing willingness to accommodate to their husbands' preferences. Furthermore, men who participate have opinions about how child rearing or housework should be done. As a husband begins to pitch in, his wife may resent his intrusion into her traditional domain.

SHARE FEELINGS   Once partners recognize their ambivalent feelings, the next step is to be open about them. Intimate sharing between mates is important; family meetings and honest group discussion with other couples can also help (Hopkins and White 1978). Again, spouses should expect to face interpersonal conflict and anger as they share; the stalled revolution can precipitate "intense and stressful negotiations" (Peterson and Gerson 1992).

EMPATHIZE   As partners share their feelings, a next step is to empathize. This can be difficult, for it is tempting instead to retaliate for past hurts. But partners who retaliate rather than empathize are trying to win or "show" the other. If couples are to maintain intimacy, they must make sure *both* partners win (see Chapter 9).

It has been suggested that as wives empathize, they try to recognize that a husband's lack of participation is very likely "due to insensitivity, not malice" (Crosby 1991, p. 167). Partly because many wives do much of the housework when their mates are not home (Thomas 1991), most husbands underestimate the

---

6. Taken from the Working Woman column by Niki Scott. Copyright 1992. Distributed by Universal Press Syndicate. Reprinted with permission. All rights reserved.

number of hours that household labor takes. It is never easy to adjust to new roles, and men especially may feel they have a lot to lose. A man who has enjoyed being emotionally cared for by his wife may feel threatened by her choosing to invest energy in school, a job, or a career. In the words of one observer, "Fear can close even the most open mind; and vague, ill-defined fears can make us all mulish. In the reactions of husbands, I detect a haunting worry about what they will lose when true gender equality arrives" (Crosby 1991, p. 162). A wife's pressing for role changes may threaten a man with the loss of authority and privilege, of a unilateral helpmate, of the "hero-at-home" status of the "good provider," and of the promise of a dependable listener (Crosby 1991, pp. 159–77). (Men gain, too, of course: They develop domestic skills, their marriage is enhanced, there is more money, and they benefit from spending time with their children [Crosby 1991]. In Hochschild's study [1989] some fathers who felt they had been emotionally deprived in relationships with their own fathers took great pleasure in creating more satisfying family relationships with and for their children.)

As husbands empathize, they need to be aware that their willingness to participate in household tasks is vitally important to wives, especially perhaps to employed wives (McHale and Crouter 1992). For one thing, wives doing a second shift are tired and need rest and some personal leisure time. They value having a "down time" due to a responsive mate's freeing them from overload or from specific tasks they dislike (Thompson 1991; Blair and Johnson 1992). On another level, a husband's sharing carries a symbolic meaning for a wife, indicating that her work is recognized and appreciated and that her husband cares (Thompson 1991). Indeed, wives are more inclined to see the division of household labor as fair when their work is obviously appreciated (Blair and Johnson 1992). A husband's participation becomes caring; his resistance, carelessness and inattentiveness (Thompson 1991, p. 187).

STRIKE AN EQUITABLE BALANCE
Researchers who studied 153 Pennsylvania couples with school children concluded the following: "Our data imply that the adjustment of individual family members, as well as harmonious family relationships, requires a *balance* among the very different and often conflicting needs and goals of different family members"

(McHale and Crouter 1992, pp. 545–46, italics in original). Once equity is habitual, calculation and constant comparison are no longer necessary; some observers point out that the balance need not be an exactly calculated 50-50 split. Such an arrangement, complete with rules and his-and-her chore lists, may seem too impersonal: People find it alienating and hurtful when others treat them according to some fixed rule rather than as individuals with particular needs and desires. We do not feel so much cared for when others' responsiveness is routinized and rule-bound (Thompson 1991, p. 188).

SHOW LOVE AND MUTUAL APPRECIATION
Once partners have committed themselves to striking a balance, they need to create ways to let each know the other is loved. Traditional role expectations were relatively rigid and limiting, but they sometimes let mates know they were cared for. When a wife sewed buttons on her husband's shirt or cooked his favorite meal, for example, he felt cared about. As spouses relinquish some of these traditional behaviors, they need to create new ways of letting each other know they care, rather than succumbing completely to the time pressures of today's hectic family life.

Many people have noted the potential of shared work and of shared provider and caregiving roles for enriching a marriage. Hochschild holds out this hope also, if the present transition from a traditional division of labor to an egalitarian one can be negotiated successfully. Chafetz (1989) predicts a period of adjustment followed by marriages that are more strongly bonded as they come to be based on similarity, rather than on difference.

In Sum

The labor force is a social invention. Traditionally, marriage has been different for men and women: The husband's job has been as breadwinner, the wife's as homemaker. These roles are changing as more and more women enter the work force. Women still remain very segregated occupationally and earn lower incomes than men, on the average.

We distinguished between two-earner and two-career marriages. In the latter, wives and husbands both earn high wages and work for intrinsic rewards. Even in such marriages, the husband's career usually has priority. Responsibility for housework falls largely on wives. Many wives would prefer shared roles, and negotiation and tension over this issue cast a shadow on many marriages. An incomplete transition to equality at work and at home affects family life profoundly.

We have emphasized that both cultural expectations and public policy affect people's options. As individuals come to realize this, we can expect pressure on public officials to meet the needs of working families by providing supportive policies: parental leave, child care, and flextime.

We have seen that paid work is not usually structured to allow time for household responsibilities and that women, rather than men, continue to adjust their time to accomplish both paid and unpaid work. "If we can't return to traditional marriage, and if we are not to despair of marriage altogether, it becomes vitally important to understand marriage as a magnet for the strains of the stalled revolution" (Hochschild 1989, p. 18). To be successful, two-earner marriages will require social policy support and workplace flexibility. But there are some things the couple themselves can keep in mind that will facilitate their management of a working-couple family. Recognition of both positive and negative feelings and open communication between partners can help working couples cope with an imperfect social world.

Household work and child care are pressure points as women enter the labor force and the two-earner marriage becomes the norm. To make it work, either the structure of work must be changed, social policy must support working families, or women and men must change their household role patterns—and very probably all three.

Key Terms

ambivalent provider couple
center care
child care
commuter marriages
coprovider couple
elder care
family day care
family-friendly workplace policies
family leave
family myths
flexible scheduling
flextime

gender strategy
good provider role
househusbands
ideological perspective
job sharing
labor force
life course solution
main/secondary provider couple
mommy track
occupational segregation
personal days
pink-collar jobs

rational investment perspective
reinforcing cycle
resource hypothesis
second shift
shift work
stalled revolution

trailing spouse
two-career marriage
two-earner marriages
two-location families
two-person single career
unpaid family work

Study Questions

1. Discuss whether or to what extent distinctions between husbands' and wives' work are disappearing.

2. What do you see as the advantages and disadvantages of men being househusbands? Discuss this from the points of view of both men and women.

3. Discuss the advantages and the disadvantages of full-time homemaking for wives, for husbands, and for children.

4. What are some differences between career wives and husbands and other categories of wives and husbands?

5. Discuss the problems inherent in two-earner and two-career families. What influence does the social environment have on these families?

6. What is meant by work–family conflict? Interview some married or single-parent friends of yours for concrete examples and for some suggestions for resolving such conflicts.

7. What is unpaid family labor, and why is it called "labor"? Discuss social policy in the United States regarding work–family conflicts.

8. How can husbands and wives negotiate decisions about job opportunities in two locations?

Suggested Readings

Beach, Betty. 1989. *Integrating Work and Family Life: The Home-Working Family.* Albany: State University of New York Press. Although less than 2 percent of American families earn income through work they do at home, you may be interested in reading about this alternative if you contemplate freelance work, a home business, or a computer link to a workplace.

Bradley, Harriet. 1989. *Men's Work, Women's Work: A Sociological History of the Sexual Division of Labour in Employment.* Minneapolis: University of Minnesota Press. Drawing on historical research and case studies from a variety of occupations from mining to shoemaking to medicine, Bradley illustrates both the wide range of women's market work and segregation within those occupations.

Caplan, Nathan, John Whitmore, and Marcella Choy. 1989. *The Boat People and Achievement in America: A Study of Family Life, Hard Work, and Cultural Values*. Ann Arbor: The University of Michigan Press. Study of work–family behaviors of refugees from Vietnam, Cambodia, and Laos, who immigrated to America in the 1970s.

Cowan, Ruth Schwartz. 1983. *More Work for Mother*. New York: Basic Books. Interesting history of household technology and why it hasn't made housework less time consuming.

Crosby, Faye J. 1991. *Juggling: The Unexpected Advantages of Balancing Career and Home for Women and Their Families*. New York: Free Press. A fairly down-to-earth how-to book of advice for juggling paid and unpaid labor today.

Gerson, Kathleen. 1985. *Hard Choices: How Women Decide About Work, Career, and Motherhood*. Berkeley: University of California Press. Interesting research on women's work and family choices.

Gerstel, Naomi and Harriet Gross. 1985. *Commuter Marriage: A Study of Work and Family*. New York: Guilford. The major study on this topic; sheds light on coresident marriages as well.

Hertz, Rosanna. 1986. *More Equal than Others: Women and Men in Dual Career Marriages*. Berkeley: University of California Press. In-depth study of some twenty career couples; explores the strengths and difficulties of two-career marriages.

Hochschild, Arlie. 1989. *The Second Shift: Working Parents and the Revolution at Home*. New York: Viking/Penguin. Powerful book on the struggles of working men and women over housework—and more; fascinating case examples that illuminate the dynamics of marriage today.

Hood, Jane C., ed. 1993. *Men, Work, and Family*. Newbury Park, CA: Sage. An interesting and readable collection of research and essays of men's roles at work and in families. Many well-known social scientists in the field are included.

Jones, Jacqueline. 1985. *Labor of Love, Labor of Sorrow: Black Women, Work, and the Family from Slavery to the Present*. New York: Basic Books. Jones is an historian, and this well-written and interesting book follows African American women's employment from slavery to the 1980s.

Ogden, Annegret S. 1986. *The Great American Housewife: From Helpmate to Wage Earner, 1776–1986*. Westport, CT: Greenwood. A well-written, well-documented historical survey, tracing the history of the housewife role from 1776 to 1986.

Romero, Mary. 1992. *Maid in the U.S.A.* New York: Routledge. A well-written, provocative and moving qualitative study of how race/ethnicity and social class impact the kind of paid work that women do. Romero explores whether women in divergent social classes are bonded by sisterhood or oppression.

Schwartz, Felice N. with Jean Zimmerman. 1992. *Breaking with Tradition: Women and Work, the New Facts of Life*. New York: Warner. Schwartz's experience of being heavily criticized for her previous article suggesting what quickly became labeled a "mommy track"—and her subsequent rethinking of some of the issues—resulted in this book for managers on workplace policy regarding work–family conflicts.

Stromberg, Ann Helton and Shirley Harkess, eds. 1988. *Working Women*, 2d ed. Palo Alto, CA: Mayfield. Articles on various types of work women do and how work affects the rest of their lives; inclusive of social diversity: class, race/ethnicity, and sexual orientation.

Suransky, Valerie Polakow. 1982. *The Erosion of Childhood*. Chicago: University of Chicago Press. Are the interests of women and children in conflict? What makes a day-care center good? Suransky explores these issues in a theoretical context, based on her observational case studies of five day-care centers.

Tilly, Louise and Joan Scott. 1978. *Women, Work, and Family*. New York: Holt, Rinehart & Winston. History of women's work and the relationship of family to work.

Family Change and Crises

With the expectation and the reality that people continue to change throughout their lives come the pressures that these individual changes exert on the family unit. External events may also require the partners to adapt. Intimate relationships may be simultaneously more meaningful and more difficult to maintain in the face of such changes.

Part Five examines how spouses and families manage change. Chapter 14 describes changing family situations, or crises, and what families can do to meet them creatively and effectively.

Chapter 15 discusses separation and divorce. Whether the decision to divorce is one partner's or both partners', divorce ends the marital commitment (though not necessarily the familial one). It is a process of separating oneself from the former spouse—emotionally, legally, socially, and psychologically. Healing after divorce involves learning to commit oneself to being single again, if only temporarily.

Chapter 16 examines the making of new commitments in remarriage. As we shall see, more and more people who have been married before are committing themselves to building a family again. Such reconstituted families are different from traditional nuclear families in some important ways. In fact, because of the special challenges in stepfamily living, couples in a remarriage may need even stronger commitment to each other than those in first marriages.

Managing Family Stress and Crises

Families are more likely to be happy when they work toward mutually supportive relationships. Nowhere does this become more apparent than in a discussion of how families manage stress and crises.

Family stress is a state of tension that arises when demands tax a family's resources. Responding to the needs of aging parents while simultaneously negotiating children's adolescent years is stressful for a family. Families experience financial pressures, like finding adequate housing on a poverty budget or financing children's education on a middle-class income. Managing full-time jobs while parenting children (discussed in Chapter 13) is a common source of family stress. Especially for women, being a member of the "sandwich generation" with both children and aging parents to care for is a source of family stress.

Family stress calls for family adjustment. In response to financial pressures, family members might adjust their budget, cutting back spending on clothing, recreation, travel, or eating out. They might cope with time pressures by reducing their responsibilities or by arranging for someone outside the immediate family to take them over. Many times, though, there is no easy way to remove pressures of time, money, or responsibilities. If solving the problem by adding resources or helpers or by subtracting responsibilities is not possible, family coping takes a more social psychological form: expressing feelings, doing things that give the family pleasure or establishing rituals that offset tension, even using humor. But constant stress can be wearing to family members and family harmony.

There is another kind of family stress that is not so much the chronic challenge of demands on a family, but rather an event that changes things. Each such event is a **crisis**—a crucial change in the course of events, a turning point, an unstable condition in affairs. A crisis is a sharper trauma to a family.

The definition of crisis encompasses three interrelated ideas:

1. Crises necessarily involve change.

2. A crisis is a turning point with the potential for either positive or negative effects, or both.

3. A crisis is a time of relative instability.

Family crises share these characteristics. They are turning points that require some change in the way family members think and act in order to meet a new situation (Hansen and Hill 1964).

When people live together over a period of time, they develop patterns of relating to one another. Family members are fairly well aware of where each fits in and what each is expected to do; the family functions smoothly (or, at least, predictably) when each member behaves according to the expectations of the others. Any change that disrupts these expectations marks the onset of a crisis. Sometimes, the event that precipitates a crisis is dramatic, unexpected, and unfortunate. The death of a child or a spouse, diagnosis of a serious illness, a breadwinner's loss of employment, a premarital pregnancy, an adoption revealed—all of these require major alterations in the way family members think, feel, and act.

Positive changes can precipitate crises, too. Suddenly getting rich or, more commonly, moving to another city or another state because of a promotion are examples.

Family **transitions**—expected or **predictable** changes in the course of family life—precipitate family stress and/or **crises.** Having a first baby or sending the youngest child off to college, for example, tax a family's resources and can bring about significant changes in family relationships and expectations. Throughout the course of family living *all* families are faced with stress, crises, and transitions.

This chapter examines how families cope with stress and crises. We'll look at family transitions, along with some ways families can face stress and crises more creatively. We'll discuss how families define or interpret stressful situations and how their definitions affect the course of a family crisis. To begin, we'll ask what precipitates family stress or crisis.

What Precipitates a Family Crisis?

Something must happen to precipitate a crisis. That something is what social scientists call a **stressor,** a precipitating event that creates stress. Stressors vary in both kind and degree, and their nature is one factor that affects how a family responds to crisis.

Types of Stressors

There are several types of stressors, as Figure 14.1 shows. Some involve the *loss of a family member* either permanently, such as through death or desertion, or temporarily, such as through hospitalization or imprisonment. A family member can be lost through divorce but can still remain a part of the family psychologically, socially, and even economically. The ambiguity of the boundaries of a postdivorce family can be stressful in itself.

A precipitating event can also involve the *addition of a member* or members to the family through birth, adoption, or marriage. Not only are in-laws (and perhaps stepparents and stepsiblings) added through marriage but also a whole array of *their* kin. Sociologist Pauline Boss (1980) described these additions and losses as family *boundary changes;* that is, family boundaries are shifting or contracting to include or exclude certain members. The addition of family members can bring into intimate social contact people who are very different from one another in values and life experience.

Sudden change in the family's income or social status can also be a stressor. Most people think of stressors as being negative, but positive changes, such as a move to a better neighborhood, a promotion, or sudden wealth or fame, can cause crisis, too. One author of this text knows a family that became wealthy suddenly when a product they had invented and manufactured on a small scale became popular nationally. This family hired a consultant to teach them how to behave in their new social circumstances. Deciding such matters as whether it would be appropriate for the children to continue to take babysitting jobs occupied a great deal of family energy.

Ongoing *unresolved conflict* among members over *familial roles* can be a stressor. As husbands' and wives' roles change, conflicts can arise over child-care responsibilities, for example, or over division of household labor. Deciding how adolescents should be disciplined

Many of the frail elderly depend upon, or even live with, family members. While care is often given with fondness and love, it also can bring stress, conflicting emotions, and great demands on time, energy, and finances.

for rules violations can bring to the surface divisive differences over parenting roles. The role of an adult child living with parents is often ambiguous and can be a source of unresolved conflict. If children of teenagers or of divorced adult children are involved, the situation becomes even more challenging.

Caring for a dependent or disabled family member is a stressor. Raising a mentally or physically disabled child is an example. (Box 14.1, "How Low-Income Mothers Care for Children with Sickle-Cell Disease," explores aspects of this stressor.) Due mainly to advancing medical technology, the number of dependent people and the severity of their disabilities have steadily increased over recent decades. For example,

Loss of a
family member

Addition of a
family member

Sudden change

Conflict over
family roles

Caring for a
disabled or
dependent
family member

Demoralizing
event

FIGURE *14.1*

I   Types of stressors.

more babies survive birth defects, and more people survive serious accidents; but they may require ongoing care and medical attention. Parents may need to see their children through bone marrow, kidney, or liver transplants, sometimes requiring several months residence at a medical center away from home. Adults with advanced AIDS may return home to be taken care of by family members. Caring for disabled family members is, of course, another stressor—for young children, who may exhibit behavior problems as a response—as well as for other adults in the household (Le Clere and Kowalewski 1994).

Finally, stressors can be *demoralizing events*—those that signal some loss of family morale. Demoralization can accompany the stressors already described. But this category also includes loss of employment, unwanted pregnancy, poverty, juvenile delinquency or criminal prosecution, scandal, family violence, mental illness, alcoholism and drug abuse, and suicide. Physical illness can be demoralizing if it carries attributions of family dysfunction (anorexia nervosa or bulimia, for example) or is socially stigmatized (AIDS, for example). Sexually transmitted diseases with less serious health implications may nevertheless present a threat to the marriage. Alzheimer's disease or head injury, in which a beloved family member seems to have become a different person, is heartbreaking.

Stressor Overload

A family can be demoralized not just by one serious, chronic problem but also by a series of small, unrelated stressors that seem to build on each other too rapidly for the members to cope effectively. Sociologist and marriage counselor Carlfred Broderick explains:

> Even small events, not enough by themselves to cause any real stress, can take a toll when they come one after another. First an unplanned pregnancy, then a move, then a financial problem that results in having to borrow several thousand dollars, then the big row with the new neighbors over keeping the dog tied up, and finally little Jimmy breaking his arm in a bicycle accident, all in three months, finally becomes too much. (Broderick 1979b, p. 352)

Even though it may be difficult to point to any single precipitating factor, an unremitting series of relatively small stressors can add up to a major demoralizing crisis. In today's economy, characterized by fewer high-paying jobs and less job security, **stressor overload** may be more common than in the past. (Appendix I addresses managing a family budget.) Characteristically, stressor overload creeps up on people without their realizing it.

In any given situation, each type of stressor will have certain characteristics.

Characteristics of Stressors

We can point to five stressor characteristics:

1. *Is the stressor expected (normative) or unexpected (nonnormative)?* A predictable life-course transition, such as having a baby, would be an expected stressor. Finding oneself involuntarily infertile, on

How Low-Income Mothers Care for Children with Sickle-Cell Disease

You may recall that in Box 2.2 we discussed three frameworks (cultural equivalent, cultural deviant, and cultural variant) for studying minority families. The *cultural variant* approach, in which minority families are studied on their own terms instead of comparing them with the majority, is favored by many scholars. University of Kansas sociologists Shirley Hill and Mary Zimmerman (1995) studied low-income, African American mothers caring for children with sickle-cell disease; their research exemplifies the cultural variant framework.

"Sickle-cell disease (SCD), a hereditary, incurable illness that primarily affects blacks, is a blood disorder that causes red blood cells to assume the shape of a sickle, thus obstructing [blood flow] and depriving bodily tissues of oxygen. [SCD patients] may experience fatigue, poor appetite, frequent pains and infections, growth retardation, incontinence, [organ damage,] stroke, paralysis, and early death." The most common symptom is a *pain crisis*—relentless, gnawing bone and joint pain.

Hill and Zimmerman wanted to know more about how mothers care for children with SCD. To find out, Hill observed patient–staff interactions at a local SCD clinic, attended SCD support group meetings, and interviewed thirty-two low-income black mothers of children with SCD. The tape-recorded interviews, for which the mothers had volunteered, were unstructured (as opposed to asking for short, specific answers to pointed questions), lasted an hour or more, and were mostly in the mothers' homes. "[The] mothers ranged in age from 21 to 72, with an average of [about 35]. Twenty-nine were biological mothers; the other three [included] an adoptive mother, a great aunt, and a grandmother. All were primarily responsible for the child's care."

To analyze their data, the researchers pored over it many times with the following three central questions in mind:

a. "What meanings did the mothers attach to having a child with SCD?

b. How did mothers manage the care of their children?

c. What factors shaped the way the mothers defined and managed SCD?"

Hill and Zimmerman found, first, that "the mothers exercised a great deal of discretion in deciding what type of care their children needed, [largely because] they rarely [received] adequate medical advice [and hence] tended to develop their caregiving

(continued)

(continued)

strategies from first-hand experience. [Much of the care required] some change in everyday practices, [such as] restricting children's activities and changing diets," in order to avoid pain crises.

Although SCD symptoms vary, there is no evidence that they are influenced by biological sex. But as the researchers continued examining their interviews and observational notes, they found that the child's gender *was* an important factor in how mothers cared for their sick youngsters. Mothers of sons with SCD saw their children as sicker than mothers of daughters; these gendered perceptions influenced their caregiving. "Compared with mothers of daughters, mothers of sons were usually more involved in caregiving [and] more likely to intervene to protect their child's health. [They also] expressed more anxiety over their child's health, [which they] were more likely to describe as fragile." Mothers with ill daughters gave them more freedom, actually encouraged them to engage in physical activities, were more likely to trust them to care for themselves, and more often expected them to tolerate their symptoms.

The authors argue that the gendered norms these mothers had for all children influenced how they treated boys and girls with SCD. Generally, low-income, African American boys are "de-fined by aggressiveness and physical activities, especially sports. Mothers of sons with SCD saw having the disease as a tragic violation of the already limited social roles available for their sons, and they responded by protecting their sons and restricting their activities."

"The protectiveness of these mothers reflects [both] the higher social value placed on males in our society [and] the growing concern" among African Americans over the survival of black males today. The authors suggest that, with "black maleness described as a 'high-risk venture'" African American parents may unwittingly have lower expectations for their sons than for their daughters and be more protective of them.

This study uses a very small sample and is located in only one area of the United States. Thus we cannot generalize Hill and Zimmerman's findings to all African American—or even to all low-income, African American—mothers caring for children with SCD. We need more studies in other areas first. Nevertheless, this study illustrates how race, social class, and gender all interact to influence our attitudes and behaviors—in this case caregiving behavior under crisis.

Source: Hill and Zimmerman 1995, pp. 45, 46, 50, 51.

the other hand, would be unexpected. As Box 14.2, "Death of a Child," points out, a child's death is probably more stressful today than in 1900 because it is more unexpected; so are the deaths of young or middle-aged adults, now becoming more prevalent due to AIDS.

2. *Is the stressor brief or prolonged?* For example, caring for a temporarily disabled family member who has had major surgery is brief compared to caring for one who is chronically ill.

3. *Is the stressor external (originating outside the family) or internal (beginning within the family)?* External stressors include such events as hurricanes and earthquakes. Internal stressors include, for example, alcoholism, chronic gambling, and family transitions. External stressors often help to solidify a family; families draw together when threatened by a tornado or blizzard, for example. Internal stressors, however, are more likely to divide and demoralize a family, primarily because members tend to

BOX **14.2**

Death of a Child

The likelihood of death in our society influences how we define a death in the family. In a society such as ours in which death is statistically infrequent, not all deaths, but only unnecessary, preventable, or "premature" ones, are highly problematic (Parsons and Lidz 1967). Under the mortality conditions that existed in this country in 1900, half of all families with three children could expect to have one die before reaching age 15. Parents defined the loss of a child, social historians believe, as an almost natural or predictable crisis and, consequently, suffered less pain than do parents today. By 1976, in contrast, the probability of a child dying was 6 in 100 (.06) (Uhlenberg 1980, p. 315). To illustrate this change, demographer Robert Wells asks readers to imagine themselves on a New Jersey village street in 1750:

> As you sit by the roadside, you notice a [funeral procession] coming toward you.... The coffin is ... that of a child. This, also, you learn is not unusual, because children commonly died during this time period. The second thing that surprises you is that the individuals following the coffin to the graveyard seem to accept their loss relatively calmly. You will later learn that since so many children die at a relatively early age, parents do not invest significant amounts of emotional energy in their children until they have survived the first five or ten years of life. In addition, their religious beliefs encourage many colonists to view death not as an ending, but as a release from earthly miseries and sins.
>
> As you watch the procession go by, you realize that about half the people in the procession are children, including several you later learn are brothers and sisters of the deceased. (Wells 1985, pp. 1–2)

Family members who lose a child today, by comparison, have not had much opportunity to witness death, dying, or grieving (Kübler-Ross 1979). The death—defined as unnatural, irrational, uncalled-for—is an uncommonly painful event, especially in a rational progress- and action-oriented society such as ours.

Loss of potential children through miscarriage, stillbirth, or infant death has the added strain of ambiguity. It is unclear whether the incipient child is a member of the family or not. Attachment to the fetus or newborn may vary greatly so that the loss may be grieved greatly or little. Add to that the generally minimal display of bereavement customary in the United States and the omission of funerals or support rituals for perinatal (birth process) loss, and "all these ambiguities mean that a family may have to cope with sharply different feelings among family members ... [and] the family as a whole may have to cope with the fact that they as a family have a very different reaction to loss than do the people around them" (Rosenblatt and Burns 1986, p. 238). In fact, a study of fifty-five instances of perinatal loss found that grief was still felt by some parents forty years later, whereas the majority reported no long-run grief and some had always defined their loss as a medical problem rather than a child death. Men and women may have different reactions to their pregnancy loss and a different sensibility about the expression of feelings (Stinson et al. 1992).

This study, because it dealt with an especially ambiguous situation, discovered a truth about death that is probably more general: "Loss through death may represent more than the person who has died" (p. 251). Here, "what is grieved may be the child, but it may also be subsequent childlessness, the absence of a desired additional child, an unpleasant medical or marital experience associated with the loss, a loss of innocence, the end to feelings of invulnerability, a loss of faith that life is fair, or something else" (p. 251).

The Compassionate Friends is a self-help group for parents who have experienced a child's death. New in-hospital or religious rituals are beginning to develop to mark perinatal or fetal loss.

blame each other—or to fight about which member is to blame—for their troubles. (Focusing on blame greatly reduces a family's ability to cope, as we shall see later in this chapter.) Some stressors are ambiguous: Unemployment may be defined either as the worker's fault or as the impersonal workings of a recession economy (Newman 1988).

4. *Does society offer concrete norms for dealing with the stressor event?* For example, our culture provides widows and widowers with fairly concrete expectations for how to behave at the funeral, but it is ambiguous with regard to how gay or cohabiting partners should behave upon a lover's death. The ambivalent social response to divorce is illustrated by the congratulatory cards now beginning to appear in drugstores. Even the death of adult siblings is culturally undefined. Friends may not take much note of a sibling's death, yet it raises complex feelings, often including "survivor guilt," a sense that it isn't "fair" for a person of comparable age to have died (Kutner 1989a).

5. *Does the stressor condition improve, remain stable, or deteriorate?* Grief usually lessens (improves) over time, for example, whereas addictions such as alcoholism or drug abuse grow progressively worse. The amount of stress caused by caring for a family member with cancer depends on whether the patient's condition improves or deteriorates.

In general, stressors are less difficult to cope with when they are expected, brief, external, defined by concrete norms, and improving. Now that we have examined stressor types and characteristics, we will look at family transitions as stressors.

Family Transitions as Stressors

The transitions that individuals make during their adult lives were discussed in Chapter 1, which points out, among other things, that partners' personal changes affect their relationship. Families as a unit evolve as well. Over the course of family living, people often marry, become parents, may divorce, may remarry, and still later make transitions to retirement and widow- or widowerhood. All of these transitions are stressors. Many are discussed in detail elsewhere in this text, and we won't repeat those discussions here. This section focuses on family transitions that occur mostly in life's second half: transitions to the sandwich generation and to the empty nest, retirement, and widow- or widowerhood.

The Sandwich Generation

Several years ago journalists and social scientists began to write about the **sandwich generation:** middle-aged (or older) individuals, usually women, who are sandwiched between the simultaneous responsibilities of caring for their dependent (sometimes young adult) children and aging parents. The sandwich generation experiences all the hectic task juggling discussed in Chapter 13 on work and family. However, stress builds as family members handle not only employment and child care but also a third category of demands: parent care.

We can expect that responsibility for attending elderly parents will become an increasingly common experience for several reasons. First, the number of elderly in our population is growing. In 1970 there were about 7½ million Americans age 75 and older; by 1994 there were over 13 million. Of those age 85 and above, there were 1.4 million in 1970, compared to 3 million in 1994 (U.S. Bureau of the Census 1995, Table 14). Projections are that by the year 2050, there will be nearly 18 million Americans over age 84 ("We the American Elderly" 1993, Figure 5).

Not just the number of elderly, but also their proportion of the total population is growing. In 1970, about 4 percent of the U.S. population was 75 or older. By 1990, the proportion was 5.5 percent. Because individuals are living longer, and because families are smaller than in the past, the ratio of middle-aged adult children to elderly parents is declining. As you can see in Figure 14.2, in 1950 there were 3 persons age 85 and older for every 100 individuals between ages 50 and 64 who might care for them. By 1990, that ratio had increased to 9 ("We the American Elderly" 1993, p. 8). In other words, there are fewer adult siblings available to share in the care of their elderly parents.

Even among the "old old" (age 85 and above), most men are married. At age 85 and over, about half of the men are married while four-fifths of women are widowed. At 65 to 74 years old, about four-fifths of men and half of women are married. One implication of these data is that most elderly men have a spouse for assistance if their health fails, while the majority of elderly women do not ("We the American Elderly" 1993, p. 7).

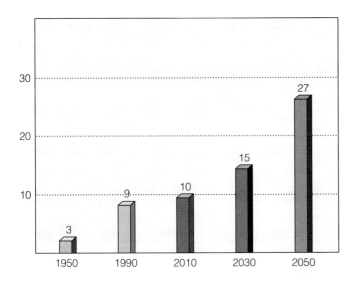

FIGURE *14.2*

Parent support ratio: 1950 to 2050. The ratio reflects persons 85 years old and over per 100 persons 50 to 64 years old. Dependency ratios for 1950 and 1990 are based on actual data. Ratios for 2010, 2030, and 2050 are projections. (Source: "We the American Elderly" 1993, U.S. Dept. of Commerce, Figure 14, pg. 8)

As people live longer, long-term chronic illness, disability, and dependency become more likely. A small minority of elderly Americans reside in nursing homes. About half of the "oldest old" living in their own homes are frail and need assistance with everyday activities. Over 90 percent of older disabled people not in institutions receive help from family; 70 percent depend exclusively on nonprofessional informal caregivers (Soldo and Agree 1988; "We the American Elderly" 1993). Relatives, usually daughters—although among Asian Americans the eldest son may be designated by custom (Kamo and Zhou 1994)—who themselves are often in their 50s and 60s, face the difficulties of providing care.

In fact, a trend today is for the dependent and disabled of all ages to be cared for at home, if possible (Freedman 1993). Families take responsibility for complicated care regimens (such as administering intravenous drugs) that have traditionally been handled only in hospitals by health care professionals. This situation places added demands on family members' time, energy, emotional commitment, and other resources. This is particularly true for the primary caregiver (Mancini and Bleiszner 1989; Glazer 1990), who often exhibits symptoms of depression (Gerstel and Gallagher 1993; Dwyer, Lee, and Jankowski 1994). With concerns about cuts in federal expenditures for Medicare (Fineman 1995), the sandwich generation may face added financial stresses.

The Postparental Period

The postparental stage of family life first came to attention through the value-laden term **empty nest.** It was assumed that mothers would feel lonely and depressed once their children had grown and left home; it is often associated with negative stereotypes of the menopausal woman.[1] Hormonal changes do occur in both women and men in middle age, and to some extent these changes affect people's self-esteem, levels of energy, and emotional stability. But the hormonal

1. The empty nest problem was highlighted in 1972 by sociologist Pauline Bart's study of women who were hospitalized for depression. She found that many of these women had invested much of their lives in being mothers and that they felt lost, unneeded, and depressed once the youngest child left home. Bart also found that women who devoted their time and energy in alternative roles or activities, such as employment, school, volunteer work, or their marital relationship, were better able to adjust (Bart 1972).

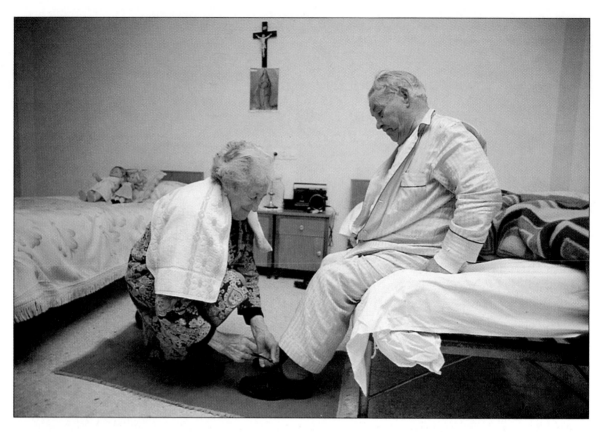

Aging and failing health can create stress both individually and within a marriage, but many spouses draw on deep reserves of affection and love to find the patience and willingness to care for the other.

changes that accompany menopause and the beginning of the empty nest were exaggerated in the 1950s and 1960s. By the late 1970s studies began to show that many women look forward to the time when their youngest child leaves home (Harkins 1978; Robertson 1978; Rubin, 1979). Changes in our society that offered middle-aged women opportunities to return to work or school made this a satisfying period of life for many. Current cohorts of women are likely to have continuous work patterns, or close to it. Indeed, now, having young-adult children who do not leave the nest as early as their parents had anticipated—or who return—can be viewed as a potential crisis.

The old stereotype of the depressed empty nest mother is also ironic because fathers have transition pains, too (Robinson and Barret 1986). This is especially true for men who are older, those who have fewer children, those who are more emotionally involved in fathering, and those who have a less satisfy-

ing marriage relationship (Lewis, Freneau, and Roberts 1979).

The postparental period is a time when couples can return their attention to each other and reorganize their lives. But these couple relationships differ from newly married ones because family history plays an important role in giving meaning to their married lives. Intergenerational and other kin contacts are usually very satisfying to middle-aged and older people. But the older generation is sometimes at a loss concerning how to maintain affectional and open communication with children and to deal with practical matters while avoiding interfering in their children's lives (Brubaker 1991).

In any case, the need to readjust the couple's relationship and roles after parenthood is a crisis situation. The outcome is more positive when parents have other meaningful roles, such as work, school, or other activities, to turn to. In fact, this is one transition with a high likelihood of a positive outcome because marital satisfaction often increases when couples have time, energy, and financial resources to invest in their couple relationship (Ishii-Kuntz and Seccombe 1989). Still, a frequent strain on couples during this period is imposed by the failing health of their aging parents. So even if their children have left home—and assuming they have no disabled dependents to care for—the psychic and economic costs of caring for one's parents may make the empty nest stage far from a period of new freedom. It may instead be a period of new family responsibilities.

Later in life, spouses may be caregivers to their mates. Although caregiving for a spouse recuperating from a hospital stay or other health condition creates stress, spouses seem patient enough and willing to provide long-term care. Whether a marriage relationship is, nevertheless, affected by the stress of caregiving varies with the age and sex of the couple, with younger wives most troubled. Some studies that examined sex differences find men to be less stressed, perhaps because their socialization prepares them to distance themselves somewhat or because they receive more help from relatives, whereas women are more burdened and depressed (Brubaker 1991). Other studies find few gender differences in stress, social support, time spent in caregiving or emotional strain, although women do experience some strain to their health (Miller 1990).

Some couples may experience divorce in middle age or later, with consequent loss of income. There may be more limited contacts with friends and kin; divorced men are especially likely to be isolated from family (Keith 1989; Brubaker 1991). Parental divorce may disrupt family rituals, especially Thanksgiving and Christmas (Pett, Lang, and Gander 1992).

It is the case during later life that morale and well-being frequently derive from nonfamily contacts—from friends and neighbors. Health is an important factor in morale in later life, and it has a substantial impact on family and other social contacts.

Retirement

Some—although not all—retired couples experience financial deprivation, finding themselves in tighter straits than are their children and grandchildren. Income usually declines by one-third to one-half upon retirement (Soldo and Agree 1988, p. 26).

Still, older people were less hard hit by the economic troubles of the past two decades. Seniors now have a poverty rate that is lower than either the overall U.S. rate or the rate for children. Cost-of-living increases built into Social Security payments did a better job of keeping pace with inflation than did earned income.

Having benefited from a rising economy during their working years and gained from the appreciated value of their homes, older people are better off, on the average, than young Americans (Peterson 1991). Serious illness, of course, can erode the savings and income of the elderly and their families. Health costs are rising, and health insurance programs, including Medicaid and Medicare, are likely to be cut back.

It should be noted also that elderly men are much better off financially than elderly women, who may receive lower Social Security payments as dependents (rather than as earners) and who generally lack pensions. Family savings may be depleted by the cost of nursing home or medical care for a spouse; women are far more likely to be the surviving spouse. Because older men are more apt to have living spouses, they are also more likely to be in a two-income family (Barringer 1992a).

Retirement represents a great change in the traditional masculine gender role. But retirement will soon be an event experienced by men who are, on the whole, less traditional in their roles. Women in their traditional roles did not mark a sharp change of activities in their 60s, but in the future the vast majority of women will take the formal step from employment to nonemployment.

We know most about the retirement experience of employed men who were unlikely to be sharing domestic roles. Accustomed to seeing himself as principal breadwinner and head of the family, the traditional retired husband faced a major role loss. A retired husband might devote more attention to family roles such as being a companionate husband and grandparent and might spend more time in homemaking tasks. Doing this can be problematic, however, for men who cling to the traditional masculine role that highly values work and achievement (Aldous 1978).

The results of one study of seventy-four husbands with multiple sclerosis are applicable here, even though the study deals with an unexpected and much less common crisis than retirement (Power 1979). All the men in the study had been their family's principal breadwinner and authority figure, but their illness upset this family organization because the husbands were no longer able to work. These men reacted in different ways. Most became "spectators" in their homes. Although they were physically able to perform many household, family, and leisure activities, they became inactive, tending to feel angry, deprived, and powerless. Some husbands remained "participants" in their families. Although they were unable to work, they found new family activities, such as shopping with their wives and working in the kitchen. These husbands felt that working had been important but was not essential to their self-esteem. They also identified with avocational and family activities and considered these equally (if not more) important to their personal satisfaction.

Although this study looked at retirement forced by illness rather than by age, many of its conclusions apply to retirement in general. The more flexible homebound men were in their outlook and role definitions, the more successful their adjustment tended to be. Retired husbands feel more useful when they take part in household activities, even traditionally feminine ones.

Retirement forces homemaking wives to adjust as well. Full-time homemakers may find it difficult to share the house that had become their exclusive territory during the day. In this situation, role flexibility for women is also important, and "both husbands and wives are happier in the aging period when they emphasize mutual help, companionship, and affection rather than trying to maintain the segregation of daily role activities characteristic of the preretirement period" (Aldous 1978, p. 204). Mutuality and continuity of sharing support seem to predict good adjustment for all couples, including traditional ones (Brubaker 1991).

There seems to be no overall relationship between retirement and marital satisfaction. Research to date suggests that simultaneous retirement or the retirement of the wife before the husband goes most smoothly in terms of the marital relationship (Lee and Shehan 1989; Brubaker 1991). Ironically, the better a couple adjusts to retirement together, the more painful may be a forthcoming transition to widow- or widowerhood.

Widowhood and Widowerhood

A final family transition is adjustment to widow- or widowerhood. Because women's life expectancy is longer and because older men remarry far more often than women do, widowhood is significantly more common in our society than is widowerhood. For in-

Retirement is a family transition that traditionally has marked a significant change in the masculine gender role. As women's roles change, retirement will increasingly mean important changes in their roles, too.

stance, 75 percent of men over age 65 live with a spouse, but only 41 percent of women do. Whereas 40 percent of women over 65 live alone, just 16 percent of men do (U.S. Bureau of the Census 1995, Table 62).[2] Widowhood is usually a permanent status for older women. Indeed, for some women, widowhood may last longer than the child-rearing stage of life.

Typically, widow- and widowerhood begin with **bereavement,** a period of mourning, followed by

2. This latter statistic points to the fact that living conditions have changed for the widowed (as well as for other unmarried elderly) over the past 100 years. In 1880, 9 percent of unmarried (widowed, divorced, and never-married) people over age 65 lived alone; in 1989, 67 percent did (Smith, 1981, p. 110; U.S. Bureau of the Census 1991b, Table 63, p. 49).

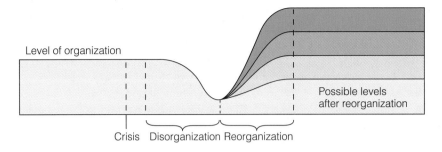

FIGURE *14.3*

I Patterns of family adaptation to crisis. (Source: Adapted from Hansen and Hill 1964, p. 810)

gradual adjustment to the new, unmarried status and to the loss. Mates who lose a partner suddenly and without warning tend to have more difficulty adjusting to the loss than do those who had warning, such as with a long illness (Newman 1979). If the spouse's dying has been prolonged, much of the mourning may have preceded the actual death, so that grieving after the death is less intense. In a study of widows in Chicago, about half said they had recovered from their husband's death within a year, although 20 percent said they had never gotten over it and did not expect to (Lopata 1973).

Bereavement manifests itself in physical, emotional, and intellectual symptoms. Recently widowed people perceived their health as declining and reported depressive symptoms (Brubaker 1991). Emotional reactions include anger, guilt, sadness, anxiety, and preoccupation with thoughts of the dead spouse. These responses tend to diminish over time and are characteristic of both men and women. Social support and activities with friends, children, and sometimes siblings facilitate adjustment.

Past research indicated that greater financial hardship ensued for widowed women than for men. Current research suggests that *both* lose ground initially, but the financial situation improves after several years, especially for the middle-aged and for those who remarry. There seems to be little difference between widows and widowers in their ultimate financial position.

Current research reveals few differences in other realms. Women more often have social support outside the family, whereas men are typically more dependent on family for support. But

men and women experience similar physical and emotional difficulties initially, but after a time seem

to cope with the loss of a spouse. . . . [B]oth establish new [single] lifestyles based on their past patterns of interaction. For both, their financial [and health] situation is related to their feelings of well-being. (Brubaker 1991, p. 233)

A spouse's death brings the conjugal unit to an end—often a profoundly painful event. Yet, "we have to learn to accept the death of others—including others we love dearly—and keep living ourselves just as we have to learn to accept our own impending deaths and to keep on living" (Marshall 1986, p. 127).

Widow- and widowerhood and the other family transitions we have been discussing are stressors that can usually be anticipated and planned for. Partners can discuss what kind of funeral each wants and make sure each knows how to carry on financially after the other's death, for example. If they still have dependent children, they can decide how these would be cared for in case both parents died at the same time. They might discuss with the children their feelings about remarriage. As we will see, anticipating and planning ahead for crises whenever possible is an important aid to meeting them effectively and creatively. First, however, we will examine the course of a family crisis.

The Course of a Family Crisis

A family crisis ordinarily follows a fairly predictable course, similar to the truncated roller coaster shown in Figure 14.3. Three distinct phases can be identified: the event that causes the crisis, the period of disorganization that follows, and the reorganizing or recovery phase after the family reaches a low point. Families have a certain level of organization before a crisis; that is, they function at a certain level of effectiveness—higher for some families, lower for others. In the period of disorganization following the crisis, family functioning declines from its initial level. Families reorganize, and after the reorganization is complete (1) they may function at about the same level as before; (2) they may have been so weakened by the crisis that they function only at a reduced level; or (3) they may have been stimulated by the crisis to reorganize in a way that makes them more effective.

At the onset of a crisis, it may seem that no adjustment is required at all. Family members may be numbed by the new or sudden stress and, in a process of denial, go about their business as if the event had not occurred.

Gradually, however, they begin to assimilate the reality of the crisis and to appraise the situation. Then the **period of disorganization** sets in.

The Period of Disorganization

At this time, family organization slumps, habitual roles and routines become nebulous and confused, and members carry out their responsibilities with less enthusiasm. Typically, and legitimately, they begin to feel angry and resentful.

Expressive relationships within the family change, some growing stronger and more supportive perhaps, and others more distant. Sexual activity, one of the most sensitive aspects of a relationship, often changes sharply and may temporarily cease. Parent–child relations may also change.

Relations between family members and their outside friends, as well as extended kin network, may also change during this phase. Some families withdraw from all outside activities until the crisis is over; as a result, they may become more private or isolated than before the crisis began. As we shall see, withdrawing from friends and kin often weakens rather than strengthens a family's ability to meet a crisis.

At the nadir, or low point, of family disorganization, conflicts may develop over how the situation should be handled. For example, in families with a seriously ill member, the healthy members are likely either to overestimate or to underestimate the sick person's incapacitation and, accordingly, to act either more sympathetically or less tolerantly than the ill member wants (Parsons and Fox 1952; Strauss and Glaser 1975). Reaching the optimal balance between nurturance and encouragement of the ill person's self-sufficiency may take time, sensitivity, and judgment.

During the period of disorganization, family members face the decision of whether to express or to smother any angry feelings they may have. As Chapter 9 points out, people can express their anger either in primarily bonding or in alienating ways. Expressing anger as blame will almost always sharpen hostilities. When family members opt to repress their anger, they risk converting it into destructive "anger insteads" or allowing it to smolder, thus creating tension and increasingly strained relations. How members cope with conflict at this point will greatly influence the family's overall level of recovery.

Recovery

Once the crisis hits bottom, things often begin to improve. Either by trial and error or (when possible) by thoughtful planning, family members usually arrive at new routines and reciprocal expectations. They are able to look past the time of crisis to envision a return to some state of normalcy and to reach some agreements about the future.

Some families do not recover intact, as today's high divorce rate illustrates. Divorce can be seen both as an adjustment to a family crisis and as a family crisis in itself (Figure 14.4).

Other families stay together, although at lower levels of organization or mutual support than before the crisis. As Figure 14.3 showed, some families remain at a very low level of recovery, with members continuing to interact much as they did at the low point of disorganization. This interaction often involves a series of circles in which one member is viewed as deliberately causing the trouble and the others blame and nag him or her to stop. This is true of many families in which one member is an alcoholic or otherwise chemically dependent, an overeater, or a chronic gambler, for example. Rather than directly expressing anger about being blamed and nagged, the offending member persists in the unwanted behavior. (Box 14.3, "Alcoholism as a Family Crisis," defines stages in the response of an alcoholic's family to the crisis. Note that it is not uncommon for recovery to be arrested at any of the early stages.)

Some families match the level of organization they had maintained before the onset of the crisis, whereas others rise to levels above what they experienced before the crisis. For example, one partner's attempted suicide might motivate both spouses to reexamine their relationship. With the help of professional men-

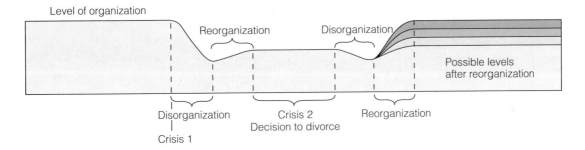

FIGURE 14.4

| Divorce as a family adjustment to crisis and as a crisis in itself.

tal health and marriage therapists, they might develop a personal marriage contract that is more supportive to both.

Reorganization at higher levels of mutual support can also result from less dramatic crises. For example, partners in midlife might view their boredom with their relationship as a challenge and revise their lifestyle to add some zest—by traveling more or planning to spend more time together rather than in activities with the whole family, for example.

Now that we have examined the course of family crises, we will turn our attention to a theory of family crisis and adaptation.

Theories of Family Crisis and Adaptation

Some years ago sociologist Reuben Hill proposed the ABC-X family crisis model, and much of what we've already noted about stressors is based on Hill's insights and research (Hill 1958; Hansen and Hill 1964). The **ABC-X model** states that A (the stressor event) interacting with B (the family's ability to cope with a crisis, their crisis-meeting resources) interacting with C (the family's appraisal of the stressor event) produces X (the crisis).

Building on this ABC-X model, Hamilton Mc-Cubbin and Joan Patterson (1983) advanced the double ABC-X model to better describe family adjustment to crises. In Hill's original model the "a" factor was the stressor event; in the **double ABC-X model,** "a" factor becomes "Aa," or "family pile-up." **Pile-up** includes not just the stressor but also previous family life (Olson and McCubbin 1983, p. 119) and hard-

ships induced by the stressor event. A parent's losing a job, for example, would be a stressor; it would soon be accompanied by hardships as bills begin to accumulate. Prior strains are the residuals of family tension that linger from unresolved stressors or that are inherent in ongoing family roles such as being a single parent or a spouse in a two-career family. Stressors and strain are connected to the family life cycle stage, as are the family resources available to cope with the stressor event (McCubbin and McCubbin 1991; Florian and Dangoor 1994).

When a family experiences a new stressor, prior strains that may have gone unnoticed come to the fore. For example, ongoing-but-ignored marital conflict may intensify when parents suddenly face caring for a child diagnosed as having a terminal illness. As another example, financial and time constraints typical of single-parent families may assume crisis-inducing importance with the addition of a stressor, such as caring for a disabled child (McCubbin 1989).

Stressors and hardships call for family coping and management skills. When family members do not have adequate resources for coping and managing—either because of prior strains or because of the character of the stressor event itself—stress emerges.

The pile-up concept of family-life stressors and strains (similar to the concept of stressor overload described earlier) is important in predicting family adjustment over the course of family life. Social scientists have hypothesized that an excessive number of life changes and strains occurring within a brief time, perhaps a year, are more likely to disrupt a family (Olson and McCubbin 1983, p. 120). Pile-up renders a

Alcoholism as a Family Crisis

When one of the adults in a family becomes an alcoholic, the course of the crisis is similar to that described in this chapter. One authority, Joan Jackson, has described seven stages in the course of alcoholism. Not all alcoholic families move through all seven stages; instead, many become mired in the early stages.

Stage 1: Attempts to Deny the Problem

Alcoholism rarely emerges full blown overnight. It is usually heralded by widely spaced incidents of excessive drinking, each of which sets off a small family crisis. The alcoholic may arrive home late with a ticket for reckless driving, for instance. Both mates try to account for such an episode as perfectly normal.

Between drinking episodes, both mates feel guilty about the alcoholic's behavior. (It is common for nonalcoholic family members to blame themselves either consciously or unconsciously for the alcoholic's drinking.) Gradually, the drinking problem and any other problems in the marriage may be sidestepped.

It takes some time for the sober spouse to realize that the drinking is neither normal nor controllable. Meanwhile, to protect themselves from embarrassment, the family may begin to eliminate outside activities and to become socially isolated.

Stage 2: Attempts to Eliminate the Problem

This stage begins when the family finally defines the drinking as not normal and tries to stop it alone. Lacking clear-cut guidelines, the nonalcoholic partner makes trial-and-error efforts. The family gradually becomes so preoccupied with finding ways to keep the alcoholic sober that they lose sight of all other family goals. Almost all thought becomes centered on alcohol. Meanwhile, family isolation peaks.

Stage 3: Disorganization

This is a stage of "What's the use?" Nothing seems effective in stabilizing the alcoholic, and efforts to change the situation become sporadic. The family gives up trying to understand the alcoholic. The children may no longer be required to show the alcoholic parent affection or respect. The sober partner, recog-

family more vulnerable to emerging from a crisis at a lower level of effectiveness (McCubbin and McCubbin 1989).

We have examined various characteristics of A, the stressor event, in the double ABC-X model. Next we will look at C, how the family defines or appraises stressor(s). We'll look at B, the family's crisis-meeting resources, after that.

Appraising the Situation

The way in which a family interprets a crisis-precipitating event may have as much or more to do with members' ability to cope as with the character of the

event itself.[3] Many events—unemployment or the accidental death of a child, for example—can be defined as either external or internal. One family might view a

3. While we are discussing the family's definition of the situation, it is important to remember the possibility that each family member experiences a stressful event in a unique way. "These unique meanings may enable family members to work together toward crisis resolution or they may prevent resolution from being achieved. That is, an individual's response to a stressor may enhance or impede the family's progress toward common goals, may embellish or reduce family cohesion, may encourage or interfere with collective efficacy. From this perspective, what is important is not the 'family's' definition of the stressor but an understanding of individual perspectives regarding stressful situations, how these perspectives relate to behavior, and the influences of members' perspectives in combination" (Walker 1985, pp. 832–33).

nizing his or her own inconsistent behavior, may become concerned about his or her sanity.

Stage 4: Attempts to Reorganize in Spite of the Problem

Here the sober partner takes sole family leadership. The alcoholic is ignored or assigned the status of a recalcitrant child and is increasingly excluded from family activities. Hostility diminishes and is replaced by feelings of pity, exasperation, and protectiveness.

The sober mate seeks assistance from public agencies and self-help groups such as Al-Anon. With an emerging new support network, the sober partner gradually regains his or her sense of worth. Some families remain at this stage indefinitely. But despite greater stabilization, subsidiary crises multiply. The alcoholic may become violent or get arrested for drunken driving. Each crisis temporarily disrupts the new family organization, but the realization that the events are caused by alcoholism often prevents the complete disruption of the family.

Stage 5: Efforts to Escape the Problem

When some major subsidiary crisis occurs—desertion or unemployment, for example—the family feels forced to take survival action. At this point many couples separate or divorce. First, however, the sober partner has to resolve the mental conflicts about deserting a sick mate. Separating from the alcoholic is further complicated by the fact that when the decision is about to be made, the alcoholic often gives up drinking for a while.

Some other events have made separation possible, however. By this time the sober partner has learned that the family runs more smoothly without the alcoholic. Taking over control has bolstered the sober mate's self-confidence, and his or her orientation has shifted from inaction to action.

Stage 6: Reorganization of the Family

Here the family reorganizes without the alcoholic. For the most part this goes relatively smoothly. The family has long ago closed ranks against the alcoholic and now feels free of the minor disruptions the drinker created in the family. Reorganization is impeded, however, if the alcoholic continues to attempt reconciliation or believes she or he must "get even" with the family for leaving or "kicking me out."

Stage 7: Reorganization with Sobriety

If the partners have not divorced by the time the alcoholic recognizes that he or she has a drinking problem and makes serious efforts to stop drinking, hope is mobilized. The family attempts to open ranks again to give the alcoholic the maximum chance for recovery.

If treatment is successful, the couple can expect many other problems to appear, however. Both mates may have unrealistic expectations; the family may harbor built-up resentment; and it may require a major adjustment for both the spouse and the children to again respect the alcoholic as a responsible adult. Gradually, perhaps through counseling or membership in groups such as Alcoholics Anonymous and Al-Anon, the difficulties may be overcome and family adjustment and reorganization with sobriety can be achieved.

Source: Jackson 1958, pp. 90–98.

husband's unemployment as the result of a contracting economy (an external event), whereas another family might see the same situation as the result of the man's lack of initiative (an internal factor). Similarly, involuntary infertility, getting AIDS, or a child's death can be interpreted as an accident or as the result of negligence or disloyalty.

Families who define a problem as their fault suffer more as individuals and also tend to provide less support than families who consider the cause to be external. For example, although it is almost impossible to distinguish degrees of anguish, parents who blame themselves for their child's developmental disability may be less able to cope with the situation than if they interpret the occurrence of the illness as beyond their control (Farber 1959; Price-Bonham and Addison 1978).

Several factors influence how family members define a crisis. One is the nature of the precipitating event itself. Another is the degree of hardship or the kind of problems the stressor creates. Temporary unemployment is less of a hardship than is a layoff in an area where there are few job prospects or a loss of a job at age 55. Growing up with a blind, deaf, or otherwise disabled parent can seem to be not only an unfair situation but also a never-ending and hopeless one (Walker 1985). A third factor that influences appraisal of a stressor event is the family's previous experience

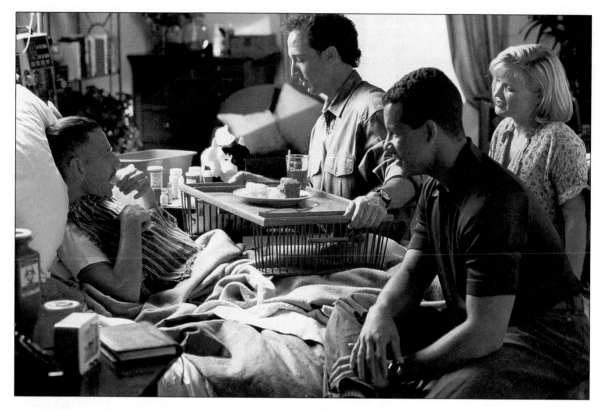

Families with strong crisis–meeting resources work together to prevent severe disharmony or disruption, as did the family and friends of this man dying of AIDS.

with crises, particularly those of a similar nature. If family members have had experience in nursing a sick member back to health, they will feel less bewildered and more capable of handling a new, similar situation. Believing from the start that a crisis is surmountable makes adjustment somewhat easier. Family members' interpretations of a crisis event shape their responses in subsequent stages of the crisis. Meanwhile, the family's crisis-meeting resources affect its appraisal of the situation.

Crisis–Meeting Resources

A family's crisis-meeting resources (B) comprise its abilities to prevent a stressor from creating severe disharmony or disruption (McCubbin and McCubbin 1989). The personal resources of each family member (for example, intelligence, problem-solving skills, physical and emotional health) are important here. At the same time, the family *as family* or family system has resources, including bonds of trust, appreciation,

and support (family harmony); sound financial management and health practices; positive communication patterns; healthy leisure activities; and overall satisfaction with the family and quality of life (Olson and McCubbin 1983, pp. 211–14; McCubbin and McCubbin 1989).

After describing stressor pile-up, Hamilton McCubbin and Marilyn McCubbin began to focus more directly on a family's crisis-meeting resources. As a result, they developed the **resiliency model of family stress, adjustment, and adaptation** (see Figure 14.5). Although this model does not use the ABC-X designations, it has the same elements: a stressor event, family appraisal of the stressor event, and the family's resources for coping with a stress, strain, or transition.

The resiliency model is a more complex model, adding an analysis of the family system and a typology of families to the basic elements. Simplifying the McCubbins' detailed analyses, which appear in various publications, we can say that family systems and types

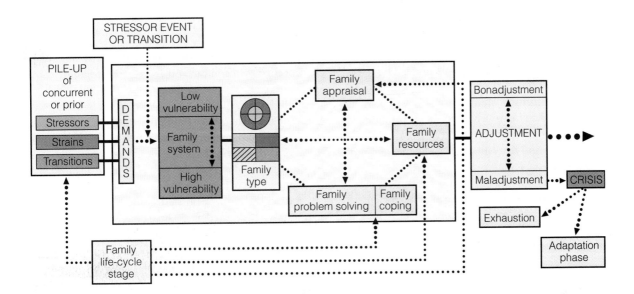

FIGURE **14.5**

The resiliency model of family stress, adjustment, and adaptation. This model elaborates on an earlier family crisis model, the double ABC–X model of family stressors and strains. Components of the double ABC–X model include: a stressor event (A); the family's appraisal of the stressor event (B); the family's crisis-meeting resources (C); the doubling of these components through pile-up; and the experience or avoidance of crisis (X).

In the resiliency model components are no longer identified by these letters. Family system characteristics and typology are added concepts, and the family's resilient adjustment or its experience of crisis have been developed in greater detail. Demands on the family from prior or concurrent stressors and strains and family life-cycle transitions have been included to better predict and understand the family's vulnerability to the stressor event. (Source: Adapted from Figure 1.2, Components of the Adjustment Phase of the Resiliency Model, McCubbin and McCubbin 1991, p. 5; see their article for more detail including other components added in a Family Adaptation Phase, Figure 1.5, p. 16)

can be low or high in vulnerability (McCubbin and McCubbin 1991). Families that cope well with stress are **strong families,** who emphasize mutual acceptance, respect, and shared values. Family members rely on one another for support. Generally accepting difficulties, they work together to solve problems with members, feeling they have input into major decisions. Strong families foster predictable family routines, rituals, and other times together (McCubbin and McCubbin 1989).

Weak families are more vulnerable to negative outcomes from crisis-provoking events than are strong families. Having a lower sense of common purpose and feeling less in control of what happens to them, they tend to cope with problems by showing diminished respect or understanding for one another. Hesitant to depend on the family for support and understanding, members may avoid one another. Weak families are also less experienced in shifting re-

sponsibilities among family members and are more resistant to compromise. There is little emphasis on family routines or predictable time together (McCubbin and McCubbin 1989).

Noting that a family's typology and the family system affect how positively it faces a crisis enables us to predict or explain its good ("bonadjustment") or bad ("maladjustment") adjustment to the stressor event. At the end point of the resiliency model, the family either successfully adapts or becomes exhausted and vulnerable to continuing crisis. The next section discusses factors that help families to meet crises creatively.

Meeting Crises Creatively

Meeting crises creatively means that after reaching the nadir in the course of the crisis, the family rises to a level of reorganization and emotional support that is equal to or higher than that which preceded the crisis. Social scientists have pointed out that most American families have some handicaps in meeting crises creatively, however.

The typical American family is under a high level of stress at all times. Providing family members with emotional security in an impersonal and unpredictable society is difficult even when things are running smoothly. Family members are trying to do this while holding jobs and managing other activities and relationships.

In addition, the average family has some weaknesses as an organization. Young and inexperienced dependents may outnumber more capable adults, and families can hardly reject their weak, ineffective, or problematic members to recruit more competent ones. Ironically, the principal function of today's family—offering a kind of unearned acceptance—makes the family less efficient in meeting crises (Parsons and Fox 1952; Hansen and Hill 1964): The family cannot tell a member to "shape up or ship out" as readily as an employer can.

Determining Factors

For some families, breaking up is the most beneficial (and perhaps the only workable) way to reorganize. Other families stay together and find ways to meet crises effectively. What differentiates families that reorganize creatively from those that do not?

A POSITIVE OUTLOOK   In times of crisis, family members make many choices, one of the most significant of which is whether to blame one member for the hardship. Casting blame, even when it is deserved, is less productive than viewing the crisis primarily as a challenge.

Put another way, choosing a positive outlook helps a person or a family to meet a crisis constructively. Electing to work toward developing more open, supportive family communication—especially in times of conflict—also helps individuals and families meet crises constructively. Families that meet a crisis with an accepting attitude, focusing on the positive aspects of their lives, do better than those that feel they have been singled out for misfortune. For example, many chronic illnesses have downward trajectories, so that both partners can realistically expect that the ill mate's health will only grow worse. Some couples are remarkably able to adjust to this, "either because of immense closeness to each other or because they are grateful for

what little life and relationship remains" (Strauss and Glaser 1975, p. 64). Ironically, it is often easier for family members to focus on the positive when they accept their own and each other's negative feelings.

SPIRITUAL VALUES AND SUPPORT GROUPS
Some researchers have found that strong religious faith is related to high family cohesiveness (Bahr and Chadwick 1985) and helps people meet crises, partly because it provides a positive way of looking at suffering. Many of the better-adjusted victims of multiple sclerosis described earlier belonged to charismatic religious groups, which acted as a source of strength (Power 1979). Self-help groups, such as Al-Anon for families of alcoholics, can also help people take a positive approach to family crises.

HIGH SELF-ESTEEM   Families meet crises more creatively when members have high self-esteem. This is because when family members focus their energies to meet a crisis, they often have to cut down on their expression of affection and emotional support (Hansen and Hill 1964, p. 806). As a result, family members must stand alone emotionally until they work out new avenues for expressing mutual support and concern, and this requires high self-esteem. Also, family members with high self-esteem are more able to deal constructively with the conflict that characterizes families at the nadir of a crisis.

OPEN, SUPPORTIVE COMMUNICATION   Families whose members interact openly and supportively meet crises more creatively. For one thing, free-flowing communication opens the way to understanding. For example, divorcing parents who interact cooperatively do their children—and themselves—a great favor. As another example, the better-adjusted husbands with multiple sclerosis believed that even though they were embarrassed when they fell in public or were incontinent, they could freely discuss these situations with their families and feel confident that their families understood (Power 1979).

Some research indicates that families who have difficulty coping with crises are far more inadequate in communicating with and supporting each other than they are in dealing with practical problems (Hansen and Hill 1964, p. 808). For example, moving can be considered a family crisis because it disrupts social networks (Lavee, McCubbin, and Patterson 1985). The

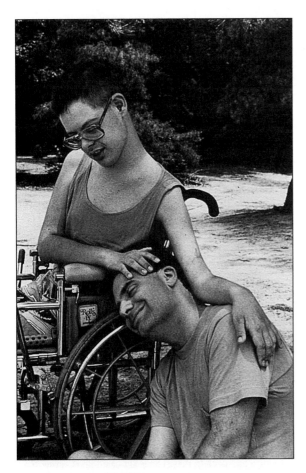

A positive outlook, spiritual values, high self-esteem, supportive communication, adaptability, informal social support—all these, along with an extended family and community resources—are factors in meeting a crisis creatively.

prolonged disruption or isolation that families experience after a move is often a result of strained relationships and inadequate supportive communication rather than of problems associated with the move itself (McAllister, Butler, and Kaiser 1973).

Knowing how to indicate the specific kind of support one needs is important at stressful times. For example, differentiating between—and knowing how to request—just listening as opposed to problem-solving discussion can help reduce misunderstandings among family members—and between family members and others as well (Perlman and Rook 1987). Box 14.4, "Communicating with the Dying," addresses the special communication problems involved in this family crisis.

ADAPTABILITY   Adaptable families are better able to respond effectively to crises. And families are more adaptable when they are more democratic and when conjugal power is fairly egalitarian (Patterson and Mc-Cubbin 1984). In families in which one member wields authoritarian power, the whole family suffers if the authoritarian leader does not make effective decisions during a crisis—and allows no one else to move into a position of leadership (Hansen and Hill 1964).

For example, a teenager may realize that his or her mother is addicted to drugs or is an alcoholic before either the mother or the father is willing to admit it. Although the whole family might benefit if the child goes to Al-Ateen, an authoritarian father might forbid the youngster to attend or refuse transportation to the meetings. In a more democratic family the youngster would be allowed to attend Al-Ateen meetings, even if the parents continued to disagree with the teenager's point of view.

Often, family power relations change in response to crisis. An ineffective leader tends to be replaced eventually by another family member—the spouse or an older child—if she or he does not solve the problem effectively (Bahr and Rollins 1971). Men who feel comfortable only as family leaders may resent their loss of power, and this resentment may continue to cause problems when the crisis is over. When wives took over all family decisions in the absence of husbands serving in the Gulf War, marital strife was sometimes the outcome upon the husband's return.

Just as traditional gender roles continue to affect most areas of family life, so, too, do they influence family crises. Social scientists Bryan Robinson and Robert Barret, in their book *The Developing Father* (1986), argue that fathers, more often than mothers, have more difficulty coping with their severely retarded children; they are especially vulnerable to social stigma. Differences between fathers and mothers in coping can cause problems in their marriage. For example, a mother's using community and social support could increase the father's stress over potential social stigma and invasion of privacy (pp. 198–99). Women who adapted to husbands' unemployment by seeking emotional support from relatives and friends found that their husbands often viewed this as disloyalty (Robertson et al. 1991).

Family adaptability in aspects other than leadership is also important. Families who can adapt their

Communicating with the Dying

Dying poses special communication problems, yet communication at this time is enormously important. Dying people, particularly older adults, often begin to prepare for death with a "life review"—evaluating and legitimating their biography or progress through life. "In the normal case, a person will ... complete the life review and reach a state of integrity." "Most old people in our society do come to accept their deaths" (Marshall 1986, pp. 140–41).

Social relations are very significant during the dying process. The life review is most successful when it is social, and those who are able to talk intimately with at least one person, often a spouse, about their awareness of death are most accepting.

Avoidance, confrontation, and reaction are three primary communication strategies that emerge in conversations with the dying. *Avoidance*, as the term implies, is an interaction with the patient that does not acknowledge death. "The motif is to carry on the interaction as if nothing were significantly different in the relationship.... In using this approach, persons try to focus on upbeat, positive topics—topics that center on current interests or news and topics that explore happy events of the patient's past" (Miller and Knapp 1986, p. 259). Dying people may become isolated or depressed when family members are willing to talk only in optimistic terms (Miller and Knapp 1986). This can be a particular problem when it is a child who is dying—and they are usually much more aware of their situation than parents may think (Christ 1982).

The confrontive style, which focuses on the impending death, may not be well received by the patient. But it is intended to be helpful, motivated by an awareness that (1) dying people may have a need to ventilate their anger or thoughts about dying, (2) patients' clarifying their illness or feelings may receive therapeutic benefits, and (3) the family member may desire to express feelings of affection or sorrow.

The reactive approach is one of listening uncritically to the dying person. "By deeming whatever topics of conversation the dying person wishes to cover as appropriate, the [family member] is free to discover the norms of interaction with the particular individual, to assess the 'needs' of the dying person, to provide emotional support through the unqualified acceptance of the person's feelings, and to initiate any conversational topics that the patient has perceived to be appropriate. This approach seeks to help the patient by giving the patient control of the conversational agenda" (Miller and Knapp 1986, p. 260).

Miller and Knapp (1986), who have done extensive research on communication with the dying based primarily on interviews with health care professionals and clergy, generally endorse a dialectical approach to these conversations; that is, rather than rely, as the professionals do, on one primary mode (usually reactive), the family member would do well to adapt the conversational strategy to the history of the family relationship, the nearness of death, the practical needs of the situation, and the emotional tone that seems intuitively appropriate at a particular time. Open communication styles have even been correlated with longer survival of terminal patients (Shapiro 1983). As with other family conversation, communication about death requires sensitivity, confidence in one's perception of the situation, and trust in one's instincts.

---

schedules and use of space, their family activities and rituals, and their connections with the outside world to the limitations and possibilities posed by the crisis will cope more effectively than families that are committed to preserving sameness.

Rituals can serve as effective tools to cope with family stress and crisis (Imber-Black, Roberts, and Whiting 1988). A study of alcoholic families found that adult children of alcoholics who came from families that had maintained family dinner and other ritu-

Many turn to their extended family in times of stress. Kin may provide emotional support, monetary support, practical help, and a strong shoulder to lean on without embarrassment.

als (or who married into families that did) were less likely to become alcoholics themselves (Bennett et al. 1987; Bennett, Wolin, and Reiss 1988; Goleman 1992a, based on an interview with psychiatrist Steven J. Wolin about his research with anthropologist Linda Bennett).

INFORMAL SOCIAL SUPPORT   It's easier to cope with crises when a person doesn't feel alone (see, for example, Turner and Avison 1985). It may go without saying that spouses do better when they feel supported by their mates (Suitor and Pillemer 1994). Families can find helpful support in times of crisis from kin, good friends, neighbors, and even acquaintances such as work colleagues. These various relationships provide a wide array of help—from lending money in financial emergencies to helping with child care to just being there for emotional support. Even continued contact with more casual acquaintances can be helpful

as often they offer useful information, along with enhancing one's sense of community (Milardo 1989). A two-year study (Gore 1978) of 100 unemployed married men recently laid off because of a plant shutdown found that those with more supportive social and kin relationships did not find jobs sooner, but they did exhibit fewer physical and mental stress symptoms.

AN EXTENDED FAMILY   Kin networks are a valuable source of support in times of crisis. As you'll recall from Chapter 2, formally extended families consist of parents and their children who live in the same household with other relatives, such as the parents or a brother or sister of one of the spouses. Just under 4 percent of American family households are extended family households (U.S. Bureau of the Census 1991b, Table 56, p. 45). Actually, this figure represents only three-generation households—that is, extended families in which three generations (parent, child, and

Extended families. *Go to chapter 2*

grandchild) live together. But other combinations of kin, such as aunts, uncles, or cousins, may live together, too.[4]

Extended families are more common among blacks (Hofferth 1984), Native Americans, and Asian Americans than among other racial and ethnic groups, and—except for Asian Americans—they are especially likely to occur when one of the parents is not residing in the household. (This seems to be a result of the greater proportion of female-headed households among blacks; a greater proportion of such households are headed by never-married women, who are especially likely to live with extended kin. Black married couples are no more likely than white married couples to live with extended kin [Hofferth 1984].) But extended families have been common among working-class whites as well, because of economic deprivation or ethnic heritage. Then, too, certain ethnic groups of any class, Asian Americans and Italian Americans, for example, may continue the extended family groupings of their homelands, with young couples residing near or with parents and aging parents living in the homes of adult children.

Even families who do not live together regularly may be organized around exchanges of goods and services among extended kin (Taylor 1986). Lower-class black communities may operate primarily as extended families, with shifting combinations of kin sharing economic resources, child-rearing responsibilities, and so forth (Stack 1974; Chatters, Taylor, and Neighbors 1989; Benin and Keith 1995). Native American families function in much the same way (Staples and Mirande 1980). And Jaime Sena-Rivera (1979) argues that the Latino family is similarly a shared-aid extended family. Caucasian working-class families have commonly had such mutual support patterns also. As

middle-class wives have begun working outside the home and families at all levels have experienced economic stress and high rates of divorce, middle-class families have begun to draw on the resources of extended kin as well (Clavan 1978). Grandparents, aunts, or other relatives may help with child rearing in two-career families. Families going through divorce often fall back on relatives for practical help and financial assistance. In other crises, kin provide a shoulder to lean on—someone who can be asked for help without causing embarrassment—which can make a crucial difference in a family's ability to recover. Indeed, with today's high divorce rate and difficult economic circumstances, adults may increasingly find that their more permanent kin ties are with their families of origin rather than with their mates. They may rely more on kin for emotional support, financial assistance, and exchange of services. Put another way, we may increasingly see a *vertical family* (attachment to parents and family of origin) as opposed to the *horizontal family* (stressing priority and permanence of the marital bond, which has characterized American society from its beginning).

Even though extended families as residential groupings represent a small proportion of family households, kin ties remain salient and may become more so. One aspect of all this that is just beginning to get research attention is reciprocal friendship and support among adult siblings (White and Riedman 1992; Horwitz 1993).

We need to be aware, though, of the dark side of helping (Fisher, Nadel, and Alagne 1982; Staub et al. 1984; Worchel 1984). Sometimes the recipient's discomfort or perceived vulnerability to control attempts on the part of the helper can cloud the relationship or discourage the person in need (Saks and Krupat 1988; Brigham 1991).

Although in most racial/ethnic groups the extended family is viewed as a resource in times of trouble, some clinicians argue that, at least in Irish American families, having the family find out about your distress may be the problem rather than the solution (Hines et al. 1992). There is some indication that Irish elderly are uncomfortable receiving financial help or personal care from adult children (Brubaker, Gorman, and Hiestand 1990). Then too, among some recent immigrant groups, such as Asians or Hispanics, expectations of the extended family may clash with the more individualistic values of

---

4. Extended families were prominent in traditional agricultural societies, for they provided a network of property ownership and support that contributed to the family's self-sufficiency. In those societies the extended family most often resulted when a young married couple moved into the home of one spouse's parents. The spouse's parents continued to own the family property until they died, and all members of the household participated in the family business or farming. Not only did these families tend to be patriarchal, but obedience to familial authority was valued more than individual advancement and growth. Furthermore, loyalty to the ties of blood kinship typically was valued more highly than intimacy between spouses.

Today's extended families may also experience tension between family loyalty and individual advancement or couple intimacy. In some of the black families studied by anthropologist Carol Stack, for example, some extended families attempted to prevent or break up marriages that pitted competition for scarce resources against the needs of kin (Stack 1974).

Americanized family members (Kamo and Zhou 1994).

COMMUNITY RESOURCES   The success with which families meet crises also depends on the community resources available to help:

> Community-based resources are all of those characteristics, competencies and means of persons, groups and institutions outside the family which the family may call upon, access, and use to meet their demands. This includes a whole range of services, such as medical and health care services. The services of other institutions in the family's . . . environment, such as schools, churches, employers, etc. are also resources to the family. At the more macro level, government policies that enhance and support families can be viewed as community resources. (McCubbin and McCubbin 1991, p. 19)

These may be particularly important for families caring for the disabled, chronically ill, or elderly in their homes (Perlman 1983; Mancini and Bleiszner 1989; McCubbin 1989). For one thing, children in today's relatively small families will find the burden of caring for aging parents much heavier than in the past when it was shared by a larger group of siblings. Divorce disrupts or weakens some family ties that might have been resources. The high rates of women's employment mean that these caregivers are not as available as in the past. Government policy and corporate initiatives to address the need to support or provide for caregivers are just beginning to develop. Still, the social support found in interpersonal ties is perhaps the single most important community resource (McCubbin and McCubbin 1991, p. 19).

One important community resource available is marriage and family counseling. Counseling can help families after a crisis occurs; it can also help when families foresee a family change or crisis. For instance, a couple might visit a counselor when expecting or adopting a baby, when deciding about work commitments and family needs, when the youngest child is about to leave home, or when the husband or wife is about to retire. Marriage counselors now offer premarital counseling to help partners anticipate some of the problems associated with the transition to married life.

Marriage counseling is not just for marriages that are in trouble but a resource that can help relationships (see Appendix H).

Whether society's resources are made available through government support for families with problems is also a factor in ability to cope.

Crisis: Disaster or Opportunity?

A family crisis is a turning point in the course of family living that requires members to change how they have been thinking and acting. We think of *crisis* as synonymous with *disaster,* but the word comes from the Greek for *decision.* Although we cannot control the occurrence of many crises, we can decide how to cope with them.

Crises—even the most unfortunate ones—have the potential for positive as well as negative effects. One therapist cites cases in which death in a family with longstanding problems precipitated counselor-supported change, enabling family members to function more effectively than they ever had functioned previously (Gelcer 1986). Whether a family emerges from a crisis with a greater capacity for supportive family interaction depends largely on how family mem-bers choose to define the crisis. A major theme of this text is that people create the kind of marriage and family they want based on the choices they make. Then, too, families whose members choose to be flexible in roles and leadership meet crises creatively.

Even though they have options and choices, however, family members do not have absolute control over their lives. Many family troubles are really the results of public issues. The serious family disorganization that results from unemployment or poverty, for example, is as much a social as a private problem. In this text we have often suggested that the tension between individualistic and family values in our society creates personal and family conflict. And the society-wide movement toward equality and greater self-actualization may spark family crises as wives and children move toward independence. Moreover, many family crises are more difficult to bear because communities lack adequate resources to help families meet them. Families must act collectively to obtain the social resources they need for effective crisis management in everyday living.

In Sum

Family crises may be expected and normative, as when a baby is born or adopted, or they may be unexpected. In either case the event that causes the crisis is called a stressor. Stressors are of various types and have varied characteristics. Generally, stressors that are expected (normative), brief, external, defined by concrete norms, and improving are less difficult to cope with.

The predictable changes of individuals and families—parenthood, midlife transitions, postparenthood, retirement, and widow- and widowerhood—are all family transitions that can be viewed as stressors. During transitions, spouses can expect their relationship to follow the course of a family crisis.

A common pattern can be traced in families' reactions to crises. Three distinct phases can be identified: the event that causes the crisis, the period of disorganization that follows, and the reorganizing or recovery phase after the family reaches a low point. The eventual level of reorganization a family reaches depends on a number of factors, including the type of stressor, the degree of stress it imposes, whether it is accompanied by other stressors, and the family's definition of the crisis situation. Various models of family crisis and reorganization try to capture this process in an analytical way, to pursue through research useful knowledge about coping with crisis.

Meeting crises creatively means resuming daily functioning at or above the level that existed before the crisis. Several factors can help: a positive outlook, spiritual values, the presence of support groups, high self-esteem, open and supportive communication within the family, adaptability, counseling, and the presence of a kin network.

Key Terms

ABC-X model	resiliency model of family
bereavement	stress, adjustment, and
crisis	adaptation
double ABC-X model	sandwich generation
empty nest	stressor
family stress	stressor overload
period of disorganization	strong families
pile-up	transitions
predictable crises	weak families

Study Questions

1. Define *crisis* and explain how family transitions fit this definition.

2. Describe the three phases of a family crisis. Use an example to illustrate each phase.

3. Differentiate among the types of stressors. How are these single events different from stressor overload?

4. What are some characteristics of stressors? Why would a stressor characterized by concrete norms be less difficult to cope with than one characterized by ambivalence?

5. Differentiate between external and internal stressors. How would they affect families differently? Which is more difficult for families to cope with? Why?

6. Discuss the factors that influence how family members appraise a crisis. How does the family's appraisal relate to the members' ability to adjust to the crisis?

7. Describe the period of disorganization in a family crisis situation. How can this phase of a crisis pull a family apart? How can it bring a family together?

8. What factors help some families recover from crisis while others remain in the disorganization phase?

9. Are family transitions, which are at least somewhat expected, easier to adjust to than unexpected crises? Why or why not?

10. Compare the transitions of sandwich generation, empty nest, retirement, and widow- or widowerhood. Are there similarities? Any differences? What factors would encourage adjustment to each transition? What factors would inhibit adjustment?

11. Discuss the advantages and the drawbacks of relying on help from an extended family.

12. What should government do to help families in crisis?

Suggested Readings

Abel, Emily K. 1991. *Who Cares for the Elderly? Public Policy and the Experiences of Adult Daughters*. Philadelphia: Temple University Press. This important question, which could not be developed in detail in our textbook, deserves this thoughtful and important book.

Adams, David W. and Eleanor J. Deveau. 1984. *Coping with Childhood Cancer*. Reston, VA: Reston Publishing of Prentice-Hall. Adams has a master's degree in social work, and

Deveau is a nurse. These authors address such issues as managing life in the hospital, the needs of a parent's other children, remission, relapse, the dying child, and grief.

Brubaker, Timothy. 1990. *Family Relationships in Later Life,* 2d ed. Newbury Park, CA: Sage. Good series of articles on all aspects of family life in middle and old age. These articles, although academic, are clearly useful to formal and informal caregivers, among others.

Davis, Fred. 1991. *Passage Through Crisis: Polio Victims and Their Families,* 2d ed. New Brunswick, NJ: Transaction. The illness studied by Davis has largely disappeared, but the dynamics of families caring for a child with a major illness have not. Because this insightful social-psychological study of families is so well-thought-of, it has been reissued thirty years after its original publication.

Earhart, Eileen and Michael J. Sporakowski. 1984. *The Family with Handicapped Members.* Special issue of *Family Relations* 33 (Jan.). Contains many useful articles based on research or clinical experience.

Figley, Charles R. and Hamilton T. McCubbin (eds.). 1983. *Stress and the Family, Vol. II: Coping with Catastrophe.* New York: Brunner/Mazel. Part of a two-volume book on stress, covering current research in this field. Stressful family situations are divided into the transitions in family stages that are widely shared (Vol. I) and crises of various sorts that only some families experience (Vol. II).

Greenfield, Josh. 1972. *A Child Called Noah.* New York: Holt, Rinehart & Winston.

———. 1978. *A Place for Noah.* New York: Holt, Rinehart & Winston.

———. 1987. *A Client Called Noah.* New York: Holt, Rinehart & Winston. Accounts of a family's emotional response and attempts at coping with an autistic child. The author is honest about negative as well as positive feelings.

Imber-Black, Evan, Janine Roberts, and Richard A. Whiting (eds.). 1988. *Rituals in Families and Family Therapy.* New York: Norton. Discusses family rituals as crisis intervention.

Jacob, Theodore. 1992. "Family Studies of Alcoholism." *Journal of Family Psychology* 5: 319–38. Review of recent research on a topic in which there is much misunderstanding and myth.

Journal of Health and Social Behavior. Academic journal that reports many studies of relationships between life events, stress, social supports, and other coping mechanisms, and physical and mental health.

Kaminer, Wendy. 1992. *I'm Dysfunctional, You're Dysfunctional: The Recovery Movement and Other Self-Help Fashions.* Reading, MA: Addison-Wesley. Although we only rarely hear criticism of self-help movements or therapy, Kaminer provides a strongly negative perspective on groups and resources that many people believe have been helpful in their lives. We need to know there is a contrary opinion.

Kleban, Morton H. 1989. "Family Help to the Elderly: Perceptions of Sons-in-Law Regarding Parent Care." *Journal of Marriage and the Family* 51(2): 303–12. Thus far a neglected research topic, this study interviewed 150 husbands whose wives were primary caregivers for an elderly mother.

Kübler-Ross, Elisabeth. 1969. *On Death and Dying.* New York: Macmillan. Now-classic conceptualization of death as a social and psychological experience. The stages of adjustment to death she identified have been applied to other losses and crises.

McCubbin, Hamilton I. and Charles R. Figley (eds.). 1983. *Stress and the Family, Vol. I: Coping with Normative Transitions.* New York: Brunner/Mazel. Part of a two-volume book on stress, covering research in this field. Stressful family situations are divided into the transitions in family stages that are widely shared (Vol. I) and crises of various sorts that only some families experience (Vol. II, Figley and McCubbin 1983).

McCubbin, Marilyn A. and Hamilton I. McCubbin. 1994. "Families Coping with Illness: The Resiliency Model of Family Stress, Adjustment, and Adaptation." Chapter 2 in *Families, Health, and Illness.* St. Louis: Mosby. Presents the newest model of family stress and coping. It is directed to families coping with illness, so it tells us much about this family stressor.

Nunnally, Elam W., Catherine S. Chilman, and Fred M. Cox. 1988. *Troubled Relationships.* Newbury Park, CA: Sage. From the "Families in Trouble" series, this volume addresses divorce, family violence, out-of-home placement of children, and troubled marital and parent–child relationships generally; includes chapters on law and public policy.

Perlman, Robert (ed.). 1983. *Family Home Care: Critical Issues for Services and Policies.* New York: Haworth. Examines caring for the chronically ill or disabled family member at home as an ongoing crisis in "an unstable triad"—the sick or dependent person's demands; family resources, and community supports.

Smith, Gregory C. 1995. *Strengthening Aging Families: Diversity in Practice and Policy.* Thousand Oaks, CA: Sage. A readable review of the literature and exploration of policy consequences and possibilities for older couples and families.

Steinem, Gloria. 1983. *Outrageous Acts and Everyday Rebellions.* New York: Holt, Rinehart & Winston. Autobiography of the well-known feminist and journalist. Steinem grew up as primary caregiver to her mentally ill mother. Her chapter on this experience conveys the impact of coping with a major family burden as a child.

Steinglass, Peter, with Linda Bennett, Steven J. Wolin, and David Reiss. 1987. *The Alcoholic Family.* New York: Basic Books. Analysis of the alcoholic family by family systems people and sociologists. This book contains a typology of families different from that presented in this text.

Tschann, Jeanne M., et al. 1989. "Resources, Stressors, and Attachment as Predictors of Adult Adjustment after Divorce: A Longitudinal Study." *Journal of Marriage and the Family* 51(4): 1033–46. Application of the double ABC-X crisis theory to adjustment after divorce.

Vine, Phillis. 1982. *Families in Pain: Children, Siblings, Spouses, and Parents of the Mentally Ill Speak Out.* New York: Pantheon. Coping with "the tangible role responsibilities of having a mentally disabled relative"; based on interviews and written by a historian with a mentally disturbed relative.

Westberg, Granger E. 1971. *Good Grief: A Constructive Approach to the Problem of Loss.* Philadelphia: Fortress Press. A small book but a classic full of information and advice on the mourning process.

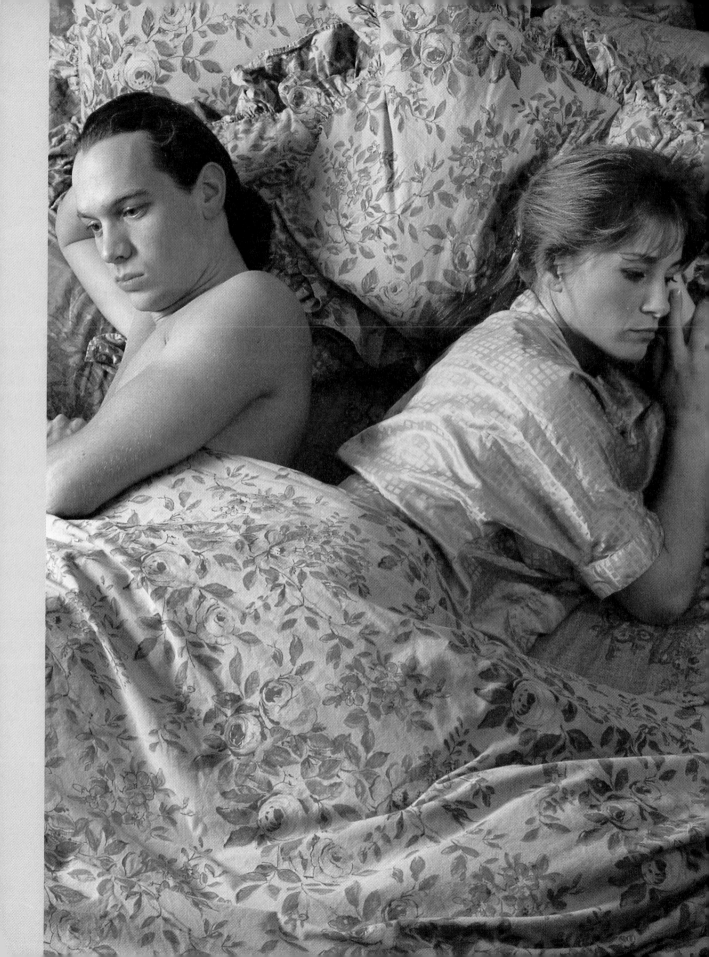

Divorce

*You cannot imagine how much we
hoped in the beginning.*
LIV ULLMANN, CHANGING

*"Marital disruption is altering the
organization of kinship in American
society."*
FRANK FURSTENBERG, JR.,
DEMOGRAPHER

Divorce has become a common experience in the United States, in all social classes, age categories, and religious and ethnic groups. In this chapter we'll look at divorce in broad terms and analyze why so many couples divorce in our society today. At the more personal level, we'll examine factors that affect people's decisions to divorce, the experience itself, and ways the experience can be made less painful and become the prelude to the future, alone or in a new marriage (remarriages are discussed in Chapter 16). We'll begin by looking at current divorce rates in the United States, among the highest in the world.

Today's High U.S. Divorce Rate

The frequency of divorce has increased sharply throughout most of the twentieth century, with dips and upswings surrounding historical events (such as the Great Depression and major wars), as Figure 15.1 shows. Before we examine the divorce rate today, we need to understand how divorce rates are reported.

How Divorce Rates Are Reported

Divorce rates are reported in many forms, some of which are more useful than others. Reports generally indicate rising rates no matter what measure is used, but the implications of these measures are frequently misunderstood.

- *Number of divorces per year.* The number of divorces recorded per year is not an accurate measure of the rate because it does not take into account the

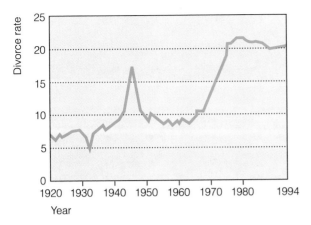

FIGURE **15.1**

Divorce rates per 1,000 married women age 15 and older in the United States, 1920–1993. (Source: U.S. National Center for Health Statistics 1990a, 1991a, 1994, p. 4)

general increase in population. There may be more divorces in a population simply because there are more people.

- *Ratio of current marriages to current divorces.* This measure is faulty because the marriages reported in the media have all taken place in the current year, whereas the divorces reported are of marriages that took place in many different years. The divorce rate then comes to depend on the marriage rate; that is, if the number of marriages goes down, the divorce rate will appear to rise, even if the number of divorces remains constant.

- *Crude divorce rate.* The crude divorce rate is the number of divorces per 1,000 population. This measure takes into account changes in size of population but includes portions of the population—children and the unmarried—not at risk for divorce.

- *Refined divorce rate.* This is the number of divorces per 1,000 married women over age 15. This measure compares the number of divorces with the total number of women eligible for divorce (adult married women) and hence is a more valid indicator of the propensity for divorce. It does not, however, predict one's chances either of divorcing over a lifetime or of divorcing at any particular age. Age-specific divorce rates (number of divorces per 1,000

married women in each age group) are available, but they do not provide an overall rate.

Ideally, a cohort of married couples would be followed over a lifetime and their rate of divorce calculated. This sort of *longitudinal study* has never been done on a large scale and is unlikely ever to be done because of the expense and the length of time necessary for collecting data and ascertaining results. Of course, any rate calculated on this basis would be applicable only to those who married in the same year because sociohistorical conditions (which affect divorce rates) would have changed over time.

On balance, the most useful and valid divorce rate appears to be the refined divorce rate, which we have presented in Figure 15.1 and in the text's discussion of divorce rates.

Our Current Divorce Rate

Between 1960 and a peak in 1979, the **refined divorce rate** rose from 9.2 divorces to 22.8 (per 1,000 married women age 15 and older) (U.S. National Center for Health Statistics 1990a). Although the divorce rate nearly doubled between 1965 and 1975, the rate of increase began to slow in the late seventies. From a 1979 peak of 22.8, the divorce rate dropped to 20.5 in 1993, lower than it had been since 1975 (U.S. National Center for Health Statistics 1994, p. 4). The same general trends in rates have occurred in all race and age groups.

Most observers conclude that the divorce rate has stabilized for the time being. One reason is that fewer people are marrying at the vulnerable younger ages. Figure 15.2 tracks divorce rates by age for both men and women in 1990. There are many reasons for the high divorce rate among teens (as Chapter 7 discusses), but most experts point to the interacting effects of economic hardship, low education, premarital pregnancy (Teti and Lamb 1989), and emotional immaturity. Although divorce rates appear to have stabilized, the fact remains that over the past twenty years they have been higher than they have ever been in our society. Moreover, demographers argue that marital dissolution, including informal or nonlegal separation, is underreported. As a result, real dissolution rates are higher than official divorce statistics indicate. Projections are that in the future about half of all marriages will last a lifetime (Bumpass, Raley, and Sweet 1995).

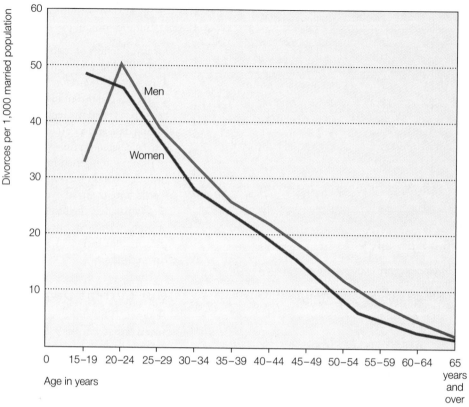

FIGURE **15.2**

Divorce rates by age for men and women, 1990. (Source: Clarke 1995b, p. 3)

The high incidence of divorce and separation is apparent in the prevalence of single-parent families. Between 1970 and 1993 the proportion of children living with one parent more than doubled while the proportion living with two parents declined by nearly 15 percentage points (Saluter 1994, p. xi). This increase in single-parent families is increasingly a result of factors other than divorce—rising unwed birthrates and greater tendency for unmarried mothers to establish independent households—but for the U.S. population as a whole, and among non-Hispanic whites in particular, the major cause is divorce (U.S. Bureau of the Census 1995, Table 75). In 1994, almost one-third (31 percent) of all family groups with a child under age 18 were headed by one parent.[1] Among whites this proportion was 25 percent, whereas among African Americans it was 65 percent, with Hispanics intermediate at 36 percent. The vast majority of all one-parent households are currently headed by women. The proportion headed by men is growing but remains small. Only about 4 percent of all family households are headed by men, and this figure varies little by race (U.S. Bureau of the Census 1995, Table 71). (See Figure 15.3.)

Projections are that one out of three white children and two out of three black children born to a marriage will experience the dissolution of their parents' marriage by the time they reach age 16 (Thornton and Freedman 1983), compared to 25–30 percent of children born in the first half of the twentieth century.[2] We'll look at reasons for this striking change in family patterns in the next section.

1. Data on the parental situation of children do not distinguish stepparents and biological parents in two-parent families.

2. It should be noted that marriages can be dissolved by death as well as divorce. Until the mid-1970s increasing divorce and decreasing mortality (death was the most common cause of marital dissolution in past times) canceled each other out, so that no overall increase in marital dissolution occurred.

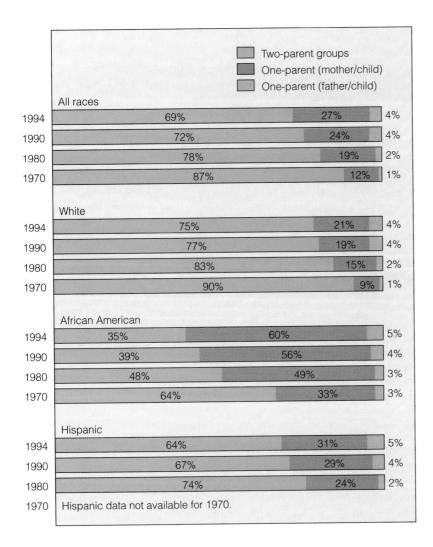

FIGURE 15.3

Composition of family groups with children, by race and Hispanic origin, 1970 to 1993.
(Source: Based on Rawlings 1994, Figure 5; U.S. Bureau of the Census 1995, Table 71)

The high divorce rate does not mean that Americans have given up on marriage; it means that they find an unhappy marriage intolerable and hope to replace it with a happier one. Meanwhile, an emerging trend is **redivorce.** Redivorces take place more rapidly than first divorces. The median time to divorce in 1988 was seven years for first marriages (U.S. Bureau of the Census 1994a, Table 143), just under six years for second marriages, and about four years for third or later marriages (U.S. National Center for Health Statistics 1991a, Table 10). Consequently, many who divorce—and their children—can expect several rapid and emotionally significant transitions in lifestyle and the family unit. They will experience stress and the need for adjustment. The stability of remarriages is addressed in greater detail in Chapter 16.

Why More Couples Are Divorcing

Various factors can bind marriages and families together: economic interdependence; legal, social, and moral constraints; and the spouses' relationship. The binding strength of some of these factors has lessened, however. We'll examine how these changes affect the divorce rate.

Given statistical projections, it is likely that by the time they are sixteen at least two and perhaps as many as four of these six boys will experience the dissolution of their parents' marriage.

Decreased Economic Interdependence

Traditionally, as we've seen, the family was a self-sufficient productive unit. Survival was far more difficult outside of families, so members remained economically bound to one another. But today, because family members no longer need each other for basic needs, they are freer to divorce than they once were.

Families are, however, still somewhat interdependent economically. As long as marriage continues to offer practical benefits, economic interdependence will help hold marriages together. The economic practicality of marriages varies according to several conditions. We'll look briefly at two: income and how it relates to divorce, and wives' employment.

DIVORCE AND INCOME   A positive relationship usually exists between marital stability and income; that is, up to a certain level, the higher the income, the less likely a couple is to divorce (Raschke 1987). A relative loss or lack of socioeconomic status seems important, too; men who are downwardly mobile—with less education than their parents—are more likely to divorce, as are those who fail to complete college. Both

the stress of living with inadequate finances and the failure to meet one's expectations for economic or educational attainment seem to contribute to marital instability (White 1990). This situation, together with the tendency of low-income groups to marry relatively early, helps explain why less well-off families have the highest rates of marital disruption, including divorce, separation, and desertion. Class differences in divorce rates, however, have been narrowing in recent years (Raschke 1987).

Although economic stress is less in high-income marriages, the higher the income, the less economic impediment there is to divorce. This relative economic independence, plus the fact that the pursuit of occupational success can erode marital intimacy, contributes to the divorce rate among high-income couples.

WIVES IN THE LABOR FORCE AND DIVORCE
Statistics show that divorce rates have risen as women's employment opportunities have increased and women have entered the labor force in large numbers. But many other changes have occurred during the same time period. How directly does women's employment affect divorce rates?

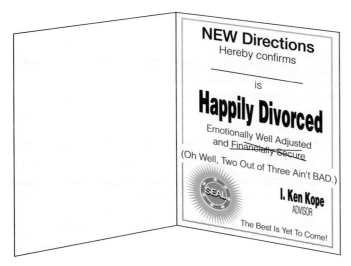

With greeting cards like this one in our everyday, or popular culture, it is clear that divorce has decreased stigma associated with it than in the past. Consumer items such as this are marketed to the divorced (and their friends), but their presence in the culture also helps to legitimate divorce.

Employed women, especially those working more hours (Greenstein 1990), do have higher divorce rates than do nonemployed women. But research indicates that wives' employment makes no difference in marital quality, which suggests an independence effect. Although it may not make marriages less happy, employment may nevertheless contribute to a divorce by giving an *unhappily* married woman the economic power, the increased independence, and the self-confidence to help her decide on divorce (Booth, Johnson, and White 1984).

Although there is no conclusive evidence that a wife's working affects marital satisfaction directly, researchers still suspect that working can have a negative indirect effect in two ways. Role overload can result in less couple time, and insufficient emotional support and household help on the part of husbands of working wives seem associated with their thoughts of divorce (Spitze 1988; Wilkie 1988). Chapter 13 addresses couples' work-family issues more fully.

The relationship between employment and divorce may run in the other direction; that is, it is also possible that women whose marriages are not going well will enter employment in anticipation of the need to be self-supporting (Spitze 1988; Wilkie 1988).

Decreased Social, Legal, and Moral Constraints

A second influence on the divorce rate has been the change in cultural values and attitudes. Although this is less true for Mexican Americans (Wagner 1988), the social constraints that once kept unhappy partners from separating operate less strongly now, and divorce is more acceptable. Divorce is not the issue it used to be in politics and corporate advancement.

No-fault divorce laws, which have eliminated legal concepts of guilt, are a symbolic representation of how our society now views divorce. The official posture of many—though not all—religions in the United States has also changed to be less critical of divorce than in the past. Fewer and fewer people now see divorce as a moral issue.

An even more basic factor has been the rise of individualistic values. Americans increasingly value personal freedom and happiness—gained, if necessary, at the cost of a marriage—over commitment to the family. They believe marriages ought to be happy if they are to continue. The definition of marriage as an emotional relationship rather than a practical one is modern (Shorter 1975; Stone, 1980). For all societies undergoing the transition from traditional to modern, divorce rates have increased.

High Expectations: The Ideal Marriage

People increasingly expect marriage to provide a happy, emotionally supportive relationship. This is an essential family function; yet high expectations for intimacy between spouses push the divorce rate upward. Research has found that couples whose expectations are more practical are more satisfied with their marriages than are those who expect more loving and expressive relationships (Troll, Miller, and Atchley 1979; see also White and Booth 1991). Indeed, many observers attribute our high divorce rate to the fact that Americans' expectations about marital happiness are *too* high. Although many couples part for serious reasons, others may do so because of unrealized (often unrealistic) expectations and general discontent.

The Changed Nature of Marriage Itself

To say that there are *no* societal constraints against divorce any more is an overstatement; nevertheless, as we have seen, they have weakened. A related issue is the changed nature of marriage itself. In Chapter 1 we argued that, due largely to our high divorce rate, marriage has been redefined as a nonpermanent—or, at best, a semi-, maybe, or hopefully permanent—relationship. We have also noted that marriage was originally perceived as a social institution for the practical purposes of economic support and responsible child rearing. But we have come to equate marriage with ongoing love, with *relationship*. Emphasis on the relationship itself over the institutional benefits of marriage results in its being defined as not necessarily permanent. As Frederick Engels, a colleague of Karl Marx and an early family theorist, noted near the turn of the twentieth century, "If only the marriage based on love is moral, then also only the marriage in which love continues." Engels went on to argue—as have some counselors and others over the past three decades—that "if affection definitely comes to an end or is supplanted by a new passionate love, separation is a benefit for both partners as well as for society" (Engels 1942, p. 73).

SELF-FULFILLING PROPHECY   Defining marriage as semipermanent becomes a self-fulfilling prophecy. Increasingly, spouses may enter the union with reservations, making "no definitive gift" of themselves. But,

"just as the person makes no definitive gift of himself, he has definitive title to nothing" (Durkheim 1951, p. 271). If partners behave as if their marriage could end, it is more likely that it will. Experts on the family who had at first remained optimistic even as divorce rates rose are now concerned about what they see as eroding family commitment (Glenn 1991).

MARITAL CONVERSATION—MORE STRUGGLE AND LESS CHITCHAT   Besides being visibly less permanent, marriage has changed in another way. No longer are the rules or normative role prescriptions for wives, husbands, or children taken for granted. Consequently, marriage entails continual negotiation and renegotiation among members about trivial matters as well as important ones. When family roles were culturally agreed on, members were likely to "share the indifferent 'intimacies' of the day" (Simmel 1950, p. 127), to engage in relatively inconsequential conversation about events outside the family. Today, however, as family roles are less precisely defined, "marital conversation is more struggle and less chitchat. . . . The conversation turns inward, to the question of defining the makeup of the family, . . . [to the] forging and fighting for identities" (Wiley 1985, pp. 23, 27; also, Chafetz 1989). One result is that increasingly, marriage and family living feels like work, and living alone may look restful by comparison.

Shifts in these areas—the economic practicality of marriage, the morality of divorce, and expectations for marital happiness—have all had the effect of expanding the alternatives to continuing a marriage.

Other Factors Associated with Divorce

So far in this section we have looked at sociohistorical and cultural factors that encourage high divorce rates. Another way to think about causes for divorce is to recognize that certain social or demographic factors are related to divorce rates. These include the following (White 1990, pp. 906–907):

- As we have already seen, *remarried mates are more likely to divorce.*

- *Cohabitation,* as discussed in Chapter 7, *is associated with higher probability of divorce.*

- *Marrying young,* also discussed in Chapter 7, *increases the likelihood of divorce.*

- *Premarital childbearing increases the risk of divorce in a subsequent marriage.* (Premarital conception, by itself, does not.)

- As discussed in Chapter 11, a married couple's having children stabilizes marriage—that is, decreases the possibility of divorce. Hence, *remaining child-free is associated with a higher likelihood of divorce.*

- A growing literature suggests that *having parents who divorced increases the likelihood of divorcing.* Reasons for this are hypothetical. It is possible that divorcing parents model divorce as a solution to marital problems, or that children of divorced parents are more likely to marry inappropriate partners. There is also evidence that children of divorced parents marry at younger ages and are more likely to experience premarital births, and both factors are associated with higher divorce rates.

- *Race is differentially associated with one's chances of divorcing.* Even among those who are similarly educated and/or earning similar incomes, African Americans are more likely than non-Hispanic whites to divorce. Some scholars have suggested that the history of slavery is an explanation inasmuch as slaves were not allowed to institutionalize marriage. But other scholars (Gutman 1976; Cherlin 1981) discount this explanation. To a significant extent, this difference in marital stability among blacks and whites remains a puzzle.

So far we have been discussing social factors that are associated with divorce. Now we will look at some of the common marital complaints voiced by divorcing couples.

Common Marital Complaints

Marital complaints given by the divorced include alcoholism, drug abuse, family violence, and, much less often, homosexuality (Myers 1989; White 1990). But knowing this tells us little about the extent to which these factors are actually associated with divorce. Indeed, it is "striking . . . that so few empirical studies of divorce take these individual complaints seriously" (White 1990, p. 908).

Meanwhile, a general (and perhaps obvious) conclusion drawn from research is that deficiencies in the emotional quality of the marriage lead to divorce (Martin and Luke 1991). In one study (Spanier and

Thompson 1987) of 210 separated people, 56 percent of the women were dissatisfied with their husband's contribution to household tasks. For 40 percent of the women and 20 percent of the men, their spouse had not lived up to their expectations as a parent. Many were unhappy with their partner as a leisure-time companion, as someone with whom to talk things over, or as a sex partner. Subsequent research finds marital stability to be positively associated with a wife's perception that household tasks are shared fairly (Greenstein 1995).

Financial and job-related problems can contribute to marital deterioration. The greatest source of strain in working- and lower-class marriages often is vulnerability to economic difficulties. Although research has discredited the idea that high income necessarily produces happy marriages (Piotrkowski, Rapoport, and Rapoport 1987), there is evidence that the husband's provision of adequate financial support is important to his and his wife's marital satisfaction, with a husband's unemployment having a definite negative effect (Aldous and Tuttle 1988). Along with obvious economic hardships, loss of identity and increased time on their hands impose stress on unemployed husbands, one result of which can be diminished displays of affection and increased hostility (Conger et al. 1990). In Spanier and Thompson's (1987) study, the following economic issues were points of conflict: amount of money one had (named by 55.6 percent of the respondents), individual's or spouse's working hours (54.1 percent), time away from home because of a job (40 percent), kind of one's or spouse's job (39 percent), and one's or spouse's work colleagues (34.6 percent). As we saw in Chapter 13, unions in which one or both partners work night or unusual shifts are more vulnerable to divorce (White and Keith 1990). Some spouses may enjoy their careers more than family life; in two-career marriages, conflicting interests and ambitions can cause problems.

In many cases, thinking about divorce is apparently precipitated by third-party involvement. In the Spanier and Thompson study, more than 60 percent of the respondents had had extramarital sexual relationships. In-laws may also have a negative influence, especially if they disapprove of the marriage (Burns 1984). Counselors suggest that some common problems—with money, sex, and in-laws, for example—are really arenas for acting out deeper conflicts, such as who will be the more powerful partner. Problems can become routine, as we saw in Chapter 8, resulting in devitalized or conflict-habituated marriages. Many spouses choose to continue in such relationships. The personal decision about divorce involves a process of balancing those alternatives against the practical and emotional satisfactions of one's present union.

Thinking About Divorce: Weighing the Alternatives

Not everyone who thinks about divorce actually gets one, of course. One assumption behind the relative love and need theory, discussed in Chapter 10, is that spouses continually compare the benefits of their union with the projected consequences of not being married. As divorce becomes a more realistic option, this same weighing of alternatives takes place.

Marital Happiness, Barriers to Divorce, and Alternatives to the Marriage

One model (Levinger 1965) derived from exchange theory (see Chapter 2) argues that spouses weigh their marital happiness against both alternatives to the marriage (possibilities for remarriage or fashioning a satisfying single life) and barriers to divorce (for example, religious beliefs against divorce, parental pressure to stay married, common friendship networks, joint economic investments, and children to consider).

There is considerable research evidence spanning the past three decades to indicate that children serve as a barrier to divorce. Because research also shows that children are likely to lessen marital happiness, we can conclude that affection for their children, along with economic and other concerns about children's welfare after divorce, discourages some parents from dissolving their marriage.

One important recent analysis (Waite and Lillard 1991) found that **firstborn** children increase marital stability through their preschool years. Other children do so only when they are very young. (Older children and those born before the marriage actually increase chances for dissolution.) "The initially stabilizing and later destabilizing effects of children combine over the course of the marriage to give parents only a modestly higher chance than childless couples of reaching their twentieth anniversary" (Waite and Lillard 1991, p. 930). These authors argue that young children's high requirements for parental time and effort (and their expense) stabilize marriages because single parenting

seems too costly compared to remaining married. As children grow older, however, they require less constant care and their presence lessens their parents' marital happiness—a situation resulting in children's later destabilization of marriages.

As Figure 15.4 shows, most divorces occur relatively early in marriage. About 8.5 percent of couples married three years get divorced, compared with about 1 percent of couples married for twenty-five years. The proportion of divorces for couples married twenty years or more has been increasing. Nevertheless, about one-third of divorces are for couples married less than five years, and almost two-thirds are for couples married less than ten years (U.S. National Center for Health Statistics 1991a, p. 4). Long marriages are less likely to end in divorce even though empirical evidence consistently shows that marital happiness declines over time. One reason for this is that barriers to divorce, such as common economic interests and friendship networks, increase over time (White and Booth 1991). One study (Black et al. 1991) that compared spouses who had initiated the divorce (the "leavers") with the "left" found that the two groups did not differ on marital satisfaction or barriers to divorce, but leavers saw more alternatives to their marriage.

Our discussion of weighing the alternatives suggests that one of the first questions married people ask themselves is whether they would be happier if they were to divorce.

"Would I Be Happier?"

This is not an easy question to answer. Deciding whether one would be happier after divorce requires one to hypothesize about future alternatives. Some people may prefer to stay single after divorce, but many partners probably weigh their chances for a remarriage. According to national surveys over the past two decades, divorced and separated individuals have lower levels of life satisfaction and a more negative general mood than married people. Divorced people also have poorer physical health than marrieds, are more often depressed (Menaghan and Lieberman 1986; Wineberg 1994), and are somewhat more inclined to suicide (Stack 1990). After controlling for socioeconomic status, these effects seem to be similar for whites and blacks (Fine, McKenry, and Chung 1992). Analysis of 6,573 respondents from the National Survey of Families and Households found that, even though factors such as income and education

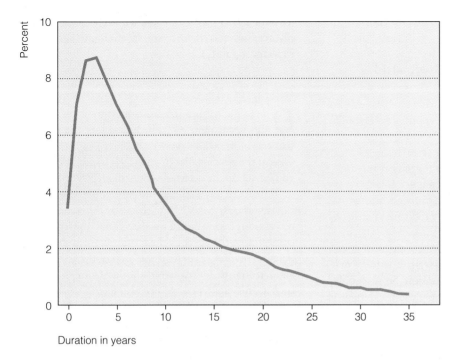

FIGURE **15.4**

Percent distribution of divorces by duration of marriage, 1990. (Source: Clarke 1995b, Figure 3, p. 4)

were more important, people in first marriages reported greater happiness and less depression than those who had ever divorced. Furthermore, people who had divorced just once were less depressed than those with a history of two or more divorces. Those currently married reported greater happiness and less depression than cohabitors, who in turn reported greater well-being on these two scores than those who lived without another adult (Kurdek 1991b).

One study of 1,755 whites in Detroit found an interesting wrinkle, however. Higher levels of depression among the divorced were *not* apparent among those who saw themselves as escaping marriages with serious, long-term problems (Aseltine and Kessler 1993). Then too, some (but not all) depression effects may mean that people who divorce are more inclined to unhappiness anyway (Gove, Style, and Hughes 1990). Counselors advise that in some cases partners might be happier trying to improve their relationship than divorcing. They warn that frequently one's motive for divorce is really to solve an *individual* problem—one that stems from a period of transition, perhaps, or

some personal inadequacies. If this is true, blaming one's partner and deciding to divorce will not be a solution unless personal growth also occurs (Counts and Sacks 1991).

Marriage can and often does provide emotional support, sexual gratification, companionship, and economic and practical benefits. But unhappy marriages do not provide all (or in some cases, any) of these benefits. Compared with unhappily married people, divorced individuals display generally better physical and emotional health and higher morale. "In short, it appears that at any particular point in time most marriages are 'good marriages' and that such marriages have a strong positive effect on well-being and that 'bad marriages' have a strong negative effect on well-being" (Gove, Style, and Hughes 1990, p. 14).

One must decide whether divorce represents a healthy step away from an unhappy relationship that cannot be satisfactorily improved or is an illusory way to solve what in reality are personal problems. Going to a marriage counselor can help partners become more aware of the consequences of divorce so that they can make this decision more knowledgeably.

Ambivalence, Vacillation, and Readiness for Divorce

As a couple moves toward divorce, they balance all of the anticipated moral, social, economic, and familial consequences against their satisfactions, securities, and unhappiness. As they weigh the alternatives, they can expect to go through periods of indecision and ambivalence. "Uncoupling," as sociologist Diane Vaughan (1986) puts it, is a long process during which former partners gradually redefine themselves and each other as single.

Once a person decides he or she wants a divorce, however, it is not always easy to wait: A sense of urgency often accompanies the decision. Some people believe that failing to act decisively will mean they are denying their own needs.

Marriage counselors advise against making quick decisions, especially if partners are motivated by feelings of hate or anger. Because of this, counselors may suggest that intensely hostile couples who say they want a divorce try a period of **structured separation,** during which they live apart for a limited period, avoid securing lawyers or getting involved in new relationships, and continue in counseling together. After such a separation, partners can usually determine more knowledgeably whether they want to divorce.[3]

Readiness for divorce implies the willingness to take responsibility for one's own contribution to the breakup, along with seeing the alternatives to the marriage as somewhat appealing; it also means feeling comfortable with the decision over an extended period of time, without extreme vacillation.

Not all divorced people wanted or were ready to end their marriage, of course. It may have been their spouses' choice. Not surprisingly, research shows that the degree of trauma a divorcing person suffers usually depends on whether that person or the spouse wanted the dissolution (Pettit and Bloom 1984), at least partly because the one feeling "left" experiences a greater loss of control (Wilder and Chiriboga 1991; but see also

Rossiter 1991). Even for those who actively choose to divorce, however, divorce and its aftermath can be unexpectedly painful.

Getting the Divorce

One of the reasons it feels so good to be engaged and newly married "is the rewarding sensation that out of the whole world, you have been selected. One of the reasons that divorce feels so awful is that you have been de-selected" (Bohannan 1970b, p. 33).

Bohannan has analyzed the contemporary American divorce experience in terms of six different facets, or "stations." He talks of the emotional, the legal, the community, the psychic, the economic, and the coparental divorce. Experience in each of these realms varies from one individual to another (some stations, such as the coparental, do not characterize every divorce). Yet they capture the complexity of the divorce experience. In this section, we will examine the first four stations listed above; the economic and coparental aspects of divorce will be explored in greater detail later in this chapter.

The Emotional Divorce

Emotional divorce involves withholding bonding emotions and communications from the relationship (Vaughan 1986), typically replacing these with alienating feelings and behavior. Partners no longer reinforce but rather undermine each other's self-esteem through endless large and small betrayals: responding with blame rather than comfort to a spouse's disastrous day, for instance, or refusing to go to a party given by the spouse's family, friends, or colleagues. As emotional divorce intensifies, betrayals become greater.

In a failing marriage, both spouses feel profoundly disappointed, misunderstood, and rejected. Because the other's very existence and presence is a symbol of failure and rejection, the spouses continually grate on each other. The couple may want the marriage to continue for many reasons—continued attachment, fear of being alone, obligations to children, the determination to be faithful to marriage vows—yet they can hurt one another as they communicate their frustration by look, posture, and tone of voice.

DIVORCE COUNSELING   For this reason, marriage counselors offer **divorce counseling,** in which partners try to negotiate various conflicts, grievances, and misunderstandings, as in marriage counseling (see, for

---

3. A danger with structured separation is that friends will define it, not as a "therapy technique" (Brooks 1985), but as the beginning of the divorce process. When friends and colleagues know about the breakup, the process often accelerates. Interestingly, one study, using National Survey of Families and Households data, has found that separated spouses of the same religion are more likely than others to reconcile (Wineberg 1994).

example, Textor 1989). The goal of this kind of therapy is to help a couple accept and adjust to the divorce so that eventually they can "use the 'crisis' of divorce as an opportunity for learning and personal growth" (Sprenkle 1989, p. 176). More specific concerns of divorce counselors are to counteract the feelings of uncooperativeness and hostility generated by the legal arrangements that accompany divorce and to help clients to help their children adjust to the loss.

The Legal Divorce

A **legal divorce** is the dissolution of the marriage by the state through a court order terminating the marriage. The principal purpose of the legal divorce is to dissolve the marriage bond so that emotionally divorced spouses can conduct economically separate lives and be free to remarry.

Two aspects of the legal divorce itself make marital breakup painful. First, divorce, like death, creates the need to grieve. But the usual divorce in court is a rational, unceremonial exchange that takes only a few minutes. And lawyers have been trained to solve problems rationally and to deal with clients in a detached, businesslike manner. Although divorcing spouses might need and want them to, very few divorce attorneys view their role as helping with the grieving process.

A second aspect of the legal divorce that aggravates the misery is the adversary system. Under our judicial system a lawyer advocates his or her client's interest only. Eager to "get the most for my client" and "protect my client's rights," opposing attorneys are not trained to and ethically are not even supposed to balance the interests of the parties and strive for the outcome that promises most mutual benefit. Also, the divorcing individuals can feel frustrated by their lack of control over a process in which the lawyers are the principals. In one study of divorced women, virtually all had complaints about their lawyers and the legal system (Arendell 1986).

DIVORCE MEDIATION **Divorce mediation** is an alternative, nonadversarial means of dispute resolution by which the couple, with the assistance of a mediator or mediators (frequently a lawyer–therapist team), negotiate the terms of their settlement of custody, support, property, and visitation issues. The couple work out a settlement best suited to the needs of their family. In the process it is hoped that they learn a pattern

of dealing with each other that will enable them to work out future disputes.

Divorce mediation is a fairly recent development in the United States, having emerged over the past twenty-five years. Research indicates that couples who use divorce mediation have less relitigation, feel more satisfied with the process and the results, and report better relationships with ex-spouses (Marlow and Sauber 1990). However, various women's task groups have noted that mediation as it is currently practiced may be biased against females, who report being labeled "unladylike" or "vindictive" if they refuse to give in on a point (Lonsdorf 1991; Woo 1992b).

NO-FAULT DIVORCE Before the 1970s the fault system predominated in divorce cases. A divorcing party had to prove she or he had grounds for divorce, such as the spouse's adultery, mental cruelty, or desertion. Just discussing grounds increased hostilities and diminished chances for cooperative negotiation: To prove grounds required a determination of which partner was guilty and which partner innocent. The one judged guilty rarely got custody of the children, and the judgment largely influenced property settlement and alimony awards. To the judge, the spouses, the friends, and the public, the one judged guilty was the loser (Fine and Fine 1994).

Beginning with California in 1970 and continuing until all states passed no-fault legislation, divorce, or marital dissolution, no longer officially requires a winner and a loser. Today, with **no-fault divorce,** a partner seeking divorce no longer has to prove grounds. Instead, a marriage is legally "dissolved" because the relationship is "irretrievably broken." No-fault divorce legally abolishes the concept of the "guilty party." But learning to think in terms of no-win/no-lose is difficult even for people in good marriages. One result, some observers suspect, is that divorcing spouses who are bent on publicly determining guilt or innocence fight the same court battle, but as a child-custody suit instead; we'll address this issue when we discuss coparental divorce. And as we shall see later in this chapter, some argue that a second, unanticipated result of no-fault statutes is that many divorced women are worse off economically.

The Community Divorce

Marriage is a public announcement to the community that two individuals have joined their lives. Marriage usually also joins extended families and friendship

Divorce affects the extended family as well as the immediate one. In some families, grandparents may lose touch with grandchildren, while in others they may become more central figures of support and stability for their grandchildren.

networks and simultaneously removes individuals from the world of dating and mate seeking. Divorce occasions reverse changes in social networks, which are sometimes disappointing or embittering, often confusing. Divorce also provides the opportunity for forming new ties.

KIN NO MORE?   Given the frequency of divorce, many extended families find themselves touched by it. Grandparents fear losing touch with grandchildren, and this does happen. Virtually all states now have statutes granting grandparents visitation rights in some circumstances (Bean, 1985–86, p. 393). These provide grandparents whose children are deceased or divorced the opportunity to go to court to seek visitation rights against a custodial parent who has refused them the opportunity to see their grandchildren. States vary as to what categories of grandparents are eligible (some laws do not include divorce situations),

procedure, and criteria for approval or disapproval. Some effort has been made to expand grandparents' rights by including them in the initial custody order and/or by entitling grandparents to pursue visitation over the opposition of still-married adult children (Marcus 1991).

In other cases, grandparents become closer to grandchildren, as adult children turn to grandparents for help or grandchildren seek emotional support (Spitze et al. 1994). Researchers and therapists have concluded that

these relationships work best when family members do not take sides in the divorce and make their primary commitment to the children.

Grandparents can play a particular role, especially if their marriages are intact: symbolic generational continuity and living proof to children that relationships can be lasting, reliable, and dependable. Grandparents also convey a sense of tradition and a special commitment to the young that extends beyond and over the parents' heads. Their encouragement, friendship, and affection has special meaning for children of divorce; it specifically counteracts the children's sense that all relationships are unhappy and transient.

For the fortunate children in our study who could rely on their extended families, the world seemed a more stable, predictable place. (Wallerstein and Blakeslee 1989, p. 111)

Women are more likely to retain in-law relationships after divorce, particularly if they had been in close contact before the divorce and if the in-law approves of the divorce (Serovich, Price, and Chapman 1991). Relationships between former in-laws are more likely to continue when children are involved. One study looked at the general character of postdivorce extended-kin relationships. This study found that in half of the cases, the kinship system included **relatives of divorce** (Johnson 1988, p. 168; see also Stacey 1990). This was most likely to occur with paternal relatives, who, of course, needed to keep in touch in order to see grandchildren living with custodial mothers. When fathers had custody, they too were likely to have contact with their former spouse's extended family (Ambert 1988). But the most commonly retained tie was between a grandmother and her former daughter-in-law. Also, grandmothers who were themselves divorced or widowed or in a remarriage were more likely to be in an expanded kin system than were those in an intact first marriage (Johnson 1988).

Still, relationships between former in-laws frequently deteriorate or vanish. In only 11 percent of the forty-eight individuals followed in a longitudinal study of postdivorce kinship did both members of the married couple maintain relations with both families, whereas in over one-third no "relations of divorce" were maintained. Proximity and time since divorce were factors. In only about 20 percent of the situations was there frequent visiting, phone contact, or mutual aid; in the remainder, relationships were characterized as "friendly" but were not very active ties (Ambert 1988).

Adult children's relationships with their own parents generally changed after the children's divorce. According to one study,

the grandmothers interviewed rarely viewed their relationship with their divorced child with the sentimentality they bestowed upon their role of grandparent; [the relationship] was characterized by underlying tension. . . . With the unscheduled life event of divorce, transitions in the relationships between divorced individuals and their parents were often regressive because a child reassumed a dependent status. Members of both generations had to revise their expectations of the other, and members of the older generation found themselves in a situation of having to give more of themselves to a child than they had expected to do at their stage of life. They were often forced into a parenting role, and this greater involvement provided more opportunity to observe and comment on their adult child's life. . . .

Even so, there was extensive contact between generations, which did not decrease over time. . . . Nevertheless, there appeared to be incongruity in their expectations. Adult children were more likely to feel that parents should be available to help them with their emotional problems than their parents felt was appropriate. Divorcing children did not want their parents to interfere in childrearing or offer unsolicited advice, while their parents felt they could voice their concerns. (Johnson 1988, pp. 190–91)

There was considerable variation in these relationships within the sample of fifty-two adult–child dyads followed over several years in Johnson's study. But most older parents espoused "modern values of personal freedom and self-fulfillment" (p. 191); that is, they did not criticize the decision to divorce from a traditional perspective.

Couples who divorce later in life and/or after long marriages often find socioemotional support from their adult children (Gander 1991), but this may be truer for mothers than for fathers (Wright and Maxwell 1991).

FRIENDS NO MORE? Important changes in one's lifestyle almost invariably mean changes in one's community of friends. When they marry, people usually replace their single friends with couple friends, and couples also change communities when they divorce. Separating from one's former community of friends and in-laws is part of the pain of divorce. Over

Fifteen Suggestions for Healing

Separation by divorce, like death, inflicts a painful emotional wound that must heal. A little book entitled *How to Survive the Loss of a Love: 58 Things to Do When There Is Nothing to Be Done* suggests some ways to facilitate the healing process.

1. Do your mourning *now*. Don't pretend, deny, cover up, or run away from the pain. Everything else can wait. The sooner you allow yourself to be with your pain, the sooner it will pass. Resisting the mourning only postpones healing, and grief can return months or even years later to haunt you.

2. Be gentle with yourself. Accept the fact that you have an emotional wound, that it is disabling, and that it will take a while before you are completely well. Treat yourself with the same care, consideration, and affection you would offer a good friend in a similar situation.

3. If possible, don't take on new responsibilities. When appropriate, let employers and coworkers know you're healing.

4. Don't blame yourself for any mistakes (real or imagined) you may have made that brought you to this loss. You can acknowledge mistakes later, when the healing process is further along.

5. Remember that it's okay to feel depressed. Let the healing process run its full course. A time of convalescence is

---

three-quarters of the women in one study reported losing friends during or after the divorce (Arendell 1986).

Divorced people often feel uncomfortable with their friends who are still married because activities are done in pairs. Conversely, a married couple may find that a friend's divorce challenges them to take another look at their own marriage, an experience that can cause them to feel anxious and uncomfortable with the divorced friend. Also, couple friends may be reluctant to become involved in a conflict over allegiances and often experience their own sense of loss and grief. A common outcome is a mutual withdrawal, during which the divorced person feels heightened loneliness.

Like many newly married people, those who are newly divorced must find new communities to replace old friendships that are no longer mutually satisfying. The initiative for change may in fact come not only from rejection or awkwardness in old friendships but also from the divorced person's finding or wanting to find friends who share with him or her the new concerns and emotions of the divorce experience. Priority

may also go to new relationships with people of the opposite sex; for the majority of divorced and widowed people, building a new community involves dating again.

The Psychic Divorce

Psychic divorce refers to the regaining of psychological autonomy through emotional separation from the personality and influence of the former spouse. In the process, one learns to feel whole and complete again and to have faith in one's ability to cope with the world. In psychic divorce, one must distance oneself from the still-loved aspects of the spouse, from the hated aspects, and "from the baleful presence that led to depression and loss of self-esteem" (Bohannan 1970b, p. 53).

Not all—and perhaps not even most—divorced people fully succeed at psychic divorce. Even for those who do, "progress is slow, if not plodding. Furthermore, the 'graph' of divorce recovery is typically jagged rather than straight, and each forward step is likely to be matched by a retreat" (Sprenkle 1989, p. 175). But counselors point out that this stage is a necessary pre-

very important. Just follow your daily routine and let yourself heal.

6. For a while, don't get involved in an all-consuming passionate romance or a new project that requires great time and energy.

7. Don't try against obvious odds to rekindle the old relationship. Futile attempts at reconciliation are painful and a waste of recuperative energy, and they slow healing and growth.

8. If you find photographs and mementos helpful to the mourning process, use them. If you find they bind you to a dead past, get rid of them. Put them in the attic, sell them, give them away, or throw them out.

9. Remember that it's okay to feel anger toward God, society, or the person who left you (even if through death). But it is *not* okay or good for you to hate yourself or to act on your anger in a destructive way. Let the anger out safely: hit a pillow, kick the bed, sob, scream. Practice screaming as loudly as you can. A car with the windows up makes a great scream chamber.

10. Use addictive prescription drugs like Valium wisely. Take them only if prescribed by your personal physician and only for a short period of time. *Don't* take them to mask your grief because your friends have grown tired of it.

11. Watch your nutrition. Take vitamins, eat good foods, and try to get plenty of rest.

12. Don't overindulge in alcohol, marijuana or other recreational chemicals, or cigarettes.

13. Pamper yourself a little. Get a manicure, take a trip, bask in the sun, sleep late, see a good movie, visit a museum, listen to music, take a long bath instead of a quick shower. As healing progresses, remember that it's okay *not* to feel depressed.

14. You might find keeping a journal or diary helpful. This way you can see your progress as you read past entries.

15. Heal at your own pace. The sadness comes and goes—though it comes less frequently and for shorter lengths of time as healing proceeds.

Source: Colgrove, Bloomfield, and McWilliams 1978.

requisite to a satisfying remarriage. Box 15.1, "Fifteen Suggestions for Healing," provides some guidelines.

To be successful, a psychic divorce requires a period of mourning. Like the death of a spouse, divorce means the end of a relationship. The experience of loss is as real as that suffered by people who are widowed. Just as a gradual process of emotional estrangement starts long before the actual legal event of divorce, the partners' emotional involvement often continues long after. This "persistence of attachment" (Weiss 1975) is real for both spouses and should be understood and addressed.

There are at least three stages in the mourning process. The first, which typically occurs before the legal divorce, is shock and denial. Sometimes, a person's inability to accept the divorce manifests itself in physical illnesses, accidents, or even suicide attempts. Eventually, however, the frustration of emotional divorce leads partners to face facts. They may say to each other, "We can't go on like this."

A second stage, characterized by anger and depression, follows this realization. These feelings often alternate, so that recently divorced people feel confused.

(Whether these feelings differ for people who experience a second or a third divorce will be a subject for future research.) Often their feelings resemble those of one ex-wife, who asked a counselor, "How come I miss so terribly someone I couldn't stand?" (Framo 1978, p. 102). Unfortunately, many former partners vent their anger in the coparental arena (Masheter 1991).

In the third stage, ex-spouses take responsibility for their own part in the demise of the relationship, forgive themselves and the mate, and proceed with their lives: The psychic divorce is then complete. As long as one views the ex-spouse as an enemy or object of ongoing anger, however, the psychic divorce has not been accomplished, because preoccupation with one's ex-spouse (Masheter 1991), along with bitterness and hate, are emotions of a continuing relationship.

Counseling can help; increasingly, communities and religious groups are providing workshops (such as "BE," or Beginning Experience) that can be helpful as well (Byrne and Overline 1992).

Deciding knowledgeably whether to divorce means weighing what we know about the consequences of

TABLE 15.1

Children Living Below Poverty Level, By Race, 1993*

	Number				Percentage			
	All races	White	Black	Hispanic	All races	White	Black	Hispanic
Children under age 18	15,727,000	9,752,000	5,125,000	3,873,000	23	18	47	41
Children under age 6 (1992)	5,600,000	2,300,000	1,800,000	1,300,000	24	14	51	44

NOTE: Data for children under age 6 are for 1992 because the U.S. Bureau of the Census deleted this information in the 1995 *Statistical Abstract of the United States.*

Sources: U.S. Bureau of the Census 1994a, Table 729; 1995, Table 747.

divorce. The next section examines the economic consequences of divorce.

The Economic Consequences of Divorce

Increasingly social scientists and policymakers worry about the economic consequences of divorce, especially for children.

Divorce, Single-Parent Families, and Poverty

As you can see from Table 15.1, poverty rates for our nation's children are discouragingly high. Nearly one-quarter of U.S. children under age 6, and 23 percent of those under 18, live in poverty. While almost half (47 percent) of African American and 41 percent of Hispanic children under age 18 live in poverty, scarcity is assuredly not simply a "minority problem." Almost one-fifth (18 percent) of non-Hispanic white children under 18 grow up in poverty as well.

Figure 15.5 gives the proportions of children under age 6 who are living in poverty, by race/ethnicity and family type. As you can see, 59 percent of all children under age 6 who reside in mother-headed, single-parent families live in poverty. This compares to 13 percent of those living in married-couple families. Among Hispanics, 69 percent of young children in mother-headed, single-parent families live in poverty, compared with 34 percent of those in married-couple

households. The relationship between racial/ethnic minority status and poverty is consistent across family types: Non-Hispanic whites are much less likely than other racial/ethnic categories to be poor. At the same time, the relationship between family type and poverty is consistent across racial/ethnic categories: Children in single-parent families, and particularly in mother-headed single-parent families, are significantly more likely to live in poverty than children in married-couple households.

While unmarried childbearing also accounts for single-parent families (particularly among African Americans), divorce remains the major cause overall (Saluter 1994, p. xii). Looking at these figures, some policymakers seem to argue that single-parent families are the major cause of poverty and hence should virtually always be discouraged. Opponents of this view argue that financial privation exists (albeit to a lesser extent) even in married-couple families and that racial/ethnic discrimination and low wages, particularly for women, account for poverty (Mauldin 1990; Lichter and Eggebeen 1994; Lichter and Landale 1995). Still others point out that for many individuals—those who have experienced family violence, for example—divorce is the lesser of two evils. Here, the primary concern is the economic betterment of single-parent families.

Meanwhile, a consistent and unequivocal research finding is that divorce is inversely related to a woman's (but not necessarily to a man's) economic status. In-

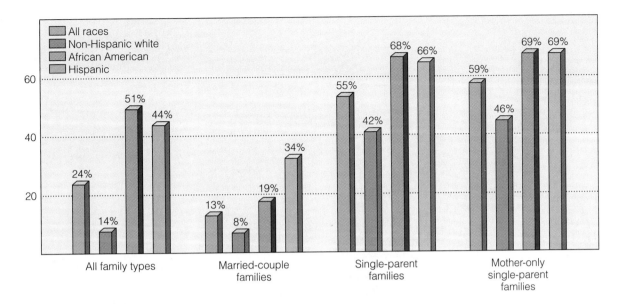

FIGURE **15.5**

| Children under age 6 below the poverty level, by family type and race, 1992. (Source: U.S. Bureau of the Census 1994a, Table 729)

deed, "the longer the period of divorce (or widow-hood) of a woman, the more she is likely to be poor" (Choi 1992, p. 40). Why is this the case?

Husbands, Wives and Economic Divorce

Upon divorce, a couple undergoes an **economic divorce** in which they become distinct economic units, each with its own property, income, control of expenditures, and responsibility for taxes, debts, and so on. Behind the idea of fair property settlement run two legal assumptions. The first is that a family is an interdependent economic unit based on the basic exchange (see Chapter 7); that is, a man could not earn the money he earns without the moral assistance and domestic services of his wife. A minority of states have community property laws based directly on this premise. In those states, family property legally belongs equally to both partners. Most states now have laws promising a divorced wife either an equitable or an equal share of the marital property.

A second assumption is that property consists of such tangible items as a house or money in the bank. Yet, except for very wealthy people, the valuable "new property" (Glendon 1981) in today's society is the earning power of a professional degree, a business or

managerial position, work experience, a skilled trade, or other "human capital." When property is legally divided in divorce, the wife may get an equal share of tangible property, such as a house or savings, but that usually does not put her on an equal footing with her former husband. An even split of the marital property may not be truly equitable if one partner has stronger earning power and benefits such as pensions but the other does not, and if the parent with custody of the children has a heavy child-support burden. Put another way, dividing property may be easy compared with ensuring that both partners and their children will have enough to live on comfortably after divorce.

About ten years ago, sociologist Lenore Weitzman addressed the plight of divorced women and their children in her a book *The Divorce Revolution* (1985). She reported a 42 percent increase in the standard of living for ex-husbands, compared with a 73 percent decline for ex-wives with minor children. Others, including sociologist Richard R. Peterson in the *American Sociological Review*, June 1996, dispute the precision of these figures (see also Smock 1993). Current research shows a 27 percent income decline for divorced women, together with a 10 percent increase in the amount ex-husbands have to spend on

Ex-husbands often complain that "she took me to the cleaners," but research and statistics show otherwise. Ex-wives' standard of living typically declines after divorce, while ex-husbands' does not.

themselves. What accounts for this situation? A fundamental reason for the income disparity between ex-husbands and their former wives is men's and women's unequal wages (Smock 1993), a situation discussed in Chapters 3 and 13. But factors associated with the divorce itself—attitudes about alimony, for example, and difficulties collecting adequate child support—make financial matters worse for women. In addition, Weitzman (1985) has argued that no-fault divorce laws are a significant cause of the postdivorce poverty of women and children.* We will look at each of these factors in some detail.

Attitudes About Alimony

Popular myth once had it that ex-wives lived comfortably on high **alimony** awards, now generally called

*As we go to press, having reanalyzed Weitzman's data, Richard R. Peterson (*American Sociological Review*, June 1996) concluded (and Weitzman concurred) that economic outcomes are essentially the same in fault and no-fault divorces.

"maintenance" or **spousal support.** (The word *alimony* comes from the Latin verb that means "to nourish." Historically, alimony rests on the assumption that the contract of marriage includes a husband's lifetime obligation to support his wife and children.) But the fact is that courts award spousal support in less than 15 percent of all divorce cases, and in the majority of cases the amount awarded is low.

This situation partially results from the assumption, beginning in the 1970s, that women and men have—or soon would have—equal earning power. Some financially dependent spouses have been awarded short-term **rehabilitative alimony,** in which the ex-husband pays his ex-wife "just enough cash to put her back on her feet, or at least in a word-processing class" (Stern 1994). But in many cases, this is not enough.

Full-time homemakers, often older, who suddenly find themselves divorced and without adequate support (called **displaced homemakers**) are particularly disadvantaged by this system. When they married, they expected the conditions and assumptions of the basic exchange to last for life. Many have few or no marketable skills, no employment record, and no pension. Nor are they in a position to pursue education or job training followed by long-term employment. The results of divorce can be devastating for them.

Some activists have argued that many wives—particularly those who left the labor force to raise children or to help with their husband's career—deserve not alimony but **entitlement:** the equivalent of severance pay for work done at home during the length of a marriage. An important basis for this argument is that the wife has contributed to her husband's employability through her domestic role, including entertaining clients and providing job assistance. Therefore, she should be paid regularly after divorce for her interest and investment in his (that is, their) continuing earning power (Weitzman 1985). Social Security provisions that now allow an ex-wife who had been married at least ten years to collect 50 percent of her ex-husband's retirement benefits (Choi 1992) are a result of political pressures to recognize this concept of entitlement. Families of women and children would be much better off after divorce, of course, if they received the child support due them.

Child Support

Child support involves money paid by the noncustodial to the custodial parent in order financially to support the children of a separated marital, cohabiting, or sexual relationship. Because mothers retain custody in 90 percent of cases—and because women are economically disadvantaged in employment—the vast majority of those ordered to pay child support are fathers.

During the 1980s the federal government made sweeping changes in child support award and collection policies and procedures. Before that, child support was negotiated as part of each individual divorce/separation agreement or paternity case. Payment was difficult to enforce, and collection of past-due amounts depended on the custodial parent's coming forward to prove delinquency.

Policy and procedural changes in child support were prompted by four separate but overlapping concerns: (1) the move during the Reagan and Bush administrations to privatize support obligations for children in an effort to lessen welfare spending; (2) concern about the growing proportion of children in female-headed, single-parent households; (3) alarm about the growing proportion of children living in poverty; and (4) public recognition of the inequitable economic consequences for women, compared with men, of divorce.

The **Child Support Amendments (1984)** to the Social Security Act, together with the **Family Support Act (1988),** did the following: (1) encouraged the establishment of paternity and consequent child support awards; (2) required states to develop numerical guidelines for determining child support awards from which judges would find it difficult to depart; (3) required periodic review of the award levels so that awards may be amended to keep up with inflation and to ensure that the noncustodial parent continues to pay an appropriate share of his or her income; and (4) enforced payments through locator services (to find nonpaying noncustodial parents), through provisions for mandatory wage withholding, and through interception of tax refunds to channel support payments to the custodial parent (Garfinkel, Oellerich, and Robins 1991). In addition, all states implemented automatic wage withholding of child support in 1994.

Child support awards have historically been and continue to be small—usually around 10 percent of the noncustodial father's income and amounting to less than half of a child's expenses. But state guidelines have been developed either according to the cost-sharing method, which calculates the award using child-rearing costs, or according to the income-sharing method, which computes the support award using one or both parents' income (Garfinkel, Oellerich, and Robins 1991). Early research on the outcomes of these policies has yielded mixed results. Some conclude that the new guidelines will significantly increase award amounts but may not necessarily lead to equivalently higher payments, so that overall child support income received should increase by about 50 percent (Garfinkel, Oellerich, and Robins 1991). It appears that automatic withholding of support payments alone increases payments by 15 to 25 percent (Teachman 1991).

Still, some evidence suggests that many lawyers and judges, influenced by their own experiences, still respond arbitrarily to child support issues (Ellis 1991). This, coupled with evidence that women may experience discrimination in divorce mediation more generally (Lonsdorf 1991; Woo 1992b), has not lessened many feminists' concerns about child support awards.

Moreover, some research suggests that the principal reason for a noncustodial parent's failure to pay is unemployment. Among families in which the absent

Women and their children experience a startling decline in their standard of living after a divorce. Although no-fault laws have made it easier to obtain a divorce, some argue that the laws tend to reduce a woman's economic leverage in receiving a favorable financial settlement.

parent has been employed during the entire previous year, payment rates are 80 percent or more. Hence,

> coercive child-support collection policies, such as automatic wage withholding, will have only limited success. Divorced fathers (perhaps in distinction to never married fathers) appear to pay quite well if they are fully employed. . . . The key to reducing poverty thus appears to be the old and unglamorous one, of solving un- and underemployment, both for the fathers and the mothers. (Braver, Fitzpatrick, and Bay 1991, pp. 184–85)

The child support issue is complex. Some noncustodial fathers provide support in ways other than money (Teachman 1991), such as child care, for example. "In some cases attempts to locate and require payments from such fathers may result in severing these ties" (Peterson and Nord 1990, p. 539). Some experts argue that fathers not involved in their chil-

dren's daily lives can never be made to pay. They propose that support of children should instead devolve upon whichever adults they are living with, whether biological, adoptive, stepparents, or other.

Two suggested alternative solutions are guaranteed child support and a children's allowance. Both are based on the principle of societywide responsibility for investing in all children. With **guaranteed child support,** used in France and Sweden, the government sends to the custodial parent the full amount of support awarded to the child, even though this sum may not have been received from the noncustodial parent. It then becomes the government's responsibility to collect the money from the parent that owes it (Salt 1991). A second alternative, a **children's allowance,** provides a government grant to all families—married or single-parent, regardless of income—based on the number of children they have. All industrialized countries except the United States have some version of a children's allowance. Policy experts argue that removing income tax deductions for children, the current practice in the United States, would help finance a children's allowance (Meyer, Phillips, and Maritato 1991).

Are No-fault Divorce Laws Responsible for Ex-Wives' and Their Children's Poverty?

Sociologist Lenore Weitzman (1985) has argued that no-fault divorce laws are a significant cause of postdivorce poverty of women and children. More recently, at least one other scholar (Parkman 1992) has argued similarly. No-fault divorce laws were intended to mitigate hostilities between divorcing partners. However, according to Weitzman and Parkman, an unintended consequence has been removal of the leverage women once had to negotiate favorable financial settlements. Because in many states divorce could not be obtained without proving fault, a husband wanting a divorce from a wife who did not was often forced to make a deal, assuming guilt and paying the price.

This process was part of the distasteful maneuvering associated with the fault-based divorce system. But Weitzman argues that, despite its negative aspects, it did result in more adequate provisions for dependent wives and children than they receive in the "equal" treatment of the no-fault system. Without the benefit of fault in a settlement—and with the changing attitudes toward traditional gender roles in which courts consciously strive to treat women and men equally—divorced women are receiving less in terms of property settlements and alimony awards than women did before no-fault. Although the abolition of fault has the desirable result of permitting either spouse to obtain a divorce (and with less legal acrimony than in the past), the practical results have been hard on mothers and children.

No-fault divorce and its equal property division depend on an assumption of equality between the sexes that does not exist in reality. Ex-wives without the education, skills, or job experience necessary to earn a reasonable income may need more than their share of tangible couple property to survive. Also, in the past a dependent wife with minor children was often given title to the family home; now it may need to be sold so that its worth can be divided. Selling the family home is particularly difficult for families with minor children, with the stress of moving and the loss of neighborhood, school, and friends. New quarters are often less adequate and may be more expensive, as monthly rent is higher than the payment on a house purchased some years ago.

Other scholars have trouble with Weitzman's assessment of the economic impact of the changeover to no-fault divorce law. For example, feminist attorney Marygold Melli (1986) argues that what Weitzman *assumes,* rather than proving, is that the years before no-fault divorce produced more favorable financial outcomes for women. Melli notes the infrequency with which alimony awards were actually made. Another criticism of Weitzman's thesis is that it presumes that husbands seek divorces and wives don't. Moreover, Melli and others see no-fault as necessary and beneficial to women, particularly those in intolerable marital situations, because divorce was frequently difficult for women to obtain under fault laws.

Much of this discussion at least indirectly addresses the impact of divorce on children. We turn now to a more direct and detailed examination of that topic.

Divorce and Children

Figure 15.6 shows the rising number of children affected by their parents' divorce through 1980 and the continuing high number in the 1990s. Estimates are that 60 percent of all children will spend some time in a single-parent family, primarily as a consequence of marital disruption (Bianchi 1990, p. 10). Furthermore, only a minority (36 percent) of children in single-parent families will experience the transition to a two-parent family within five years (44 percent of white children and 23 percent of black children; Bumpass and Sweet 1989).

How do separation and divorce affect children? Experts disagree, and the situation may be changing as well, so we cannot give a definitive answer to this question. It also depends on what comparison is being made. Children's self-concepts are affected not so much by family structure as by the quality of familial relationships. Living in an intact family characterized by unresolved tension and alienating conflict can cause as great or greater emotional stress and lower self-esteem in children than living in a supportive single-parent family (Peterson and Zill 1986; Kline, Johnston, and Tschann 1991).

Nevertheless, the divorce experience is both psychologically stressful and, in most cases as we have seen, financially disadvantageous for children.

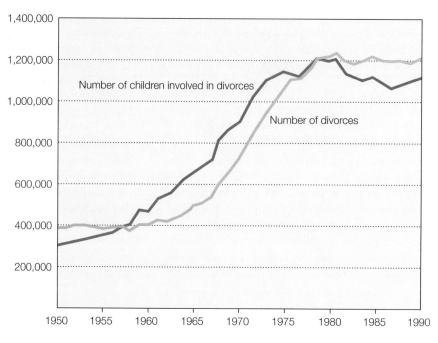

FIGURE *15.6*

Estimated number of divorces and children involved in divorce, United States, 1950–1990. (Source: Clarke 1995b, Figure 1, p. 2)

The Various Stresses for Children of Divorce

During and for a period of time after divorce, children typically feel guilty, depressed, and anxious (Hetherington 1973; Wallerstein and Kelly 1980). In the only longitudinal study of children's postdivorce adjustment, psychologists Judith Wallerstein and Joan Kelly interviewed all of the members of some sixty families with one or more children in counseling at the time of the parents' separation in 1974. Wallerstein and her colleagues reinterviewed children at one year, two years, five years, ten years, and in some cases, fifteen years later. Children appeared worst in terms of their psychological adjustment at one year after separation, many having declined significantly since their parents' separation. By two years postdivorce, households had generally stabilized. At five years, many of the 131 children seemed to have come through the experience fairly well; 63 percent were either in excellent or reasonably good psychological health. Another 37 percent, however, were not coping well, with anger

playing a significant part in the emotional life of 23 percent. (In Figures 15.7 and 15.8, children's drawings suggest the ways the children have discovered to accommodate their parents' divorce.)

By 1984, after following her sample of children of divorce for ten years, Wallerstein and colleagues found the majority to be approaching economic self-sufficiency, enrolled in educational programs, and in general, to be responsible young adults. Still, the overall impression left by Wallerstein's research is one of loss. Children may lose fathers, who become disinterested and detached; they may lose mothers, who are overwhelmed by the task of supporting the family and managing a household alone and who either see little chance of happiness for themselves or else are busy pursuing their "second chance" (Wallerstein and Blakeslee 1989).

One of Wallerstein's major findings is that girls, who seemed, as in other studies, far better adjusted than boys in the early postdivorce years, evidenced a "sleeper effect," which surfaced as they reached adoles-

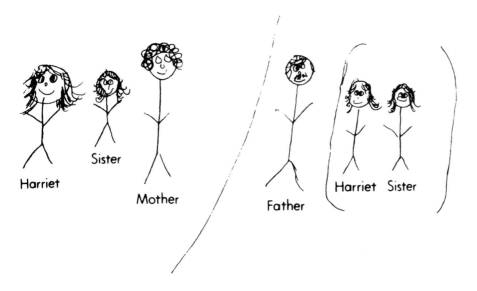

FIGURE **15.7**

This drawing reveals the creative coping of a child whose parents are divorcing. She has figured out a way to include her father as well as keep within the bounds of reality as she knows it. (Source: Isaacs, Montalvo, and Abelsohn 1986, p. 280)

cence and young adulthood. They were beset with "lingering sorrow" and seemed hesitant about marriage and childbearing. Said one of Wallerstein's respondents, "How can you expect commitment when anyone can change his mind?"

Children of divorce experience the loss of their identity as a member of an intact family, along with the loss of daily interaction with one of their parents. They get less help with homework (Astone and McLanahan 1991), and at school and elsewhere they may suffer from the cultural stigma of being from a "broken home" (Amato 1991), although this is lessening, and schools are instituting programs to help children cope with their parents' divorce (Peres and Pasternack 1991).

CHILDREN OF REDIVORCE   In some cases, but certainly not all, a new stepparent can prove a stabilizing and supportive influence. Counselors—who are now beginning to see children who have experienced divorce more than once—encourage divorced stepparents to maintain contact with their former stepchildren. Wallerstein found that half of the children in her study had experienced a second divorce of one or both

parents. According to Clifford Sager, a psychiatrist and director of family psychiatry at the Jewish Board of Family and Children's Services in New York, the dangers to the emotional health of both the stepchild and the stepparent when a remarriage breaks up have been overlooked—"woefully underestimated." In Sager's words,

we have given too little thought to what kind of values we are demonstrating when we just walk away from these relationships or cut them off. The child is left with the fear that nobody can be trusted and perhaps he wasn't worth the love anyway. The stepparents must realize that after making all kinds of motions about being a caring love-parent or friend they can't suddenly disappear. They have a moral obligation. It's something the biological parent must understand too. (in Brooks 1984b, p. 46)

LESS MONEY AND LESS EDUCATION   Children whose parents have divorced will more than likely have noticeably less money to live on, and the chances are fairly good that they will suffer economic depriva-

FIGURE **15.8**

The sister of the drawer of Figure 15.7 used a jagged line to separate her father from the rest of the family. (Source: Isaacs, Montalvo, and Abelsohn 1986, p. 281)

tion. This is especially significant because some of the negative impact of divorce can be attributed to economic deprivation (Garfinkel and McLanahan 1986; McLanahan and Booth 1989).

We may think here in terms of children at the lower end of the socioeconomic scale, but one researcher found considerable deprivation among middle-class children compared to what they could have expected had their parents remained married (Wallerstein and Blakeslee 1989).

Because divorce settlements seldom include arrangements for how parents will pay for children's college education—and because family savings are often eroded by the costs involved with divorcing—financing the high costs of college for children of divorced parents can be especially problematic. Some children of divorced parents who would otherwise have attended college may find that they cannot afford it, whereas others who would have gone away to college may find themselves attending a local school.

Wallerstein, who followed her sample into young adulthood, was surprised at the extent of their educational downward mobility. Sixty percent of the study children were likely to receive less education than their fathers; 45 percent, less than their mothers. This phenomenon is connected to the financial marginality of most divorced mothers and the defection of many fa-

thers (Stern 1994; Amato, Rezac, and Booth 1995). Even divorced fathers who had retained close ties, who had the money or could save it, and who ascribed importance to education, seemed to feel less obligated to support their children through college.

> Children of divorce feel less protected economically; unlike children from intact families, whose parents usually continue to support them through college and sometimes even beyond, the children of divorce face an abrupt, premature end to an important aspect of their childhood. . . . [Moreover,] nowhere is it writ that psychological or emotional support stops at age eighteen, but many children of divorce cannot help but feel that when child support stops, something else stops in the social contract between parent and child. (Wallerstein and Blakeslee 1989, p. 157)

Along with this may come a favoring of boys over girls. In Wallerstein's sample, fathers who helped with college were twice as likely to help sons.

Because the Wallerstein study is based on a small sample of children once in counseling, it has been challenged by studies with more representative samples and that reach somewhat less pessimistic conclusions. The absence of a comparison group to tell us how children in intact homes are faring these days also

While many children do adapt to their parents' divorce, they often express a great sense of loss, no matter how rationally a divorce is handled. Holidays, in particular, can be stressful for divorced families.

renders the study inconclusive (although whenever possible, Wallerstein referred to other studies including relevant national sample data). Box 15.2 on page 493, "How It Feels When Parents Divorce," points to many issues raised in this section.

SURVEY VERSUS CLINICAL FINDINGS

One analysis (Dawson 1991) from a nationally representative sample of 17,110 children under age 18 found that those living with single mothers or stepfathers were more likely to have school, behavioral, and health problems than were those living with both biological parents. There is evidence that adolescents in single-parent families are at greater risk for substance abuse and premarital sex (Flewelling and Bauman 1990; Needle, Su, and Doherty 1990). The latter may result from these teenagers' more liberal attitudes toward nonmarital sex as a result of their divorced parents' dating behavior (Whitbeck, Simons, and Kao 1994).

Meanwhile, sociologists Paul Amato and Bruce Keith have conducted two important, systematic reviews and statistical analyses of virtually all studies of children's and young adults' postdivorce adjustment through the 1980s. After analyzing ninety-two studies of children, Amato and Keith (1991a) concluded that parental divorce is associated with negative outcomes in academic achievement, conduct, psychological adjustment, self-esteem, and social relations. A similar review of thirty-three studies found that adults who had experienced parental divorce as children, compared with those from continuously intact families, have poorer psychological adjustment, lower socioeconomic attainment, and greater marital instability (Amato and Keith 1991b). "The cumulative picture that emerges from the evidence suggests that parental divorce (or some factor connected with it) is associated with lowered well-being among both children and adult children of divorce" (Amato 1993, p. 23).

Still, the differences are small, and smaller today than in the past, when divorce was less common. Clinical studies (such as Wallerstein's, previously described) found stronger negative effects than studies using representative samples. "This is due to the fact that a great deal of variability is present among children of divorce, with some experiencing problems and others adjusting well or even showing improvements in behavior" (Amato 1993, p. 23). Overall, Amato and Keith concluded that, at least for effects-of-divorce research, "clinical studies should not be generalized to the larger community" (1991b, p. 55).

REASONS FOR NEGATIVE EFFECTS OF DIVORCE ON CHILDREN

Amato (1993) has summarized five theoretical perspectives found in the literature concerning the reasons for negative outcomes among children of divorced parents. These are as follows:

1. The **life stress perspective** assumes that, just as divorce is known to be a stressful event for adults, it must also be so for children. Furthermore, divorce is not one single event but a process of associated

Research shows many possible explanations for negative effects of divorce on children. Meanwhile, some studies suggest that children may accept their parents' divorce as a desirable alternative to ongoing family conflicts.

events (Morrison and Cherlin 1995)—moving, changing schools, giving up pets, loss of contact with grandparents and other relatives—that may be distressing to children. Reminiscent of our discussion of stressor pile-up and overload in Chapter 14, this perspective holds that an accumulation of negative stressors results in problems for children of divorce.

2. The **parental loss perspective** assumes that a family with both parents living in the same household is the optimal environment for children's development. Both parents are important resources, providing children emotional support, practical assistance, information, guidance, and supervision, as well as modeling social skills like cooperation, negotiation, and compromise.

Accordingly, the absence of a parent from the household is problematic for children's socialization.

3. The **parental adjustment perspective** notes the importance of the custodial parent's psychological adjustment. Supportive and appropriately disciplining parents facilitate their children's well-being. However, the stress of divorce impairs a parent's child-rearing skills, with probable negative consequences for children.

4. The **economic hardship perspective** assumes that economic hardship brought about by marital dissolution is primarily responsible for the problems faced by children whose parents divorce (Entwisle and Alexander 1995).

BOX 15.2

How It Feels When Parents Divorce

In the following excerpts, five children of divorce tell their own stories. As you will see, they talk about issues raised in this chapter.

Zach, Age 13

Even though I live with my Dad and my sister lives with my Mom, my parents have joint custody, which means we can switch around if we feel like it. I think that's the best possible arrangement because if they ever fought over us, I know I would have felt I was like a check in a restaurant—you know, the way it is at the end of a meal when two people are finished eating and they both grab for the check and one says, "I'll pay it," and the other one says, "No, this one is mine," and they go back and forth, but secretly neither one really wants it, they just go on pretending until someone finally grabs it, and then that one's stuck....

My parents knew they couldn't live together, but they also knew it was nobody's fault. It was as if they were magnets—as if when you turn them the opposite way they can't touch.... Neither of them ever blamed the other person, so they worked it out the best they could—for their sakes and ours, too.

Nevertheless, it's very sad and confusing when your parents are divorced. I think I was five when they separated....

When my parents first split up, it affected me a lot.... I got real fat and my grades went way down, so I went to a psychologist. She made me do a lot of things which seemed dumb at the time—like draw pictures and answer lots of silly questions.... My school work suffered because I was so distracted thinking about my situation that I couldn't listen very well, and for a long time I didn't work nearly as hard as I should have. Everyone told me I was an underachiever, and my parents tell me I still am, but I don't think so. What I do think is that I am a lot more independent—a go-out-and-do-it-yourself person....

I've heard about kids who are having all these problems because their parents are getting divorced, but I can't understand what the big deal is. I mean, it's upsetting, sure, but just because your parents are separated it doesn't mean you're going to lose anybody.... It's not something I talk about very much. Most of my friends would rather talk about MTV than talk about divorce.

Ari, Age 14

When my parents were married, I hardly ever saw my Dad because he was always busy working. Now that they're divorced, I've gotten to know him more because I'm with him every weekend. And I really look forward to the weekends because it's kind of like a break—it's like going to Disneyland because there's no set schedule, no "Be home by five-thirty" kind of stuff. It's open. It's free. And my father is always buying me presents.

My Mom got remarried and divorced again, so I've gone through two divorces so far. And my father's also gotten remarried—to someone I don't get along with all that well. It's all made me feel that people shouldn't get married—they should just live together and make their own agreement. Then, if things get bad, they don't have to get divorced and hire lawyers and sue each other. And, even more important, they don't have to end up hating each other.

I'd say that the worst part of the divorce is the money problem. It's been hard on my Mom because lots of times she can't pay her bills, and it makes her angry when I stay with my father and he buys me things. She gets mad and says things like, "If he can buy you things like this, then he should be able to pay me." And I feel caught in the middle for two reasons: First, I can't really enjoy whatever my Dad does get for me, and second, I don't know who to believe. My Dad's saying, "I don't really owe her any money," and my Mom's saying he does. Sometimes I fight for my Mom and sometimes I fight for my Dad, but I wish they'd leave me out of it completely.

Jimmy, Age 10

My mother divorced my father when I was two and a half. My Mom took me to live with her, but she couldn't cope, and so she

(continued)

(continued)

was going to put me up for adoption. There was a big scene because my Dad didn't want that to happen, and so I went to live with him ... for the next three years—until the first "snatch." ...

One weekend while I was visiting [my mother], she and Don [my mother's boyfriend] packed up their car with a lot of stuff and we all piled in. I kept asking her, "Where are we going?" And finally she said, "Take a wild guess." I guessed a couple of places and finally I said Florida because that was the one place she had talked about a lot—she had some relatives who lived there. I didn't have any idea how far away Florida was, so I asked her how long it would take us to get there, and she said about two days. All I could think of was, "Wow! We're going to Florida!" I remember asking her, "What about Daddy?" And she just said, "Don't worry about it."

After we got there she put me in school, but since she didn't want to send for my school records, she started me in first grade instead of putting me in second grade, where I belonged. And I had to have all three of my shots over again, which I hated. The reason she couldn't write for my school records is because then my father might find out where I was and she didn't want that to happen.

School was really easy for me and I'd just sit there being bored. It seems like all we did was learn the alphabet, which I already knew, and play with blocks and do dumb things like that. I didn't have to stay for long, though, because one day my mother decided we were going to move without paying our next rent. We packed up all our belongings, climbed out the window, got in the car, and drove off. We slept in the car that night, and the next day we found a nice apartment. I went to another school, which was better because at least I was in second grade, where I belonged, but I didn't stay there for very long, either. One morning the principal came into my classroom and told me to pick up my stuff—to take some crayons and paper—because some man had come to pick me up. The man told me he was a detective and showed me a picture of my father. ... Before I knew what was happening we were driving off in his car. ...

That night [my father and I] went out to eat with the detective, and the next day my father said he had two surprises for me. The first surprise was that we were going bowling, and later on he said that the second surprise was that we were all going to go on a big airplane. That's how I got to take my first plane ride—which was a lot of fun! The three of us flew back to New York and then went to my Dad's house—my old home—and that night we had a big party with lots of balloons and cake. Everyone was there—all my father's relatives and friends came and everyone was hugging me. The next week, my father ... won custody again, and this time he actually got to keep me because my mother didn't even try to get me back. It's hard to believe, but I haven't seen her since then, which is about four years now. At first I was glad because I was so happy to settle down and be in a good school, but then I started feeling sad, too, because I missed my mother. I still miss her.

---

5. The **interparental conflict perspective** holds that conflict between parents prior to, during, and after the divorce is responsible for the lowered well-being of children of divorce. Although the literature offers only modest support for the first four perspectives stated, research strongly supports this fifth perspective (Amato 1993). Many studies indicate that much of what appear to be negative results for children of divorce are probably not simply the results of divorce per se but also of parental and family conflict before, during, and after the divorce (Smith 1990; Amato and Keith 1991b; Dawson 1991; Morrison and Cherlin 1995).

FAMILY RELATIONS VERSUS FAMILY STRUCTURE It is important to note that the interparental

Caleb, Age 7

My parents aren't actually divorced yet. But they're getting one soon. They stopped living together when I was one and a half, and my Dad moved next door. Then, when I was five, he moved to Chicago, and that hurt my feelings because I realized he was really leaving and I wouldn't be able to see him every day. My father's an artist, and when he lived next door to us in New York, I used to go to his studio every day and watch him when he was welding. I had my own goggles and tools, and we would spend many an hour together. I remember when I first heard the bad news that he was moving away, because I almost flipped my lid. My father said he would be divorcing my Mom but that he wouldn't be divorcing me and we'd still see each other a lot—but not as often. I started crying then and there, and ever since then I've been hoping every single second that he'd move back to New York and we'd all live together again. I don't cry much anymore because I hold it back, but I feel sad all the same.

I get to visit my father quite often. And Shaun. He's my collie. My cat lives in New York with me and Mom. Whenever I talk with Daddy on the phone I can hear Shaun barking in the background. The hardest thing for me about visiting my father is when I have to leave, and that makes me feel bad—and mad—inside. I still wish I could see him every day like I did when I was little. It's hard to live with just one person, because you don't have enough company, though my Mom has lots of great baby-sitters and that helps a little.

Tito, Age 11

It seems like my parents were always fighting. The biggest fight happened one night when we were at a friend's house. Mommy was inside the house crying, and Daddy was out on the sidewalk yelling and telling my mother to come down, and my little sister, Melinda, and I were outside with a friend of my father's. We were both crying because we were so frightened. Then Daddy tried to break the door down, so Mommy came downstairs. And then the police cars came and Daddy begged Mommy to stay quiet and not say anything and to give him another chance, but she was so unhappy that she got into one of the cars.... I was only four but I remember everything.

We stayed with our cousin for about two months, and during this time I saw my father whenever he visited us at my grandmother's house.... I was always happy to see him, but sometimes it made me feel sad, too, because I would look forward to our visits so much, and then when we were together it could never be as perfect as I was hoping it would be. He was still so angry at Mommy's leaving him that it was hard for him to feel anything else for anybody....

About the time of the divorce I started to get into fights with other kids, and my mother got worried. She thought I must be feeling very angry and having a hard time expressing my feelings, so she took me to a therapist.... We got really close and he'd talk to me about my problems with my Dad. This went on for about two years, and during that time he helped me realize that the divorce was better for me in the long run because our home was more relaxed and there wasn't so much tension in the air.

The other thing that happened around this time was that my mother found out about an organization called Big Brothers, where I could have another male figure in my life.... They paired me off with a guy named Pat Kelly, and we've been getting together every weekend for a couple of years.... Pat and I do a lot of things like play baseball or video games and eat hot dogs. But the best thing we do is talk—like when I do something good in school I can tell him, and if I feel sad I can talk about that, too. His parents got divorced when he was twelve, and so we have a lot of the same feelings.

Chapter 1 argues that people's families of orientation and social class influence their life options. How is that point evident in these children's stories? Overall, do you see these stories as hopeful, as dismaying, or both? Why?

Source: Krementz 1984.

---

conflict explanation (described above) finds family relations and communication patterns more important to children's well-being than any particular family form or structure (Wenk et al. 1994). As an example, one survey analysis of 1,291 Michigan adolescents found that teens in intact families who described their fathers as not interested in them had lower self-esteem than did those in mother-headed, single-parent families (Clark and Barber 1994). Perhaps what matters, the authors surmise,

are not only adolescents' perceptions of how parents think and behave but adolescents' accounts to themselves of why parents behave as they do. A lack of paternal interest following divorce, when it occurs, could conceivably be attributed to one or more divorce-related factors, such as a father's wish to avoid conflictual situations with his ex-wife. It may be that in two-parent families, a low level of paternal interest is harder for an adolescent to "explain away" or view as beyond his or her control than would be the case for an adolescent whose parents are divorced. (p. 613)

Our overall conclusion is that in choosing to divorce, parents are not necessarily enhancing their children's development—although they may be preventing more damaging experiences—but they are not necessarily condemning them to a disastrous and disappointing life either.

Even Wallerstein, whose research presents the most negative view of children's postdivorce adjustment, says:

When people ask whether they should stay married for the sake of the children, I have to say, "Of course not." All our evidence shows that children turn out less well adjusted when exposed to open conflict, where parents terrorize or strike one another, than do children from divorced families. And while we lack systematic studies comparing unhappily married families and divorced families, I do know that it is not useful to provide children with a model of adult behavior that avoids problem solving and that stresses martyrdom, violence, or apathy. A divorce undertaken thoughtfully and realistically can teach children how to confront serious life problems with compassion, wisdom, and appropriate action. . . . Our findings do not support those who would turn back the clock. (Wallerstein and Blakeslee 1989, p. 305)

From studies that reach different conclusions about overall outcome, we can still learn much that is potentially useful about what postdivorce circumstances are most beneficial to children's development and what pitfalls to avoid. Wallerstein and Blakeslee consider a good mother–child (or, custodial parent–child) bond to be the most significant factor (see also Tschann et al. 1989). Another highly important factor in children's adjustment to divorce is the divorced parents' relationship with each other.

FIGHTING THROUGH THE CHILDREN   In some divorces, "children are handed a hunting license by one parent to go out and take shots at the other" (Wallerstein and Blakeslee 1989, p. 188). In the

Wallerstein study, more than half of 8- and 12-year-old children were asked by one parent to spy on the other—to go through bureau drawers and the like. Even though married partners may be angry and feel vindictive, they do not usually draw children openly into the spousal wars in the way that ex-spouses do.

Visitation is one frequent arena of parental disputes: The child isn't ready to go when visitation time starts or the visiting parent brings the child home late, for instance. Visitation can be an arena in which ex-spouses continue the basic conflicts they experienced before the divorce (Kelly and Wallerstein 1990). Counselors urge ex-spouses to work out irritations and conflict directly with each other rather than through their children. Parents need to communicate as openly as possible about their divorce *without* speaking negatively about the ex-spouse. Counselors encourage the attitude "We had our problems, but that doesn't mean he (or she) is bad or that I'm bad." Some jurisdictions sponsor "visitation support groups," in which parents and stepparents discuss their problems with others, or provide court-supported conciliation of visitation disputes.

Experts agree that adjusting to divorce is easier for children *and* parents when former spouses cooperate (Brody and Forehand 1990; Ahrons and Miller 1993; Bronstein et al. 1994). The child's maintaining ties with the noncustodial parent can be important also (Kline, Johnston, and Tschann 1991). We turn now to custody issues.

Custody Issues

About 53 percent of couples who divorce have children under age 18 (U.S. National Center for Health Statistics 1991a, Table 4). A basic issue for them is determining which parent will take **custody,** or assume primary responsibility for making decisions about the children's upbringing and general welfare. Legal issues surrounding custody, visitation rights, and (sometimes) child support payments in *re*divorces (that is, divorcing stepfamilies) are just now beginning to emerge.

As it is formalized in divorce decrees, child custody is usually an extension of the basic exchange: Divorced fathers have legal responsibility for financial support while divorced mothers continue the physical, day-to-day care of their children. Fathers were automatically given custody in the United States until the nineteenth century, but from the beginning of the twenti-

One of the more difficult and recurrent moments in life after divorce is the transfer of the child from one parent to the other. It is easier for both children and parents when former spouses cooperate over custody and financial issues.

eth century until recently, courts have presumed that mother custody was virtually always in the child's best interest (Grych and Fincham 1992). Today custody is awarded to mothers in about 90 percent of cases, and most often this has been agreed upon by the couple or the mother has custody by default.

With worsening economic prospects and changing attitudes and gender roles, however, a few mothers are voluntarily relinquishing custody and more fathers are seeking custody. Today, a small but growing number of fathers are awarded both custody and child support from ex-wives. The odds of father custody are slightly higher when the children are older, especially when the eldest child is male and when the father is the plaintiff in the divorce (Fox and Kelly 1995).

Under current laws a father and a mother who want to retain custody have theoretically equal chances, and judges try to assess the relationship between each parent and the child. However, because mothers are typically the ones who have physically cared for the child, and because many judges still have traditional attitudes about gender, some courts give preference to mothers (Robinson and Barret 1986, p. 87). Generally, studies have found nothing to preclude father custody or to prefer it (Rosenthal and Keshet 1980).

In a study of noncustodial mothers based on interviews with over 500 mostly white women in 44 states, over 90 percent reported that the process of becoming noncustodial was stressful. Most commonly, money (30 percent), child's choice (21 percent), difficulty in handling the children (12 percent), avoidance of moving the children (11 percent), and self-reported instability or problems (11 percent) were given as reasons. Only 9 percent reported losing their children in a court battle or ceding custody to avoid a custody fight (Greif and Pabst 1988, p. 88). Box 15.2 includes one child's story (Jimmy's) in which the father eventually won custody, presumably following the mother's financial and perhaps emotional instability.

| A small but growing number of fathers are seeking, and are awarded, custody.

Judith Fisher, who also studied noncustodial mothers, believes that women should not relinquish custody because they feel inadequate in comparison to their successful husbands. At the same time, she strongly supports the freedom of men and women to make choices about custody—including the woman's choice to live apart from her children—without guilt or stigma. She urges

> (1) the negation of the unflattering stereotypes of noncustody mothers; (2) sensitivity to the demands society places on mothers (as opposed to parents) and to the needs of all family members; (3) supportiveness of the woman's choice when it appears to have been well thought out; and (4) . . . prohibitions against blaming the mother when others (the children, the children's father, the courts) decide that the children should live apart from her. (Fisher 1983, p. 357)

Children living with fathers typically have more contact and are emotionally closer to noncustodial parents, that is, their mothers (Greif and Pabst 1988; Lewin 1990a).

THE VISITING PARENT   Almost all of the research and discussion to date on visiting parents have been about fathers, but some research suggests that the same findings apply to visiting mothers (Spanier and Castro 1979, p. 249). Fathers without custody experience a sense of extreme loss. There is some evidence that, ironically, more emotionally involved fathers may cope with this loss by visiting their children *less* often than fathers with less emotional involvement before the divorce (Kruk 1991). Evidence suggests this can also be true for stepfathers and men who have cohabited with women and their children (Brooks 1984b; Dullea 1987b).

Early visits often display what has been called the Disneyland Dad syndrome, in which the father indulges his child to make the visit as happy as possible. As time goes by, this phase generally passes. Often fathers become increasingly frustrated with their loss of influence over the children's upbringing, and they let themselves drift away from the children. Visiting typically declines over time (Seltzer 1991a) and often becomes episodic. This lack of divorced father–child contact persists into and throughout the child's adulthood, eventually reducing the probability that a divorced father can think of his children as potential sources of support in times of need (Amato 1994; Aquilino 1994b; Cooney 1994; Lawton, Silverstein, and Bengston 1994).

Research is inconclusive regarding the relationship of a father or father figure to a child's overall development and well-being (Hawkins and Eggebeen 1991). In some cases, such as verbal, physical, or sexual abuse, father contact may actually be damaging to children (King 1994). But studies generally indicate that a father's decreased visiting can be painful for children. In the Wallerstein study, fathers visited their children with more regularity than other research has found, but they were often psychologically detached. In one case, "Almost always, there would be other adults around or adult activities planned. Carl watched hundreds of hours of television at his father's house, feeling more and more alone and removed from his earlier visions of family life" (Wallerstein and Blakeslee 1989, p. 79). Children often forgave geographically distant fathers who did not appear frequently but were very hurt by those nearby fathers who rarely visited. Some fathers managed to retain a fantasy image of themselves and their children relating happily in the face of a quite different reality.

Other studies have found fathers to have vanished even more thoroughly. A study by Furstenberg and Harris followed 1,000 children from 1976 to 1987 (and included children of unwed as well as divorced parents). Forty-two percent had not seen their fathers in the previous year, and over half had never been in the father's home. "Men regard marriage as a package deal. They cannot separate their relations with their children from their relations to their former spouse. When that relationship ends, the paternal bond usually withers within a few years too" (Furstenberg, in Lewin 1990a, p. A18; Furstenberg and Cherlin 1991).

CHILD SNATCHING   At the other extreme is **child snatching:** kidnapping one's children from the other parent. A surprisingly high number of noncustodial parents engage in child snatching. A 1988–89 survey of over 10,000 families with children found a rate of reported child snatching that, projected to the U.S. population, would mean some 163,200 children are kidnapped each year by noncustodial parents who took them to another state and/or intended to keep them (Finkelhor, Hotaling, and Sedlak 1991).

Child snatching is frightening and confusing for the child, can be physically dangerous, and is usually detrimental to the child's psychological development (Wallerstein and Blakeslee 1989). Yet for years the snatching of a child by a biological parent was not legally considered kidnapping; in most states it was a misdemeanor called *custodial interference.* Now, however, due to the passage of the federal uniform child custody acts, states recognize out-of-state custody decrees and do more to find and extradite offenders. Child snatching is an extreme act, but it points up the frustration involved in arrangements regarding sole custody and visiting parents. An alternative is joint custody.

JOINT CUSTODY   When parents live close to each other and when both are committed to the arrangement, joint custody or shared living arrangements, or both, can bring the experiences of both parents closer together, providing advantages to each. In **joint custody,** both divorced parents continue to take equal responsibility for important decisions regarding the child's general upbringing.

There are two variations in joint custody agreements. One is joint legal and physical custody, in which parents or children move periodically (usually weekly, biweekly, or monthly) to reside with the children. The second variation, recommended by some policy experts (see, for example, Furstenberg and Cherlin 1991), is shared *legal custody*—in which both parents still have the right to participate in important decisions and retain a symbolically important legal authority—with *physical custody* (that is, residential care of the child) going to one parent. Parents with higher incomes and education are more likely to have joint custody (Seltzer 1991b).

There are several advantages to arrangements for joint legal and physical custody. (Table 15.2 lists advantages and disadvantages of joint custody from a father's perspective.) Shared custody gives children the chance for a more realistic and normal relationship with each parent (Arditti and Keith 1993). The parents of Zach, whose story is included in Box 15.2, have joint custody; this may be one reason for his conclusion that "just because your parents are separated it doesn't mean you're going to lose anybody." Furthermore, both parents may feel they have the opportunity to pass their own beliefs and values on to their children. In addition, neither parent is overloaded with sole custodial responsibility and its concomitant loss of personal freedom.

Despite its possibilities, joint custody arrangements remain limited in number and experimental in character. Many attorneys and judges, along with some

TABLE *15.2*

Advantages and Disadvantages of Joint Custody from a Father's Perspective

Advantages	Disadvantages
Fathers can have more influence on the child's growth and development—a benefit for men and children alike.	Children lack a stable and permanent environment, which can affect them emotionally.
Fathers are more involved and experience more self-satisfaction as parents.	Children are prevented from having a relationship with a "psychological parent" as a result of being shifted from one environment to another.
Parents experience less stress than sole-custody parents.	Children have difficulty gaining control over and understanding of their lives.
Parents do not feel as overburdened as sole-custody parents.	Children have trouble forming and maintaining peer relationships.
Generally, fathers and mothers report more friendly and cooperative interaction in joint custody than in visitation arrangements, mostly because the time with children is evenly balanced and agreement exists on the rules of the system.	Long-term consequences of joint custody arrangements have not been systematically studied.
Joint custody provides more free social time for each single parent.	
Relationships with children are stronger and more meaningful for fathers.	
Parental power and decision making are equally divided, so there is less need to use children to barter for more.	

Source: Robinson and Barret 1986, p. 89.

child psychiatrists, continue to favor sole custody. One reason is that ex-spouses may be no more able to agree on child-related decisions at this point than they were able to make joint decisions while married (Coysh et al. 1989).

One study (Maccoby, Depner, and Mnookin 1990) of divorced California families compared co-parental interaction among parents with joint physical custody (170 families) to those with mother custody (420 families) and to those with father custody (74 families). Eighteen months after the divorce, just one-fifth of parents with father custody and one-fourth of those with mother custody described their postdivorce relationship as cooperative. More—but still only one-third—of those with joint custody said their co-parental relationship was cooperative. Of the families with joint custody, nearly 50 percent said they discussed their children once a week or more, but only 34 percent said they consciously tried to coordinate

rules about bedtime, TV, or homework in the two households.

Another difficulty with joint physical custody is today's high rate of geographic mobility. When ex-mates move to separate regions, it would be difficult, and very likely harmful, for the children to divide the year between two different schools and communities. Even without the problems of geographic mobility, some children who have experienced joint custody report feeling "torn apart," particularly as they get older (Simon 1991). Although some youngsters appreciate the contact with both parents and even the "change of pace" (Krementz 1984, p. 53), others don't. The following account from 11-year-old Heather shows both sides:

The way it works now is we switch houses every seven days—on Friday night at five-thirty. At first we tried switching every three days and that was

Maintaining the parent–child bond is significant in a child's adjustment to divorce. Shared custody provides a framework wherein a child can maintain relationships with both mother and father.

crazy, and then we tried five days and that was still too confusing. . . . And switching on school nights was awful. Friday nights are perfect because we don't have to worry about homework, and if Matthew [her brother] and I both have after-school activities it doesn't matter because my Dad can take me and Mom can drive Matt. Another reason it works out well is because each of our parents likes to help us with our homework and when we stay with them for a week at a time it's easier for them to keep up with what we're doing. . . .

And as long as they're divorced, I don't see any alternative because it wouldn't seem right to live with either parent a hundred percent of the time and only see the other one on weekends. But switching is definitely the biggest drag in my life— like it's just so hard having two of everything. My rooms are so ugly because I never take the time to decorate them—I can't afford enough posters and I don't bother to set up my hair stuff in a special way because I know that I'll have to take it right back down and bring it to the next house. Now I'm thinking that I'll try to make one room my real room and have the other one like camping out. I can't buy two of everything, so I might as well have one good room that's really mine. (Krementz 1984, pp. 76–78)

Some feminists oppose joint custody because they believe ex-husbands might use it to intervene excessively in the lives of their former wives (Brophy 1985). Also, joint custody is expensive. Each parent must maintain housing, equipment, toys, and often a separate set of clothes for the children and must sometimes pay for travel between homes. Mothers, more than fathers, would find it difficult to maintain a family household without child support, which is often not awarded when custody is shared.

Research does not support the presumption that joint custody is best for children of divorced parents, but neither does it indicate that it is more harmful than traditional sole custody; several studies have found no difference between joint and sole custody (Coysh et al. 1989; Kline et al. 1989).

Joint custody parents recognize that they are in the minority, and as yet experts are unclear on what advice to give. School-age children seem to adjust better than preschoolers, perhaps because their more sophisticated cognitive skills allow them to place the frequent changes in routine in perspective (Steinman 1981). One thing is clear: *Court-ordered* joint custody opposed by one or both parents does have more negative outcomes than do other forms of custody (Johnston, Kline, and Tschann 1988).

This discussion of mothers' and fathers' custody issues suggests that being divorced is in many ways a very different experience for men and women. Just as sexually segregated or traditional roles can lead to "his" and "her" marriages (Bernard 1982), so also can they result in "his" and "her" divorces.

Her and His Divorce

We have seen that gender roles diminish communication and understanding between women and men. Perhaps nowhere is the lack of understanding more evident than in the debate over which partner—the ex-wife or the ex-husband—is the victim of divorce. Both are victims. The first year after divorce is especially stressful for both ex-spouses. Divorce wields a blow to each one's self-esteem. Both feel they have failed as spouses and, if there are children, as parents. They question their ability to get along well in a remarriage. Yet each has particular difficulties that are related to societal expectations.

Her Divorce

Ex-wives often feel helpless and physically unattractive. Women who were married longer, particularly those oriented to traditional gender roles, may feel like "nobodies," having lost the identity associated with their husband's status. Getting back on their feet is particularly difficult for women in this group. Older women face real disappointments. They have few op-

portunities for meaningful career development and limited opportunities to remarry (Choi 1992).

Divorced mothers who retain sole custody of their children also experience difficulties. They often undergo severe overload as they attempt to provide not only for their financial self-support but also for the day-to-day care of their children (Campbell and Moen 1992). Their difficulties are aggravated by discrimination in hiring, promotion, and salaries, by the high cost of child care, and often by their less extensive work experience and training. All in all, custodial mothers frequently feel very much alone as they struggle with money, scheduling, and discipline problems, often passing up chances to meet people and, even if they want to, to marry again. Objective difficulties are reflected in decreased psychological well-being (Doherty, Su, and Needle 1989).

Researchers have painted a rather gloomy picture of women's life after divorce. Of course, many divorced women do go on to fulfilling single lives or happy remarried ones.

Meanwhile, the ex-husband may seem to be reaping all of the benefits with no unpleasant consequences. How accurate is this picture of the husband's divorce?

His Divorce

In fact, men take divorce much harder than many people—particularly their ex-wives—may think (Myers 1989). Ex-husbands' anger, grief, and loneliness may be aggravated by the traditional male gender role, which discourages them from sharing their pain with other men (Myers 1989). Many become more socially active to compensate.

But they go on missing their families and children. At least one therapist maintains that divorced, noncustodial fathers have more radical readjustments to make in their lifestyles than do custodial mothers. In return for the responsibilities and loss of freedom associated with single parenthood, custodial mothers escape much of the loneliness the loss of family status might otherwise offer and are somewhat rewarded by social approval for rearing their children (see Myers 1989).

Noncustodial fathers often retain the financial obligations of fatherhood while experiencing few of its joys. Whether it takes place in the children's home, the father's residence, or at some neutral spot, visitation is typically awkward and superficial. And the man may worry that if his ex-wife remarries, he will lose even more influence over his children's upbringing. For many individuals, parenthood plays an important role in adult development. "Removed from regular contact with their children after divorce, many men stagnate" (Wallerstein and Blakeslee 1989, p. 143).

Many ex-husbands feel shut out and lost because women may control access to emotional nurturance and may also have a stronger influence on children (Thomas and Forehand 1993). Yet for most women, men still hold the keys to economic security, and ex-wives suffer financially more than do ex-husbands. This interrelationship also applies to women's and men's separate experiences in divorce. The fact is that both men's and women's grievances could be alleviated by eliminating the economic discrimination toward women and the gender role expectations that create husbands' and wives' divorces. This leads us to some consideration of how divorce affects the families of the future.

Forming Families: The Next Generation

We have talked about the general impact of divorce on children, but what do we know specifically about how a parental divorce affects adult children's married and family lives? Sociologists have paid particular attention to the effects of parental divorce on two broad areas regarding adult children's family lives: (1) the quality of intergenerational relationships between adult children and their divorced parents, and (2) the marital stability for adult children of divorced parents.

Adult Children of Divorced Parents and Intergenerational Relationships

There is evidence that adult children of divorced parents have probably come to accept their parents' divorce as a desirable alternative to ongoing family conflict (Amato and Booth 1991). Nevertheless, several recent studies point to one conclusion: Ties between adult children and their parents are generally weaker (less close or supportive) when the parents are highly conflicted and "divorce prone," even if they are still together (Booth and Amato 1994b), or when the parents are divorced (Aquilino 1994a, 1994b; Cooney 1994; White 1994; Furstenberg, Hoffman, and Shrestha 1995; Lye et al. 1995; Marks 1995). The effect for divorced parents is stronger for fathers, who

"How was your lonely little dinner, sir?"

were usually the noncustodial parent, but the relationship has been found for mothers as well.

Sociologist Lynn White (1994) analyzed data from 3,625 National Survey of Families and Households respondents to examine the long-term consequences of childhood family divorce for adults' relationships with their parents. Using a broad array of indicators of family solidarity—relationship quality, contact frequency, and perceived and actual social support (doing favors, lending and giving money, feeling that one can call on the parent for help in an emergency—White found that adults raised by single parents reported lowered solidarity with them. They saw their parents less often, had poorer quality relationships, felt less able to count on parents for help and emotional support, and actually received less support. White found these negative effects to be stronger regarding, but not limited to, the noncustodial parent (usually the father).

In another study, sociologist William Aquilino (1994a) analyzed National Survey of Families and Households data from 3,281 young adults between ages 19 and 34 who grew up in intact families and had therefore lived with both biological parents from birth to age 18. Aquilino separated his sample into three categories: those whose parents were still married, those whose parents had divorced (about 20 percent of the sample), and those whose parent had been widowed. Aquilino found that parental divorce did not reduce the amount of help adult children gave their parents, but it did reduce the amount of help and money that the divorced parents gave their sons (though not their daughters). Furthermore, Aquilino found that even when parents divorce after the child is 18, the divorce seems negatively to affect the quality of their relationship. Children of divorced parents were in contact with their parents less often and reported lower relationship quality overall. These findings applied to both mothers and fathers, although the effect was much stronger for fathers. At the conclusion of still another analysis of different national survey data from adults, about 20 percent of whom had experienced the divorce of their parents, demographer Frank Furstenberg and his colleagues went so far as to pronounce that "marital disruption is altering the organization of kinship in American society. When men relinquish ties to their children during childhood, they rarely resume those ties in later life" (Furstenberg, Hoffman, and Shrestha 1995, p. 330).

Generally, evidence suggests that adult children of divorced parents feel less obligation to remain in contact with them and are less likely to receive help from them or to provide help to them. Social scientists (Lye et al. 1995) have posited four reasons for these findings:

1. Children raised in divorced, single-parent families may have received fewer resources from their custodial parents than did their friends in intact families, and thus they may feel less obliged to reciprocate.

2. Strain in single-parent families, deriving from the custodial parent's emotional stress and/or economic hardship, may weaken subsequent relations between adult children and their parents.

3. The reciprocal obligations of family members in different generations may be less clear in single-parent, postdivorce families—a situation that may also tend to weaken intergenerational relations in later life.

4. Adult children raised in divorced, single-parent families may still be angry, feeling that their parents failed to fulfill their obligation to provide a stable, two-parent household.

Marital Stability for Adult Children of Divorced Parents

Studies show that children of divorced parents do, in fact, have higher divorce rates than children from stable households. And research has documented other negative effects for children of divorced parents. On average, children of divorced parents acquire less education, marry sooner, and are more likely to cohabit and/or to have children before marriage (Keith and Finlay 1988; McLanahan and Bumpass 1988). All these factors are associated with the increased likelihood of divorce (Bumpass, Martin, and Sweet 1991; Thornton 1991).

But not all studies find parental divorce to be the long-term family handicap that we might expect. Although adult children whose parents divorced may be more negative about their families of origin, they seem little different from children of intact marriages in their goals and attitudes toward marriage and families—only perhaps a little more realistic (Amato and Booth 1991). They value marriage, believe that it should be forever, that partners should be monogamous, and that one's truly important relationships are in families. At the same time, they are "aware of [marriage's] limitations and tolerant toward its alternatives" (Amato 1988, pp. 453, 460). Moreover, although adult children of divorced parents express more accepting attitudes toward divorce, adult children from intact families that were characterized by conflict express this attitude also (Amato and Booth 1991).

Graham Spanier, a former president of the National Council on Family Relations, took as his theme the resilience of children and families in unpromising circumstances. "Many children who experience abuse, family disruption, or poverty nevertheless reach adulthood, against great odds, with a strong commitment to family life" (1989, p. 3). "Some widely held beliefs about the negative impact on children of divorce or parental absence may be overstated, if not wrong" (p. 10). Spanier also suggests that individuals can move on from difficult childhoods to create a satisfying family life:

Hope stems from the evidence that much childhood trauma has the prospect of being left behind; that some of the setbacks of childhood are short-lived; that we should be skeptical about any intellectual conviction about the superiority of the intact nuclear family; that growing insight into protective favors may help us help children to negotiate risk situations better; that individual and family resilience, or hardiness, exists and can be learned or enhanced; that . . . [responsible individuals in extended family networks] can forge continuity out of chaos; and that the intergenerational dynamics of the future may dictate stronger familial bonds. (Spanier 1989, p. 10)

It would be nice to conclude here. At the same time, family scholars and policymakers increasingly express concern that the rising number and proportion of single-parent families are causally related to rising childhood poverty rates—and this seems especially true for whites (Eggebeen and Lichter 1991). Both optimistic and pessimistic views of the impact of divorce on children are grounded in current research, and there seems no way of reconciling the diverse perspectives one encounters in the sociology of divorce.

We turn in Chapter 16 to a consideration of a next step for many divorced people: remarrying.

In Sum

Divorce rates have risen sharply in this century, and divorce rates in the United States are now the highest in the world. In the past decade, however, they have begun to level off.

Reasons why more people are divorcing than in the past have to do with changes in society: economic interdependence and legal, moral, and social constraints are lessening; expectations for intimacy are increasing; and expectations for permanence are declining. People's personal decisions to divorce, or to redivorce, involve weighing marital complaints—most often problems with communication or the emotional quality of the relationship—against the possible consequences of divorce. Two consequences that receive a great deal of consideration are how the divorce will affect children, if there are any, and whether it will cause serious financial difficulties.

The divorce experience is almost always far more painful than people expect. Bohannan has identified six ways in which divorce affects people. These six stations of divorce are the emotional divorce, the legal divorce, the community divorce, the psychic divorce, the economic divorce, and the coparental divorce. The psychic divorce involves a healing process that individuals must complete before they can fully enter new intimate relationships.

The economic divorce is typically more disastrous for women than for men, and this is especially so for custodial mothers. Over the past fifteen years, child support policies have undergone sweeping changes, which are only now beginning to result in evaluation research.

Researchers have proposed five possible theories to explain negative effects of divorce on children. These include the life stress perspective, the parental loss perspective, the parental adjustment perspective, the economic hardship perspective, and the interparental conflict perspective. The interparental conflict perspective, which asserts that conflict between parents before, during, and after divorce is responsible for children's lowered well-being, is most strongly supported by research.

Husbands' and wives' divorce experiences, like husbands' and wives' marriages, are different. Both the overload that characterizes the wife's divorce and the loneliness that often accompanies the husband's divorce, especially when there are children, can be lessened in the future by more androgynous settlements. Divorce counseling can help make the experience less painful. Joint custody offers the opportunity of greater involvement by both parents, although its impact is still being evaluated. So also is the effect of parents' divorce on children's marital prospects.

Key Terms

alimony/spousal support	interparental conflict
children's allowance	perspective
child snatching	joint custody
child support	legal divorce
Child Support Amendments	life stress perspective
(1984)	no-fault divorce
custody	parental adjustment
displaced homemakers	perspective
divorce counseling	parental loss perspective
divorce mediation	psychic divorce
economic divorce	readiness for divorce
economic hardship	redivorce
perspective	refined divorce rate
emotional divorce	rehabilitative alimony
entitlement	relatives of divorce
Family Support Act (1988)	structured separation
guaranteed child support	

Study Questions

1. What three factors bind marriages and families together? How have these factors changed, and how has the divorce rate been affected?

2. Discuss the relationship of economics to marital stability.

3. Two reasons for the high divorce rate in the United States are the myth of the ideal relationship and the rise of individualistic values. Can we promote these and at the same time support the present institution of marriage?

4. What are the consequences of divorce for children? When is divorce a better alternative for the children than maintaining the marriage? A worse alternative?

5. Describe the six stations of divorce, and speculate on how each station might be made less difficult.

6. Describe the five perspectives that explain the possible negative effects of divorce on children. Do any of these theories seem plausible? Why or why not?

7. How is "his" divorce different from "her" divorce? How are these differences related to societal expectations?

8. Differentiate between joint custody and shared living arrangements. How can these narrow the gap between his and her divorces and help lessen the pain of divorce for some couples with children?

9. Discuss family policy measures with regard to child support and women's and children's poverty. Is the support of America's children ultimately the responsibility of their parents, or of society in general? How do unemployment and gender/racial job discrimination affect questions of children's poverty?

10. Discuss the impact of parental divorce on adult children. What are some research findings in this regard? How might individuals make personal, knowledgeable choices that might result in improved consequences?

11. When, in your opinion, would divorce be the lesser of two evils? When would it not?

Suggested Readings

Ahrons, Constance. 1994. *The Good Divorce: Raising Your Family Together When Your Marriage Comes Apart.* New York: HarperCollins. Written for a general rather than professional audience, this book is based on the author's BiNuclear Family Study of family functions after divorce. Readable presentation of divorce research with a gender theme and hopeful outlook. *Each* ex-spouse (and their children) in ninety-six families was interviewed at the 1-year, 3- and 5-year points postdivorce. Useful perspective, much information and advice for those who find themselves in family relationships that cross the divorce barrier.

Arendell, Terry. 1986. *Mothers and Divorce: Legal, Economic and Social Dilemmas.* Berkeley: University of California Press. A broad look at the situation of divorcing and postdivorce women through the eyes of those interviewed by Arendell. Most found their present circumstances not what they had anticipated. The book's theme is the disadvantage, especially economic, of divorced women.

———. 1995. *Men and Divorce.* Thousand Oaks, CA: Sage. Qualitative research on men's experiences of divorce. Based on interviews with seventy-five men, with integration of expert sources in law, medicine, and mental health.

Buxton, Amity P. 1991. *The Other Side of the Closet: The Coming-Out Crisis for Straight Spouses.* Santa Monica, CA: IBS Press. An examination of divorce as a result of a partner's coming out as gay. Sympathetic to the straight partner.

Fineman, Martha Albertson. 1991. *The Illusion of Equality: The Rhetoric and Reality of Divorce Law Reform.* Chicago: University of Chicago Press. Challenges the imputed benefits of divorce law reform, contending that "no-fault" divorce and related changes have disadvantaged women economically and as parents.

Furstenberg, Frank F. and Andrew J. Cherlin. 1991. *Divided Families: What Happens to Children When Parents Part.* Cambridge, MA: Harvard University Press. Argues for public policy directed at helping custodial parents function better and shielding children from parental conflict after divorce.

Gold, Lois. 1992. *Between Love and Hate.* New York: Plenum. Written by a social worker, this book is mainly a guide for divorcing adults on "facing the crisis of separation constructively"; discusses healing, custody, conflict, negotiating, mediation, and other relationship issues of families involved in divorce or separation; gives practical advice, exercises, and rituals for such tasks as letting go and forgiving.

Haynes, John M. and Gretchen L. Haynes. 1989. *Mediating Divorce.* San Francisco: Jossey-Bass. A casebook of strategies for legal family negotiations in divorce mediation. Written primarily for professional mediators, this book explores the key assumptions of divorce mediation, along with various strategies and skills.

Johnson, Colleen Leahy. 1988. *Ex-Familia: Grandparents, Parents, and Children Adjust to Divorce.* New Brunswick, NJ: Rutgers University Press. Divorcing families handle their kin relations in varying ways, some of which work, others of which do not. This anthropological study of intergenerational relations expands our knowledge about divorce, pointing out that the extended family must make adjustments too.

Kelly, Joan B. and Judith S. Wallerstein. 1990. *Surviving the Breakup.* New York: Basic Books. This book includes some advice for coparenting after divorce, based on research and the best interest of the children.

Kitson, Gay C. 1992. *Portrait of Divorce: Adjustment to Marital Breakdown.* New York: Guilford. A scholarly and readable book resulting from Kitson and colleagues' longitudinal studies of divorcing families, conducted from the mid-1970s through the late 1980s.

Myers, Michael F. 1989. *Men and Divorce.* New York: Guilford. This book, written by a therapist, is based on the premises that men have a lot of difficulty coping with divorce and that many divorcing men fall between the cracks, receiving little or insufficient helpful attention. Among others, topics include the newly separated man, abandoned husbands, divorced fathers, gay divorcing men, and common themes (for example, intimacy, control, sexuality, grief, sexism, and loneliness) in treatment.

Office of Child Support Enforcement. 1989. *Handbook on Child Support Enforcement.* U.S. Department of Health and Human Services. This government pamphlet explains procedures in child support collection, such as establishing paternity, locating an absent parent, and establishing and enforcing a support order. Available free by writing: Handbook, Dept. 628M, Consumer Information Center, Pueblo, CO 81009.

Parkman, Allen M. 1992. *No-Fault Divorce: What Went Wrong?* Boulder, CO: Westview. Analyzes the aftermath of no-fault divorce laws in the United States, much in the tradition of Weitzman's *The Divorce Revolution* (1985).

Riley, Glenda. 1991. *Divorce: An American Tradition.* New York: Oxford University Press. An historical analysis of American values, attitudes, and behaviors regarding divorce since the establishment of the United States. Argues that traditional American values of independence, freedom, and individualism are much in keeping with high divorce rates.

Sugarman, Stephen D. and Herma Hill Kay (eds.). 1990. *Divorce Reform at the Crossroads.* New Haven: Yale University Press. A collection of serious readings on social policy regarding divorce and child support by leading authors in the field.

Textor, Martin (ed.). 1989. *The Divorce and Divorce Therapy Handbook.* Northvale, NJ: Jason Aronson Inc. A collection of readings on divorce therapy, including topics such as marital therapy and the divorcing family, mediating child custody disputes, school interventions with children of divorced parents, group therapy for divorcing or divorced adults, single-parent families, and stepfamilies.

Teyber, Edward. 1985. *Helping Your Children Through Divorce: A Compassionate Guide for Parents.* New York: Pocket. Focuses on children's concerns and offers fairly realistic guidelines for parents. (There are many good books written for children on this topic available in bookstores and libraries.)

Vaughan, Diane. 1986. *Uncoupling: Turning Points in Intimate Relationships.* New York: Oxford University Press. Uses sociological theory to analyze the process of separating from a partner; describes the process of coming apart based on interviews.

Wallerstein, Judith and Sandra Blakeslee. 1989. *Second Chances: Men, Women, and Children a Decade After Divorce.* New York: Ticknor and Fields. Latest report on a longitudinal study of children's adjustment to divorce, which has much to say about mothers and fathers, as well; includes guidelines about the tasks of divorcing parents and their children.

Ware, Chi. 1984. *Sharing Parenthood After Divorce.* New York: Bantam. Thoughtful book on how divorced couples can work together as parents.

Remarriages

Marrying again after being divorced or widowed is an alternative that many Americans are choosing. Today remarriages make up a significant proportion of all marriages. They are now common enough that congratulatory greeting cards to remarrying parents and others are available, a fairly recent development.

Until about twenty-five years ago, research and advice concerning remarriages was sparse, but more recently, social scientists, counselors, and journalists have given remarriage considerable attention. Some of their findings have shown that many remarried people consider themselves very happy. They are glad they chose to remarry and satisfied with their marital relationships and their lives. At the same time, remarriages are often beset with special challenges. More and more, counselors and agencies are helping remarried couples and families deal with these challenges.

This chapter explores remarriages today. We will discuss choosing a remarriage partner, noting some social factors that influence that choice. We'll examine happiness and stability in remarriage, as well as the fact that no cultural model exists to guide interactions among the often complex networks of kin. We'll focus on two particularly difficult problem areas in remarriages: stepchildren and finances. Finally, we will explore the writing of a remarriage agreement. We'll begin with an overview of remarriage in the United States today.

Remarriage: Some Basic Facts

Remarriages (marriages in which at least one partner had previously been divorced or widowed) are increasingly frequent in the United States today, compared to the early decades of the twentieth century. The remarriage rate rose sharply during World War II, peaking as the war ended. During the 1950s both the divorce and remarriage rates declined and remained relatively low until the 1960s, when they began to rise again. The remarriage rate peaked again in about 1966 but has declined since then (U.S. National Center for Health Statistics 1991a).

Although the divorce rate declined along with the remarriage rate in the late 1940s, the situation is not the same today. The divorce rate continued to climb after 1966 and has leveled off at a high point. But the remarriage rate has declined steadily since 1966. The decline is due in part to the economic uncertainties of recent years, which discourage people, particularly divorced men who may already be paying child support, from assuming financial responsibility for a new instant family.[1]

Nevertheless, approximately two-thirds of the people who divorce each year remarry eventually (U.S. National Center for Health Statistics 1991c, Table 4). According to the latest figures available, in 1988 remarriages comprised 46 percent of all marriages, compared with 31 percent in 1970 (U.S. Bureau of the Census 1995, Table 143). As Figure 16.1 shows, one-fifth of all marriages are between two formerly divorced people (Clarke 1995a, p. 4). Some divorce lawyers and counselors, however, report having begun to see "regulars"—people who divorce, remarry, redivorce, then remarry and perhaps redivorce yet again (Leerhsen 1985). But the vast majority of remarriages are second marriages.

The average divorced person who remarries does so within less than three years after divorce (U.S. National Center for Health Statistics 1991b, p. 5). Rapidity—and overall probability—of remarriage varies, however, for women of different ethnic origins. About 44 percent of white women remarry within three years of a divorce or spouse's death, compared with 20 per-

---

1. Some states, concerned about child support, passed legislation designed to prevent the remarriage of people whose child support was not paid up. But the Supreme Court ruled in *Zablocki v. Redhail* (1978) that marriage was so fundamental a right it could not be abridged in this way. In fact, the law had led to increased living together without marriage rather than increased compliance with child support orders.

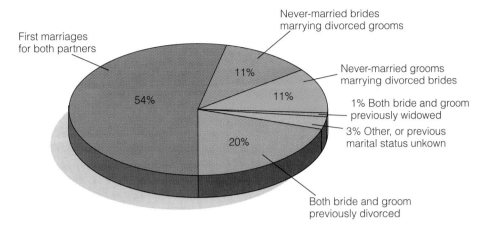

FIGURE 16.1

| U.S. marriages by previous marital status of bride and groom, 1990. (Source: Clarke 1995a, p. 4)

cent of African American women and 23 percent of Latinas. Analogous percentages for five years after dissolution are 53 percent for whites, 25 percent for blacks, and 30 percent for Latinas (U.S. Bureau of the Census 1995, Table 147).

Single-parent and remarried families may still be stereotyped as not as normal or functional as first-marriage, nuclear families (Ganong, Coleman, and Mapes 1990). But there is no question that the increased incidence of divorce and remarriage has led to somewhat

| Many marriages today are remarriages, a fact that greeting card companies have come to acknowledge.

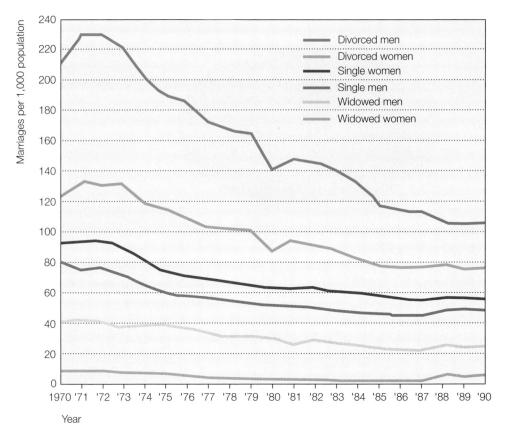

FIGURE **16.2**

Marriage rates of never-married, divorced, and widowed men and women, 1970–1990. Marriage rates presented here are number of marriages per 1,000 population in the specified marital status category. (Sources: U.S. National Center for Health Statistics 1991b, Table 5; Clarke 1995a, Table 6)

greater cultural tolerance of these families. Fewer and fewer Americans view divorce and remarriage as a moral issue. Remarriages have always been fairly common in the United States, but until the twentieth century almost all remarriages followed widowhood. As recently as the 1920s a majority of remarrying brides and grooms had been widowed; 91 percent of remarrying brides and grooms are divorced. Today, the largest proportion of the remarried population are divorced people who have married other divorced people. Remarriage is much more likely to occur, age for age, among divorced women than among widowed women. In fact, divorced men are more than twice as likely to marry again than never-married men. Figure

16.2 contrasts the marriage rates of divorced and widowed men and women with those of never-married people and illustrates change over time.

Remarriage and Children's Living Arrangements

One result of the significant number of remarriages today is that more Americans are living with other people's children—and more children are living with other than their biological parents. About 40 percent of remarried households contain one or more stepchildren under age 18 (Glick 1989). Sixteen percent of children in married-couple households with children are stepchildren, most living with their biological

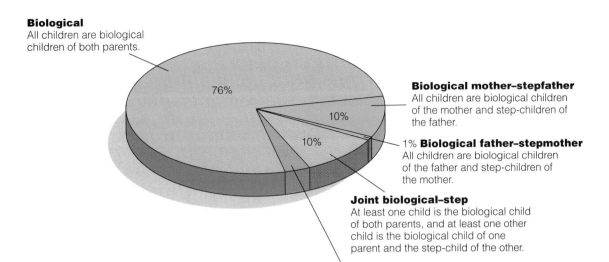

Biological
All children are biological children of both parents.

76%

Biological mother–stepfather
All children are biological children of the mother and step-children of the father.

10%

10%

1% **Biological father–stepmother**
All children are biological children of the father and step-children of the mother.

Joint biological–step
At least one child is the biological child of both parents, and at least one other child is the biological child of one parent and the step-child of the other.

3% Other, including adoptive

FIGURE **16.3**

| Married–couple households with children, 1990. (Source: Adapted from U.S. Bureau of the Census 1994a, Table 78)

mother and stepfather (Moorman and Hernandez 1989). As you can see from Figure 16.3, in about three-quarters (76 percent) of married-couple households with children, all the children are the biological offspring of both parents. In another 10 percent of these households, the children are the biological offspring of the mother and stepchildren of the father. In another 10 percent of these households, at least one child is a biological child of one parent and a stepchild of the other parent, while at least one other child (assumedly born after the parents' remarriage) is the biological child of both parents (U.S. Bureau of the Census 1994a, Table 78).

Figure 16.3 points up the fact that stepfamilies are not all alike: A stepparent can be mother or father (although usually it is the father), and a stepfamily may or may not contain biological children of both remarried parents. Moreover, in some stepfamilies both parents are stepparents to their spouse's biological children. (This situation is not directly illustrated in Figure 16.3.)

Figure 16.3 looks at how married-couple households are distributed. Another, related way to think about how remarriage has affected the family experiences of children is presented in Table 16.1. In 1993,

71 percent of all American children under 18 (77 percent of whites, 36 percent of African Americans, and 65 percent of Hispanics) were living with two parents (U.S. Bureau of the Census 1994a, Table 80). Table 16.1 shows how the living arrangements of those children who are residing with two parents are distributed, by race and Hispanic origin. As you can see from the table, the proportion of children living with both biological parents has declined some since 1980 while the percentage of children living with a stepfather has increased. These related trends are not surprising, given current high divorce rates followed by remarriages. Among African American children who live in two-parent families, nearly 64 percent reside with both biological parents and another 31 percent live with their biological mother and a stepfather. These figures compare with 81.5 percent and 14.6 percent, respectively, for all races taken together. The higher proportion of African American children living with a stepfather reflects the higher divorce rate among African Americans as well as the lower life expectancy for African American men.

The remainder of this chapter explores what family life is like for children—and their parents—who are

TABLE *16.1*

Percentage of Children Under Age 18 Living with Biological and Step Married-Couple Parents by Race and Hispanic Origin of Mother, 1980 and 1990

	All races[1]		White		African American		Hispanic origin[2, 3]
	1980	1990	1980	1990	1980	1990	1990
Biological mother and father	83.7	81.5	84.7	83.0	71.5	63.6	81.1
Biological mother and stepfather	11.3	14.6	10.3	13.2	23.2	31.3	15.3
Biological father and stepmother	1.5	1.3	1.6	1.4	1.2	1.0	0.8
Other, including adoptive and unknown	3.5	2.6	3.4	2.4	4.1	4.1	2.8

[1]Includes other races, such as Asians and Native Americans, not shown separately here.

[2]Persons of Hispanic origin may be of any race.

[3]Data for Hispanics is not available for 1980.

Source: U.S. Bureau of the Census 1995, Table 77.

living in stepfamilies. We begin with a question about courtship. Do people choose partners differently the second (or third) time around?

Choosing Partners the Next Time: Variations on a Theme

Courtship before remarriage may differ in many respects from courtship before first marriage. It may proceed much more rapidly, with the people involved viewing themselves as mature adults who know what they are looking for—or it may be more cautious, with the partners needing time to recover from, or being wary of repeating, their previous marital experience. It is likely to have an earlier, more open sexual component—which may be hidden from the children through a series of complex arrangements. It may include both outings with the children and evenings at home as partners seek to recapture their accustomed domesticity.

Although courtship for remarriage has not been a major topic for research, Canadian sociologists Roy Rodgers and Linda Conrad have begun to remedy that. After reviewing the literature on divorce, stepfamilies, and family structure, they have posed several hypotheses for future testing. Generally, the hypotheses point up the complicated interrelationships between courting parents, their respective children, and their ex-spouses. For example, if one ex-spouse begins courtship while the other does not, conflict between the former spouses may escalate. Moreover, the non-courting partner may try to interfere with the ex-spouse's new relationship. A custodial ex-wife might do this by sabotaging her ex-husband's time with the children; ex-husbands might do this by threatening physical violence or by withholding—or legally changing—financial support. Not only ex-spouses but children, too, may react negatively to a parent's dating. Generally, "the more the custodial parent's new partner displaces the child as a source of emotional support for the parent, the greater the probability of a negative reaction of the child to the new partner" and the more problems are likely to arise (Rodgers and Conrad 1986, p. 771).

All of this suggests that courtship toward remarriage differs from that for first marriages in important ways. Nevertheless, the basic structure of the remarriage process has much in common with initial marriage. We'll look at two significant factors that we first

TABLE 16.2

Women's Median Family Income in Various Marital Statuses

	Women in First Marriages	Women Who Later Divorced or Separated	Remarried Women
Whites			
Separated/divorced in the late 1960s–mid-1970s	$25,381 (N = 430)	$14,720 (N = 284)	$27,498 (N = 141)
Separated/divorced in the 1980s	$24,020 (N = 416)	$13,712 (N = 258)	$24,801 (N = 158)
African Americans			
Separated/divorced in the late 1960s–through the mid-1970s	$18,086 (N = 226)	$9,167 (N = 195)	$17,121 (N = 30)
Separated/divorced in the 1980s	$16,988 (N = 133)	$8,971 (N = 110)	$28,722 (N = 23)

NOTE: N indicates the number of women in each category.

Source: Adapted from Smock 1993, p. 359.

examined in Chapter 7 with regard to first marriages: the traditional exchange and homogamy.

The Traditional Exchange in Remarriage

Jessie Bernard (1982) pointed out that as a whole, married life tends to place greater stresses on women who become mothers and, at the same time, reduce stress for men. This would suggest that ex-husbands more than ex-wives may want to remarry.

On the other hand, economists and other social scientists point out that ex-wives have much to gain financially by being remarried. In fact, most women's standard of living does not improve after divorce unless they remarry. One extensive longitudinal study (Chapter 2 explains longitudinal studies) by Pamela Smock (1993) used three large national data sets to monitor the financial well-being of women as they married, divorced, and then remarried over the years. Table 16.2 presents findings from this research. As you can see from the table, the situations of women who separated or divorced during the 1980s is essentially the same as those who did so in the late 1960s and 1970s. In constant 1987 dollars, the median annual

income of married white women before they divorced was approximately $24,000–$25,000. With divorce or separation, annual income dropped by more than 40 percent—to approximately $14,000 ($14,720 for those divorcing in the 1960s and 1970s, and $13,712 for those doing so in the 1980s). With remarriage, the women's annual family income returned to what it had been prior to their divorce. This same pattern was found for African American women, with the exception that overall their annual family earnings were lower than whites. Note, however, that black women who remarried in the 1980s had annual family incomes higher than their white counterparts ($28,722 compared with $24,801). This situation may reflect the somewhat higher proportion of African American wives in the labor force. Chapter 15 addresses the economic plight of divorced women in more detail. Recalling that discussion and integrating it with Smock's findings, we can conclude, as have other social scientists, that, "Given current policies, the economic well-being of mothers and children is strikingly dependent on marriage and remarriage" (Duncan and Rodgers 1987, p. 178).

Meanwhile, women typically have more to gain

Cartoon by Jan Eliot from Roz Warren, [Ed.] *Men Are from Detroit, Women Are from Paris*, p. 98. Reprinted by permission of Universal Press Syndicate.

financially from being remarried than men do. The following account by a single male in his late 20s illustrates this fact:

> She was eyeing me and eyeing my house as a nice place to live with her son. It was the first thing she did. We went to my house one night and she says, "Boy, you've really got a nice backyard and a nice house here." And I'm thinking, "Why is she saying that?" It didn't click at the time, but she told this buddy of mine that her plan was to move in here with her kid—whatever his name was. Jason, that was the name, Jason. I could not stand that kid at all. So when I found this out, that I was her meal ticket, I thought, "Hea-a-a-vy." And that was the last time I saw her. She had great plans for me, but then I thought "Nah." (personal interview)

From the discussion in Chapter 7 of the marriage market, the reader will remember that in recent years the exchange has become more lopsided than it was in the past. Both homemaking services and sex are more available to single men, whereas women, especially older women and those who have been out of the labor force for a number of years, often still find it difficult to get good-paying jobs and support themselves. The scales are tipped even more in men's favor in the remarriage market, in which women's remarriage rate

is less than half of men's. In part, that is a consequence of the very low remarriage rate of widows, for whom few partners are available in later life. But even comparing only divorced people, men's remarriage rates are substantially higher, particularly after age 30 (see Figure 16.2). Two factors influence the exchange to work against women's remarriage.

CHILDREN AND REMARRIAGE   One factor that works against women in the remarriage market is the presence of dependent children. As we saw in Chapter 15, the woman usually retains custody of children from a previous marriage. As a result, a prospective second husband may look on her family as a financial—and also an emotional and psychological—liability. Furthermore, according to one qualitative study of thirty single-parent white women employed outside the home, the independent and assertive attitudes and behaviors necessary for single parenthood are contradictory to those they assumed they should exhibit in order to remarry (Quinn and Allen 1989). Then too, children are often strongly loyal to the first family and may oppose or have strong reservations about the divorce for as long as five years afterward. In general, the more children a woman has, the less likely she is to remarry (Glick and Lin 1986b).

AGE AND REMARRIAGE   A second factor, age, works against women in several ways. Figure 16.4 shows the remarriage rates for divorced women in six age categories. Although the pattern of remarriage since 1960 has been similar for all age categories, the remarriage rate for younger women has been consistently higher. As discussed in Chapters 6 and 7, women live longer, on the average, than men do: by age 65 there are only 85 men for every 100 women (see Table 7.1). The *double standard of aging* works against women in the remarriage market. In our society, women are considered to be less physically attractive with age, and they may also be less interested in fulfilling traditional gender role expectations.

Age works against both women's and men's remarriage in another way. Older people often face considerable social opposition to remarriage—both from restrictive pension and Social Security regulations and from their friends and children. Grown children may feel that remarriage is inappropriate, or may be concerned about the biological parent's continued interest in them or, ultimately, the estate.

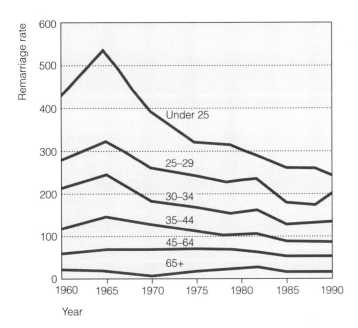

FIGURE *16.4*

Remarriage rates per 1,000 divorced and widowed women over age 15, by age at remarriage, 1960–1990. (Source: Clarke 1995a, Table 6)

Homogamy in Remarriage

Homogamy has traditionally been a second important factor influencing marriage choices. We saw in Chapter 7 that homogamy can be important in both the choice of a first marriage partner and the degree to which couples live happily ever after. Does it affect second marriages in the same way? The answer to that question is, on the whole, no. Older people, particularly those who are widowed, are likely to remarry homogamously. Sometimes the new partner is someone who reminds them of their first spouse or is someone they've known for years. This rule does not apply to middle-aged or younger people who choose second marriage partners.

Choosing a remarriage partner differs from making a marital choice the first time inasmuch as there is a smaller pool of eligibles with a wider range of any given attribute. As prospective mates move from their late 20s into their 30s, they affiliate in occupational circles and interest groups that assemble people from more diverse backgrounds. There is also some evidence that divorced people tend more toward heterogamy than nondivorced people the first time

around and that they simply accentuate this tendency when they remarry. As a result, remarriages are less homogamous than first marriages with partners varying in age, educational background (U.S. National Center for Health Statistics 1990b), and religious background. (The age difference between spouses is about four years in remarriages of divorced people and five years for widowed people, compared to about two years for first marriages [U.S. National Center for Health Statistics 1990b].) Some observers note that homogamy increases the likelihood for marital stability and point to this preference for and increase in heterogamy as a partial explanation for the fact that the divorce rate is higher for second than for first marriages. The next section discusses stability and happiness in remarriage.

Happiness and Stability in Remarriage

As pointed out elsewhere in this text, marital happiness and stability are not the same. *Marital happiness* refers to the quality of the marital relationship

Drawing by Leo Cullum; © 1989 The New Yorker Magazine, Inc.

"Well, the children are grown up, married, divorced, and remarried. I guess our job is done."

whether or not it is permanent; *stability* refers simply to the duration of the union. We'll look at both ways of evaluating remarriage.

Happiness in Remarriage

Researchers consistently find little difference in marital happiness between first and later marriages (White and Booth 1985b; Vemer et al. 1989; Glenn 1990, 1991). In fact, not only is there little difference generally in marital satisfaction, but there also appears to be little difference in marital quality for remarriages of men compared to those of women; stepfathers compared to stepmothers; double remarriages (both parties were previously married) compared to single ones; or simple compared to complex stepfamily households (with both parties bringing children to the marriage). The effect of stepchildren on marital satisfaction is minimal, with one national sample study finding a slight negative effect (in unions where both spouses are remarried) (White and Booth 1985b), whereas an analysis of a number of studies found a minimal positive effect (Vemer et al. 1989). Despite the challenges peculiar to remarriages (discussed later in this chapter), remarriages tend to be as happy as first marriages (Coleman and Ganong 1990, pp. 931–32).

Some writers have suggested that older, more experienced mates in remarriages are likely to be more mature than when they began their first marriages. (In 1990 the mean age at remarriage was 35.3 for brides and 38.7 for grooms, compared with 24.6 and 26.5 for first marriages [Clarke 1995a, Table 10].) Counselors often add that people who ended troubled first marriages through divorce often are still experiencing personal conflicts, which they must resolve before they can expect to succeed in a second marriage. Consequently, many counselors advise waiting at least two years after divorce before entering into another serious relationship.

NEGATIVE STEREOTYPES AND REMARRIEDS' HAPPINESS   To use the family ecology theoretical perspective, one way that remarital satisfaction is influenced by the wider society is through the negative stereotyping of remarriages (Ganong, Coleman, and Mapes 1990), with such potentially harmful myths as, "A stepfamily can never be as good as a family in which children live with both natural parents" (Kurdek and Fine 1991, p. 567). In one study of thirty-one white, middle-class spouses in stepfather families, researchers found, especially among the wives, that believing in none or very few of these negative myths and having high optimism about the remarriage were related to high family, marital, and personal satisfaction (Kurdek and Fine 1991). One study (Kurdek 1989a) that compared relationship quality between re-

As Americans live longer, many of them are finding that love and intimacy can occur at any age. People tend to choose partners differently the second time around and, although remarriage rates are relatively for those over 65, some do find happiness in a late remarriage.

marrieds and first marrieds found that high satisfaction with social support from friends and families of origin, together with high expressiveness in partners, were more important to marital satisfaction than whether the union was a first marriage or a remarriage.

The Stability of Remarriages

Even though remarriages are about as likely to be happy as intact first marriages, they differ somewhat in the likelihood of their stability, with remarriages about 20 percent more likely to end in divorce than first marriages (Martin and Bumpass 1989). One projection predicts eventual convergence of divorce rates for first and subsequent marriages (Norton and Moorman 1987). In fact, some scholars consider the current divergence between divorce rates for first and subsequent marriages to be relatively minimal (Furstenberg 1987). By 1988, husbands and wives who had jointly remarried at ages 25–44 in 1972 had lower divorce rates than did those who were first married as teenagers (Clarke and Wilson 1994). "This indicates that marrying at a young age was a stronger determinant of divorce than was a previous marriage of either or both spouses" (Wilson and Clarke 1992, p. 123).

Nevertheless, remarriages *are* somewhat less likely to be stable. Furthermore, people tend to get a subsequent divorce in less time than it took them to obtain their initial divorce—about six years for a second and four years for a third divorce versus eight years for a first divorce.

There are several reasons for the lower overall stability of remarriages. First, people who divorce in the first place are disproportionately from lower middle- and lower-class groups, which generally have higher divorce rates.

Second, people who remarry after divorce are, as a group, more accepting of divorce (White and Booth 1985b; Booth and Edwards 1992) and may have already demonstrated that they are willing to choose divorce as a way of resolving an unsatisfactory marriage.

Third, remarrieds receive less social support from their families of origin and are generally less integrated with parents and in-laws (Booth and Edwards 1992), a situation that can act as a barrier to divorce.

Fourth, remarriages present some special stresses on a couple, stresses that are not inherent in first marriages. Our culture has not yet fully evolved norms or traditions that provide remarried partners and their families with models for appropriate behavior. As we gain experience with the special challenges of remarriage, which are discussed throughout this chapter, remarriages may become more stable.

Perhaps the most significant factor in the comparative instability of remarriages is the presence of stepchildren. Sociologists Lynn White and Alan Booth explored the puzzling question of why remarried couples are as happy as couples in their first marriages, yet have higher divorce rates. They interviewed a national sample of over 2,000 married people (under age 55) in 1980 and reinterviewed four-fifths of them in 1983. During that interval, **double remarriages,** in which both partners had been married before, were twice as likely to have broken up as those of people in their first marriage. **Single remarriages,** in which only one partner had been previously married, did not differ significantly from first marriages in their likelihood of divorce.

Double remarriage increased the probability of divorce by 50 percent over what it would otherwise have been, and having stepchildren in the home increased it an additional 50 percent.[2] Although the presence of stepchildren did not make so much difference in single remarriages, double remarriages involving stepchildren have a very elevated risk of dissolution. Of course, these are the most complex family types in a society that has not yet developed a normative structure for stepfamilies.

Nevertheless, White and Booth found that marital quality does not vary greatly between first marriages and single or double remarriages. It is satisfaction with family life generally that is affected. Individuals in families with stepchildren are more likely to report that they have problems with the children, that they are unhappy with the way the spouse gets along with their children, that they would enjoy living apart from their children, that marriage has a negative effect on their relationships with their children, and that, if they had it to do over again, they would not remarry. "Since they report only modestly lower marital happiness, we interpret this evidence that the stepfamily, rather than the marriage, is stressful. . . . These data suggest that . . . if it were not for the children these marriages would be stable. The partners manage to be relatively happy despite the presence of stepchildren, but they nevertheless are more apt to divorce because

of child-related problems" (White and Booth 1985b, p. 696).

Remarriage and Well-Being

We often compare first and subsequent marriages, but another comparison involves remarriage and continued singlehood as lifestyles. A longitudinal study of 180 men and women divorced in Pennsylvania and followed for two and a half years found that those who had not remarried were as well off on eight measures of well-being as were those who had remarried. "Many researchers and clinicians, as well as those who experience divorce, assume that remarriage following divorce may be a significant—if not the most significant—alteration in social circumstance influencing enhanced well-being" and is the culturally preferred outcome (Spanier and Furstenberg 1982, p. 709). In one study (Weingarten 1985) remarrieds saw themselves as happier than the divorced; another study (Mitchell 1983) found remarrieds' overall well-being to exceed that of the divorced. When studies are reviewed as a group (Coleman and Ganong 1990), there seems to be virtually no difference in well-being in terms of satisfaction with one's life and health, suicide propensity, self-esteem, emotional balance, psychosomatic symptoms, and changes in habits between remarried people and those who continued to be single, whether living with another adult or not. In Spanier and Furstenberg's research, there was also no difference between men and women in the various categories, nor did presence or absence of children affect well-being.

The researchers found that people were not very accurate in predicting at the time of the divorce whether or not they would remarry. "Only 22 percent of those who had been eager to remarry had done so, compared with 30 percent who had said they were reluctant and 19 percent of those who said they would probably never remarry" (Spanier and Furstenberg 1982, p. 714). Nor was remarriage related to age, income, occupational status, education, religion, religiosity, or gender within the group, or to presence or absence of children. Partners who had initiated the divorce were more likely to have remarried, a situation that the researchers attributed to greater readiness—because they had begun adjusting to the idea of divorce earlier—rather than to greater psychological health or development of a potential marriage relationship before the divorce. Among remarried people,

---

2. In presenting their findings, White and Booth take into account the fact that divorce occurs earlier in remarriages.

TABLE *16.3*

Major Structural Characteristics of Three American Family Patterns

Stepfamilies	Nuclear (Intact) Families	Single-Parent Families
Biological parent is elsewhere.	Both biological parents are present.	Biological parent is elsewhere.
Virtually all members have recently sustained a primary relationship loss.		All members have recently sustained a primary relationship loss.
An adult couple is in the household.	An adult couple is in the household.	
Relationship between one adult (parent) and child predates the marriage.	Spousal relationship predates parental ones.	Parental relationship is the primary family relationship.
Children are members in more than one household.	Children are members in only one household.	Children may be members in more than one household.
One adult (stepparent) is not legally related to a child (stepchild).	Parents and child(ren) are legally related.	Parent and child(ren) are legally related.

Source: Visher and Visher 1979.

Spanier and Furstenberg found that the quality of the remarriage was related to well-being, and they suggested that variations among the remarried may account for the lack of difference between remarried and single people (Spanier and Furstenberg 1987).

It appears from this research that there are many routes to postmarital happiness. It is encouraging to learn that there are many ways of making a life after divorce and that those who do not choose or have no opportunity to remarry can build a satisfying life. Given the sex ratio in the older years, many older women in particular will not find partners available. Individuals, both men and women, may come to prefer a single lifestyle.

Remarried Families: A Normless Norm

When neither spouse in a remarriage has children, the couples' union is usually very much like a first marriage. This is probably also true for the estimated 18 percent of remarried parents who have all their children after the remarriage (see Glick 1989,

p. 25). But when at least one spouse has children from a previous marriage, family life often differs sharply from that of first marriages. A primary reason for this is the fact that our society offers members of **remarried families** no *cultural script,* or set of socially prescribed and understood guidelines for relating to each other or for defining responsibilities and obligations.

Although society tends to broadly apply to second marriages the rules and assumptions of first marriages, these rules often ignore the complexities of remarried families and leave many questions unanswered. Three areas in which these shortcomings are most apparent are within the remarried families, in relationships with kin, and in family law.

Characteristics of Remarried Families

The complexity of stepfamily structure, which affects family relationships, is illustrated in Table 16.3. It compares stepfamilies with the original nuclear family and with the single-parent family, which generally is a stage between marriages and remarriages. The repeated transitions from one family structure to another create a prolonged period of upheaval and stress. Yet sufficient adjustment to a single-parent household is likely to have occurred, so that a new adaptation to a two-parent remarried family is all the more difficult. A very tight emotional bond often exists in single-parent families.

The unique characteristics of stepfamilies were pointed out more than a decade ago by Emily Visher and John Visher (1979):

1. There is a biological parent outside the stepfamily unit and an adult of the same sex as the absent parent in the household.

2. Most children in stepfamilies hold membership in two households, with two sets of rules.

3. The role models for stepparents are poorly defined.

4. The fact that remarried families come together from diverse historical backgrounds accentuates the need for tolerance of differences. For example, a remarried mother's biological children may be used to saying grace before meals while her stepchild(ren) are not.

5. Step-relationships are new and untested and not a given as they are in intact families. Even when the new groups are in tune with each other, there is never the comfort of *knowing* that there is a bond of caring and love. Outward signals and signs are continuously needed in many stepfamilies to show that caring and loving really exist.

6. The children in stepfamilies have at least one extra set of grandparents.

Despite these common characteristics, remarried families are of various types. The simplest is one in which a divorced or widowed spouse with one child remarries a never-married childless spouse. In the most complex, both remarrying partners bring children in from previous unions and also have a mutual child or children together. If a remarriage is followed by redivorce and a subsequent remarriage, the new remarried family structure is even more complex. Moreover,

ex-spouses remarry, too, to persons who have spouses by previous marriages, and who also have mutual children of their own. This produces an extraordinarily complicated network of family relationships in which adults have the roles of parent, stepparent, spouse, and ex-spouse; some adults have the role of custodial parent and others have the role of noncustodial, absent parent. The children all have roles as sons or daughters, siblings, residential stepsiblings, nonresidential stepsiblings, residential half-siblings, and nonresidential half-siblings. There are two subtypes of half-sibling roles: those of children related by blood to only one of the adults, and the half-sibling role of the mutual child. Children also have stepgrandparents and ex-stepgrandparents as well as grandparents. (Beer 1989, p. 8)

Kin Networks in the Remarried Family

Relationships with kin outside the immediate remarried family are complex and uncharted as well. Two decades ago, one American anthropologist observed this of the American scene:

Many people are married to people who have been married to other people who are now married to still others to whom the first parties may not have been married, but to whom somebody has likely been married. (Tiger 1978, p. 14)

The observation describes a large segment of American families today, and the complexity it portrays is striking in several respects. One is the fact that our language has not yet caught up with the proliferation of new family roles. As family members separate and then join new families formed by remarriage, the new kin do not so much *replace* as *add* to kin from the first marriage (White and Riedmann 1992). What are the new relatives to be called? There may be stepparents, stepgrandparents, and stepsiblings, but what, for instance, does a child call the new wife that her or his noncustodial father has married? Or if a child alternates between the new households or remarried parents in a joint custody arrangement, what does he or she call "home," and where is his or her "family"? According to sociologist William Beer, "I began my in-depth interviewing of one stepfamily with a question about who the family members were. The response was a confused look and the question, 'Well, when do you mean?' " (1989, p. 9).

Anthropologist Paul Bohannan suggests a new term to define another previously unnamed relationship, the person one's former spouse remarries. He

This family portrait is of a mother and stepfather of two full sisters, along with a baby son from the new union. The remarried family structure, which is complex and has many unique characteristics, has no accepted cultural script. When all members are able to work throughtfully together, adjustment to a new family life can be easier.

calls this person one's **quasi-kin** (Bohannan 1970a). But he and other observers point to another way in which our culture has not yet caught up with the needs of the new remarried family: We have few mutually accepted ways of dealing with these new relationships. Interaction takes place between these families, often on a regular basis (about 60 percent of divorced parents' children have ongoing relationships with their noncustodial parent), yet there are no set ways of dealing with quasi-kin.

It does appear that in many cases there is little meaningful interaction between former spouses. "The popular impression that many divorced couples are unable to disengage is not substantiated. Indeed, we found just the opposite to be true. The overwhelming majority completely severed their ties and in doing so abandoned the prospect of collaborative child care" (Spanier and Furstenberg 1987, p. 57; see also Lewin 1990a).

The Binuclear Family Research Project collected data from ninety-eight pairs of Wisconsin families at one, three, and five years after divorce. At three years, only 9 percent of divorced spouses continued to engage in **coparenting**—shared decision making and

parental supervision in such areas as discipline and schoolwork or shared holidays and recreation. But conflict and mutual support had not declined. Satisfaction declined for fathers, but not for mothers. Tentative data indicate that the coparental relationship deteriorated if one parent remarried, especially the man.

Biological parents and their new partners had high rates of coparenting, with 62 percent of stepmothers and 73 percent of stepfathers reporting joint involvement in seven of ten areas of child rearing. Researchers Ahrons and Wallisch (1987) pose, but do not answer, the question of whether stepparents are "substitute" or "alternative" parents, but other research suggests that it is often the former (Seltzer and Bianchi 1988; Wallerstein and Blakeslee 1989). Cultural norms, on the other hand, do not clearly indicate which role the stepparent should play.

Because of the cultural ambiguity of remarried family relationships, social scientist Andrew Cherlin calls the remarried family an "incomplete institution" (1978). A second symptom of its incompleteness is the lack of the legal definitions for roles and relationships in this institution.

Family Law and the Remarried Family

Because family law assumes that marriages are first marriages, there are no legal provisions for several remarried family problems: balancing husbands' financial obligations to their spouses and children from current and previous marriages; defining a wife's obligations to husbands and children from the new and the old marriages; or reconciling the competing claims of current and former spouses for shares of the estate of a deceased spouse (Weitzman 1981; Beer 1989, pp. 10–11). Research suggests, incidentally, that remarried people may be reluctant to commit all of their economic resources to a second marriage and take care to protect their individual interests and those of their biological children (Fishman 1983).

Legal regulations concerning incest are also inadequate for stepfamilies. In all states, marriage and sexual relations are prohibited between people closely related by blood. Many states have found that these restrictions do not cover sexual relations or marriage between family members not related by blood—between stepsiblings or between a child and a stepparent, for example—and have moved to modify their laws to protect children in remarried families from abuse. Margaret Mead pointed out that incest taboos serve the important function of allowing children to develop affection for and identification with other family members without risking sexual exploitation.

In some states stepparents do not have the authority to see the school records of stepchildren or make medical decisions for them. The preservation of stepparent–stepchild relations when death or divorce severs the marital tie is also an issue. Visitation rights (and corresponding support obligation) of stepparents is only just beginning to be legally clarified. When a biological parent dies, the absence of custodial preference for stepparents over extended kin or even foster placement may result in children being removed from a home in which they had close psychological ties to a stepparent. Such ties may have taken a long time to develop.

Stepparenting: A Challenge in Remarriage

For most first-married couples, the biggest problems are immaturity, sexual difficulties, and personal lack of readiness for marriage. For a remarried spouse, in contrast, stepchildren and finances present the greatest challenges.

Many studies have found children in stepfamilies to be just as happy and well adjusted as children in intact families, or nearly so (see, for example, Furstenberg 1987; Pasley and Ihinger-Tallman 1988). One important study on a national sample of black and white couples concluded that stepchildren are not necessarily associated with more frequent marital conflict. Moreover, the negative impact of stepchildren declines with the length of the remarriage (MacDonald and DeMaris 1995). Case Study 16.1 "My (Remarried) Family," illustrates these points. But even though stepchildren may not cause undue family conflict and typically turn out to be well adjusted, being a stepparent (or married to one) often poses problems for the marriage.

Nevertheless, remarried couples often go on to have children together. About half of women who remarry have children, usually within two years after the wedding. This is more likely to occur if the woman has no children or only one from the first marriage, and it is also more common among white couples.

Because remarried women with two or more children are not so likely to have another child, researchers infer that the assumption that partners have children to validate their relationship is not true. Rather, women seem to be claiming adult status in becoming mothers, or simply completing a family of desired size (Wineberg 1990).

Demographer Howard Wineberg expresses concern about the impact on remarriage of having a child so early in what is bound to be a complex adjustment. Yet, the children a couple have from their new marriage seem to be associated with both increased happiness and stability (White and Booth 1985b). According to Cherlin:

> This is what we would expect, since children from a previous marriage expand the family across households and complicate the structure of family roles and relationships. But children born into the new marriage bring none of these complications. Consequently, only children from a previous marriage should add to the special problems of families in remarriage. (1978, p. 645)

My (Remarried) Family

The following essay was written for a marriage and family course by a young college student named David.

I'm writing this paper about my family. My family is made up of my family that I reside with and then my dad, stepmother, and half-sister that I visit. My family I reside with is who I consider my real family. We are made up of five girls and three boys, a cat and a dog, my mom and stepfather. The oldest is Harry, then down the line goes Diane, Barbara Ann, Kathy, Debi, Mel, Sharon, and myself. My sister Debi and I are the only kids from my mom's original marriage, so as you can see my mom was taking a big step facing six new kids. (My mom has guts!)

I still consider my dad "family," but I don't come into contact with him that much now. I would like to concentrate on my new "stepfamily," but I really don't like that word for it. My family is my family. My brothers and sisters are *all* my brothers and sisters whether they are step or original. My stepfather, although I don't call him "dad," is my father. My grandparents, step or original, are my grandparents. I can honestly say I love them all the same.

It all began when I was four. I don't remember much about my parents' divorce. The one real memory I have is sleeping with my dad downstairs while my sister slept upstairs with my mom....

My mom met my stepfather, Harry, through mutual friends who went to our church. I was in first grade, and I don't really remember much about their dating. All I knew was either these strange kids came over to my house or I went over to theirs. They were married in March 1976. I was five. At the reception I got to see all my relatives, old and new. When my parents left for their honeymoon, I began to cry. My aunt did a good job of consoling me.

After my parents got married, we moved. We didn't move into my stepfather's and his kids' house but to a new house altogether for everyone. I still live there. I love it there and probably will live in that area all my life.

I remember when we first went to look at the house. We live on a golf course. Mel, my brother (step), and sister (step), and I went out on this big sand trap to get a look around. The next

(continued)

Some Reasons Stepparenting Is Difficult

One study investigating children's contact with absent parents found such contact to be significantly less likely when the biological parent had been replaced by a stepparent or adoptive parent. "The interpretation [of the data] implies that noncustodial parents discard ties to their biological children with divorce, and ties are replaced through remarriage" as men become stepparents in another family. "Children tend to have only two or fewer parents at a time" (Seltzer and Bianchi 1988, p. 674).

Overall, research has found little difference in self-esteem, problem behaviors, or intelligence between children in stepfamilies and those living with both biological parents (Coleman and Ganong 1990, p. 928). But there are special difficulties associated with bringing families together under the same roof. Stepsiblings may not get along because they resent sharing their room or possessions, not to mention their parent (Beer 1989, 1992). In Lilian Messinger's early study, couples reported a diverse array of problems. For instance, ties with the noncustodial parent created a triangle effect that made the spouse's previous marriage

(continued)

thing I remember is getting pushed off and swallowing sand and not being able to breathe. I don't know if this was just an older brother thing or a new stepbrother thing. In any event, Mel came over and helped me get my breath back and all was fine. That was just the first of our many childhood altercations. There were lots of other fights over things from broken toys to slap shots in the face.

Well, we moved into our new house, and I had to change schools. I had to leave all my old friends and make some new ones. I did make new friends, and what was neat was that our two football teams played each other every year. So I got to play against my old friends with my new friends. When I got into high school, we were all united again. So I had gotten a new family, a new house, and a new school with new friends. I would have to say it was a major life transition.

As I grew up there were some tough times. I got picked on, but the youngest always does. But as I got older, it got to be less and less. And by the time I made it into high school, my long-time adversary, Mel, was beginning to be my best friend. I owe him a lot. During my childhood, my stepfather was extremely busy with work, and my dad was seldom around. It was Mel who taught me how to ride a bike and play basketball, baseball, and football. He also watched out for the younger kids, me and Sharon. I owe him a lot and would do anything for him.

Both my parents agreed about most issues (discipline, for example), and this led to smooth communication between our parents and us kids. The only thing detrimental I can think of that came out of being in a large family was that I had poor study habits as a young child. The problem was my stepfather can watch TV, listen to the radio, and prepare a balance sheet all at the same time. So he let his kids listen to the radio or watch TV while they did their homework. My mom didn't agree with this, but it was too hard for her to enforce not watching TV with so many kids already used to doing it. . . .

I love all my brothers and sisters very much and would do anything for them. It is neat to see them get married and have kids. I have three nieces and five nephews. It is very exciting to get the whole family together. A lot of my family, nuclear and extended, live in our area. My grandparents, aunts, and uncles, cousins, three sisters and their kids all live within twenty minutes of our house. Having roots and strong family connections are two things I'm very thankful for. These are things I received from my stepfather because if I lived with my dad, I would be on what I consider his nomadic journey: he moves about every three years.

I think I'm very lucky. I had a solid upbringing and relatively few problems. I owe my stepfather a lot. He has given me clothes and food, taught me important lessons, instilled in me a good work ethic, and seen to it that I get a good education—all the way through college. Without him a lot of these things would not be possible; I am grateful for everything he has done for me. I'm happy with the way things turned out, and I love my family dearly.

What might be a reason that David does not like to apply the term stepfamily *to his own family? How does David's essay illustrate some potential problems specific to stepfamilies? Do you think David's parents have a stable marriage? Why or why not? Divorce and remarriage can be thought of as transitions and/or crises, the subject matter of Chapter 14. In what ways does David's essay illustrate meeting crises creatively, as discussed in that chapter? If you live in a stepfamily, how does David's experience compare with your own?*

seem "more real" than the second union; the children, upset after visits with the noncustodial parent, were required to make a major adjustment that made life difficult for everyone else; and the natural parents sometimes feel caught between loyalties to the biological child and their desire to please their partner (Messinger 1976). More specifically, three major problem areas can be identified in remarried families:

financial burdens, role ambiguity, and negative feelings of the children, who often don't want the new family to work.

FINANCIAL STRAINS

Observers point out that the problems that burden remarriages frequently begin with the previous divorce. This is particularly evident in the case of finances. Frequently, money problems arise because of obligations left over from first marriages. Remarried husbands may end up financially responsible for children from their first marriages *and* for their stepchildren. In many states an ex-wife's alimony (though not child support) automatically ends when she remarries. And one study has found that remarriage by a custodial mother prompts sizable reductions in child support from her ex-husband (Hill 1992). Stepfamilies have lower incomes than do other married couples, and they are less educated as well (White and Booth 1985b; U.S. Bureau of the Census 1989c).

Even though disproportionately more second wives are employed outside their homes than are first wives, remarried husbands report feeling caught between the often impossible demands of their former family and their present one. Some second wives also feel resentful about the portion of the husband's income that goes to his first wife to help support his children from that marriage. Or a second wife may feel guilty about the burden of support her own children place on their stepfather.

The emotional impact of the previous divorce also causes money problems. Among Messinger's respondents, some women reported stashing money away in case of a second divorce, and some men refused to revise their wills and insurance policies for the same reason. For many couples, money became a sensitive issue that neither partner talked about (Messinger 1976). Partners in remarriage are less likely to think and act in terms of "our" money (Fishman 1983).

ROLE AMBIGUITY

Another basic problem in remarried families is that roles of stepchild and stepparent are neither defined nor clearly understood (Schwebel, Fine, and Renner 1991). As one stepfamily expert explains,

> the role of the stepparent is precarious; the relationship between a stepparent and stepchild only exists in law as long as the biological parent and stepparent are married. If the biological parent dies, the

stepparent instantly loses any legal claim to custody over his [or her] stepchild(ren), and custody reverts back to the surviving biological parent. The only way for a person to cement his or her legal tie to stepchildren is by adopting them, but this requires the cooperation of the noncustodial parent, which may not always be forthcoming. (Beer 1989, p. 11)

Legally, the stepparent is a nonparent with no prescribed rights or duties. Indeed, the term *stepparent* originally meant a person who replaced a dead parent, not an acquired quasi-kin who becomes an additional parent.

Uncertainties arise when the role of parent is shared between stepparent and the noncustodial natural parent (Beer 1992). What Bohannan pointed out over twenty-five years ago remains true today: Stepparents aren't "real" parents, but "the culture so far provides no norms to suggest how they are different" (Bohannan 1970a, p. 119). Some studies indicate that stepparents are less involved as parents (Thomson, McLanahan, and Curtin 1992) and that spouses of stepparents expect this (Giles-Sims 1984). Meanwhile, other studies find biological mothers dissatisfied when stepfathers are not very involved (Furstenberg 1987, in Pasley and Ihinger-Tallman 1988). Not surprisingly, relatively low role ambiguity has been associated with higher remarital satisfaction, especially for wives, and with greater parenting satisfaction, especially for stepfathers (Kurdek and Fine 1991).

One result of role ambiguity is that society seems to expect acquired parents and children to love each other in much the same way as biologically related parents and children do. In reality, however, this is not often the case. Interviews with remarried couples often reveal guilt about the lack of positive feelings (or even the presence of negative feelings) toward their partners' children. Delia Ephron, author of a book on remarriage based on personal experience, says that discipline was a consistent source of family conflict:

> I thought my husband wasn't being strict enough, but what I found out was that stepfathers and stepmothers all think the real parent of the kids isn't being strict enough.
>
> When you're a stepparent, you think you're an unbiased observer—when really you're an unbiased observer with a grudge, because you're an outsider and the very thing that's making you "unbiased" is something you resent. (in Bennetts 1986, p. 10-E)

Similarly, stepchildren often don't react to their parent's new spouse as though he or she were the "real" parent. The irony of expecting "real" parent–child love between acquired kin is further compounded by the fact that stepparents are not generally expected to be "equal" in disciplining or otherwise controlling their stepchildren (Beer 1992).

Moreover, adolescent stepchildren may have considerable family power (Smith 1992). Our discussion of power in Chapter 10 focused on marital power, but children are a force to be reckoned with in any family. Giles-Sims' research on stepfamilies found mothers to have more power on both major and everyday decisions than did either stepfathers or adolescents. But in 12 percent of the families, adolescents had more power than either parent, particularly in everyday decision making. "It is possible that the disruptive influence of adolescents on remarriage [citations omitted] is related to their power position" (Giles-Sims and Crosbie-Burnett 1989, p. 1076). Surprisingly, *better* coparental relations between the ex-spouses were associated with greater adolescent power. Perhaps active contact with the noncustodial parent gave an adolescent the option of an alternative home. Those adolescents with alternative resources did have more power; their power was diminished by stepfathers' support of the family. It was also less when stepfathers had children from their previous marriage. Giles-Sims speculates that this indicates an experience with parenting that makes the stepfather more effective and more authoritative. Adolescents also had less power in stepfamilies of greater longevity, reflecting either the breakup of the most troubled families or a gradual consolidation of the new stepfamily, which is evident over time in some (though not all) research (Pasley and Ihinger-Tallman 1988). Of course, children's attitudes as well as their power can have an impact on the new marriage.

STEPCHILDREN'S HOSTILITY   A third reason for the difficulty in stepparent–child relationships lies in children's lack of desire to see them work. Children often harbor fantasies that their original parents will reunite. Wallerstein and colleagues' longitudinal research showed that five years after divorce, 30 percent of children or adolescents still strongly disapproved of their parents' divorce, and another 42 percent had reservations (Wallerstein and Kelly 1980). Children who want their natural parents to remarry may feel that sabotaging the new relationship can help achieve that

goal. In the case of remarriage after widowhood, children may have idealized, almost sacred, memories of the parent who died and may not want another to take his or her place. As a result, stepchildren can prove hostile adversaries. This is especially true for adolescents.

For a young teenager, a parent's remarriage may prove more difficult to accept than was the divorce (Krementz 1984). At puberty, when children are discovering their own sexuality, it is remarkable how conservative they can expect their parents to be with regard to *their* sexuality. Adolescence can be a trying time for parents; teens can be impatient, self-centered, and argumentative. They can be especially distrustful, suspicious, and resentful toward a new parent, and they are sometimes verbally critical of a stepparent's goals, values, or personal characteristics. Anger displacement, described in Chapter 9, may also play a part in the stepparent's difficulty. Many adolescents blame their parents or themselves, or both, because the first marriage broke up. The stepparent becomes a convenient scapegoat for their hostilities.

Sociologists Lynn White and Alan Booth's (1985b) analysis of stability in remarriages, discussed earlier, considered one additional point: that family tension can be resolved by the child's rather than the partner's exit from the home. Speculating that children might be moved out by sending them to live with the other parent or forcing them to become independent, White and Booth found that indeed teenage children in stepfamilies leave home at significantly lower ages than do teens in intact families. For example, of teens age 16 and over in the first wave of the study, 65 percent of stepchildren were no longer living at home three years later, compared to 44 percent of children in intact families. White and Booth were working from the perspective of the married parents; qualitative data from Wallenstein's study of divorced families suggest that stepchildren often feel excluded and out of place in their new household, shut out by the tight bond between the remarried couple. They frequently report a sense of rejection as they perceive the largest share of the parent's attention and energy going to the new marriage (Wallerstein and Blakeslee 1989). The following, from a student essay, helps to illustrate this process:

I grew up with my biological mother and stepfather without much happiness. As I grew up, I never felt my stepfather ever showed me any love nor

Conflicting expectations concerning a stepmother's role can make it stressful. When stepparents can ignore the myths and negative images of the role and maintain optimism about the remarriage, they are more likely to have high family, marital, and personal satisfaction.

spent any time with me and I feel he influenced my mother to act the same. I can still remember only feeling like a financial burden to this man, and perhaps this is why I moved out of their house at 19 as opposed to 22 or 23 or after college like many of my friends.

Although virtually all stepchildren and stepparents are uncomfortable with some aspects of their family role, certain difficulties are more likely to trouble stepmothers, and others are more common to stepfathers. We'll look at each of these roles and the problems associated with them in more detail.

The Stepmother Trap

The role of stepmother is thought by clinicians and parents to be more difficult than that of stepfather (Ihinger-Tallman and Pasley 1987, pp. 100–103). Some research by sociologists points to the same conclusion (White and Riedmann 1992; Aquilino 1994b; White 1994). One important reason for this is a contradiction in expectations for the role of stepmother. Other explanations for the greater difficulty of the *residential* stepmother role include the fact that stepmother families, more than stepfather families, may

begin after difficult custody battles and/or have a history of particularly troubled family relations. Then too, a remarried wife, more often than a remarried husband, may suddenly find herself a full-time stepmother, even though she had never anticipated this (Ihinger-Tallman and Pasley 1987, pp. 100–101). Phyllis Raphael (1978) refers to it as the **stepmother trap:** On the one hand, society seems to expect romantic, almost mythical loving relationships between stepmothers and children (Smith 1990). On the other hand, stepmothers are seen and portrayed as cruel, vain, selfish, competitive, and even abusive (remember Snow White's, Cinderella's, and Hansel and Gretel's). Stepmothers are accused of giving preferential treatment to their own children. As a result, writes Lucile Duberman, in our society "a stepmother must be exceptional before she is considered acceptable. No matter how skillful and patient she is, all her actions are suspect" (1975, p. 50). Consequently, stepmothers tend to be more stressed, anxious, and depressed than other mothers (Santrock and Sitterle 1987) and also more stressed than stepfathers (Pasley and Ihinger-Tallman 1988).

Conflicting expectations seem to come with the stepmother role, and relations between stepmothers and daughters seem particularly troubled. Some researchers find that stepmothers behave more negatively toward stepchildren than do stepfathers, and children in stepmother families seem to do less well in terms of their behavior (Pasley and Ihinger-Tallman 1988). Yet, other studies indicate that stepmothers can have a positive impact on stepchildren. Observational research found that children interacted in a more socially mature manner with stepmothers than with their biological fathers. "Stepmothers were much more likely to be highly active participants in the lives of children than were their stepfather counterparts. . . . Children from stepmother families perceived their relationship with their stepparent as more positive than children from stepfather families" (Santrock and Sitterle 1987, p. 287; see also Smith 1990, pp. 118–32).

Still, some stepmothering situations often make the role especially complicated. For example, special problems accompany the role of part-time or weekend stepmother when women are married to noncustodial fathers who see their children regularly. The part-time stepmother may try to establish a loving relationship with her husband's children only to be openly rejected, or she may feel left out by the father's ongoing relationship with his offspring. One study of women who were dating divorced fathers found that the woman would help the man with house and child care in an effort to free more of his time for intimacy and social life with her. To her frustration, however, the man often spent the extra free time with his children (Rosenthal and Keshet 1978).

Part-time stepmothers can feel left out not only by the father's relationship with his children but also by his continued relationship with his ex-wife. Noncustodial fathers can spend long hours on the telephone with their ex-wives discussing their children's school problems, orthodontia, illnesses, and even household maintenance and repairs.

One stepmother summed up the feelings of many others in her role: "I heard my stepson on the phone to his mother say to her, 'I love you,' and it hurt so much! You think you've come to terms with the whole situation, and then some tiny thing can make you feel completely thrown out again, after [several] years" (in Smith 1990, p. 62). (Box 16.1, "Some Stepparenting Tips," contains some advice that may make stepparenting easier.)

Stepfathering

Men who decide to marry a woman with children come to their new responsibilities with varied emotions. The motivations may be far different from those that make a man assume responsibility for his biological children. "I was really turned on by her," said one stepfather of his second wife. "Then I met her kids." This sequence is fairly common, and a new husband may have both positive and negative reactions, ranging from admiration to fright to contempt. Data from the 1987–88 National Survey of Families and Households show that about half (55 percent) of stepfathers find it somewhat or definitely true that "having stepchildren is just as satisfying as having your own children" (Marsiglio 1992b, p. 204).

Research into the stepfather role suggests two contrasting conclusions. The first is that stepchildren tend to be well adjusted and to get along with their stepfathers as well as other children do with their natural fathers. The second is that stepfathers tend to view themselves as less effective than natural fathers view themselves (Beer 1992). Stepfathers' relatively low appraisal of their own performance may reflect the special problems and discomforts they experience in the stepfathering role. We'll look at a few of these.

ODD MAN OUT IN AN IN-GROUP   A new stepfather typically enters a household headed by a mother. When a mother and her children make up a single-parent family, the woman tends to learn autonomy and self-confidence, and her children may do more work around the house and take more responsibility in family decisions than do children in two-parent households. These are positive developments, but to enter such a family, a man must work his way into a closed group (Robinson and Barret 1986, pp. 118–43). For one thing, the mother and children share a common history, one that does not yet include a new stepfather.

Living arrangements can cause a new husband to feel like the odd man out. One stepfather moved into his wife's house because she didn't want to move her children to a new school and neighborhood. But, he said, he "never felt at home there. . . . I'd sit someplace, and then I'd move someplace else. I didn't have a place there. I couldn't find a place there for the first few months" (in Bohannan and Erickson 1978, p. 59). In fact, stepfathers do have less power relative to

Some Stepparenting Tips

Preparing to Live in Step

In a stepfamily at least three (and usually more) individuals find themselves struggling to form new familial relationships while still coping with reminders of the past. Each family member brings to the situation expectations and attitudes that are as diverse as the personalities involved. The task of creating a successful stepfamily, as with any family, will be easier for all concerned if each member tries to understand the feelings and motivations of the others as well as his or her own.

It is important to discuss the realities of living in a stepfamily before the marriage, when problems that are likely to arise can be foreseen and examined theoretically. If you are contemplating entering a steprelationship, here are some points to consider:

1. Plan ahead! Some chapters of Parents Without Partners conduct "Education for Remarriage" workshops. Contact your local chapter or write to Parents Without Partners.

2. Examine your motives and those of your future spouse for marrying. Get to know him or her as well as possible under all sorts of circumstances. Consider the possible impact of contrasting lifestyles.

3. Discuss the modifications that will be required in bringing two families together. Compare similarities and differences in your concepts of child rearing.

4. Explore with your children the changes remarriage will bring: new living arrangements, new family relationships, the effect on their relationship with their noncustodial parent.

5. Give your children ample opportunity to get to know your future spouse well. Consider your children's feelings, but don't allow them to make your decision about remarriage.

6. Discuss the disposition of family finances with your future spouse. An open and honest review of financial assets and

responsibilities may reduce unrealistic expectations and resultant misunderstandings.

7. Understand that there are bound to be periods of doubt, frustration, and resentment.

Living in Step

Any marriage is complex and challenging, but the problems of remarriage are more complicated because more people, relationships, feelings, attitudes, and beliefs are involved than in a first marriage. The two families may have differing roles, standards, and goals. Because its members have not shared past experiences, the new family may have to redefine rights and responsibilities to fit the individual and combined needs.

Time and understanding are key allies in negotiating the transition from single-parent to stepfamily status. Consideration of the following points may ease the transition process:

1. Let your relationship with stepchildren develop gradually. Don't expect too much too soon—from the children or yourself. Children need time to adjust, accept, and belong. So do parents.

2. Don't try to replace a lost parent; be an additional parent. Children need time to mourn the parent lost through divorce or death.

3. Expect to deal with confusing feelings—your own, your spouse's, and the children's. Anxiety about new roles and relationships may heighten competition among family members for love and attention; loyalties may be questioned. Your children may need to understand that their relationship with you is valued but different from that of your relationship with your spouse and that one cannot replace the other. You love and need them both, but in different ways.

4. Recognize that you may be compared with the absent partner. Be prepared to be tested, manipulated, and chal-

(continued)

(continued)

lenged in your new role. Decide, with your mate, what is best for your children and stand by it.

5. Understand that stepparents need support from natural parents on child-rearing issues. Rearing children is tough; rearing someone else's is tougher.

6. Acknowledge periods of cooperation among stepsiblings. Try to treat stepchildren and your own with equal fairness.

Communicate! Don't pretend that everything is fine when it isn't. Acknowledge problems immediately and deal with them openly.

7. Admit that you need help if you need it. Don't let the situation get out of hand. Everyone needs help sometimes. Join an organization for stepfamilies; seek counseling.

Source: U.S. Department of Health, Education, and Welfare 1978.

---

adolescent stepchildren when they move into the mother–child home (Giles-Sims and Crosbie-Burnett 1989).

THE HIDDEN AGENDA The new husband–father may feel out of place not only because of his different background but also because he has a different perspective on family life. After years of living as a single-parent family, for instance, both the mother and children are likely to have developed a heightened concern over the fairness of chore allocation. A newcomer, especially if he assumes the traditional male role in a two-earner remarriage, may draw complaints that he is not contributing enough. This last bone of contention is part of another source of potential conflict: the hidden agenda.

The **hidden agenda** is one of the first difficulties a stepfather encounters: The mother or her children, or both, may have expectations about what the stepfather will do but may not think to give the new husband a clear picture of what those expectations are. The stepfather may have a hidden agenda of his own. For example, he may see his new stepchildren as unruly and decide they need discipline. Or he may find that after years of privacy, a bustling house full of children disrupts his routine.

A part of the stepchildren's hidden agenda, as we

have seen, involves the extent to which they will let the new husband play the father role. Children may be adamant in their distaste for or jealousy of the stepfather, or they may be ready and anxious to accept the stepfather as a new daddy. This last is particularly true of young children.

Stepfathers tend to be more distant and detached than stepmothers (Santrock and Sitterle 1987). "It has been suggested by clinicians that such detached parenting behaviors are essential for improved relationships of stepparent and stepchild, especially during the early years of remarriage"(Pasley and Ihinger-Tallman 1988, p. 210), and stepfathers with such a parenting style do have better relationships with their stepchildren (citing Hetherington 1987). Young adult children may be mature enough to think of the new addition to the family primarily as their mother's husband rather than as a stepfather.

The hidden agendas of mother, children, and stepfather also involve supposedly simple matters of everyday living, such as food preferences, personal space, and the division of labor. Problems can arise when a meat-and-potatoes person joins a gourmet-dinner household, for example.

Discipline is likely to be a particularly tricky aspect of both the children's and the parents' hidden agendas.

A few problem areas are notable. There are now two parents rather than one to establish house rules and to influence children's behavior, but the parents may not agree. A second problem can be the holdover influence of the biological father. To the new father, there may sometimes seem to be *three* parents instead of two—especially if the noncustodial father sees the children regularly—with the biological father wielding more influence than the stepfather. A third problem can be the development of children's responsibility and participation in decision making in single-parent families. The children may be unwilling to go back to being "children"—that is, dependent on and subject to adult direction. The new parent may view them as spoiled and undisciplined rather than mature.

A stepfather can react to these difficulties in finding a place in a new family in one of four ways. First, the stepfather can be driven away. Second, he may take control, establishing himself as undisputed head of the household, and force the former single-parent family to accommodate to his preferences. Third, he may be assimilated into a family with a mother at its head and have relatively little influence on the way things are done. And fourth, the stepfather, his new wife, and her children can all negotiate new ways of doing things (Isaacs, Montalvo, and Abelsohn 1986, pp. 248–64). This is the most positive alternative for everyone, and it is best achieved by writing a personal remarriage agreement.

Writing a Personal Remarriage Agreement

Reasons for writing a personal marriage agreement were discussed in Chapter 8. To varying degrees, these reasons also apply to remarriage agreements. But additional reasons make it useful to negotiate a personal remarriage agreement or at least to talk over the issues involved. One reason is that remarriages are often beset with specific problems. Many times partners do not foresee these problems and are therefore not prepared to deal with them. (In many locations, prospective spouses can participate in remarriage preparatory courses to alert remarrying couples to common problems and to help them find ways to discuss inevitable conflicts.)

A second reason for negotiating a remarriage agreement is that partners who marry after divorce may want to try harder to make their marriage work but may be reluctant to fully commit themselves emotionally. Divorce and the struggles of the marriage prior to divorce can leave scars. When not openly acknowledged, past failure, rejection, loss, and guilt can undermine a new intimate relationship without either partner understanding what is happening. One way to counteract this situation is to share negative feelings about oneself, then to try to negotiate a relationship in which each partner feels as secure, as positive about himself or herself, and as comfortable as possible.

A third, related reason for writing a remarriage agreement is that second-time spouses may feel the conflict taboo (described in Chapter 9) even more than do spouses in first marriages. In their desire to make their marriage work, divorced and remarried spouses may feel too battle-scarred to open a can of worms. Accordingly, they may gloss over differences that need airing and resolution—differences they may not have hesitated to fight about in their first marriages. As Chapter 9 points out, avoiding conflicts is a serious mistake.

Finally, remarriage agreements are important because, as we have seen, society has not yet evolved an effective cultural model for these complex relationships. Unless they discuss their expectations, new mates are likely to be unrealistic. Furthermore, the legal issues involved in remarriage are best given careful consideration before the marriage. Inheritance is one: Do you want your money and property to go to your new spouse or your children? What does the law in your state permit on this? Obligation of a stepfather to support stepchildren is another important issue.

Custodial parents may want their new spouse to adopt the children. This generally involves a waiver of parental rights by the biological parent, who may not wish to do so. Recent court cases have addressed this issue; such termination of parental rights is extremely rare. Children above a certain age, perhaps 14, may or must give their own consent to stepparent adoption in some states.

Rights of stepparents to visitation or even custody of a stepchild in the event of death or divorce from the child's parent is a crucial issue, when so many people become closely attached to stepchildren. Law in this area—case law and legislation—is rapidly changing. Although it seems unlikely—and with reason—that biological parents will be legally replaced by stepparents, it is important to indicate in a will or other statement that the biological parent would like his/her children's relationship with a stepparent preserved through visitation, if that is the case.

The most general point here is that people who remarry should not assume that they can do some of these things without checking with a lawyer. Individuals may also want to be become active in relevant public policy areas that concern them.

Chapter 8 suggests some questions to address when creating a personal marriage agreement; the majority of these apply to second marriages as well. Remarriage agreements, like other marriage agreements, should be revised as situations and partners change.

Divorce and remarriage, the subject of two of the chapters in Part Five, have become commonplace in our society. Knowledgeably choosing to both divorce and remarry requires anticipating the consequences of those choices.

In Sum

Although remarriages have always been fairly common in the United States, patterns have changed. Remarriages are far more frequent now than they were earlier in this century, and they follow divorce more often than widowhood. The courtship process by which people choose remarriage partners has similarities to courtship preceding first marriages, but the basic exchange often weighs more heavily against older women, and homogamy tends to be less important.

Second marriages are usually about as happy as first marriages, but they tend to be slightly less stable. An important reason is the lack of a cultural script. Relationships in immediate remarried families and with kin are often complex, yet there are virtually no social prescriptions and few legal definitions to clarify roles and relationships.

The lack of cultural guidelines is clearest in the stepparent role. Stepparents are often troubled by financial strains, role ambiguity, and stepchildren's hostility. Marital happiness and stability in remarried families are greater when the couple have strong social support, high expressiveness, a positive attitude about the remarriage, low role ambiguity, and little belief in negative stereotypes and myths about remarriages or stepfamilies. Personal remarriage agreements can help

to establish an understanding where few social norms exist.

Key Terms

coparenting	remarriages
double remarriages	remarried families
hidden agenda	single remarriages
quasi-kin	stepmother trap

Study Questions

1. Discuss the similarities and differences between courtship before remarriage and courtship before first marriage.

2. How is the basic exchange in remarriage tipped against women? Compare this with the basic exchange in first marriages.

3. How might remarriage rates be different if divorced women were not at such an economic disadvantage, compared with married women?

4. Why are remarriages somewhat less stable than first marriages even though they are about as likely to be happy as are intact first marriages?

5. The remarried family has been called an incomplete institution. What does this mean? How does this affect the people involved in a remarriage? Include a discussion of kin networks and family law. Do you think this situation is changing?

6. What evidence can you gather from observation and/or your own personal experience to show that stepfamilies: (a) may be more culturally acceptable today than in the past, and (b) remain negatively stereotyped as not as functional or as normal as first-marriage, nuclear families?

7. What are some problems faced by both stepmothers and stepfathers? What are some problems particularly faced by stepfathers? Why might the role of stepmother be more difficult than that of stepfather? How might these problems be resolved or alleviated?

8. What are some reasons for writing a personal remarriage agreement? Is it more or less important to write a remarriage agreement than a first marriage agreement? Why? Discuss some topics that could be important to consider in a remarriage agreement but not in a first marriage agreement.

Suggested Readings

Beer, William R. 1989. *Strangers in the House: The World of Stepsiblings and Half-Siblings*. New Brunswick, NJ: Transaction. This readable book by a sociologist explores the experiences of stepsiblings and half-siblings and includes topics such as sexuality between stepsibs and changes in birth and age order when a stepfamily begins.

————. 1992. *American Stepfamilies*. New Brunswick, NJ: Transaction. This book addresses the complex structure of remarried families, compares them to first-married families, and then addresses issues concerning raising children in stepfamilies.

Einstein, Elizabeth. 1982. *The Stepfamily: Living, Loving, and Learning*. New York: Macmillan. This award-winning (award from the American Psychological Association) book chronicles the development stages of stepfamily living through experiences of the author (a stepchild and a stepmother), interviews with about fifty other stepfamilies, and discussions with professionals who work with stepfamilies.

Ephron, Delia. 1986. *Funny Sauce: Us, the Ex, the Ex's New Mate, the New Mate's Ex, and the Kids*. New York: Viking. How to live in stepfamilies, based on personal experience.

Ganong, Lawrence H. and Marilyn Coleman. 1994. *Remarried Family Relationships*. Thousand Oaks, CA: Sage. Authors have done significant research on remarriage and stepfamilies. Part of a series on "close relationships," this book presents a comprehensive view of remarried families and includes recent research.

Ihinger-Tallman, Marilyn and Kay Pasley. 1987. *Remarriage*. Newbury Park, CA: Sage. Comprehensive, concise review of research; intended for the classroom but also useful to the general reader.

Journal of Family Issues. 1992. (June, vol. 13, no. 2), Lynn K. White (ed.) Full issue devoted to remarriage.

Maglin, Nan Bauer and Nancy Schneidewind. 1989. *Women and Stepfamilies: Voices of Anger and Love*. Philadelphia: Temple University Press. Personal essays and scholarly articles by and about stepmothers, stepdaughters, stepgrandmothers, adult steprelations, and others.

Mahoney, Margaret M. 1994. *Stepfamilies and the Law*. Ann Arbor: University of Michigan Press. Up-to-date reference on this area of family law. Things are changing as to stepparents' rights and responsibilities in law.

Martin, Don. 1992. *Stepfamilies in Therapy: Understanding Systems, Assessment, and Intervention*. San Francisco: Jossey-Bass. Written for family therapists or therapy students, this book combines material such as that in this chapter with counseling perspectives and techniques.

Pasley, Kay and Marilyn Ihinger-Tallman eds. 1987. *Remarriage and Stepparenting: Current Research and Theory*. New York: Guilford. Important volume.

————., eds. 1994. *Stepparenting: Issues in Theory, Research, and Practice*. Westport, CT: Greenwood. Addressed to an academic audience, this volume reports research, practice, and theory development regarding stepfamilies. Includes articles by most active researchers in this area.

Robinson, Bryan E. and Robert L. Barret. 1986. *The Developing Father*. New York: Guilford. Includes a good chapter on stepfathers.

Rosenberg, Maxine B. 1990. *Talking About Stepfamilies*. New York: Bradbury. Children in stepfamilies tell about the positives and negatives of daily life in their own words.

Smith, Donna. 1990. *Stepmothering*. New York: St. Martin's Press. Emphatic, qualitative study of what being a stepmother involves and feels like. The final chapter is called "The Worth of a Stepmother."

Afterword

We have worked with four themes throughout this text: (1) Personal decisions (always trade-offs) must be made throughout the life course. (2) In making those decisions, we are all influenced by the culture and society in which we find ourselves. (3) Our U.S. society is changing, with more ethnic diversity and decreased marital and family permanence. (4) As our society continues to change—and as we make personal decisions in this context—what we choose feeds back into society, resulting in yet more, or ongoing, change.

Families and Society

Throughout this book we have pointed to the mutual influence of family and society. The larger society affects individuals and families by the economy's effect on the material base of family life; the societal values that pervade schools and other institutions; social definitions of appropriate family forms, gender roles, and sexual morality; the extent to which reproduction is encouraged or discouraged and child rearing is esteemed; whether and how family roles are

supported in the workplace; and in many other ways. Society can be generally favorable toward families or not. It can take them for granted or give them support and encouragement. It can encourage some families and discourage others.

If people are to shape the kinds of family living they want, they need to reach beyond their own marriages and families to get involved in electoral politics and federal and state legislation. One's role as a family member, as much as one's role as a citizen, has come to require participation in public policy decisions.

A Change from Institution to Relationship

As the majority of American families turn from emphasizing the utilitarian benefits of marriage (such as financial support, household maintenance, and child rearing) to focusing on intimate, supportive relationships, loving becomes central to marriage and family life. Sex becomes less a physical act for procreation and more a potential source of pleasure and bonding. Individuals choose marriage partners not so much for their basic exchange values as for unique personal qualities.

With the transition to marriage as a relationship comes an emphasis on communicating effectively and supportively. Contemporary families exist mainly because members see that being together is, in itself, good. One outcome of this attitude, ironically, is a high divorce rate. As families move from utilitarian arrangements to intrinsic relationships, expectations for intimacy rise, and divorce becomes available as an option.

The family continues to function as a supplier of practical needs. Families are no longer self-sufficient economic units, but they channel and mediate the practical, health, and educational needs of family members—needs that, in turn, are handled by entities such as stores, hospitals, and schools. The larger society greatly influences how families help members meet their practical needs and, to a great extent, members' capacity for shaping a supportive familial environment.

Choices Throughout Life

Making decisions about one's marriage and family begins in early adulthood and lasts into old age. People repeatedly make decisions about how to balance the need for individuality with that for commitment and togetherness. All individuals today, whether married or single, female or male, must determine the balance between caring for and about themselves and caring for and about others—their spouses and partners, parents and children, sisters, brothers, and family friends, the independent and the dependent, their own generation and future generations. From these ties come much of the responsibility in and meaning of life.

Some Final Thoughts About Family Values and Social Policy

Family values have been much in the news during the past several years. A problem with all the debate about family values, however, is that seldom do people define what they mean by "family values." Now that you've completed a course on marriages and families, why not list what you think of as family values? You can do it on this page. We've given you room for ten, although you may not think of that many. If you think of more than ten, never mind the lines and keep going.

WHAT I THINK OF AS FAMILY VALUES

1. _____
2. _____
3. _____
4. _____
5. _____
6. _____
7. _____
8. _____
9. _____
10. _____

THINKING CRITICALLY ABOUT FAMILY VALUES
Now that you've listed what you think of as family values, let's think critically about them for a few minutes. First, let's take each value on your list and ask whether it is a *family* value or an *individual* value. Because we expect people to learn individual values in families, we often mistake individual values for family values. A family value would be one that pertains specifically to family structure and/or relationships. An example might be sexual fidelity to one's partner or mate. An individual value would be one that you expect individuals to exhibit whether they live in families or not. Examples are honesty or courtesy. Now go back to your list and indicate whether each of your entries is a family or an individual value.

Finally, ask yourself whether each of the values you have listed requires a particular family structure (a two-parent heterosexual family, for example) in order to endure. Could other family structures also foster these values? If yes, why? If no, why not?

FAMILY VALUES AND SOCIAL POLICY   Now let's think about what can be done to promote the values you have listed. As a first step, think of something the federal government could do to encourage each value on your list. Imagine that money is no object. Next, take into account that the United States has a serious budget deficit and a goal of balancing the budget. What could the federal government do that would *not* cost money?

Now let's think about what your state government might do to promote each value. With money? Without money? Next let's think of some things your neighborhood or community might do to encourage each value on your list. And finally, why not think of some things that you yourself can choose to do that would foster each of these values?

If you have time in or outside class, you might share your ideas with classmates or with your family members and friends. We, your authors, would be interested in your thoughts as well. You can send them to us c/o Wadsworth Publishing Company, Belmont, CA, 94002. Families are important! Hence the choices we make about them, in our own lives and as a society, could not be more vital.

APPENDIX

Human Sexual Anatomy

If you are to understand sexual relations between individuals, you need to be aware of both the anatomy and physiology of sex and the attitudes and emotions that shape people's feelings about their own sexuality and that of others. In Appendix A we will consider the first of these elements, the anatomy and physiology of sex. We will look at female and male sexual anatomy and describe the **genitalia,** or external reproductive parts, and then the internal reproductive systems of each sex.

Female Genital Structures

The external genitalia of a woman are technically referred to as the **vulva.** The vulva is composed of the following structures:

- the **mons veneris,** or pubic mound; an area of fatty tissue above the pubic bone

- the **labia majora** (Latin for "greater lips"; the singular form of *labia* is *labium*), two rounded folds of skin; and, within them, the **labia minora** (or "lesser lips")

- **prepuce,** or clitoral hood: a fold of skin that covers the clitoris when it is not erect and is formed where the labia minora join

- the **clitoris,** which consists of an internal shaft composed of **erectile tissue**—tissue that becomes engorged with blood during arousal, causing it to increase in size—and a **glans,** a highly sensitive tip, about the size of a pea

- the **urethra,** the opening through which urine passes from the bladder to the outside

- the **vestibule,** or entryway to the vagina

- the **perineum,** the area between the vestibule and the anus (the opening from the rectum and bowel)

- the **hymen,** a ring of tissue that partially covers the vaginal opening. The hymen contains small blood vessels that may bleed the first time the tissue is broken: at first intercourse, first insertion of a tampon, during masturbation, or as a result of some accidental injury

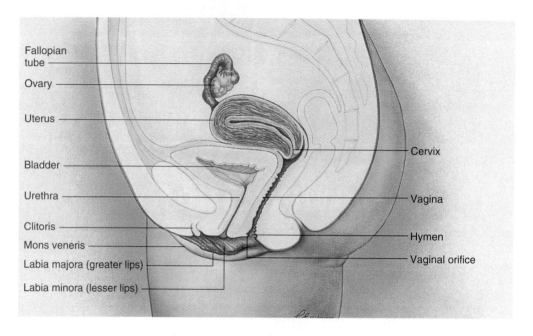

FIGURE *A.1*

| Female urogenital system.

Labels on figure:
Fallopian tube
Ovary
Uterus
Bladder
Urethra
Clitoris
Mons veneris
Labia majora (greater lips)
Labia minora (lesser lips)
Cervix
Vagina
Hymen
Vaginal orifice

The main internal structures of the female reproductive system are the vagina, the cervix, the uterus, the fallopian tubes, and the ovaries (see Figure A.1.)

The **vagina** is the passageway from the uterus to the external genitalia. It is a potential space within a woman's body. Usually, the vaginal walls touch one another; but the vagina is elastic and capable of opening wide enough to allow a baby to pass through during birth. Such stretching would be extremely painful if the vagina had the same number of nerve endings as many of the structures of the vulva have. Therefore, the vagina is not so sensitive to feeling.

At the top of the vagina is the **cervix:** the neck of the uterus (*cervix* means "neck" in Latin). The **uterus,** or womb, is a cavity whose purpose is to cradle a fetus until birth. Leading from the uterus are two passageways, called **fallopian tubes,** that connect a woman's uterus with her ovaries.

Ovaries are female **gonads,** or sex glands. Women have two ovaries, one on each side of their bodies. They produce reproductive cells (**ova,** or eggs) and two female sex hormones, estrogen and progesterone.

Ordinarily, the ovaries alternate in producing one ovum per month, in a process called **ovulation.**[1] The egg, barely visible, travels along the fallopian tubes to the uterus.

In preparing to receive the egg, the lining of the uterus, called the **endometrium,** thickens with a layer of tissue and blood (see Figure A.2). This tissue can nourish an embryo during the early stages of pregnancy, if the egg becomes fertilized during its passage from the ovaries. When fertilization does not occur, the egg and the unused endometrial tissue and blood are discarded during **menstruation.**[2]

Male Genital Structures

The external male genitalia are the **penis** and the scrotum (see Figure A.3). Like the female clitoris, the penis is composed of an erectile shaft and a sensitive tip,

1. Sometimes, women produce more than one egg at a time. This is one way that twins or multiple children are conceived.

2. Menstruation occurs in monthly cycles, ranging from about twenty-one to thirty-five days. Travel, anxiety, or change in diet can make periods more or less frequent.

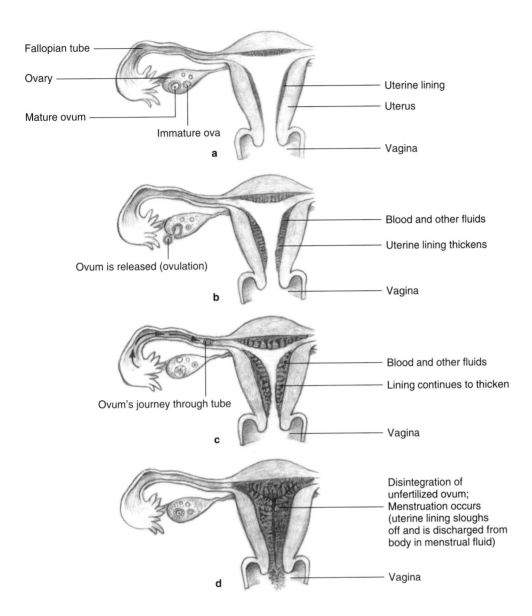

Fallopian tube

Ovary

Mature ovum

Immature ova

a

Uterine lining

Uterus

Vagina

Ovum is released (ovulation)

b

Blood and other fluids

Uterine lining thickens

Vagina

Ovum's journey through tube

c

Blood and other fluids

Lining continues to thicken

Vagina

Disintegration of unfertilized ovum; Menstruation occurs (uterine lining sloughs off and is discharged from body in menstrual fluid)

Vagina

d

FIGURE $A.2$

The menstrual cycle: (a) During the early part of the cycle, an ovum matures in an ovary; the endometrium, or uterine lining, begins to thicken. (b) About fourteen days after the onset of the last menstruation, a mature ovum is released; the endometrium is thick and spongy. (c) The ovum travels through one of the fallopian tubes; blood and other fluids engorge the uterine lining. (d) If the ovum is not fertilized, the endometrium breaks down and sloughs off in a form of bleeding (menstruation).

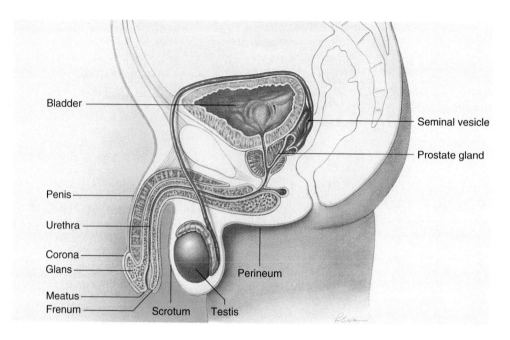

FIGURE *A.3*

I   Male urogenital system.

or glans. The glans is especially sensitive to touch at the **corona:** a crownlike ridge at its base. If a male has not been circumcised, the glans is covered by a thin membrane, the **foreskin,** when his penis is not erect (in circumcision, this foreskin is removed).[3] On the side of the penis, which rests against the scrotum, is the **frenum,** the place where the foreskin is or was connected to the penis. The frenum, even more than the corona, is sensitive to tactile stimulation.

When a man is not sexually aroused, his penis is flaccid. When erect, penises vary somewhat in size and are usually about six inches in length and about an inch and a half in diameter. The **urethra** runs through the penis and carries male reproductive cells

and urine, though never at the same time.[4] The opening at the tip of the penis is called the **meatus,** Latin for "passage."

Behind the penis hangs a sac, the **scrotum,** which holds the two male gonads, the **testicles.** One testicle is usually lower than the other. Testicles, sometimes called *testes,* are the male counterpart to the female ovaries. They produce the male reproductive cells, called **sperm,** as well as male hormones such as testosterone. Unlike ovaries, however, testicles are external structures. That is because they must be maintained at a temperature lower than body temperature in order to produce living sperm. Between the scrotum and the anal opening is an area called the **perineum.** As in women, this area is sensitive to the touch.

The internal male reproductive structures are also shown in Figure A.3. Above the testicles, near the internal surface of the rectal walls, are two glands, the **seminal vesicles** and the **prostate.** These glands produce **semen,** the milky fluid that carries the sperm through the urethra and out the meatus. There are usually between 200 million and 500 million sperm in a teaspoonful of semen. Sperm are ejaculated,

3. In our society, circumcision is usually performed shortly after a male baby's birth. There are probably no significant differences in cancer rates or other medical or sexual problems that can be related to circumcision (Haas and Haas 1993, p. 99). For some, circumcision is an important cultural or religious ritual. Beginning in the 1980s, the circumcision of male babies has been challenged by those who point to a medical risk as well as psychological trauma to the infant (Konner 1988).

4. A man's urethra cannot carry urine while his penis is erect because erection automatically blocks the opening from his bladder to his urethra.

or ejected, during the rhythmic contractions of **orgasm.**

If they are ejected into a woman's vagina, sperm move toward her fallopian tubes. Sperm can live in the fallopian tubes from two to five days. If one sperm cell fuses with a female's egg, fertilization occurs and a fetus is conceived. There are a number of methods that can be used to prevent conception, and these are discussed in Appendix D.

The structures described above make up the male and female reproductive systems. Their reproductive functions are discussed in Appendix C. Appendix B describes the physiology of human sexual response and sexual expression.

Human Sexual Response

Appendix A describes human sexual anatomy; here we examine the physiology of sexual response.

The Four Stages of Human Sexual Response

Through carefully controlled laboratory observation over a period of eleven years, sex researchers William Masters and Virginia Johnson (1966) recorded in detail the bodily changes that take place as a consequence of **sexual arousal:** the awakening, stirring up, or excitement of sexual desires and feelings in either ourselves or others. Masters and Johnson described four phases of human sexual response: excitement, plateau, orgasm, and resolution. These phases characterize both men's and women's responses and take place in sex with partners of the same or the opposite sex. Specific stimulation and sexual movements may vary with the presence or absence of a partner and the sex of the partner, as well as individual preference and spontaneity on any given occasion. However, the underlying physiological response is the same.

EXCITEMENT   When people begin to feel sexually aroused, they enter the **excitement phase** of sexual re-

sponse. Many forms of stimuli—fantasy, sights, sounds, smells, touches—can cause sexual excitement.[1] Women and men share several responses during the excitement phase, including an increase in blood pressure and pulse rate and faster breathing. There is a heightened feeling in and awareness of the genitals. This is caused by engorgement, or **congestion,** of the genital blood vessels, which causes the affected tissue to swell and, often, coloration to deepen. Another effect of excitement is **myotonia:** increased muscle tension, especially in the abdominal region and in the long muscles of the arms and legs. A measleslike rash, called a *sex flush,* appears on the abdomen and chest in about 75 percent of sexually excited women and 25 percent of aroused men.

In women, sexual excitement is marked by the onset of vaginal lubrication, or "sweating" of fluid from the inner walls of the vagina. In men, sexual excitement is characterized by erection of the penis, caused by congestion.

---

1. For a detailed review of men's and women's responses to sexual stimulation, see Baron and Byrne 1984.

The excitement phase can be stopped intentionally by removing the sexual stimulus. It can also be stopped or interrupted unintentionally through distractions, such as babies crying, phones ringing, or changes in lighting and temperature, or through feelings of anxiety or guilt. Once interrupted, the excitement phase can be resumed.

PLATEAU   The **plateau phase** involves an intensification of processes begun during the excitement phase, with several marked bodily changes. The color of the penile glans and the labia minora becomes a deeper red or reddish-purple. There is increased tension in both involuntary and voluntary muscles. Pelvic thrusting, which begins voluntarily, grows more rapid and becomes involuntary, especially among men. Heart rates may nearly double, and blood pressure continues to rise. If the sex flush appeared on a woman during the excitement phase, much of her body will now be flushed. A man may now show the first signs of a sex flush, which begins under his rib cage and spreads over his chest, neck, and face.

In men the corona becomes more swollen. Several drops of fluid, which is not semen but which can contain some sperm cells, may emerge from the meatus.[2] Late in this phase a woman's clitoris pulls deeply underneath the clitoral hood. This, along with the marked change in color in the labia minora, is evidence of her impending orgasm.

The plateau phase may be intentionally prolonged by decreasing the stimulation, returning to the excitement phase, and then increasing stimulation. If stimulation is withdrawn and not restored, sexual tensions will decrease only very gradually. This can be an uncomfortable process, with feelings of fullness and pressure in the pelvis, cramps, lower back pain (Masters and Johnson 1973, p. 119), and general physical and emotional frustration.

ORGASM   In **orgasm,** or climax, sexual tension reaches its peak and is suddenly discharged. This ex-

tremely pleasurable and totally involuntary response, the **orgasmic phase,** lasts a few seconds and is accompanied by pronounced physiological changes. Heart and pulse rates peak. Breathing becomes deeper and faster than in the plateau phase, so that an individual may sometimes momentarily experience a shortage of oxygen. The senses of smell, taste, hearing, sight, and feeling (except for genital sensation) are temporarily diminished. The sex flush is brightest at this point. Muscles in the neck, legs, arms, buttocks, and abdomen may contract spasmodically, and hand and foot muscles often contract strongly. Involuntary rhythmic contractions in the vagina and penis also occur, though their strength varies from person to person and from orgasm to orgasm. In adult men, orgasm is almost always accompanied by **ejaculation,** the rhythmic discharge of seminal fluid containing sperm. Once these contractions begin, a man cannot voluntarily stop ejaculation.

Men normally experience a single orgasm; women, however, can be **multiorgasmic,** experiencing several orgasms successively during one sexual encounter. About 15 percent of women regularly have multiple orgasms. In women who experience multiple orgasms, each successive orgasm is often more intense than the preceding one.

RESOLUTION   During the **resolution phase** of sexual activity, partners' bodies return to their unstimulated state. The genitals resume normal size and color; the sex flush disappears, muscles relax, and erect nipples soften. Heart rate, blood pressure, and respiratory rates revert to normal.

Return of the penis to its flaccid state begins quickly, then proceeds more slowly. During this stage men experience a **refractory period:** a time during which they cannot become sexually aroused. The refractory period usually lasts at least twenty minutes and may be considerably longer, particularly as a man grows older. In women the resolution phase lasts about ten to fifteen minutes and occasionally as long as a half hour. During this time women may remain sexually aroused and, with continued or renewed stimulation, can experience subsequent orgasms.

2. If this fluid is discharged while the penis is in the vagina, a woman can be impregnated. Thus, interrupting intercourse before ejaculation (coitus interruptus) is not a reliable birth control method.

APPENDIX C

Conception, Pregnancy, and Childbirth

A female's ovaries alternate in releasing one egg, or ovum, each month, in a process called **ovulation.** Ovulation takes place about fourteen days before a menstrual period; thus a woman's most fertile time is usually midway between menstrual periods, when the ovum is traveling through the fallopian tube to the uterus.

Conception

When sperm enter a female's vagina during coitus, they move into the fallopian tubes and can live there from two to five days. **Conception** takes place upon **fertilization,** or the joining of the sperm cell with the ovum. If this takes place in the fallopian tube, the fertilized egg, or **zygote,** moves down to the uterus, where it embeds itself in the thickened lining, or endometrium (see Figure C.1), a process called **implantation.** Until an umbilical cord is formed during about the fifth week, the endometrial tissue provides nourishment for the developing fetus.

Pregnancy

The fertilization and implantation processes just described take place during the **germinal period,** or first two weeks of pregnancy. During this early period, the woman usually isn't aware that she is pregnant. By the fourth week, however, she may begin to notice some changes.

The first signs a woman often notices are a cessation of menstruation (because the endometrial tissue will not be sloughed off), nausea (a physical reaction to the zygote's embedding itself in the uterine wall), changes in the size and fullness of the breasts, darkened coloration of the **areolae** around the nipples, fatigue, and frequency of urination, a result of pressure on the bladder from the expanding uterus. Not all of these signs, including nausea and cessation of menstruation, are always present, so a woman who suspects she is pregnant should have a pregnancy test even if not all classical signs of pregnancy are present.

THE EMBRYONIC STAGE   The **embryonic stage** of pregnancy lasts from the second until about the eighth week. During this stage the head, skeletal system, heart, and digestive system begin to form. Also during this time a sac of salty, watery fluid called **amniotic fluid** surrounds the fetus to cushion and protect it. In later stages of pregnancy, doctors can detect

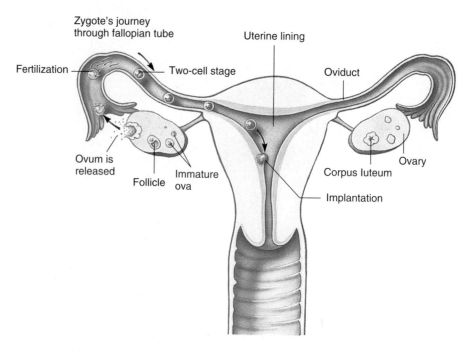

FIGURE *C.1*

| Ovulation, fertilization, and the germinal period of pregnancy.

some fetal defects by withdrawing a tiny portion of this amniotic fluid through the mother's abdomen with a syringe and testing it in a laboratory. (See the section "Amniocentesis and Other Prenatal Testing" later in this appendix.) During this period, the **placenta,** which holds the fetus in place inside the uterus and functions in nourishment, develops. (The placenta will be discharged in the final stage of childbirth.)

THE FETAL STAGE   The **fetal period** of development lasts from about eight weeks until birth. This span, including the germinal and embryonic periods, is often broken down into three three-month trimesters. During the fetal period, the organs and structural system that budded during the embryonic stage refine themselves and grow. Some of the changes that take place up to fifteen weeks are illustrated in Figure C.2.

In the third month, the facial features become differentiated. The lips take shape, the nose begins to stand out, and the eyelids are formed, although they remain fused. The fingers and toes are well developed, and fingernails and toenails are forming.

During the fourth month most of the fetus's bones have formed, although they are still soft cartilage and will not be completely hardened into bone until many years after birth.

In the fifth month, the fetal heartbeat can be heard through a stethoscope. Around this time, too, the **quickening**—the first fetal movements apparent to the mother herself—progresses from a mild fluttering to solid kicks against the side of the mother's abdomen. Any nausea the mother may have experienced usually disappears by now, and she is in the most comfortable period of her pregnancy.

In the sixth month, the fetus grows to a foot in length and about twenty ounces in weight. The fetus now has eyelashes, it can open and close its eyes and may even learn to suck its thumb. By the end of this month, its essential anatomy and physiology are almost complete; further development consists largely of an increase in size and refinement and stabilization of the organs' functions. A fetus born or aborted at this time is likely to emerge alive and may live several hours. Survival beyond that will require constant medical attention, and the chances for survival are slim.

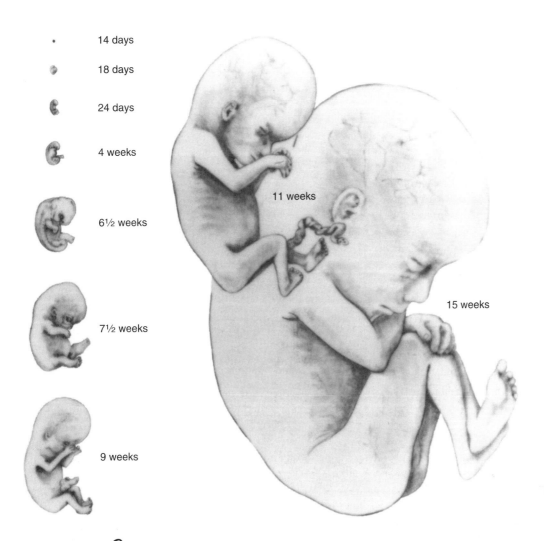

14 days

18 days

24 days

4 weeks

6½ weeks

7½ weeks

9 weeks

11 weeks

15 weeks

FIGURE $C.2$

| Prenatal development.

By seven months the fetus weighs about two and a half pounds. If it is born now, it will have a 50 percent chance of survival with the aid of specialized attention and equipment. A baby born in the eighth month of pregnancy has a very good chance of survival, because its development is virtually complete.

In the eighth and ninth months of pregnancy, the fetus grows very rapidly, gaining an average of a half pound per week. At this time, the mother is likely to feel generally healthy but may also be uncomfortable because of the crowding in her expanding uterus and because weight increases may disrupt her equilibrium and her ability to get around. Toward the end of preg-

nancy, the fetus usually changes its position so that the head is in the lower part of the uterus. This marks the beginning of preparation for birth.

Monitoring Fetal Development

Recent years have seen extraordinary scientific advances in monitoring fetal development. Here we look at two such advances—ultrasound and amniocentesis—and the newer techniques of chorionic villus sampling and alpha-fetoprotein testing.

ULTRASOUND   In ultrasound, sound waves are bounced off the abdomen of the pregnant woman to determine the shape and position of the fetus. Some obstetricians already use it in every pregnancy; others prefer to use it only when clearly indicated because of the absence of long-term studies. Doctors who use it at about the fifth month say that it helps to predict the date of birth within two weeks, that it can detect twins 90 percent of the time, and that it shows whether the fetus is maturing as it should. Ultrasound can also reveal several different kinds of birth defects—especially malformations of the skeleton—early enough for a legal abortion if parents choose to have one.

The use of ultrasound has implications beyond diagnosis, however. The fact that sonograms permit prospective parents to do something they have never been able to do before—observe the fetus—is pushing back parental bonds to before birth (Powledge 1983, p. 38). Patients often ask for copies of the Polaroid fetal snapshots. As one genetic counselor said, "It seems that with sonography there's really that *connection,* especially if the fetus happens to move. . . . I guess seeing that image creates a stronger bond" (in Powledge 1983, p. 38). Ultrasound imaging may be the most common technique for prenatal diagnosis; amniocentesis is another.

AMNIOCENTESIS AND OTHER PRENATAL TESTING

Amniocentesis and other prenatal testing provide information to parents about their risk of having a child with a birth defect—that is, a condition substantially lowering the quality of life or leading to premature illness and death. Common concerns of parents are Down's syndrome and spina bifida, a neural-tube defect in which the spinal covering fails to close, commonly leading to severe mental and physical disability and early death.

In amniocentesis, a physician inserts a needle through the abdominal wall into the uterus, withdrawing a small amount of amniotic fluid. Cells and other substances that the fetus has cast off float in this fluid, which technicians can examine for clues to fetal health. Among other things, the cells in the sampling are cultivated and checked for evidence of possible birth defects, particularly Down's syndrome. When doctors suspect that a woman might give birth to a child with a particular disorder—often because she carries a recessive gene for this disorder or because she has already given birth to a child with the disorder—scientists can examine the fluid for other conditions as well, including nearly 100 rare genetic diseases. As women postpone childbearing to older ages, they have

more concern about the risk of such birth defects as Down's syndrome; the risk increases with age. Pregnant women over 35 are usually advised to have amniocentesis.

Experts on prenatal diagnosis note that an increasing number of women under 35 are asking for prenatal testing. These are prospective parents who feel that they could not give a disabled child the necessary time and attention. For them, prenatal diagnosis is simply a logical extension of family planning (Powledge 1983). Some physicians are opposed to providing these tests to women under 35, believing that the risk of testing outweighs the risk of having a baby with a birth defect. Others point out that despite the relatively low risk and because there are so many more births to younger women, 80 percent of Down's syndrome children are born to mothers under 35 (Kolata 1987).

When performed by experienced medical personnel, amniocentesis appears reasonably safe, but the technique is not without risks. Hazards include spontaneous abortion and risk of premature birth, with fetal damage occurring in slightly under 1 percent of cases. Moreover, amniocentesis must take place during the second trimester when sufficient amniotic fluid is present. This timing is a major drawback. The prospect of a second-trimester abortion is more emotionally troubling, and a later abortion heightens the physical risk to the woman.

The newer techniques of chorionic villus sampling (CVS) and alpha-fetoprotein testing can provide information earlier in pregnancy, from the fetal membrane and a sample of the mother's blood (respectively), but carry some risks and uncertainty as well. CVS seems to cause miscarriages in about 2.4 percent of cases. Recently, some research has reported increased risk of fetal deformity (Gilbert 1993). Another problem with the alpha-fetoprotein test is that it is suggestive but not definitive. Doctors fear that with frequent use of these tests, parents will unnecessarily abort normal fetuses in many cases, whereas in others major defects will not have been identified (Kolata 1987). Social scientists working in this field find parents to be troubled by CVS-based decisions as well as those related to amniocentesis (Seals 1990).

Virtually all testing programs include genetic counselors who can advise parents as to the significance of their test results and help them work through their decisions and their emotions. Because detection of an abnormal fetus gives prospective parents the chance to knowledgeably choose abortion, antiabortion groups have objected strenuously to prenatal screening and genetic counseling. Although amniocentesis is very much established as a normal part of prenatal care for those at risk, and newer forms are becoming so, genetic testing and the accompanying abortion decision remain ethically and personally difficult choices. We turn now to the childbirth process.

Childbirth

The process of childbirth takes place in three stages: labor, delivery, and afterbirth (Figure C.3).

LABOR **Labor** is the process by which the baby is propelled from the mother's body through a series of contractions of the muscles of the uterus. Labor usually begins with mild contractions, at intervals of about fifteen to twenty minutes. The contractions increase steadily over the first phase of labor (usually from six to eighteen hours for the first birth, shorter for subsequent births); they also increase in intensity and duration until by the end of labor each contraction lasts a minute or more.

During labor some other changes usually take place. The cervix dilates from its normal size (about one-eighth inch) to approximately four inches in preparation for the baby's passage. A second occurrence is the expulsion of a bloody plug (sometimes called *show*) from the base of the uterus through the vagina. During pregnancy the plug helped prevent infectious bacteria from entering the uterus through the cervix. And third, the amniotic membrane (often called the *bag of waters*) ruptures, and amniotic fluid flows from the vagina. Show and breakage of waters are usually signs of imminent delivery. Together with these, full dilation of the cervix marks the beginning of the second, or delivery, stage of childbirth.

DELIVERY The second phase of childbirth is the **delivery** of the baby. This phase extends from the time the cervix is completely dilated until the fetus is expelled—a process that may last from less than twenty minutes to (rarely) more than ninety minutes. The mother can often speed the birth process at this stage by tightening the muscles in her diaphragm, abdomen, and back so that the uterine muscles are aided in pushing the baby through the cervix. Her active participation at this point may also help reduce pain.

When the baby appears at the vaginal opening (*crowning*), its head usually turns so that the back of its skull emerges first, as is shown in Figure C.3. After

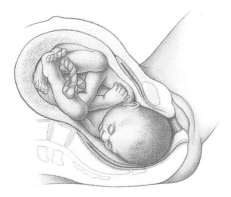

a

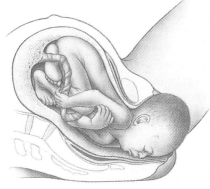

b

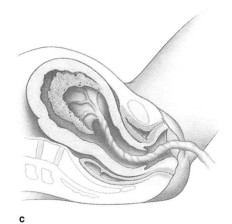

c

FIGURE *C.3*

Events in the childbirth process. **(a)** early stages of labor; cervix is dilating; baby's head starts to turn; **(b)** baby's head begins to emerge; **(c)** afterbirth.

the head emerges, the infant usually turns again to find the path of least resistance. This kind of delivery, in which the baby's skull emerges first, occurs in about 95 percent of births. The remaining 5 percent of deliveries are more difficult: If the baby's buttocks, shoulder, foot, or face emerge first (**breech presentation**), the baby will not be able to take as compact a shape as it passes through the vagina.

Oversized babies (the average newborn weighs 7½ pounds) can also cause problems because the baby's head must pass between the bones of the mother's pelvic arch. If the baby is too large, or if the mother's or baby's physical condition makes the stress of childbirth dangerous, a physician may decide to deliver the child by cesarean, or Caesarean, section (so called after Julius Caesar, who was supposedly born in this way). A **cesarean section** is a surgical operation in which a physician makes an incision in the mother's abdomen and uterine wall to remove the infant.[1]

Another source of complications may be weak uterine contractions (often caused by anesthetics). If contractions are not strong enough to expel the baby, a physician may use forceps—tongs that fit around the baby's head—to draw the baby out through the vagina. This procedure is risky, however, for the inaccurate placement of forceps, along with the force necessary to pull the infant free, may cause disfigurement or brain damage.

AFTERBIRTH   The third and final stage of childbirth takes place between two and twenty minutes after delivery. It consists of the expulsion of the **afterbirth:** the placenta, the amniotic sac, and the remainder of the umbilical cord.

---

1. An unprecedented 25 percent of the nation's infants were born by cesarean section in 1988, compared to 4.5 percent in 1965 (U.S. Bureau of the Census 1989a). Since then, the percentage of cesarean births has declined, but only slightly, to 22.8 percent in 1993 (U.S. Bureau of the Census 1995, Table 97). Reasons offered for this high rate include the now common practice of fetal monitoring, which can trigger unnecessary intervention, physicians' concern about liability if they fail to take all precautionary measures, and the possibility that for-profit hospitals subtly encourage cesarean births to maximize profits (Winslow 1991b).

Of course, difficult births may require surgical intervention, and there is no question that cesareans often are life-saving procedures for both mothers and infants. But experts express alarm at the high rates of cesarean births because they are riskier for mothers and infants. A cesarean section also deprives a mother of the experience of normal childbirth. Cost is also a factor; in 1988, the cost of a cesarean delivery was, on the average, $1,700 more than that of a vaginal delivery. An important public health goal is reducing the number of cesarean deliveries. In a departure from past practice, doctors now encourage even women who have had one cesarean birth to try vaginal deliveries for subsequent births (Rothman 1984; Leary 1988a; Ryan 1988; Brody 1989a).

NATURAL CHILDBIRTH   Hospitalization and a doctor's assistance have contributed greatly over the years to the sharp decrease in infant and maternal death during childbirth. The overwhelming majority, 99.1 percent, of babies born are delivered in hospitals (U.S. National Center for Health Statistics 1989). An emerging issue is that, due to high costs and insurance company policies, mothers and newborns are sent home too soon (Begley 1995). The reliance on hospitals and doctors, has also given rise to a negative reaction against the treatment of childbirth as a medical problem rather than a natural event (Rothman, 1982, 1989). Many physicians, nurses, and expectant mothers prefer not to rely on forceps, heavy anesthesia, or other often unnecessary procedures. They feel that more natural methods of delivery are more emotionally satisfying to both the mother and father and are often better for the infant (Korte 1995).

Infants born under heavy sedation are less responsive and alert; they also have somewhat reduced chances for surviving a medical emergency. The parent–child bond is less easily established. Alternate forms of birthing include natural childbirth, in which the use of anesthesia is minimized and the father's presence and participation are encouraged. The baby may be given to the mother for nursing or affectionate contact even before the umbilical cord is tied.

A recent development, still not legally or administratively permissible in all states, is the use of midwives rather than physicians as professional birth attendants.[2] Midwives are paraprofessionals specially trained to assist mothers in the birth process; they provide both emotional support and professional expertise and recognize that the baby belongs to the family, not to the medical establishment (Korte 1995). Although the numbers of births attended by midwives, both in and out of hospitals, are still relatively small, they have increased substantially in recent years. In 1993, 196,228 babies (4.8 percent of all babies born that year) were delivered by midwives, over eight times the number in 1975. Just over 25,000 babies were born in a residence, 8,000 of them delivered by a person other than a physician or certified nurse-midwife (U.S. National Center for Health Statistics 1989; Ventura et al. 1995, Table 38, p. 71).

---

2. Midwifery by registered nurses (often with a requirement of supervision or collaboration with a physician) is legal in all fifty states, but midwives who aren't nurses and largely do deliveries at home are not legally recognized in most states (Nazario 1990a).

Contraceptive Techniques

Contraceptives, techniques and methods to prevent conception, can be divided into three groups. One group uses various chemical substances or mechanical devices, or both. These include the pill, the IUD or intrauterine device, the diaphragm, vaginal foam, cream, or jellies, vaginal suppositories, the contraceptive sponge, the cervical cap, Norplant, and Depo-Provera (all used by females), and the condom or rubber (used by males). Chemical substances alter the biochemistry of the body, whereas mechanical devices form barriers between the sperm and ovum. Figure D.1 displays some types of contraceptive devices. They are described in Table D.1.

A second method of contraception is the surgical sterilization of either the male (vasectomy) or the female (usually by tubal ligation). A **vasectomy** involves tying the tubes between the testicles (where sperm is produced) and the penis (through which the seminal fluid is ejaculated). The procedure can be done in a doctor's office and is safe. Following a vasectomy the male will be able to have erections, enjoy sex, and ejaculate as before the sterilization, but he will not be able to cause pregnancies because there will be no sperm in his ejaculate. A **tubal ligation** involves cutting or scar-ring the fallopian tubes between a woman's ovaries and her uterus so that eggs cannot pass into the tubes to be fertilized. Tubal ligation must be done in a hospital and is more expensive than a vasectomy, but it is also safe. Unlike the techniques in the first group, which are generally reversible when one decides to have children, sterilization is a one-time procedure that is virtually 100 percent effective and usually permanent. Microsurgical techniques to restore fertility have been developed but are not always successful. Thus, individuals should be certain about their decision to give up the capacity to have children before undergoing sterilization. About 30 percent of U.S. women have been surgically sterilized (U.S. Bureau of the Census 1995, Table 107).

The third type of conception control avoids all surgery, chemicals, and devices. Instead, it is based on controlling sexual behavior. One such method, periodic abstinence, often called natural family planning or fertility awareness, depends on the couple's awareness of the woman's ovulation cycle: The couple refrains from intercourse for several days before and after the woman ovulates. Ovulation may be detected

(cont. on p. 565)

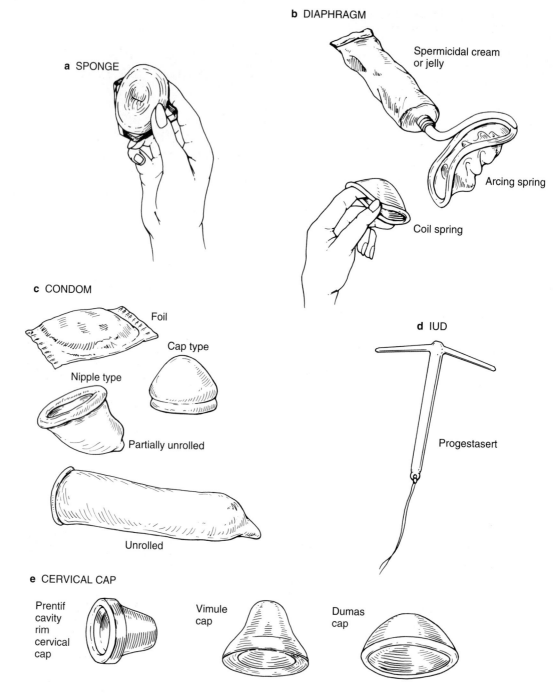

a SPONGE

b DIAPHRAGM

Spermicidal cream or jelly

Arcing spring

Coil spring

c CONDOM

Foil

Cap type

Nipple type

Partially unrolled

Unrolled

d IUD

Progestasert

e CERVICAL CAP

Prentif cavity rim cervical cap

Vimule cap

Dumas cap

FIGURE *D.1*

Some types of contraceptive devices: a the sponge, which contains spermicide that is released in the woman's vagina; b a diaphragm—spermicidal cream is squeezed into the cup and around the rim before the diaphragm is inserted in the vagina; c condoms, placed over the man's penis; d an IUD, an intrauterine device, which is inserted into the uterus by a clinician; e a cervical cap, similar to the diaphragm but inserted farther into the vagina, covering the cervix.

f LAPAROSCOPY AND TUBAL
   STERILIZATION TECHNIQUES

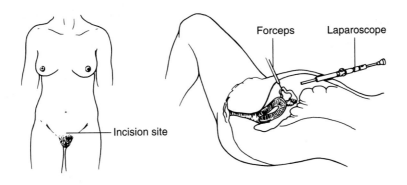

Incision site

Forceps   Laparoscope

Fallopian tubes tied

Pomeroy method   Healed

Fallopian tubes cut

Irving method   Healed

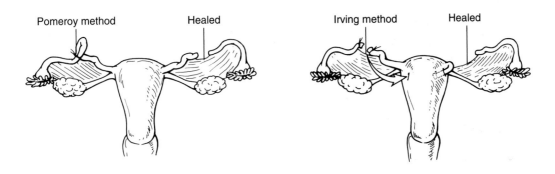

g VASECTOMY

Incision made

Vas deferens
divided

Incisions closed
after procedure

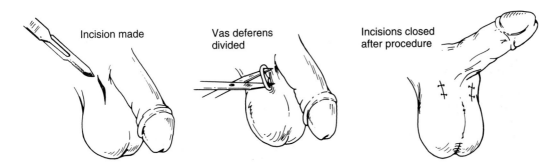

FIGURE **D.1** (continued)

Surgical sterilization: f For women, the surgical procedure includes the use of a laparoscope, a
long slender optical instrument that is inserted through a small incision in the abdominal wall
and is used to visualize the interior of the abdominal cavity. The fallopian tubes are then either
tied (Pomeroy technique) or cut (Irving method) to prevent eggs from passing and being fertilized. g For men, a vasectomy involves tying or cutting the vas deferens, the sperm-carrying
tubes from the testicles to the penis, so that sperm are not ejaculated.

A Comparison of Birth Control Methods

The Pill

What This Is

A series of pills that a woman takes once each day for a month. At the end of the month, she starts a new package of pills. The pills are made of hormones, much like those a woman's body makes to control her menstrual cycle.

How It Works

Most kinds of birth control pills keep a woman's ovaries from releasing eggs. Other kinds of pills change the lining of her uterus, or change the mucus in her cervix (the opening into her uterus).

How Well It Works

The pill is the one of best temporary methods of birth control for women in the United States. If 100 women use the pill for one year, 3 will get pregnant. A woman who forgets even one pill has a chance of getting pregnant, so women who always remember to take their daily pill are better protected.

Main Advantages

The pill is very effective, and easy to use. Women don't have to think about it while they are having sex. Women who use it can have more regular menstrual periods, with less cramps, blood loss, and anemia. They can also have less premenstrual tension, fewer ectopic pregnancies, less pelvic inflammatory disease, acne and rheumatoid arthritis. They have some protection from noncancerous breast tumors, ovarian cysts, and cancer of the ovary or endometrium.

Who Can Use It

Most women can use the pill unless: they are over 35 and smoke more than fifteen cigarettes a day, or have had blood clots, inflamed veins, serious liver disease, unexplained vaginal bleeding, cancer of the breast or uterus, and certain other tumors. Some conditions, like diabetes or high blood pressure, could be affected by the pill. Unless you are over 35 and a heavy smoker, the pills available today are safer than pregnancy.

How It Is Used

Take your pills as directed, the same time each day, without skipping any. When you want to get pregnant, stop taking the pill and use another birth control method until your periods are regular. After having a baby, talk with your doctor about going back on the pill, especially if you breast-feed your baby. You will need regular medical checkups, and any new clinician you see should be told you are taking the pill. You don't need to take a vacation from the pill.

Possible Problems

Like all kinds of medicine, the pill can cause problems for some women. Minor problems, that usually go away in a few months, include: breast tenderness, nausea, vomiting, weight change, or spotting between periods. If a woman is taking another kind of medicine along with the pill, she may not have as much protection from pregnancy, and needs to talk about this with her clinician. Women on the pill are slightly more likely than other women to have blood clots, strokes, heart attacks if you are over 40, or liver tumors. Chances of these problems increase as a woman gets older, or if she smokes or has other problems like high blood pressure, higher levels of some blood fats, or diabetes. To learn more details, talk with your clinician, and read the information that comes with your pills.

Warning Signals

Call your clinician right away if you have any of these signs: unusual swelling or pain in your legs; yellow color in your eyes or skin; pain in your abdomen, chest or arms; shortness of breath; severe headache; severe depression; blurred or double vision, or other eye problems.

Norplant

What This Is

A set of six capsules placed under the skin of a woman's arm.

How It Works

A hormone in the capsules leaks slowly into the woman's body to keep her ovaries from producing eggs. It also changes the mucus in her cervix, the opening to her uterus.

How Well It Works

The capsules contain five years worth of hormone, and give 99 percent protection from pregnancy. A woman who weighs more than 150 pounds may have slightly less protection.

Main Advantages

Norplant is very effective. Once it is in place in her body, the woman does not have to think about using birth control. It does not contain any estrogen, so it does not have side effects from that hormone. Once it is removed, the woman is fertile again.

Who Can Use It

Most women can use it unless: they are pregnant, they have active blood clotting problems, abnormal bleeding from the vagina, liver disease or tumors, or breast cancer.

How It Is Used

A clinician will insert the six capsules under the skin of a woman's upper arm. They need to be removed by a clinician after the five years of protection are up, or if the woman wants to get pregnant or has side effects. **The woman should continue to get a checkup and Pap smear test for cancer each year, even if the capsules are still in place.**

Possible Problems

Most women have irregular menstrual bleeding during the first months of use. A few women may have headaches or other problems, and their clinician can decide whether or not these are related to Norplant. Insertion requires minor surgery, and some women report complications with removal.

Warning Signals

Call your clinician right away if you have: very heavy bleeding from the vagina; severe pain in the lower abdomen; pain, pus or bleeding where the capsules were inserted; loss of a capsule; severe headaches; or miss a period after having regular menstrual cycles.

Depo-Provera Injections

What This Is

A contraceptive injection of long-acting progestins lasting three months or more.

How It Works

Like the pill and Norplant, Depo-Provera is a hormone that keeps a woman's body from producing eggs.

How Well It Works

Studies have shown that approximately 1 out of every 400 women who use Depo-Provera for a year will become pregnant; this is a protection rate of 99.75 percent. One reason for the high effectiveness of Depo-Provera is that each injection actually provides more than three months of protection. Hence the woman has a four- to six-week grace period during which she can be late for her next shot but still be protected.

Main Advantages

Depo-Provera is very effective. Once injected, a woman need not think about birth control for about three months. Like Norplant, Depo-Provera does not contain estrogen and so does not have side effects from that hormone.

Who Can Use It

Depo-Provera is indicated for women who have developed estrogen-related complications while taking the pill and for women who want no more children but do not choose to be surgically sterilized. Lactating women can use it. Women with sickle cell disease should not use it, nor should any woman with undiagnosed abnormal genital bleeding. Although women do not appear to have long-term post–Depo-Provera infertility problems, the return of fertility does appear to be delayed until the effects of the injection(s) have elapsed. Hence some programs choose not to provide it to women planning to become pregnant in the future. Other programs offer it to women who have three or more children.

How It Is Used

A woman goes to a doctor or clinic to be injected every three months. A woman should use an additional contraceptive method for two weeks after her first injection.

Possible Problems

Excessive vaginal bleeding and amenorrhea (the stopping of periods) are the most frequent reasons women give for discontinuing Depo-Provera. Bleeding is often heavier during periods, which increases the risk of developing anemia. Breakthrough bleeding (vaginal bleeding between regular periods) can occur. Pregnancy may result if a woman in her 40s misinterprets prolonged amenorrhea as menopause, assumes she is sterile, and discontinues contraception. Depo-Provera users occasionally report decreased libido, depression, headaches, dizziness, weight gain, and allergic reactions. There is some concern about the increased possibility of breast cancer after Depo-Provera use for 6 years or longer. Many women experience post–Depo-Provera infertility lasting six to twelve months. However, fertility returns in 80 percent of women within one year after injections are stopped. There is no evidence that the delay in the return of fertility leads to long-term infertility.

Warning Signals

Breakthrough bleeding should be checked by a physician because it can also be indicative of illness. Excessive bleeding also indicates that a woman should be checked, particularly for anemia. Especially in the first month of using

(continued)

Depo-Provera Injections

Depo-Provera, a woman who stops having periods should have this situation checked in order to rule out other causes, including pregnancy. Feelings of depression, headaches, and dizziness should be checked by a physician.

Intrauterine Device (IUD)

What This Is

A small device made of plastic. Some contain copper, or a hormone. A clinician chooses the right type for a woman, and inserts it into her uterus. Some can stay there for four years, but others must be changed each year.

How It Works

IUDs prevent a woman's egg from being fertilized by the man's sperm, and change the lining of her uterus.

How Well It Works

If 100 women use the IUD for a year, 4 will get pregnant. Most pregnancies happen in the first year of use. Women who check their IUD strings regularly, or couples who use condoms and foam during the woman's most fertile time, have better protection.

Main Advantages

A woman with an IUD does not have to think about using her birth control method every time she has sex. The IUD does not affect a woman's hormones.

Who Can Use It

Women who have had a baby, have only one faithful sex partner, and have never had PID (pelvic inflammatory disease). Other possible reasons for not using the IUD include certain conditions of the woman's reproductive system, active or recent infection of her tubes or ovaries, very heavy periods, abnormal vaginal bleeding, anemia, or a history of tubal pregnancy. People who are allergic to copper, or are having diathermy heat treatments, can't use copper IUDs. A clinician will evaluate your history if you have had certain illnesses, such as heart disease, and you should have a pregnancy test to make sure you are not pregnant before the IUD is inserted.

How It Is Used

A clinician will insert the IUD into a woman's uterus, usually during her menstrual period. This may be uncomfortable while it is being done. A string from the IUD hangs down into the woman's vagina, and she can feel it after her menstrual period to make sure the IUD is still in place. A woman should have a checkup within three months after the IUD is inserted, and then every year. A woman who wants to get pregnant needs to have a clinician remove the IUD.

Possible Problems

Most women have few serious problems, or none at all. Some have more cramping, especially after the IUD is inserted, bleeding between periods, or heavier and longer periods. The IUD may come out, and the woman can get pregnant. Rarely, a woman will get pregnant with the IUD still inside her uterus. If you have signs of pregnancy, get examined right away, because the IUD should be taken out as soon as possible if you are pregnant. This reduces the chance of serious infection, miscarriage, or premature delivery. Some pregnancies to women with an IUD in place are ectopic (outside the uterus), and this requires surgery. Infection of a woman's tubes or ovaries is more common in IUD users than in other women, especially if they have had these infections before, or have had more than one sexual partner, or a sexual partner who may have picked up an infection from someone else. Such an infection can increase the risk of a woman having a tubal pregnancy, cause her to become sterile, or, seldom, require a hysterectomy. An infection that is not treated could be fatal. Rarely, the IUD punctures the wall of the woman's uterus, usually while it is being inserted. If this happens, surgery may be needed to remove it.

Warning Signals

Call your clinician right away if: you can't feel your IUD string; you have severe cramping or increasing pain in your lower abdomen; you have pain or bleeding during sex; you have unexplained fever and/or chills; you have increased, or bad-smelling discharge from your vagina; you have a missed, late, or light period.

Diaphragm or Cervical Cap

What This Is

A soft rubber barrier in a woman's vagina, used with a contraceptive cream or jelly.

How It Works

The diaphragm or cervical cap is put into a woman's vagina before intercourse. It covers the entrance to her uterus, and the cream or jelly stops the man's sperm from moving. The diaphragm can be put in the woman's vagina 6 hours ahead of intercourse, and left in for 24 hours. The cervical cap can be left in her vagina for up to 48 hours.

How Well It Works

If 100 women use a diaphragm or a cervical cap for a year, 18 will get pregnant. Women get better protection by checking each time they have intercourse to make sure that the diaphragm or cap covers the cervix, and also by having their partner use a condom.

Main Advantages

Once you learn how to use them, putting them in is easy. A woman can put it in at bedtime, or her partner can help her. If it is in place, neither the man or woman should be able to feel it. The contraceptive cream or jelly used with the diaphragm or the cervical cap also helps to protect against some infections, including the AIDS virus.

Who Can Use It

Diaphragms can be used by most women when they are not menstruating. They are not recommended for women with poor muscle tone in their vaginas, a sagging uterus, or vaginal obstructions. *Cervical caps* can be worn by most women when they are not menstruating, and by some women who can't wear a diaphragm. However, some women can't be fitted with the sizes of cervical caps now available. The cervical cap can be harder to fit, and harder to put in and remove, than the diaphragm.

How It Is Used

A woman must be fitted with the right size of diaphragm or cervical cap. She will be shown how to put it in, and take it out. She should always use contraceptive cream or jelly along with it. The diaphragm or cap should be in place every time a couple has sex. (If the woman is menstruating, the couple can use contraceptive foam and a condom instead.) A woman should check to see if the size is still right after she has a full-term pregnancy, a miscarriage or abortion past the first three months of pregnancy, pelvic surgery, or a weight gain or loss of more than 10 pounds.

Possible Problems

Most women have no side effects. Some women who use the diaphragm are prone to get bladder infections. A mild allergic reaction to rubber, cream or jelly may occur. A woman with very short fingers may need an inserter to use the diaphragm, and may not be able to use the cervical cap. If the woman is on top during sex, her diaphragm may be pushed out of place. If the cervical cap is worn more than three days, or if the woman has an infection, an unpleasant odor can result. Women can check their diaphragm or cap for weak spots or holes by holding it up to the light.

Warning Signals

A woman should call her clinician if she has any discomfort while the diaphragm or cervical cap is in place, if it does not stay in place, or if she has irritation or itching in the genital area, frequent bladder infections, or an unusual discharge from her vagina.

Over-the-Counter Birth Control Methods for Women

What This Is

These are birth control methods you can buy without a prescription. They all contain chemicals called spermicides that paralyze the man's sperm cells, but do not harm a woman's vagina. Always look for the word *contraception* or *contraceptive* on the package to be sure you are getting a birth control method.

How It Works

The spermicide is put into various foams, creams, jellies, suppository capsules or films that are put deep into a woman's vagina before she has intercourse. Those that are not liquids melt into liquids to put the sperm-stopping chemicals at the opening to the woman's uterus. The *contraceptive sponge* is moistened and put deep in the woman's vagina before she has intercourse. It contains the sperm-stopping chemicals, and has a nylon loop so it is easy to take out. The *foams* help block the opening to the woman's uterus with bubbles, and also contain the sperm-stopping chemicals. The *creams, jellies, suppositories,* and *films* melt into a thick liquid that coats the woman's vagina and contains the sperm-stopping chemicals. *Contraceptive sponges* are barriers that cover the opening to the woman's uterus and stop the man's sperm from entering it, and they also release the sperm-stopping chemicals.

How Well It Works

The sponge has a 6–28 percent failure rate. If 100 woman use the *contraceptive sponge* for a year, about 18 will get pregnant. If they use one of the other over-the-counter methods for a year, about twenty will get pregnant. **However, if women use one of these methods, and their partners use a condom, only half as many will get pregnant.**

(continued)

Over-the-Counter Birth Control Methods for Women

The methods work better if they are used correctly. Each package has detailed instructions about using the method.

Main Advantages

These are easy to buy in most drugstores and many supermarkets, or from your family planning clinic. You don't need a prescription. The sperm-stopping chemicals in these methods also help protect you against some sexually transmitted infections, including AIDS. Once you learn how to use them, they are easy to use, and either you or your partner can put them into your vagina.

Who Can Use It

Almost any woman can use the *foams, creams, jellies, suppositories,* and *films.* Almost any woman who can use a tampon can use the *contraceptive sponge,* but if she has short fingers it may be hard to put it in her vagina properly. **Women who have had Toxic Shock Syndrome should not use the sponge.** Don't use the sponge any time you are bleeding from your vagina, including during your menstrual period.

How It Is Used

The *contraceptive foams, creams, jellies, suppositories,* and *films* must be in a woman's vagina before she has intercourse. They are only effective for one hour. After that time, if a woman is having intercourse, she must put more of her method into her vagina. **Each package has directions to follow.** A woman should not douche, or wash out her vagina, for at least 6 hours after intercourse, because this will wash away the sperm-stopping chemicals. The *contraceptive sponge* needs to be moistened with at least 2 tablespoons of clean water in order to make the sperm-killing chemicals active. Fold it in half, away from the loop on the bottom. This make it narrow so it is easy to slide into your vagina as far back as you can reach. When you let go, it should spring back into shape and cover the opening to your uterus. You can slide your fingers around it to be sure it is where it belongs, and feel the loop to be sure it is facing the opening to your vagina. You can leave the sponge in your vagina up to 24 hours, but to make sure all the sperm are stopped, it needs to be left in at least 6 hours after you have intercourse. Do not douche while the sponge is in your vagina. To take it out, reach in and pull gently on the nylon loop. Throw it away. Don't use it again, and don't flush it down the toilet. If you have any problems using it, ask your clinician.

Possible Problems

Some men and women are irritated by the sperm-stopping chemicals in these birth control methods. Trying another brand may help. Some women find these methods to be messy. A few women are allergic to the sponge. **If a woman cannot remove the sponge from her vagina, or if it breaks into pieces, she should call her clinician or go to a hospital emergency room to have it taken out.** Women who use the *contraceptive sponge* may have a greater chance of getting **Toxic Shock Syndrome,** a condition also associated with the use of some tampons. To keep this risk down, never leave a sponge in your vagina more than 24 hours, never use the sponge when there is bleeding from your vagina, and don't use it after childbirth, miscarriage or abortion until your clinician says it is OK.

Warning Signals

Women who use the *contraceptive sponge* should call their clinician if they are having itching or irritation in or around their vagina, or a bad smell or unusual discharge from their vagina. The signs of **Toxic Shock Syndrome** are: fever of 101 degrees or over, skin rash that looks like sunburn, vomiting, diarrhea, fatigue, and aching muscles. See a doctor right away if these happen to you.

The Male Condom

What This Is

A sheath of rubber or animal tissue that a man can wear over his penis during intercourse.

How It Works

The condom catches the semen that comes out of a man's penis before, during and after he ejaculates, or comes. This keeps his sperm from getting into the woman's vagina. Rubber (latex) condoms also help protect against some infections, including the virus that causes AIDS.

How Well It Works

If 100 women have partners who use condoms for one year, about 11–14 will get pregnant. If the woman uses an over-

The Male Condom

the-counter birth control method like the foams, creams and jellies, they have better protection from pregnancy and from infections.

Main Advantages

A man who uses a condom can take care of preventing pregnancy, and protect himself and his partner from spreading infections. There are no side effects, except for the very few people who are allergic to the rubber or chemicals added to the condom. Condoms are easy to buy. They are also a good method to use as a backup birth control method in an emergency, or to give additional protection to another kind of birth control. Some men find that wearing a condom helps them keep an erection longer. Couples can even use a condom while the woman is pregnant, to protect against sexually transmitted infections.

How It Is Used

After a man's penis is erect (hard), but before it has any contact with the woman's genital area, either the man or the woman can put the condom on his penis. Hold the rolled-up condom on the tip of the man's penis. Pinch the air out of the end of the condom, so that there is about $^1/_2$ inch of it left to catch his semen when it comes out. Pull down his foreskin, if necessary, and roll the condom down over his erect penis, all the way down to his pubic hair. Smooth out any air bubbles. If you use a lubricant, use only water-based products like KY-Jelly, or women's over-the-counter birth control creams or jellies. Lubricants like vase-line, that contain oil, can weaken the rubber and make the condom break. After the man ejaculates (comes), hold the rim of the condom against the man's penis as he pulls it out of the woman's vagina. This way, as the man's penis gets soft, the condom is less likely to slip off and spill semen into the woman's vagina. Throw the used condom away, and use a fresh condom each time.

If the condom breaks while the couple is having sex, the man should pull his penis out of the woman's vagina right away, take off the broken condom, and put on a new one. The woman can put birth control foam, cream or jelly into her vagina right away to try to stop any sperm that may have spilled from the broken condom.

Possible Problems

If the condom is not put onto the man's penis correctly, leaving space at the end where his semen can go, it is more likely to break. People must be careful to hold the condom while the man pulls his penis out of the woman's vagina, so that his semen does not spill. Some couples think that using the condom breaks the mood of lovemaking. Other couples enjoy putting it on. Some men say they have less feeling when they wear a condom. Others say that using a condom helps them keep their erection longer.

Warning Signals

A man can learn how it feels when a condom breaks during intercourse. If he needs to, he can practice wearing a condom while he masturbates, and breaking it to see how it feels.

The Female Condom

What This Is

A lubricated, polyurethane pouch about 7 inches long, with flexible rings at both ends and inserted into the vagina like a diaphragm. The inner ring fits behind the pubic bone and the outer ring remains outside the body.

How It Works

The condom is inserted into and lines the vagina to prevent sperm from going into the uterus. The plastic pouch extends outside the vagina to cover the outer lips. Each condom is used only once. Most women insert it up to twenty minutes before intercourse.

How Well It Works

The female condom is 79–95 percent effective, with a failure rate of 5–21 percent. The condom is most effective when used with a spermicide.

Main Advantages

There are no major health concerns with use of the female condom. No physical exam is needed before using it, and it can be bought over the counter in a drugstore without a doctor's prescription. Besides protecting against pregnancy, the female condom can protect against STDs/HIV. Another advantage for women is that it puts control of contraception and STD/HIV protection directly in their hands, with no need to possibly cajole a male partner to use a condom.

Possible Problems

The condom can be difficult to insert, may cause minor irritation and/or discomfort, and can break if not stored and used properly. The outer ring can be pushed into the vagina during intercourse, causing pain or other problems. The male partner's penis can slip to the side of the condom, so that the condom does not properly collect the sperm or protect against STD/HIV.

(continued)

The Morning After Pill

What This Is

The morning after pill does not provide ongoing protection against pregnancy and is intended and used only as an emergency contraceptive technique. Usually two hormone pills need to be taken within 72 hours after a single act of unprotected intercourse. The hormone pills appear to prevent a fertilized egg either from entering the uterus or from attaching itself to the uterus in order to begin growth and development.

How It Works

A highly sensitive urine pregnancy test is done to be sure that the woman is not already pregnant. If not, one hormone pill is taken within 72 hours of unprotected intercourse. Another dose is taken 12 hours after the first dose. A follow-up pregnancy test is done three weeks later.

How Well It Works

The morning after pill is between 98.4 and 99.4 percent effective, with a failure rate of 0.6 to 1.6 percent.

Main Advantages

This is a form of emergency contraception. It is also a method that can be used for rape victims.

Possible Problems

This is intended for emergency contraception only and cannot be used if a woman has had more than one act of unprotected intercourse during her monthly menstrual cycle. Minor side effects include possible nausea, bloated feeling, tender breasts, and mood changes. Use may temporarily change a woman's menstrual cycle. Generally, the medical risks are similar to those associated with birth control pills.

Fertility Awareness

What This Is

A woman can watch changes in her body to see when she is fertile (able to get pregnant).

How It Works

There is a period of about eight days in a woman's menstrual cycle when she is most likely to get pregnant if she has intercourse. Some women keep records of their temperature each morning, before they get out of bed. A woman can also check the mucus in her vagina, which changes at the time when her ovary releases an egg. A woman using fertility awareness must pay attention to her body signs each day, and keep careful records. During the woman's most fertile time, the couple can avoid having sex. People who use this method need to take training to learn how to watch the changes in the woman's body, and to understand what they mean.

How Well It Works

If 100 couples use this method for a year, there will be about 24 pregnancies. People who are very careful about keeping records, and about avoiding sex during the woman's fertile time, have better results.

Main Advantages

People who use this method don't need to use pills, or IUDs, with their side effects. The thermometers and charts needed to use it are easy to get. This method is highly approved by some religious groups.

Who Can Use It

Any woman can use fertility awareness if she is in good health, so that her body's signs of fertility are clear, and if she has been carefully taught how to check her body and keep good records. This method works best if a woman has regular menstrual cycles. It also works best if she has only one sexual partner, who works together with her to make it succeed. Most women who succeed check both their temperature and their vaginal mucus, instead of depending on just one symptom of fertility.

How It Is Used

The woman checks her body each day, following the directions she got from her instructor, and keeps careful records. Normally cloudy, tacky mucus from her vagina will become clear and slippery and will stretch between the fingers as the woman's egg is released from her ovary. She and her partner avoid having sex during the times her records show she might be fertile. If she has a problem understanding her records, she can call her instructor for help.

Possible Problems

Many things, such as a fever, or not sleeping, can change a woman's temperature. Other things, like an infection, or using a medicine or douche in her vagina, can change her vaginal mucus. If these signs are changed, she won't be able to tell her fertile time, and the couple needs to avoid sex for her entire menstrual cycle. Using this method takes a lot of dedication and self-control from both the woman and the man. People can get frustrated when they go a long time without having sex. Even when this method is used correctly, people do not have as much protection from pregnancy as they would have if they used other birth control methods.

Sources: Basic table provided courtesy of Planned Parenthood Federation of America, Inc. *The Methods of Birth Control,* 1993; Additional information on Depo-Provera is from Hatcher et al. 1992.

by observing changes in the woman's cervical mucus and by monitoring changes in basal body temperature. The effectiveness of this technique, described in Table D.1, depends on how correctly and diligently it is used.

Another method, withdrawal, is not included in Table D.1. It depends on the male's withdrawing his penis from the woman's vagina before he experiences orgasm. This technique is not effective: The male is tempted not to withdraw; and even if he does, the few drops of seminal fluid that are emitted before orgasm may contain sperm, making it possible for the woman to become pregnant.

In order to choose which alternative or alternatives to use,[1] people need to consider how each method works, how effective it is, its advantages and disadvantages, its side effects, its health implications for the user, and its long-term effects on the ability to have children. Of the methods discussed, oral contraceptives and sterilization are the most frequently used, as Table D.1 shows.

We have not yet mentioned one means of controlling fertility: **abortion.** Abortion differs from the methods already discussed because it does not prevent conception but terminates the development of the fetus after conception.[2]

---

1. Combining two (or more) methods—such as diaphragm and spermicidal cream—can often increase effectiveness greatly.

2. A form of fertility regulation termed *menstrual extraction* is difficult to classify as contraception or abortion. In menstrual extraction, the contents of the uterus are suctioned and scooped out at about the time of the expected menstrual period, whether or not a pregnancy has occurred. If performed regularly and under the assumption that there is no pregnancy, it can be viewed as a measure for health or convenience (avoiding debilitating periods), not as abortion. In effect, however, it would abort a zygote and may be performed with that purpose in mind.

ℰ

High-Tech Fertility

High technology has come to conception and pregnancy as modern science continues to develop new fertilization techniques. Among the less dramatic and more common are new drugs and microscopic surgical procedures. More dramatic and widely publicized high-tech fertilization methods include artificial insemination, in vitro fertilization, surrogate motherhood, and embryo transfers.

Artificial Insemination

This procedure may be indicated when a wife is presumably fertile but her husband is not. In **artificial insemination,** a physician injects sperm into a woman's vagina when she is ovulating.

AIH—ARTIFICAL INSEMINATION BY HUSBAND

In cases in which the husband's sperm count is low, the physician may accumulate several of the husband's ejaculations (which are preserved by refrigeration) so that the greater quantity of semen introduced into the vagina overcomes the low sperm count.

AID—ARTIFICIAL INSEMINATION BY DONOR

Should AIH fail, sperm from a donor other than the husband can be obtained and either mixed with the husband's sperm or used alone. When a donor's sperm is used, the physician may attempt to match the donor's physical characteristics with those of the husband. The practice of mixing the donor's and the husband's sperm makes for the possibility of fertilization by the husband's sperm—a possibility that can be psychologically important to the couple.

Often without disclosing it to others, couples and individuals choose artificial insemination far more widely than people may realize (Isaacs and Holt 1987). In 1989 the procedure was responsible for 65,000 births in the United States; an estimated 20,000 infants are born annually in the United States as a result of AID (Halpern 1989c, p. 149). The procedure is relatively simple (compared to those described below) and may even be done by an individual or couple at home.

In Vitro Fertilization

With **in vitro fertilization (IVF),** a fetus is conceived outside a woman's body (in a laboratory dish or jar) but develops within a woman's uterus. The process can be used when a woman with diseased or blocked fallopian tubes wants to give birth. An egg is removed from her surgically, fertilized in the laboratory with sperm from her husband (or, if her husband is infertile, from another donor), and implanted in her uterus after two days as a multicelled embryo. Pregnancy and childbirth follow the natural pattern. In more and more cases, fertilized embryos are frozen in liquid nitrogen for implantation later. Success rate is about 20 percent; cost is $6,000 to $10,000 for a single cycle of IVF (Brant, Springen, and Rogers 1995).

GIFT   In a related process known as **gamete intrafallopian transfer (GIFT),** eggs are collected from the ovaries and put into a catheter outside the body. Sperm are put into the same catheter but are kept apart from the eggs by an air bubble. The eggs and sperm are then placed in the woman's fallopian tubes. If fertilization subsequently occurs, the fertilized egg then travels to the uterus as is the case in a normal pregnancy. Success rate is about 28 percent; cost is $6,000 to $10,000 per attempt (Brant, Springen, and Rogers 1995).

ZIFT   Another related process is **zygote intrafallopian transfer (ZIFT).** In this procedure the fertilized egg is implanted in the woman's fallopian tubes after only one day, as a zygote (still a single cell) (Elmer-Dewitt 1991b).

The first in vitro baby was born in England in 1978. By 1992 there were 250 IVF clinics in the United States (Shellenbarger 1992g). It was estimated in 1989 that one cycle of IVF offered couples a 6 percent chance of having a baby (Halpern 1989c, p.

150). Success rate is about 24 percent; cost is $8,000 to $10,000 per attempt (Brant, Springen, and Rogers 1995).

Surrogate Motherhood

When a woman cannot carry a child to term and her husband is fertile and wants a child biologically his own, they can turn to a **surrogate mother.** Here a husband fathers a child with another woman by artificial insemination; this woman, the surrogate mother, carries the child to term and then turns the baby over to the couple. Note that the term *surrogate,* or substitute, is inaccurate because she is in fact the child's biological mother.

Embryo Transfers

In addition to this form of regular surrogacy, in which the surrogate mother is also the genetic mother of the child, is the newer pure surrogacy involving women who contribute no egg but carry the infant to term. Here an **embryo or ovum transfer** has occurred, in which fertilized eggs (not the receiving woman's) are transferred into her uterus.

Embryo transfer can occur when a woman who will rear the child cannot herself carry it to term; her fertilized eggs are transferred to the uterus of a gestational or pure surrogate mother who will carry the fetus to term and then deliver and relinquish the infant. Embryo transfer can also occur when the mother who will carry the infant plans to keep and rear it but cannot or does not want to (for example, because of a genetically transmitted disease) use her own eggs and genetic material ("First 'Ovum Transfer' Baby Born" 1984, p. 7).

Sexually Transmitted Diseases

AIDS has taken center stage, but other **sexually transmitted diseases** (**STDs**) threaten comfort, health, reproductive capacity, and sometimes life. The effects of STDs may extend to infants born to infected mothers. Public health officials have become very aware in recent years of the serious impact of STDs formerly thought to be relatively trivial. Early treatment can usually prevent the most severe outcomes.

About 12 million Americans become infected with at least one STD annually. It is estimated that 40 million Americans have at least one STD, and that of these, 30 million have herpes (CDC Hotline Statistical Information, November 1992).

STDs can be grouped according to different organizing principles: Diseases may be classified by the bacterium or virus that produces them, or they may be classified by their symptomatic effects. Such terms as vaginitis, urethritis, enteritis, and pelvic inflammatory disease refer to inflammations or infections of the vagina, urinary tract, digestive tract, and abdomen, respectively.

The Division of Sexually Transmitted Diseases at the Centers for Disease Control (CDC) in Atlanta is a central resource for information and research on STDs. The toll-free CDC STD Hotline number is 1-800-227-8922.

CDC guidelines recommend these principles regardless of the STD being treated:

1. Refrain from sex while infectious and while under treatment.

2. Inform sex partners so they can be treated.

3. Use condoms to prevent future transmission, as many STDs tend to recur.

Continue treatment as long as recommended and return for follow-up visits. Successful treatment of STDs can require a great deal of trial and error on the physician's part and considerable patience from the patient.

Because many STDs can be asymptomatic—that is, have no visible symptoms to indicate the presence of the disease—it is important for people at risk to have frequent checkups with knowledgeable physicians. "At risk" in this context probably means anyone with multiple sex partners or partners who themselves have a history of multiple partners or other current

Disease	Symptoms and Complications	Treatment
Chancroid	Genital ulcers; enlarged, sometimes painful lymph nodes in the groin.	Antibiotics.
Chlamydia	Urinary tract, rectal, or vaginal infections. These infections can cause pelvic inflammatory disease in women, which can lead to infertility and is potentially life-threatening.	Tetracycline or other antibiotics.
Cytomegalovirus	Frequently asymptomatic. Severe impact on people with impaired immunological systems, who may develop gastrointestinal problems or blindness. Possible retardation or death of newborn infants.	Contact CDC for current developments.
Enteric infections	Inflammation of the intestinal tract with symptoms of abdominal pain and cramping, diarrhea, fever, nausea, and vomiting, of varying severity. Can lead to systemic infections or severe dehydration. Commonly called "gay bowel" because it is associated with sexual practices common to gays. Patients may have a history of frequent oral–genital or oral–anal contact or a recent sex partner with intestinal illness.	Dietary regimen; medical treatment appropriate to enteric infections generally.
Genital warts	Warts, sometimes painful, on genital organs or rectum. Papilloma virus that causes them has been linked to cancer.	Surgical, chemical, or cryogenic (freezing of tissue) removal.
Gonorrhea	Genital irritation, discharge, painful urination. Can promote pelvic inflammatory disease in women, which can, in turn, lead to sterility. Sterility possible in men. Newborns of infected mothers at risk.	Tetracycline, penicillin, or other antibiotics. Penicillin-resistant strains have developed, making gonorrhea riskier than it had been and turning treatment into a search for the right antibiotic.
Granuloma inguinale	Genital or rectal ulcers. In the United States, more common in southern and southwestern states.	Antibiotics.

relationships. Individuals whose general health makes them vulnerable to infection are also at risk. For the truly cautious individual, any sexual relationship that is *not* long-term and *not* known to be monogamous could be considered risky. Some people, particularly women who have regular gynecological checkups any-

way, might simply wish to make STD screening a part of their health care.

Table F.1 describes the symptoms, complications, and treatment of STDs other than AIDS. AIDS is discussed at length in Chapter 5.

Disease	Symptoms and Complications	Treatment
Hepatitis B	This viral disease contracted through sexual contact or infected blood or needles is similar to hepatitis A, a liver disease contracted through contact with infected people or their urine or feces. Symptoms are general, such as nausea, vomiting, pain, malaise, loss of appetite, and jaundice but can be asymptomatic. Can cause short-term liver inflammation or chronic diseases such as cirrhosis and liver cancer.	None, other than good diet and rest, the common treatment for hepatitis. There is a vaccine for hepatitis B, recommended especially for overseas travelers.
Herpes (genital)	The genital lesions of herpes are similar to the cold sores that appear on the mouth. May be asymptomatic, but symptoms often take the form of a general malaise (tiredness, depression, low energy, not feeling right). Frequently recurs, but first episode is usually the worst. Can cause blindness, hearing problems, or death of infants born to infected mothers.	Acyclovir capsules or ointment can alleviate symptoms.
Lymphogranuloma venereum	Genital or rectal infection. Enlarged lymph nodes in the groin; pain, fever, and inflammation.	Tetracycline.
Syphilis	A genital sore, which disappears, followed in a few weeks by fever, swollen lymph nodes, and rash, which also recedes (but syphilis may be asymptomatic). May be in remission for some years, but arrival of last stage can damage heart, nervous system, brain, and other organs. In infected pregnant women, can cause stillbirth or birth defects.	Antibiotics. Early diagnosis is important; curable in early stages, but later-stage syphilis cannot be successfully treated.
Vaginitis	*Trichomonas,* bacterium, fungus, other infectious organisms, allergens, or physical irritation can cause inflammation, discharge, and itching of the vagina and genital area. Male sex partners may develop urethritis (inflammation of the urinary tract) or penile lesions. Vaginitis from some causes can result in an oral infection in newborn infants.	Physicians treat this common and persistent condition with various medications and also need to treat infected people's partners to avoid reinfection. Women's health books such as *The New Our Bodies, Ourselves* (Boston Women's Health Collective 1992) provide useful advice on proper diet, clothing, hygiene, and so on, which can help prevent some forms of vaginitis.

Sources: U.S. Centers for Disease Control, "Sexually Transmitted Disease Summary, 1990"; U.S. Department of Health and Human Services, "1993 Sexually Transmitted Diseases Treatment Guidelines."

Sexual Dysfunctions and Therapy

Therapists distinguish six sexual dysfunctions: premature ejaculation, retarded ejaculation, and impotence among men; and general female sexual dysfunction, female orgasmic dysfunction, and vaginismus among women (see Table G.1).

Sexual dysfunctions, along with the more general situation of decreased sexual desire, can be related to certain physical disabilities and chronic diseases, such as heart disease, diabetes, hepatitis, and hyperthyroidism (Levay, Sharpe, and Kagle 1981). Some surgical and chemotherapy treatments for cancer can cause sexual dysfunctions (Levay, Sharpe, and Kagle 1981). Also, many drugs are related to sexual difficulties. These include drugs for hypertension and some heart conditions, antidepressants, antianxiety drugs, most narcotics (for example, heroin, morphine, codeine, methadone), and LSD, marijuana, and cocaine (Horwith and Imperato-McGinley 1983, pp. 191–92).

Premature Ejaculation

Premature ejaculation, the inability to control the ejaculatory reflex voluntarily, is one of the most common male sexual complaints. A man might ejaculate after several minutes of foreplay or just after entering his partner's vagina. In contrast, a man who has good ejaculatory control can continue to engage in sex play while in a highly aroused state.

The essential problem is not how quickly the man ejaculates but his inability to control the reflex. One way therapists deal with premature ejaculation is to teach a couple an exercise through which the man can gradually learn to control his orgasm. Therapists report that in most cases they have treated, premature ejaculation eventually disappears (Kaplan 1983; Nadelson and Marcotte 1983, pp. 32–33).

Retarded Ejaculation

A man afflicted with **retarded ejaculation,** or ejaculatory inhibition, cannot trigger orgasm. Clinicians once thought that retarded ejaculation was relatively rare, but it now appears to be prevalent, at least in its mildest form.

In mild form, ejaculatory inhibition is confined to specific anxiety-producing situations, such as when a man is with a new partner or when he feels guilty about the sexual encounter. In more severe cases, a

TABLE G.1

Common Sexual Dysfunctions

Dysfunction	Symptoms	Usual Treatment
Premature ejaculation	Inability of a man to control ejaculatory reflex	Repeated stimulation to the point just before ejaculation
Retarded ejaculation	Inability of a man to trigger orgasm; may be situational or a general dysfunction	Sexual exercises combined with therapeutic counseling; temporary avoidance of intercourse and use of other means to elicit ejaculation
Impotence (erectile inhibition)	Inability of a man to produce or maintain an erection	Sexual exercises combined with therapeutic counseling; focus shifted away from performance aspect of sexual interaction
General female sexual dysfunction	Inability of a woman to derive erotic pleasure from sexual stimulation	Education about arousal techniques; creation of relaxed, sensuous environment free from pressure to have intercourse
Female orgasmic dysfunction	Difficulty of a woman in reaching orgasm	Focus on helping woman learn to reach climax by herself, then with husband in sexual exercises not initially aimed at intercourse
Vaginismus	Involuntary contraction of vaginal walls that prevents coitus	Correction of possible physical conditions; counseling plus exercises to recondition musculature

NOTE: Treatment is preceded by a complete physical examination to identify any physiological causes for the disturbance.

man may seldom experience orgasm during intercourse but may be able to achieve it by masturbation or by a partner's fondling or oral stimulation.

Once physical or drug-related causes are ruled out, treatment consists of marriage counseling sessions along with a series of progressive sexual exercises designed to relieve the man of his fears about intercourse. The rate of success with therapy is fairly high (Kaplan 1983; Nadelson and Marcotte 1983, p. 21).

Impotence

A man suffering from **impotence**, or erectile inhibition, is unable to produce or maintain an erection. Although he may become aroused in a sexual encounter and want to have intercourse, he cannot. Clinicians

and researchers estimate that half the male population has experienced at least temporary impotence.

Physicians estimate that between 40 and 60 percent of the impotence cases they see have physical causes ("Male Impotence" 1985). In some cases, chronic impotence is successfully treated by surgically inserting an inflatable implant into the penis and scrotum.

Although impotence can sometimes be a sign of deeper psychological problems, it is more often the result of situational causes, such as a fear of sexual failure, pressures created by an excessively demanding partner, or guilt. Unfortunately, our society tends to equate the capacity to have an erection with adult masculinity, so that even transient impotence may cause a man to feel anxious, frustrated, and humiliated. In impotence, as in other sexual dysfunctions, the anxiety produced by one otherwise insignificant and transitory failure can initiate a downward spiral in which anxiety retards sexual responsiveness, leading to more anxiety about performance, less sexual success, and so on.

Depression or relationship discord, or both, often accompany impotence, and these symptoms must be at least somewhat relieved before a therapist can treat the impotence itself. Therefore, therapists combine sexual exercises at home with therapeutic counseling.

The exercises are designed to free the man from pressures to perform and let him simply enjoy his sexual feelings. Essentially, the couple is instructed to caress each other during sexual play but *not* to have intercourse. Permission to enjoy himself without having to perform allows the man to relax without worrying whether his body will respond. Paradoxically, the more he relaxes, the more likely his body *is* to respond. This same philosophy lies behind much of the treatment for female sexual dysfunction.

General Female Sexual Dysfunction

Women who experience **general female sexual dysfunction** derive little if any erotic pleasure from sexual stimulation. Some women have never experienced erotic pleasure; others have at one time but no longer do. Often they enjoyed petting before marriage but became unable to respond when intercourse was the expected goal of sex.

Besides giving the couple basic information, therapists encourage them to create a relaxed, sensuous atmosphere at home, one that allows for the natural unfolding of sexual responses. In one exercise, the couple take turns caressing each other, but they do not progress to sexual intercourse and orgasm. Freed from the pressure to have intercourse, a woman can often experience erotic sensations, and the couple can gradually build on this sensation of pleasure until they are eventually ready for intercourse.

Female Orgasmic Dysfunction

Difficulty in reaching orgasm, or **female orgasmic dysfunction,** is the most common sexual complaint among women. About 20 percent experience orgasm infrequently or not at all (Konner 1990; see also Laumann et al. 1994). A few women cannot reach a climax under any circumstances. More often, a woman can reach orgasm, but only under specific conditions.

Kaplan points out that many women with this dysfunction enjoy sex; they just "get stuck" at the plateau phase and cannot proceed to a climax (Masters and Johnson 1970; Masters, Johnson, and Kolodny 1994). As in impotence, anxiety about performance can further inhibit a woman's sexual responsiveness.

Treatment for women who have never experienced orgasm usually begins by focusing on the woman alone. The therapist asks the woman to masturbate at home alone, stressing that the environment should be free from distractions and interruptions. Another approach for women who have never experienced orgasm is group education. Women meet together to learn about their bodies; they are then encouraged to masturbate at home until they become familiar and confident with their own response cycles (Masters, Johnson, and Kolodny 1994). Research confirms the effectiveness of both masturbation and various forms of talk therapy, including general couple therapy (Konner 1990).

Once a woman can stimulate herself to climax, her partner enters the treatment program. The couple is told to make love as usual, except that after the man ejaculates, he stimulates his partner to orgasm. The woman is told to be utterly selfish, not to monitor her progress toward orgasm but to simply enjoy her sensations (Masters, Johnson, and Kolodny 1994). Women are cautioned that watching one's own response to see if it's "right"—that is, headed toward orgasm—tends to inhibit physical responsiveness and to contribute to tension that sometimes develops into long-term sexual problems (Masters and Johnson 1970; Masters, Johnson, and Kolodny 1994). Rather, each partner is to enjoy the pleasurable sensations produced by the caresses of the partner.

This treatment is helpful in letting couples see beyond the myth of the simultaneous orgasm—the erroneous idea that true love or really great sex means that both partners must reach orgasm at the same time. Sometimes partners do climax simultaneously, but not usually. The belief that they *should* can leave the woman, who is typically slower to become aroused, frustrated; it may even encourage her to fake it. It may be better to take turns in being pleasured to orgasm.

DIRECT CLITORAL VERSUS VAGINAL STIMULATION   One reason it may be better to take turns is that many women report they do not reach orgasm through vaginal stimulation in intercourse, despite the earlier myth that vaginal orgasms are more mature than clitoral orgasms. As Appendix A points out, there is a much greater concentration of nerve endings in the clitoral prepuce than in the vagina itself. One possible pattern is for the husband to stimulate his wife's clitoris until she reaches orgasm and then to enter her vagina to attain his own climax.

One sex therapist estimates that of the 90 percent of women who have experienced orgasm, only about half do so regularly during intercourse. Most experts and therapists see the need for direct clitoral stimulation as normal to female sexuality. The best strategy, it would seem, would be for the individual woman to be aware and make her partner aware of her own response pattern.

Some women never achieve orgasm with a partner even with direct clitoral stimulation, although they are able to climax by masturbating. Typically, this situation reflects a woman's anxiety, ambivalence, or anger about the relationship. Treating this type of inorgasm typically involves individual or marital therapy or both.

Vaginismus

Vaginismus is relatively rare. A woman with this dysfunction is anatomically normal, but whenever her partner attempts to penetrate her vagina, the vaginal muscles involuntarily contract so that intercourse is impossible. Typically, vaginismic women are, at least unconsciously, afraid of vaginal penetration and intercourse.

After any physical conditions have been corrected, therapists treat vaginismus by seeking in counseling sessions to uncover the basis for the woman's fear of vaginal entry. Then progressive exercises are used to recondition the muscles at the entrance to the vagina. The length of the treatment varies, but therapists report excellent results (Kaplan, 1974).

In all of the dysfunctions we have described, a common thread is the emotional climate of a couple's relationship. This emotional relationship may be a cause of the dysfunction or it may be affected by the dysfunction or both. Although therapists help a couple overcome their immediate sexual difficulties, they also try to help partners recognize and avoid alienating practices that can become obstacles to mutually pleasurable sex.

Sex Therapy

In traditional approaches to treating sexual dysfunctions, therapists looked for subtle and profound psychological sources, such as unresolved emotional conflicts from childhood or severe marital power struggles. These causes and therapies still exist, but today most therapists focus on more immediate and obvious reasons for the dysfunction: anxieties about sexual failure or fear that the partner expects too much or that the partner will reject sexual advances. These fears create various sexual defenses, introduce conscious control into lovemaking, and thus inhibit people from abandoning themselves to the experience. One important feature of Masters and Johnson's (1970) therapy is its attempt to remove performance pressure by insisting that the couple *not* strive for orgasm or even have intercourse but rather focus on all-over body pleasure and pleasuring.

Masters and Johnson laid down some ground rules for sex therapy in their book *Human Sexual Inadequacy* (1970). They said that therapists should work in male–female teams and with both partners in the relationship. They stressed that the team should be comfortable with their own sexuality and nonjudgmental about the full range of human sexual activity. Since then, some therapists—Helen Singer Kaplan, head of the Sex Therapy and Education Program in New York, for example—have successfully treated couples without a cotherapist. However, many contemporary therapists follow the Masters and Johnson guidelines. Moreover, therapists are violating professional ethics if they become sexually involved with clients.

Therapy normally begins with a physical and psychological examination of both partners. Although it is commonly held that only about 10 percent of sexual dysfunctions have a physiological basis, the last decade has witnessed a significant return to physical treatments, such as hormone-injection treatment and/or penile implant.

A legitimate therapist will give a couple a clear picture of what to expect during treatment and will probably make a therapeutic contract with them that clearly establishes the couple's responsibility for their treatment.

One way to check therapists' qualifications is to find out whether they belong to either EAST or AASEC. EAST (Eastern Association of Sex Therapists) is a group of approximately 100 legitimate practitioners. AASEC (American Association of Sex Educators and Counselors) has developed a certification program that requires academic training, supervised therapy experience, and a written examination. In the absence of membership in either association, therapists are more likely to be legitimate if they are accountable to a community agency, teaching hospital, medical school, or university. Some states license or certify therapeutic professionals of various kinds.

If a couple cannot find a therapist with these qualifications, or if costs are prohibitive, there are other choices, such as, for example, biofeedback or hypnosis—but the success rates for these is difficult to determine. Group therapy for sex problems has developed. This approach is especially effective for women, in all-women's groups, who may not respond as well to couples therapy if power issues are an underlying problem and the male tries to dominate the sessions. A choice for couples is to recognize that sexual problems often reflect the general relationship and to seek help from a qualified marriage counselor.

Sex therapy, as with therapy or social action in other areas of life, has often been in advance of scientific research. This does not necessarily mean that therapy programs should not be developed or that people should not seek sex therapy; however, we need to keep in mind the limits of our certainty about it. Increasingly in recent years, individuals have chosen medical, rather than psychological, treatment for some sexual dysfunctions, particularly impotence.

Marriage and Close Relationship Counseling

Marriage and close relationship counseling is a professional service dedicated both to helping individuals, couples, and families gain insight into the actually or potentially troublesome dynamics of their relationship(s) and to teaching clients more effective and supportive communication techniques. Experts have suggested that couples or families should visit a counselor when communication is typically hostile or conflict goes unresolved, when they cannot figure out how to resolve difficulties themselves, when a partner is thinking of leaving a committed relationship, or when a problem in the relationship appears to be linked to a personality disorder in one or more family members (such as chronic drinking, drug abuse, severe depression, or deep feelings of insecurity and inadequacy). But counseling is also appropriate—and perhaps more effective—as a preventive technique, undertaken at the onset of a family crisis or when a couple or family sees a potentially troublesome transition ahead. Today people go to counselors for help in working through premarital and engagement issues, as well as cohabitation, marriage, divorce, remarriage, and stepfamily issues.

Qualifications of Counselors

The qualifications of marriage counselors vary. A counselor who is a member of the **American Association for Marital and Family Therapy (AAMFT)** has a graduate degree (in either medicine, law, social work, psychiatry, psychology, human development and family studies, or the ministry) in addition to special training in marriage or family therapy or both and at least three years of clinical training and experience under a senior counselor's supervision.

Not all those who practice counseling are so well qualified, however, and some have, in fact, taken on the responsibility of training themselves. About half the states license marriage counselors. Some require counselors to pass oral and written tests in order to practice. Many states, however, require no license at all. The safest way to choose a counselor is to select one who belongs to the AAMFT. AAMFT has a toll-free number for referrals in your area: 1-800-374-AMFT (1-800-374-2638). Or, send a self-addressed, stamped envelope to AAMFT, 1133 15th St. N.W., Washington, DC 20005. Personal references from friends may also be helpful.

It is important to have a counselor you like, trust, and feel is sympathetic to you. It is also important that the counselor respect your religious and personal values. If after three or four sessions you do not feel comfortable with the counselor or don't believe she or he is effective, it might be a good idea to try someone else (Ambroz 1995).

Marriage Counseling Approaches

Marriage/close relationship counseling can be either a short- or long-term arrangement. A difficulty might be cleared up in a few weekly sessions. In other cases, counseling might last a year or more. Or a couple or family might work through a problem in a few visits, then quit with the understanding that they'll return if conflicts once more begin to go unresolved. In all cases, counseling should have definite goals and should aim at termination instead of becoming an indefinite program.

Qualified counselors have widely varying approaches to their work; couples and families would do well to inquire about this before engaging a counselor. For example, counselors whose primary training is in psychiatry or psychoanalysis may view problems in relationships as the result of at least one family member's personal neurosis. Such counselors would believe that restoring each individual to emotional health is the first and most important step in improving the relationship and so would probably suggest seeing each family member individually.

Another approach to marriage counseling is **conjoint marital counseling.** Instead of counseling one partner at a time, the counselor sees husband and wife together. In such an approach, counselors help partners learn to interact more constructively. A related approach is family systems therapy, in which as many family members as possible—sometimes even extended kin—are engaged together in therapy.

All we've said about counseling is based on the presumption that partners are willing to cooperate. It is entirely possible, however, that one's partner may not be willing. Something no counselor can or will attempt to accomplish is changing a person to a partner's liking without active cooperation from all involved (Ambroz 1995).

Managing a Family Budget

Americans think of their spending power in terms of how much money they make. But they can increase their real income—the amount of goods and services their money will buy—as much as 10 percent just by planning (Smith 1989). And there are other good reasons to budget:

- Budgeting encourages people to stop and think— to make conscious decisions about what goods and services they really want—before they spend.

- Budgeting makes it easier to adjust irregular income (whether through commissions, fluctuating profits, or seasonal employment) to regular expenses. Planning to reserve money in good months can help even things out when income falls.

- Budgeting helps when family income increases or decreases and when the amount families need to spend on certain things changes. For example, neglecting to plan ahead during an expansive period (such as just before retirement) can accustom a family to a lifestyle it may be surprised to find it can't sustain during the lean years that follow. Bud-

geting also helps during inflation and recession as real income falls.

- Budgeting helps people discover and plug leaks in their expenditures. Records can show where the money goes and help people avoid spending carelessly. Using a budget helps families pinpoint just how much of their income is **discretionary income:** uncommitted money that can be spent as the family pleases. Being aware of that figure helps in making decisions about spending.[1]

- Budgeting encourages two-paycheck partners to decide whether the second wage-earner's salary is to be considered permanent or temporary.

- Finally, budgeting encourages family members to reexamine their personal and family goals and values and to negotiate and cooperate about spending.

---

1. Don't feel you're in the minority, by the way, if you find yourself without discretionary money. According to one government-business report, only about one-third of American households have money left over after paying for a comfortabled standard of living ("One-third" 1985).

Budgeting is one avenue to renegotiating a couple's personal marriage agreement. Because it helps get important issues into the open, budgeting can minimize power politics in a family and thereby help increase intimacy.

Steps in Planning a Budget

Here are some general tips for planning a budget. You can adapt them to your own interests and needs.

1. *Assess the situation.* You must first know where you stand. Prepare a balance sheet showing all of your financial assets and liabilities.

2. *Set your goals.* Decide on your priorities and goals, whether they may be paying off your college loan in two years, saving for a house down payment, or simply paying off your credit-card debt.

3. *Estimate your income.* Make as accurate an estimate as you can, including such income as salaries, gifts, commissions, and bonuses. Don't be overly optimistic about raises or projected commissions and profits. Unrealistic estimates can result in overspending.

4. *Compile a spending inventory.* To get an accurate picture of current expenses, keep track of everything you spend for at least a week. Your spending habits—and knowledge of them—may surprise you. If you're skeptical and don't see a reason to keep track, be aware that some overweight people insist they never eat; when they keep a daily record of the calories they take in, they are amazed. In a similar fashion, you may find, for example, that you spend $25 on lunches, not the $10 you had thought.

5. *Prepare the budget.* You might create a budget form similar to Figure I.1. To adapt it, you might omit some items and substitute others. Write in intended expenses for each item for every month.

6. *Follow through.* This involves keeping daily records of expenditures in a convenient notebook or in an account book bought for that purpose. At the end of each month, you should total daily expenditure figures for each item, then compare them with the amounts planned in step 4. If January had more food expenditures than were budgeted, either this expenditure would have to be cut back in February or more money allocated for food by taking it from some other category. If excessive expenses show up in successive months, you may decide to eliminate some expensive purchases. If expenditures for one account consistently fall below what was planned, the budget can be revised.

About Saving

Being able to save regularly is highly unlikely (if not impossible) for low-income and poverty-level families. But middle-income people need to consider it seriously. Financial advisers recommend regularly saving 10 percent of your gross income. If that seems too difficult, start with less—even just 1 percent—and work up (Smith 1989). Employed people can usually save automatically by setting up payroll deduction plans at work. In this case, a fixed percentage of your salary is regularly deposited into a credit union or bank savings account. The value here is that the amount is deducted *before* you receive your check so that after a few paychecks you may not miss it. Another way to save automatically is to have your bank regularly withdraw a fixed amount from your checking account and deposit it into a savings account.

Individuals or families can save for a particular item—perhaps for a house down payment or a college education for the children, or for a short-term consumer purchase such as a VCR or a vacation. Families should also gradually establish an emergency fund (the old "saving for a rainy day" maxim). When possible, emergency funds ought to equal three to six months' living expenses. Remember that you'll need to retrieve money in an emergency fund quickly; for example, don't put it in a ten-year government bond.

Besides an emergency fund, saving toward retirement has become increasingly important, and people need to begin early. Today's workers cannot count on Social Security or even typical corporate pension plans to maintain their standard of living in retirement. Therefore, it makes sense to begin retirement saving as soon as you can. "Early savers have a silent partner that does half the work," according to one financial planner. "It's called compounding" (in Luciano 1989, p. 85). For example, a 25-year-old who saves $2,000 each year for 10 years at 8 percent compounded annual interest will have $198,000 at age 60. But if a 45-year-old began the same savings plan, he or she would have just $43,000 at age 60 (Luciano 1989, p. 85).

There are ways to cut your taxes and save toward retirement at the same time. Supplemental Retirement Accounts (SRAs) are tax-deferred savings plans available to those with employment income whose companies participate. Interest accrued on these savings is

	January	February	March	April	May	June	July	August	September	October	November	December
TOTAL INCOME												
Breadwinner A												
Breadwinner B												
EXPENDITURES												
"His" personal money												
"Her" personal money												
Family expenses												
1. Food												
2. Housing												
3. Household operation												
4. Household furnishings												
5. Clothing												
Father												
Mother												
Child 1												
Child 2												
6. Transportation												
Automobile												
7. Taxes												
Income (withholding)												
Social Security												
Property												
8. Personal care (shampoo, deodorant, cosmetics, etc.)												
9. Medical care												
10. Dental care												
11. Insurance (life, disability)												
12. Savings												
13. Vacation and recreation												
14. Newspapers, magazines, postage												
15. Education												
16. Dues												
17. Contributions												
18. Gifts												
19. Miscellaneous												

FIGURE *I.1*

Sample budget form for one- or two-paycheck families.
(Source: Adapted from Gordon and Lee 1977, p. 369)

not taxed until withdrawn in retirement. In some cases, contributions may be tax deductible, depending on your income and marital status. Keogh is a retire- ment savings plan for people with self-employment income. Contributions are tax deductible—up to 20 percent of your net self-employment income or a particular dollar figure, whichever is less—and earnings are tax deferred until withdrawn (Luciano 1989, p. 85). At the least, you should ask at your bank for details. But you might want to consult an investment adviser about the best strategy for maximizing the long-term payoff from your savings.

About Credit

In our country, credit cards and time purchases have become a way of life in recent years. Yet credit costs money, and it should be considered a purchase in its own right. Financial writer Sylvia Porter (1976) has offered some right and wrong reasons for borrowing. Right reasons include establishing a household or beginning a family, making major purchases, taking advantage of seasonal sales, financing college or other educational expenses, and genuine emergencies.

Wrong reasons include buying to boost morale, to increase one's status, or on impulse. People misuse credit when they use it to maintain an adequate cash reserve. Another misuse is financing purchases against an uncertain but hoped-for raise or future financial windfall.

SOURCES OF CREDIT   People borrow money from many different sources. Some turn to their parents or, less often, to brothers and sisters; in such cases, interest is likely to be low. Sometimes, relatives help each other by putting up security or cosigning bank loans. Family borrowers need to consider psychological as well as monetary costs of intrafamilial transactions.

Another source of credit is public lending agencies: credit unions, banks, finance companies, and pawnshops. Interest rates vary widely among the different types of agencies. One of the most highly advertised sources of credit is the small loan company, which specializes in lending to borrowers who have little or no security. These sources serve a purpose, for without them many people with no security or poor credit ratings could not legally borrow money. But because of the risk small loan companies take, they charge extremely high interest rates and are best avoided in favor of other sources of credit.

Another way to borrow money is by arranging retail **installment financing,** in which people contract with a dealer to pay for major purchases over a period of time. The dealer receives the markup on the merchandise and the interest on the unpaid balance. When making major purchases, people should study installment contracts with care and ask questions about anything they don't understand. Find out how much the item will *really* cost—that is, the price of the item plus the hidden cost of financing. Compare interest rates and consider whether a personal bank loan would be a less expensive way of paying for the item.

It is important to shop for credit as carefully as for any other major purchase. Furthermore, people should try to pay off one installment obligation before taking on a new one. Moreover, consumers need to remember that using credit cards is the same as borrowing money, and credit-card purchases need to be budgeted the same way cash purchases are. Statistics on credit misuse are high. A good rule is to limit borrowing so that debt payments, excluding mortgage payments, account for no more than between 15 and 20 percent of take-home pay.

FINANCIAL OVEREXTENSION AND WHAT TO DO ABOUT IT   Optimally, families pay all their installment and other loans monthly, along with other regular expenses, and also set some money aside for savings. Some advisers suggest that a family is overextended if they have less than three months' take-home pay in savings for emergencies.

If an individual or a couple is financially overextended, the first step in solving the problem is to make a conscious choice to change things. Plan a budget jointly with a spouse or other family member who shares the budget responsibility. Look for places to cut expenses. Resolve not to use credit cards or take out new loans until your situation has improved. If necessary, cut up all your credit cards—the first action credit counselors take for many of their clients (Klein 1989). Ask your creditors if they will agree to spread payments over a longer period. Usually, they will be willing to work out some temporary arrangement.

If you feel it's necessary, consult a credit-counseling service. Write the National Foundation for Consumer Credit or check the yellow pages of your phone book. Credit-counseling services associated with the national foundation provide free budget advice and charge a nominal fee for debt management. Often this involves working out a reduced-payment arrangement with your creditors. It is important to make sure you're dealing with a legitimate credit counselor, though. Some unscrupulous people advertising themselves as credit counselors may charge such high fees that the fees will outweigh the benefit of any services they offer—and push you over the brink of indebtedness.

Some people aim to solve their credit problems by taking out a **consolidation loan:** a single loan large enough to cover all outstanding debts. Because the consolidation loan is repaid over an extended period of time, the monthly payments are smaller than they would be if a couple were making separate payments on a variety of shorter-term bills. The total interest, however, is greater. A problem with consolidation loans is that people tend to use them as a stopgap measure rather than changing their buying habits. "A consolidation loan is fine if you get religion," said one bank official. The problem is that many people don't reform. "Unless you're a paragon of self-discipline, resist this temptation" (Klein 1989). Most families end up owing money again within about six months because they continue to charge on their paid-up credit cards.

Some people file for bankruptcy as a means of getting on their financial feet again. Indeed, the number of U.S. bankruptcies has more than doubled since 1985 ("Credit-Card Role" 1989). A recent federal law makes declaring bankruptcy easier than it used to be, but declaring bankruptcy will seriously impair a credit rating and will remain on the credit record for up to ten years.[2]

---

2. The Fair Debt Collection Practices Act provides protection for over-extended consumers. Among other things, the law prohibits independent collection agencies from making abusive or threatening phone calls, calling before 8 A.M. or after 9 P.M., or contacting friends or employees except to verify where you work or live. Direct any questions or complaints to the Fedceral Trade Com mission in Washington, D.C., or to a local Legal Aid Society.

Glossary

A

AAMFT American Association for Marital and Family Therapy. Membership requires a graduate degree in medicine, law, social work, psychiatry, psychology, or the ministry; special training in marriage or family therapy or both; and at least three years of clinical training and experience under a senior counselor's supervision.

ABC-X model A model of family crisis in which A (the stressor event) interacts with B (the family's resources for meeting a crisis) and with C (the definition the family formulates of the event) to produce X (the crisis). See also *double ABC-X model, strains, resiliency model, stressor.*

abortion The expulsion of the fetus or embryo from the uterus either naturally (spontaneous abortion or miscarriage) or medically (induced abortion).

abstinence The standard that maintains that regardless of the circumstances, nonmarital intercourse is wrong for both women and men. Many religions espouse abstinence as a moral imperative. See also *double standard, permissiveness with affection, permissiveness without affection.*

acquaintance rape Forced or unwanted sexual contact between acquaintances. See also *date rape.*

A-frame relationship A relationship style (symbolized by the capital letter A) in which partners have a strong couple identity but little self-esteem; therefore, they are dependent on each other rather than interdependent. See also *H-frame relationship, M-frame relationship.*

afterbirth The placenta, amniotic sac, and the remainder of the umbilical cord, all of which are expelled during delivery after the birth of the baby.

agape The love style that emphasizes unselfish concern for the beloved in which one attempts to fulfill the other's needs even when that means some personal sacrifice. See also *eros, ludus, mania, pragma, storge.*

age expectations Societal expectations about how people should think and behave because of their age. For instance: Women over ____ (you fill in the blank) should not wear bikinis; men over ____ should not wear tank tops.

agreement reality Knowledge based on agreement about what is true.

AIDS (acquired immune deficiency syndrome) A sexually transmitted disease involving breakdown in the manufacture of white blood cells, which resist viruses, bacteria, and fungi.

alienating fight tactics Tactics in fighting that tend to create distance between intimates: They don't resolve conflict; they increase it. See also *gaslighting, gunnysacking, kitchen-sink fight.*

alimony/spousal support Derived from the Latin verb meaning "to nourish" or "to give food to"; the traditional condition in which the breadwinning husband undertook the obligation to support his dependent wife and children even after divorce. With the advent of no-fault divorce laws, the term *alimony* was replaced by "spousal support" or "maintenance." The change in terminology reflects a departure from the obligation of lifetime support, which was premised on wifely economic dependency in a lifelong contractual relationship. Spousal support is presumed to be temporary, a practical means of meeting a spouse's economic needs during a period of readjustment.

ambivalent provider couple The wife's income is essential to the family, but her responsibility as an economic provider is not acknowledged.

amniotic fluid Salty, watery fluid that surrounds the developing fetus within the mother's uterus, cushioning and protecting it; also called *amnion.*

androgyny The social and psychological condition in which individuals can think, feel, and behave in ways that express both instrumental and expressive character traits. Androgyny is the combination of both masculine and feminine qualities in one individual. Androgynous people will probably be better equipped to deal with industrial society. Moreover, androgynous partners may find greater intimacy in marriage than do more traditional spouses.

anger "insteads" Ways people deal with their anger rather than expressing it directly. Some substitutes for open anger are overeating, boredom, depression, physical illness, and gossip.

archival family function The creating, storing, preserving, and passing on of particular objects, events, or rituals that family members consider relevant to their personal identities and to maintaining the family as a unique existential reality or group.

areolae The pigmented areas of the breasts surrounding the nipples.

artificial insemination A process in which a physician injects sperm into a woman's vagina when she is ovulating in an attempt to achieve pregnancy.

assignment The designation of gender identity to a hermaphrodite shortly after birth. The person is then treated accordingly.

attachment disorder An emotional disorder in which a person defensively shuts off the willingness or ability to make future attachments to anyone.

attribution Assigning or attributing character traits to other people. Attributions can be positive or negative, such as when a spouse is told he or she is an interesting or a boring person.

autocratic discipline Discipline of children that places the power of determining rules and limits entirely in the hands of the parent, with little or no input from the growing child. See also *democratic discipline, laissez-faire discipline.*

B

battered woman syndrome A circumstance in which a battered woman feels incapable of making any change in her way of living.

bereavement A period of mourning after the death of a loved one.

bonding fighting Fighting that brings intimates closer together rather than leaving them just as far apart or pushing them even farther apart. See also *alienating fight tactics.*

borderwork Interaction rituals are based on and reaffirm boundaries and differences between girls and boys.

breech presentation In childbirth, a delivery in which the baby's buttocks, shoulder, foot, or face emerge first. This makes for a more difficult delivery than the vertex presentation.

C

case study A written summary and analysis of data gained by psychologists, psychiatrists, counselors, and social workers when working directly with individuals and families. Case studies are often used as sources in scientific investigation.

center care Group child care provided in day-care centers for a relatively large number of children.

cervix At the top of the vagina in the female, the neck of the uterus.

cesarean section A surgical operation in which a physician makes an incision in the mother's abdomen and uterine wall to remove the infant. Named after Julius Caesar, who was supposedly delivered this way.

checking-it-out A communication or fighting technique in which a person asks the other whether her or his perceptions of the other's feelings or thoughts are correct.

child abuse Overt acts of aggression against a child, such as excessive verbal derogation, beating, or inflicting physical injury. Sexual abuse is a form of physical child abuse. See also *emotional child abuse or neglect.*

child care As defined by policy researchers, the full-time care and education of children under age 6, care before and after school and during vacations for older children, and overnight care when employed parents must travel.

child neglect Failure to provide adequate physical or emotional care for a child. See also *emotional child abuse or neglect.*

children's allowance A type of child support that provides a government grant to all families—married or single-parent, regardless of income—based on the number of children they have.

child snatching Kidnapping one's own children from the other parent after divorce.

child support Money paid by the noncustodial to the custodial parent to financially support children of a separated marital, cohabiting, or sexual relationship.

Child Support Amendments (1984) Amendments to the Social Security Act that (1) encouraged establishment of paternity and consequent child support awards, (2) required states to develop numerical guidelines for determining child support awards, (3) required periodic review of award levels, and (4) enforced payments by various means.

Chodorow's theory of gender A theory of gender socialization that combines psychoanalytic ideas about identification of children with parents with an awareness of what those parents' social roles are in our society.

choosing by default Making semiconscious or unconscious choices when one is not aware of all the possible alternatives or when one pursues the path of least resistance. From this perspective, doing nothing about a problem or issue, or making no choice, is making a serious choice—that is, the choice to do nothing.

choosing knowledgeably Making choices and decisions after (1) recognizing as many options or alternatives as possible; (2) recognizing the social pressures that can influence personal choices; (3) considering the consequences of each alternative; and (4) becoming aware of one's own values.

clitoris Part of the female genitalia: female erectile tissue, consisting of an internal shaft and a tip, or glans, which contains concentrations of nerve endings and is highly sensitive.

closed adoption An adoption process in which the adoptive and biological families have no communication and do not know one another's names. See also *open adoption, semi-open adoption.*

closed marriage A synonym for static marriage.

coercive power One of the six power bases, or sources of power. This power is based on the dominant person's ability and willingness to punish the partner either with psychological-emotional or physical violence, or with more subtle methods of withholding affection.

cohabitation Living together in an intimate, sexual relationship without traditional, legal marriage. Sometimes referred to as living together or marriage without marriage. Cohabitation can be a courtship process or an alternative to legal marriage, depending on how partners view it.

coitus Sexual intercourse.

commitment (to intimacy) The determination to develop relationships in which experiences cover many areas of personality, problems are worked through, conflict is expected and seen as a normal part of the growth process, and there is an expectation that the relationship is basically viable and worthwhile.

commitment (Sternberg's triangular theory of love) The short-term decision that one loves someone and the long-term aspect—the commitment to maintain that love; one dimension of the triangular theory of love.

common law marriage A legal concept whereby cohabiting partners are considered legally married if certain requirements are met, such as showing intent to enter into a marriage and living together as husband and wife for a certain period of time. Most states have dropped common law marriage, but cohabiting relationships may sometimes have a similar effect on property ownership and rights in children.

commune A group of adults and perhaps children who live together, sharing aspects of their lives. Some communes are group marriages, in which members share sex; others are communal families, with several monogamous couples, who share everything except sexual relations, and their children.

commuter marriage A marriage in which the two partners live in different locations and commute to spend time together. Also called two-location family.

conception The moment in which an ovum (egg) joins with a sperm cell and a fetus begins to develop.

conflict-habituated marriage A marital relationship characterized by ongoing tension and unresolved conflict. See also *devitalized, passive-congenial, total, vital marriage.*

conflict perspective Theoretical perspective that emphasizes social conflict in a society and within families. Power and dominance are important themes.

conflict taboo The cultural belief that conflict and anger are wrong, which thereby discourages people from dealing with these negative emotions openly.

congestion The engorgement of blood vessels, which causes the affected tissues to swell. Occurs to the genitals during the excitement phase of sexual response.

conjoint marital counseling Counseling in which the counselor sees the husband and wife together rather than one at a time.

conjugal kin The term used to define relationships acquired through marriage (spouses and in-laws), as opposed to consanguineous relatives (parents and grandparents), who are blood related. See also *consanguineous kin.*

conjugal power Power exercised between spouses.

consanguineous kin The term used to define blood-related kin (parents and grandparents), as opposed to conjugal kin (spouses and in-laws), who are related through marriage. See also *conjugal kin.*

consensual validation The process whereby people depend on others, especially significant others, to help them affirm their definitions of and attitudes and feel-

ings about reality. Consensual validation is important in modern society because social reality is no longer just taken for granted as the way things are.

consolidation loan A single loan large enough to cover all outstanding debts.

consummate love A complete love, in terms of Sternberg's triangular theory of love, in which the components of passion, intimacy, and commitment come together.

continuum of social attachment A conception of social attachment developed by Catherine Ross that emphasizes the quality of the attachment and its relationship to happiness or depression. Her research found that singles are not all socially unattached, isolated, or disconnected, and that people who are in relationships that are unhappy are more depressed than people who are alone and without a partner.

contraceptives Techniques and devices that prevent conception.

coparenting Shared decision making and parental supervision in such areas as discipline and schoolwork or shared holidays and recreation.

coprovider couple Both partners have a primary responsibility for the family's economic support.

corona A crownlike ridge at the base of the glans on the penis.

courtly love Popular during the twelfth century and later, courtly love is the intense longing for someone other than one's marital partner—a passionate and sexual longing that ideally goes unfulfilled. The assumptions of courtly love influence our modern ideas about romantic love.

courtship The process whereby a couple develops a mutual commitment to marriage.

crisis A crucial change in the course of events, a turning point, an unstable condition in affairs.

crude birthrate The number of births per thousand population. See also *total fertility rate.*

cultural deviant A theoretical framework that emphasizes those features that minority families exhibit that distinguish them from white mainstream families. These qualities are viewed as negative or pathological.

cultural equivalent A theoretical framework that emphasizes those features that minority families have in common with mainstream white families.

cultural variant A theoretical approach that calls for making contextually relevant interpretations of minority families. Minority families are studied on their own terms (as opposed to favorably or unfavorably with regard to mainstream white families) and comparisons are made within those groups.

custodial grandparent A parent of a divorced, custodial parent.

custody Primary responsibility for making decisions about the children's upbringing and general welfare.

D

date rape Forced or unwanted sexual contact between people who are on a date. See also *acquaintance rape.*

dating A form of courtship in which, through a series of appointed meetings, an exclusive relationship between two people often evolves.

delivery The second phase (after labor) of childbirth, lasting from the time the cervix is completely dilated until the fetus is expelled.

democratic discipline A parenting attitude and method in which all family members have some input into family and discipline decisions; whenever possible, rules are discussed ahead of time, with both children and parents participating. See also *autocratic discipline, laissez-faire discipline.*

dependence The general reliance on another person or on several others for continuous support and assurance, coupled with subordination to that other. A dependent partner probably has low self-esteem and is having illegitimate needs met by the partner on whom he or she is dependent. See also *independence, interdependence.*

developmental model of child rearing Popularized by many child-rearing experts, this view sees the child as an extremely plastic organism with virtually unlimited potential that the parent is called on to tap and encourage. The model ignores or deemphasizes parents' personal rights. See also *interactive perspective.*

devitalized marriage A marital relationship in which a couple have lost the original zest, intimacy, and meaningfulness that were once a part of their relationship. See also *conflict-habituated marriage, passive-congenial marriage, total marriage, vital marriage.*

discretionary income Uncommitted income that people or families can spend as they please.

displaced homemaker A full-time housewife who, through divorce or widowhood, loses her means of economic support.

displacement A passive-aggressive behavior in which a person expresses anger with another by being angry at or damaging people or things the other cherishes. See also *passive-aggression.*

disrupted adoptions Adoptions in which the child to be adopted is returned to the agency before the adoption is legally final. See also *dissolved adoptions.*

dissolved adoptions Adoptions in which the adopted child is returned after the adoption is final. See also *disrupted adoptions.*

divorce counseling Counseling in which partners go together to negotiate conflicts, grievances, and misunderstandings. The goal of divorce counseling is to help a couple separate with a minimum of destructiveness to themselves and to their children.

divorce mediation A nonadversarial means of dispute resolution by which the couple, with the assistance of a mediator or mediators (frequently a lawyer–therapist team), negotiate the terms of their settlement of custody, support, property, and visitation issues.

divorce readiness See *readiness for divorce.*

domestic partners Two people who have chosen to share one another's lives in an intimate and committed relationship.

dominant dyad A centrally important twosome that symbolizes the culture's basic values and kinship obligations. In white, middle-class America, the husband–wife dyad is expected to take precedence over any others.

double ABC-X model A variation of the ABC-X model of family crises that emphasizes the impact of unresolved prior crises. In this model, A becomes Aa and represents not only the current stressor event but also family pile-up, or residual strains from prior crises.

double message See *mixed message.*

double remarriage A remarriage in which both partners were previously married.

double standard The standard according to which premarital sex is more acceptable for males than for females. Reiss subdivided the double standard into orthodox and transitional. See also *abstinence, permissiveness with affection, permissiveness without affection.*

double standard of aging A sociological concept describing the situation in which men are not considered old or sexually ineligible as early in their lives as women are. Americans view older men as distinguished and older women as just plain old.

dowry A sum of money or property brought to the marriage by the female.

dual-career family Family in which both partners have a strong commitment to the lifetime development of careers.

dual-earner family Family in which the wife as well as the husband is employed, but her work is not viewed as a lifetime career. Her work is considered secondary in importance and in psychological involvement to her family role, and it does *not* serve as an important source of identity. Also called "two-paycheck marriage." See also *two-earner marriage*.

E

economic divorce The separation of the couple into separate economic units, each with its own property, income, control of expenditures, and responsibility for taxes, debts, and so on.

economic hardship perspective One of the five theoretical perspectives concerning the negative outcomes among children of divorced parents. From the economic hardship perspective, the marital dissolution is primarily responsible for the problems faced by children. See also *life stress perspective, parental adjustment perspective, parental loss perspective, interparental conflict perspective*.

ejaculation The rhythmic discharge of seminal fluid containing sperm from the penis during orgasm.

elder abuse Overt acts of aggression toward the elderly, in which the victim may be physically assaulted, emotionally humiliated, purposefully isolated, or materially exploited.

elder care The provision of assistance with daily living activities for an elderly relative who is chronically frail, ill, or disabled.

elder neglect Acts of omission in the care and treatment of the elderly.

embryonic stage The period of pregnancy after the first two weeks (germinal period) until about eight weeks, during which the fetal head, skeletal system,

heart, and digestive system begin to form. See also *fetal period*.

embryo transplant The implantation of a fertilized egg, donated by a fertile woman, into an infertile woman. See also *ovum transplant*.

emotion A strong feeling arising without conscious mental or rational effort, such as joy, reverence, anger, fear, love, or hate. Emotions are neither bad nor good and should be accepted as natural. People can and should learn to control what they *do* about their emotions.

emotional child abuse or neglect A parent or other caregiver's being overly harsh and critical, failing to provide guidance, or being uninterested in a child's needs.

emotional divorce Withholding any bonding emotions and communication from the relationship, typically replacing these with alienating feelings and behavior.

empty nest An old (some say outdated) sociological term referring to the postparental stage of family life. With the contracting economy, the trend to stay single longer, and the high divorce rate, more and more parents are complaining that their nests won't seem to empty.

endogamy Marrying within one's own social group. See also *exogamy*.

endometrium The lining of the uterus, which thickens with a layer of tissue and blood in order to nourish an embryo should an egg become fertilized. If no egg is fertilized, the endometrial tissue and blood are discarded during menstruation.

entitlement In divorce, the equivalent of severance pay for work done at home during the length of the marriage, paid to a wife upon divorce instead of traditional alimony.

erectile tissue Genital tissue that becomes engorged with blood during sexual arousal, causing it to increase in size. In women, erectile tissue composes the clitoris; in men, the penis.

eros The love style characterized by intense emotional attachment and powerful sexual feelings or desires. See also *agape, ludus, mania, pragma, storge*.

exchange theory Theoretical perspective that sees relationships as determined by the exchange of resources and the reward–cost balance of that exchange. This theory predicts that people tend to marry others whose social class, education, physical attractiveness, and even self-esteem are similar to their own.

excitement phase One of the four phases of sexual arousal described by Masters and Johnson. The excitement phase begins when people begin to feel sexually aroused and is characterized by increased breathing, blood pressure, and pulse rates; vasocongestion of the penis and clitoris; and vaginal lubrication in women. See also *orgasmic phase, plateau phase, resolution phase.*

exogamy Marrying a partner from outside one's own social group. See also *endogamy.*

experiential reality Knowledge based on personal experience.

experiment One tool of scientific investigation, in which behaviors are carefully monitored or measured under controlled conditions; also called "laboratory observation."

expert power One of the six power bases, or sources of power. This power stems from the dominant person's superior judgment, knowledge, or ability.

expressive character traits Such traits as warmth, sensitivity to the needs of others, and the ability to express tender feelings, traditionally associated with women. Expressive character traits and roles complement instrumental character traits.

expressive sexuality The view of human sexuality in which sexuality is basic to the humanness of both women and men; all individuals are free to express their sexual selves; and there is no one-sided sense of ownership.

extended family Family including relatives besides parents and children, such as aunts or uncles. See also *nuclear family.*

external stressor An event that precipitates a family crisis and that originates from outside the family itself. See also *internal stressor.*

F

fallopian tubes The tubes that connect a female's uterus with her ovaries. Named after sixteenth-century Italian anatomist Gabriel Fallopius, who first described them.

familism The valuing of traditional family living; also, cherishing family values, such as family togetherness, cohesiveness, and loyalty.

family Any sexually expressive or parent–child or other kin relationship in which people live together with a commitment, in an intimate interpersonal relationship. Family members see their identity as importantly attached to the group, which has an identity of its own. Families today take several forms: single-parent, remarried, dual-career, communal, homosexual, traditional, and so forth. See also *extended family, nuclear family.*

family cohesion That intangible emotional quality that holds groups together and gives members a sense of common identity.

family day care Child care provided in a day-care worker's home.

family development perspective Theoretical perspective that gives attention to changes in the family over time.

family ecology perspective This theoretical perspective explores how a family influences and is influenced by the environments that surround it. A family is interdependent first with its neighborhood, then with its social-cultural environment, and ultimately with the physical-biological environment. All parts of the model are interrelated and influence one another.

family-friendly workplace policies Workplace policies that are supportive of employee efforts to combine family and work commitments.

family functions What families do for members of society. Family functions include responsible reproduction and child rearing; providing economic support for members, including health maintenance; and providing emotional security in an impersonal society.

family leave A leave of absence from work granted to family members to care for new infants, newly adopted children, ill children, aging parents, or to meet similar family needs or emergencies.

family life cycle Stages of family development defined by the addition and subtraction of family members, children's stages, and changes in the family's connection with other social systems.

family myths Versions of reality that obscure a core truth in orer to manage a family's tension.

family of orientation The family in which an individual grows up. Also called "family of origin."

family of procreation The family that is formed when an individual marries and has children.

family policy All the actions, procedures, regulations, attitudes, and goals of government that affect families.

family power Power among family members.

family stress State of tension that arises when demands tax a family's resources.

Family Support Act (1988) Legislation with the same purposes as the 1984 Child Support Amendments.

family systems theory An umbrella term for a wide range of specific theories. This theoretical perspective examines the family as a whole. It looks to the patterns of behavior and relationships within the family, in which each member is affected by the behavior of others. Systems tend toward equilibrium and will react to change in one part by seeking equilibrium by restoring the old system or creating a new one.

family values Values that focus on the family group as a whole and on maintaining family identity and cohesiveness. See also *familism.*

fecundity The technical term for biological reproductive capacity.

feedback The communication technique of stating in different words what another person has said or revealed nonverbally.

female orgasmic dysfunction A female sexual dysfunction in which a woman becomes sexually aroused but cannot reach orgasm.

feminist perspective Feminist theories are conflict theories. The primary focus of the feminist perspective is that male dominance in families and society is oppressive to women. The mission of this perspective is to end this oppression of women (or related pattern of subordination based on social class, race/ethnicity, age, or sexual orientation) by developing knowledge that confronts this disparity. See also *conflict perspective.*

fertility In everyday language, the ability to reproduce biologically. See also *fecundity.*

fertilization The joining of an ovum (egg) with a sperm cell.

fetal period The period of pregnancy lasting from about eight weeks until birth. See also *embryonic stage, germinal period.*

flexible marriage One that allows and encourages partners to grow and change both as individuals and in the relationship. A synonym is *open marriage.*

flexible monogamy The condition in which a couple allows each other some degree of sexual freedom with outside partners but also expects that each will remain the other's primary sexual partner. See also *primariness, sexual exclusivity.*

flexible scheduling A type of employment scheduling that includes scheduling options such as job sharing or flextime, among others. See also *flextime, personal days.*

flextime A policy that permits an employee some flexibility to adjust working hours to suit family needs or personal preference.

foreskin A thin membrane covering the glans of the penis at birth, which is sometimes removed by circumcision.

frenum The place where the foreskin is or was connected to the penis.

functional requisites Basic requirements or needs that must be fulfilled in order for a society to persist.

G

gamete intrafallopian transfer (GIFT) An in vitro fertilization procedure in which eggs are collected from the ovaries, put into a catheter outside the body, and then placed (along with sperm) inside a woman's fallopian tubes.

gaslighting An alienating tactic in which one partner chips away at the other's perception of her- or himself and at the other's definitions of reality. Gaslighting can literally drive people insane. See also *consensual validation.*

gender Attitudes and behavior associated with and expected of the two sexes. The term *sex* denotes biology; *gender* refers to social role.

gendered The way every aspect of people's lives and relationships are influenced by gender, by whether we are male or female.

gender roles Masculine and feminine prescriptions for behavior. The masculine gender role demands instrumental character traits and behavior, whereas the feminine gender role demands expressive character traits and behavior. Traditional gender roles are giving way to androgyny, but they're by no means gone.

gender stereotypes Strongly held exaggerations or overgeneralizations about gender.

gender strategy A way of working through everyday family situations that takes into account an individual's beliefs and deep feelings about gender roles, as well as her or his employment and other nonfamily commitments.

general female sexual dysfunction A female sexual dysfunction in which a woman experiences very little if any erotic pleasure from sexual stimulation.

general sexual unresponsiveness See *general female sexual dysfunction.*

genitalia The external reproductive parts of women and men.

germinal period The first two weeks of pregnancy. See also *embryonic stage, fetal period.*

getting together A courtship process different from dating, in which groups of women and men congregate at a party or share an activity.

glans The sensitive tip at the end of the clitoris in women and of the penis in men.

gonads Sex glands, or glands secreting sex hormones: the ovaries in women, the testicles in men.

good provider role A specialized masculine role that emerged in this country in about the 1830s and that emphasizes the husband as the only or the primary economic provider for his family.

group marriage A family form in which several couples share a household and sexual relations. See also *commune.*

guaranteed child support Type of child support, used in France and Sweden, in which the government sends to the custodial parent the full amount of support awarded to the child.

gunnysacking An alienating fight tactic in which a person saves up, or gunnysacks, grievances until the sack gets too heavy and bursts, and old hostilities pour out.

H

habituation The decreased interest in sex that results from the increased accessibility of a sexual partner and the predictability in sexual behavior with that partner over time.

hermaphrodite A person whose genitalia cannot be clearly identified as either female or male at birth. Attending physicians assign a sex identity to hermaphrodites at birth or shortly thereafter.

heterogamy Marriage between partners who differ in race, age, education, religious background, or social class. Compare with *homogamy.*

heterosexism The taken-for-granted system of beliefs, values, and customs that places superior value on heterosexual behavior (as opposed to homosexual) and denies or stigmatizes nonheterosexual relations. This tendency also sees the heterosexual or straight family as standard.

heterosexuals Individuals who prefer sexual partners of the opposite sex.

H-frame relationship Relationships that are structured like a capital H: Partners stand virtually alone, each self-sufficient and neither influenced much by the other. An example would be a devitalized, dual-career marriage. See also *A-frame relationship, M-frame relationship.*

hidden agenda Associated with stepfathers who, with remarriage, assume functioning single-parent families of a mother and her children. The family may have a hidden agenda, or assumptions and expectations about how the stepfather will behave—expectations and assumptions that are often not passed on to the new stepfather.

HIV/AIDS Human immunodeficiency virus, the virus that causes AIDS.

holistic view of sex The view that conjugal sex is an extension of the whole marital relationship, which is not chopped into compartments, with sex reduced to a purely physical exchange.

homogamy Marriage between partners of similar race, age, education, religious background, and social class. See also *heterogamy.*

homophobia Fear, dread, aversion to, and often hatred of homosexuals.

homosexuals People who prefer same-sex partners.

hormones Chemical substances secreted into the bloodstream by the endocrine glands.

househusband A man who takes a full-time family care role, rather than being employed; the male counterpart to a housewife.

hymen In the female, a ring of tissue that partly covers the vaginal opening.

hypergamy A marriage in which a person gains social rank by marrying someone of higher rank.

I

identity A sense of inner sameness developed by individuals throughout their lives: They know who they are throughout their various endeavors and pursuits, no matter how different these may be.

ideological perspective A perspective used in explaining the division of household labor in families, which points to the impact of cultural expectations on household labor.

illegitimate needs Needs that arise from feelings of self-doubt, unworthiness, and inadequacy. Loving partners cannot fill each other's illegitimate needs no matter how much they try. One fills one's illegitimate needs best by personally working to build one's self-esteem. A first step might be doing something nice for oneself rather than waiting for somebody else to do it.

imaging Choosing to look and behave in ways that one imagines one's partner will consider attractive. An imaging partner complements a romanticizing one. Imaging is the opposite of authenticity.

implantation The process in which the fertilized egg, or zygote, embeds itself in the thickened lining of the uterus.

impotence A sexual dysfunction in which a male is unable to produce or maintain an erection.

Incest Sexual relations between related individuals.

independence Self-reliance and self-sufficiency. To form lasting intimate relationships, independent people must choose to become interdependent.

independents Type of marriage (in Fitzpatrick's typology) characterized by liberal ideology of marriage, egalitarian gender roles, moderate interdependency, and acceptance of conflict.

individualistic values Values that encourage self-fulfillment, personal growth, doing one's own thing, autonomy, and independence. Individualistic values can conflict with family values.

informational power One of the six power bases, or sources of power. This power is based on the persuasive content of what the dominant person tells another individual.

installment financing Contracting with a dealer to pay for major purchases over a period of time.

institution See *social institution.*

instrumental character traits Traits that enable one to accomplish difficult tasks or goals—for example, ra-

tionality, leadership. Traditionally, people thought men were born with instrumental character traits. See also *expressive character traits.*

interactional pattern A pattern noted by sociologist Jesse Bernard among middle-class marriages (as opposed to the parallel relationship pattern among working-class marriages). See also *parallel relationship pattern.*

interactionist perspective Theoretical perspective that focuses on internal family dynamics; the ongoing action and response of family members to each other.

interactive influence of biology and society Gender-linked characteristics are interactively influenced by both biology (nature) and society (nurture).

interactive perspective A perspective of parenting that considers the influence between parent and child reciprocal rather than flowing from parent to child.

intercourse The insertion of the penis into the vagina, also called coitus.

interdependence A relationship in which people who have high self-esteem make strong commitments to each other, choosing to help fill each other's legitimate, but not illegitimate, needs.

internalize The process of making a cultural belief, value, or attitude one's own. When internalized, an attitude becomes a part of us and influences how we think, feel, and act. Internalized attitudes become valued emotionally and therefore are difficult to change. When people do begin to change them, they can expect to go through a period of ambivalence.

internal stressor An event that precipitates a family crisis and that originates from within the family. See also *external stressor.*

interparental conflict perspective One of the five theoretical perspectives concerning the negative outcomes among children of divorced parents. From the interparental conflict perspective, the conflict between parents prior to, during, and after the divorce is responsible for the lowered well-being of the children of divorce. This perspective is strongly supported by research on the effects of divorce on children. See also *life stress perspective, parental loss perspective, parental adjustment perspective, economic hardship perspective.*

interracial marriages Unions between partners of the white, African American, Asian, or Native American races with a spouse outside their own race.

intimacy Committing oneself to a particular other and honoring that commitment in spite of some personal sacrifices while sharing one's inner self with the other. Intimacy requires interdependence.

intimacy (Sternberg's triangular theory of love) Close, connected, and bonded feelings in loving relationships, including sharing oneself with the loved one; one dimension of the triangular theory of love.

intrinsic marriage A marriage in which the emphasis is on the intensity of feelings about each other and the centrality of the spouse's welfare in each mate's scale of values. Intrinsic marriages are primary, intimate relationships. See also *utilitarian marriage.*

in vitro fertilization (IVF) A process in which a baby is conceived outside a woman's body, in a laboratory dish or jar, but develops within the woman's uterus.

involuntary infertility The condition of wanting to conceive and bear a child but being physically unable to do so.

involuntary stable singles Older divorced, widowed, and never-married people who wanted to marry or remarry, but have not found a mate and have come to accept being single as a probable life situation.

involuntary temporary singles Singles who would like, and expect, to marry. These can be younger never-marrieds who do not want to be single and are actively seeking mates, as well as somewhat older people who had not previously been interested in marrying but are now seeking mates.

J

jealousy Emotional pain, anger, and uncertainty arising when a valued relationship is threatened or perceived to be threatened. Research has shown that men and women experience and react differently to jealousy.

job sharing Two people sharing one job. See also *flexible scheduling.*

joint custody A situation in which both divorced parents continue to take equal responsibility for important decisions regarding their child's general upbringing.

K

kin scripts framework A theoretical framework for studying ethnic minority families that includes three culturally relevant family concepts: kin-work, kin-

time, and kin-scription. Kin scripts theorizing helps to make theory and research less biased and more relevant to minority families. In addition, it can be applied to mainstream white families.

kitchen-sink fight An alienating type of fight that does not focus on specific, here-and-now issues: Everything but the kitchen sink gets into the battle.

L

labia majora Two rounded folds of skin, the external lips in the female genitalia.

labia minora Latin for "lesser lips"; two folds of tissue with the labia majora in the female genitalia.

labor In childbirth, the process by which the baby is propelled from the mother's body.

labor force A social invention that arose with the industrialization of the nineteenth century when people characteristically became wage earners, hiring out their labor to someone else.

la familia In Hispanic families, "the family" means the extended family as well as the nuclear family.

laissez-faire discipline Discipline in which parents let children set their own goals, rules, and limits. See also *autocratic discipline, democratic discipline.*

legal divorce The dissolution of a marriage by the state through a court order terminating the marriage.

legitimate needs Needs that arise in the present rather than out of the deficits accumulated in the past.

legitimate power One of the six power bases, or sources of power. This power stems from the more dominant individual's ability to claim authority, or the right to request compliance.

leveling Being transparent, authentic, and explicit about how one truly feels, especially concerning the more conflictive or hurtful aspects of an intimate relationship. Among other things, leveling between intimates implies self-disclosure and commitment (to intimacy).

life course solution A policy proposal designed to ease work/family tension in which policymakers and employers would rethink the traditionally accepted pattern of education, then full-time employment, then retirement.

life spiral An alternative to the life cycle model of the adult life course; the life spiral model stresses the incorporation by individuals of both traditional and alternative roles throughout their lifetime.

life stress perspective One of the five theoretical perspectives concerning the negative outcomes among children of divorced parents. From the life stress perspective, divorce involves the same stress for children as for adults, and divorce is not one single event but a process of stressful events—moving, changing schools, and so on. See also *parental loss perspective, parental adjustment perspective, economic hardship perspective, interparental conflict perspective.*

longitudinal study One technique of scientific investigation, in which researchers study the same individuals or groups over an extended period of time, usually with periodic surveys.

looking-glass self The concept that people gradually come to accept and adopt as their own the evaluations, definitions, and judgments of themselves that they see reflected in the faces, words, and gestures of those around them.

love A deep and vital emotion resulting from significant need satisfaction, coupled with a caring for and acceptance of the beloved, and resulting in an intimate relationship. Love may make the world go 'round, but it's a lot of work too.

love style A distinctive character or personality that loving or lovelike relationships can take. One social scientist has distinguished six: agape, eros, ludus, mania, pragma, and storge.

ludus The love style that focuses on love as play and on enjoying many sexual partners rather than searching for one serious relationship. This love style emphasizes the recreational aspect of sexuality. See also *agape, eros, mania, pragma, storge.*

M

main/secondary provider couple Husband provides family's primary economic support; wife contributes to the family income, but her earnings are seen as inessential. She takes primary responsibility for homemaking.

male dominance The cultural idea of masculine superiority; the idea that men exercise the most control and influence over society's members.

mania The love style that combines strong sexual attraction and emotional intensity with extreme jealousy and moodiness, in which manic partners alternate between euphoria and depression. See also *agape, eros, ludus, pragma, storge.*

manipulating Seeking to control the feelings, attitudes, and behavior of one's partner or partners in underhanded ways rather than by assertively stating one's case.

marital rape A husband's compelling a wife against her will to submit to sexual contact that she finds offensive.

marriage gradient The sociological concept that men marry down in educational and occupational status and women marry up. As a result, never-married women may represent the cream of the crop, whereas never-married men may be the bottom of the barrel. The marriage gradient is easy to observe in the way American couples match their relative heights and ages, with the men slightly taller and older.

marriage market The sociological concept that potential mates take stock of their personal and social characteristics and then comparison shop or bargain for the best buy (mate) they can get.

marriage premise By getting married, partners accept the responsibility to keep each other primary in their lives and to work hard to ensure that their relationship continues. See *primariness.*

martyring Giving others more than one receives in return to maintain relationships. Martyrs often punish the person to whom they are martyring by letting her or him know "just how much I put up with."

masturbation Self-stimulation to provide sexual pleasure.

meatus The opening at the tip of the penis.

Medicaid A federally mandated and funded program that funds health care for low-income U.S. citizens.

Medicare A federally mandated and funded program that funds health care for elderly U.S. citizens.

menstruation The (about monthly) process of discarding an unfertilized ovum, unused tissue, and blood through the vaginal opening.

M-frame relationship Relationship based on couple interdependence. Each partner has high self-esteem, but they mutually influence each other and experience loving as a deep emotion. See also *A-frame relationship, H-frame relationship.*

mixed message Two simultaneous messages that contradict each other; also called a double message. For example, society gives us mixed messages regarding family values and individualistic values and about premarital sex. People, too, can send mixed messages, as when a partner says, "Of course I always like to talk with you" while turning up the TV.

modern family Throughout most of this century, the traditional nuclear family—husband, wife, and children living in one household. See also *postmodern family, nuclear family.*

modern society Social scientists refer to advanced industrial societies as modern because these tend to encourage individualism, social and personal change, and flexibility.

mommy track Term applied to Felice Schwartz's proposal that corporations should develop a separate career ladder for women who wish to work fewer hours so that they may have more time for their families. In return, they would not expect to advance as far or as fast as full-time career women (and men).

monogamy The sexually exclusive union of a couple.

mons veneris The female pubic mound, an area of fatty tissue above the pubic bone.

multiorgasmic Capable of experiencing several successive orgasms during one sexual encounter.

myotonia Increased muscle tension, often as a result of sexual arousal.

N

narcissism Concern chiefly or only with oneself, without regard for the well-being of others. Narcissism is selfishness, not self-love. People with high self-esteem care about and respect themselves *and* others. Narcissistic, or selfish, people, on the other hand, have low self-esteem, are insecure, and therefore worry unduly about their own well-being and very little about that of others.

natural family A traditional cultural view in which the normal, typical, or right family consisted only of two monogamous heterosexual parents and their children. Husbands were primary breadwinners; wives were homemakers. See also *nuclear family.*

naturalistic observation A technique of scientific investigation in which a researcher lives with a family or social group or spends extensive time with them, carefully recording their activities, conversations, gestures, and other aspects of everyday life.

neo-local residence pattern The socially approved practice and cultural expectation that newlyweds will leave their parents' homes and set up their own. (Other residence patterns in other cultures are matrilocal and patrilocal, in which newlyweds reside with the bride's or groom's family, respectively.)

neutralizing power The practice of weakening a powerful person's control by refusing to cooperate.

new Christian right Also called the Moral Majority or the Religious Right, this loose coalition of religious fundamentalists and political conservatives believes that American government and social institutions must be made to operate according to what they see as Christian principles.

no-fault divorce The legal situation in which a partner seeking a divorce no longer has to prove grounds. Virtually all states now have no-fault divorce.

noncustodial grandparent A parent of a divorced, noncustodial parent.

no-power A situation in which partners are equally able to influence each other and, at the same time, are not concerned about their relative power vis-à-vis each other. No-power partners negotiate and compromise instead of trying to win.

nuclear family A family group comprising only the wife, the husband, and their children. See also *extended family, natural family.*

O

occupational segregation The distribution of men and women into substantially different occupations. Women are overrepresented in clerical and service work, for example, whereas men dominate the higher professions and the upper levels of management.

open adoption An adoption process in which there is direct contact between the biological and adoptive parents, ranging from one meeting before the child is born to lifelong friendship. See also *closed adoption, semi-open adoption.*

open marriage A synonym for flexible marriage, in which partners allow and encourage each other to grow and change over the years. Open marriages need not be sexually open; they can, by mutual agreement, be sexually exclusive.

opportunity costs (of children) The economic opportunities for wage earning and investments that parents forgo when rearing children.

orgasm The climax in human sexual response during which sexual tension reaches its peak and is suddenly discharged. In men, ejaculation almost always accompanies orgasm.

orgasmic phase The third of four progressive phases of sexual arousal described by Masters and Johnson. The orgasmic phase is characterized by extremely pleasurable sexual sensations and by involuntary rhythmic contractions in the vagina and penis. See also *excitement phase, plateau phase, resolution phase.*

ova Plural of ovum.

ovaries Two female gonads, or sex glands, that produce reproductive cells called ova, or eggs.

overpermissiveness (in child rearing) Allowing a child's undesirable acts, such as hitting or biting a parent in anger, as expressions of feelings.

ovulation The process by which the ovary produces an ovum, or egg. Usually, the two ovaries alternate, so that only one ovulates each month.

ovum An egg produced by the female ovary. Usually the two ovaries alternate in producing one ovum each month in a process called ovulation. The plural of ovum is ova.

ovum transplant A process in which a fertilized egg is implanted into an infertile woman. See also *embryo transplant.*

P

paradoxical pregnancy The concept that the more guilty and disapproving a sexually active woman is about premarital sex, the less likely she is to use contraceptives regularly, if at all.

parallel relationship pattern A pattern noted by sociologist Jesse Bernard among working-class marriages (as opposed to the interactional pattern among middle-class marriages) in which the husband was expected to be a hard-working provider and the wife a good housekeeper and cook. See also *interactional pattern.*

parental adjustment perspective One of the five theoretical perspectives concerning the negative outcomes among children of divorced parents. From the parental adjustment perspective, the parent's child-rearing skills are impaired as a result of the divorce, with probable negative consequences for the children. See also *life stress perspective, parental loss perspective, economic hardship perspective, interparental conflict, perspective.*

parental loss perspective One of the five theoretical perspectives concerning the negative outcomes among children of divorced parents. From the parental loss perspective, divorce involves the absence of a parent from the household, which deprives children of the optimal environment for their emotional, practical, and social support. See also *life stress perspective, parental adjustment perspective, economic hardship perspective, interparental conflict perspective.*

parenting alliance The relationship between mothers and fathers or gay and lesbian parents as they coparent children, whether or not they are married or living together.

passion (Sternberg's triangular theory of love) The drives that lead to romance, physical attraction, sexual consummation, and so on in a loving relationship; one dimension of the triangular theory of love.

passive-aggression Expressing anger at some person or situation indirectly, through nagging, nitpicking, or sarcasm, for example, rather than directly and openly. See also *displacement, sabotage.*

passive-congenial marriage A marital relationship in which spouses accent things other than emotional closeness; unlike devitalized partners, passive-congenial spouses have always done so. See also *conflict-habituated, devitalized, total, vital marriage.*

patriarchal sexuality The view of human sexuality in which men own everything in the society, including women and women's sexuality, and males' sexual needs are emphasized while females' needs are minimized.

penis The penis and the scrotum together make up the external male genitalia. The penis is composed of an erectile shaft and a sensitive tip, or glans.

perineum In the female, the area between the vestibule to the vagina and the anus. In the male, the area between the scrotum and the anus.

period of disorganization That period in a family crisis, after the stressor event has occurred, during which family morale and organization slump and habitual roles and routines become nebulous.

permissiveness (in child rearing) Setting limits without admonishing a child for feelings, wishes, or appropriate childish behavior such as getting wet at the beach or dirty at a picnic.

permissiveness with affection The standard that permits premarital sex for women and men equally, provided they have a fairly stable, affectionate relationship. See also *abstinence, double standard, permissiveness without affection.*

permissiveness without affection The standard that allows premarital sex for women and men regardless of how much stability or affection there is in their relationship. Also called the "recreational standard." See also *abstinence, double standard, permissiveness with affection.*

personal days Unpaid days off from work with no questions asked.

personal marriage agreement An articulated, negotiated agreement between partners about how each will behave in many or all aspects of the marriage. Personal marriage contracts need to be revised as partners change. A synonym is relationship agreement.

personal power Power exercised over oneself.

pile-up Concept from family stress and crisis theory that refers to the accumulation of family stressors and prior hardships.

pink-collar jobs Low-status, low-pay jobs still reserved primarily for women. The pink-collar job ghetto includes secretaries, beauticians, clerks, domestic workers, bookkeepers, waitresses, and so forth.

placenta Tissue and membrane that hold the fetus in place inside the uterus and function in nourishment; discharged in childbirth.

plateau phase The second phase of sexual arousal, during which the bodily changes begun during the excitement phase intensify and pelvic thrusting, which begins voluntarily, grows more rapid and becomes involuntary, especially among men. See also *excitement phase, orgasmic phase, resolution phase.*

pleasure bond The idea, from Masters and Johnson's book by the same name, that sexual expression between intimates is one way of expressing and strengthening the emotional bond between them.

pleasuring Spontaneously doing what feels good at the moment during a sexual encounter; the opposite of spectatoring.

polygamy A marriage system in which a person takes more than one spouse. Polygyny describes one man with multiple wives, while a marriage of a woman with plural husbands is termed polyandry.

pool of eligibles A group of individuals who, by virtue of background or birth, are most likely to make compatible marriage partners.

postmodern family As a result of progressively increasing family diversity, today's family has little or no objective reference to a particular structure. In the postmodern perspective, there is tremendous variability of family forms in contemporary society, leading some theorists to conclude that the concept of "family" no longer has any objective meaning.

power The ability to exercise one's will. Personal power, or autonomy, is power exercised over oneself. Social power is the ability to exercise one's will over others.

power politics Power struggles between spouses in which each seeks to gain a power advantage over the other; the opposite of a no-power relationship.

pragma The love style that emphasizes the practical, or pragmatic, element in human relationships and involves the rational assessment of a potential (or actual) partner's assets and liabilities. See also *agape, eros, ludus, mania, storge.*

predictable crisis Fairly predictable transitions over the course of family living that can be considered crises or critical opportunities for creatively reorganizing a family's values, attitudes, roles, or relationships.

premature ejaculation A sexual dysfunction in which a man is unable to control his ejaculatory reflex voluntarily.

prepuce Part of the female genitalia; the fold of skin that sometimes covers the clitoris, formed where the labia minora join; the clitoral hood.

primariness Commitment to keeping one's partner the most important person in one's life. See also *commitment (to intimacy).*

primary group A group, usually relatively small, in which there are close, face-to-face relationships. The family and a friendship group are primary groups. See also *secondary group.*

primary parent Parent who takes full responsibility for meeting child's physical and emotional needs by providing the major part of the child's care directly and/or managing the child's care by others.

principle of least interest The postulate that the partner with the least interest in the relationship is the one who is more apt to exploit the other. See also *relative love and need theory.*

private adoption Also called independent adoptions, private adoptions are those arranged directly between adoptive parent(s) and the biological mother, usually through an attorney. See also *public adoption.*

programmatic postponers Couples who postpone parenthood, so as to arrive at late first-time parenthood, by a deliberate, self-conscious process of mutual intention, negotiation, and planning.

pronatalist bias A cultural attitude that takes having children for granted.

prostate A male internal reproductive organ that, along with the seminal vesicles, produces semen.

psychic divorce Regaining psychological autonomy after divorce; emotionally separating oneself from the personality and influence of the former spouse.

psychic intimacy The sharing of people's minds and feelings. Psychic intimacy may or may not involve sexual intimacy.

public adoption Adoptions that take place through licensed agencies that place children in adoptive families. See also *private adoption.*

Q

quasi-kin Anthropologist Paul Bohannan's term for the person one's former spouse remarries. The term is also used more broadly to refer to former in-laws and other former and added kin resulting from divorce and remarriage.

quickening The first fetal movements apparent to the pregnant woman.

R

rapport talk In Deborah Tannen's terms, this is conversation engaged in by women aimed primarily at gaining or reinforcing rapport or intimacy. See also *report talk.*

rational investment perspective Theory arguing that couples attempt to maximize the family economy by trading off between time and energy investments in paid market work and unpaid household labor.

readiness for divorce A state combining a willingness to take responsibility for one's own contribution to the breakup, seeing the alternatives to the marriage as somewhat appealing, and feeling comfortable with the decision to divorce without extreme vacillation over an extended period of time.

recreational standard See *permissiveness without affection.*

redivorce An emerging trend in U.S. society. Redivorces take place more rapidly than first divorces so that many who divorce (and their children) can expect several rapid and emotionally significant transitions in lifestyle and family unit.

referent power One of the six power bases, or sources of power. This power is based on the less dominant person's emotional identification with the more dominant individual.

refined divorce rate Number of divorces per thousand married women over age 15.

refractory period A time after orgasm during which a man cannot become sexually aroused; it usually lasts at least twenty minutes and may be considerably longer, particularly in older men.

rehabilitative alimony A plan under which a man pays his former wife support for a limited period after their divorce while she goes to school or otherwise gets retraining and finds a job.

reinforcing cycle A cycle regarding women's earnings and paid and unpaid family work in which cultural expectations and persistent discrimination result in employed males receiving higher average earnings than women employed full-time, and hence in women's doing more unpaid family work.

relationship agreement See *personal marriage agreement.*

relative love and need theory Theory of conjugal power that holds that the spouse with the least to lose if the marriage ends is the more powerful in the relationship.

relatives of divorce Kinship ties established by marriage, but retained after the marriage is dissolved. For example, the relationship of a former mother-in-law and daughter-in-law.

remarriages Marriages in which at least one partner has already been divorced or widowed. Remarriages are becoming increasingly common for Americans.

remarried family A family consisting of a husband and wife, at least one of whom has been married before, and one or more children from the previous marriage of either or both spouses. There are more remarried families in the United States today, and they usually result from divorce and remarriage.

report talk In Deborah Tannen's terms, this is conversation engaged in by men aimed primarily at conveying information. See also *rapport talk*.

representative sample Survey samples that reflect, or represent, all the people about whom social scientists want to know something.

resiliency model of family stress, adjustment, and adaptation Complex model of family stress and adaptation developed from the double ABC-X model.

resolution phase The final phase of sexual arousal described by Masters and Johnson, during which partners' bodies return to their unstimulated state. See also *excitement phase, plateau phase, orgasmic phase*.

resource hypothesis Hypothesis by Blood and Wolfe that because conjugal power was no longer distributed according to sex, the relative power between wives and husbands would result from their relative resources (for example, age, education, job skills) as individuals.

retarded ejaculation A sexual dysfunction in which a man, although sexually aroused, cannot trigger orgasm.

reward power One of the six power bases, or sources of power. This power is based on an individual's ability to give material or nonmaterial gifts and favors.

role Expected behavior associated with a particular social position. For example, "mother" and "father" are family positions and we expect mothers and fathers to act in certain ways.

role making Improvising a course of action and fitting it to that of others. In role making we use our acts to alter the traditional expectations and obligations associated with a role.

role taking Playing at the expected behavior associated with a social position. This term is also used to mean imaginatively placing oneself in another's position to better understand the other's perspective.

romanticizing Imagining or fabricating many qualities of a person the romanticizer wants to love, as well as many sentimentalized aspects of loving, marriage, parenthood, sexual expression, and so forth.

S

sabotage A passive-aggressive action in which a person tries to spoil or undermine some activity another has planned. Sabotage is not always consciously planned. See also *passive-aggression*.

safer sex The use of latex condoms and the selective limiting of number of sex partners as a precaution against contracting or spreading HIV/AIDS.

sandwich generation Middle-aged (or older) individuals, usually women, who are sandwiched between the simultaneous responsibilities of caring for their dependent children (sometimes young adults) and aging parents.

scapegoating A negative family interaction behavior in which one family member is consistently blamed for everything that goes wrong in the household.

scientific investigation The systematic gathering of information—using surveys, experiments, naturalistic observation, and case studies—from which it often is possible to generalize with a significant degree of predictability.

scrotum A sac behind the penis that holds the two male gonads or sex glands, the testicles.

secondary group A group, often large, characterized by distant, practical relationships. An impersonal society is characterized by secondary groups and relations. See also the opposite, *primary group*.

second shift Sociologist Arlie Hochschild's term for the domestic work that employed women must perform after coming home from a day on the job.

selective reduction Also called "selective termination," a process involved in reproductive technology in which some (but not all) fetuses in multiple pregnancies (resulting from ovulation-stimulating fertility drugs or the GIFT procedure) are selectively aborted, usually in the first trimester.

self-concept The basic feelings people have about themselves, their abilities, and their worth; how people think of or view themselves.

self-disclosure Letting others see one as one really is. Self-disclosure demands authenticity.

self-esteem Feelings and evaluations people have about their own worth.

self-identification theory of gender A theory of gender socialization, developed by psychologist Lawrence

Kohlberg, that begins with a child's categorization of self as male or female. The child goes on to identify sex-appropriate behaviors in the family, media, and elsewhere, and to adopt those behaviors.

self-sufficient economic unit A situation in which family members cooperatively produce what they consume. The American rural family of the eighteenth century tended to be a self-sufficient economic unit; virtually no American family is one today.

semen The milky fluid that carries the sperm through the urethra and out the meatus.

seminal vesicles Internal male reproductive glands that, with the prostate, produce semen, the milky fluid that carries the sperm through the urethra and out the meatus.

semi-open adoption An adoption process in which biological and adoptive families exchange personal information, such as letters or photographs, but they do not have direct contact. See also *closed adoption, open adoption.*

sensuality A more general term than *sexuality,* conveying a general awareness and readiness to respond to experiences from all of the sense organs.

separates Type of marriage (in Fitzpatrick's typology) characterized by ambivalence toward family, conservative gender roles, little interdependence, and avoidance of conflict.

sex the different chromosomal, hormonal, and anatomical components of males and females at birth.

sex ratio The ratio of men to women in a given society or subgroup of society.

sexual abuse A form of child abuse that involves forced, tricked, or coerced sexual behavior—exposure, unwanted kissing, fondling of sexual organs, intercourse, rape, and incest—between a young person and an older person.

sexual arousal The process of awakening, stirring up, or exciting sexual desires and feelings in ourselves or others.

sexual exclusivity Expectations for strict monogamy in which a couple promise or publicly vow to have sexual relations only with each other. See also *flexible monogamy.*

sexual intercourse The insertion of the penis into the vagina. Also called *coitus.*

sexual intimacy A level of interpersonal interaction in which partners have a sexual relationship. Sexual intimacy may or may not involve psychic intimacy.

sexually open marriage A marriage agreement in which spouses agree that each may have openly acknowledged sexual relationships with others while keeping the marriage relationship primary.

sexually transmitted diseases (STDs) Highly contagious diseases transmitted from one person to another through sexual contact. There are more than a dozen STDs, but the most serious are AIDS, syphilis, gonorrhea, and herpes simplex virus, type 2 (HSV-2).

sexual orientation The attraction an individual has for a sexual partner of the same or opposite sex.

sexual responsibility The assumption by each partner of responsibility for his or her own sexual response.

shared parenting Mother and father (or two homosexual parents) who both take full responsibility as parents.

shift work As defined by the Bureau of Labor Statistics, any work schedule in which more than half an employee's hours are before 8 A.M. or after 4 P.M.

significant others People whose opinions about one are very important to one's self-esteem. Good friends are significant others, as are family members.

single Any person who is divorced, widowed, or never-married.

single-parent family A family consisting of a never-married, divorced, or widowed parent and biological or adopted children. The majority of single-parent families in the United States today are headed by women and result from divorce.

single remarriage A remarriage in which only one of the partners was previously married.

social institution A system of patterned and predictable ways of thinking and behaving—beliefs, values, attitudes, and norms—concerning important aspects of people's lives. The five major social institutions are family, religion, government or politics, economics, and education.

socialization The process by which society influences members to internalize attitudes, beliefs, values, and expectations.

social learning theory According to this theory, children learn gender roles as they are taught by parents, schools, and the media.

social power The ability to exercise one's will over others.

Social Security Social insurance whose chief purpose is financial support of the elderly. Based on employer–employee contributions, it is available at a minimum age of 62 to formerly employed persons or their spouses. Payments to current beneficiaries far exceed their actual contributions. Other Social Security programs provide support to minor children whose parents have died and to the disabled.

spectatoring A term Masters and Johnson coined to describe the practice of emotionally removing oneself from a sexual encounter in order to watch oneself and see how one is doing. See also *pleasuring*.

sperm Male reproductive cells.

stalled revolution As used by sociologist Arlie Hochschild, the juxtaposition of women's entry into the paid labor force but without men's doing more unpaid family work.

static (closed) marriage A marriage that does not change over the years and does not allow for changes in the partners. Static marriage partners rely on their formal, legal bond to enforce permanence and sexual exclusivity. Static marriages are more inclined to become devitalized than are flexible marriages.

stepmother trap The conflict between two views: Society sentimentalizes the stepmother's role and expects her to be unnaturally loving toward her stepchildren but at the same time views her as a wicked witch.

Sternberg's triangular theory of love Psychologist Robert Sternberg's formulation of love, in which a variety of types of love are constructed from the three basic dimensions of passion, intimacy, and commitment.

storge An affectionate, companionate style of loving. See also *agape, eros, ludus, mania, pragma*.

stressor A precipitating event that causes a crisis—it is often a situation for which the family has had little or no preparation. See also *ABC-X model, double ABC-X model*.

stressor overload A situation in which an unremitting series of small crises adds up to a major crisis.

strong families Families that emphasize mutual acceptance, respect, and shared values; members rely on each other for emotional support.

structural antinatalism The structural, or societal, condition in which bearing and rearing children is discouraged either overtly or—as may be the case in the United States—covertly through subtle economic discrimination against parents.

structured separation A limited period during which spouses live apart, avoid securing lawyers for a divorce, avoid getting involved in new relationships, and continue in marriage counseling together.

structure-functional perspective Theoretical perspective that looks to the functions that institutions perform for society and the structural form of the institution.

subfecundity Also called secondary infertility, the condition in which parents who have at least one child have difficulty having additional children.

suppression of anger Repression—the involuntary, unconscious blocking of painful thoughts, feelings, or memories from the conscious mind—as it applies to anger. Repressed feelings of anger often come out in other ways, such as overeating and feeling bored or depressed.

surrogate mother A woman who carries within her uterus a developing fetus for a couple who cannot conceive and carry an infant naturally. The surrogate mother delivers the infant, then turns it over to the couple.

survey A technique of scientific investigation using questionnaires or brief face-to-face interviews or both. An example is the United States census. See also *longitudinal study*.

SVR (stimulus-values-roles) The three-stage filtering sequence couples undergo to determine whether they are appropriately matched. The first stage involves physical attraction. The second stage assesses values consensus on a range of issues. The final stage explores role compatibility, wherein couples test and negotiate how they will play out their marital roles.

swinging A marriage agreement in which couples exchange partners in order to engage in purely recreational sex.

symbiotic relationship A relationship based on the mutual meeting of illegitimate needs. See also *legitimate needs*.

symbolic interaction theory Perspective on family life that takes as its starting point the meaning behavior has for family members and the interplay between self-concept and family interaction.

systems theory Theoretical perspective that examines the family as a whole. It looks to the patterns of behavior and relationship within the family, in which each member is affected by the behavior of others. Systems tend toward equilibrium and will react to change in one part by seeking equilibrium by restoring the old system or creating a new one.

T

testicles Male gonads, or sex glands, which hang in the scrotum behind the penis and produce male reproductive cells (sperm) and the hormone testosterone.

theoretical perspective A way of viewing reality, or a lens through which analysts organize and interpret what they observe. Researchers on the family identify those aspects of families that are of interest to them, based on their own theoretical perspective.

theory of complementary needs Theory developed by social scientist Robert Winch suggesting that we are attracted to partners whose needs complement our own. In the positive view of this theory, we are attracted to others whose strengths are harmonious with our own so that we are more effective as a couple than either of us would be alone.

three-phase cycle of violence A relationship–violence cycle in which (1) tension builds between two parties; (2) the situation escalates, exploding in a violent episode; and (3) the violent person becomes contrite, apologizing and treating the other lovingly. Predictably, the cycle continually repeats itself.

total fertility rate For a given year, the number of births that women would have over their reproductive lifetimes if all women at each age had babies at the rate for each age group that year.

total marriage A marital relationship in which partners are intensely bound together psychologically; an intrinsic rather than a utilitarian marriage. Total relationships are similar to vital relationships but are more multifaceted. See also *conflict-habituated, devitalized, passive-congenial, vital marriage.*

traditionals Type of marriage (in Fitzpatrick's typology) characterized by conventional ideology of marriage, conservative gender roles, interdependency, and conflict on serious issues only.

trailing spouse The spouse of a relocated employee who moves with him or her.

transitions Predictable changes in the course of family life.

tubal ligation Cutting or scarring the fallopian tubes between a woman's ovaries and uterus so that eggs cannot pass into the tubes to be fertilized.

two-career marriage See *dual-career family.*

two-earner marriage Both partners are in the paid labor force.

two-location families See *commuter marriage.*

two-person single career The situation in which one spouse, usually the wife, encourages and participates in the other partner's career without direct recognition or personal remuneration.

two-stage marriage An alternative to more formal dating proposed by the late anthropologist Margaret Mead. Americans would first enter into individual marriages involving no children but a serious though not necessarily lifelong commitment. Couples who were compatible in individual marriages might choose to move into the second stage, parental marriage, which would presume lifelong commitment and the ability to cooperatively support and care for a child or children.

U

unpaid family work The necessary tasks of attending to both the emotional needs of all family members and the practical needs of dependent members, such as children or elderly parents, and maintaining the family domicile.

urethra The opening in women and men through which urine passes from the bladder to the outside.

uterus A cavity inside the female in which a fetus grows until birth; also called womb.

utilitarian marriage A union begun or maintained for primarily practical purposes. See also *intrinsic marriage.*

V

vagina The passageway in the female from the uterus to the outside; the birth canal.

vaginismus A sexual dysfunction in which anatomically normal vaginal muscles involuntarily contract whenever a sex partner attempts penetration, so that intercourse is impossible.

vasectomy Tying the tubes (vas deferens) between the testicles and the penis so that sperm will not be included in a man's ejaculation of semen.

vestibule Entryway to the vagina.

violence between intimates Murders, rapes, robberies, or assaults committed by spouses, ex-spouses, boyfriends, or girlfriends.

vital marriage A marital relationship in which partners are intensely bound together psychologically; an intrinsic, as opposed to a utilitarian, marriage. See also *conflict-habituated, devitalized, passive-congenial, total marriage.*

voluntary stable singles Singles who are satisfied to have never married, divorced people who do not want to remarry, cohabitants who do not intend to marry, and those whose lifestyles preclude marriage, such as priests and nuns.

voluntary temporary singles Younger never-marrieds and divorced people who are postponing marriage or remarriage. They are open to the possibility of marriage, but searching for a mate has a lower priority than do other activities, such as career.

vulva The female external genitalia.

W

weak families Families that have a low sense of common purpose, feel in little control over what happens to them, and tend to cope with problems by showing diminished respect and/or understanding for each other.

wheel of love An idea developed by Ira Reiss in which love is seen as developing through a four-stage, circular process, including rapport, self-revelation, mutual dependence, and personality need fulfillment.

Z

zygote A fertilized ovum (egg). See also *embryonic stage, fetal period, germinal period.*

zygote intrafallopian transfer (ZIFT) An in vitro fertilization process in which fertilized eggs are implanted in a woman's fallopian tubes while the fertilized egg is still a single cell, or zygote.

References

A

Abbey, Antonia, Frank M. Andrews, and L. Jill Halman. 1991. "Gender's Role in Responses to Infertility." *Psychology of Women Quarterly* 15:295–316.
———. 1992. "Infertility and Subjective Well-Being: The Mediating Roles of Self-Esteem, Internal Control, and Interpersonal Conflict." *Journal of Marriage and the Family* 54 (2) (May): 408–17.

Abel, Emily K. 1991. *Who Cares for the Elderly? Public Policy and the Experiences of Adult Daughters.* Philadelphia: Temple University Press.

"Abortion as a Contraceptive." 1988. *Newsweek,* June 13, p. 71.

Acker, Michele and Mark H. Davis. 1992. "Intimacy, Passion, and Commitment in Adult Romantic Relationships: A Test of the Triangular Theory of Love." *Journal of Social and Personal Relationships* 9:21–50.

Adams, Karen L. and Norma C. Ware. 1995. "Sexism and the English Language: The Linguistic Implications of Being a Woman." Pp. 331–346 in *Women: A Feminist Perspective.* 5th ed., edited by Jo Freeman. Mountain View, CA: Mayfield.

Adler, Jerry. 1993. "Sex in the Snoring '90s." *Newsweek.* (Apr.) 26: 55–57.

Adler, Nancy E., Henry P. David, Brenda N. Major, Susan H. Roth, Nancy F. Russo, and Gail E. Wyatt. 1990. "Psychological Responses After Abortion," *Science* 248:41–44.

Ahlander, Nancy Rollins and Kathleen Slaugh Bahr. 1995. "Beyond Drudgery, Power, and Equity: Toward an Expanded Discourse on the Moral Dimensions of Housework in Families." *Journal of Marriage and the Family* 57 (1) (Feb.): 54–68.

Ahlburg, Dennis A. and Carol J. De Vita. 1992. *New Realities of the American Family. Population Bulletin* 47 (2). Washington, DC: Population Reference Bureau.

Ahrons, Constance R. and Richard B. Miller. 1993. "The Effect of the Post-Divorce Relationship on Paternal Involvement: A Longitudinal Analysis." *American Journal of Orthopsychiatry* 63 (3) (July): 462–79.

Ahrons, Constance R. and Lynn Wallisch. 1987. "Parenting in the Binuclear Family: Relationships Between Biological and Stepparents." Pp. 225–56 in

Remarriage and Stepparenting, edited by Kay Pasley and Marilyn Ihinger-Tallman. New York: Guilford.

Ahuvia, Aaron C. and Mara B. Adelman. 1992. "Formal Intermediaries in the Marriage Market: A Typology and Review." *Journal of Marriage and the Family* 54 (May): 452–63.

"AIDS Front: Good News and Grim News." 1995. *U.S. News & World Report.* Feb. 13.

Ainsworth, M. D. S. 1973. "The Development of Infant–Mother Attachment." In *Review of Child Development Research,* vol. III, edited by B. M. Caldwell and H. N. Ricuiti. Chicago: University of Chicago Press.

Aldous, Joan. 1978. *Family Careers: Developmental Change in Families.* New York: Wiley.

———. 1990. "Specification and Speculation Concerning the Politics of Workplace Family Policies." *Journal of Family Issues* 11 (4) (Dec.): 355–67.

Aldous, Joan and Wilfried Dumon, eds. 1980. *The Politics and Programs of Family Policy.* Notre Dame, IN: University of Notre Dame Press.

Aldous, Joan, Lawrence Marsh, and Scott Trees. 1985. "Families and Inflation." Paper prepared for presentation at the 80th Annual Meeting of the American Sociological Association, Aug. 26–30.

Aldous, Joan and Robert C. Tuttle. 1988. "Unemployment and the Family." Pp. 17–41 in *Employment and Economic Problems: Families in Trouble,* vol. 1, edited by Catherine S. Chilman, Elam W. Nunnally, and Fred M. Cox. Newbury Park, CA: Sage.

Alexander, Pamela C., Sharon Moore, and Elmore R. Alexander III. 1991. "What Is Transmitted in the Intergenerational Transmission of Violence?" *Journal of Marriage and the Family* 53 (3) (Aug.): 657–68.

Alexander, Suzanne. 1991. "More Working Mothers Opt for Flexibility of Operating a Franchise from Home." *The Wall Street Journal,* Jan. 13, p. B1.

Allen, Katherine R. 1993. "The Dispassionate Discourse of Children's Adjustment to Divorce." *Journal of Marriage and the Family* 55 (1) (Feb.): 46–49.

Allen, Katherine R. and David H. Demo. 1995. "The Families of Lesbians and Gay Men: A New Frontier in Family Research." *Journal of Marriage and the Family* 57 (1) (Feb.): 111–27.

Allen, Robert L. and Paul Kivel. 1994. "Men Changing Men." *Ms.* Sept./Oct.: 50–53.

Allen, W. 1978. "The Search for Applicable Theories of Black Family Life." *Journal of Marriage and the Family* 40 (1):117–31.

Allgeier, E. R. 1983. "Sexuality and Gender Roles in the Second Half of Life." Pp. 135–37 in *Changing Boundaries: Gender Roles and Sexual Behavior,* edited by Elizabeth Rice Allgeier and Naomi B. McCormick. Palo Alto, CA: Mayfield.

Almquist, Elizabeth M. 1995. "The Experiences of Minority Women in the United States: Intersections of Race, Gender, and Class." Pp. 573–606 in *Women: A Feminist Perspective.* 5th ed., edited by Jo Freeman. Mountain View, CA: Mayfield.

Almond, Brenda and Carole Ulanowsky. 1990. "HIV and Pregnancy." *Hastings Center Report* (Mar./Apr.): 16–21.

Altman, I. and W. W. Haythorn. 1965. "Interpersonal Exchange in Isolation." *Sociometry* 28:411–26.

Altman, Irwin and Dalmas A. Taylor. 1973. *Social Penetration: The Development of Interpersonal Relations.* New York: Holt, Rinehart & Winston.

Alvardo, Donna. 1992. "Children Having Children." *San Jose Mercury News,* July 1, pp. 1D, 6D.

Alvarez, Lizette. 1995. "Pint-Size Interpreters of World for Parents." *New York Times,* Oct. 1, p. A16.

Alwin, Duane, Philip Converse, and Steven Martin. 1985. "Living Arrangements and Social Integration." *Journal of Marriage and the Family* 47:319–34.

Amato, Paul R. 1988. "Parental Divorce and Attitudes Toward Marriage and Family Life." *Journal of Marriage and the Family* 50:453–61.

———. 1991. "The 'Child of Divorce' as a Person Prototype: Bias in the Recall of Information About Children in Divorced Families." *Journal of Marriage and the Family* 53 (1) (Feb.): 59–70.

———. 1993. "Children's Adjustment to Divorce: Theories, Hypotheses, and Empirical Support." *Journal of Marriage and the Family* 55 (1) (Feb.): 23–28.

———. 1994. "Father–Child Relations, Mother–Child Relations, and Offspring Psychological Well-Being in Early Adulthood." *Journal of Marriage and the Family* 56 (4) (Nov.): 1031–42.

Amato, Paul R. and Alan Booth. 1991. "The Consequences of Divorce for Attitudes Toward Divorce and Gender Roles." *Journal of Family Issues* 12 (3) (Sept.): 306–22.

Amato, Paul R. and Bruce Keith. 1991a. "Parental Divorce and the Well-Being of Children: A Meta-analysis." *Psychological Bulletin* 110:26–46.

———. 1991b. "Parental Divorce and Adult Well-Being: A Meta-analysis." *Journal of Marriage and the Family* 53 (1) (Feb.): 43–58.

Amato, Paul R., Sandra J. Rezac, and Alan Booth. 1995. "Helping Between Parents and Young Adult Offspring: The Role of Parental Marital Quality, Divorce, and Remarriage." *Journal of Marriage and the Family* 57 (2) (May): 363–74.

Ambert, Anne-Marie. 1988. "Relationships with Former In-laws After Divorce: A Research Note." *Journal of Marriage and the Family* 50:679–86.

———. 1994. "A Qualitative Study of Peer Abuse and Its Effects: Theoretical and Empirical Implications." *Journal of Marriage and the Family* 56 (1) (Feb.): 119–30.

Ambroz, Juliann R. 1995. "Keeping Love Alive: How Couples Counseling Can Work for You." *Mothering* (Fall): 75–80.

"American Indians by the Numbers." 1992. *New York Times,* Feb. 26.

Ames, Katherine. 1992. "Domesticated Bliss." *Newsweek.* March 23.

American Psychological Association. Public Interest Directorate. 1987. *Research Review: The Psychological Sequelae of Abortion.* Washington, DC: American Psychological Association.

"Americans Became More Committed to Both Sides of Debate over Abortion During the Past Decade." 1990. *Family Planning Perspectives* 22:40.

Ames, Katherine. 1992. "Domesticated Bliss." *Newsweek,* Mar. 23.

Anderson, Margaret L. 1988. *Thinking About Women: Sociological Perspectives on Sex and Gender* 2d ed. New York: Macmillan.

Andrews, Bernice and Chris R. Brewin. 1990. "Attributions of Blame for Marital Violence: A Study of Antecedents and Consequences." *Journal of Marriage and the Family* 52 (3) (Aug.): 757–67.

"And Two Babies Make Just the Right Number." 1991. *The Wall Street Journal,* Dec. 18.

Annas, George J. 1991. "Crazy Making: Embryos and Gestational Mothers." *Hastings Center Report* (Jan./Feb.): 35–38.

Anson, Ofra. 1989. "Marital Status and Women's Health Revisited: The Importance of a Proximate Adult." *Journal of Marriage and the Family* 51:185–94.

Anthony, E. James and Bertram J. Cohler, eds. 1987. *The Invulnerable Child.* New York: Guilford.

Anti-Defamation League of B'nai Brith. 1981. "The American Story: The Hernandez Family." New York: Anti-Defamation League. Video.

Antill, J. K. 1987. "Parents' Beliefs and Values about Sex Roles, Sex Differences, and Sexuality." Pp. 294–328 in *Sex and Gender,* edited by P. Shaver and C. Hendrick. Newbury Park, CA: Sage.

Aquilino, William S. 1994a. "Later Life Parental Divorce and Widowhood: Impact on Young Adults' Assessment of Parent–Child Relations." *Journal of Marriage and the Family* 56(4) (Nov.): 908–22.

———. 1994b. "Impact of Childhood Family Disruption on Young Adults' Relationships with Parents." *Journal of Marriage and the Family* 56 (2) (May): 295–313.

Arafat, I. and Betty Yorburg. 1973. "On Living Together Without Marriage." *Journal of Sex Research* 9:21–29.

Arditti, Joyce A. and Timothy Z. Keith. 1993. "Visitation Frequency, Child Support Payment, and the Father–Child Relationship Postdivorce." *Journal of Marriage and the Family* 55 (3) (Aug.): 699–712.

Arendell, Terry. 1986. *Mothers and Divorce: Legal, Economic, and Social Dilemmas.* Berkeley: University of California Press.

Ariès, Phillipe. 1962. *Centuries of Childhood: A Social History of Family Life.* New York: Knopf.

Arond, Miriam and Samuel L. Pauker, M.D. 1987. *The First Year of Marriage.* New York: Warner.

Aronson, Jane. 1992. "Women's Sense of Responsibility for the Care of Old People: But Who Else Is Going to Do It?" *Gender & Society* 6 (1) (Mar.):8–29.

Aseltine, Robert H. Jr., and Ronald C. Kessler. 1993. "Marital Disruption and Depression in a Community Sample." *Journal of Health and Social Behavior* 34 (Sept.): 237–51.

Astone, Nan Marie and Sara S. McLanahan. 1991. "Family Structure, Parental Practices and High School Completion." *American Sociological Review* 56 (June): 309–20.

Atkin, David J., Jay Moorman, and Carolyn A. Lin. 1991. "Ready for Prime Time: Network Series Devoted to Working Women in the 1980s." *Sex Roles* 25 (11/12):677–83.

Atkinson, Alice M. 1992. "Stress Levels of Family Day Care Providers, Mothers Employed Outside the Home, and Mothers at Home." *Journal of Marriage and the Family* 54 (2) (May): 379–86.

————. 1994. "Rural and Urban Families' Use of Child Care." *Family Relations* 43: 16–22.

Atkinson, Maxine P. and Stephen P. Blackwelder. 1993. "Fathering in the 20th Century." *Journal of Marriage and the Family* 55 (3) (Nov.): 975–86.

Auerbach, Judith D. 1990. "Employer-Supported Child Care as a Women-Responsive Policy." *Journal of Family Issues* 11 (4) (Dec.): 384–400.

Axinn, William G. and Arland Thornton. 1992. "The Relationship Between Cohabitation and Divorce: Selectivity or Causal Influence?" *Demography* 29 (3) (Aug.): 357–74.

————. 1993. "Mothers, Children, and Cohabitation: The Intergenerational Effects of Attitudes and Behavior." *American Sociological Review* 58 (2) (Apr.): 233–45.

B

Babbie, Earl. 1992. *The Practice of Social Research,* 6th ed. Belmont, CA: Wadsworth.

————. 1995. *The Practice of Social Research,* 7th ed. Belmont, CA: Wadsworth.

Bach, George R. and Ronald M. Deutsch. 1970. *Pairing.* New York: Avon.

Bach, George R. and Peter Wyden. 1970. *The Intimate Enemy: How to Fight Fair in Love and Marriage.* New York: Avon.

Bachrach, Christine, Patricia F. Adams, Soledad Sambrano, and Kathryn A. London. 1990. "Adoption in the 1980's." U.S. National Center for Health Statistics, Advance Data, No. 181, Jan. 5.

Bachrach, Christine A., Kathy Shepherd Stolley, and Kathryn A. London. 1992. "Relinquishment of Premarital Births: Evidence from National Survey Data." *Family Planning Perspectives* 24 (1) (Jan./Feb.): 27–32.

Bahr, Howard M. and Bruce A. Chadwick. 1985. "Religion and Family in Middletown, USA." *Journal of Marriage and the Family* 47 (2):407–14.

Bahr, Stephen J. and Boyd C. Rollins. 1971. "Crisis and Conjugal Power." *Journal of Marriage and the Family* 33:360–67.

Baker, James N. 1989. "Lesbians: Portrait of a Community." *Newsweek,* Mar. 12, p. 24.

Baldwin, James. 1988. "A Talk to Teachers." Pp. 3–12 in *Multicultural Literacy,* edited by Rick Simonson and Scott Walker. St. Paul, MN: Graywolf Press.

Barber, Brian K. 1994. "Cultural, Family, and Personal Contexts of Parent–Adolescent Conflict." *Journal of Marriage and the Family* 56 (2) (May): 375–86.

Bardwick, Judith M. 1971. *Psychology of Women: A Study of Biocultural Conflicts.* New York: Harper & Row.

Barker-Benfield, G. J. 1975. *The Horrors of the Half-Known Life: Male Attitudes Toward Women and Sexuality in Nineteenth-Century America.* New York: Harper & Row.

Barnett, Rosalind C. 1994. "Home-to-Work Spillover Revisited: A Study of Full-Time Employed Women in Dual-Earner Couples." *Journal of Marriage and the Family* 56 (3) (Aug.): 647–56.

Barnett, Rosalind C. and Grace K. Baruch. 1987. "Social Roles, Gender, and Psychological Distress." Pp. 122–43 in *Gender and Stress,* edited by Rosalind C. Barnett, Lois Biener, and Grace K. Baruch. New York: Free Press.

Barnett, Rosalind C., Nancy L. Marshall, and Joseph H. Pleck. 1992. "Men's Multiple Roles and Their Relationship to Men's Psychological Distress." *Journal of Marriage and the Family* 54 (2) (May): 358–67.

Baron, Robert A. and Donn Byrne. 1984. *Social Psychology,* 4th ed. Boston: Allyn & Bacon.

Barrett, Paul M. 1992. "High Court Prepares to Take Up Abortion Case Almost Guaranteed to Add to Political Confusion." *The Wall Street Journal,* Apr. 15, p. A22.

Barringer, Felicity. 1992a. "Among Elderly, Men's Prospects Are the Brighter." *New York Times,* Nov. 10.

————. 1992b. "A Census Disparity for Asians in the U.S." *New York Times,* Sept. 20.

————. 1992c. "Rate of Marriage Continues Decline." *New York Times,* July 19.

Barron, James. 1989. "Homosexuals See 2 Decades of Gains, But Fear Setbacks." *New York Times,* June 25.

Bart, Pauline. 1972. "Depression in Middle-Aged Women." Pp. 163–86 in *Women in Sexist Society: Studies in Power and Powerlessness,* edited by Vivian Gornick and Barbara K. Moran. New York: New American Library.

Barth, Richard P. and Marianne Berry. 1988. *Adoption and Disruption: Rates, Risks, and Responses.* New York: Aldine.

Bartholet, Elizabeth. 1993. *Adoption and the Politics of Parenting.* Boston: Houghton Mifflin.

Basow, Susan. 1991. "The Hairless Ideal: Women and Their Body Hair." *Psychology of Women Quarterly* 15:83–96.

Basow, Susan H. 1992. *Gender: Stereotypes and Roles.* 3d ed. Pacific Grove, CA: Brooks/Cole.

Baucom, Donald H. and Paige K. Besch. 1985. "Personality Processes and Individual Differences." *Journal of Personality and Social Psychology* 48 (5):1218–26.

Baumrind, Diana. 1971. "Current Patterns of Parental Authority." *Developmental Psychology Monographs* 4:1–102.

———. 1978. "Parental Disciplinary Patterns and Social Competence in Children." *Youth and Society* 9:239–76.

———. 1994. "The Social Context of Child Maltreatment." *Family Relations* 43 (4) (Oct.): 360–68.

Beach, Frank A. 1977. "Cross-Species Comparisons and the Human Heritage." Pp. 296–316 in *Human Sexuality in Four Perspectives,* edited by Frank A. Beach. Baltimore: Johns Hopkins University Press.

Bean, Kathleen S. 1985–86. "Grandparent Visitation: Can the Parent Refuse?" *Journal of Family Law* 24:393–449.

Beattie, Melody. 1987. *Codependent No More: How to Stop Controlling Others and Start Caring for Yourself.* New York: Harper/Hazelden.

Beck, Melinda. 1988a. "Miscarriages." *Newsweek,* Aug. 15, pp. 46–52.

———. 1988b. "Willing Families, Waiting Kids." *Newsweek,* Sept. 12, p. 64.

———. 1990. "Trading Places." *Newsweek,* July 16, pp. 48–54.

Becker, Gary S. 1991. *A Treatise on the Family.* 2d ed. Cambridge, MA: Harvard University Press.

Becker, Gay. 1990. *Healing the Infertile Family.* New York: Bantam.

Bedard, Marcia E. 1992. *Breaking with Tradition: Diversity, Conflict, and Change in Contemporary Families.* Dix Hills, NY: General Hall.

Beer, William R. 1989. *Strangers in the House: The World of Stepsiblings and Half Siblings.* New Brunswick, NJ: Transaction.

———. 1992. *American Stepfamilies.* New Brunswick, NJ: Transaction.

Begley, Sharon. 1995. "Deliver, Then Depart." *Newsweek,* July 10, p. 62.

Begley, Sharon and Daniel Glick. 1994. "The Estrogen Complex." *Newsweek,* March 21: 76–77.

"Behind Jobless Figures, A Rise in Poor Children." 1991. *New York Times,* Nov. 26.

Belcastro, Philip A. 1985. "Sexual Behavior Differences Between Black and White Students." *The Journal of Sex Research* 21 (1):56–67.

Belkin, Lisa. 1989. "Bars to Equality of Women Seen as Eroding Slowly." *New York Times,* Aug. 20.

Bell, Alan P., Martin S. Weinberg, and Sue Kiefer Hammersmith. 1981. *Sexual Preference: Its Development in Men and Women.* Bloomington: University of Indiana Press.

Bell, Richard Q. 1974. "Contributions of Human Infants to Caregiving and Social Interaction." Pp. 11–19 in *The Effect of the Infant on Its Care Giver. Origins of Behavior Series,* vol. 1, edited by Michael Lewis and Leonard A. Rosenblum. New York: Wiley.

Bellah, Robert N., Richard Madsen, William M. Sullivan, Ann Swidler, and Steven M. Tipton. 1985. *Habits of the Heart: Individualism and Commitment in American Life.* Berkeley and Los Angeles: University of California Press.

Belsky, Jay. 1990. "Parental and Nonparental Child Care and Children's Socioemotional Development: A Decade in Review." *Journal of Marriage and the Family* 52 (4) (Nov.): 885–903.

———. 1991. "Parental and Non-parental Child Care and Children's Socioemotional Development." Pp. 122–40 in *Contemporary Families: Looking Forward, Looking Back,* edited by Alan Booth. Minneapolis: National Council on Family Relations.

Belsky, Jay and Michael J. Rovine. 1990. "Patterns of Marital Change Across the Transition to Parenthood: Pregnancy to Three Years Postpartum." *Journal of Marriage and the Family* 52:5–19.

Bem, Sandra. 1975. "Androgyny vs. the Tight Little Lives of Fluffy Women and Chesty Men." *Psychology Today* 9:58–62.

Bem, S. L. 1981. "Gender Schema Theory: A Cognitive Account of Sex Typing." *Psychological Review* 88:354–64.

Benenson, Harold. 1984. "Women's Occupation and Family Achievement in the U.S. Class System: A Critique of the Dual-Career Family Analysis." *British Journal of Sociology* 35:19–41.

Bengtson, Vern L. and Dale Dannefer. 1987. "Families, Work and Aging: Implications of Disordered Cohort Flow for the Twenty-First Century." Pp. 256–89 in *Health in Aging: Sociological Issues and Policy Directions,* edited by Russell A. Ward and Sheldon S. Tobin. New York: Springer.

Benin, Mary and Verna M. Keith. 1995. "The Social Support of Employed African American and Anglo Mothers." *Journal of Family Issues* 16(3) (May): 275–97.

Bennett, Linda A., Steven J. Wolin, and David Reiss. 1988. "Deliberate Family Process: A Strategy for Protecting Children of Alcoholics." *British Journal of Addiction* 83:821–29.

Bennett, Linda A., Steven J. Wolin, David Reiss, and Martha A. Teitelbaum. 1987. "Couples at Risk for Transmission of Alcoholism: Protective Influences." *Family Process* 26:111–29.

Bennett, Neil G., David E. Bloom, and Cynthia K. Miller. 1995. "The Influence of Nonmarital Childbearing on the Formation of First Marriages." *Demography* 32 (1) (Feb.): 47–62.

Bennetts, Leslie. 1986. "Stepparenting." *Omaha World-Herald,* Dec. 7.

Benson, R. C. 1983. *Handbook of Obstetrics and Gynecology.* Los Altos, CA: Lange Medical Publishers.

Berardo, Felix M. 1985. "Age Heterogamy in Marriage." *Journal of Marriage and the Family* 47:553–66.

Berg, Barbara. 1984. "Early Signs of Infertility." *Ms.,* May, pp. 68ff.

Berger, Peter L., Brigitte Berger, and Hansfried Kellner. 1973. *The Homeless Mind: Modernization and Consciousness.* New York: Random House.

Berger, Peter L. and Hansfried Kellner. 1970. "Marriage and the Construction of Reality." Pp. 49–72 in *Recent Sociology No. 2,* edited by Hans Peter Dreitzel. New York: Macmillan.

Berger, Peter L. and Thomas Luckmann. 1966. *The Social Construction of Reality: A Treatise in the Sociology of Knowledge.* New York: Doubleday.

Berk, Richard A. and Phyllis J. Newton. 1985. "Does Arrest Really Deter Wife Battery? An Effort to Replicate the Findings of the Minneapolis Spouse Abuse Experiment." *American Sociological Review* 50:253–62.

Berk, Richard A., Phyllis J. Newton, and Sarah Fenstermaker Berk. 1986. "What a Difference a Day Makes: An Empirical Study of the Impact of Shelters for Battered Women." *Journal of Marriage and the Family* 48:481–90.

Berk, Sarah Fenstermaker. 1985. *The Gender Factory: The Apportionment of Work in American Households.* New York: Plenum.

———. 1988. "Women's Unpaid Labor: Home and Community." Pp. 287–302 in *Women Working.* 2d ed., edited by Ann Helton Stromberg and Shirley Harkess. Mountain View, CA: Mayfield.

Bernard, Jessie. 1964. "The Adjustment of Married Mates." Pp. 675–739 in *The Handbook of Marriage and the Family,* edited by Harold T. Christensen. Chicago: Rand McNally.

———. 1982 [1972]. *The Future of Marriage,* 2d ed. New York: Bantam.

———. 1986. "The Good-Provider Role: Its Rise and Fall." Pp. 125–44 in *Family in Transition: Rethinking Marriage, Sexuality, Child Rearing, and Family Organization.* 5th ed., edited by Arlene S. Skolnick and Jerome H. Skolnick. Boston: Little, Brown.

Berner, R. T. 1992. *Parents Whose Parents Were Divorced.* New York: Haworth Press.

Besharov, Douglas J. 1993. "Truth and Consequences: Teen Sex." *The American Enterprise.* Jan./Feb.: 52–59.

Beutel, Ann M. and Margaret Mooney Marini. 1995. "Gender and Values." *American Sociological Review* 60 (3) (June): 436–49.

Bianchi, Suzanne. 1990. *America's Children: Mixed Prospects. Population Bulletin* 45 (1). Washington, DC: Population Reference Bureau.

Billy, John O. G., Karin L. Brewster, and William R. Grady. 1994. "Contextual Effects on the Sexual Behavior of Adolescent Women." *Journal of Marriage and the Family* 56 (2) (May): 387–404.

Bird, Chole E. and Catherine E. Ross. 1993. "Houseworkers and Paid Workers: Qualities of the Word and Effects on Personal Control." *Journal of Marriage and the Family* 55 (Nov.): 913–25.

"Births to Women in Their Late 30s Have Risen Among Both Blacks and Whites Since 1980." 1992. *Family Planning Perspectives* 24:91–92.

Bishop, Katherine. 1989. "San Francisco Grants Recognition to Partnerships of Single People." *New York Times,* July 8.

Black, Leora E., Matthew M. Eastwood, Douglas H. Sprenkle, and Elaine Smith. 1991. "An Exploratory Analysis of Leavers Versus Left as It Relates to

Levinger's Social Exchange Theory of Attractions, Barriers, and Alternative Attractions." *Journal of Divorce and Remarriage* 15 (1/2):127–40.

Blackman, Julie. 1989. *Intimate Violence: A Study of Injustice.* New York: Columbia University Press.

Blair, Sampson Lee and Michael P. Johnson. 1992. "Wives' Perceptions of the Fairness of the Division of Household Labor: The Intersection of Housework and Ideology." *Journal of Marriage and the Family* 54 (3) (Aug.): 570–81.

Blair, Sampson Lee and Daniel T. Lichter. 1991. "Measuring the Division of Household Labor: Gender Segregation of Housework Among American Couples." *Journal of Family Issues* 12 (1) (Mar.): 91–113.

Blaisure, Karen and Katherine R. Allen. 1995. "Feminists and the Ideology and Practice of Marital Equality." *Journal of Marriage and the Family* 57 (1) (Feb.): 5–19.

Blake, Judith. 1974. "Can We Believe Recent Data on Birth Expectations in the U.S.?" *Demography* 11:25–44.

Blankenhorn, David. 1995. *Fatherless America: Confronting Our Most Urgent Social Problem.* New York: Basic Books.

Blauner, Robert. 1992. "The Ambiguities of Racial Change." Pp. 54–65 in *Race, Class, and Gender: An Anthology,* edited by Margaret L. Andersen and Patricia Hill Collins. Belmont, CA: Wadsworth.

Blee, Kathleen M. and Ann R. Tickamyer. 1995. "Racial Differences in Men's Attitudes About Women's Gender Roles." *Journal of Marriage and the Family* 57 (1) (Feb.): 21–30.

Blood, Robert O., Jr., and Donald M. Wolfe. 1960. *Husbands and Wives: The Dynamics of Married Living.* New York: Free Press.

Blumberg, Rae Lesser and Marion Tolbert Coleman. 1989. "A Theoretical Look at the Gender Balance of Power in the American Couple." *Journal of Family Issues* 10:225–50.

Blumstein, Philip and Pepper Schwartz. 1983. *American Couples: Money, Work, Sex.* New York: Morrow.

Bly, Robert. 1990. *Iron John: A Book About Men.* Reading, MA: Addison-Wesley.

Bogert, Carroll. 1994. "Bringing Back Baby." *Newsweek,* Nov. 21, pp. 78–79.

Bohannan, Paul. 1970a. "Divorce Chains, Households of Remarriage, and Multiple Divorces." Pp. 113–23 in *Divorce and After,* edited by Paul Bohannan. New York: Doubleday.

———. 1970b. "The Six Stations of Divorce." Pp. 29–55 in *Divorce and After,* edited by Paul Bohannan. New York: Doubleday.

Bohannan, Paul and Rosemary Erickson. 1978. "Stepping In." *Psychology Today* 12:53–54.

Bollier, David. 1982. *Liberty and Justice for Some.* New York: Frederick Ungar.

Booth, Alan. 1977. "Wife's Employment and Husband's Stress: A Replication and Refutation." *Journal of Marriage and the Family* 39:645–50.

Booth, Alan and Paul R. Amato. 1994a. "Parental Gender Role Nontraditionalism and Offspring Outcomes." *Journal of Marriage and the Family* 56 (4) (Nov.): 865–77.

———. 1994b. "Parental Marital Quality, Parental Divorce, and Relations with Parents." *Journal of Marriage and the Family* 56 (1) (February): 21–34.

Booth, Alan and James Dabbs. 1992. "Testosterone and Men's Marriages." Paper presented at the 1992 Annual Meeting of the Population Association of America. Denver, Apr. 30–May 2.

Booth, Alan and John Edwards. 1985. "Age at Marriage and Marital Instability." *Journal of Marriage and the Family* 47 (1):67–75.

———. 1992. "Starting Over: Why Remarriages Are More Unstable." *Journal of Family Issues* 13 (2) (June): 179–94.

Booth, Alan and David Johnson. 1988. "Premarital Cohabitation and Marital Success." *Journal of Family Issues* 9:255–72.

Booth, Alan, David R. Johnson, Ann Branaman, and Alan Sico. 1995. "Belief and Behavior: Does Religion Matter in Today's Marriage?" *Journal of Marriage and the Family* 57 (3) (Aug.): 661–71.

Booth, Alan, David R. Johnson, and Lynn K. White. 1984. "Women, Outside Employment, and Marital Instability." *American Journal of Sociology* 90 (3):567–83.

Borello, Gloria M. and Bruce Thompson. 1990. "A Note Regarding the Validity of Lee's Typology of Love." *Journal of Psychology* 124:639–44.

Boss, Pauline G. 1980. "Normative Family Stress: Family Boundary Changes Across the Lifespan." *Family Relations* 29:445–52.

Boss, Pauline G., William J. Doherty, Ralph LaRossa, Walter R. Schumm, and Suzanne K. Steinmetz, eds. 1993. *Sourcebook of Family Theories and Methods: A Contextual Approach.* New York: Plenum.

Bossard, James H. and E. S. Boll. 1943. *Family Situations.* Philadelphia: University of Pennsylvania Press.

Boston Women's Health Book Collective. 1992. *The New Our Bodies, Ourselves.* New York: Simon & Schuster.

"Both Sexes Are Drawn to Working at Home." 1990. *The Wall Street Journal,* May 24.

Boulding, Elise. 1976. "Familial Constraints on Women's Work Roles." Pp. 95–117 in *Women and the Workplace,* edited by Martha Blaxall and Barbara Reagan. Chicago: University of Chicago Press.

Bouton, Katherine. 1987. "Fertility and Family." *Ms.,* Apr., p. 92.

Bower, Bruce. 1995. "Depression: Rates in Women, Men . . . and Stress Effects Across Sexes." *Science News* 147 (22) (June 3): 346.

Bowers v. *Hardwick.* 1986. 478 U.S. 186, 92 L.Ed.2d 140, 106 S.Ct. 2841.

Bowlby, John. 1969. *Attachment and Loss.* New York: Basic Books.

Boyer, Debra and David Fine. 1992. "Sexual Abuse as a Factor in Adolescent Pregnancy and Child Maltreatment." *Family Planning Perspectives* 24 (1) (Jan./Feb.): 4–11.

Boyer, King David, Jr. 1981. "Changing Male Sex Roles and Identities." Pp. 158–165 in *Men in Difficult Times,* edited by Robert A. Lewis. Englewood Cliffs, NJ: Prentice-Hall.

Boyum, Lisa Ann and Ross D. Parke. 1995. "The Role of Family Emotional Expressiveness in the Development of Children's Social Competence." *Journal of Marriage and the Family* 57 (3) (Aug.): 593–608.

Bradsher, Keith. 1989. "Young Men Pressed to Wed for Success." *New York Times,* Dec. 13.

———. 1990. "Modern Tale of Woe: Being Left at the Altar." *New York Times,* Mar. 7.

Branden, Nathaniel. 1980. *The Psychology of Romantic Love.* Los Angeles: J. P. Tarcher.

———. 1988. "A Vision of Romantic Love." Pp. 218–31 in *The Psychology of Love,* edited by Robert J. Sternberg and Michael L. Barnes. New Haven, CN: Yale University Press.

———. 1994. *Six Pillars of Self-Esteem.* New York: Bantam.

Brandt, Anthony. 1982. "Avoiding Couple Karate: Lessons in the Marital Arts." *Psychology Today* 16:38–43.

Brant, Martha, Karen Springen, and Adam Rogers. 1995. "The Baby Myth." *Newsweek,* Sept. 4, pp. 38–47.

Braver, Sanford L., Pamela J. Fitzpatrick, and R. Curtis Bay. 1991. "Noncustodial Parent's Report of Child Support Payments." *Family Relations* 40 (2) (Apr.): 180–85.

Bray, Rosemary. 1992. "So How Did I Get Here?" *New York Times Magazine,* Nov. 8, pp. 34–35, 38–40, 42.

Brayfield, April. 1995. "Juggling Jobs and Kids: The Impact of Employment Schedules on Fathers' Caring for Children." *Journal of Marriage and the Family* 57 (2) (May): 321–32.

Brazelton, T. Berry. 1989a. "Bringing Up Baby: A Doctor's Prescription for Busy Parents." *Newsweek,* Feb. 13, pp. 68–69.

———. 1989b. "Working Parents." *Newsweek,* Feb. 13, pp. 66–69.

Brigham, John C. 1991. *Social Psychology.* 2d ed. New York: HarperCollins.

Brines, Julie. 1994. "Economic Dependency, Gender, and the Division of Labor at Home." *American Journal of Sociology* 100 (3) (Nov.): 652–88.

Broderick, Carlfred B. 1979a. *Couples: How to Confront Problems and Maintain Loving Relationships.* New York: Simon & Schuster.

———. 1979b. *Marriage and the Family.* Englewood Cliffs, NJ: Prentice-Hall.

Brody, Charles J. and Lala Carr Steelman. 1985. "Sibling Structure and Parental Sex-Typing of Children's Household Tasks." *Journal of Marriage and the Family* 47 (2) (May): 265–73.

Brody, Gene H. and Rex Forehand. 1990. "Interparental Conflict, Relationship with the Noncustodial Father, and Adolescent Post-Divorced Adjustment." *Journal of Applied Psychology* 11 (2) (April–June): 312–36.

Brody, Gene H., Zolinda Stoneman, Douglas Flor, and Chris McCrary. 1994. "Religion's Role in Organizing Family Relationships: Family Process in Rural, Two-Parent African American Families." *Journal of Marriage and the Family* 56 (4) (Nov.): 878–88.

Brody, Jane E. 1988. "Widespread Abuse of Drugs by Pregnant Women Is Found." *New York Times,* Aug. 30.

———. 1989a. "Personal Health: Research Casts Doubt on Need for Many Caesarean Births as Their Rate Soars." *New York Times,* July 27.

———. 1989c. "Who's Having Sex? Data Are Obsolete, Experts Say." *New York Times,* Feb. 18.

———. 1991. "Better Conduct? Train Parents, Then Children." *New York Times,* Dec. 3.

Broman, Clifford L. "Race Differences in Marital Well-Being." *Journal of Marriage and the Family* 55 (3) (Aug.): 724–32.

Bronstein, Phyllis, Miriam F. Stoll, JoAnn Clauson, Craig Abrams, and Maria Briones. 1994. "Fathering After Separation or Divorce: Factors Predicting Children's Adjustment." *Family Relations* 43: 469–79.

Brooks, Andrée. 1984a. "Child-Care Homes Divide Communities." *New York Times,* July 19.

———. 1984b. "Stepparents and Divorce: Keeping Ties to Children." *New York Times,* July 29.

———. 1985. "Trial Separation as a Therapy Technique." *New York Times Style,* Sept. 6.

Brophy, Julia. 1985. "Child Care and the Growth of Power: The Status of Mothers in Custody Disputes." Pp. 97–116 in *Women in Law: Explorations in Law, Family, and Sexuality,* edited by Julia Brophy and Carol Smart. London: Routledge & Kegan Paul.

Brubaker, Ellie, Mary Anne Gorman, and Michele Hiestand. 1990. "Stress Perceived by Elderly Recipients of Family Care." Pp. 267–81 in *Family Relationships in Later Life.* 2d ed., edited by Timothy H. Brubaker. Newbury Park, CA: Sage.

Brubaker, Timothy H. 1991. "Families in Later Life: A Burgeoning Research Area." Pp. 226–48 in *Contemporary Families: Looking Forward, Looking Back,* edited by Alan Booth. Minneapolis: National Council on Family Relations.

Brush, Lisa D. 1990. "Violent Acts and Injurious Outcomes in Married Couples: Methodological Issues in the National Survey of Families and Households." *Gender and Society* 4:56–67.

Bryant, S. and Demian. 1990. "National Survey Results of Gay Couples in Long-Lasting Relationships." *Partners: Newsletter for Gay & Lesbian Couples* (May/June): 1–16.

Bubolz, Margaret M. and M. Suzanne Sontag. 1993. "Human Ecology Theory." Pp. 419–48 in *Sourcebook of Family Theories and Methods: A Contextual Approach,* edited by Pauline G. Boss, William J. Doherty, Ralph LaRossa, Walter R. Schumm, and Suzanne K. Steinmetz. New York: Plenum.

Buchanan, Patrick J. 1983. "Nature Exacts Awful Penalty from Gays." *Omaha World-Herald,* May 25, p. 36.

Bulcroft, Richard A. and Kris A. Bulcroft. 1993. "Race Differences in Attitudinal and Motivational Factors in the Decision to Marry." *Journal of Marriage and the Family* 55 (2) (May): 338–55.

Bumpass, Larry L. 1990. "What's Happening to the Family? Interactions Between Demographic and Institutional Change." *Demography* 27: 483–98.

Bumpass, Larry L., Teresa Castro Martin, and James E. Sweet. 1991. "The Impact of Family Background and Early Marital Factors on Marital Disruption." *Journal of Family Issues* 12 (1) (Mar.): 22–42.

Bumpass, Larry L., R. K. Raley, and J. Sweet. 1995. "The Changing Character of Stepfamilies: Implications of Cohabitation and Nonmarital Childbearing." *Demography* 32 (3) (Aug.): 425–36.

Bumpass, Larry L., and James A. Sweet. 1989. "Children's Experience in Single-Parent Families: Implications of Cohabitation and Marital Transitions." *Family Planning Perspectives* 21:256–61.

———. 1995. "Cohabitation, Marriage, Nonmarital Childbearing, and Union Stability: Preliminary Findings from NSFH2." Population Association of America Annual Meeting, San Francisco, April.

Bumpass, Larry L., James A. Sweet, and Andrew Cherlin. 1991. "The Role of Cohabitation in Declining Rates of Marriage." *Journal of Marriage and the Family* 53 (4) (Nov.): 913–27.

Bunker, Barbara B., Josephine M. Zubek, Virginia J. Vanderslice, and Robert W. Rice. 1992. "Quality of Life in Dual-Career Families: Commuting Versus Single-Residence Couples." *Journal of Marriage and the Family* 54 (3) (May): 399–407.

Bureau of Justice Assistance. 1993. *Family Violence: Interventions for the Justice System.* Washington, DC: U.S. Department of Justice.

Burchell, R. Clay. 1975. "Self-Esteem and Sexuality." *Medical Aspects of Human Sexuality* (Jan.): 74–90.

Burgess, Ernest and Harvey Locke. 1953 [1945]. *The Family: From Institution to Companionship.* New York: American Book Co.

Burnley, Cynthia S. 1987. "The Impact of Emotional Support for Single Women." *Journal of Aging Studies* 1 (3) (Fall): 253–64.

Burns, A. 1984. "Perceived Causes of Marriage Breakdown and Conditions of Life." *Journal of Marriage and the Family* 46:551–62.

Burns, A. and R. Homel. 1989. "Gender Division of Tasks by Parents and Their Children." *Psychology of Women Quarterly* 13:113–25.

Burns, David D. 1989. *The Feeling Good Handbook.* New York: Penguin.

Buscaglia, Leo. 1982. *Living, Loving, and Learning.* New York: Holt, Rinehart & Winston.

Buunk, Bram. 1982. "Strategies of Jealousy: Styles of Coping with Extramarital Involvement of the Spouse." *Family Relations* 31:13–18.

———. 1991. "Jealousy in Close Relationships: An Exchange Theoretical Perspective." Pp. 148–77 in *The Psychology of Jealousy and Envy,* edited by Peter Salovey. New York: Guilford.

Buxton, Amity Pierce. 1991. *The Other Side of the Closet: The Coming-Out Crisis for Straight Spouses.* Santa Monica, CA: IBS Press.

———. 1992. "Don't Forget the Abandoned Spouses of Gays." *New York Times,* Aug. 15.

Buzawa, Eve S. and Carl G. Buzawa. 1990. *Domestic Violence: The Criminal Justice Response.* Newbury Park, CA: Sage.

———. 1992. *Domestic Violence: The Criminal Justice Response.* Westport, CN: Auburn House.

Byrne, Donn E. 1977. "A Pregnant Pause in the Sexual Revolution." *Psychology Today* 11:67–68.

Byrne, Robert C. and Harry M. Overline. 1992. "A Study of Divorce Adjustment Among Paraprofessional Group Leaders and Group Participants." *Journal of Divorce and Remarriage* 17 (1/2):171–92.

Byrne, W. 1994. "The Biological Evidence Challenged." *Scientific American,* May.

C

Caldera, Y. M., A. C. Huston, and M. O'Brien. 1989. "Social Interactions and Play Patterns of Parents and Toddlers with Feminine, Masculine, and Neutral Toys." *Child Development* 60:70–76.

Call, Vaughn, Susan Sprecher, and Pepper Schwartz. 1995. "The Incidence and Frequency of Marital Sex in a National Sample." *Journal of Marriage and the Family* 57 (3) (Aug.): 639–52.

Campbell, Angus A. 1975. "The American Way of Mating: Marriage, Si; Children, Maybe." *Psychology Today* 8:37–43.

Campbell, Marian L. and Phyllis Moen. 1992. "Job–Family Role Strain Among Employed Single Mothers of Preschoolers." *Family Relations* 41 (2) (Apr.): 205–11.

Campbell, Susan. 1991. "Male Day-Care Workers Face Prejudice." *Omaha World-Herald,* July 14.

Canavan, Margaret M., Walter J. Meyer, III, and Deborah C. Higgs. 1992. "The Female Experience of Sibling Incest." *Journal of Marital and Family Therapy* 18 (2) (Apr.): 129–42.

Cancian, Francesca M. 1985. "Gender Politics: Love and Power in the Private and Public Spheres." Pp. 253–64 in *Gender and the Life Course,* edited by Alice S. Rossi. New York: Aldine.

———. 1987. *Love in America: Gender and Self-Development.* New York: Cambridge University Press.

Caplow, Theodore, Howard M. Bahr, Bruce A. Chadwick, Reuben Hill, and Margaret Holmes Williamson. 1982. *Middletown Families: Fifty Years of Change and Continuity.* Minneapolis: University of Minnesota Press.

Carey v. *Population Services International.* 1977. 431 U.S. 678, 52 L.Ed.2d 675, 97 S.Ct. 2010.

Cargan, Leonard and Matthew Melko. 1982. *Singles: Myths and Realities.* Newbury Park, CA: Sage.

Carlson, Bonnie. 1990. "Adolescent Observers of Marital Violence." *Journal of Family Violence* 5 (4): 285–99.

Carlson, Margaret. 1990. "Abortion's Hardest Cases." *Time,* July 9, pp. 22–25.

———. 1994. "Old Enough to Be Your Mother." *Time,* Jan. 10, p. 41.

Carol, Arthur. 1983. "Pay Gaps: Worry Not for Women But Humdrum Households." *Wall Street Journal,* Aug. 12.

Carter, B. 1991. "Children's TV, Where Boys Are King." *New York Times,* May 1, pp. A1, C18.

Carver, Karen Price and Jay D. Teachman. 1993. "Female Employment and First Union Dissolution in Puerto Rico." *Journal of Marriage and the Family* 55 (3) (Aug.): 686–98.

Casey v. *Planned Parenthood of Southeastern Pennsylvania.* 1992.

Cate, Rodney M. and Sally A. Lloyd. 1992. *Courtship.* Newbury Park, CA: Sage.

Cazenave, N. A. 1984. "Race, Socioeconomic Status, and Age: The Social Context of American Masculinity." *Sex Roles* 11:639–56.

Celis, William, 3d. 1992. "Hispanic Youths Quitting School at a High Rate." *New York Times,* Oct. 14.

Centers for Disease Control. 1993a. "Facts About HIV/AIDS and Race/Ethnicity." Nov. Atlanta: U.S. Department of Health and Human Services, Centers for Disease Control.

———. 1993b. "Condoms and Their Use in Preventing HIV Infection and Other STDs." July 30. Atlanta: U.S. Department of Health and Human Services, Centers for Disease Control.

———. 1994a. "Facts About Recent Trends in Reported U.S. AIDS Cases." Aug. Atlanta: U.S. Department of Health and Human Services, Centers for Disease Control.

———. 1994b. "Facts About Adolescents and HIV/AIDS." Dec. Atlanta: U.S. Department of Health and Human Services, Centers for Disease Control.

Chafetz, Janet Saltzman. 1988. "The Gender Division of Labor and the Reproduction of Female Disadvantage: Toward an Integrated Theory." *Journal of Family Issues* 9:108–31.

———. 1989. "Marital Intimacy and Conflict: The Irony of Spousal Equality." Pp. 149–56 in *Women: A Feminist Perspective,* 4th ed., edited by Jo Freeman. Mountain View, CA: Mayfield.

"Changes in Sexual Permissiveness." 1992. *GSS News* (General Social Survey) No. 6 (Sept.):4–6.

Chapman, Audrey B. 1988. "Male–Female Relations." Pp. 190–200 in *Black Families.* 2d ed., edited by Harriette P. McAdoo. Newbury Park, CA: Sage.

"Charge in Fetus's Death Spawns Battle over Drug-Using Mothers." 1992. *The Wall Street Journal,* June 18.

Charny, Israel. 1974. "Marital Love and Hate." In *Violence in the Family,* edited by Suzanne K. Steinmetz and Murray A. Straus. New York: Dodd, Mead.

Chartrand, Sabra. 1992. "Parents Recall Ordeal of Prosecuting in Artificial-Insemination Fraud Case." *New York Times,* Mar. 15, p. 10.

Chase-Lansdale, P. Lindsay, Jeanne Brooks-Gunn, and Roberta L. Palkoff. 1991. "Research and Programs for Adolescent Mothers: Missing Links and Future Promises." *Family Relations* 40 (4) (Oct.): 396–403.

Chatters, Linda M., Robert Joseph Taylor, and Harold W. Neighbors. 1989. "Size of Informal Helper Network Mobilized During a Serious Personal Problem Among Black Americans." *Journal of Marriage and the Family* 51 (3):667–76.

Chen, Kathy. 1995. "Women Pack Up After Fractious Meeting." *The Wall Street Journal,* Sept. 18, p. B1.

Cherlin, Andrew J. 1978. "Remarriage as Incomplete Institution." *American Journal of Sociology* 84:634–50.

———. 1981. *Marriage, Divorce, Remarriage.* Cambridge, MA: Harvard University Press.

Cherlin, Andrew and Frank F. Furstenberg. 1986. *The New American Grandparent: A Place in the Family, A Life Apart.* New York: Basic Books.

Chilman, Catherine. 1978. "Families of Today." Paper presented at the Building Family Strengths Symposium, University of Nebraska, Lincoln, May.

Chilman, Catherine Street. 1993. "Hispanic Families in the United States." Pp. 141–63 in *Family Ethnicity: Strength in Diversity,* edited by Harriette Pipes McAdoo. Newbury Park, CA: Sage.

Chira, Susan. 1991. "Poverty's Toll on Health Is Plague of U.S. Schools." *New York Times,* Oct. 5.

———. 1994a. "Push to Revamp Ideal for American Fathers." *New York Times,* June 19.

———. 1994b. "Role of Parents Diminishes as Pupils Age, Report Says." *New York Times,* Sept. 5.

Chodorow, Nancy. 1978. *The Reproduction of Mothering: Psychoanalysis and the Sociology of Gender.* Berkeley: University of California Press.

Choi, Namkee G. 1992. "Correlates of the Economic Status of Widowed and Divorced Elderly Women." *Journal of Family Issues* 13 (1) (Mar.): 38–54.

Chojnack, Joseph T. and W. Bruce Walsh. 1990. "Reliability and Concurrent Validity of the Sternburg Triangular Love Scale." *Psychological Reports* 67: 219–24.

Chow, E. N. 1985. "The Acculturation Experience of Asian American Women." Pp. 238–51 in *Beyond Sex Roles,* 2d ed., edited by A. Sargent. St. Paul, MN: West.

Christ, Grace. 1982. "Dis-synchrony of Coping Among Children with Cancer, Their Families, and the Treating Staff." Pp. 85–96 in *Psychosocial Interventions in Chronic Pediatric Illness,* edited by Adolph E. Christ and Kalmen Flamenhaft. New York: Plenum.

Christensen, Kathleen E. and Graham L. Staines. 1990. "Flextime: A Viable Solution to Work/Family Conflict?" *Journal of Family Issues* 11 (4) (Dec.): 455–76.

Cicerelli, Victor G. 1990. "Family Support in Relation to Health Problems of the Elderly." Pp. 212–28 in *Family Relations in Later Life.* 2d ed., edited by Timothy H. Brubaker. Newbury Park, CA: Sage.

Clark, Jennifer and Bonnie L. Barber. 1994. "Adolescents in Postdivorce and Always-Married Families: Self-Esteem and Perceptions of Fathers' Interest." *Journal of Marriage and the Family* 56 (3) (Aug.): 608–14.

Clarke, Sally C. 1995a. "Advance Report of Final Marriage Statistics, 1989 and 1990." *Monthly Vital Statistics Report* 43 (12), July 14. U.S. Department of Health and Human Services, National Center for Health Statistics.

Clarke, Sally C. 1995b. "Advance Report of Final Divorce Statistics, 1989 and 1990." *Monthly Vital Statistics Report* 43 (9). Supp., March 22.

Clarke, Sally C. and Barbara F. Wilson. 1994. "The Relative Stability of Remarriages: A Cohort Approach Using Vital Statistics." *Family Relations* 43:305–10.

Clatworthy, Nancy M. 1975. "Living Together," Pp. 67–89 in *Old Family/New Family,* edited by Nona Glazer-Malbin. New York: Van Nostrand.

Clavan, Sylvia. 1978. "The Impact of Social Class and Social Trends on the Role of Grandparent." *Family Coordinator* 27:351–57.

Clinton, Hillary Rodham. 1990. "In France, Day Care Is Every Child's Right." *New York Times,* Apr. 7.

Cole, Thomas. 1983. "The 'Enlightened' View of Aging." *Hastings Center Report* 13:34–40.

Coleman, M. and L. H. Ganong. 1990. "Remarriage and Stepfamily Research in the 1980s: Increased Interest in an Old Family Form." *Journal of Marriage and the Family* 52:925–40.

Colgrove, Melba, Harold Bloomfield, and Peter McWilliams. 1978. *How to Survive the Loss of a Love: 58 Things to Do When There Is Nothing to Be Done.* New York: Bantam.

Collins, J. A., J. B. Garner, E. H. Wilson, W. Wrixon, and R. F. Casper. 1984. "A Proportional Hazard's Analysis of the Clinical Characteristics of Infertile Couples." *American Journal of Obstetrics and Gynecology* 148:527–32.

Collins, Patricia Hill. 1991. *Black Feminist Thought: Knowledge, Consciousness, and the Politics of Empowerment.* New York: Routledge.

Collins, W. A. 1990. "Parent–Child Relationships in the Transition to Adolescence: Continuity and Change in Interaction, Affect and Cognition." Pp. 85–106 in *From Childhood to Adolescence: A Transitional Period? Advances in Adolescent Development,* vol. 2, edited by R. Montemayor, G. R. Adams, and T. P. Gullotta. Newbury Park, CA: Sage.

Coltrane, Scott. 1990. "Birth Timing and the Division of Labor in Dual-Earner Families: Exploratory Findings and Suggestions for Further Research." *Journal of Family Issues* 11:157–81.

Coltrane, Scott and Masako Ishii-Kuntz. 1992. "Men's Housework: A Life Course Perspective." *Journal of Marriage and the Family* 54 (2) (Feb.): 43–57.

Comfort, Alex. 1972. *The Joy of Sex.* New York: Crown.

Commission on Lesbian and Gay Concerns of the American Psychological Association. 1991. "Avoiding Heterosexual Bias in Language." *American Psychologist* 46:973–74.

Committee on the Judiciary, United States Senate. 1992. "Violence Against Women: A Week in the Life of America." Oct. Washington, DC: U.S. Government Printing Office.

Conger, Rand D., Glen H. Elder, Jr., Frederick O. Lorenz, Katherine J. Conger, Ronald L. Simons, Les B. Whitbeck, Shirley Huck, and Janet N. Melby. 1990. "Linking Economic Hardship to Marital Quality and Instability." *Journal of Marriage and the Family* 52 (3) (Aug.): 643–56.

Connor, Michael E. 1986. "Some Parenting Attitudes of Young Black Fathers." Pp. 159–68 in *Men in Families,* edited by Robert A. Lewis and Robert E. Salt. Newbury Xark, CA: Sage.

Connor, Steve. 1994. "Downward Spiral in Quality of Sperm." *San Francisco Chronicle,* Aug. 28, p. A2.

Cooley, Charles Horton. 1902. *Human Nature and the Social Order.* New York: Scribner's.

———. 1909. *Social Organization.* New York: Scribner's.

Cooney, Rosemary, Lloyd H. Rogler, Rose Marie Hurrel, and Vilma Ortiz. 1982. "Decision Making in Intergenerational Puerto Rican Families." *Journal of Marriage and the Family* 44:621–31.

Cooney, Teresa M. 1994. "Young Adults' Relations with Parents: The Influence of Recent Parental Divorce." *Journal of Marriage and the Family* 56 (1) (Feb.): 45–56.

Cooney, Teresa M. and Dennis P. Hogan. 1991. "Marriage in an Institutionalized Life Course: First

Marriage Among American Men in the Twentieth Century." *Journal of Marriage and the Family* 53:178–90.

Cooney, Teresa M., and Peter Uhlenberg. 1991. "Changes in Work-Family Connections Among Highly Educated Men and Women." *Journal of Family Issues* 12 (1) (Mar.): 69–90.

Coontz, Phyllis D., and Judith A. Martin. 1988. "Understanding Violent Mothers and Fathers: Assessing Explanations Offered by Mothers and Fathers for the Use of Control Punishment." Pp. 77–90 in *Family Abuse and Its Consequences: New Directions in Research,* edited by Gerald T. Hotaling, David Finkelhor, John T. Kirkpatrick, and Murray A. Straus. Newbury Park, CA: Sage.

Coontz, Stephanie. 1992. *The Way We Never Were: American Families and the Nostalgia Trap.* New York: Basic Books.

Cooper, Helene. 1992. "Tennessee Court Refuses to Give Custody of 7 Embryos to Mother." *The Wall Street Journal,* June 2, p. B8.

Coopersmith, Stanley. 1967. *The Antecedents of Self-Esteem.* San Francisco: Freeman.

Copenhaver, Stacey and Elizabeth Grauerholz. 1991. "Sexual Victimization Among Sorority Women: Exploring the Link Between Sexual Violence and Institutional Practices." *Sex Roles* 24 (1/2):31–41.

Cose, Ellis. 1994a. "Truths About Spouse Abuse." *Newsweek,* Aug. 8, p. 49.

———. 1994b. "The Year of the Father." *Newsweek,* Oct. 31, p. 61.

———. 1995. "Black Men and Black Women." *Newsweek,* June 15, pp. 66–69.

Coston, John. 1992. "More Work, Less Play Is Rule of the Day." *The Wall Street Journal,* Feb. 14, p. A9.

Counts, Robert M. and Anita Sacks. 1991. "Profiles of the Divorce Prone: The Self-Involved Narcissist." *Journal of Divorce and Remarriage* 15 (1/2):51–74.

"Couple Sues a County Agency for Barring a Transracial Adoption." 1990. *The Wall Street Journal,* Nov. 5.

Coverman, Shelly. 1985. "Explaining Husbands' Participation in Domestic Labor." *Sociological Quarterly* 26 (1):81–97.

Cowan, Alison Leigh. 1989. "Trends in Pregnancies Challenge Employees." *New York Times,* Apr. 17.

Cowan, C. P. and P. A. Cowan. 1992. *When Partners Become Parents.* New York: Basic Books.

Cowan, Ruth Schwartz. 1983. *More Work for Mother: The Ironies of Household Technology from the Open Hearth to the Microwave.* New York: Basic Books.

Cowley, Geoffrey. 1995. "Ethics and Embryos." *Newsweek,* June 12, pp. 66–67.

Coysh, William S., Janet R. Johnston, Jeanne M. Tschann, Judith S. Wallerstein, and Marsha Kline. 1989. "Parental Postdivorce Adjustment in Joint and Sole Physical Custody Families." *Journal of Family Issues* 10:52–71.

Craig, Stephen. 1992. "The Effect of Television Day Part on Gender Portrayals in Television Commercials: A Content Analysis." *Sex Roles* 26 (5/6):197–211.

Cramer, James C. 1995. "Racial and Ethnic Differences in Birthweight: The Role of Income and Financial Assistance." *Demography* 32 (2) (May): 231–47.

"Credit-Card Role in Bankruptcies Debated." 1989. UPI, in *Omaha World-Herald,* Oct. 8.

Critelli, Joseph W., Emilie J. Myers, and Victor E. Loos. 1986. "The Components of Love: Romantic Attraction and Sex Role Orientation." *Journal of Personality* 54:354–76.

Crohan, Susan E. 1992. "Marital Happiness and Spousal Consensus on Beliefs About Marital Conflict: A Longitudinal Investigation." *Journal of Social and Personal Relationships* 9:89–102.

Crosby, Faye J. 1991. *Juggling: The Unexpected Advantages of Balancing Career and Home for Women and Their Families.* New York: The Free Press.

Crosby, John F. 1985 [1976]. *Illusion and Disillusion: The Self in Love and Marriage.* 2d, 3d eds. Belmont, CA: Wadsworth.

———. 1991. *Illusion and Disillusion: The Self in Love and Marriage.* 4th ed. Belmont, CA: Wadsworth.

Crossen, Cynthia. 1990. "Baby Ads Spark Debate over Ethics." *The Wall Street Journal,* Dec. 26.

Cuber, John and Peggy Harroff. 1965. *The Significant Americans.* New York: Random House. (Published also as *Sex and the Significant Americans,* Baltimore: Penguin, 1965.)

Cunningham-Burley, Sarah. 1987. "The Experience of Grandfatherhood." Pp. 91–105 in *Reassessing Fatherhood: New Observations on Fathers and the Modern Family,* edited by Charles Lewis and Margaret O'Brien. Newbury Park, CA: Sage.

"Cupid's Arrows Strike at Random." 1981. *St. Louis Post-Dispatch,* Feb. 15, p. 2-B.

Cutright, Phillips and Herbert L. Smith. 1988. "Intermediate Determinants of Racial Differences in 1980 U.S. Nonmarital Fertility Rates." *Family Planning Perspectives* 20:119–23.

D

Dahms, Alan M. 1976. "Intimacy Hierarchy." Pp. 85–104 in *Process in Relationship: Marriage and Family.* 2d ed., edited by Edward A. Powers and Mary W. Lees. New York: West.

Daniels, Pamela and Kathy Weingarten. 1980. "Postponing Parenthood: The Myth of the Perfect Time." *Savvy Magazine,* May, pp. 55–60.

D'Antonio, William V. and Joan Aldous, eds. 1983. *Families and Religions: Conflict and Change in Modern Society.* Newbury Park, CA: Sage.

D'Antonio, William V., James Davidson, Dean Hoge, and Ruth Wallace. 1989. *American Catholic Laity in a Changing Church.* Kansas City, MO: Sheed and Ward.

Darling, Carol A. 1987. "Family Life Education." Pp. 815–34 in *Handbook of Marriage and the Family,* edited by Marvin B. Sussman and Suzanne K. Steinmetz. New York: Plenum.

Darling, Carol A., David J. Kallen, and Joyce E. VanDusen. 1984. "Sex in Transition, 1900–1980." *Journal of Youth and Adolescence* 13 (5):385–99.

David, Deborah S. and Robert Brannon, eds. 1976. *The Forty-Nine Percent Majority; The Male Sex Role.* Reading, MA: Addison-Wesley.

Davidson, Bernard, Jack Balswick, and Charles Halverson. 1983. "Affective Self-Disclosure and Marital Adjustment: A Test of Equity Theory." *Journal of Marriage and the Family* 45:93–102.

Davis, Angela. 1989. *Women, Culture, and Politics.* New York: Random House.

Davis, Fred. 1991 [1963]. *Passage Through Crisis: Polio Victims and Their Families.* 2d ed. New Brunswick, NJ: Transaction.

Davis, Bob and Dana Milbank. 1992. "If the U.S. Work Ethic Is Fading, Alienation May Be Main Reason." *The Wall Street Journal,* Feb. 7, pp. A1, A5.

Dawson, Deborah A. 1991. "Family Structure and Children's Health and Well-Being: Data from the 1988 National Health Interview Survey on Child Health." *Journal of Marriage and the Family* 53 (3) (Aug.): 573–84.

DeBuono, Barbara A., Stephen H. Zinner, Maxim Daamen, and William M. McCormack. 1990. "Sexual Behavior of College Women in 1975, 1986, and 1989." *New England Journal of Medicine* 222:821–25.

de Lama, George. 1994. "Hawaii May Lead Way on Same-Sex Marriage." *Chicago Tribune,* May 15, p. I-17.

De Lisi, R. and L. Soundranayagam. 1990. "The Conceptual Structure of Sex Stereotypes in College Students." *Sex Roles* 23:593–611.

DeMaris, Alfred and William MacDonald. 1993. "Premarital Cohabitation and Marital Instability: A Test of the Unconventionality Hypothesis." *Journal of Marriage and the Family* 55 (2) (May): 399–407.

DeMaris, Alfred, Meredith D. Pugh, and Erika Harman. 1992. "Sex Differences in the Accuracy of Recall of Witnesses of Portrayed Dyadic Violence." *Journal of Marriage and the Family* 54 (2) (May): 335–45.

DeMause, Lloyd. 1975. "Our Forebears Made Childhood a Nightmare." *Psychology Today* 8:85–88.

D'Emilio, John, and Estelle B. Freedman. 1988. *Intimate Matters: A History of Sexuality in America.* New York: Harpers.

Derlega, Valerian J., Sandra Metts, Sandra Petronio, and Stephen T. Margulis. 1993. *Self-Disclosure.* Newbury Park, CA: Sage.

de Santis, Marie. 1990. "Hate Crimes Bill Excludes Women." *Off Our Backs,* June.

DeVault, Marjorie L. 1991. *Feeding the Family: The Social Organization of Caring as Gendered Work.* Chicago: University of Chicago Press.

Dilworth-Anderson, Peggye, Linda M. Burton, and Eleanor Boulin Johnson. 1993. "Reframing Theories for Understanding Race, Ethnicity, and Families." Pp. 627–46 in *Sourcebook of Family Theories and Methods: A Contextual Approach,* edited by Pauline G. Boss, William J. Doherty, Ralph LaRossa, Walter R. Schumm, and Suzanne K. Steinmetz. New York: Plenum.

Dion, Karen K. and Kenneth L. Dion. 1988. "Romantic Love: Individual and Cultural Perspectives." Pp. 264–89 in *The Psychology of Love,* edited by Robert J. Sternberg and Michael L. Barnes. New Haven, CN: Yale University Press.

———. 1991. "Psychological Individualism and Romantic Love." *Journal of Personality and Social Psychology* 6:17–33.

Dobash, Russell P., R. Emerson Dobash, Margo Wilson, and Martin Daly. 1992. "The Myth of Sexual Symmetry in Marital Violence." *Social Problems* 39 (1) (Feb.): 71–91.

Doherty, William, J., Susan Su, and Richard Needle. 1989. "Marital Disruption and Psychological Well-Being: A Panel Study." *Journal of Family Issues* 10:72–85.

Doniger, Nancy. 1992. "Single Mothers: The Changing Mix." *New York Times*, Oct. 5, p. A-16.

Dorfman, Andrea. 1989. "Alcohol's Youngest Victims." *Time*, Aug. 28, p. 60.

Dorris, Michael. 1989. *The Broken Cord*. New York: Harper & Row.

Doudna, Christine. 1981. "Where Are the Men for the Women at the Top?" Pp. 21–34 in *Single Life*, edited by Peter J. Stein. New York: St. Martin's Press.

Douglas, Susan. 1995. "Sitcom Women: We've Come a Long Way. Maybe." *Ms.* (Nov./Dec.): 76–80.

Doyle, J. 1989. *The Male Experience*. Dubuque, IA: William C. Brown.

Duberman, Lucile. 1975. *The Reconstituted Family: A Study of Remarried Couples and Their Children*. Chicago: Nelson-Hall.

DuBois, David L., Susan K. Eitel, and Robert D. Felner. 1994. "Effects of Family Environment and Parent–Child Relationships on School Adjustment During the Transition to Early Adolescence." *Journal of Marriage and the Family* 56 (2) (May): 405–14.

Dullea, Georgia. 1987a. "AIDS and Divorce: A New Legal Arena." *New York Times*, Sept. 21.

———. 1987b. "Divorces Spawn Confusion Over Stepparents' Rights." *Omaha World-Herald*, Mar. 18.

———. 1987c. "False Child-Abuse Charges Seem to Be on the Increase." *New York Times*, reprinted in *Omaha World-Herald*, Jan. 19.

———. 1987d. "When Illness Destroys a Marriage." *New York Times*, Sept. 21.

———. 1988a. "Gay Couples' Wish to Adopt Grows, Along with Increasing Resistance." *New York Times*, Feb. 7.

———. 1988b. "Prenuptial Stress over the Contract: It's Sign or Stay Single." *New York Times*, July 6.

Duncan, Greg and Willard Rodgers. 1987. "Single-Parent Families: Are Their Economic Problems Transitory or Persistent?" *Family Planning Perspectives* 19:171–78.

Dunning, Jennifer. 1986. "Women and AIDS: Discussing Precautions." *New York Times*, Nov. 3.

Durkheim, Emile. 1933 [1893]. *Division of Labor in Society*, translated by George Simpson. New York: Macmillan.

———. 1951. *Suicide*, translated by John A. Spaulding and George Simpson. Glencoe, IL: Free Press.

Dutton, Donald G. and James J. Browning. 1988. "Concern for Power, Fear of Intimacy, and Aversive Stimuli for Wife Assault." Pp. 163–75 in *Family Abuse and Its Consequences: New Directions in Research*, edited by Gerald T. Hotaling, David Finkelhor, John T. Kirkpatrick, and Murray A. Straus. Newbury Park, CA: Sage.

Duvall, E. M. 1957. *Family Development*. Philadelphia: Lippincott.

Dwyer, Jeffrey W., Gary R. Lee, and Thomas B. Jankowski. 1994. "Reciprocity, Elder Satisfaction, and Caregiver Stress and Burden: The Exchange of Aid in the Family Caregiving Relationship." *Journal of Marriage and the Family* 56 (1) (Feb.): 35–43.

Dwyer, Jeffrey W. and Karen Seccombe. 1991. "Elder Care as Family Labor: The Influence of Gender and Family Position." *Journal of Family Issues* 12 (2) (June): 229–47.

E

Eargle, Judith. 1990. *Household Wealth and Asset Ownership:1988*. U.S. Bureau of the Census. Household Economic Studies Series P-70, No. 22. Washington, DC: Government Printing Office.

Easterlin, Richard. 1987. *Birth and Fortune: The Impact of Numbers on Personal Welfare*. 2d rev. ed. Chicago: University of Chicago Press.

Easterlin, Richard A. and Eileen M. Crimmins. 1985. *The Fertility Revolution: A Supply-Demand Analysis*. Chicago: University of Chicago Press.

Eaton, William J. 1992. "House Panel Approves Abortion-Rights Bill." *San Francisco Chronicle*, July 1, p. A4.

Eberstadt, Nicholas. 1992. "America's Infant Mortality Problem: Parents." *The Wall Street Journal*, Jan. 20, p. A12.

Economic Report of the President. 1989. Washington: U.S. Government Printing Office. Jan.

Edwards, John N. 1987. "Changing Family Structure and Youthful Well-Being." *Journal of Family Issues* 8 (4):355–72.

———. 1991. "New Conceptions: Biosocial Innovations and the Family." *Journal of Marriage and the Family* 53:349–60.

Eggebeen, David J. and Daniel T. Lichter. 1991. "Race, Family Structure, and Changing Poverty Among American Children." *American Sociological Review* 56 (Dec.): 801–17.

Ehrbar, Al. 1993. "Pride of Progress." *The Wall Street Journal,* Mar. 16, Pp. A1, A11.

Ehrenreich, Barbara. 1983. *The Hearts of Men: American Dreams and the Flight from Commitment.* Garden City, NY: Anchor/Doubleday.

Ehrensaft, Diane. 1990. *Parenting Together: Men and Women Sharing the Care of Their Children.* Urbana: University of Illinois Press.

Eisenstadt v. *Baird.* 1972. 405 U.S. 398.

Eitzen, D. Stanley and Maxine Baca Zinn. 1992. "Work and Economic Transformation." Pp. 178–82 in *Race, Class, and Gender,* edited by Margaret Andersen and Patricia Hill Collins. Belmont, CA: Wadsworth.

Elder, Glen. 1974. *Children of the Great Depression.* Chicago: University of Chicago Press.

Elder, Glen H., Jr. 1977. "Family History and the Life Course." *Journal of Family History* 2:279–304.

Elkin, Larry M. 1994. *First Comes Love, Then Comes Money.* New York: Doubleday.

Ellis, Walter L. 1991. "The Effects of Background Characteristics of Attorneys and Judges on Decision Making in Domestic Relations Court: An Analysis of Child Support Awards." *Journal of Divorce and Remarriage* 16 (1/2):107–20.

Elmer-Dewitt, Philip. 1991a. "How Safe Is Sex?" *Time,* Nov. 25, p. 72.

———. 1991b. "Making Babies." *Time,* Sept. 30, pp. 56–63.

Elson, John. 1989. "The Rights of Frozen Embryos." *Time,* July 24, p. 63.

Engels, Friedrich. 1942 [1884]. *The Origin of the Family, Private Property, and the State.* New York: International.

England, Paula and George Farkas. 1986. *Households, Employment and Gender: A Social, Economic and Demographic View.* New York: Aldine.

Entwisle, Doris R. and Karl L. Alexander. 1995. "A Parent's Economic Shadow: Family Structure Versus Family Resources as Influences on Early School Achievement." *Journal of Marriage and the Family* 57 (2) (May): 399–409.

Epstein, Cynthia Fuchs. 1988. *Deceptive Distinctions: Sex, Gender, and the Social Order.* New Haven and New York: Yale University Press and Russell Sage Foundation.

Epstein, Norma, Lynda Evans, and John Evans. 1994. "Marriage." Pp. 115–25 in *Encyclopedia of Human Behavior,* vol. 3, edited by V. S. Ramachandran. New York: Academic.

Erickson, Nancy S. 1991. "Battered Mothers of Battered Children: Using Our Knowledge of Battered Women to Defend Them Against Charges of Failure to Act." *Current Perspectives in Psychological, Legal, and Ethical Issues,* vol. 1A: Children and Families: Abuse and Endangerment: 197–218.

Erickson, Rebecca J. 1993. "Reconceptualizing Family Work: The Effect of Emotion Work on Perceptions of Marital Quality." *Journal of Marriage and the Family* 55 (3) (Nov.): 888–900.

Erikson, Erik. 1979. Interview in *Omaha World-Herald,* Aug. 5.

Ettelbrick, Paula L. 1989. "Since When Is Marriage a Path to Liberation?" *OUT/LOOK, National Gay and Lesbian Quarterly* 6 (Fall).

Ettner, Susan L. 1995. "The Impact of 'Parent Care' on Female Labor Supply Decisions." *Demography* 32 (1) (Feb.): 63–80.

Etzkowitz, Henry and Peter Stein. 1978. "The Life Spiral: Human Needs and Adult Roles." *Alternative Lifestyles* 1:434–46.

Eyer, Diane E. 1992. *Mother–Infant Bonding: A Scientific Fiction.* New Haven, CN: Yale University Press.

Eyre, Linda and Richard Eyre. 1994. *Teaching Your Children Joy. Teaching Your Children Responsibility. Teaching Your Children Values.* 3 books. New York: Simon & Schuster, Fireside Books.

Evans, Sara. 1978. *Personal Politics.* New York: Knopf.

F

Faber, Adele and Elaine Mazlish. 1982. *How to Talk So Kids Will Listen and Listen So Kids Will Talk.* New York: Avon.

Falbo, T. 1976. "Does the Only Child Grow Up Miserable?" *Psychology Today* 9:60–65.

Faludi, Susan. 1991. *Backlash.* New York: Crown.

Farber, Bernard. 1959. "Effects of a Severely Mentally Retarded Child on Family Integration." *Monographs of the Society for Research in Child Development* 24 (71):80–94.

Fassinger, Polly. 1990. "The Meanings of Housework for Single Parents: Insights into Gender Strategies." Paper presented at the annual meeting of the Midwest Sociological Society, Chicago, Apr. 13.

"The Fate of the Frozen Embryos." 1992. *New York Times,* June 6, p. 14.

Felson, Richard B. and Natalie Russo. 1988. "Parental Punishment and Sibling Aggression." *Social Psychology Quarterly* 51 (1): 11–18.

Ferguson, Susan J. 1995. "Marriage Timing of Chinese American and Japanese American Women." *Journal of Family Issues* 16 (3) (May): 314–43.

Ferree, Myra Marx. 1990. "Beyond Separate Spheres: Feminism and Family Research." *Journal of Marriage and the Family* 52:866–84.

———. 1991. "The Gender Division of Labor in Two-Earner Marriages: Dimensions of Variability and Change." *Journal of Family Issues* 12 (2) (June): 158–80.

Fiese, Barbara H., Karen A. Hooker, Lisa Kotary, and Janet Schwagler. 1993. "Family Rituals in the Early Stages of Parenthood." *Journal of Marriage and the Family* 55 (3) (Aug.): 633–42.

Fine, Mark A. and David R. Fine. 1994. "An Examination and Evaluation of Recent Changes in Divorce Laws in Five Western Countries: The Critical Role of Values." *Journal of Marriage and the Family* 56 (2) (May): 249–63.

Fine, Mark A., Patrick C. McKenry, and Hyunsook Chung. 1992. "Post-Divorce Adjustment of Black and White Single Parents." *Journal of Divorce and Remarriage* 17 (3/4):121–34.

Fineman, Howard. 1994. "Clinton's Values Blowout." *Newsweek,* Dec. 19, Pp. 24–27.

———. 1995. "Mediscare." *Newsweek,* Sept. 18, Pp. 38–43.

Fineman, Martha Albertson and Roxanne Mykitiuk. 1994. *The Public Nature of Private Violence: The Discovery of Domestic Abuse.* New York: Routledge.

Finkelhor, David and Larry Baron. 1986. "High Risk Children." Pp. 60–88 in *A Sourcebook on Child Sexual Abuse,* edited by David Finkelhor. Newbury Park, CA: Sage.

Finkelhor, David, Gerald Hotaling, and Andrea Sedlak. 1991. "Children Abducted by Family Members: A National Household Survey of Incidence and Episode Characteristics." *Journal of Marriage and the Family* 53 (3) (Aug.): 805–17.

Finkelhor, David and Karl Pillemer. 1988. "Elder Abuse: Its Relationship to Other Forms of Domestic Violence." Pp. 244–54 in *Family Abuse and Its Consequences: New Directions in Research,* edited by Gerald T. Hotaling, David Finkelhor, John T. Kirkpatrick, and Murray A. Straus. Newbury Park, CA: Sage.

Finkelhor, David and Kersti Yllo. 1985. *License to Rape: Sexual Abuse of Wives.* New York: Holt.

"First 'Ovum Transfer' Baby Born." 1984. *Chronicle of Higher Education,* Feb. 15, pp. 7, 9.

Fisher, Jeffrey D., Arie Nadler, and S. Alagne. 1982. "Recipient Reactions to Aid." *Psychological Bulletin* 91:25–54.

Fisher, Judith L. 1983. "Mothers Living Apart from Their Children." *Family Relations* 32:351–57.

Fisher, Lucy R. 1986. *Linked Lives: Adult Daughters and Their Mothers.* New York: Harper.

Fishman, Barbara. 1983. "The Economic Behavior of Stepfamilies." *Family Relations* 32:359–66.

Fishman, Katherine Davis. 1992. "Problem Adoptions." *The Atlantic Monthly* 270 (3) (Sept.): 37–69.

Fitzpatrick, Mary Anne. 1988. *Between Husbands and Wives: Communication in Marriage.* Newbury Park, CA: Sage.

Flaks, D. K., I. Ficher, F. Masterpasqua, and G. Joseph. 1995. "Lesbians Choosing Motherhood: A Comparative Study of Lesbians and Heterosexual Parents and Their Children." *Developmental Psychology* 31 (1): 105–14.

Flanery, James Allen. 1992. "Siblings May Suffer from Fetal Abuse." *Omaha World-Herald,* June 17.

Flewelling, Robert L. and Karl E. Bauman. 1990. "Family Structure as a Predictor of Initial Substance Use and Sexual Intercourse in Early Adolescence." *Journal of Marriage and the Family* 52 (1) (Feb.): 171–81.

Florian, Victor and Nira Dangoor. 1994. "Personal and Familial Adaptation of Women with Severe Physical Disabilities: A Further Validation of the Double ABCX Model." *Journal of Marriage and the Family* 56 (3) (Aug.): 735–46.

Flynn, Clifton P. 1990. "Relationship Violence by Women: Issues and Implications." *Family Relations* 39 (2) (Apr.):194–98.

———. 1994. "Regional Differences in Attitudes Toward Corporal Punishment." *Journal of Marriage and the Family* 56 (May): 314–24.

Foderaro, Lisa W. 1988. "In Age of AIDS, Blind Dating Is Back." *New York Times,* Jan. 21.

Folk, Karen and Yunae Yi. 1994. "Piecing Together Child Care with Multiple Arrangements: Crazy Quilt or Preferred Pattern for Employed Parents of Preschool Children?" *Journal of Marriage and the Family* 56 (3) (Aug.): 669–80.

Foote, Nelson. 1954. "Sex as Play." *Social Problems* 1:159–63.

Ford, Clellan A. and Frank A. Beach. 1971. "Human Sexual Behavior in Perspective." Pp. 155–71 in *Family in Transition: Rethinking Marriage, Sexuality, Child Rearing, and Family Organization,* edited by Arlene S. Skolnick and Jerome H. Skolnick. Boston: Little, Brown.

"Foreign-Born Mothers Fertility-Rate Leaders." 1987. *Omaha World-Herald,* Jan. 6.

Forgatch, Marion S. 1989. "Patterns and Outcome in Family Problem Solving: The Disrupting Effect of Negative Emotion." *Journal of Marriage and the Family* 51:115–24.

Forrest, Jacqueline Darroch. 1987. "Unintended Pregnancy Among American Women." *Family Planning Perspectives* 19:76–77.

Forste, Renata, Koray Tanfer, and Lucky Tedrow. 1995. "Sterilization Among Currently Married Men in the United States, 1991." *Family Planning Perspectives* 27 (3) (May/June): 100–107, 122.

Fossett, Mark A. and K. Jill Kiecolt. 1993. "Mate Availability and Family Structure Among African Americans in U.S. Metropolitan Areas." *Journal of Marriage and the Family* 55 (2) (May): 288–302.

Fox, Freer L. and Robert F. Kelly. 1995. "Determinants of Child Custody Arrangements at Divorce." *Journal of Marriage and the Family* 57 (3) (Aug.): 693–708.

Fox-Genovese, E. 1991. *Feminism Without Illusions: A Critique of Individualism.* Chapel Hill: University of North Carolina Press.

Fracher, Jeffrey and Michael S. Kimmel. 1992. "Hard Issues and Soft Spots: Counseling Men About Sexuality." Pp. 438–50 in *Men's Lives.* 2d ed. edited by Michael S. Kimmel and Michael A. Messner. New York: Macmillan.

Framo, James L. 1978. "The Friendly Divorce." *Psychology Today* 11:77–80, 99–102.

Francoeur, Robert T. 1987. "Human Sexuality." Pp. 509–34 in *Handbook of Marriage and the Family,* edited by Marvin B. Sussman and Suzanne K. Steinmetz. New York: Plenum.

———, ed. 1994. *Taking Sides: Clashing Views on Controversial Issues in Human Sexuality.* 4th ed. New York: Dushkin.

Franklin II, Clyde W. 1986. "Black Male–Black Female Conflict: Individually Caused and Culturally Nurtured." Pp. 106–13 in *The Black Family; Essays and Studies,* edited by Robert Staples. Belmont, CA: Wadsworth.

———. 1988. *Men and Society.* Chicago: Nelson-Hall.

———. 1992. "Hey, Home—Yo, Bro": Friendship Among Black Men." Pp. 201–14 in *Men's Friendships: Research on Men and Masculinities,* edited by Peter M. Nardi. Newbury Park, CA: Sage.

Freedman, Samuel G. 1986. "New York Offers Help to Unmarried Fathers." *New York Times,* reprinted in *Omaha World-Herald,* Dec. 7.

Freedman, Vicki A. 1993. "Kin and Nursing Home Lengths of Stay: A Backward Recurrence Time Approach." *Journal of Health and Social Behavior* 34 (June): 138–52.

Freeman, William J. 1994. "Letters." *U.S. News & World Report,* Nov. 14, p. 8.

French, J. R. P., and Bertram Raven. 1959. "The Basis of Power." In *Studies in Social Power,* edited by D. Cartwright. Ann Arbor: University of Michigan Press.

Friedan, Betty. 1963. *The Feminine Mystique.* New York: Dell.

———. 1978. "Where Are Women in 1978?" *Cosmopolitan,* Aug., pp. 196, 206–11.

Friedman, Joel, Marcia M. Boumil, and Barbara Ewert Taylor. 1992. *Date Rape: What It Is, What It Isn't, What It Does to You, What You Can Do About It.* Deerfield Beach, FL: Health Communications.

Fromm, Erich. 1956. *The Art of Loving.* New York: Harper & Row.

Frye, Marilyn. 1992. "Lesbian 'Sex.'" Pp. 109–19 in *Essays in Feminism 1976–1992,* edited by Marilyn Frye. Freedom, CA: Crossing Press.

Fullilove, Mindy Thompson, Robert E. Fullilove, Katherine Haynes, and Shirley Gross. 1990. "Black Women and AIDS Prevention: A View Toward Understanding the Gender Roles." *Journal of Sex Research* 27:47–64.

Furstenberg, Frank F., Jr. 1987. "The New Extended Family: The Experience of Parents and Children after Remarriage." Pp. 42–61 in *Remarriage and Stepparenting,* edited by Kay Pasley and Marilyn Ihinger-Tallman. New York: Guilford.

———. 1991. "As the Pendulum Swings: Teenage Childbearing and Social Concern." *Family Relations* 40:127–38.

———. 1992. "Good Dads—Bad Dads: Two Faces of Fatherhood." Pp. 342–62 in *Families in Transition.* 7th ed., edited by Arlene S. and Jerome H. Skolnick. New York: Harper.

Furstenberg, Frank F., Jr., J. Brooks-Gunn, and S. Philip Morgan. 1987. *Adolescent Mothers in Later Life.* New York: Cambridge University Press.

Furstenberg, Frank F., Jr. and Andrew J. Cherlin. 1991. *Divided Families: What Happens to Children When Parents Part.* Cambridge, MA: Harvard University Press.

Furstenberg, Frank F., Jr., Saul D. Hoffman, and Laura Shrestha. 1995. "The Effect of Divorce on Intergenerational Transfers: New Evidence." *Demography* 32 (3) (Aug.): 319–33.

Furstenberg, Frank F., Jr. and Mary Elizabeth Hughes. 1995. "Social Capital and Successful Development Among At-Risk Youth." *Journal of Marriage and the Family* 57 (3) (Aug.): 580–92.

G

Gable, Sara, Jay Belsky, and Keith Crnic. 1992. "Marriage, Parenting, and Child Development: Progress and Prospects." *Journal of Family Psychology* 5:276–94.

Gagnon, John H. 1977. *Human Sexualities.* Glenview, IL: Scott, Foresman.

Galen, Michele. 1994. "White, Male, and Worried." *Business Week,* Jan. 31, pp. 50–55.

Galinsky, Ellen. 1986. "Family Life and Corporate Policies." Pp. 109–45 in *In Support of Families,* edited by Michael W. Yogman and T. Berry Brazelton. Cambridge, MA: Harvard University Press.

Galinsky, Ellen and Peter J. Stein. 1990. "The Impact of Human Resource Policies on Employees: Balancing Work/Family Life." *Journal of Family Issues* 11 (4) (Dec.): 368–83.

Gallagher, Janet. 1987. "Prenatal Invasions and Interventions: What's Wrong with Fetal Rights." *Harvard Women's Law Journal* 9:9–58.

"Gallup Poll Shows More Americans Say Premarital Sex Is Wrong, Reversing Trend of Last 20 Years." 1988. *Family Planning Perspectives* 20:180–81.

Gander, Anita Moore. 1991. "After the Divorce: Familial Factors that Predict Well-Being for Older and Younger Persons." *Journal of Divorce and Remarriage* 15 (1/2):175–92.

Ganong, Lawrence H. and Marilyn Coleman. 1992. "Gender Differences in Expectations of Self and Future Partner." *Journal of Family Issues* 13 (1) (Mar.): 55–64.

Ganong, Lawrence H., Marilyn Coleman, and Dennis Mapes. 1990. "A Meta-analytic Review of Family Structure Stereotypes." *Journal of Marriage and the Family* 52 (2) (May): 287–97.

Gans, Herbert J. 1982 [1962]. *The Urban Villagers: Group and Class in the Life of Italian-Americans,* updated and expanded edition. New York: Free Press.

Garbarino, James. 1977. "The Human Ecology of Child Maltreatment: A Conceptual Model for Research." *Journal of Marriage and the Family* 39:721–35.

Gardner, Richard A. 1993. "Modern Witch Hunt— Child Abuse Charges." *The Wall Street Journal,* Feb. 22, p. A-10.

Garfinkel, Irwin and Sara S. McLanahan. 1986. *Single Mothers and Their Children: A New American Dilemma.* Washington, DC: Urban Institute Press.

Garfinkel, Irwin, Donald Oellerich, and Philip K. Robbins. 1991. "Child Support Guidelines: Will They Make a Difference?" *Journal of Family Issues* 12 (4) (Dec.): 404–29.

Garrett, Patricia, John Ferron, Nicholas Ng'Andu, Donna Dryant, and Gloria Harbin. 1994. "A Structural Model for the Development Status of Young Children." *Journal of Marriage and the Family* 56 (1) (Feb.): 147–63.

"Gay Rights: What's the Law?" 1994. *New York Times,* Apr. 25, p. C12.

Geist, William E. 1987. "In AIDS Era, a New Club for Singles." *New York Times,* Apr. 15.

Gelcer, Esther. 1986. "Dealing with Loss in the Family Context." *Journal of Family Issues* 7:315–36.

Gelles, Richard J. 1994. "Ten Risk Factors." *Newsweek,* July 4, p. 29.

Gelles, Richard J. and Jon R. Conte. 1990. "Domestic Violence and Sexual Abuse of Children: A Review of Research in the Eighties." *Journal of Marriage and the Family* 52 (4) (Nov.): 1045–58.

Gelles, Richard J. and Murray A. Straus. 1988. *Intimate Violence: The Definitive Study of the Causes and Consequences of Abuse in the American Family.* New York: Simon & Schuster.

Gelman, David. 1990. "A Is for Apple, P Is for Shrink." *Newsweek,* Dec. 24, pp. 64–66.

———. 1992. "Born or Bred?" *Newsweek,* Feb. 24, pp. 46–57.

Gerhard, Susan. 1994. "Cybersex: The New Abstinence." *San Francisco Bay Guardian,* Aug. 24, pp. 23–24.

Geronimus, Arline T. 1991. "Teenage Childbearing and Social and Reproductive Disadvantage: The Evolution of Complex Questions and the Demise of Simple Answers." *Family Relations* 40 (4) (Oct.): 463–71.

Gerson, Kathleen. 1985. *Hard Choices: How Women Decide About Work, Career, and Motherhood.* Berkeley: University of California Press.

———. 1993. *No Man's Land: Men's Changing Commitments to Family and Work.* New York: HarperCollins, Basic Books.

Gerstel, Naomi and Sally K. Gallagher. 1993. "Kinkeeping and Distress: Gender, Recipients of Care, and Work–Family Conflict." *Journal of Marriage and the Family* 55 (3) (Aug.): 598–607.

Gibbs, Nancy. 1989. "The Baby Chase." *Time,* Oct. 9, pp. 86–89.

———. 1991. "Marching Out of the Closet." *Time,* Aug. 19, pp. 14–15.

Gibbs, Nancy R. 1993. "Bringing Up Father." *Time,* June 28, pp. 53–61.

Gilbert, Susan. 1993. "The Waiting Game." *New York Times Magazine.* April 25: 68, 71–72, 92.

Giles-Sims, Jean. 1984. "The Stepparent Role: Expectations, Behavior, and Sanctions." *Journal of Family Issues* 5:116–30.

Giles-Sims, Jean and Margaret Crosbie-Burnett. 1989. "Adolescent Power in Stepfather Families: A Test of Normative Resource Theory." *Journal of Marriage and the Family* 52:1065–78.

Gilgun, Jane E. 1995. "We Shared Something Special: The Moral Discourse of Incest Perpetrators." *Journal of Marriage and the Family* 57 (2) (May): 265–81.

Gillespie, Dair. 1971. "Who Has the Power? The Marital Struggle." *Journal of Marriage and the Family* 33:445–58. Reprinted in *Women: A Feminist Perspective,* edited by Jo Freeman. Mountain View, CA: Mayfield.

Gilmore, David D. 1990. *Manhood in the Making: Cultural Concepts of Masculinity.* New Haven, CT: Yale University Press.

Ginott, Haim G. 1965. *Between Parent and Child.* New York: Macmillan.

Gittelson, Natalie. 1984. "American Jews Rediscover Orthodoxy." *New York Times Magazine,* Sept. 9, pp. 40ff.

Glaser, Danya and Stephen Frosh. 1988. *Child Sexual Abuse.* Chicago: Dorsey.

Glass, Jennifer. 1992. "Housewives and Employed Wives: Demographic and Attitudinal Change, 1972–1986." *Journal of Marriage and the Family* 54 (Aug.): 559–69.

Glazer, Nona. 1990. "The Home as Workshop: Women as Amateur Nurses and Medical Care Providers." *Gender and Society* 4:479–99.

Glendon, Mary Ann. 1981. *The New Family and the New Property.* Toronto: Butterworths.

———. 1987. *Abortion and Divorce in Western Law.* Cambridge, MA: Harvard University Press.

———. 1989. *The Transformation of Family Law: State, Law, and Family in the United States and Western Europe.* Chicago: University of Chicago Press.

Glenn, Norval D. 1982. "Interreligious Marriage in the United States: Patterns and Recent Trends." *Journal of Marriage and the Family* 44:555–68.

———. 1987. "Continuity Versus Change, Sanguineness Versus Concern: Views of the American Family in the Late 1980s." Introduction to a special issue of the *Journal of Family Issues* 8:348–54.

———. 1990. "Quantitative Research on Marital Quality in the 1980s: A Critical Review." *Journal of Marriage and the Family* 52 (Nov.): 818–31.

———. 1991. "The Recent Trend in Marital Success in the United States." *Journal of Marriage and the Family* 53 (2) (May): 261–70.

———. "A Plea for Objective Assessment of the Notion of Family Decline." *Journal of Marriage and the Family* 55 (3) (Aug.): 542–44.

Glenn, Norval D. and Sara McLanahan. 1982. "Children and Marital Happiness: A Further Specification of the Relationship." *Journal of Marriage and the Family* 44:63–72.

Glenn, N. D. and Charles N. Weaver. 1988. "The Changing Relationship of Marital Status to Reported Happiness." *Journal of Marriage and the Family* 50:317–24.

Glick, Paul C. 1988. "Fifty Years of Family Demography: A Record of Social Change." *Journal of Marriage and the Family* 50 (4):861–73.

———. 1989. "Remarried Families, Stepfamilies, and Stepchildren: A Brief Demographic Profile." *Family Relations* 38 (Jan.): 24–27.

Glick, Paul C. and Arthur J. Norton. 1979. "Marrying, Divorcing, and Living Together in the U.S. Today." *Population Bulletin* 32 (5). Washington, DC: Population Reference Bureau.

Glick, Paul C. and Graham Spanier. 1980. "Married and Unmarried Cohabitation in the United States." *Journal of Marriage and the Family* 42:19–30.

Glick, Paul C. and Sung-Ling Lin. 1986a. "More Young Adults Are Living With Their Parents: Who Are They?" *Journal of Marriage and the Family* 48:107–12.

———. 1986b. "Recent Changes in Divorce and Remarriage." *Journal of Marriage and the Family* 48 (4):737–47.

Glick, Peter. 1991. "Trait-Based and Sex-Based Discrimination in Occupational Prestige, Occupational Salary, and Hiring." *Sex Roles* 25 (5/6):351–78.

"Gloria: Women Are Winning." 1992. Associated Press, in *Omaha World-Herald,* July 18.

Goffman, Erving. 1959. *The Presentation of Self in Everyday Life.* Garden City, NY: Doubleday.

Gold, Steven J. 1993. "Migration and Family Adjustment: Continuity and Change Among Vietnamese in the United States." Pp. 300–14 in *Family Ethnicity: Strength in Diversity,* edited by Harriette Pipes McAdoo. Newbury Park, CA: Sage.

Goldberg, Herb. 1987. *The Inner Male: Overcoming Roadblocks to Intimacy.* New York: New American Library.

Goldberg, Steven. 1973. *The Inevitability of Patriarchy.* New York: Morrow.

Goldner, Virginia. 1993. "Feminist Theories." Pp. 623–25 in *Sourcebook of Family Theories and Methods: A Contextual Approach,* edited by Pauline G. Boss, William J. Doherty, Ralph LaRossa, Walter R. Schumm, and Suzanne K. Steinmetz. New York: Plenum.

Goldscheider, Frances and Calvin Goldscheider. 1994. *"Leaving and Returning Home in 20th Century America." Population Bulletin* 48 (4) (Mar.).

Goldscheider, Frances K. and Calvin Goldscheider. 1989. "Family Structure and Conflict: Nest-leaving Expectations of Young Adults and Their Parents." *Journal of Marriage and the Family* 51:87–97.

Goldstein, Arnold P., Harold Keller, and Diane Erne. 1985. *Changing the Abusive Parent.* Champaign, IL: Research Press.

Goleman, Daniel. 1985. "Patterns of Love Charted in Studies." *New York Times,* Sept. 10.

———. 1986. "Two Views of Marriage Explored: His and Hers." *New York Times,* Apr. 1.

———. 1988a. "Adding the Sounds of Silence to the List of Health Risks." *New York Times,* Aug. 4.

———. 1988c. "The Lies Men Tell Put Women in Danger of AIDS." *New York Times,* Aug. 14.

———. 1989. "For a Happy Marriage, Learn How to Fight." *New York Times,* Feb. 21.

———. 1990a. "Aggression in Men: Hormone Levels Are a Key." *New York Times,* July 17.

———. 1990b. "Anger over Racism Is Seen as a Cause of Blacks' High Blood Pressure." *New York Times,* Apr. 24.

———. 1992a. "Family Rituals May Promote Better Emotional Adjustment." *New York Times,* Mar. 11.

———. 1992b. "Gay Parents Called No Disadvantage." *New York Times,* Dec. 21.

"Goodbye to the 'Condom Queen.'" 1994. *Newsweek,* Dec. 19, p. 26.

Goode, William J. 1959. "The Theoretical Importance of Love." *American Sociological Review* 24:38–47.

———. 1963. *World Revolution and Family Patterns.* New York: Free Press.

———. 1982. "Why Men Resist." Pp. 131–50 in *Rethinking the Family: Some Feminist Questions,* edited by Barrie Thorne and Marilyn Yalom. New York: Longmans.

Goodman, Ellen. 1987. "M Is for Money, Not Mother." Boston Globe Newspaper Company/Washington Post Writers Group. In *The Daily Nebraskan,* Feb. 17.

Gordon, Leland J. and Stewart M. Lee. 1977. *Economics for Consumers.* New York: Van Nostrand.

Gordon, Linda. 1988. *Heroes of Their Own Lives: The Politics and History of Family Violence.* New York: Viking.

Gordon, Thomas. 1970. *Parent Effectiveness Training.* New York: Wyden.

Gore, Susan. 1978. "The Effect of Social Support in Moderating the Health Consequences of Unemployment." *Journal of Health and Social Behavior* 19:157–65.

Gore, Susan, and Thomas W. Mangione. 1983. "Social Roles, Sex Roles, and Psychological Distress:

Additive and Interactive Models of Sex Differences." *Journal of Health and Social Behavior* 24:300–12.

Gorman, Christine. 1991. "Returning File Against AIDS." *Time,* June 24, p. 44.

———. 1992. "Sizing Up the Sexes." *Time,* Jan. 20, pp. 42–51.

———. 1993. "When AIDS Strikes Parents." *Time,* Nov. 18, pp. 76–77.

Gottfried, Adele E. and Allen W. Gottfried, eds. 1994. *Redefining Families: Implications for Children's Development.* New York: Plenum.

Gottlieb, Annie. 1979. "The Joyful Marriage." *Redbook,* Nov., pp. 29, 194–96.

Gottman, John M. 1979. *Marital Interaction: Experimental Investigations.* New York: Academic.

Gottman, John M. and L. J. Krotkoff. 1989. "Marital Interaction and Satisfaction: A Longitudinal View." *Journal of Consulting and Clinical Psychology* 57:47–52.

Gove, Walter R., and Robert D. Cruchfield. 1982. "The Family and Juvenile Delinquency." *The Sociological Quarterly* 23:301–19.

Gove, Walter R., Carolyn Briggs Style, and Michael Hughes. 1990. "The Effect of Marriage on the Well-Being of Adults." *Journal of Family Issues* 11 (1) (Mar.): 4–35.

Grant, Linda, Layne A. Simpson, Xue Lan Rong, and Holly Peters-Golden. 1990. "Gender, Parenthood, and Work Hours of Physicians." *Journal of Marriage and the Family* 52 (2) (Feb.): 39–49.

Grauerholz, Elizabeth. 1987. "Balancing the Power in Dating Relationships." *Sex Roles* 17:563–71.

Gray, John. 1995. *Mars and Venus in the Bedroom.* New York: HarperCollins.

Gray-Little, Bernadette. 1982. "Marital Quality and Power Processes Among Black Couples." *Journal of Marriage and the Family* 44:633–46.

Greeley, Andrew. 1989. "Protestant and Catholic." *American Sociological Review* 54:485–502.

Greenblatt, Cathy Stein. 1983. "The Salience of Sexuality in the Early Years of Marriage." *Journal of Marriage and the Family* 45:289–99.

Greenfield, Sidney J. 1969. "Love and Marriage in Modern America: A Functional Analysis." *Sociological Quarterly* 6:361–77.

Greenhouse, Steven. 1992. "Income Data Show Years of Erosion for U.S. Workers." *New York Times,* Sept. 7.

Greenstein, Theodore N. 1990. "Marital Disruption and the Employment of Married Women." *Journal of Marriage and the Family* 52 (3) (Aug.): 657–76.

———. 1995. "Gender Ideology, Marital Disruption, and the Employment of Married Women." *Journal of Marriage and the Family* 57 (1) (Feb.): 31–42.

Gregg, Gail. 1986. "Putting Kids First." *New York Times Magazine,* Apr. 13, pp. 47ff.

Greif, Geoffrey and Mary S. Pabst. 1988. *Mothers Without Custody.* Lexington, MA: Heath.

Grimes, Ronald L. 1995. *Marrying and Burying: Rites of Passage in a Man's Life.* Boulder, CO: Westview.

Grimm-Thomas, Karen and Maureen Perry-Jenkins. 1994. "All in a Day's Work: Job Experiences, Self-Esteem, and Fathering in Working-Class Families." *Family Relations* 43:174–81.

Gringlas, Marcy and Marsha Weinraub. 1995. "The More Things Change . . . Single Parenting Revisited." *Journal of Family Issues* 16 (1) (Jan.): 29–52.

Griswold, Robert L. 1993. *Fatherhood in America: A History.* New York: Basic Books.

Griswold v. *Connecticut.* 1965. 381 U.S. 479, 14 L.Ed.2d 510, 85 S.Ct. 1678.

Gross, Harriet Engel. 1980. "Dual-Career Couples Who Live Apart: Two Types." *Journal of Marriage and the Family* 42:567–76.

Gross, Jane. 1991b. "A Milestone in the Fight for Gay Rights: A Quiet Suburban Life." *New York Times,* June 30.

———. 1991c. "More Young Single Men Clinging to Apron Strings." *New York Times,* June 16.

———. 1991d. "New Challenge of Youth: Growing Up in a Gay Home." *New York Times,* Feb. 11.

———. 1992a. "Collapse of Inner-City Families Creates America's New Orphans." *New York Times,* Mar. 29.

———. 1992b. "Divorced, Middle-Aged and Happy: Women, Especially, Adjust to the 90's." P. 10 in *The New York Times, Themes of the Times.* Englewood Cliffs, NJ: Prentice-Hall.

———. 1994. "After a Ruling, Hawaii Weighs Gay Marriages." *New York Times,* Apr. 25, pp. A1, C12.

Grubb, W. Norton and Robert H. Wilson. 1989. "Sources of Increasing Inequality in Wages and Salaries, 1960–80." *Monthly Labor Review* 112:3–13.

Grych, John H. and Frank D. Fincham. 1992. "Interventions for Children of Divorce: Toward Greater

Integration of Research and Action." *Psychological Bulletin* 111 (3): 434–54.

Guelzow, Maureen G., Gloria W. Bird, and Elizabeth H. Koball. 1991. "An Exploratory Path Analysis of the Stress Process for Dual-Career Men and Women." *Journal of Marriage and the Family* 53 (1) (Feb.): 151–64.

Gutis, Philip S. 1989b. "Family Redefines Itself, and Now the Law Follows." *New York Times,* May 28.

———. 1989c. "What Is a Family? Traditional Limits Are Being Redrawn." *New York Times,* Aug. 31.

Gutman, Herbert G. 1976. *The Black Family in Slavery and Freedom, 1750–1925.* New York: Pantheon.

Guyon, Janet. 1992. "Inequality in Granting Child-Care Benefits." *The Wall Street Journal,* Oct. 23, pp. A1, A5.

Guy-Sheftall, B. and P. Bell-Scott. 1989. "Finding a Way: Black Women Students and the Academy." Pp. 47–56 in *Educating the Majority: Women Challenge Tradition in Higher Education,* edited by C. S. Pearson, D. L. Savlik, and J. G. Touchton. New York: American Council on Education/Macmillan.

H

Haas, Kurt and Adalaide Haas. 1993. *Understanding Sexuality.* 3d ed. St. Louis: Mosby.

Haas, Linda. 1990. "Gender Equality and Social Policy: Implications of a Study of Parental Leave in Sweden." *Journal of Family Issues* 11 (4) (Dec.): 401–23.

Hacker, Andrew, ed. 1983. *U/S: A Statistical Portrait of the American People.* New York: Viking.

Hagar, Laura. 1995. "Why Has AIDS Education Failed?" *Express: The East Bay's Weekly,* June 23, pp. 1, 10–16.

Hagestad, G. 1986. "The Family: Women and Grandparents as Kin Keepers." Pp. 141–60 in *Our Aging Society,* edited by A. Pifer and L. Bronte. New York: Norton.

Haines, James and Margery Neely. 1987. *Parents' Work Is Never Done.* Far Hills, NJ: New Horizon Press.

Hale-Benson, J. E. 1986. *Black Children: Their Roots, Culture, and Learning Styles.* Provo, UT: Brigham Young University Press.

Hale, Christiane B. 1990. *Infant Mortality: An American Tragedy.* Population Reference Bureau, Inc. Publication # 18, Apr.

Hallenbeck, Phyllis N. 1966. "An Analysis of Power Dynamics in Marriage." *Journal of Marriage and the Family* 28:200–203.

Haller, Max. 1981. "Marriage, Women, and Social Stratification: A Theoretical Critique." *American Journal of Sociology* 86:766–95.

Halpern, Sue. 1989a. "AIDS: Rethinking the Risk." *Ms.,* May, pp. 80–87.

———. 1989b. "And Baby Makes Three." *Ms.,* Jan.–Feb., p. 151.

———. 1989c. "Infertility: Playing the Odds." *Ms.,* Jan.–Feb., pp. 147–56.

Hamachek, Don E. 1971. *Encounters with the Self.* New York: Holt, Rinehart & Winston.

———. 1992. *Encounters with the Self,* 4th ed. New York: Holt, Rinehart & Winston.

Hampton, Robert L., ed. 1991. *Black Family Violence.* Lexington, MA: D.C. Heath/Lexington.

Hampton, Robert L., and Eli H. Newberger. 1988. "Child Abuse Incidence and Reporting by Hospitals: Significance of Severity, Class and Race." Pp. 212–23 in *Coping with Family Violence: Research and Policy Perspectives,* edited by Gerald T. Hotaling, David Finkelhor, John T. Kirkpatrick, and Murray A. Straus. Newbury Park, CA: Sage.

Hanawalt, Barbara. 1986. *The Ties that Bound: Peasant Families in Medieval England.* New York: Oxford University Press.

Hansen, Donald A. and Reuben Hill. 1964. "Families Under Stress." Pp. 782–819 in *The Handbook of Marriage and the Family,* edited by Harold Christensen. Chicago: Rand McNally.

Hansen, Gary L. 1985. "Perceived Threats and Marital Jealousy." *Social Psychology Quarterly* 48 (3):262–68.

Hansson, Robert O., R. Eric Nelson, Margaret D. Carver, David H. NeeSmith, Elsie M. Dowling, Wesla L. Fletcher, and Peter Suhr. 1990. "Adult Children with Frail Elderly Parents: When to Intervene?" *Family Relations* 39:153–58.

Haraway, Donna. 1989. *Primate Visions: Gender, Race, and Nature in the World of Modern Science.* New York: Routledge, Chapman and Hall.

Hardcastle, James R. 1995. "Singles in Northern Virginia's Condos." *New York Times,* Aug. 20, p. Y25.

Hardesty, C. and J. Bokemeier. 1989. "Finding Time and Making Do: Distribution of Household Labor in Nonmetropolitan Marriages." *Journal of Marriage and the Family* 51:253–67.

Hardy, Janet B., Anne K. Duggan, Katya Masnyk, and Carol Pearson. 1989. "Fathers of Children Born to Young Urban Mothers." *Family Planning Perspectives* 21 (4) (July/Aug.): 159–87.

Harkins, Elizabeth B. 1978. "Effects of Empty Nest Transition on Self-Report of Psychological and Physical Well-Being." *Journal of Marriage and the Family* 40:549–56.

Harry, Joseph. 1983. "Gay Male and Lesbian Relationships." Pp. 216–34 in *Contemporary Families and Alternative Lifestyles: Handbook on Research and Theory*, edited by Eleanor D. Macklin and Roger H. Rubin, Newbury Park, CA: Sage.

———. 1990. "A Probability Sample of Gay Males." *Journal of Homosexuality* 19:89–104.

Hatcher, Robert A., Felicia Stewart, James Trussell, Deborah Kowal, Felicia Guest, Gary K. Stewart, and Willard Cates. 1992. *Contraceptive Technologies 1990–1992,* 15th rev. ed. New York: Irvington.

Hatfield, Julie. 1985. "Adolescents Have Need to Try Their Wings." *Boston Globe*, reprinted in *Omaha World-Herald*, May 6.

Haveman, Robert H. and Barbara Wolfe. 1994. *Succeeding Generations: On the Effects of Investments in Children.* New York: Russell Sage Foundation.

Hawke, Sharryl and David Knox. 1978. "The One-Child Family: A New Life-style." *Family Coordinator* 27:215–19.

Hawkins, Alan J. and David J. Eggebeen. 1991. "Are Fathers Fungible? Patterns of Coresident Adult Men in Maritally Disrupted Families and Young Children's Well-being." *Journal of Marriage and the Family* 53 (4) (Nov.): 958–72.

Hayden, Dolores. 1981. *The Grand Domestic Revolution: A History of Feminist Designs for American Homes, Neighborhoods, and Cities.* Cambridge, MA: MIT Press.

Hayes, Arthur S. 1991. "Courts Concede the Sexes Think in Unlike Ways." *The Wall Street Journal,* May 28, pp. B1, B5.

Hayes, Cheryl D., ed. 1987. *Risking the Future: Adolescent Sexuality, Pregnancy, and Childbearing,* vol. 1. Washington, DC: National Academy Press.

Hayghe, Howard. 1982. "Dual Earner Families: Their Economic and Demographic Characteristics." Pp. 27–40 in *Two Paychecks*, edited by Joan Aldous. Newbury Park, CA: Sage.

Heaton, Tim B. 1984. "Religious Homogamy and Marital Satisfaction Reconsidered." *Journal of Marriage and the Family* 46:729–33.

Heaton, Tim B., S. L. Albrecht, and T. K. Martin. 1985. "The Timing of Divorce." *Journal of Marriage and the Family* 47:631–39.

Heaton, Tim B. and Edith L. Pratt. 1990. "The Effects of Religious Homogamy on Marital Satisfaction and Stability." *Journal of Family Issues* 11 (2) (June): 191–207.

Hedges, Larry V. and Amy Nowell. 1995. "Sex Differences in Mental Test Scores, Variability, and Numbers of High-Scoring Individuals." *Science* 269 (5220) (July 7): 41–46.

Heiss, Jerold. 1991. "Gender and Romantic Love Roles." *The Sociological Quarterly* 32:575–92.

Helgeson, Vicki S. 1994. "Relation of Agency and Communion to Well-Being: Evidence and Potential Explanations." *Psychological Bulletin* 116 (3) (Nov.): 412–29.

Hendrick, Clyde and Susan S. Hendrick. 1989. "Research on Love: Does It Measure Up?" *Journal of Personality and Social Psychology* 56:784–94.

Henley, Nancy and Jo Freeman. 1989. "The Sexual Politics of Interpersonal Behavior." Pp. 457–69 in *Women: A Feminist Perspective*, 4th ed., edited by Jo Freeman. Mountain View, CA: Mayfield.

Henley, Nancy and J. Freeman. 1995. "The Sexual Politics of Interpersonal Behavior." Pp. 79–91 in *Women: A Feminist Perspective*. 5th ed. Mountain View, CA: Mayfield.

Henshaw, Stanley K. 1987. "Characteristics of U.S. Women Having Abortions, 1982–1983." *Family Planning Perspectives* 19:5–9.

———. 1990. "Induced Abortion: A World Review, 1990." *Family Planning Perspectives* 22:76–89.

———. 1995. "Factors Hindering Access to Abortion Services." *Family Planning Perspectives* 27 (2) (Mar./Apr.): 54–59.

Henshaw, Stanley K., Jacqueline Darroch Forrest, and Jennifer Van Vort. 1987. "Abortion Services in the United States, 1984 and 1985." *Family Planning Perspectives* 19:63–70.

Herbert, Tracy Bennett, Roxane Cohen Silver, and John H. Ellard. 1991. "Coping with an Abusive Relationship: I. How and Why Do Women Stay?" *Journal of Marriage and the Family* 53 (2) (May): 311–25.

Herek, Gregory M. and Kevin T. Berrill, eds. 1992. *Hate Crimes: Confronting Violence Against Lesbians and Gay Men.* Newbury Park, CA: Sage.

Hernandez, Pedro M., Andrea H. Beller, and John W. Graham. 1995. "Changes in the Relationship Between Child Support Payments and Educational Attainment of Offspring, 1979–1988." *Demography* 32 (2) (May): 249–260.

Hershey, Robert D., Jr. 1989. "Unemployment Rate Drops to 4.9%, Lowest Mark Since '73." *New York Times,* Apr. 8.

Hertz, Rosanna. 1986. *More Equal than Others: Women and Men in Dual Career Marriages.* Berkeley: University of California Press.

Hetherington, E. Mavis. 1973. "Girls Without Fathers." *Psychology Today* 6:47–52.

———. 1987. "Family Relations Six Years After Divorce." Pp. 185–205 in *Remarriage and Stepparenting,* edited by Kay Pasley and Marilyn Ihinger-Tallman. New York: Guilford.

Hevesi, Dennis. 1994. "U.S. Court Faults a New York List on Child Abusers." *New York Times,* Mar. 6, p. A-1.

Hewlett, Sylvia Ann. 1986. *A Lesser Life: The Myth of Women's Liberation in America.* New York: Morrow.

Higginbotham, Evelyn Brooks. 1992. "African-American Women's History and the Metalanguage of Race." *Signs: Journal of Women in Culture and Society* 17 (2) (Winter): 251–74.

Hill, Martha S. 1992. "The Role of Economic Resources and Remarriage in Financial Assistance for Children of Divorce." *Journal of Family Issues* 13 (2) (June): 158–78.

Hill, Reuben. 1958. "Generic Features of Families Under Stress." *Social Casework* 49:139–50.

Hill, Shirley A. and Mary K. Zimmerman. 1995. "Valiant Girls and Vulnerable Boys: The Impact of Gender and Race on Mothers' Caregiving for Chronically Ill Children." *Journal of Marriage and the Family* 57 (1) (Feb.): 43–53.

Hiller, Dana V. and William W. Philliber. 1986. "The Division of Labor in Contemporary Marriage: Expectations, Perceptions, and Performance." *Social Problems* 33 (3):191–201.

Hilts, Philip J. 1990b. "Panel Says Government Is Not Leading AIDS Fight." *New York Times,* Apr. 25.

———. 1992. "How Abortion Pill Works." *The Wall Street Journal,* July 24, p. A7.

Hinds, Michael DeCourcy. 1992. "Graduates Facing Worst Prospect in Last 2 Decades." *New York Times,* May 19.

Hines, Paulette Moore, Nydia Garcia-Preto, Monica McGoldrick, Rhea Almeida, and Susan Weltman. 1992. "Intergenerational Relationships Across Cultures." *Families in Society: The Journal of Contemporary Human Services* 73 (June): 323–84.

Hochschild, Arlie R. 1983. *The Managed Heart: Commercialization of Human Feeling.* Berkeley: University of California Press.

———. 1989. *The Second Shift: Working Parents and the Revolution at Home.* New York: Viking/Penguin.

Hochswender, Woody. 1990. "For Today's Fathers, Their Holiday Seems a Bit Set in Its Ways." *New York Times,* June 17.

Hock, Ellen, M. Therese Gnezda, and Susan L. McBride. 1984. "Mothers of Infants: Attitudes Toward Employment and Motherhood Following Birth of the First Child." *Journal of Marriage and the Family* (May): 425–31.

Hofferth, Sandra L. 1984. "Kin Networks, Race, and Family Structure." *Journal of Marriage and the Family* 46:791–806.

Hofferth, Sandra L. and Cheryl D. Hayes, eds. 1987. *Risking the Future: Adolescent Sexuality, Pregnancy, and Childbearing,* vol. 2. Washington, DC: National Academy Press.

Hoffman, Lois W. and F. Ivan Nye, eds. 1974. *Working Mothers.* San Francisco: Jossey-Bass.

Hogan, Dennis P. and Nan Marie Astone. 1986. "The Transition to Adulthood." *Annual Review of Sociology* 12:109–30.

Hogue, Carol J. Rowland, Willard Cates, Jr., and Christopher Tietze. 1983. "Impact of Vacuum Aspiration on Future Childbearing: A Review." *Family Planning Perspectives* 15:119–26.

Holloway, Lynette. 1994. "A Grandmother Fights for Her 2d Generation." *New York Times,* Dec. 12, p. 59.

Holmes, Ivory H. 1983. *The Allocation of Time by Women Without Family Responsibilities.* Lanham, MD: University Press of America.

Holmes, Steven A. 1990. "Day Care Bill Marks a Turn Toward Help for the Poor." *New York Times,* Jan. 25.

Hondagneu-Sotelo, Pierrette. 1992. "Overcoming Patriarchal Constraints: The Reconstruction of

Gender Relations Among Mexican Immigrant Women and Men." *Gender and Society* 6 (3) (Sept.): 393–415.

Hood, Jane C. 1986. "The Provider Role: Its Meaning and Measurement." *Journal of Marriage and the Family* 48:349–59.

Hopfensperger, Jean. 1990. "A Day for Courting, Hmong Style." *Minneapolis Star Tribune,* Nov. 24, p. 1B.

Hopkins, June and Priscilla White. 1978. "The Dual-Career Couple: Constraints and Supports." *Family Coordinator* 27:253–59.

Hopper, Joseph. 1993. "The Rhetoric of Motives in Divorce." *Journal of Marriage and the Family* 55 (4) (Nov.): 801–13.

Horwith, Melvin and Julianne Imperato-McGinley. 1983. "The Medical Evaluation of Disorders of Sexual Desire in Males and Females." Pp. 183–95 in *The Evaluation of Sexual Disorders: Psychological and Medical Aspects,* edited by Helen Singer Kaplan. New York: Brunner/Mazel.

Horwitz, Allan V. 1993. "Adult Siblings as Sources of Social Support for the Seriously Mentally Ill: A Test of the Serial Model." *Journal of Marriage and the Family* 55 (3) (Aug.): 623–32.

Hotovy, Steven. 1991. "Adopting the Older Child: A Personal Reflection." *America,* May 18, pp. 541–49.

Houseknect, Sharon K. 1987. "Voluntary Childlessness." Pp. 369–418 in *Handbook of Marriage and the Family,* edited by Marvin B. Sussman and Suzanne K. Steinmetz. New York: Plenum.

Howard, J. A. 1988. "Gender Differences in Sexual Attitudes: Conservatism or Powerlessness?" *Gender and Society* 2:103–14.

Howard, Marion and Judith Blamey McCabe. 1990. "Helping Teenagers Postpone Sexual Involvement." *Family Planning Perspectives* 22:21–26.

Hrdy, Sara Blaffer. 1981. *The Woman That Never Evolved.* Cambridge, MA: Harvard University Press.

Hsu, F. L. K. 1971. *Kinship and Culture.* Chicago: Aldine.

Huber, Joan. 1973. "From Sugar and Spice to Professor: Ambiguities in Identity Transformation." Pp. 125–35 in *Academic Women on the Move,* edited by Alice S. Rossi and Ann Calderwood. New York: Russell Sage Foundation.

———. 1986. "Trends in Gender Stratification, 1970–1985." *Sociological Forum* 1:476–95.

———. 1989. "A Theory of Gender Stratification." Pp. 110–19 in *Feminist Frontiers II,* edited by Laurel Richardson and Verta Taylor. New York: Random House.

Huber, Joan and Glenna Spitze. 1983. *Sex Stratification.* New York: Academic.

Hudson, Margaret F. 1986. "Elder Mistreatment: Current Research." Pp. 125–66 in *Elder Abuse: Conflict in the Family,* edited by Karl A. Pillemer and Rosalie S. Wolf. Dover, MA: Auburn.

Hueschler, Anne (YWCA: Women Against Violence). 1989. Personal communication, Apr. 15.

Hughes, Diane, Ellen Galinsky, and Anne Morris. 1992. "The Effects of Job Characteristics on Marital Quality: Specifying Linking Mechanisms." *Journal of Marriage and the Family* 54 (1) (Feb.): 31–42.

Hughes, Kathleen A. 1991. "Pregnant Professionals Face Subtle Bias at Work as Attitudes Toward Them Shift." *The Wall Street Journal,* Feb. 6, pp. B1, B6.

Hughes, Michael and Walter R. Gove. 1989. "Explaining the Negative Relationship Between Social Integration and Mental Health: The Case of Living Alone." Paper presented at the annual meeting of the American Sociological Association, San Francisco, Aug.

Hunt, Janet G. and Larry L. Hunt. 1977. "Dilemmas and Contradictions of Status: The Case of the Dual-Career Family." *Social Problems* 24:407–16.

———. 1986. "The Dualities of Careers and Families: New Integrations or New Polarizations?" Pp. 275–89 in *Family in Transition: Rethinking Marriage, Sexuality, Child Rearing, and Family Organization.* 5th ed., edited by Arlene S. Skolnick and Jerome H. Skolnick. Boston: Little, Brown.

Hunt, Morton. 1974. *Sexual Behavior in the Seventies.* Chicago: Playboy Press.

Huston, A. C., V. C. McLoyd, and C. G. Coll. 1994. "Children and Poverty: Issues in Contemporary Research." *Child Development* 65 (2):275–82.

Hutchinson, Earl Ofari. 1992. *Black Fatherhood: The Guide to Male Parenting.* Los Angeles: Middle Passage Press.

Hutchison, Ray and Miles McNall. 1994. "Early Marriage in a Hmong Cohort." *Journal of Marriage and the Famly* 56 (3) (Aug.): 579–90.

Hyde, Barbara L. and Margaret Texidor. 1994. "A Description of the Fathering Experience Among Black Fathers." Pp. 157–164 in *The Black Family: Essays and Studies.* 5th ed., edited by Robert Staples. Belmont, CA: Wadsworth.

Hyde, Janet Shibley, Elizabeth Fennema, and Susan Lamon. 1990. "Gender Differences in Mathematics Performance: A Meta-analysis." *Psychological Bulletin* 106:139–55.

Hyde, Janet S. and Elizabeth Ashby Plant. 1995. "Magnitude of Psychological Gender Differences: Another Side to the Story." *The American Psychologist* 50 (3) (Mar.): 159–62.

I

Ihinger-Tallman, Marilyn and Kay Pasley. 1987. *Remarriage.* Newbury Park, CA: Sage.

Imber-Black, Evan, Janine Roberts, and Richard A. Whiting, eds. 1988. *Rituals in Families and Family Therapy.* New York: Norton.

Ingoldsby, Bron B. and Suzanna Smith. 1995. *Families in Multicultural Perspective.* New York: Guilford.

Ingrassia, Michele. 1995. "Ordered to Surrender." *Newsweek,* Feb. 6: 44–45.

Isaacs, Marla Beth, Braulio Montalvo, and David Abelsohn. 1986. *The Difficult Divorce: Therapy for Children and Families.* New York: Basic Books.

Isaacs, Stephen L. and Renee J. Holt. 1987. *Redefining Procreation: Facing the Issues. Population Bulletin* 42. Washington, DC: Population Reference Bureau.

Ishii-Kuntz, Masako and Karen Seccombe. 1989. "The Impact of Children upon Social Support Networks Throughout the Life Course." *Journal of Marriage and the Family* 51 (3):777–90.

Island, David and Patrick Letellier. 1991. *Men Who Beat the Men Who Love Them: Battered Gay Men and Domestic Violence.* New York: Haworth.

J

"J" (Joan Garrity). 1977. *Total Loving.* New York: Simon & Schuster.

Jackson, James S., Wayne R. McCullough, and Gerald Gurin. 1988. "Family, Socialization, Environment, and Identity Development in Black Americans." Pp. 242–56 in *Black Families,* edited by Harriette Pipes McAdoo. Newbury Park, CA: Sage.

Jackson, Joan J. 1958. "Alcoholism and the Family." *Annals of the American Academy of Political and Social Science* 315:90–98.

Jasso, Guillermina. 1985. "Marital Coital Frequency and the Passage of Time: Estimating the Separate Effects of Spouses' Ages and Marital Duration, Birth and Marriage Cohorts, and Period Influences." *American Sociological Review* 50:224–41.

Jelen, Ted G. and Marthe A. Chandler. 1994. *Abortion Politics in the United States and Canada.* Westport, CT: Praeger.

Jendrick, Margaret Platt. 1993. "Grandparents Who Parent Their Grandchildren: Effects on Lifestyle." *Journal of Marriage and the Family* 55 (3) (Aug.): 609–21.

Joesch, Jutta M. 1991. "The Effect of the Price of Child Care on AFDC Mothers' Paid Work Behavior." *Family Relations* 40 (2) (Apr.): 161–66.

———. 1994. "Children and the Timing of Women's Paid Work After Childbirth: A Further Specification of the Relationship." *Journal of Marriage and the Family* 56 (2) (May): 429–40.

Johann, Sara Lee. 1994. *Domestic Abusers: Terrorists in Our Homes.* Springfield, IL: Thomas.

John, Daphne, Beth Anne Shelton, and Kristen Luschen. 1995. "Race, Ethnicity, Gender, and Perceptions of Fairness." *Journal of Family Issues* 16 (3) (May): 357–79.

Johnson, Clifford M., Leticia Mirande, Arloc Sherman, and James D. Weill. 1991. *Child Poverty in America.* Washington, DC: Children's Defense Fund.

Johnson, Colleen L. 1975. "Authority and Power in Japanese-American Marriage." Pp. 182–96 in *Power in Families,* edited by Ronald E. Cromwell and David H. Olson. Newbury Park, CA: Sage.

———. 1988. *Ex Familia: Grandparents, Parents, and Children Adjust to Divorce.* New Brunswick, NJ: Rutgers University Press.

Johnson, Dirk. 1990. "Chastity Organization: Starting Over in Purity." *New York Times,* Jan. 28.

———. 1991. "Polygamists Emerge from Secrecy, Seeking Not Just Peace but Respect." *New York Times,* Apr. 9.

Johnson, Michael. P. 1995. "Patriarchal Terrorism and Common Couple Violence: Two Forms of Violence Against Women." *Journal of Marriage and the Family* 57 (2) (May): 283–94.

Johnson, Tanya. 1986. "Critical Issues in the Definition of Elder Maltreatment." Pp. 167–96 in *Elder Abuse: Conflict in the Family,* edited by Karl A. Pillemer and Rosalie S. Wolf. Dover, MA: Auburn House.

Johnson, Walter R. and D. Michael Warren. 1994. *Inside the Mixed Marriage: Accounts of Changing Attitudes, Patterns, and Perceptions of Cross-Cultural and Interracial Marriages.* New York: University Press of America.

Johnston, Janet R., Marsha Kline, and Jeanne M. Tschann. 1988. "Ongoing Post-Divorce Conflict in Families Contesting Custody: Does Joint Custody and Frequent Access Help?" Paper presented at the annual meeting of the American Orthopsychiatric Association, San Francisco, Mar. Cited in Judith Wallerstein and Sandra Blakeslee, *Second Chances* (New York: Ticknor and Fields, 1989), p. 321.

Jones, Ann. 1994. "Where Do We Go from Here?" *Ms.* (Sept./Oct.): 56–63.

Joseph, Elizabeth. 1991. "My Husband's Nine Wives." *New York Times,* May 23.

Jourard, Sidney M. 1976. "Reinventing Marriage." Pp. 231–37 in *Process in Relationship: Marriage and Family.* 2d ed., edited by Edward A. Powers and Mary W. Lees. New York: West.

———. 1979. "Marriage Is for Life." Pp. 230–37 in *Choice and Challenge: Contemporary Readings in Marriage.* 2d ed., edited by Carl E. Williams and John F. Crosby. Dubuque, IA: Wm. C. Brown.

"Judge Backs Marine Who Sued the U.S. over AIDS Infection." 1990. *New York Times,* Apr. 28.

Judis, John B. 1994. "Crosses to Bear: The Many Faces of the Religious Right." *The New Republic* 211 (11) (Sept. 12): 21–26.

Julian, Teresa W., Patrick C. McKenry, and Mary W. McKelvey. 1994. "Cultural Variations in Parenting." *Family Relations* 43:30–37.

K

Kadushin, A. and Judith A. Martin. 1981. *Child Abuse: An Interactional Event.* New York: Columbia University Press.

Kalish, Susan. 1994. "Rising Costs of Raising Children." *Population Today* 22 (7/8) (July/Aug.): 4–5.

Kalmijn, Matthijs. 1991. "Status Homogamy in the United States." *American Journal of Sociology* 97 (2) (Sept.): 496–523.

Kalmuss, Debra. 1992. "Adoption and Black Teenagers: The Viability of a Pregnancy Resolution Strategy." *Journal of Marriage and the Family* 54 (3) (Aug.): 485–95.

Kalmuss, Debra, Andrew Davidson, and Linda Cushman. 1992. "Parenting Expectations, Experiences, and Adjustment to Parenthood: A Test of the Violated Expectations Framework." *Journal of Marriage and the Family* 54 (3) (Aug.): 516–26.

Kalmuss, Debra, Pearila B. Namerow, and Ursula Bauer. 1992. "Short-Term Consequences of Parenting Versus Adoption Among Young Unmarried Women." *Journal of Marriage and the Family* 54 (1) (Feb.): 80–90.

Kalof, Linda and Timothy Cargill. 1991. "Fraternity and Sorority Membership and Gender Dominance Attitudes." *Sex Roles* 25 (7/8):417–23.

Kamo, Yoshinori and Min Zhou. 1994. "Living Arrangements of Elderly Chinese and Japanese in the United States." *Journal of Marriage and the Family* 56 (3) (Aug.): 544–58.

Kann, Mark E. 1986. "The Costs of Being on Top." *Journal of the National Association for Women Deans, Administrators, and Counselors* 49:29–37.

Kantor, David and William Lehr. 1975. *Inside the Family: Toward a Theory of Family Process.* San Francisco: Jossey-Bass.

Kantrowitz, Barbara. 1987. "The Year of Living Dangerously." *Newsweek on Campus,* Apr., pp. 12–21.

———. 1989. "Advocating a 'Mommy Track.' " *Newsweek,* Mar. 13, p. 45.

———. 1992. "Teenagers and AIDS." *Newsweek,* Aug. 3, pp. 45–49.

Kaplan, Helen Singer. 1974. "No-Nonsense Therapy for Six Sexual Malfunctions." *Psychology Today* 8:132–38.

———. 1983. *The Evaluation of Sexual Disorders: Psychological and Medical Aspects.* New York: Brunner/Mazel.

Kaplan, Marion A., ed. 1985. *The Marriage Bargain: Women and Dowries in European History.* New York: Harrington Park Press.

Karen, R. 1994. *Becoming Attached: Unfolding the Mystery of the Infant–Mother Bond and Its Impact on Later Life.* New York: Warner.

Karoly, Lynn A. and Gary Burtless. 1995. "Demographic Change, Rising Earnings Inequality, and the Distribution of Personal Well-Being." *Demography* 32 (3) (Aug.): 379–405.

Kassel, V. 1966. "Polygyny after 60." *Geriatrics* 21:214–18.

Kassorla, Irene. 1973. *Putting It All Together.* New York: Brut Publications; Hawthorn/Dutton.

Katz, Stan J. and Aimee E. Liu. 1988. *False Love and Other Romantic Illusions: Why Love Goes Wrong and How to Make It Right.* New York: Ticknor & Fields.

Keith, Carolyn. 1995. "Family Caregiving Systems: Models, Resources, and Values." *Journal of Marriage and the Family* 57 (1) (Feb.): 179–89.

Keith, Pat M. 1989. *The Unmarried in Later Life.* Newbury Park, CA: Sage.

Keith, Verna M. and Barbara Finlay. 1988. "The Impact of Parental Divorce on Children's Educational Attainment, Marital Timing, and Likelihood of Divorce." *Journal of Marriage and the Family* 51:797–809.

Kelly, Joan Berlin and Judith S. Wallerstein. 1990. *Surviving the Breakup.* New York: Basic Books.

Kelly, James R. 1990. "The Koop Report and a Better Politics of Abortion." *America* 162 (21):542–46.

———. 1991. "Abortion: What Americans Really Think and the Catholic Challenge." *America* 165 (13): 310–16.

Kelly, Mary E. 1990. "Choice and Constraint: Nurturing Behavior in Women and Men." Paper presented at the annual meeting of the Midwest Sociological Society, Chicago, Apr. 13.

Kempe, C. Henry, Frederic N. Silverman, Brandt F. Steele, William Droegemueller, and Henry K. Silver. 1962. "The Battered-Child Syndrome." *Journal of the American Medical Assn.* 181 (July 7): 17–24.

Kephart, William. 1971. "Oneida: An Early American Commune." Pp. 481–92 in *Family in Transition: Rethinking Marriage, Sexuality, Child Rearing, and Family Organization,* edited by Arlene S. Skolnick and Jerome H. Skolnick. Boston: Little, Brown.

Kern, Louis J. 1981. *An Ordered Love: Sex Roles and Sexuality in Victorian Utopias—the Shakers, the Mormons, and the Oneida Community.* Chapel Hill: University of North Carolina Press.

Kilborn, Peter T. 1990. "Tales from the Digital Treadmill." *New York Times,* June 3.

———. 1992b. "Lives of Unexpected Poverty in Center of a Land of Plenty." *New York Times,* July 7.

———. 1992d. "Sad Distinction for the Sioux: Homeland Is No. 1 in Poverty." *New York Times,* Sept. 29.

Kilbourne, Jean. 1994. "'Gender Bender' Ads: Same Old Sexism." *New York Times.* May 15, p. F-13.

Kimmel, Michael S. 1989. "From Pedestals to Partners: Men's Responses to Feminism." Pp. 581–94 in *Women: A Feminist Perspective.* 4th ed., edited by J. Freeman. Mountain View, CA: Mayfield.

———. 1995a. "Misogynists, Masculinist Mentors, and Male Supporters: Men's Responses to Feminism." Pp. 561–72 in *Women: A Feminist Perspective.* 5th ed., edited by Jo Freeman. Mountain View, CA: Mayfield.

———. 1995b. *Manhood in America: A Cultural History.* Berkeley: University of California Press.

King, Valerie. 1994. "Variation in the Consequences of Nonresident Father Involvement for Children's Well-Being." *Journal of Marriage and the Family* 56 (3) (Nov.): 963–72.

King, Valerie and Glen H. Elder, Jr. 1995. "American Children View Their Grandparents: Linked Lives Across Three Rural Generations." *Journal of Marriage and the Family* 57 (1) (Feb.): 165–78.

Kingsbury, Nancy and John Scanzoni. 1993. "Structural-Functionalism." Pp. 195–217 in *Sourcebook of Family Theories and Methods: A Contextual Approach,* edited by Pauline G. Boss, William J. Doherty, Ralph LaRossa, Walter R. Schumm, and Suzanne K. Steinmetz. New York: Plenum.

Kingston, Paul W. 1990. "Illusions and Ignorance About the Family-Responsive Workplace." *Journal of Family Issues* 11 (4) (Dec.): 438–54.

Kinsey, Alfred, Wardell B. Pomeroy, and Clyde E. Martin. 1948. *Sexual Behavior in the Human Male.* Philadelphia: Saunders.

———. 1953. *Sexual Behavior in the Human Female.* Philadelphia: Saunders.

Kitano, Harry and Roger Daniels. 1995. *Asian Americans: Emerging Minorities.* 2d ed. Englewood Cliffs, NJ: Prentice-Hall.

Kite, M. E. and K. Deaux. 1987. "Gender Belief Systems: Homosexuality and the Implicit Inversion Theory." *Psychology of Women Quarterly* 11:83–96.

Kitson, Gay C. 1992. *Portrait of Divorce: Adjustment to Marital Breakdown.* New York: Guilford.

Klass, Perri. 1989. "AIDS: The Youngest Victims." *New York Times Magazine,* June 18, pp. 34–35, 56–58.

Kleban, Morton H., Elaine M. Brody, Claire B. Schoonover, and Christine Hoffman. 1989. "Family Help to the Elderly: Perceptions of Sons-in-law Regarding Parent Care." *Journal of Marriage and the Family* 51 (2):303–12.

Klebanov, Pamela Kato, Jeanne Brooks-Gunn, and Greg Duncan. 1994. "Does Neighborhood and Family Poverty Affect Mothers' Parenting, Mental Health, and Social Support?" *Journal of Marriage and the Family* 56 (2) (May): 441–55.

Klein, Robert J. 1989. "Cutting Your Debt." Pp. 16–19 in *Money Guide: Basics of Personal Finance,* edited by *Money Magazine.* Spring.

Kline, Jennie, Zena Stein, Mervyn Susser, and Dorothy Warburton, eds. 1986. "Induced Abortion and the Chromosomal Characteristics of Subsequent Miscarriages (Spontaneous Abortions)." *American Journal of Epidemiology* 123:1066–79.

Kline, Marsha, Janet R. Johnston, and Jeanne M. Tschann. 1991. "The Long Shadow of Marital Conflict: A Model of Children's Postdivorce Adjustment." *Journal of Marriage and the Family* 53 (2) (May): 297–309.

Kline, Marsha, Jeanne M. Tschann, Janet R. Johnston, and Judith Wallerstein. 1989. "Children's Adjustment in Joint and Sole Physical Custody Families." *Developmental Psychology* 25:430–38.

Klitsch, Michael. 1995. "Still Waiting for the Contraceptive Revolution." *Family Planning Perspectives* 27 (6) (Nov./Dec.): 246–53.

Knox, David H., Jr. 1975. *Marriage: Who? When? Why?* Englewood Cliffs, NJ: Prentice-Hall.

Kobren, Gerri. 1988. "Older First-Time Moms Better Educated, Secure." *Omaha-World Herald,* Dec. 30.

Koch, Joanne and Lew Koch. 1976. "A Consumer's Guide to Therapy for Couples." *Psychology Today* 9:33–38.

Kohlberg, Lawrence. 1966. "A Cognitive-Developmental Analysis of Children's Sex-Role Concepts and Attitudes." Pp. 82–173 in *The Development of Sex Differences,* edited by Eleanor E. Maccoby. Palo Alto, CA: Stanford University Press.

Kohn, M. L. 1977. *Class and Conformity: A Study in Values.* Chicago: University of Chicago Press.

Kolata, Gina. 1987. "Tests of Fetuses Rise Sharply Amid Doubts." *New York Times,* Sept. 22.

———. 1988. "Researchers Report Miscarriages Occurring in 31% of Pregnancies." *New York Times,* July 27.

———. 1989c. "Growing Movement Seeks to Help Women Infected with AIDS Virus." *New York Times,* May 4.

———. 1989d. "Lesbian Partners Finding the Means to Be Parents." *New York Times,* Jan. 30.

Konner, Melvin. 1988. "The Aggressors." *New York Times Magazine,* Aug. 14, pp. 33ff.

———. 1990. "Women and Sexuality." *New York Times Magazine,* Apr. 29, pp. 24, 26.

Koop, C. Everett. n.d. *Surgeon General's Report on Acquired Immune Deficiency Syndrome.* Washington, DC: U.S. Department of Health and Human Services.

Korman, Sheila. 1983. "Nontraditional Dating Behavior: Date-Initiation and Date-Expense-Sharing Among Feminists and Nonfeminists." *Family Relations* 32:575–81.

Korman, Sheila and G. Leslie. 1982. "The Relationship between Feminist Ideology and Date-Expense-Sharing to Perceptions of Sexual Aggression in Dating." *Journal of Sex Research* 18:114–29.

Kornfein, Madeleine, Thomas S. Weisner, and Joan C. Martin. 1979. "Women into Mothers: Experimental Family Life-Styles." Pp. 259–91 in *Women into Wives: The Legal and Economic Impact of Marriage,* edited by Jane Roberts Chapman and Margaret Gates. Newbury Park, CA: Sage.

Korte, Diana. 1995. "Midwives on Trial." *Mothering* (Fall): 52–63.

Koss, Mary P., Christine A. Gidycz, and Nadine Wisniewski. 1987. "The Scope of Rape: Incidence and Prevalence of Sexual Aggression and Victimization in a National Sample of Higher Education Students." *Journal of Consulting and Clinical Psychology* 55 (2):162–70.

Kraft, Joan Marie and James E. Coverdill. 1994. "Employment and the Use of Birth Control by Sexually Active Single Hispanic, Black, and White Women." *Demography* 31 (4) (Nov.): 593–602.

Kramarow, Ellen A. 1995. "The Elderly Who Live Alone in the United States: Historial Perspectives on Household Change." *Demography* 32 (3) (Aug.): 335–52.

Krementz, Jill. 1984. *How It Feels When Parents Divorce.* New York: Knopf.

Kreppner, Kurt. 1988. "Changes in Parent–Child Relationships with the Birth of the Second Child." *Marriage and Family Review* 12 (3–4):157–81.

Krieger, Lisa M. 1995. "AIDS Loses Urgency in Nation's List of Worries." *San Francisco Examiner,* Jan. 29, pp. A1, A9.

Krokoff, Lowell J. 1987. "The Correlates of Negative Affect in Marriage: An Exploratory Study of Gender Differences." *Journal of Family Issues* 8:111–35.

Kruk, Edward. 1991. "Discontinuity Between Pre- and Post-Divorce Father–Child Relationships: New Evidence Regarding Paternal Disengagement." *Journal of Divorce and Remarriage* 16 (3/4):195–227.

Kruttschnitt, Candace, Jane D. McLeod, and Maude Dornfeld. 1994. "The Economic Environment of Child Abuse." *Social Problems* 41 (2) (May): 299–315.

Ku, Leighton, Freya L. Sonenstein, and Joseph H. Pleck. 1995. "When We Use Condoms and Why We Stop." *Population Today* (March): 3.

Kübler-Ross, Elisabeth. 1979. *On Death and Dying.* New York: Macmillan.

Kuhn, Manfred. 1955. "How Mates Are Sorted." In *Family, Marriage, and Parenthood,* edited by Howard Becker and Reuben Hill. Boston: Heath.

Kurdek, Lawrence A. 1989a. "Relationship Quality for Newly Married Husbands and Wives: Marital History, Stepchildren, and Individual-Difference Predictors." *Journal of Marriage and the Family* 51 (4) (Nov.): 1053–64.

———. 1989b. "Relationship Quality in Gay and Lesbian Cohabiting Couples: A 1-Year Follow-up Study." *Journal of Social and Personal Relationships* 6:35–59.

———. 1991. "The Relations Between Reported Well-being and Divorce History, Availability of a Proximate Adult, and Gender." *Journal of Marriage and the Family* 53 (1) (Feb.): 71–78.

Kurdek, Lawrence A. 1994. "Areas of Conflict for Gay, Lesbian, and Heterosexual Couples: What Couples Argue About Influences Relationship Satisfaction." *Journal of Marriage and the Family* 56 (4) (Nov.): 923–34.

———. 1995. "Predicting Change in Marital Satisfaction from Husbands' and Wives' Conflict Resolution Styles." *Journal of Marriage and the Family* 57 (1) (Feb.): 153–64.

Kurdek, Lawrence A. and Mark A. Fine. 1991. "Cognitive Correlates of Satisfaction for Mothers and Stepfathers in Stepfather Families." *Journal of Marriage and the Family* 53 (3) (Aug.): 565–72.

Kutner, Lawrence. 1988a. "Parent & Child: Parents Change: Children Must Too." *New York Times,* Mar. 24.

———. 1988b. "Parent and Child: Working at Home; or, The Midday Career Change." *New York Times,* Dec. 8.

———. 1989a. "Parent & Child: 'The Forgotten Mourners': Help for Adults Whose Sibling Has Died." *New York Times,* Dec. 21.

———. 1989b. "Parent & Child: When Mom and Dad Do Not Agree on Discipline, the Child May Take Advantage of the Confusion." *New York Times,* Sept. 21.

———. 1990a. "Parent & Child: Chasms of Pain and Growth: When Generation Gaps Are Too Wide to Leap." *New York Times,* Mar. 8.

———. 1990b. "Parent & Child: Confronting Mother and Father About Their Battles and the Effect It Has on the Family." *New York Times,* Mar. 29.

———. 1990c. "Parent & Child: Money Matters, for Some as Confidential as Sex, Are Often A Stressful Subject." *New York Times,* Jan. 11.

———. 1990d. "Parent & Child: When a Child Marries: The Pride and the Pitfalls." *New York Times,* May 10.

———. 1991. "Parent & Child: First Comes the Baby, Then Anger and Frustration When Not All Goes According to Expectations." *New York Times,* Nov. 7.

L

Lacayo, Richard. 1989a. "Whose Life Is It?" *Time,* May 1, pp. 20–24.

———. 1989b. "Nobody's Children." *Time,* Oct. 9, pp. 91–95.

Lackey, Chad and Kirk R. Williams. 1995. "Social Bonding and the Cessation of Partner Violence Across Generations." *Journal of Marriage and the Family* 57 (2) (May): 295–305.

Ladner, Joyce A. 1977. "Mixed Families: White Parents and Black Children." *Society* (Sept./Oct.):70–78.

Laing, Ronald D. 1971. *The Politics of the Family.* New York: Random House.

Lally, Catherine F. and James W. Maddock. 1994. "Sexual Meaning Systems of Engaged Couples." *Family Relations* 43:53–60.

Lamanna, Mary Ann. 1977. "The Value of Children to Natural and Adoptive Parents." Ph.D. dissertation, Department of Sociology, University of Notre Dame.

Lambert, Wade. 1991. "Discrimination Afflicts People with HIV." *The Wall Street Journal,* Nov. 19, pp. B1, B9.

Langan, P. and C. Innes. 1986. *Preventing Domestic Violence Against Women.* Bureau of Justice Statistics. Washington, DC: Department of Justice.

Langhinrichsen-Rohling, Jennifer, Natalie Smutzler, and Dina Vivian. 1994. "Positivity in Marriage: The Role of Discord and Physical Aggression Against Wives." *Journal of Marriage and the Family* 56 (1) (Feb.): 69–79.

Langley, Patricia. 1988. Reproductive Technology Fact Sheet. Unpublished.

LaRossa, Ralph and Donald C. Reitzes. 1993. "Symbolic Interactionism and Family Studies." Pp. 135–63 in *Sourcebook of Family Theories and Methods: A Contextual Approach,* edited by Pauline G. Boss, William J. Doherty, Ralph LaRossa, Walter R. Schumm, and Suzanne K. Steinmetz. New York: Plenum.

Larson, R. and M. Ham. 1993. "Stress and 'Storm and Stress' in Early Adolescence: The Relationship of Negative Events with Dysphoric Affect." *Developmental Psychology* 29:130–40.

Lasch, Christopher. 1977. *Haven in a Heartless World: The Family Besieged.* New York: Basic Books.

———. 1980. *The Culture of Narcissism.* New York: Warner.

Laslett, Peter. 1971. *The World We Have Lost: England Before the Industrial Age.* 2d ed. New York: Scribner's.

Lauer, Jeanette and Robert Lauer. 1985. "Marriages Made to Last." *Psychology Today,* June.

Laumann, Edward O., John H. Gagnon, Robert T. Michael, and Stuart Michaels. 1994. *The Social Organization of Sexuality: Sexual Practices in the United States.* Chicago: University of Chicago Press.

Lauritzen, Paul. 1990. "What Price Parenthood?" *Hastings Center Report* (Mar./Apr.): 38–46.

Lavee, Yoav, Hamilton I. McCubbin, and Joan M. Patterson. 1985. "The Double ABCX Model of Family Stress and Adaptation: An Empirical Test by Analysis of Structural Equations with Latent Variables." *Journal of Marriage and the Family* 47 (4):811–26.

Lawson, Carol. 1990. "Fathers, Too, Are Seeking a Balance Between Their Families and Careers." *New York Times,* Apr. 12.

Lawton, Leora, Merril Silverstein, and Vern Bengston. 1994. "Affection, Social Contact, and Geographic Distance Between Adult Children and Their Parents." *Journal of Marriage and the Family* 56 (1) (Feb.): 57–68.

Leach, Penelope. 1994. *Children First.* New York: Knopf.

Leary, Warren E. 1988. "Experts Caution Against Repeated Caesareans and Recommend Natural Deliveries." *New York Times,* Oct. 27.

———. 1991. "Hypertension Among Blacks Tied to Bias, Poverty, and Diet." *New York Times,* Feb. 6.

LeClere, Felicia B. and Brenda M. Kowalewski. 1994. "Disability in the Family: The Effects on Children's Well-Being." *Journal of Marriage and the Family* 56 (2) (May): 457–68.

Lee, Gary R., Julie K. Netzer, and Raymond T. Coward. 1994. "Filial Responsibility Expectations and Patterns of Intergenerational Assistance." *Journal of Marriage and the Family* 56 (3) (Aug.): 559–65.

Lee, Gary R., Karen Seccombe, and Constance L. Shehan. 1991. "Marital Status and Personal Happiness: An Analysis of Trend Data." *Journal of Marriage and the Family* 53:839–44.

Lee, Gary R. and Constance L. Shehan. 1989. "Retirement and Marital Satisfaction." *Journal of Gerontology* 44:S226–30.

Lee, John Alan. 1973. *The Colours of Love.* Toronto: New Press.

———. 1981. "Forbidden Colors of Love: Patterns of Gay Love." Pp. 128–39 in *Single Life: Unmarried Adults in Social Context,* edited by Peter J. Stein. New York: St. Martins.

Leerhsen, Charles. 1985. "Reading, Writing and Divorce." *Newsweek,* May 13, p. 74.

Leibowitz, Arleen and Jacob Alex Klerman. 1995. "Explaining Changes in Married Mothers' Employment over Time." *Demography* 32 (3) (Aug.): 365–78.

Leland, John. 1995. "Bisexuality." *Newsweek,* July 17, pp. 44–50.

Leland, Nancy Lee, Donna Petersen, Mary Braddock, and Greg Alexander. 1995. "Variations in Pregnancy Outcomes by Race among 10–14-Year-Old Mothers in the United States." *Public Health Reports* 110 (1) (Jan.-Feb.): 53–58.

LeMasters, E. E. and John DeFrain. 1989. *Parents in Contemporary America: A Sympathetic View,* 5th ed. Belmont, CA: Wadsworth.

Lemonick, Michael D. 1994. "Not So Fertile Ground." *Time,* Sept. 19, pp. 68–69.

Lennon, Mary Clare and Sarah Rosenfield. 1994. "Relative Fairness and the Division of Housework:

The Importance of Options." *American Journal of Sociology* 100 (2) (Sept.): 506–31.

Lenz, Elinor and Barbara Myerhoff. 1985. *The Feminization of America.* Los Angeles: Tarcher (St. Martin's).

Levay, Alexander N., Lawrence Sharpe, and Arlene Kagel. 1981. "The Effects of Physical Illness on Sexual Functioning." Pp. 169–90 in *Sexual Problems in Medical Practice,* edited by Harold I. Lief, M.D. Monroe, WI: American Medical Association.

LeVay, S. and D. H. Hamer. 1994. "Evidence for a Biological Influence in Male Homosexuality." *Scientific American* (May).

Levinger, George. 1965. "Marital Cohesiveness and Dissolution: An Integrative Review." *Journal of Marriage and the Family* 27:19–28.

Levinson, David. 1989. *Family Violence in Cross-Cultural Perspective.* Newbury Park, CA: Sage.

Levy, Barrie, ed. 1991. *Dating Violence: Young Women in Danger.* Seattle,: Seal Press.

Levy-Shiff, R. 1994. "Individual and Contextual Correlates of Marital Change Across the Transition to Parenthood." *Developmental Psychology* 30: 591–601.

Lewin, Tamar. 1989a. "Family or Career? Choose, Women Told." *New York Times,* Mar. 8.

———. 1989b. "Small Tots, Big Biz." *New York Times Magazine,* Jan. 29, pp. 30–31, 89–92.

———. 1990a. "Father's Vanishing Act Called Common Drama." *New York Times,* June 4.

———. 1992. "Rise in Single Parenthood Is Reshaping U.S." *New York Times,* Oct. 5.

Lewis, Robert A., Phillip J. Freneau, and Craig L. Roberts. 1979. "Fathers and the Postparental Transition." *Family Coordinator* 28:514–20.

L'Hommedieu, Elizabeth. 1991. "Walking Out on the Boys." *Time,* July 8, pp. 52–53.

Libby, Roger W. 1976. "Social Scripts for Sexual Relationships." In *Sexuality Today and Tomorrow,* edited by Sol Gordon and Roger W. Libby. North Scituate, MA: Duxbury.

Lichter, Daniel T., Robert N. Anderson, and Mark D. Hayward. 1995. "Marriage Markets and Marital Choice." *Journal of Family Issues* 16 (4) (July): 412–31.

Lichter, Daniel T. and David J. Eggebeen. 1994. "The Effect of Parental Employment on Child Poverty." *Journal of Marriage and the Family* 56 (3) (Aug.): 633–45.

Lichter, Daniel T. and Nancy S. Landale. 1995. "Parental Work, Family Structure, and Poverty Among Latino Children." *Journal of Marriage and the Family* 57 (2) (May): 346–54.

Lichter, Daniel T., Felicia B. LeClere, and Diane K. McLaughlin. 1991. "Local Marriage Markets and the Marital Behavior of Black and White Women." *American Journal of Sociology* 96 (4) (Jan.): 843–67.

Liebow, Elliot. 1967. *Tally's Corner.* Boston: Little, Brown.

Lin, Chien and William T. Liu. 1993. "Relationships Among Chinese Immigrant Families." Pp. 271–86 in *Family Ethnicity: Strength in Diversity,* edited by Harriette Pipes McAdoo. Newbury Park, CA: Sage.

Linton, Sally. 1971. "Woman the Gatherer: Male Bias in Anthropology." In *Women in Cross-Cultural Perspective,* edited by Lenore Jacobs. Champaign-Urbana: University of Illinois Press.

Lipman, Joanne. 1993. "The Nanny Trap." *The Wall Street Journal,* April 14, pp. A1, A8.

Lipmen-Blumen, Jean. 1984. *Gender Roles and Power.* Englewood Cliffs, NJ: Prentice-Hall.

Lips, Hilary M. 1995. "Gender-Role Socialization: Lessons in Femininity." Pp. 128–48 in *Women: A Feminist Perspective.* 5th ed., edited by Jo Freeman. Mountain View, CA: Mayfield.

Loeber, Rolf and Magda Stouthamer-Loeber. 1986. "Family Factors as Correlates and Predictors of Juvenile Conduct Problems and Delinquency." Pp. 29–149 in *Crime and Justice,* vol. 7, edited by Michael Tonry and Normal Morris. Chicago: University of Chicago Press.

London, Kathryn A. 1991. "Advance Data Number 194: Cohabitation, Marriage, Marital Dissolution, and Remarriage: United States, 1988." U.S. Department of Health and Human Services: Vital and Health Statistics of the National Center, Jan. 4.

Lonsdorf, Barbara J. 1991. "The Role of Coercion in Affecting Women's Inferior Outcomes in Divorce: Implications for Researchers and Therapists." *Journal of Divorce and Remarriage* 16 (1/2):69–106.

Lopata, Helena Znaniecki. 1973. "Living Through Widowhood." *Psychology Today* 7:87–92.

Losh-Hesselbart, Susan. 1987. "Development of Gender Roles." Pp. 535–64 in *Handbook of Marriage and the Family,* edited by Marvin B. Sussman and Suzanne K. Steinmetz. New York: Plenum.

Loving v. *Virginia.* 1967. 388 U.S. 1, 87 S.Ct. 1817, 18 L.Ed.2d 1010.

Lowinsky, Naomi Ruth. 1992. *Stories from the Motherline: Reclaiming the Mother-Daughter Bond, Finding Our Feminine Souls.* Los Angeles: Jeremy P. Tarcher.

Lublin, Joann S. 1992. "Spouses Find Themselves Worlds Apart as Global Commuter Marriages Increase." *The Wall Street Journal,* Aug. 19, pp. B1, B6.

Luciano, Lani. 1989. "Saving for Retirement." Pp. 84–87 in *Money Guide: Basics of Personal Finance,* edited by *Money Magazine.* Spring.

Luker, Kristin. 1984. *Abortion and Politics of Motherhood.* Berkeley: University of California Press.

———. 1990. Interview by Margo Adler on "Morning Edition." National Public Radio, May 14.

Luster, Tom, Robert Boger, and Kristi Hannan. 1993. "Infant Affect and Home Environment." *Journal of Marriage and the Family* 55 (3) (Aug.): 651–61.

Luster, Tom, Kelly Rhoades, and Bruce Haas. 1989. "The Relation Between Parental Values and Parenting Behavior: A Test of the Kohn Hypothesis." *Journal of Marriage and the Family* 51:139–47.

Luster, Tom and Stephen A. Small. 1994. "Factors Associated with Sexual Risk-Taking Behaviors Among Adolescents." *Journal of Marriage and the Family* 56 (3) (Aug.): 622–32.

Lye, Diane N., Daniel H. Klepinger, Patricia Davis Hyle, and Anjanette Nelson. 1995. "Childhood Living Arrangements and Adult Children's Relations with Their Parents." *Demography* 32 (2) (May): 261–80.

Lynd, Robert S. and Helen Merrell Lynd. 1929. *Middletown: A Study in American Culture.* New York: Harcourt & Brace.

Lytton, H. and D. M. Romney. 1991. "Parents' Differential Socialization of Boys and Girls: A Meta-analysis." *Psychological Bulletin* 109:267–96.

M

Maccoby, Eleanor E., Charlene E. Depner, and Robert H. Mnookin. 1990. "Coparenting in the Second Year after Divorce." *Journal of Marriage and the Family* 52 (1) (Feb.): 141–55.

Maccoby, Eleanor E. and Carol Nagy Jacklin. 1974. *The Psychology of Sex Differences.* Stanford, CA: Stanford University Press.

Maccoby, Eleanor E. and J. A. Martin. 1983. "Socialization in the Context of the Family: Parent–Child Interaction." Pp. 1–101 in *Handbook of Child Psychology,* vol. 4, *Socialization, Personality, and Social Development,* edited by E. Mavis Hetherington. New York: Wiley.

MacDonald, William L. and Alfred DeMaris. 1995. "Remarriage, Stepchildren, and Marital Conflict: Challenges to the Incomplete Institutionalization Hypothesis." *Journal of Marriage and the Family* 57 (2) (May): 387–98.

Macklin, Eleanor D. 1983. "Nonmarital Heterosexual Cohabitation: An Overview." Pp. 49–74 in *Contemporary Families and Alternative Lifestyles: Handbook on Theory and Research,* edited by Eleanor D. Macklin and Roger H. Rubin. Newbury Park, CA: Sage.

———. 1987. "Nontraditional Family Forms." Pp. 317–53 in *Handbook of Marriage and the Family,* edited by Marvin B. Sussman and Suzanne K. Steinmetz. New York: Plenum.

———. 1988. "AIDS: Implications for Families." *Family Relations* 37 (Apr.): 141–49.

Maddock, J. W., M. J. Hogan, A. L. Antonove, and M. S. Matskovsky, eds. 1993. *Peristroika and Family Life: Post-USSR and US Perspectives.* New York: Guilford.

Majors, Richard G. and Janet M. Billson. 1992. *Cool Pose: The Dilemmas of Black Manhood in America.* Lexington, MA: Heath.

Makepeace, James M. 1981. "Courtship Violence Among College Students." *Family Relations* 30:97–102.

Malcolm, Andrew H. 1991. "Helping Grandparents Who Are Parents Again." *New York Times,* Nov. 19.

"Male Contraceptive OK'd for Tests." 1990. *Omaha World-Herald,* Jan. 4.

"Male Impotence Has Emotional, Physical Roots." 1985. *Los Angeles Daily News,* in *Omaha World-Herald,* Oct. 20.

Mancini, Jay A. and Rosemary Bleiszner. 1989. "Aging Parents and Adult Children: Research Themes in Intergenerational Relations." *Journal of Marriage and the Family* 51 (2):275–90.

Manke, Beth, Brenda L. Seery, Ann C. Crouter, and Susan M. McHale. 1994. "The Three Corners of Domestic Labor: Mothers,' Fathers,' and Children's Weekday and Weekend Housework." *Journal of Marriage and the Family* 56 (3) (Aug.): 657–68.

Manning, Wendy D. 1993. "Marriage and Cohabitation Following Premarital Conception." *Journal of Marriage and the Family* 55 (3) (Nov.): 839–50.

Marcus, Amy Dockser. 1990. "Medical, Social Changes May Spur Courts to Reformulate Definition of Parenthood." *The Wall Street Journal,* Sept. 18, pp. B1, B13.

———. 1991. "Grandparents Turn to the Courts to Seek Permission to Visit Their Grandchildren." *The Wall Street Journal,* June 5, p. B1.

Mare, Robert D. 1991. "Five Decades of Educational Assortative Mating." *American Sociological Review* 56 (Feb.): 15–32.

Marecek, Jeanne. 1995. "Gender, Politics, and Psychology's Ways of Knowing." *The American Psychologist* 50 (3) (Mar.): 162–64.

Margolin, Leslie. 1989. "Gender and the Prerogatives of Dating and Marriage: An Experimental Assessment of a Sample of College Students." *Sex Roles* 20 (1/2): 91–101.

Marks, James D. 1992. "A Victory for the New American Family." *New York Times,* Feb. 1.

Marks, Nadine F. 1995. "Midlife Marital Status Differences in Social Support Relationships with Adult Children and Psychological Well-Being." *Journal of Family Issues* 16 (1) (Jan.): 5–28.

Marks, Stephen R. 1989. "Toward a Systems Theory of Marital Quality." *Journal of Marriage and the Family* 51:15–26.

Marlow, Lenard and S. Richard Sauber. 1990. *The Handbook of Divorce Mediation.* New York: Plenum.

Marshall, Susan E. 1995. "Keep Us on the Pedestal: Women Against Feminism in Twentieth-Century America." Pp. 547–60 in *Women: A Feminist Perspective.* 5th ed., edited by Jo Freeman. Mountain View, CA: Mayfield.

Marshall, Victor. 1986. "A Sociological Perspective on Aging and Dying." Pp. 125–46 in *Later Life: The Social Psychology of Aging,* edited by Victor M. Marshall. Newbury Park, CA: Sage.

———. 1992b. "Stepfathers with Minor Children Living at Home." *Journal of Family Issues* 13 (2) (June): 195–214.

———. 1993. "Adolescent Males' Orientation Toward Paternity and Contraception." *Family Planning Perspectives* 25 (1) (Jan./Feb.): 22–31.

Marsiglio, William and Denise Donnelly. 1991. "Sexual Relations in Later Life: A National Study of Married Persons." *Journal of Gerontology: Social Sciences* 46 (6): 5338–44.

Martin, April. 1993. *The Lesbian and Gay Parenting Handbook: Creating and Raising Our Families.* New York: HarperCollins.

Martin, C. L. 1989. "Children's Use of Gender-Related Information in Making Social Judgments." *Developmental Psychology* 25:80–88.

Martin, Douglas. 1990. "Odd Man Out: Helping Big Boys Feel Like Fathers." *New York Times,* Apr. 7.

Martin, Peter and Landy Luke. 1991. "Divorce and the Wheel Theory of Love." *Journal of Divorce and Remarriage* 15 (1/2):3–22.

Martin, Philip and Elizabeth Midgley. 1994. *Immigration to the United States: Journey to an Uncertain Destination. Population Bulletin* 49 (2) (Sept.). Washington DC: Population Reference Bureau.

Martin, Teresa Castro and Larry Bumpass. 1989. "Trends in Marital Disruption." *Demography* 26:37–52.

Marvin v. *Marvin.* 1976. 18 Cal.3d 660, 134 Cal.Rptr. 815, 557 P.2d 106.

Masheter, Carol. 1991. "Postdivorce Relationships Between Ex-spouses: The Roles of Attachment and Interpersonal Conflict." *Journal of Marriage and the Family* 53 (1) (Apr.): 103–10.

Maslow, Abraham H. 1943. "A Theory of Human Motivation." *Psychological Review* 50:370–96.

Mason, Mary Ann. 1988. *The Equality Trap.* New York: Simon & Schuster.

Masters, William H. and Virginia E. Johnson. 1966. *Human Sexual Response.* Boston: Little, Brown.

———. 1970. *Human Sexual Inadequacy.* Boston: Little, Brown.

———. 1973. "Orgasm, Anatomy of the Female." In *The Encyclopedia of Sexual Behavior,* edited by Albert Ellis and Albert Abarbanel. New York: Aronson.

———. 1976. *The Pleasure Bond: A New Look at Sexuality and Commitment.* New York: Bantam.

Masters, William H., Virginia E. Johnson, and Robert C. Kolodny. 1994. *Heterosexuality.* New York: HarperCollins.

Mattessich, Paul and Reuben Hill. 1987. "Life Cycle and Family Development." Pp. 437–69 in *Handbook of Marriage and the Family,* edited by Marvin B. Sussman and Suzanne K. Steinmetz. New York: Plenum.

Matthews, Ralph and Anne Martin Matthews. 1986. "Infertility and Involuntary Childlessness: The

Transition to Nonparenthood." *Journal of Marriage and the Family* 48:641–49.

Mauldin, Teresa A. 1990. "Women Who Remain Above the Poverty Level in Divorce: Implications for Family Policy." *Family Relations* 39 (2) (Apr.): 141–46.

May, Rollo. 1969. *Love and Will.* New York: Norton.

———. 1975. "A Preface to Love." Pp. 114–19 in *The Practice of Love,* edited by Ashley Montagu. Englewood Cliffs, NJ: Prentice-Hall.

McAdoo, John L. 1988. "The Roles of Black Fathers in the Socialization of Black Children." Pp. 257–69 in *Black Families.* 2d ed., edited by Harriette Pipes McAdoo. Newbury Park, CA: Sage.

McAllister, Ronald J., Edgar W. Butler, and Edward J. Kaiser. 1973. "The Adaptation of Women to Residential Mobility." *Journal of Marriage and the Family* 35:197–204.

McCarthy, Kate. 1991. "Recovering from Harassment." *Minnesota Monthly,* Aug., pp. 28–92.

McClain, Leanita. 1992. "The Middle Class Black's Burden." Pp. 120–22 in *Race, Class, and Gender: An Anthology,* edited by Margaret L. Andersen and Patricia Hill Collins. Belmont, CA: Wadsworth.

McClelland, David C. 1986. "Some Reflections on the Two Psychologies of Love." *Journal of Personality* 54:334–53.

McCormick, Naomi B. and Clinton J. Jessor. 1983. "The Courtship Game: Power in the Sexual Encounter." Pp. 64–86 in *Changing Boundaries: Gender Roles and Sexual Behavior,* edited by Elizabeth Rice Allgeier and Naomi B. McCormick. Mountain View, CA: Mayfield.

McCormick, Richard A., S.J. 1992. "Christian Approaches: Catholicism." Paper presented at the Surrogate Motherhood and Reproductive Technologies Symposium, Creighton University, Jan. 13.

McCubbin, Hamilton I. and Marilyn A. McCubbin. 1991. "Family Stress Theory and Assessment: The Resiliency Model of Family Stress, Adjustment and Adaptation." Pp. 3–32 in *Family Assessment Inventories for Research and Practice.* 2d ed., edited by Hamilton I. McCubbin and Anne I. Thompson. Madison: University of Wisconsin, School of Family Resources and Consumer Services.

McCubbin, Hamilton I. and Joan M. Patterson. 1983. "Family Stress and Adaptation to Crisis: A Double ABCX Model of Family Behavior." Pp. 87–106 in *Family Studies Review Yearbook,* vol. 1, edited by David H. Olson and Brent C. Miller. Newbury Park, CA: Sage.

McCubbin, Marilyn A. 1989. "Family Stress and Family Strengths: A Comparison of Single- and Two-Parent Families with Handicapped Children." *Research in Nursing and Health* 12:101–10.

McCubbin, Marilyn A. and Hamilton I. McCubbin. 1989. "Theoretical Orientations to Family Stress and Coping." Pp. 3–43 in *Treating Families Under Stress,* edited by Charles Figley. New York: Brunner/Mazel.

McCutcheon, A. L. 1988. "Denominations and Religious Intermarriage: Trends Among White Americans in the Twentieth Century." *Review of Religious Research* 29:213–27.

McDaniel, Antonio. 1994. "Historical Racial Differences in Living Arrangements of Children." *Journal of Family History* 19 (1): 57–77.

McDonald, Gerald W. and Marie Withers Osmond. 1980. "Jealousy and Trust: Unexplored Dimensions of Social Exchange Dynamics." Paper presented at the National Council on Family Relations Workshop on Theory Construction and Research Methodology, Portland, OR, Oct. 21.

McGill, Michael E. 1985. *The McGill Report on Male Intimacy.* New York: Holt, Rinehart & Winston.

McGrew, W. C. 1981. "The Female Chimpanzee as a Human Evolutionary Prototype." Pp. 35–74 in *Woman the Gatherer,* edited by Frances Dahlberg. New Haven, CN: Yale University Press.

McHale, Susan M., W. T. Bartko, Ann C. Crouter, and M. Perry-Jenkins. 1990. "Children's Housework and Psychological Functioning: The Mediating Effects of Parents' Sex-Role Behaviors and Attitudes." *Child Development* 61:1413–26.

McHale, Susan M. and Ann C. Crouter. 1992. "You Can't Always Get What You Want: Incongruence Between Sex-Role Attitudes and Family Work Roles and Its Implications for Marriage." *Journal of Marriage and the Family* 54 (3) (Aug.): 537–47.

McLanahan, Sara S. and Karen Booth. 1989. "Mother-Only Families: Problems, Prospects, and Politics." *Journal of Marriage and the Family* 51:557–80.

McLanahan, Sara S. and Larry Bumpass. 1988. "Intergenerational Consequences of Family Disruption." *American Journal of Sociology* 94:130–52.

McLaughlin, S. D., D. L. Manninen, and L. D. Winges. 1988. "The Consequences of the Relinquishment Decision Among Adolescent Mothers." *Social Work* 33:320–24.

McLeod, Jane D. 1995. "Social and Psychological Bases of Homogamy for Common Psychiatric Disorders." *Journal of Marriage and the Family* 57 (1) (Feb.): 201–14.

McRoy, Ruth G., Harold D. Grotevant, and Louis A. Zurcher, Jr. 1988. *Emotional Disturbance in Adopted Adolescents: Origins and Development.* New York: Praeger.

Mead, George Herbert. 1934. *Mind, Self, and Society.* Chicago: University of Chicago Press.

Mead, Margaret. 1949. *Male and Female: A Study of the Sexes in a Changing World.* New York: Morrow.

———. 1966. "Marriage in Two Steps." *Redbook,* July.

Meilander, Gilbert. 1992. "Christian Approaches: Protestantism." Paper presented at the Surrogate Motherhood and Reproductive Technologies Symposium, Creighton University, Jan. 13.

Melli, Marygold S. 1986. "Constructing a Social Problem: The Post-Divorce Plight of Women and Children." *American Bar Foundation Research Journal* 1986:759–72.

Menaghan, Elizabeth and Morton Lieberman. 1986. "Changes in Depression Following Divorce: A Panel Study." *Journal of Marriage and the Family* 48:319–28.

Mendola, Mary. 1980. *The Mendola Report: A New Look at Gay Couples.* New York: Crown.

Mensch, Barbara and Denise B. Kandel. 1992. "Drug Use as a Risk Factor for Premarital Teen Pregnancy and Abortion in a National Sample of Young White Women." *Demography* 29 (3) (Aug.): 409–29.

Meredith, William H., Douglas A. Abbott, Mary Ann Lamanna, and Gregory Sanders. 1989. "Rituals and Family Strengths: A Three Generation Study." *Family Perspectives* 23:75–83.

Merson, Julie. 1995. "Eggs for Sale." *The San Francisco Bay Guardian,* May 10, pp. 14–15.

Messinger, Lilian. 1976. "Remarriage Between Divorced People with Children from Previous Marriages: A Proposal for Preparation for Remarriage." *Journal of Marriage and Family Counseling* 2:193–200.

Messner, Michael A. 1992. *Power at Play: Sports and the Problem of Masculinity.* Boston: Beacon.

Meyer, Daniel R., Elizabeth Phillips, and Nancy L. Maritato. 1991. "The Effects of Replacing Income Tax Deductions for Children with Children's Allowances." *Journal of Family Issues* 12 (4) (Dec.):467–91.

Meyer, Thomas J. 1984. " 'Date Rape': A Serious Campus Problem That Few Talk About." *Chronicle of Higher Education,* Dec. 5, pp. 1, 12.

Michaels, Marguerite and James Willwerth. 1989. "How America Has Run Out of Time." *Time,* Apr. 24, pp. 58–67.

Milano, Elyce and Stephen F. Hall. 1986. "Sex-Roles in Dating: Paying vs. Putting Out." Unpublished paper.

Milardo, Robert M. 1989. "Theoretical and Methodological Issues in the Identification of the Social Networks of Spouses." *Journal of Marriage and the Family* 51 (1):165–74.

Miller, Baila. 1990. "Gender Differences in Spouse Caregiver Strain: Socialization and Role Explanations." *Journal of Marriage and the Family* 52:311–21.

Miller, Richard B. and Jennifer Glass. 1989. "Parent–Child Attitude Similarity Across the Life Course." *Journal of Marriage and the Family* 51:991–97.

Miller, Vernon D., and Mark L. Knapp. 1986. "Communication Paradoxes and the Maintenance of Living Relationships with the Dying." *Journal of Family Issues* 7:255–76.

Mills, C. Wright. 1973 [1959]. *The Sociological Imagination.* New York: Oxford University Press.

Milner, Jan. 1990. "Strangers in a Strange Land: Housewives in the 1980's." Paper presented at the annual meeting of the Midwest Sociological Society, Chicago, Apr. 14.

Mitchell, Juliet and Ann Oakley, eds. 1986. *What Is Feminism?* New York: Pantheon.

Mitchell, K. 1983. "The Price Tag of Responsibility: A Comparison of Divorced and Remarried Mothers." *Journal of Divorce* 6:33–42.

Moen, Phyllis. 1992. *Women's Two Roles: A Contemporary Dilemma.* New York: Auburn House.

Mohr, James. 1978. *Abortion in America: The Origins and Evolution of National Policy, 1800–1900.* New York: Oxford University Press.

Molyneux, Guy. 1995. "Losing by the Rules." *Los Angeles Times,* Sept. 3, pp. M1, M3.

Money, John and Anke A. Ehrhardt. 1974. *Man and Woman, Boy and Girl: Differentiation and Dimorphism of Gender Identity from Conception to Maturity.* New York: New American Library/Mentor.

Moore, Kristin A., Christine Winquist Nord, and James L. Peterson. 1989. "Nonvoluntary Sexual Activity Among Adolescents." *Family Planning Perspectives* 21 (3) (May–June): 110–14.

Moore, Kristin A., Margaret C. Simms, and Charles L. Betsey. 1986. *Choice and Circumstance: Racial Differences in Adolescent Sexuality and Fertility.* New Brunswick, NJ: Transaction.

Moore, Teresa. 1993. "Pain of Color-Coded Child Rearing." *San Francisco Chronicle.*, Feb. 15, p. B3.

Moorman, Jeanne E. and Donald J. Hernandez. 1989. "Married Couple Families with Step, Adopted, and Biological Children." *Demography* 26:267–78.

Morganthau, Tom. 1992. "Losing Ground." *Newsweek,* Apr. 6, pp. 20–22.

Morrison, Donna R. and Andres J. Cherlin. 1995. "The Divorce Process and Young Children's Well-Being: A Prospective Analysis." *Journal of Marriage and the Family* 57 (3) (Aug.): 800–12.

Morrow, Lance. 1992. "Family Values." *Time,* Aug. 31, pp. 22–27.

Mosher, William D. 1990. "Fecundity and Infertility in the United States, 1965–1988." *Advance Data.* 192. U.S. Department of Health and Human Services.

Mosher, William D. and William F. Pratt. 1985. "Fecundity and Infertility in the United States, 1965–82." U.S. National Center for Health Statistics, Advance Data, No. 104, Feb. 11.

———. 1990. "Contraceptive Use in the United States, 1973–88." U.S. National Center for Health Statistics, Advance Data, No. 182, Mar. 20.

———. 1993. "AIDS-Related Behavior Among Women 15–44 Years of Age: United States, 1988 and 1990." *Advance Data* No. 239. Dec. 22. Hyattsville, MD: U.S. Department of Health and Human Services, National Center for Health Statistics.

"Most Americans Remain Opposed to Abortion Ban and Continue to Support Women's Right to Decide." 1984. *Family Planning Perspectives* 16:233–34.

Muller, Chandra. 1995. "Maternal Employment, Parent Involvement, and Mathematics Achievement Among Adolescents." *Journal of Marriage and the Family* 57 (1) (Feb.): 85–100.

Murdock, George. 1949. *Social Structure.* New York: Free Press.

Murnen, Sarah K., Annette Perot, and Don Byrne. 1989. "Coping with Unwanted Sexual Activity: Normative Responses, Situational Determinants, and Individual Differences." *Journal of Sex Research* 26:85–106.

Murphy, John E. 1988. "Date Abuse and Forced Intercourse Among College Students." Pp. 285–96 in *Family Abuse and Its Consequences: New Directions in Research,* edited by Gerald T. Hotaling, David Finkelhor, John T. Kirkpatrick, and Murray A. Straus. Newbury Park, CA: Sage.

Murstein, Bernard I. 1980. "Mate Selection in the 1970s." *Journal of Marriage and the Family* 42:777–92.

———. 1986. *Paths to Marriage.* Newbury Park, CA: Sage.

Myers, Michael F. 1989. *Men and Divorce.* New York: Guilford.

N

Nadelson, Carol C. and David B. Marcotte. 1983. *Treatment Interventions in Human Sexuality.* New York: Plenum.

Nanji, Azim A. 1993. "The Muslim Family in North America." Pp. 229–42 in *Family Ethnicity: Strength in Diversity,* edited by Harriette Pipes McAdoo. Newbury Park, CA: Sage.

Nasar, Sylvia. 1992b. "However You Slice the Data the Richest Did Get Richer." *New York Times,* May 11.

———. 1992c. "More College Graduates Taking Low-Wage Jobs." *New York Times,* Aug. 7.

Nason, Ellen M. and Margaret M. Poloma. 1976. *Voluntarily Childless Couples: The Emergence of a Variant Life Style.* Newbury Park, CA: Sage.

National Gay & Lesbian Task Force (NGLTF) flyer. n.d. Washington DC: National Gay & Lesbian Task Force.

National Institute of Mental Health. 1985. *Plain Talk About Wife Abuse.* Rockville, MD: U.S. Department of Health and Human Services.

"Navaho Tribal Court Supported by Utah Court in Adoption Case." 1988. *New York Times,* Dec. 15.

Nazario, Sonia L. 1990a. "Midwifery Is Staging Revival as Demand for Prenatal Care, Low-Tech Births Rises." *The Wall Street Journal,* Sept. 25.

———. 1990b. "Identity Crisis: When White Parents Adopt Black Babies, Race Often Divides." *The Wall Street Journal,* Sept. 20.

———. 1992. "Schools Teach the Virtues of Virginity." *The Wall Street Journal,* Feb. 20, pp. B1, B5.

Needle, Richard H., S. Susan Su, and William J. Doherty. 1990. "Divorce, Remarriage, and Adolescent Substance Use: A Prospective Longitudinal Study." *Journal of Marriage and the Family* 52 (1) (Feb.):157–69.

Neisen, Joseph H. 1990. "Heterosexism: Redefining Homophobia for the 1990s." *Journal of Gay and Lesbian Psychotherapy* 1: 21–35.

Nelson, L. J. and J. Cooper. 1990. *Children's Reactions to Success and Failure with Computers.* Manuscript submitted for publication. [cited in Basow 1992]

Nemy, Enid. 1991. "More Singles Jilt the City for the Suburbs." *New York Times,* May 9.

Ness, Carol. 1992. "Abortion Curbs Upheld." *San Francisco Chronicle,* June 29, pp. A1, A12.

Nevid, Jeffrey S. 1995. *Human Sexuality in a World of Diversity.* 2d ed. Boston: Allyn & Bacon.

Newcomb, Paul R. 1979. "Cohabitation in America: An Assessment of Consequences." *Journal of Marriage and the Family* 41:597–603.

"New Fathers: Trend or Phenomenon." 1991. *Family Affairs* 4, p. 13.

Newman, Katherine S. 1988. *Falling from Grace: The Experience of Downward Mobility in the American Middle Class.* New York: Random House.

Newman, Louis. 1992. "Jewish Approaches." Paper presented at the Surrogate Motherhood and Reproductive Technologies Symposium, Creighton University, Jan. 13.

Newman, Robert. 1979. "For Women Only: Warning of Widowhood Eases Adjustment Period." United Feature Syndicate, in *Omaha World-Herald,* Apr. 11.

Nieves-Squires, S. 1991. *Hispanic Women: Making Their Presence on Campus Less Tenuous.* Washington, DC: Association of American Colleges, Project on the Status and Education of Women.

Nock, Steven L. 1995. "A Comparison of Marriages and Cohabiting Relationships." *Journal of Family Issues* 16 (1) (Jan.): 53–76.

Noller, Patricia and Mary Anne Fitzpatrick. 1991. "Marital Communication in the Eighties." Pp. 42–53 in *Contemporary Families: Looking Forward, Looking Back,* edited by Alan Booth. Minneapolis: National Council on Family Relations.

Norton, Arthur J. and Jeanne E. Moorman. 1987. "Current Trends in Marriage and Divorce Among American Women." *Journal of Marriage and the Family* 49:3–14.

O

Obejas, Achy. 1994. "Women Who Batter Women." *Ms.* (Sept./Oct.): 53.

O'Boyle, Thomas F. 1990. "Fear and Stress in the Office Take Toll." *The Wall Street Journal,* Nov. 6, pp. B1, B16.

O'Brien, Patricia. 1980. "How to Survive the Early Years of Marriage." Pp. 51–54 in *Marriage and Family 80/81,* edited by Robert H. Walsh and Ollie Pocs. Guilford, CT: Dushkin.

O'Connell, Martin and David E. Bloom. 1987. *Juggling Jobs and Babies: America's Child Care Challenge.* Occasional Paper No. 12 (Feb.). Washington, DC: Population Reference Bureau.

Olson, David H. and Hamilton I. McCubbin. 1983. *Families: What Makes Them Work.* Newbury Park, CA: Sage.

O'Neil, Robin and Ellen Greenberger. 1994. "Patterns of Commitment to Work and Parenting: Implications for Role Strain." *Journal of Marriage and the Family* 56 (1) (Feb.): 101–118.

O'Neill, Nena and George O'Neill. 1972. *Open Marriage: A New Life Style for Couples.* New York: M. Evans.

———. 1974. *Shifting Gears: Finding Security in a Changing World.* New York: M. Evans.

"One-Third Have Discretionary Money." 1985. *Omaha World-Herald,* Dec. 1.

Oppenheimer, Valerie Kincaide. 1974. "The Life Cycle Squeeze: The Interaction of Men's Occupational and Family Life Cycles." *Demography* 11:227–45.

———. 1994. "Women's Rising Employment and the Future of the Family in Industrial Societies." *Population and Development Review* 20 (2) (June): 293–342.

Orbuch, Terri L. and Lindsay Custer. 1995. "The Social Context of Married Women's Work and Its Impact on Black Husbands and White Husbands." *Journal of Marriage and the Family* 57 (2) (May): 333–45.

Orenstein, Peggy. 1994. *School Girls: Young Women, Self-Esteem, and the Confidence Gap.* New York: Doubleday.

Oropesa, R. S., Daniel T. Lichter, and Robert N. Anderson. 1994. "Marriage Markets and the Paradox of Mexican American Nuptiality." *Journal of Marriage and the Family* 56 (4) (Nov.): 889–907.

Ortega, Suzanne T., Hugh P. Whitt, and J. Allen Williams. 1988. "Religious Homogamy and Marital Happiness." *Journal of Family Issues* 9:224–39.

Otten, Alan L. 1990. "Who Cares for Kids Depends on Their Status." *The Wall Street Journal*, Nov. 29, p. B1.

Overall, Christine. 1990. "Selective Termination of Pregnancy and Women's Reproductive Autonomy." *Hastings Center Report* (May/June): 6–11.

P

Padavic, Irene. 1992. "White-Collar Work Values and Women's Interest in Blue-Collar Jobs." *Gender & Society* 6 (2) (June): 215–30.

Paden, Shelley L. and Cheryl Buehler. 1995. "Coping with Dual-Income Lifestyle." *Journal of Marriage and the Family* 57 (1) (Feb.): 101–10.

Paludi, M. A., ed. 1990. *Ivory Power: Sexual Harassment on Campus*. Albany: State University of New York Press.

Parke, Ross D. and Ronald G. Slaby. 1983. "The Development of Aggression." Pp. 547–641 in *Handbook of Child Psychology*, vol. 4, *Socialization, Personality, and Social Development*, edited by E. Mavis Hetherington. New York: Wiley.

Parkman, Allen M. 1992. *No-Fault Divorce: What Went Wrong?* Boulder, CO: Westview.

Parsons, Talcott. 1943. "The Kinship System of the Contemporary United States." *American Anthropologist* 45:22–38.

Parsons, Talcott and Robert F. Bales. 1955. *Family, Socialization, and Interaction Process*. Glencoe, IL: Free Press.

Parsons, Talcott and Renee Fox. 1952. "Illness, Therapy, and the Modern American Family." *Journal of Social Issues* 8:31–44.

Parsons, Talcott and Victor Lidz. 1967. "Death in American Society." Pp. 133–70 in *Essays on Self-Destruction*, edited by Edwin Scheidman. New York: Science House.

Pasley, Kay and Marilyn Ihinger-Tallman. 1988. "Remarriage and Stepfamilies." Pp. 204–21 in *Variant Family Forms*, edited by Catherine S. Chilman, Elam W. Nunnally, and Fred M. Cox. Newbury Park, CA: Sage.

Passell, Peter. 1991. "Chronic Poverty, Black and White." *New York Times*, Mar. 6.

Patterson, Charlotte. 1992. "Children of Lesbian and Gay Parents." *Child Development* 63:1025–42.

Patterson, Gerald R. 1982. *Coercive Family Process*. Eugene, OR: Castalia.

Patterson, Joan M. and Hamilton I. McCubbin. 1984. "Gender Roles and Coping." *Journal of Marriage and the Family* 46 (1):95–104.

Pear, Robert. 1991. "5.5 Million Children in U.S. Are Hungry, a Study Finds." *New York Times*, Mar. 27.

Pearlin, Leonard I. 1975. "Status Inequality and Stress in Marriage." *American Sociological Review* 40:344–57.

Peck, M. Scott, 1978. *The Road Less Traveled: A New Psychology of Love, Traditional Values and Spiritual Growth*. New York: Simon & Schuster.

Pence, E. and M. Paymar. 1993. *Education Groups for Men Who Batter: The Duluth Model*. New York: Springer.

Peplau, Letitia A. 1981. "What Homosexuals Want in Relationships." *Psychology Today* 15:28–38.

Peplau, Letitia A. and Susan Miller Campbell. 1989. "The Balance of Power in Dating and Marriage." Pp. 121–37 in *Women: A Feminist Perspective*. 4th ed., edited by Jo Freeman. Mountain View, CA: Mayfield.

Peplau, Letitia A. and Steven L. Gordon. 1983. "The Intimate Relationships of Lesbians and Gay Men." Pp. 226–44 in *Changing Boundaries: Gender Roles and Sexual Behavior*, edited by Elizabeth Rice Allgeier and Naomi B. McCormick. Palo Alto, CA: Mayfield.

Peres, Yochanan and Rachel Pasternack. 1991. "To What Extent Can the School Reduce the Gaps Between Children Raised by Divorce and Intact Families?" *Journal of Divorce and Remarriage* 15 (3/4):143–58.

Perlman, Daniel and Karen S. Rook. 1987. "Social Support, Social Deficits, and the Family." Pp. 17–44 in *Family Processes and Problems: Social Psychological Aspects*. Applied Social Psychology Annual, vol. 7, edited by Stuart Oskamp. Newbury Park, CA: Sage.

Perlman, Robert, ed. 1983. *Family Home Care: Critical Issues for Services and Policies*. New York: Haworth Press.

"Perot Backers Like Underdog: Buchanan Tops Republican Field." 1995. *Los Angeles Times*, Aug. 13, pp. 1A, 14A.

Perry-Jenkins, Maureen and Ann C. Crouter. 1990. "Men's Provider Role Attitudes: Implications for

Household Work and Marital Satisfaction." *Journal of Family Issues* 11:136–56.

Perry-Jenkins, Maureen and Karen Folk. 1994. "Class, Couples, and Conflict: Effects of the Division of Labor on Assessments of Marriage in Dual-Earner Families." *Journal of Marriage and the Family* 56 (1) (Feb.): 165–80.

Peters, Marie Ferguson. 1988. "Parenting in Black Families with Young Children." Pp. 228–41 in *Black Families*. 2d ed., edited by Harriette Pipes McAdoo. Newbury Park, CA: Sage.

Peters, Marie F. and Harriette P. McAdoo. 1983. "The Present and Future of Alternative Lifestyles in Ethnic American Cultures." Pp. 288–307 in *Contemporary Families and Alternative Lifestyles: Handbook on Research and Theory*, edited by Eleanor D. Macklin and Roger H. Rubin. Newbury Park, CA: Sage.

Peterson, Gary W. and Boyd C. Rollins. 1987. "Parent–Child Socialization." Pp. 471–507 in *Handbook of Marriage and the Family*, edited by Marvin B. Sussman and Suzanne K. Steinmetz. New York: Plenum.

Peterson, James L. and Christine Winquist Nord. 1990. "The Regular Receipt of Child Support: A Multistep Process." *Journal of Marriage and the Family* 52 (2): (May): 539–51.

Peterson, James L. and Nicholas Zill. 1986. "Marital Disruption, Parent–Child Relationships, and Behavioral Problems in Children." *Journal of Marriage and the Family* 48:295–307.

Peterson, Oliver. 1991. "Many Older People Are Richer Than Other Americans." *New York Times*, Oct. 30.

Peterson, Richard R. 1989. *Women, Work, and Divorce*. New York: State University of New York Press.

Peterson, Richard R. and Kathleen Gerson. 1992. "Determinants of Responsibility for Child Care Arrangements Among Dual-Earner Couples." *Journal of Marriage and the Family* 54 (3) (Aug.): 527–36.

Peterson, Sharyl Bender and Traci Kroner. 1992. "Gender Biases in Textbooks for Introductory Psychology and Human Development." *Psychology of Women Quarterly* 16:17–36.

Pett, Marjorie A., Nancy Lang, and Anita Gander. 1992. "Late-Life Divorce: Its Impact on Family Rituals." *Journal of Family Issues* 13:526–53.

Pettit, Ellen J. and Bernard L. Bloom. 1984. "Whose Decision Was It? The Effects of Initiator Status on Adjustment to Marital Disruption." *Journal of Marriage and the Family* 46 (3):587–95.

Phillips, Linda R. 1986. "Theoretical Explanations of Elder Abuse: Competing Hypotheses and Unresolved Issues." Pp. 197–217 in *Elder Abuse: Conflict in the Family*, edited by Karl A. Pillemer and Rosalie S. Wolf. Dover, MA: Auburn House.

Pietropinto, Anthony and Jacqueline Simenauer. 1977. *Beyond the Male Myth: What Women Want to Know About Men's Sexuality, A National Survey*. New York: Times Books.

Pillemer, Karl A. 1986. "Risk Factors in Elder Abuse: Results from a Case-Control Study." Pp. 239–64 in *Elder Abuse: Conflict in the Family*, edited by Karl A. Pillemer and Rosalie S. Wolf. Dover, MA: Auburn House.

Pina, Darlene L. and Vern L. Bengtson. 1993. "The Division of Household Labor and Wives' Happiness: Ideology, Employment, and Perceptions of Support." *Journal of Marriage and the Family* 55 (4) (Nov.): 901–12.

Pines, Maya. 1981. "Only Isn't Lonely (or Spoiled or Selfish)." *Psychology Today* 15:15–19.

Piotrkowski, Chaya S., Robert N. Rapoport, and Rhona Rapoport. 1987. "Families and Work." Pp. 251–83 in *Handbook of Marriage and the Family*, edited by Marvin B. Sussman and Suzanne K. Steinmetz. New York: Plenum.

Pleck, Joseph H. 1977. "The Work–Family Role System." *Social Problems* 24:417–27.

———. 1985. *Working Wives/Working Husbands*. Newbury Park, CA: Sage.

———. 1992. "Prisoners of Manliness." Pp. 98–107 in *Men's Lives*. 2d ed, edited by Michael S. Kimmel and Michael A. Messner. New York: Macmillan.

Polakow, Valerie. 1993. *Single Mothers and Their Children in the Other America*. Chicago: University of Chicago Press.

Pollitt, Katha. 1990. " 'Fetal Rights': A New Assault on Feminism." *The Nation*, Mar., pp. 409–18.

Pomerleau, A., D. Bolduc, G. Malcuit, and L. Cossetts. 1990. "Pink or Blue: Environmental Stereotypes in the First Two Years of Life." *Sex Roles* 22:359–67.

Popenoe, David. 1993. "American Family Decline, 1960–1990: A Review and Appraisal." *Journal of Marriage and the Family* 55 (3) (Aug.): 527–55.

Population Information Program. The Johns Hopkins University. 1984. "After Contraception: Dispelling Rumors About Later Childbearing." *Population Reports,* Series 1, No. 28.

Population Reference Bureau. 1992. "1992 World Population Data Sheet of the Population Reference Bureau, Inc." Washington, DC: Population Reference Bureau.

Porter, Sylvia. 1976. *Sylvia Porter's Money Book.* New York: Avon.

Porterfield, Ernest. 1982. "Black-American Intermarriages in the United States." Pp. 17–34 in *Intermarriages in the United States,* edited by Gary Crester and Joseph J. Leon. New York: Haworth.

Potuchek, Jean L. 1992. "Employed Wives' Orientation to Breadwinning: A Gender Theory Analysis." *Journal of Marriage and the Family* 54 (3) (Aug.): 548–58.

"Poverty Looms for Many Women Who Take Time Off from Jobs." 1991. *The Wall Street Journal,* Feb. 5.

Power, Paul W. 1979. "The Chronically Ill Husband and Father: His Role in the Family." *Family Coordinator* 28:616–21.

Powledge, Tabitha M. 1983. "Windows on the Womb." *Psychology Today* 17:37–42.

Pozzetta, George E., ed. 1991. *Immigrant Family Patterns: Demography, Fertility, Housing, Kinship, and Urban Life.* New York: Garland.

"Pregnant Woman Can't Be Prosecuted for Transmitting Crack to Baby." 1991. *The Wall Street Journal,* July 18.

"Pregnant Woman Draws Paint-Sniffing Sentence." 1992. *Omaha World-Herald,* Feb. 12, p. 13.

Presser, Harriet B. 1988. "Shift Work and Child Care Among Young Dual-Earner American Parents." *Journal of Marriage and the Family* 50:133–48.

Prial, Dunstan. 1988. "Domestic Violence Data Change Police Strategy." *Omaha World-Herald,* Dec. 27.

Price-Bonham, Sharon, and Susan Addison. 1978. "Families and Mentally Retarded Children: Emphasis on the Father." *Family Coordinator* 27:221–30.

Pruchino, Rachel, Christopher Burant, and Norah D. Peters. 1994. "Family Mental Health: Marital and Parent–Child Consensus as Predictors." *Journal of Marriage and the Family* 56 (3) (Aug.): 747–58.

Putka, Gary. 1990. "Effort to Teach Teens About Homosexuality Advances in Schools." *The Wall Street Journal,* June 12, pp. A1, A9.

Q

Quarm, Daisy. 1984. "Sexual Inequality: The High Costs of Leaving Parenting to Women." Pp. 187–208 in *Women in the Workplace: Effects on Families,* edited by Kathryn M. Borman, Daisy Quarm, and Sarah Gideonse. Norwood, NJ: Ablex.

Quindlen, Anna. 1990. "Men at Work." *New York Times,* Feb. 18.

Quinn, Peggy and Katherine R. Allen. 1989. "Facing Challenges and Making Compromises: How Single Mothers Endure." *Family Relations* 28 (Oct.): 390–95.

Qvortrup, M. B., G. Sgritta, and H. Wintersberger, eds. 1994. *Childhood Matters: Social Theory, Practice, and Politics.* Vienna: European Centre for Social Welfare Policy and Research.

R

Raabe, Phyllis Hutton. 1990. "The Organizational Effects of Workplace Family Policies." *Journal of Family Issues* 11 (4) (Dec.): 477–91.

Rae, Scott B. 1994. *The Ethics of Commercial Surrogate Motherhood: Brave New Families?* Westport, CT: Praeger.

Ragone, Helena. 1994. *Surrogate Motherhood: Conception in the Heart.* Oxford: Westview.

Ransford, H. Edward and Jon Miller. 1983. "Race, Sex, and Feminist Outlooks." *American Sociological Review* 48:46–59.

Raphael, Phyllis. 1978. "The Stepmother Trap." *McCall's,* Feb., pp. 188–94.

Raschke, Helen C. 1987. "Divorce." Pp. 597–624 in *Handbook of Marriage and the Family,* edited by Marvin B. Sussman and Suzanne K. Steinmetz. New York: Plenum.

Raven, Bertram, Richard Centers, and Arnoldo Rodrigues. 1975. "The Bases of Conjugal Power." Pp. 217–32 in *Power in Families,* edited by Ronald E. Cromwell and David H. Olson. Beverly Hills: Sage.

Ravo, Nick. 1993. "With a New Kind of Housing, Togetherness Is Built Right In." *New York Times,* Feb. 25.

Rawlings, Steve W. 1994. *Household and Family Characteristics: March 1993.* U.S. Bureau of the Census, Current Population Reports, P20-477. Washington, DC: U.S. Government Printing Office.

Raymond, Janice G. 1989. "The International Traffic in Women: Women Used in Systems of Surrogacy and Reproduction." *Reproductive and Genetic Engineering* 2 (1):51–57.

———. 1991. "Women as Wombs." *Ms.* 1 (6), May/June, pp. 28–33.

Reese, Thomas J. 1992. "Bishops Meet at Notre Dame." *America,* July 4–11, pp. 4–6.

Register, Cheri. 1991. *"Are Those Kids Yours?": American Families with Children Adopted from Other Countries.* New York: Free Press.

Reid, John. 1982. *Black America in the 1980s. Population Bulletin* 37. Washington, DC: Population Reference Bureau.

Reiss, David, Sandra Gonzalez, and Norman Kramer. 1986. "Family Process, Chronic Illness, and Death: On the Weakness of Strong Bonds." *Archives of General Psychiatry* 43:795–804.

Reiss, I. L., and G. L. Lee. 1988. *The Family System in America.* 4th ed. New York: Holt, Reinhart & Winston.

Reiss, Ira L. 1976. *Family Systems in America.* 2d ed. Hinsdale, IL: Dryden.

———. 1986. *Journey into Sexuality: An Exploratory Voyage.* Englewood Cliffs, NJ: Prentice-Hall.

Renzetti, Claire M. 1992. *Violent Betrayal: Partner Abuse in Lesbian Relationships.* Newbury Park, CA: Sage.

Resnick, M. 1984. "Studying Adolescent Mothers' Decision Making About Adoption and Parenting." *Social Work* 29:4–10.

Reynolds v. *United States.* 1878. 98 U.S. 145, 25 L.Ed. 244.

Richardson, Brenda Lane. 1988. "Working Women Do Go Home Again." *New York Times,* Apr. 20.

Richardson, Laurel Walum. 1985. *The New Other Woman: Contemporary Single Women in Affairs with Married Men.* New York: Free Press.

Ridley, Carl, Dan J. Peterman and Arthur W. Avery. 1978. "Cohabitation: Does It Make for a Better Marriage?" *Family Coordinator* 27:129–36.

Rigdon, Joan E. 1991. "Exploding Myth: Asian-American Youth Suffer a Rising Toll for Heavy Pressure." *The Wall Street Journal,* July 10.

Riley, Glenda. 1991. *Divorce: An American Tradition.* New York: Oxford University Press.

Riordan, Teresa and Sue Kirchhoff. 1995. "Women on the Hill: Can They Make a Difference?" *Ms.* (Jan./Feb.): 85–89.

Rimer, Sara. 1988a. "Child Care at Home: 2-Women, Complex Roles." *New York Times,* Dec. 26.

———. 1988b. "Women, Jobs and Children: A New Generation Worries." *New York Times,* Nov. 27.

Robertson, Elizabeth B., Glen H. Elder, Jr., Martie L. Skinner, and Rand D. Conger. 1991. "The Costs and Benefits of Social Support in Families." *Journal of Marriage and the Family* 53:403–16.

Robertson, Joan F. 1978. "Women in Mid-life: Crisis, Reverberations, and Support Networks." *Family Coordinator* 27:375–82.

Robinson, Bryan E. and Robert L. Barret. 1986. *The Developing Father: Emerging Roles in Contemporary Society.* New York: Guilford.

Robinson, Bryan, Patsy Skeen, and Lynda Walters. 1987. "The AIDS Epidemic Hits Home." *Psychology Today* (Apr.): 48–52.

Robinson, Ira, B. Ganza, S. Katz, and E. Robinson. 1991. "Twenty Years of the Sexual Revolution, 1965–1985: An Update." *Journal of Marriage and the Family* 53:216–20.

Robson, Ruthann and S. E. Valentine. 1990. "Lov(h)ers: Lesbians as Intimate Partners and Lesbian Legal Theory." *Temple Law Review* 63: 511–41.

Rodgers, Joseph Lee and David C. Rowe. 1990. "Adolescent Sexual Activity and Mildly Deviant Behavior." *Journal of Family Issues* 11 (3) (Sept.): 274–93.

Rodgers, Roy H. and Linda M. Conrad. 1986. "Courtship for Remarriage: Influences on Family Reorganization After Divorce." *Journal of Marriage and the Family* 48:767–75.

Rodgers, Roy H. and James M. White. 1993. "Family Development Theory." Pp. 225–54 in *Sourcebook of Family Theories and Methods: A Contextual Approach,* edited by Pauline G. Boss, William J. Doherty, Ralph LaRossa, Walter R. Schumm, and Suzanne K. Steinmetz. New York: Plenum.

Rodman, Hyman. 1991. "Should Parental Involvement Be Required for Minors' Abortions?" *Family Relations* 40 (2) (Apr.): 155–60.

Rogers, David. 1995. "House Votes to Ban Late-Term Abortion in Challenges to Supreme Court Decision." *The Wall Street Journal,* Nov. 2.

Rogers, Patrick. 1994. "Surviving the Second Wave." *Newsweek,* Sept. 19, pp. 50–51.

Rogers, Susan M. and Charles F. Turner. 1991. "Male–Male Sexual Contact in the U.S.A.: Findings from Five Sample Surveys, 1970–1990." *The Journal of Sex Research* 28 (4):491–519.

Romano, Lois and Jacqueline Trescott. 1992. "Love in Black & White." *Redbook,* Feb., pp. 88–94.

Romero, Mary. 1992. *Maid in the U.S.A.* New York: Routledge.

Rosemond, John. 1991a. "Remembering ABCs Will Keep Parent out of Homework Game." *Omaha World-Herald,* Oct. 2.

———. 1991b. "Too-Helpful Parents Make Children Feel Helpless." *Omaha World-Herald,* Oct. 1.

Rosenberg, Elinor B. 1992. *The Adoption Life Cycle: The Children and Their Families Through the Years.* New York: Free Press.

Rosenblatt, Paul C. and Linda Hammer Burns. 1986. "Long-Term Effects of Perinatal Loss." *Journal of Family Issues* 7:237–54.

Rosenthal, A. M. 1987. "Individual Ethics and the Plague." *New York Times,* May 28.

Rosenthal, Andrew. 1990. "President Signs Law on Hate Crimes." *New York Times,* Apr. 24.

Rosenthal, Elisabeth. 1991. "As More Tiny Infants Live, Choices and Burdens Grow." *New York Times,* Sept. 29.

Rosenthal, Kristine and Harry F. Keshet. 1978. "The Not Quite Stepmother." *Psychology Today* 12:82–86.

———. 1980. *Fathers Without Partners.* New York: Rowman & Littlefield.

Ross, Catherine E. 1987. "The Division of Labor at Home." *Social Forces* 65 (3):816–33.

———. 1995. "Reconceptualizing Marital Status as a Continuum of Social Attachment." *Journal of Marriage and the Family* 57 (1) (Feb.): 129–40.

Ross, Catherine, John Mirowsky, and Patricia Ulrich. 1983. "Distress and the Traditional Female Role." *American Journal of Sociology* 89:670–82.

Rossi, Alice, ed. 1973. *The Feminist Papers.* New York: Bantam.

———., ed. 1994. *Sexuality Across the Life Course.* Chicago: University of Chicago Press.

Rossi, Alice S. 1968. "Transition to Parenthood." *Journal of Marriage and the Family* 30:26–39.

———. 1984. "Gender and Parenthood." *American Sociological Review* 49:1–19.

Rossiter, Amy B. 1991. "Initiator Status and Separation Adjustment." *Journal of Divorce and Remarriage* 15 (1/2):141–56.

Rothblum, Esther and Ellen Cole, eds. 1989. *Loving Boldly: Issues Facing Lesbians.* New York: Harrington Park.

Rothman, Barbara Katz. 1982. *In Labor: Women and Power in the Birthplace.* New York: Norton.

———. 1984. "The Meaning of Choice in Reproductive Technology." Pp. 1–33 in *Test-Tube Women,* edited by Rita Arditti, Renate Duelli Klein, and Shelley Minden. London: Routledge & Kegan Paul.

———. 1989. *Recreating Motherhood: Ideology and Technology in a Patriarchal Society.* New York: Norton.

Rowland, Mary. 1994. "Hurdles for Unmarried Partners." *New York Times,* May 22.

"RU-486 . . . Again." 1991. *Hastings Center Report.* July–Aug., p. 45.

Rubin, Lillian B. 1976. *Worlds of Pain: Life in the Working-Class Family.* New York: Basic Books.

———. 1979. *Women of a Certain Age: The Midlife Search for Self.* New York: Harper & Row.

———. 1983. *Intimate Strangers: Men and Women Together.* New York: Harper & Row.

———. 1990. *Erotic Wars: What Happened to the Sexual Revolution?* New York: Farrar, Straus, and Giroux.

———. 1992. *Worlds of Pain: Life in the Working-Class Family.* New York: Basic Books.

———. 1994. *Families on the Fault Line: America's Working Class Speaks About the Family, the Economy, Race, and Ethnicity.* New York: HarperCollins.

Rubiner, Betsy. 1994. "The Hidden Damage." *The Des Moines Register,* Oct. 9, p. 2E.

Rucci v. *Rucci.* 1962. 23 Conn. Supp. 221, 181A 2d, 125 S.Ct.

Rudolph, Barbara. 1989. "Adrift in the Doldrums." *Time,* July 31, pp. 32–34.

Ruefli, Terry, Olivia Yu, and Judy Barton. 1992. "Brief Report: Sexual Risk Taking in Smaller Cities." *The Journal of Sex Research* 29 (1):95–108.

Ruggles, Steven. 1994. "The Origins of African-American Family Structure." *American Sociological Review* 59 (Feb.): 136–51.

Rundblad, Georganne. 1990. "Talking Stones: An Examination of the Effects of Declining Mortality on Expression of Sentiment." Paper presented at the annual meeting of the Midwest Sociological Society, Chicago, Apr. 13.

Russell, Diana E. H. 1982. *Rape in Marriage.* New York: Macmillan.

Ryan, Barbara and Eric Plutzer. 1989. "When Married Women Have Abortions: Spousal Notification and Marital Interaction." *Journal of Marriage and the Family* 51:41–50.

Ryan, Kenneth J. 1988. "Giving Birth in America, 1988." *Family Planning Perspectives* 20:298–301.

S

Sachs, Andrea. 1990. "When the Lullaby Ends." *Time,* June 4, p. 82.

Sack, Kevin. 1991. "Court Rejects Visiting Rights for Former Lesbian Partner." *New York Times,* May 31.

Sadker, M., and D. Sadker. 1986. "Sexism in the Classroom: From Grade School to Graduate School." *Phi Delta Kappan* (Mar.): 512–15.

Safilios-Rothschild, Constantina. 1967. "A Comparison of Power Structure and Marital Satisfaction in Urban Greek and French Families." *Journal of Marriage and the Family* 29:345–52.

———. 1970. "The Study of Family Power Structure: A Review 1960–1969." *Journal of Marriage and the Family* 32:539–43.

———. 1983. "Toward a Social Psychology of Relationships." Pp. 306–12 in *Family in Transition: Rethinking Marriage, Sexuality, Child Rearing, and Family Organization,* 4th ed., edited by Arlene S. Skolnick and Jerome H. Skolnick. Boston: Little, Brown.

Saks, Michael J. and Edward Krupat. 1988. *Social Psychology and Its Applications.* New York: Harper & Row.

Salholz, Eloise. 1990. "The Future of Gay America." *Newsweek,* Mar. 12, pp. 20–25.

———. 1993. "For Better or For Worse." *Newsweek,* May 24, p. 69.

Salt, Robert. 1991. "Child Support in Context: Comments on Rettig, Christensen, and Dahl." *Family Relations* 40 (2) (Apr.): 175–78.

Saltzman, Amy. 1991. "Trouble at the Top." *U.S. News & World Report,* June 17, pp. 40–48.

Saluter, Arlene F. 1994. *Marital Status and Living Arrangements: March 1993.* U.S. Bureau of the Census, Current Population Reports, Series P20-478. Washington, DC: U.S. Government Printing Office.

Samuelson, Robert J. 1992. "How Our American Dream Unraveled." *Newsweek,* Mar. 2, pp. 32–39.

———. 1995. "RIP: The War on Poverty." *Newsweek,* Oct. 9, p. 59.

Sandalow, Marc. 1995. "Clinton Says GOP Putting Kids at Risk." *The San Francisco Chronicle,* Mar. 4, pp. A1, A15.

Sanders, Cheryl. 1992. "Gender and Racial Perspectives: The Black Community." Paper presented at the Surrogate Motherhood and Reproductive Technologies Symposium, Creighton University, Jan. 13.

Santrock, John W. and Karen A. Sitterle. 1987. "Parent–Child Relationships in Stepmother Families." Pp. 273–99 in *Remarriage and Stepparenting,* edited by Kay Pasley and Marilyn Ihinger-Tallman. New York: Guilford.

Satir, Virginia. 1972. *Peoplemaking.* Palo Alto, CA: Science and Behavior Books.

Scanzoni, J. and W. Marsiglio. 1993. "New Action Theory and Contemporary Families." *Journal of Family Issues* 14:105–32.

Scanzoni, John H. 1970. *Opportunity and the Family.* New York: Free Press.

———. 1972. *Sexual Bargaining: Power Politics in the American Marriage.* Englewood Cliffs, NJ: Prentice-Hall.

Scarf, Maggie. 1995. *Intimate Worlds: Life Inside the Family.* New York: Random House.

Schafer, Robert B. and Patricia M. Keith. 1981. "Equity in Marital Roles Across the Family Life Cycle." *Journal of Marriage and the Family* 43:359–67.

Schmalz, Jeffrey. 1993. "Poll Finds an Even Split on Homosexuality's Cause." *New York Times: Themes of the Year,* Fall, pp. 1, 11.

Schmidt, William E. 1990. "Valentine in a Survey: Fidelity Is Thriving." *New York Times,* Feb. 12.

Schmitt, Eric. 1992. "Wall of Silence Impedes Inquiry into a Rowdy Navy Convention." *New York Times,* June 14, pp. 1, 20.

———. 1993. "In Fear, Gay Soldiers Marry for Camouflage." *New York Times,* July 12, p. A7.

Schnittger, Maureen H. and Gloria W. Bird. 1990. "Coping Among Dual-Career Men and Women Across the Family Life-Cycle." *Family Relations* 39 (2) (Apr.): 199–205.

Schoen, Robert. 1992. "First Unions and the Stability of First Marriages." *Journal of Marriage and the Family* 54 (May): 281–84.

Schoen, Robert and Robin M. Weinick. 1993. "Partner Choice in Marriages and Cohabitations." *Journal of Marriage and the Family* 55 (2) (May): 408–14.

Schor, Juliet B. 1991. *The Overworked American: The Unexpected Decline of Leisure.* New York: Basic Books.

Schrof, Joannie M. 1994. "Sex in America." *U.S. News & World Report,* Oct. 17, pp. 74–81.

Schwartz, Felice N. 1989a. "Management Women and the New Facts of Life." *Harvard Business Review* 67:65–76.

———. 1989b. "The 'Mommy Track' Isn't Anti-Woman." *New York Times,* Mar. 22.

———. 1992. *Breaking with Tradition: Women and Work, the New Facts of Life.* New York: Warner.

Schwebel, Andrew I., Mark A. Fine, and Maureen A. Renner. 1991. "A Study of Perceptions of the Stepparent Role." *Journal of Family Issues* 12 (1) (Mar.): 43–57.

Scott, Joseph W. 1980. "Black Polygamous Family Formulation." *Alternative Lifestyles* 3:41–64.

———. 1991. "From Teenage Parenthood to Polygamy: Case Studies in Black Family Formation." Pp. 278–88 in *The Black Family: Essays and Studies,* edited by Robert Staples. Belmont, CA: Wadsworth.

Scott, Niki. 1992. "Irate Husband Vents His Side of the Story." *Working Woman,* Apr. 13.

Seager, Joni and Ann Olson. 1986. *Women in the World: An International Atlas.* New York: Simon & Schuster.

Seals, Brenda. 1990. Personal communication.

Seccombe, Karen. 1991. "Assessing the Costs and Benefits of Children: Gender Comparisons Among Childfree Husbands and Wives." *Journal of Marriage and the Family* 53 (1) (Feb.): 191–202.

Segal, Lynne. 1994. *Straight Sex: Rethinking the Politics of Pleasure.* Berkeley: University of California Press.

Segers, Mary C. and Timothy A. Byrnes, eds. 1995. *Abortion Politics in American States.* New York: M. E. Sharpe.

Segura, Denise A. and Beatriz M. Pesquera. 1995. "Chicana Feminisms: Their Political Context and Contemporary Expressions." Pp. 617–31 in *Women: A Feminist Perspective.* 5th ed., edited by Jo Freeman. Mountain View, CA: Mayfield.

Seidler, Victor J. 1992. "Rejection, Vulnerability, and Friendship." Pp. 15–34 in *Men's Friendships: Research on Men and Masculinities,* edited by Peter M. Nardi. Newbury Park, CA: Sage.

Seidman, Steven. 1991. *Romantic Longings: Love in America, 1830–1980.* New York: Routledge.

Seltzer, Judith A. 1991a. "Relationships Between Fathers and Children Who Live Apart: The Father's Role After Separation." *Journal of Marriage and the Family* 53 (1) (Feb.): 79–101.

———. 1991b. "Legal Custody Arrangements and Children's Economic Welfare." *American Journal of Sociology* 96:895–929.

Seltzer, Judith A. and Suzanne M. Bianchi. 1988. "Children's Contact with Absent Parents." *Journal of Marriage and the Family* 50:663–77.

Sena-Rivera, Jaime. 1979. "Extended Kinship in the United States: Competing Models and the Case of La Familia Chicana." *Journal of Marriage and the Family* 41:121–29.

Sennett, Richard and Jonathan Cobb. 1974. *The Hidden Injuries of Class.* New York: Random House.

"Sense of Identity, Melting Pot in U.S. Portrait." 1985. *New York Times,* reprinted in *Omaha World-Herald,* Feb. 24.

Serovich, Julianne M., Sharon J. Price, and Steven F. Chapman. 1991. "Former In-Laws as a Source of Support." *Journal of Divorce and Remarriage* 17 (1/2):17–26.

"Sex Education in Classroom Found to Bring Few Changes." 1989. *New York Times,* Mar. 16.

Shahan, Lynn. 1981. *Living Alone and Liking It!: A Complete Guide to Living on Your Own.* New York: Stratford Press (dist. by Harper & Row).

Shamim, Ishrat and Quamrul Ahsan Chowdhury. 1993. *Homeless and Powerless: Child Victims of Sexual Exploitation.* Dhaka, Bangladesh: University of Dhaka Press.

Shapiro, Johanna. 1983. "Family Reactions and Coping Strategies in Response to the Physically Ill or Handicapped Child: A Review." *Social Science and Medicine* 17:913–31.

Shapiro, Laura. 1990. "Guns and Dolls." *Newsweek,* May 26, pp. 56–65.

Sharff, Jagna Wojcicka. 1981. "Free Enterprise and the Ghetto Family." *Psychology Today* (Mar.): 41–47.

Shaver, Katherine. 1991. "Retirees Decide to Mind Their Own Businesses." *The Wall Street Journal,* July 18, pp. B1–B2.

Sheehan, Constance L., E. Wilbur Bock, and Gary R. Lee. 1990. "Religious Heterogamy, Religiosity, and Marital Happiness: The Case of Catholics." *Journal of Marriage and the Family* 52 (Feb.): 73–79.

Shellenbarger, Sue. 1991a. "Child-Care Setups Still Fall Apart Often." *The Wall Street Journal,* Sept. 26.

———. 1991b. "Companies Team Up to Improve Quality of Their Employees' Child-Care Choices." *The Wall Street Journal*, Oct. 17, pp. B1, B4.

———. 1991c. "Leaving Infants for Work Boosts Child-Care Costs." *The Wall Street Journal*, July 22.

———. 1991d. "More Job Seekers Put Family Needs First." *The Wall Street Journal*, Nov. 15, pp. B1, B6.

———. 1991e. "Work & Family: Business Travel's Toll on Parents Grows." *The Wall Street Journal*, Nov. 13.

———. 1991f. "Work & Family: Firms' On-Site Day Care Proves Too Expensive." *The Wall Street Journal*, Aug. 16, p. B1.

———. 1991g. "Work & Family: Men Become Evasive About Family Demands." *The Wall Street Journal*, Aug. 16, p. B1.

———. 1992a. "Ads Urge Careful Shopping for Day Care." *The Wall Street Journal*, July 6, pp. B1, B18.

———. 1992b. "Averting Career Damage from Family Policies." *The Wall Street Journal*, June 24, p. B1.

———. 1992c. "Concerns Seek Help from Public Agencies." *The Wall Street Journal*, Sept. 2.

———. 1992d. "Employees Take Pains to Make Flextime Work." *The Wall Street Journal*, Aug. 18.

———. 1992e. "Family Issues Hit Home with Single Fathers." *The Wall Street Journal*, May 27.

———. 1992f. "Indicators of Quality in Day Care Worsen." *The Wall Street Journal*, May 6, p. B1.

———. 1992g. "Infertile Employees Seek Firms' Support." *The Wall Street Journal*, May 12, pp. B1, B5.

———. 1992h. "Work & Family: Employers Try to See If Family Benefits Pay." *The Wall Street Journal*, Apr. 3.

———. 1992i. "Work & Family: Flexible Policies May Slow Women's Careers." *The Wall Street Journal*, Apr. 22.

———. 1992j. "Work & Family: States Create Patchwork of Family-Leave Laws." *The Wall Street Journal*, July 28.

———. 1993. "So Much Talk, So Little Action." *The Wall Street Journal*, June 21.

Shellenbarger, Sue and Cathy Trost. 1992a. "Annual List of Family-Friendly Firms Is Issued by Working Mother Magazine." *The Wall Street Journal*, Sept. 22, pp. A2, A4.

———. 1992b. "Partnership of 109 Companies Aims to Improve Care Nationwide for Children and the Elderly." *The Wall Street Journal*, Sept. 11, p. A12.

Shelton, Beth Anne. 1990. "The Distribution of Household Tasks: Does Wife's Employment Status Make a Difference?" *Journal of Family Issues* 11 (2) (June): 115–35.

———. 1992. *Women, Men and Time.* New York: Greenwood.

Shelton, Beth Anne and Daphne John. 1993. "Ethnicity, Race, and Difference: A Comparison of White, Black, and Hispanic Men's Household Labor Time." Pp. 131–50 in *Men, Work, and Family*, edited by Jane C. Hood. Newbury Park, CA: Sage.

Sherman, Lawrence W. 1992. *Policing Domestic Violence: Experiments and Dilemmas.* New York: Free Press.

Sherman, Lawrence W. and Richard A. Berk. 1984. "Deterrent Effects of Arrest for Domestic Assault." *American Sociological Review* 49:261–72.

Sherman, Suzanne, ed. 1992. *Lesbian and Gay Marriage: Private Commitments, Public Ceremonies.* Philadelphia: Temple University Press.

Shields, Nancy M. and Christine R. Hanneke. 1988. "Multiple Sexual Victimization: The Case of Incest and Marital Rape." Pp. 255–69 in *Family Abuse and Its Consequences: New Directions in Research*, edited by Gerald T. Hotaling, David Finkelhor, John T. Kirkpatrick, and Murray A. Straus. Newbury Park, CA: Sage.

Shihadeh, Edward S. 1991. "The Prevalence of Husband-Centered Migration: Employment Consequences for Married Mothers." *Journal of Marriage and the Family* 53 (2) (May): 432–44.

Shorter, Edward. 1975. *The Making of the Modern Family.* New York: Basic Books.

Shostak, Arthur. 1987. "Singlehood." Pp. 355–67 in *Handbook of Marriage and the Family*, edited by Marvin B. Sussman and Suzanne K. Steinmetz. New York: Plenum.

Shostak, Arthur and Gary McLouth. 1984. *Men and Abortion: Lessons, Losses, and Love.* New York: Praeger.

Simmel, Georg. 1950. *The Sociology of Georg Simmel*, translated and edited by Kurt H. Wolfe. Glencoe, IL: Free Press.

Simon, Barbara Levy. 1987. *Never Married Women.* Philadelphia: Temple University Press.

Simon, Rita. 1993. "Should White Families Be Allowed to Adopt African American Children?" *Health* (July/Aug.): 22.

Simon, Rita J. 1990. "Transracial Adoptions Can Bring Joy: Letters to the Editor." *The Wall Street Journal,* Oct. 17.

Simon, Rita James and Howard Altstein. 1987. *Transracial Adoptees and Their Families: A Study of Identity and Commitment.* New York: Praeger.

Simon, Stephanie. 1991. "Joint Custody Loses Favor for Increasing Children's Feeling of Being Torn Apart." *The Wall Street Journal,* July 15, p. B1.

Simons, Ronald L., Christine Johnson, Jay Beaman, and Rand D. Conger. 1993. "Explaining Women's Double Jeopardy: Factors that Mediate the Association Between Harsh Treatment as a Child and Violence by a Husband." *Journal of Marriage and the Family* 55 (3) (Aug.): 713–23.

Simons, Ronald. L., Christine Johnson, and Rand D. Conger. 1994. "Harsh Corporal Punishment Versus Quality of Parental Involvement as an Explanation of Adolescent Maladjustment." *Journal of Marriage and the Family* 56 (3) (Aug.): 591–607.

Simons, Ronald L. and Les B. Whitbeck. 1991. "Sexual Abuse as a Precursor to Prostitution and Victimization Among Adolescent and Adult Homeless Women." *Journal of Family Issues* 12 (3) (Sept.):361–79.

Singer, Bennett L. and David Deschamps, eds. 1994. *Gay & Lesbian Stats.* New York: New Press.

Singer v. *Hara.* 1974. 11 Wash. App. 247, 522 P 2d 1187.

Singh, Gopal K., T. J. Mathews, Sally C. Clarke, Trina Yannicos, and Betty L. Smith. 1995. "Annual Summary of Births, Marriages, Divorces, and Deaths: United States, 1994." *Monthly Vital Statistics Report* 43 (13). National Center for Health Statistics, U.S. Department of Health and Human Services. Oct. 23.

Skolnick, Arlene S. 1978. *The Intimate Environment: Exploring Marriage and Family,* 2d ed. Boston: Little, Brown.

Slater, Philip. 1976. *The Pursuit of Loneliness: American Culture at the Breaking Point.* 2d ed. Boston: Beacon.

Small, Stephen A. and Donell Kerns. 1993. "Unwanted Sexual Activity Among Peers During Early and Middle Adolescence: Incidence and Risk Factors." *Journal of Marriage and the Family* 55 (3) (Nov.): 941–52.

Small, Stephen A. and Tom Luster. 1994. "Adolescent Sexual Activity: An Ecological, Risk-Factor Approach." *Journal of Marriage and the Family* 56 (1) (Feb.): 181–92.

Small, Stephen A. and Dave Riley. 1990. "Toward a Multidimensional Assessment of Work Spillover into Family Life." *Journal of Marriage and the Family* 52:51–61.

Smith, Daniel Scott. 1981. "Historical Change in the Household Structure of the Elderly in Economically Developed Societies." In *Aging,* edited by James C. March. New York: Academic.

Smith, Donna. 1990. *Stepmothering,* New York: St. Martin's Press.

Smith, Gregory C. 1995. *Strengthening Aging Families: Diversity in Practice and Policy.* Thousand Oaks, CA: Sage.

Smith, Lee. 1988. "Who Is the Family of the Adolescent Mother?" Panel presentation at the annual meeting of the National Council on Family Relations, New Orleans, Nov. Custom Audio Tape No. 8866. Bridgeport, IL.

Smith, Marguerite T. 1989. "Saving." Pp. 12–15 in *Money Guide: Basics of Personal Finance,* edited by *Money Magazine.* New York: Time Inc. Magazines.

Smith-Rosenberg, Carroll. 1975. "The Female World of Love and Ritual: Relations Between Women in Nineteenth Century America." *Signs* 1:1–29.

Smith, Thomas A. 1992. "Family Cohesion in Remarried Families." *Journal of Divorce and Remarriage* 17 (1/2):49–66.

Smock, Pamela J. 1993. "The Economic Costs of Marital Disruption for Young Women over the Past Two Decades." *Demography* 30 (3) (Aug.): 353–71.

Smolowe, Jill. 1991. "He Had Been Punished Enough." *Time,* May 13, p. 54.

———. 1992. "Politics: The Feminist Machine." *Time,* May 4, pp. 34–36.

Soldo, Beth J. and Emily M. Agree. 1988. *America's Elderly. Population Bulletin* 43 (3). Washington, DC: Population Reference Bureau.

Somers, Marsha D. 1993. "A Comparison of Voluntarily Childfree Adults and Parents." *Journal of Marriage and the Family* 55 (3) (Aug.): 643–50.

Sonkin, Daniel Jay, Del Martin, and Lenore E. Auerbach Walker. 1985. *The Male Batterer: A Treatment Approach.* New York: Singer.

Sontag, Susan. 1976. "The Double Standard of Aging." Pp. 350–66 in *Sexuality Today and Tomorrow,* edited by Sol Gordon and Roger W. Libby. North Scituate, MA: Duxbury.

South, Scott J. 1993. "Racial and Ethnic Differences in the Desire to Marry." *Journal of Marriage and the Family* 55 (2) (May): 357–70.

———. 1995. "Do You Need to Shop Around?" *Journal of Family Issues* 16 (4) (July): 432–49.

South, Scott J. and Kim M. Lloyd. 1992a. "Marriage Markets and Non-Marital Fertility in the U.S." *Demography* 29:247–64.

———. 1992b. "Marriage Opportunities and Family Formation: Further Implications of Imbalanced Sex Ratios." *Journal of Marriage and the Family* 54 (May): 440–51.

Spanier, Graham B. 1976. "Measuring Dyadic Adjustment: New Scales for Assessing the Quality of Marriage and Similar Dyads." *Journal of Marriage and the Family* 38:15–28.

———. 1989. "Bequeathing Family Continuity." *Journal of Marriage and the Family* 51:3–13.

Spanier, Graham B. and Robert F. Castro. 1979. "Adjustment to Separation and Divorce: An Analysis of 50 Case Studies." *Journal of Divorce* 2:241–53.

Spanier, Graham B. and Frank F. Furstenberg, Jr. 1982. "Remarriage After Divorce: A Longitudinal Analysis of Well-Being." *Journal of Marriage and the Family* 44:709–20.

———. 1987. "Remarriage and Reconstituted Families." Pp. 419–34 in *Handbook of Marriage and the Family,* edited by Marvin B. Sussman and Suzanne K. Steinmetz. New York: Plenum.

Spanier, Graham B. and Linda Thompson. 1987. *Parting: The Aftermath of Separation and Divorce,* updated edition. Newbury Park, CA: Sage.

Sperling, Susan. 1991. "Baboons with Briefcases: Feminism, Functionalism, and Sociobiology in the Evolution of Primate Gender." *Signs: Journal of Women in Culture and Society* 17 (11):1–27.

Spiro, Melford. 1956. *Kibbutz: Venture in Utopia.* New York: Macmillan.

Spitze, Glenna. 1988. "Women's Employment and Family Relations: A Review." *Journal of Marriage and the Family* 50:585–618.

Spitze, Glenna, John R. Logan, Glenn Deane, and Suzanne Zerger. 1994. "Adult Children's Divorce and Intergenerational Relationships." *Journal of Marriage and the Family* 56 (2) (May): 279–93.

Spitze, Glenna and Russell Ward. 1995. "Household Labor in Intergenerational Households." *Journal of Marriage and the Family* 57 (2) (May): 355–61.

Sprecher, Susan. 1989. "Premarital Sexual Standards for Different Categories of Individuals." *Journal of Sex Research* 26:232–48.

Sprecher, Susan and Kathleen McKinney. 1993. *Sexuality.* Newbury Park, CA: Sage.

Sprecher, Susan, Kathleen McKinney, and Terri Orbuch. 1987. "Has the Double Standard Disappeared?: An Experimental Test." *Social Psychology Quarterly* 50:24–31.

Sprecher, Susan, Kathleen McKinney, Robert Walsh, and Carrie Anderson. 1988. "A Revision of the Reiss Premarital Sexual Permissiveness Scale." *Journal of Marriage and the Family* 50:821–28.

Sprecher, Susan, Quintin Sullivan, and Elaine Hatfield. 1994. "Mate Selection Preferences: Gender Differences Examined in a National Sample." *Journal of Personality and Social Psychology* 66 (6) (June): 1074–81.

Sprenkle, Douglas H. 1989. "The Clinical Practice of Divorce Therapy." Pp. 171–95 in *The Divorce and Divorce Therapy Handbook,* edited by Martin Textor. Northvale, NJ: Jason Aronson.

Stacey, Judith. 1990. *Brave New Families: Stories of Domestic Upheaval in Late Twentieth Century America.* New York: Basic Books.

———. 1993. "Good Riddance to 'The Family': A Response to David Popenoe." *Journal of Marriage and the Family* 55 (3) (Aug.): 545–47.

Stacey, William A., Lonnie R. Hazlewood, and Anson Shupe. 1994. *The Violent Couple.* Westport, CN: Praeger.

Stack, Carol B. 1974. *All Our Kin: Strategies for Survival.* New York: Harper & Row.

Stack, Steven. 1990. "New Micro-level Data on the Impact of Divorce on Suicide, 1959–1980: A Test of Two Theories." *Journal of Marriage and the Family* 52 (1) (Feb.):119–27.

———. 1994. "The Effect of Geographic Mobility on Premarital Sex." *Journal of Marriage and the Family* 56 (1) (Feb.): 204–8.

Stalcup, Brenda, ed. 1995. *Human Sexuality: Opposing Viewpoints.* San Diego: Greenhaven.

Stanley v. *Illinois.* 1972. 405 U.S. 645, 92 S.Ct. 1208, 31 L.Ed.2d 551.

Staples, Robert. 1972. "The Sexuality of Black Women." *Sexual Behavior* 2:4–15.

———. 1981. "Black Singles in America." Pp. 40–51 in *Single Life,* edited by Peter J. Stein. New York: St. Martin's.

———. 1985. "Changes in Black Family Structure: The Conflict Between Family Ideology and Structural Conditions." *Journal of Marriage and the Family* 47:1005–13.

———. 1991. "Changes in Black Family Structure: The Conflict Between Family Ideology and Structural Conditions." Pp. 28–36 in *The Black Family: Essays and Studies.* 4th ed., edited by Robert Staples. Belmont, CA: Wadsworth.

———. 1994. *The Black Family: Essays and Studies.* 5th ed. Belmont, CA: Wadsworth.

Staples, Robert and Leanor Boulin Johnson. 1993. *Black Families at the Crossroads: Challenges and Prospects.* San Francisco: Jossey-Bass.

Staples, Robert and Alfredo Mirande. 1980. "Racial and Cultural Variations Among American Families: A Decennial Review of the Literature on Minority Families." *Journal of Marriage and the Family* 42:887–903.

Staub, Ervin, Daniel Bar-Tal, Jerzy Karylowski, and Janusz Reykowski. 1984. *Development and Maintenance of Prosocial Behavior: International Perspectives on Positive Morality.* New York: Plenum.

Stein, Peter J. 1976. *Single.* Englewood Cliffs, NJ: Prentice-Hall.

———, ed. 1981. *Single Life: Unmarried Adults in Social Context.* New York: St. Martin's.

Steinberg, L. 1990. "Interdependency in the Family: Autonomy, Conflict, and Harmony in the Parent–Adolescent Relationship." In *At the Threshold: The Developing Adolescent,* edited by S. Feldman and G. Elliot. Cambridge, MA: Harvard University Press.

Steinem, Gloria. 1992. *Revolution from Within: A Book of Self-Esteem.* Boston: Little, Brown.

Steinman, Susan B. 1981. "The Experience of Children in a Joint Custody Arrangement: A Report of a Study." *American Journal of Orthopsychiatry* 51:403–14.

Steinmetz, Suzanne K. 1977. *The Cycle of Violence: Assertive, Aggressive, and Abusive Family Interactions.* New York: Praeger.

Stephen, Timothy D. 1985. "Fixed Sequence and Circular-Causal Models of Relationship Development: Divergent Views on the Role of Communication in Intimacy." *Journal of Marriage and the Family* 47:955–63.

Stephens, William N. 1963. *The Family in Cross-Cultural Perspective.* New York: Holt, Rinehart & Winston.

Stern, Gabriella. 1991. "Young Women Insist on Career Equality, Forcing the Men in Their Lives to Adjust." *The Wall Street Journal,* Sept. 16, pp. B1, B3.

Stern, Linda. 1994. "Money Watch: Divorce." *Newsweek,* June 6, p. 58.

Sternberg, Robert J. 1988a. "Triangulating Love." Pp. 119–38 in *The Psychology of Love,* edited by Robert J. Sternberg and Michael L. Barnes. New Haven, CN: Yale University Press.

———. 1988b. *The Triangle of Love: Intimacy, Passion, Commitment.* New York: Basic Books.

Stets, Jan E. 1990. "Verbal and Physical Aggression in Marriage." *Journal of Marriage and the Family* 52 (2) (May): 501–14.

———. 1991. "Cohabiting and Marital Aggression: The Role of Social Isolation." *Journal of Marriage and the Family* 53 (3) (Aug.): 669–80.

Stevens, Amy and Milo Geyelin. 1990. "Surrogate Mother Denied Parental Rights." *The Wall Street Journal,* Oct. 23.

Stewart, Mary White. 1984. "The Surprising Transformation of Incest: From Sin to Sickness." Paper presented at the annual meeting of the Midwest Sociological Society, Chicago, Apr. 18.

Stinnett, Nick. 1979. "In Search of Strong Families." Pp. 23–30 in *Building Family Strengths: Blueprints for Action,* edited by Nick Stinnett, Barbara Chesser, and John DeFrain. Lincoln: University of Nebraska Press.

———. 1983. "Strong Families." Pp. 27–38 in *Prevention in Family Services,* edited by David R. Mace. Newbury Park, CA: Sage.

Stinson, Kandi M., Judith N. Lasker, Janet Lohmann, and Lori J. Toedter. 1992. "Parents' Grief Following Pregnancy Loss: A Comparison of Mothers and Fathers." *Family Relations* 41:218–23.

Stoddard, Thomas B. 1989. "Why Gay People Should Seek the Right to Marry." *OUT/LOOK, National Gay and Lesbian Quarterly* 6 (Fall).

Stone, Lawrence. 1980. *The Family, Sex, and Marriage in England, 1500–1800.* New York: Harper & Row.

Stout, Hilary. 1992. "Adequacy of Spending on AIDS Is an Issue Not Easily Resolved." *The Wall Street Journal,* Apr. 22, pp. A1, A6.

Straus, Murray. 1964. "Power and Support Structures of the Family in Relation to Socialization." *Journal of Marriage and the Family* 26:318–26.

Straus, Murray A. and Richard Gelles. 1986. "Societal Change and Change in Family Violence from 1975 to 1985 as Revealed by Two National Surveys." *Journal of Marriage in the Family* 48:465–79.

———. 1988. "How Violent Are American Families? Estimates from the National Family Violence Resurvey and Other Studies." Pp. 14–36 in *Family Abuse and Its Consequences: New Directions in Research,* edited by Gerald T. Hotaling, David Finkelhor, John T. Kirkpatrick, and Murray A. Straus. Newbury Park, CA: Sage.

Straus, Murray A., Richard J. Gelles, and Suzanne K. Steinmetz. 1980. *Behind Closed Doors: Violence in the American Family.* New York: Doubleday.

Strauss, Anselm and Barney Glaser. 1975. *Chronic Illness and the Quality of Life.* St. Louis: Mosby.

Stubblefield, P. G., R. R. Monson, S. C. Schoenbaum, C. E. Wolfson, D. J. Cookson, and K. C. Ryan. 1984. "Fertility After Induced Abortion: A Prospective Follow-up Study." *Obstetrics and Gynecology* 62:186.

"Study Sees Little Distress After Abortion." 1990. *New York Times,* Apr. 6.

"Suicides Tied with AIDS on the Rise." 1990. *Omaha World-Herald,* Apr. 5.

Suitor, J. Jill and Karl Pillemer. 1994. "Family Caregiving and Marital Satisfaction: Findings from a 1-Year Panel Study of Women Caring for Parents with Dementia." *Journal of Marriage and the Family* 56 (3) (Aug.): 681–90.

Sullivan, Andrew. 1995. *Virtually Normal: An Argument About Homosexuality.* New York: Knopf.

Sullivan, Ronald. 1992. "Judge Says Lesbian Can Adopt Companion's Child." *New York Times,* Jan. 31.

Suro, Roberto. 1992. "Generational Chasm Leads to Cultural Turmoil for Young Mexicans in U.S." *New York Times,* Jan. 20.

Surra, Catherine A. 1990. "Research and Theory on Mate Selection and Premarital Relationships in the 1980s." *Journal of Marriage and the Family* 52 (Nov.): 844–65.

"Surrogacy for Pay Is Banned." 1992. *The Wall Street Journal,* June 29.

Swartz, Mimi. 1992. "Love and Hate at Texas A & M." *Texas Monthly,* Feb., pp. 64–71.

Sweet, James, Larry Bumpass, and Vaughn Call. 1988. *The Design and Content of the National Survey of Families and Households* (Working Paper NSFH-1). Madison: University of Wisconsin, Center for Demography and Ecology.

Swim, Janet K. 1994. "Perceived Versus Meta-analytic Effect Sizes: An Assessment of the Accuracy of Gender Stereotypes." *Journal of Personality and Social Psychology* 66 (1) (Jan.): 21–37.

Szasz, Thomas S. 1976. *Heresies.* New York: Doubleday/Anchor.

T

Taffel, Selma. 1987. "Characteristics of American Indian and Alaska Native Births: United States, 1984." *Monthly Vital Statistics Report* 36 (3), Suppl., U.S. National Center for Health Statistics, June 19.

Tan, Amy. 1989. *The Joy Luck Club.* New York: Ivy Books.

Tanfer, Koray and Lisa A. Cubbins. 1992. "Coital Frequency Among Single Women: Normative Constraints and Situational Opportunities." *The Journal of Sex Research* 29 (2):221–50.

Tannen, Deborah. 1990. *You Just Don't Understand.* New York: Morrow.

Tanouye, Elyse. 1992a. "Abortion Pill Battle Moves to State Level." *The Wall Street Journal,* Apr. 6, pp. B1, B9.

———. 1992b. "Abortion-Rights Forces Plan to Pursue Return of Pills Despite Justices' Ruling." *The Wall Street Journal,* July 20.

Tavris, Carol. 1992. *The Mismeasure of Woman.* New York: Simon & Schuster.

Taylor, Robert J., Linda M. Chatters, M. Belinda Tucker, and Edith Lewis. 1990. "Developments in Research on Black Families: A Decade Review." *Journal of Marriage and the Family* 52 (Nov.): 993–1014.

———. 1991. "Developments in Research on Black Families: A Decade Review." Pp. 274–96 in *Contemporary Families Looking Forward, Looking Back,* edited by Alan Booth. Minneapolis: National Council on Family Relations.

Taylor, Robert Joseph. 1986. "Receipt of Support from Family Among Black Americans: Demographic and Familial Differences." *Journal of Marriage and the Family* 48:67–77.

Teachman, Jay D. 1983. "Early Marriage, Premarital Fertility, and Marital Dissolution." *Journal of Family Issues* 4:105–26.

———. 1991. "Who Pays? Receipt of Child Support in the United States." *Journal of Marriage and the Family* 53 (3) (Aug.): 759–72.

Terrelonge, Pauline. 1995. "Feminist Consciousness and Black Women." Pp. 607–16 in *Women: A Feminist Perspective.* 5th ed., edited by Jo Freeman. Mountain View, CA: Mayfield.

Teti, Douglas M. and Michael Lamb. 1989. "Socioeconomic and Marital Outcomes of Adolescent Marriage, Adolescent Childbirth, and Their Co-occurrence." *Journal of Marriage and the Family* 51:203–12.

Textor, Martin R. 1989. *The Divorce and Divorce Therapy Handbook.* Northvale, NJ: Jason Aronson.

Thomas, Alexander, Stella Chess, and Herbert G. Birch. 1968. *Temperament and Behavior Disorders in Children.* New York: New York University Press.

Thomas, Amanda and Rex Forehand. 1993. "The Role of Paternal Variables in Divorced and Married Families." *American Journal of Orthopsychiatry* 63 (1) (Jan.): 154–68.

Thomas, Darwin and Jean Edmondson Wilcox. 1987. "The Rise of Family Theory: A Historical-Critical Analysis." Pp. 103–24 in *Handbook of Marriage and the Family,* edited by Marvin B. Sussman and Suzanne K. Steinmetz. New York: Plenum.

Thomas, Rich. 1991. "Middle-Class Blessings: Another View." *Newsweek,* Nov. 4, p. 25.

Thomas, Veronica G. 1990. "Determinants of Global Life Happiness and Marital Happiness in Dual-Career Black Couples." *Family Relations* 39 (2) (Apr.): 174–78.

Thompson, Linda. 1991. "Family Work: Women's Sense of Fairness." *Journal of Family Issues* 12 (2) (June): 181–96.

Thompson, Linda and Alexis J. Walker. 1991. "Gender in Families." Pp. 76–102 in *Contemporary Families: Looking Forward, Looking Back,* edited by Alan Booth. Minneapolis: National Council on Family Relations.

Thomson, Elizabeth and Ugo Colella. 1992. "Cohabitation and Marital Stability: Quality or Commitment?" *Journal of Marriage and the Family* 54 (May): 259–67.

Thomson, Elizabeth, Sara S. McLanahan, and Roberta Braun Curtin. 1992. "Family Structure, Gender, and Parental Socialization." *Journal of Marriage and the Family* 54 (2) (May): 368–78.

Thorne, Barrie. 1992. "Girls and Boys Together . . . But Mostly Apart: Gender Arrangements in Elementary School." Pp. 108–23 in *Men's Lives.* 2d ed., edited by Michael S. Kimmel and Michael A. Messner. New York: Macmillan.

Thornton, Arland. 1991. "Influence of the Marital History of Parents on the Marital and Cohabitational Experiences of Children." *American Journal of Sociology* 96 (4) (Jan.): 868–94.

Thornton, Arland and Deborah Freedman. 1983. "The Changing American Family." *Population Bulletin* 38. Washington, DC: Population Reference Bureau.

Thurman, Judith. 1982. "The Basics: Chodorow's Theory of Gender." *Ms.* (Sept.): 35–36.

Tiger, Lionel. 1969. *Men in Groups.* New York: Vintage.

———. 1978. "Omnigamy: The New Kinship System." *Psychology Today* 12:14.

Tilly, Louise A. and Joan W. Scott. 1978. *Women, Work, and Family.* New York: Holt, Rinehart & Winston.

Tolstoy, Leo. 1911. [1869] *War and Peace.* New York: Dutton.

Torres, Aida and Jacqueline Darroch Forrest. 1988. "Why Do Women Have Abortions?" *Family Planning Perspectives* 20:169–76.

Toufexis, Anastasia. 1990. "Sex Lives and Videotape." *Time,* Oct. 20, p. 104.

———. 1991. "Innocent Victims." *Time,* May 13, pp. 56–60.

Treas, Judith. 1995. "Older Americans in the 1990s and Beyond." *Population Bulletin* 50 (2) (May).

Tribe, Lawrence H. 1990. *Abortion: The Clash of Absolutes.* New York: Norton.

Triedman, Kim. 1989. "A Mother's Dilemma." *Ms.,* July–Aug., pp. 59–64.

Troiden, Richard R. 1988. *Gay and Lesbian Identity: A Sociological Analysis.* New York: General Hall.

Troll, Lillian E. 1985. "The Contingencies of Grandparenting." Pp. 135–50 in *Grandparenthood,* edited by Vern L. Bengtson and Joan F. Robertson. Newbury Park, CA: Sage.

Troll, Lillian E., Sheila J. Miller, and Robert C. Atchley. 1979. *Families in Later Life.* Belmont, CA: Wadsworth.

Troost, Kay Michael and Erik Filsinger. 1993. "Emerging Biosocial Perspectives on the Family." Pp. 677–710 in *Sourcebook of Family Theories and*

Methods: A Contextual Approach, edited by Pauline G. Boss, William J. Doherty, Ralph LaRossa, Walter R. Schumm, and Suzanne K. Steinmetz. New York: Plenum.

Trost, Cathy. 1992. "To Cut Costs and Keep the Best People, More Concerns Offer Flexible Work Plans." *The Wall Street Journal,* Feb. 18, p. B1.

Trost, Cathy and Carol Hymowitz. 1990. "Careers Start Giving in to Family Needs." *The Wall Street Journal,* June 18, p. B1.

Trussell, James. 1988. "Teenage Pregnancy in the United States." *Family Planning Perspectives* 20:262–72.

Trzcinski, Eileen and Matia Finn-Stevenson. 1991. "A Response to Arguments Against Mandated Parental Leave: Findings from the Connecticut Survey of Parental Leave Policies." *Journal of Marriage and the Family* 53 (2) (May): 445–60.

Tschann, Jeanne M., Janet R. Johnston, Marsha Kline, and Judith S. Wallerstein. 1989. "Family Process and Children's Functioning During Divorce." *Journal of Marriage and the Family* 51 (2) (May): 431–44.

Tucker, Judith E., ed. 1993. *Arab Women: Old Boundaries, New Frontiers.* Bloomington: Indiana University Press.

Turner, Barbara F. and Catherine Adams. 1988. "Reported Change in Adult Sexual Activity over the Adult Years." *Journal of Sex Research* 25:289–303.

Turner, Ralph H. 1976. "The Real Self: From Institution to Impulse." *American Journal of Sociology* 81 (5):989–1016.

Turner, R. Jay and William R. Avison. 1985. "Assessing Risk Factors for Problem Parenting: The Significance of Social Support." *Journal of Marriage and the Family* 47 (4):881–92.

U

Udry, J. Richard. 1974. *The Social Context of Marriage.* 3d ed. Philadelphia: Lippincott.

———. 1994. "The Nature of Gender." *Demography* 31 (4) (Nov.): 561–73.

Uhlenberg, Peter. 1980. "Death and the Family." *Journal of Family History* 5 (Fall): 313–20.

U.S. Bureau of the Census. 1986. *Statistical Abstract of the United States, 1986.* Washington, DC: U.S. Government Printing Office.

———. 1988. "Households, Families, Marital Status, and Living Arrangements, March 1988: Advance Report." *Current Population Reports,* Series P-20, No. 432. Washington, DC: U.S. Government Printing Office.

———. 1989a. *Statistical Abstract of the United States,* 109th ed. Washington, DC: U.S. Government Printing Office.

———. 1989b. "Fertility of American Women: June 1988." *Current Population Reports,* Population Characteristics Series P-20, No. 436. Issued May 1989.

———. 1989c. *Stepchildren and Their Families.* Washington, DC: U.S. Bureau of the Census.

———. 1991a. "Population Profile of the United States 1991." *Current Population Reports,* Special Studies, Series P-23, No. 173. Washington, DC: U.S. Government Printing Office.

———. 1991b. *Statistical Abstract of the United States 1991,* 111th ed. Washington, DC: U.S. Government Printing Office.

———. 1993. "Money Income of Households, Families and Persons in the United States: 1992." *Current Population Reports,* Series P-60. No. 184. Washington, DC: U.S. Government Printing Office.

———. 1994a. *Statistical Abstract of the United States.* Washington, DC: U.S. Government Printing Office.

———. 1994b. "How We're Changing: Demographic State of the Nation 1995." *Current Population Reports,* Special Studies Series P-23, No. 188. Washington, DC: U.S. Government Printing Office.

———. 1995. *Statistical Abstract of the United States.* Washington, DC: U.S. Government Printing Office.

U.S. Bureau of Labor Statistics. 1986. "Employment in Perspective: Women in the Labor Force, Third Quarter 1986." Report 733. Washington, DC: U.S. Department of Labor.

———. 1990. "Employment in Perspective: Women in the Labor Force, First Quarter 1990." Report 786. Washington, DC: U.S. Bureau of Labor Statistics.

———. 1994. *Employment in Perspective: Women in the Labor Force.* Report 872. Washington, DC: U.S. Department of Labor.

———. 1995a. *The Employment Situation: December 1994.* Washington, DC: U.S. Department of Labor.

———. 1995b. *Employment in Perspective: Women in the Labor Force.* Report 889. Washington, DC: U.S. Department of Labor.

U.S. Centers for Disease Control. 1992. *HIV/AIDS Surveillance: First Quarter Edition.* (Apr.) Atlanta: U.S. Department of Health and Human Services, Centers for Disease Control.

———. 1993. "1993 Sexually Transmitted Diseases Treatment Guidelines." Atlanta: U.S. Department of Health and Human Services, Centers for Disease Control.

U.S. Department of Health, Education, and Welfare. 1975. *Child Abuse and Neglect: Volume I, An Overview of the Problem.* Publication #(OHD) 75-30073. Washington, DC: U.S. Government Printing Office.

U.S. Department of Health, Education, and Welfare, National Institute of Mental Health. 1978. *Yours, Mine, and Ours: Tips for Stepparents.* Washington, DC: U.S. Government Printing Office.

U.S. Department of Justice. 1994. "Violence Between Intimates." *Bureau of Justice Statistics Selected Findings: Domestic Violence.* Nov. No. NCJ-149259. Washington, DC: U.S. Department of Justice, Office of Justice Programs.

U.S. National Center for Health Statistics. 1984. "Advance Report of Final Natality Statistics, 1982." *Monthly Vital Statistics Report* 33(6), Suppl. Sept. 28.

———. 1985. "Advance Report of Final Natality Statistics, 1983." *Monthly Vital Statistics Report* 34(6), Suppl. Sept. 20.

———. 1987. "Advance Report of Final Natality Statistics, 1985." *Monthly Vital Statistics Report* 36(4), Suppl., July 17.

———. 1989. "Advance Report of Final Natality Statistics, 1987." *Monthly Vital Statistics Report* 38(3), Suppl., June 29.

———. 1990a. "Advance Report of Final Divorce Statistics, 1987." *Monthly Vital Statistics Report* 38(12), Suppl. 1, May 15.

———. 1990b. "Advance Report of Final Marriage Statistics, 1987." *Monthly Vital Statistics Report* 38(12), Suppl., Apr. 3.

———. 1990c. "Advance Report of Final Natality Statistics, 1988." *Monthly Vital Statistics Report* 39(4), Suppl., Aug. 15.

———. 1991a. "Advance Report of Final Divorce Statistics, 1988." *Monthly Vital Statistics Report* 39(12), Suppl. 2, May 21.

———. 1991b. "Advance Report of Final Marriage Statistics, 1988." *Monthly Vital Statistics Report* 40(4), Suppl., Aug. 26.

———. 1991c. "Annual Summary of Births, Marriages, Divorces, and Deaths: United States, 1990." *Monthly Vital Statistics Report,* 39(13), Aug. 18.

———. 1994. "Annual Summary of Births, Marriages, Divorces, and Deaths: United States, 1993." *Monthly Vital Statistics Report* 42 (13). Hyattsville, MD: Public Health Service.

"U.S. Scraps Study of Teen-age Sex." 1991. *New York Times,* July 25.

V

Valdivieso, Rafael and Cary Davis. 1988. *U.S. Hispanics: Challenging Issues for the 1990s. Population Bulletin* 17, in *Population Trends and Public Policy Series,* Dec.

Van den Haag, Ernest. 1974. "Love or Marriage." Pp. 134–42 in *The Family: Its Structures and Functions.* 2d ed., edited by Rose Laub Coser. New York: St. Martin's.

Vander Zanden, James W. 1981. *Social Psychology.* 2d ed. New York: Random House.

VanLear, C. Arthur. 1992. "Marital Communication Across the Generations: Learning and Rebellion, Continuity and Change." *Journal of Social and Personal Relationships* 9:103–23.

Vannoy, Dana. 1991. "Social Differentiation, Contemporary Marriage, and Human Development." *Journal of Family Issues* 12:251–67.

Vannoy, Dana and William W. Philliber. 1992. "Wife's Employment and Quality of Marriage." *Journal of Marriage and the Family* 54 (2) (May): 387–98.

Vannoy-Hiller, Dana and William W. Philliber. 1989. *Equal Partners: Successful Women in Marriage.* Newbury Park, CA: Sage.

Vaughan, Diane. 1986. *Uncoupling: Turning Points in Intimate Relationships.* New York: Oxford University Press.

Vazquez-Nuttall, E., I. Romero-Garcia, and B. DeLeon. 1987. "Sex Roles and Perceptions of Femininity and Masculinity of Hispanic Women: A Review of the Literature." *Psychology of Women Quarterly* 11:409–25.

Vega, William A. 1990. "Hispanic Families in the 1980s: A Decade of Research." *Journal of Marriage and the Family* 52 (4) (Nov.): 1015–24.

Vemer, Elizabeth, Marilyn Coleman, Lawrence H. Ganong, and Harris Cooper. 1989. "Marital Satisfaction in Remarriage: A Meta-analysis." *Journal of Marriage and the Family* 51:713–25.

Ventura, Stephanie J. 1988. "Births of Hispanic Parentage, 1985." *Monthly Vital Statistics Report* 36 (11), Suppl., Feb. 26.

Ventura, Stephanie J., Joyce A. Martin, Selma M. Taffel, T. J. Mathews, and Sally C. Clarke. 1995. "Advance Report of Final Natality Statistics, 1993." National Center for Health Statistics, *Monthly Vital Statistics Report* 44 (3). Suppl., Sept. 21.

Viscott, David, 1976. *How to Live with Another Person.* New York: Random House.

Visher, Emily B. and John S. Visher. 1979. *Stepfamilies: A Guide to Working with Stepparents and Stepchildren.* New York: Brunner/Mazel.

Vogel, Ezra F. and Norman W. Bell. 1960. "The Emotionally Disturbed Child as Family Scapegoat." Pp. 382–97 in *Modern Introduction to the Family,* edited by Norman W. Bell and Ezra F. Vogel. Glencoe, IL: Free Press.

W

Wade, Nicholas. 1994. "How Men and Women Think." *New York Times Magazine,* June 12, p. 32.

Wadman, Meredith K. 1992. "Mothers Who Take Extended Time Off Find Their Careers Pay a Heavy Price." *The Wall Street Journal,* July 16, pp. B1–B2.

Wagner, R. M. 1988. "Changes in Extended Family Relationships for Mexican American and Anglo Single Mothers." Pp. 158–73 in *Minority and Ethnic Issues in the Divorce Process,* edited by C. A. Everett. New York: Haworth Press.

Waite, Linda J. 1995. "Does Marriage Matter?" *Demography* 32 (4) (November): 483–507.

Waite, Linda J., Frances Kobrin Goldscheider, and Christina Witsberger. 1986. "Nonfamily Living and the Erosion of Traditional Family Orientations Among Young Adults." *American Sociological Review* 51:541–54.

Waite, Linda J. and Lee A. Lillard. 1991. "Children and Marital Disruption." *American Journal of Sociology* 96 (4) (Jan.): 930–53.

Waldman, Steven and Lincoln Caplan. 1994. "The Politics of Adoption." *Newsweek,* Mar. 21, pp. 64–65.

Walker, Lenore E. 1988. "The Battered Woman Syndrome." Pp. 139–48 in *Family Abuse and Its Consequences: New Directions in Research,* edited by Gerald T. Hotaling, David Finkelhor, John T. Kirkpatrick, and Murray A. Straus. Newbury Park, CA: Sage.

Walker, Lou Ann. 1985. "When a Parent Is Disabled." *New York Times,* June 20.

Wallace, Michele. 1995. "For Whom the Bell Tolls: Why America Can't Deal with Black Feminist Intellectuals." Village Voice Literary Supplement. November: 19–24.

Wallace, Pamela M. and Ian H. Gotlib. 1990. "Marital Adjustment During the Transition to Parenthood: Stability and Predictors of Change." *Journal of Marriage and the Family* 52:21–29.

Waller, Willard. 1937. "The Rating and Dating Complex." *American Sociological Review* 2:727–34.

———. 1951. *The Family: A Dynamic Interpretation.* New York: Dryden. (Revised by Reuben Hill.)

Wallerstein, Judith and Sandra Blakeslee. 1989. *Second Chances: Men, Women, and Children a Decade After Divorce.* New York: Ticknor & Fields.

———. 1995. *The Good Marriage: How and Why Love Lasts.* Boston: Houghton Mifflin.

Wallerstein, Judith and Joan Kelly. 1980. *Surviving the Break-Up: How Children Actually Cope with Divorce.* New York: Basic Books.

Wallis, Claudia. 1989. "Onward, Women!" *Time,* Dec. 4, pp. 80–89.

Walster, Elaine H. 1965. "The Effects of Self-Esteem on Romantic Liking." *Journal of Experimental and Social Psychology* 1:184–97.

Walster, Elaine, E. Aronson, D. Abrams, and L. Rottman. 1966. "Importance of Physical Attractiveness in Dating Behavior." *Journal of Personality and Social Psychology* 4:508–16.

Walster, Elaine and G. William Walster. 1978. *A New Look at Love.* Reading, MA: Addison-Wesley.

Ward, Angela. 1988. "A Feminist Mystique." *Newsweek,* Sept. 12, pp. 8–9.

Warner, Gary A. 1992. "Statistics Needed to Determine Scope of Military Rape Problem." *The Orange County Register,* reprinted in *Omaha World-Herald,* July 13, p. 8.

Warshaw, Robin. 1995. "I Never Called It Rape." Pp. 253–58 in *Debating Sexual Correctness,* edited by Adele M. Stan. New York: Delta.

Watson, K. 1986. "Birth Families: Living with the Adoption Decision." *Public Welfare,* pp. 5–10.

Watson, Roy E. L. and Peter W. DeMeo. 1987. "Premarital Cohabitation Versus Traditional Courtship and Subsequent Marital Adjustment: A Replication and Follow-Up." *Family Relations* 36:193–97.

Watson, Russell. 1984. "What Price Day Care?" *Newsweek,* Sept. 10, pp. 14–21.

Weber, Max. 1948. *The Theory of Social and Economic Organization,* edited by Talcott Parsons. New York: Free Press.

Webster-Stratton, Carolyn. 1992. "Individually Administered Video Parent Training: Who Benefits?" *Cognitive Therapy and Research* 16:31–52.

Webster v. *Reproductive Health Services.* 1989. 851 F.2d 1071.

Weeks, Jeffrey. 1985. *Sexuality and Its Discontents: Meanings, Myths, and Modern Sexualities.* London: Routledge & Kegan Paul.

Weeks, John R. 1992. *Population: An Introduction to Concepts and Issues.* 5th ed. Belmont, CA: Wadsworth.

Weigert, Andrew and Ross Hastings. 1977. "Identity Loss, Family and Social Change." *American Journal of Sociology* 28:1171–85.

Weinberg, Martin S. and Colin J. Williams. 1988. "Black Sexuality: A Test of Two Theories." *Journal of Sex Research* 25:197–218.

Weingarten, Helen R. 1985. "Marital Status and Well-Being: A National Study Comparing First-Married, Currently Divorced, and Remarried Adults." *Journal of Marriage and the Family* 47:653–62.

Weiss, Robert S. 1975. *Marital Separation: Managing After a Marriage Ends.* New York: Basic Books.

———. 1987. "Men and Their Wives' Work." Pp. 109–21 in *Spouse, Parent, Worker: On Gender and Multiple Roles,* edited by F. J. Crosby. New Haven, CT: Yale University Press.

Weitzman, Lenore J. 1981. *The Marriage Contract: Spouses, Lovers, and the Law.* New York: Free Press.

———. 1985. *The Divorce Revolution: The Unexpected Social and Economic Consequences for Women and Children in America.* New York: Free Press.

Wellington, Alison J. 1994. "Accounting for the Male/Female Wage Gap Among Whites: 1976 and 1985." *American Sociological Review.* 56 (6) (Dec.): 839–49.

Wells, Robert V. 1985. *Uncle Sam's Family: Issues in and Perspectives on American Demographic History.* Albany: State University of New York Press.

Wenk, DeeAnn and Patricia Garrett. 1992. "Having a Baby: Some Predictions of Maternal Employment Around Childbirth." *Gender & Society* 6 (1) (Mar.): 49–65.

Wenk, DeeAnn, Constance L. Hardesty, Carolyn S. Morgan, and Sampson Lee Blair. 1994. "The Influence of Parental Involvement on the Well-Being of Sons and Daughters." *Journal of Marriage and the Family* 56 (1) (Feb.): 229–34.

Werner, Emmy E. 1992. "The Children of Kauai: Resilience and Recovery in Adolescence and Adulthood." *Journal of Adolescent Health* 13:262–68.

Werner, Emmy E. and Ruth S. Smith. 1992. *Overcoming the Odds: High Risk Children from Birth to Adulthood.* Ithaca, NY: Cornell University Press.

"We the American Asians." 1993. U.S. Bureau of the Census, Department of Commerce. Washington, DC. Issued Sept.

"We the American Elderly." 1993. U.S. Bureau of the Census, Department of Commerce. Washington, DC. Issued Sept.

"We the American Hispanics." 1993. U.S. Bureau of the Census, Department of Commerce. Washington, DC. Issued Nov.

Weston, Kath. 1991. *Families We Choose: Lesbians, Gays, Kinship.* New York: Columbia University Press.

Wexler, Richard. 1991. "Child-Abuse Hysteria Snares Innocent Victims." *The Wall Street Journal,* June 4.

Whitbeck, Les. B., Danny R. Hoyt, and Shirley M. Huck. 1993. "Family Relationship History, Contemporary Parent–Grandparent Relationship Quality, and the Grandparent–Grandchild Relationship." *Journal of Marriage and the Family* 55 (4) (Nov.): 1025–35.

Whitbeck, Les B., Ronald L. Simons, and Meei-Ying Kao. 1994. "The Effects of Divorced Mothers' Dating Behaviors and Sexual Attitudes on the Sexual Attitudes and Behaviors of Their Adolescent Children." *Journal of Marriage and the Family* 56 (3) (Aug.): 615–21.

White, Carol Wayne. 1995. "Toward an Afra-American Feminism." Pp. 529–46 in *Women: A Feminist Perspective.* 5th ed., edited by Jo Freeman. Mountain View, CA: Mayfield.

White, Gregory L. and Paul E. Mullen. 1989. *Jealousy: Theory, Research, and Clinical Strategies.* New York: Guilford.

White, Jack E. 1993. "Growing Up in Black and White." *Time,* May 17, pp. 48–49.

White, Lynn K. 1987. "Freedom Versus Constraint: The New Synthesis." *Journal of Family Issues* 8 (4):468–70.

———. 1990. "Determinants of Divorce: A Review of Research in the Eighties." *Journal of Marriage and the Family* 52 (Nov.): 904–12.

———. 1994. "Growing Up with Single Parents and Stepparents: Long-Term Effects on Family Solidarity." *Journal of Marriage and the Family* 56 (4) (Nov.): 935–48.

White, Lynn K. and Alan Booth. 1985a. "The Transition to Parenthood and Marital Quality." *Journal of Family Issues* 6 (4):435–49.

———. 1985b. "The Quality and Stability of Remarriages: The Role of Stepchildren." *American Sociological Review* 50:689–98.

———. 1991. "Divorce Over the Life Course: The Role of Marital Happiness." *Journal of Family Issues* 12 (1) (Mar.): 5–21.

White, Lynn K., Alan Booth, and John N. Edwards. 1986. "Children and Marital Happiness." *Journal of Family Issues* 7 (2):131–47.

White, Lynn and Bruce Keith. 1990. "The Effect of Shift Work on the Quality and Stability of Marital Relations." *Journal of Marriage and the Family* 52:453–62.

White, Lynn K. and Agnes Riedmann. 1992. "When the Brady Bunch Grows Up: Step/Half- and Full-sibling Relationships in Adulthood." *Journal of Marriage and the Family* 54 (1) (Feb.): 197–208.

Whiting, B. and C. P. Edwards. 1988. *Children of Different Worlds: The Formation of Social Behavior.* Cambridge, MA: Harvard University Press.

Whitman, David and Dorian Friedman. 1994. "The White Underclass." *U.S. News & World Report,* Oct. 17, pp. 40–53.

Whittaker, Terri. 1995. "Violence, Gender and Elder Abuse: Towards a Feminist Analysis and Practice." *Journal of Gender Studies* 4 (1): 35–45.

Whyte, Martin King. 1990. *Dating, Mating, and Marriage.* New York: Aldine de Gruyter.

Wilder, H. B. and David A. Chiriboga. 1991. "Who Leaves Whom: The Importance of Control." Pp. 224–47 in *Divorce: Crisis, Challenge or Relief?* edited by David A. Chiriboga and Linda S. Catron. New York: New York University Press.

Wiley, Norbert F. 1985. "Marriage and the Construction of Reality: Then and Now." Pp. 21–32 in *The*

Psychosocial Interior of the Family. 3d ed., edited by Gerald Hantel. Hawthorne, NY: Aldine.

Wilkie, Jane Riblett. 1988. "Marriage, Family Life, and Women's Employment." Pp. 149–66 in *Women Working.* 2d ed., edited by Ann Helton Stromberg and Shirley Harkess. Mountain View, CA: Mayfield.

———. 1991. "The Decline in Men's Labor Force Participation and Income and the Changing Structure of Family Economic Support." *Journal of Marriage and the Family* 53 (1) (Feb.): 111–22.

Wilkinson, Doris. 1993. "Family Ethnicity in America." Pp. 15–59 in *Family Ethnicity: Strength in Diversity,* edited by Harriette Pipes McAdoo. Newbury Park, CA: Sage.

William Petschek National Jewish Family Center of the National Jewish Committee. 1986. "Intermarriage." *Newsletter* 6 (1).

Williams, Kirk R. 1992. "Social Sources of Marital Violence and Deterrence: Testing an Integrated Theory of Assaults Between Partners." *Journal of Marriage and the Family* 54 (3) (Aug.): 620–29.

Williams, Lena. 1989. "Teen-Age Sex: New Codes Amid Old Anxiety." *New York Times,* Feb. 27.

Williams, Linda S. 1992. "Adoption Actions and Attitudes of Couples Seeking In Vitro Fertilization." *Journal of Family Issues* 13 (1) (Mar.): 99–113.

Williams, Norma. 1990. *The Mexican American Family: Tradition and Change.* Dix Hills, NY: General Hall.

Willie, Charles Vert. 1986. "The Black Family and Social Class." Pp. 224–31 in *The Black Family: Essays and Studies,* edited by Robert Staples. Belmont, CA: Wadsworth.

Wilson, Barbara Foley and Sally Cunningham Clarke. 1992. "Remarriages: A Demographic Profile." *Journal of Family Issues* 13 (2) (June): 123–41.

Wilson, William Julius. 1987. *The Truly Disadvantaged: The Inner City, The Underclass, and Public Policy.* Chicago: University of Chicago Press.

Winch, Robert F. 1958. *Mate Selection: A Study of Complementary Needs.* New York: Harper & Row.

Wineberg, Howard. 1990. "Childbearing After Remarriage." *Journal of Marriage and the Family* 52:31–38.

———. 1994. "Marital Reconciliation in the United States: Which Couples Are Successful?" *Journal of Marriage and the Family* 56 (1) (Feb.): 80–88.

Wingert, Pat and Steven Waldman. 1995. "Did Washington Hate Gays? Fighting to Take Homosexual

History Out of the Closet." *Newsweek,* Oct. 16, p. 81.

Winslow, Ron. 1990. "Study Is Good News for People Waiting to Have Children." *The Wall Street Journal,* Mar. 8.

———. 1991. "C-Sections Tied to Economic Factors in Study." *The Wall Street Journal,* Jan. 2.

Wolf, Rosalie S. 1986. "Major Findings from Three Model Projects on Elderly Abuse." Pp. 218–38 in *Elder Abuse: Conflict in the Family,* edited by Karl A. Pillemer and Rosalie S. Wolf. Dover, MA: Auburn House.

Women's Research and Education Institute. 1990. "The American Woman 1990–91." In "Bias, Tradition Seen in Pay Lag of U.S. Women." *Omaha World-Herald,* Apr. 26.

"Women's Summit in Ireland Left Many Delegates Unsatisfied." 1992. *Los Angeles Times,* reprinted in *Omaha World-Herald,* July 13, p. 4.

Woo, Junda. 1992a. "Adoption Suits Target Agencies for Negligence." *The Wall Street Journal,* July 9, pp. A1, A6.

———. 1992b. "Mediation Seen as Being Biased Against Women." *The Wall Street Journal,* Aug. 4, pp. B1, B9.

Woodman, Sue. 1995. "How Teen Pregnancy Has Become a Political Football." *Ms.* (Jan./Feb.): 90–91.

Worchel, Stephen. 1984. "The Darker Side of Helping: The Social Dynamics of Helping and Cooperation." Pp. 379–418 in *Development and Maintenance of Prosocial Behavior: International Perspectives on Positive Morality,* edited by E. Staub et al. New York: Plenum.

Wright, Carol L. and Joseph W. Maxwell. 1991. "Social Support During Adjustment to Later-Life Divorce: How Adult Children Help Parents." *Journal of Divorce and Remarriage* 15 (3/4):21–48.

Wright, Gwendolyn. 1981. *Building the Dream: A Social History of Housing in America.* New York: Pantheon.

Wright, Robert. 1994. *The Moral Animal: Evolutionary Psychology and Everyday Life.* New York: Pantheon.

Wu, Zheng. 1995. "The Stability of Cohabitation Relationships: The Role of Children." *Journal of Marriage and the Family* 57 (1) (Feb.): 231–36.

Wulf, Steve. 1995. "Generation Excluded." *Newsweek,* Oct. 23, p. 86.

Wyatt, Gail Elizabeth and Gloria Johnson Powell, eds. 1988. *Lasting Effects of Child Sexual Abuse.* Newbury Park, CA: Sage.

Wyman, Hastings, Jr. 1994. "When the Love Boat Sails for Honolulu." *The Washington Blade,* Mar. 9, p. 35.

Y

Yankelovich, Daniel A. 1981. *New Rules: Searching for Self-Fulfillment in a World Turned Upside Down.* New York: Random House.

Yarrow, Andrew L. 1987. "Older Parents' Child: Growing Up Special." *New York Times,* Jan. 26.

Yllo, Kersti and Michele Bograd, eds. 1988. *Feminist Perspectives on Wife Abuse.* Newbury Park, CA: Sage.

Yoachum, Susan and Louis Freedberg. 1995. "Boxer Finds Herself in Media Spotlight." *San Francisco Chronicle,* Sept. 8, p. A14.

Yoshihama, Mieko, Asha L. Parekh, and Doris Boyington. 1991. "Dating Violence in Asian/Pacific Communities." Pp. 184–95 in *Dating Violence: Young Women in Danger,* edited by Barrie Levy. Seattle: Seal Press.

Young-Hee, Yoon and Linda J. Waite. 1994. "Converging Employment Patterns of Black, White, and Hispanic Women: Return to Work After First Birth." *Journal of Marriage and the Family* 56 (Feb.): 209–17.

"Young Indians Prone to Suicide." 1992. *New York Times,* Mar. 25.

Z

Zabin, Laurie Schwab, Rebeca Wong, Robin M. Weinick, and Mark R. Emerson. 1992. "Dependency in Urban Black Families Following the Birth of an Adolescent's Child." *Journal of Marriage and the Family* 54 (3) (Aug.): 496–507.

Zablocki v. *Redhail.* 1978. 434 U.S. 374, 54 L.Ed.2d 618, 98 S.Ct. 673.

Zahn-Waxler, Carolyn, ed. 1995. *Sexual Orientation and Human Development.* Special issue of *Developmental Psychology,* vol. 31.

Zelizer, Viviana K. 1985. *Pricing the Priceless Child: The Changing Social Value of Children.* New York: Basic Books.

Acknowledgments

This page constitutes an extension of the copyright page. We have made every effort to trace the ownership of all copyrighted material and to secure permission from copyright holders. In the event of any question arising as to the use of any material, we will be pleased to make the necessary corrections in future printings. Thanks are due to the following authors, publishers, and agents for permission to use the material indicated.

Page 18, Figure 1.2: Figure adapted from Shifting Gears, by Nena O'Neill and George O'Neill, p. 167, 1974. Copyright © 1974 by Nena O'Neill and George O'Neill. Reprinted by permission of the publisher, M. Evans and Company, New York, NY; **Page 29,** Figure 2.1: Adapted from "Human Ecology Theory," by Margaret M. Bubolz and M. Suzanne Sontag, 1993, pp. 432. In Pauline G. Boss, William J. Doherty, Ralph La Rossa, Walter R. Schumm & Suzanne K. Steinmetz, [Eds.], *Sourcebook of Family Theories and Methods: A Contexual Approach.* Copyright © 1993 Plenum Publishers, Inc. Adapted by permission; **Pages 71–72:** Excerpt from *Hard Choices: How Women Decide about Work, Career and Motherhood,* by Kathleen Gerson, 1985, pp. 18–19. University of California Press. Copyright © 1985 The Regents of the University of California. Reprinted by permission; **Pages 75–76:** Excerpt from "The Costs of Being on Top," by Mark E. Kann, *Journal of the National Association for Women Deans, Administrators, and Counselors, 49* (Summer). Copyright © 1986 by the National Association for Women Deans, Administrators, and Counselors. Reprinted by permission; **Pages 77–78:** Taken from the *Working Woman* column by Niki Scott. Copyright 1992 Dist. by Universal Press Syndicate. Reprinted with permission. All rights reserved; **Page 90,** Figure 4.1: Adapted from "Triangulating Love," by Robert J. Sternberg, 1988, fig. 6.1, p. 121. In Robert J. Sternberg and Michael L. Barnes, [Eds.] *The Psychology of Love.* Copyright © 1988 Yale University Press. Adapted by permission; **Pages 94–95:** Excerpt from *Honoring the Self,* by Nanthaniel Branden, pp. 3–4, 1983. Jeremy P. Tarcher, Inc. Copyright © 1983 by Nathaniel Branden. Reprinted by permission of the author; **Page 95,** Box 4.1: Excerpt from *Encounters with the Self,* by Don E. Hamachek, copyright © 1971 by Holt, Rinehart & Winston, Inc. [pp. 248–51] Reprinted by permission of the publisher; **Page 96,** Box 4.2: Excerpt from *Encounters with the Self,* by Don E. Hamachek, copyright © 1971 by Holt, Rinehart & Winston, Inc. [pp. 248–51] Reprinted by permission of the publisher; **Page 99,** Figure 4.2: Adapted from *The Family System in America,* 4th ed., by Ira L. Reiss and G. L. Lee, 1988, p. 103. Copyright © 1988 by Holt, Rinehart & Winston. Reprinted by permission; **Page 100,** Figure 4.3: Adapted from "Interpersonal Exchange in Isolation," by Irwin Altman and William H. Haythorn, 1965, *Sociometry, 28,* 411–26, figure 3. American Sociological Association; **Page 112,** Table 5.1: "How the Public Views Gay Issues," from "Poll Finds an Even Split on Homosexuality's Cause," by Jeffry Schmalz, March 5, 1993. Copyright © 1993 by The New York Times. Reprinted by permission; **Page 116,** Box 5.1: Excerpt from "Lesbian Sex," by Marilyn Frye, 1992, pp. 109–119. In Marilyn Frye, [Ed.] *Essays in Feminism 1976-1992.* Copyright © 1992 by Marilyn Frye. Reprinted by permission of Crossing Press, Freedom, CA; **Pages 118–119:** Excerpt from *School Girls: Young Women, Self-Esteem and the Generation Gap,* by Peggy Orenstein, 1994, p. 61. Copyright © 1994 by Peggy Orenstein. Reprinted by permission of Doubleday, a division of Bantam Doubleday Dell; **Page 122,** Figure 5.1: From "The Incidence and Frequency of Marital Sex in a National Sample," by Vaughn Call, Susan Sprecher & Pepper Schwartz, 1995, pp. 646, *Journal of Marriage and the Family, 57 [3]* (August), 639–652. Copyright © 1995 by National Council on Family Relations. Reprinted by permission of the authors and the National Council on Family Relations, 3989 Central Avenue, N.E. Suite 550, Minneapolis, MN 55421; **Page 126:** Excerpt from *The Significant Americans* by John F. Cuber and Peggy Haroff, [p. 136]. Copyright © 1965 by John F. Cuber. Used by permission of the publisher, Dutton, an imprint of New American Library, a division of Penguin Books USA Inc.; **Page 149,** Table 6.1: From "Racial and Ethnic Differences in the Desire to Marry," by Scott J. South, 1993, *Journal of Marriage and the Family, 55 (2),* 357–370. Copyright © 1993 by the National Council on Family Relations, 3989 Central Avenue, N.E., Suite 550, Minneapolis, MN 55421. Reprinted by permission; **Page 150,** Figure 6.5: From "Leaving and Returning Home in 20th Century America," by Frances Goldscheider and Calvin Goldscheider, 1994, pp. 15, 17, 20, *Population Bulletin, 48 [4]* [March]. Population Reference Bureau. Reprinted by permission; **Page 152,** Box 6.1: Reprinted by permission of Judith Harazin; **Page 156,** Table 6.2: Excerpt from Peter J. Stein [Ed.] *Single Life: Unmarried Adults in Social Context,* 1981, p. 11. Copyright © 1981 by St. Martin's Press. Reprinted with permission of the publisher; **Page 170,** Table 6.6: From "Marital Status and Personal Happiness: An Analysis of Trend Data," by Gary

R. Lee, Karen Seccombe & Constance L. Shehan, 1991, p. 841, *Journal of Marriage and the Family, 53, (4),* 839-844. Copyright © 1991 by National Council on Family Relations. Reprinted by permission of the authors and the National Council on Family Relations, 3989 Central Avenue, N.E. Suite 550, Minneapolis, MN 55421; **Page 172:** Excerpt from *Never Married Women,* by Barbara Levy Simon, 1987, pp. 31–32. Copyright © 1987 by Temple University Press. Reprinted by permission of Temple University Press; **Page 174:** Excerpt from *Never Married Women,* by Barbara Levy Simon, 1987, pp. 71. Copyright © 1987 by Temple University Press. Reprinted by permission of Temple University Press; **Page 190,** Box 7.1: Excerpts from "Formal Intermediaries in the Marriage Market: A Typology and Review," by Aaron C. Ahuvia and Mara B. Adelman, 1992, *Journal of Marriage and the Family,* 54, 2, 452–463. Copyright © 1992 by the National Council on Family Relations. Reprinted by permission of the authors and the National Council on Family Relations, 3989 Central Avenue, N.E. Suite 550, Minneapolis, MN 55421; **Page 195,** Table 7.3: From "Black-American Intermarriages in the United States," by Ernest Porterfield, 1992, pp. 17–34. In Gary Cretser and Joseph J. Leon, [Eds.] *Intermarriages in the United States.* Copyright © 1982 by Haworth Press, Inc. Reprinted by permission. All rights reserved; **Page 198:** Excerpts from *Pairing* by George R. Bach and Ronald M. Deutsch, 1970, pp. 46–47. Reprinted by permission of Wyden Books; **Page 200,** Box 7.2: Pi Kappa Phi Fraternity; Meyer, T. J., 1984, pp. 1, 12; From "Statement of Position on Sexual Abuse" and reproduction of poster reprinted by permission of Pi Kappa Phi Fraternity; *The Chronicle of Higher Education,* 1986; Koss, Gidycz, and Wisneiwski 1987; **Page 203:** Excerpts from "The Courtship Game: Power in The Sexual Encounter," by N. B. McCormick, and C. J. Jessor, 1983, pp. 66–67, 85. In E. R. Allegier and N. B. McCormick [Eds.], *Changing Boundaries: Gender Roles and Sexual Behavior.* Mayfield Publishing Company. Reprinted by permission; **Page 207,** Table 7.4: From "Cohabitation: Does it Make for a Better Marriage?," by Carl A. Ridley, Dan J. Peterman & Arthur W. Avery, 1978, *Family Coordinator,* 27:2, 129–136. Copyright © 1978 by the National Council on Family Relations. Reprinted by permission of the authors and the National Council on Family Relations, 3989 Central Avenue, N.E. Suite 550, Minneapolis, MN 55421; **Page 209,** Table 7.5: Excerpts from "Cohabitation and Marital Stability: Quality or Comitment?," by Elizabeth Thomson and Ugo Colella, 1992, *Journal of Marriage and the Family, 54,* 2, 259–267. Copyright © 1992 by the National Council on Family Relations. Reprinted by permission of the authors and the National Council on Family Relations, 3989 Central Avenue, N.E. Suite 550, Minneapolis, MN 55421; **Page 210,** Box 7.3: Excerpt reprinted with the permission of Simon & Schuster from *Total Loving,* by "J". Copyright © 1977 by Joan Garrity; **Pages 222–224:** Excerpts from *The Significant Americans* by John F. Cuber and Peggy Haroff., [p. 50, 55, 135–136] Copyright © 1965 by John F. Cuber. Used by permission of the publisher, Dutton, an imprint of New American Library, a division of Penguin Books USA Inc.; **Page 255:** Excerpt reprinted with the permission of Simon & Schuster, from *Couples: How to Confront Problems and Maintain Loving Relationships,* by Carlfred B. Broderick, pp. 40-41. Copyright © 1979 by Dr. Carlfred B. Broderick; **Page 274:** Excerpts from *American Couples,* by Philip Blumstein and Pepper W. Schwartz, 1983, p. 141. Copyright © 1983 by Philip Blumstein and Pepper W. Schwartz. Reprinted by permission of William Morrow & Company, Inc.; **Page 278:** Excerpt from *Gender Roles and Power,* by Jean Lipman-Blumen, 1984, pp. 30–31. Copyright © 1984 by Jean Lipman-Blumen. Reprinted by permission of Prentice-Hall, Englewood Cliffs, NJ; **Page 279:** Excerpts from *American Couples,* by Philip Blumstein and Pepper W. Schwartz, 1983, pp. 152–3. Copyright © 1983 by Philip Blumstein and Pepper W. Schwartz. Reprinted by permission of William Morrow & Company, Inc.; **Page 280:** Excerpts from *American Couples,* by Philip Blumstein and Pepper W. Schwartz, 1983, p. 146. Copyright © 1983 by Philip Blumstein and Pepper W. Schwartz. Reprinted by permission of William Morrow & Company, Inc.; **Page 281,** Box 10.2: Excerpt reprinted with the permission of Simon & Schuster, from *Couples: How to Confront Problems and Maintain Loving Relationships,* by Carlfred B. Broderick, pp. 117–23. Copyright © 1979 by Dr. Carlfred B. Broderick; **Page 285:** Excerpt from "Why Men Resist," by William J. Goode, 1982, p. 140. In Barrie Thorne and Marilyn Yalom, [Eds.] *Rethinking the Family: Some Feminist Questions.* Longman Publishing Group. Reprinted by permission of Marilyn Yalom, Stanford University; **Page 288,** Table 10.2: From "How Violent Are American Families," by Murry A. Straus and Richard J. Gelles, 1988, pp. 18–19. In Gerald A. Hotaling [Ed.] *Family Abuse and Its Consequences: New Directions in Research.* Copyright © 1988 by Sage Publications. Reprinted by permission of Sage Publications, Inc.; **Page 309:** Excerpt from *Uncle Sam's Family: Issues and Perspectives on American Democracy,* by Robert V. Wells, 1985, pp 47–50. Copyright © 1985 by the State University of New York Press. Reprinted by permission; **Page 329,** Table 11.1: From Gallup Survey, 1991. Reprinted by permission of Gallup Poll News Service; **Page 337,** Table 11.2: Reproduced with the permission of The Alan Guttmacher Institute from Christine A. Bachrach, Kathy Shepherd Stolley & Kathryn A. London, "Relinquishment of Premarital Births: Evidence from National Survey Data," *Family Planning Perspectives, Vol. 24,* No. 1, January/February 1992; **Page 341:** Excerpt from "Adopting the Older Child: A Personal Reflection," by Steven Hotovy, *America,* May 18, 1991. Reprinted with permission of America Press, Inc., 106 West 56th Street, New York, NY 10019. Copyright © 1991 All rights reserved; **Page 357,** Figure 12.1: From "Single Mothers: The Changing Mix," by Nancy Doniger, October 5, 1992, p. A-16. Copyright © 1992 by *The New York Times* Company. Reprinted by permission; **Page 358,** Table 12.1: Adapted from "However You Slice the Data the Richest Did Get Richer," ["Two Views of Family Income"] by Sylvia Nasar, May 11, 1992, p. C-5. Copyright © 1992 by *The New York Times* Company. Reprinted by permission; **Page 362,** Figure 12.4: Reprinted from Martha R. Burt, *Over the Edge: The Growth of Homelessness in the 1980s,* [p. 35] © 1991 the Russell Sage Foundation. Used with permission of the Russell Sage Foundation; **Page 366:** Excerpts from "A Talk to the Teachers," in *The Price of the Ticket: Collected Nonfiction, 1948–1985,* by James Baldwin, St. Martin's Press, 1985. Reprinted by permission of the James Baldwin Estate; **Page 375,** Box 12.1: Excerpts reprinted with permission of Rawson Associates/Scribner, an imprint of Simon & Schuster from *How to Talk So Kids Will Listen and Listen So Kids Will Talk* by Adele Faber and Elaine Mazlish. Copyright © 1980 by Adele Faber and Elaine Mazlish; **Page 381,** Figure 12.5: From *The New American Grandparent: A Place in the Family,* by Andrew J. Cherlin and Frank F. Furstenberg, Jr. Copyright © 1986 by Basic Books, Inc. Reprinted by permission of A. J. Cherlin; **Page 393,** Table 13.1: Adapted from *Women, Men and Time,* by Beth Ann Shelton, 1992, pp. 39. Greenwood Press/an imprint of Greenwood Publishing Group. Westport, Ct. Copyright © 1992 by Beth Ann Shelton. Reprinted by permission; **Pages 393–394:** From "Men at Work," by Anna Quindlen, February 18, 1990, p. 19. Copyright © 1990 by *The New York Times* Company. Reprinted by permission; **Page 404,** Table 13.2: Adapted from *Women, Men and Time,* by Beth Ann Shelton, 1992, pp. 67–68, Tables 4.2 and 4.3. Greenwood Press/an imprint of Greenwood Publishing Group. Westport, Ct. Copyright © 1992 by Beth Ann Shelton. Reprinted by permission; **Page 405,** Table 13.3: Adapted from "Ethnicity, Race, and Difference: A Comparison of White, Black, and Hispanic Men's Household Labor Time," by Beth Ann Shelton and Daphne John, 1993, pp. 139, 141, Tables 7.1 and 7.2. In Jane C. Hood [Ed.] *Men, Work and Family.* Copyright © 1993 by Sage Publications. Adapted by permission of Sage Publications, Inc.; **Page 406:** Excerpt from *The Second Shift,* by Arlie Hochschild and Ann Machung, p. 9. Copyright © 1989 by Arlie Hochschild. Used by permission of Viking Penguin, a division of Penguin Books USA Inc.; **Page 409:** Excerpt from *The Second Shift,* by Arlie Hochschild and Ann Machung, pp. 9–10. Copyright © 1989 by Arlie Hochschild. Used by permission of Viking Penguin, a division of Penguin Books USA Inc.; **Page 411,** Box 13.1: From "Bringing Up Baby: A Doctor's Prescription for Busy Parents," by T. Berry Brazelton, M.D., in *Newsweek,* February 13, 1989, pp. 68–69. Reprinted by permission of T. Berry Brazelton; **Page 412:** Excerpt from *The Second Shift,* by Arlie Hochschild and Ann Machung, p. 31. Copyright © 1989 by Arlie Hochschild. Used by permission of Viking Penguin, a division of Penguin Books USA Inc.; **Page 413,** Box 13.2: Excerpts from "Child Care at Home: 2-Women, Complex Roles," by Sara Rimer, December 26, 1988. Copyright © 1988 by *The New York Times* Company. Reprinted by permission; **Page 413,** Box 13.2: Excerpt from "The Nanny Trap," by Joanne Lipman, April 14, 1993, pp. A1, A8. Reprinted by permission of the *Wall Street Journal.* Copyright © 1993 Dow Jones & Company Inc. All rights reserved worldwide; **Page 415:** Excerpts from *More Equal than Others: Women and Men in Dual-Career Marriages,* by Rosanna Hertz. Copyright © 1986 The Regents of the University of California. Reprinted by permission of the University of California Press; **Page 416:** Excerpt from "Spouses Find Themselves Worlds Apart as Global Commuter Marriages Increase," by Joann S. Lublin, 1992, pp. B1, B6. Reprinted by permission of the *Wall Street Journal.* Copyright © 1992 Dow Jones & Company Inc. All rights reserved worldwide; **Page 421:** Excerpt from "The 'Mommy Track' Isn't Anti-Woman," by Felice N. Schwartz, March 22, 1989, Op Ed page. Copyright © 1989 by *The New York Times* Company. Reprinted by permission; **Page 426,** Box 13.4: Excerpt from *The Second Shift,* by Arlie Hochschild and Ann Machung. Copyright © 1989 by Arlie Hochschild. Used by permission of Viking Penguin, a division of Penguin Books USA Inc.; **Pages 427–428:** Taken from the *Working Woman* column by Niki Scott. Copyright 1992 Dist. by Universal Press Syndicate. Reprinted with permission. All rights reserved; **Page 438:** Excerpt from *Marriage and the Family,* by Carlfred B. Broderick, p. 353. Reprinted by permission of Prentice-Hall, Inc. Englewood, NJ; **Page 441,** Box 14.2: Excerpt from *Uncle Sam's Family: Issues and Perspectives on American Democracy,* by Robert V. Wells, pp. 1–2. Copyright © 1985 by the State University of New York Press. Reprinted by permission; **Page 447,** Figure 14.3: Adapted from "Families Under Stress" by Donald A. Hansen and Reuben Hill, p. 810. In Harold Christensen (Ed.), *The Handbook of Marriage and the Family.* Copyright © 1964 by Rand McNally & Company. Reprinted by permission; **Page 450,** Box 14.3: Abridgement of "Alcoholism and the Family," by Joan J. Jackson, pp. 90–98, 1958,

Annals of the American Academy of Political and Social Science, 315. Copyright © 1958 by the American Academy of Political and Social Science. Reprinted by permission; **Page 453,** Figure 14.5: Adapted from "Family Stress Theory and Assessment," by H. I. McCubbin and M. A. McCubbin, figure 1.5, p. 16, 1991. In H. I. McCubbin and Annie I. Thompson, *Family Assessment Inventories for Research and Practice,* University of Wisconsin–Madison. Adapted by permission; **Page 459:** Excerpt from "Family Stress Theory and Assessment," by H. I. McCubbin and M. A. McCubbin, p. 19, 1991. In H. I. McCubbin and Annie I. Thompson, *Family Assessment Inventories for Research and Practice,* University of Wisconsin–Madison. Adapted by permission; **Page 468,** Figure 15.2: From "Advance Report of Final Divorce Statistics, 1989 and 1990," by Sally C. Clarke, 1995, *Monthly Vital Statistics Report,* 43 (9), March 22, p. 3. U.S. Department of Health and Human Services, National Center for Health Statistics; **Page 475,** Figure 15.4: From "Advance Report of Final Divorce Statistics, 1989 and 1990," by Sally C. Clarke, 1995, *Monthly Vital Statistics Report,* 43 (9), March 22, p. 4. U.S. Department of Health and Human Services, National Center for Health Statistics; **Pages 478–479:** Excerpt from *Second Chances: Men, Women and Children a Decade After Divorce,* by Judith S. Wallerstein and Sandra Blakeslee, 1989, p. 111. Copyright © 1989 by Judith S. Wallerstein and Sandra Blakeslee. Reprinted by permission of Ticknor & Fields/Houghton Mifflin Co. All rights reserved; **Page 479:** Excerpts from *Ex Familia: Grandparents, Parents, and Children Adjust to Divorce,* by Colleen Leahy Johnson, 1988, pp. 190–191. Copyright © 1988 by Rutgers, The State University. Used with permission of Rutgers University Press; **Page 480,** Box 15.1: Abridged from *How to Survive the Loss of Love: 58 Things To Do When There is Nothing to Be Done,* by Melba Colgrove, Harold Bloomfield & Peter McWilliams, 1978. Copyright © 1976. Used by permission of Bantam Books; **Page 488,** Figure 15.6: From "Advance Report of Final Divorce Statistics, 1989 and 1990," by Sally C. Clarke, 1995, *Monthly Vital Statistics Report,* 43 (9), March 22, p. 2. U.S. Department of Health and Human Services, National Center for Health Statistics; **Page 489,** Figure 15.7: Adapted from *The Difficult Divorce: Therapy for Children and Families,* by Marla Beth Isaacs, Braulio Montalvo & David Abelsohn, 1986, p. 280. Copyright © 1986 by Basic Books, Inc. Reprinted by permission; **Page 490,** Figure 15.8: Adapted from *The Difficult Divorce: Therapy for Children and Families,* by Marla Beth Isaacs, Braulio Montalvo & David Abelsohn, 1986, p. 281. Copyright © 1986 by Basic Books, Inc. Reprinted by permission; **Page 490:** Excerpt from *Second Chances: Men, Women and Children a Decade After Divorce,* by Judith S. Wallerstein and Sandra Blakeslee, 1989, p. 157. Copyright © 1989 by Judith S. Wallerstein and Sandra Blakeslee. Reprinted by permission of Ticknor & Fields/Houghton Mifflin Co. All rights reserved; **Page 493,** Box 15.2: Excerpts from *How It Feels When Parents Divorce,* by Jill Krementz, 1984. Copyright © 1984 by Jill Krementz. Reprinted by permission of Alfred A. Knopf, Inc.; **Page 496:** Excerpt from *Second Chances: Men, Women and Children a Decade After Divorce,* by Judith S. Wallerstein and Sandra Blakeslee, 1989, p. 305. Copyright © 1989 by Judith S. Wallerstein and Sandra Blakeslee. Reprinted by permission of Ticknor & Fields/Houghton Mifflin Co. All rights reserved; **Pages 500–501:** Excerpts from *How It Feels When Parents Divorce,* by Jill Krementz, pp. 76–78, 1984. Copyright © 1984 by Jill Krementz. Reprinted by permission of Alfred A. Knopf, Inc.; **Page 503:** Excerpt from *Second Chances: Men, Women and Children a Decade After Divorce,* by Judith S. Wallerstein and

Photo Credits

Name Index

Subject Index

Emotional child abuse or neglect, 296
Emotional divorce, 476
Emotional interdependence, 96–97
Emotional maturity and marriage, 213–214
Emotional security, family as, 32–35
Emotions, 84
Empathy, 97–98
  with partners, 428–429
Employee benefits, 391
Empty nest stage
  shared parenting and, 353
  as stressor, 443–445
Ending relationships, 210–211
Endogamy, 188
Endometriosis, 331
Endometrium, 541
English-speaking and power, 272
Enteric infections, 570
Entitlement after divorce, 485
Environment, family ecology perspective and, 28
Equality, conjugal power and, 278
Equal Pay Act of 1963, 398
Equal Rights Marriage Fund, 233
Equity, martyring and, 93
Erectile tissue, 541
Eros, 91
Escape, marriage for, 215
Eskimo fertility rates, 311
Esteem. See Self-esteem
Estrangement, 283
Ethics, reproduction technology and, 333
Ethnicity. See also African Americans; Asian Americans; Hispanics; Immigration; Native Americans
  adoption of minority children, 339–340
  bonding fighting and, 255
  conjugal power and, 276
  distribution of population, 7
  English-speaking and power, 272
  extended family and, 458–459
  HIV/AIDS and, 129–130
  homogamy and, 188
  housework and, 404–405
  impact of, 12
  lifestyle and, 42
  naturalistic observation of racial/ethnic communities, 45
  parenting and, 363–368

parenting images and, 350
single persons and, 145
studying ethnic families, 40–41
Europe, dependent care in, 423
Evading fights, 251–252
Evolutionary psychologists, 108
Exchange theory, 37–38, 183–185
  fair exchange in marriage, 189–190
  traditional exchange, 185
Excitement phase, 547–548
Exogamy, 188
Experience
  perceptions and, 41–42
  scientific investigation and, 42–43
Experiential reality of families, 41
Experimental groups, 45
Experiments, 44–45
Expert power, 271
Expressive character traits, 52
Expressive sexuality, 109
Extended family, 34, 220. See also Grandparents
  crises and, 457–458
  custody of children and, 479
  divorce and, 478
  ethnicity and, 221–222
  period of disorganization after crisis, 448
External stressors, 440, 442
Extramarital affairs, 226–230, 473–474
  effects of, 229–230
  jealousy, 230
  reasons for, 227, 229
  recovering from, 230, 232

Fair Debt Collection Practices Act, 585
Fallopian tubes, 541
Families. See also Extended family; Nuclear family; Parenting; Relatives
  archival family function, 20
  belonging in, 19–20
  boundary changes, 437
  life cycle of, 30
  myths, family, 425
  one-child families, 317–318
  of orientation, 220
  policy, family, 29–30
  of procreation, 220
  size of, 307–308
  society and families, 537
  stalled revolution, 406
  structural characteristics of, 521

Familism. See Family values
The Family: From Institution to Companionship (Burgess & Locke), 4
Family and Medical Leave Act of 1993, 420, 423
Family cohesion, 263–266
Family crises. See Crises
Family day care, 418
Family decline, 20–21, 34
Family development perspective, 30
Family disorganization, 34
Family ecology perspective, 28–30, 340
Family-friendly workplace policies, 423
Family law. See Legal issues
Family leave, 420
Family myths, 425, 426–427
Family stress. See Stress
Family Support Act (1988), 485
Family systems theory, 38–39
Family values, 18–22
Fathers. See also Parenting
  child support, 485–487
  custody issues, 496–502
  full-time fathers, 352
  good dad/bad dad images, 354
  images of, 350
  retarded children and, 455
  role of, 353–354
  in stepfamilies, 530, 532–533
  as visiting parent, 400
Fear, wife abuse and, 291
Feedback
  checking-it-out process, 256
  conflict and, 256
Femininities, 54
Feminism, 72–75
  child support, concerns about, 485
  contraceptive responsibility, 319
  dating and, 199
  joint custody, 501–502
  options and, 72–75
  perspective of, 39–40
  shared parenting and, 351
Fertility awareness, 555, 564–565
Fertility rate, 306–307
  among African Americans, 308–309
  among Asian Americans, 310–311
  among Hispanics, 309–310
  among Native Americans, 310–311
  differential fertility rates, 308–311

education and, 310
trends in U.S., 307–311
Fertilization, 549
Fetal alcohol syndrome, 320
Fetus
  fetal stage of pregnancy, 550–551
  monitoring development of, 551–555
Fictive kin, 220
Field settings for experiments, 45
Fight evading, 251–252
Finances
  bereavement and, 447
  children of divorce and, 489–490
  divorce and, 473
  family budget, management of, 581–585
  HIV/AIDS and, 134
  overextension, resources for, 584–585
  stepfamilies, financial strain in, 527
First baby, arrival of, 30
First marriages, 209
First-married sex, 124
First meeting, attraction at, 196
First-trimester abortions, 329
Flexible marriage, 238–239
  contracting for, 239–242
Flexible scheduling, 420–421
Flextime, 420
Foams, contraceptive, 561–562
Folk concept of family, 248
Foraging societies, 62
Foreskin, 544
Foster Grandparents, 377
Free floating singles, 155
Frenum, 544
Freud, Sigmund, 60, 391
Friendships
  of black men, 176
  divorce and, 479–480
  marriage agreement provisions for, 241
  marriage changing, 236
  men's friendships, 172
  opposite-sex friends, marriage and, 231
  secondary group, 4
  single persons and, 169
Full-time fathers, 352
Functions of family, 30–35

Games and gender, 68
Gamete intrafallopian transfer (GIFT), 333, 336, 568
Gaslighting, 248
Gay men. See also Homosexuality

Househusbands, 394
Housework, 403–404
  divorce and, 473
  husbands sharing, 393
  race/ethnicity and, 404–405
  rational investment perspective
    on, 407–408
  resource hypothesis on,
    407–408
  satisfaction and sharing,
    425–426
  second shift, 405–407
  sharing housework, 403–
    404
  in two-earner couples,
    246–248
  why women do, 407–408
*How to Survive the Loss of a Love:
  58 Things to Do When There
  Is Nothing to Be Done*
  (Bloomfield &
  McWilliams), 480–481
Hsiao, 221, 222
Human Sexual Inadequacy
  (Masters & Johnson),
  576–577
Human Sexual Response (Masters
  & Johnson), 135
Hunting and gathering societies,
  62
Husband abuse, 293
*Husbands and Wives: The
  Dynamics of Married Living*
  (Blood & Wolfe), 271
Hyde Amendment, 328
Hymen, 541
Hypergamy, 194
Hysterotomy, 330

Ideal marriage expectation, 472
Identity
  in family, 36
  homosexuality and
    homophobia, 113–114
Ideological perspective and
  housework, 407
Illegitimate needs, 85
Ill family members, 437–440
Illiteracy, 360
Imaging *vs.* intimacy, 198
Immigration
  conjugal power, 275–276
  English-speaking and power,
    272
  family ties and, 14
Impersonality, 84
Implantation, 549
Impotence, 574–575
Incest, 296
  stepfamilies, 524

Income. *See also* Finances
  divorce and, 470, 515
  gender and, 59–60
  middle-income families, 9
  of single persons, 157–159
Independent adoptions,
  338–339
Independent couples, 263
Individualistic values, 20
Individual marriage, 204–205
Industrial Revolution, 390
Industrial societies, 62–63
Infants
  parenting infants. *See* Infants
  working parents and, 410
Infertility, 331–337. *See also*
  Artificial insemination; In
  vitro fertilization (IVF)
  embryo transfers, 568
  emotional burden of, 332
  GIFT (gamete intrafallopian
    transfer), 333, 336, 568
  inequality issues and,
    335–336
  personal choice and,
    336–337
  surrogate mothers, 334, 568
  treatment of, 332–333
  ZIFT (zygote intrafallopian
    transfer), 568
Informational power, 271
Installment financing, 584
Institution
  child-rearing institution,
    family as, 33
  defined, 30–31
Instrumental character traits, 52
Insurance
  cohabitation and, 165
  HIV/AIDS and, 134
Intelligence. *See* IQs
Interactional pattern of marriage,
  222
Interactionist perspective, 35–37
  of child rearing, 373
  on human sexuality, 108
Interclass marriages, 192–193
Interdependence, 96–97
Internalizing cultural attitudes,
  64–65
Internal stressors, 440, 442
Internet, sexual talk on, 116
Interparental conflict perspective,
  494
Interracial adoptions, 339–340
Interracial marriages, 193–194
Interreligious marriage, 190–192
Interview effects of interviewer,
  44
Interview questions, 43

Intimacy, 87–88, 89–90
  imaging *vs.*, 198
  in intrinsic marriage, 222
  time for, 126–128
Intimacy reduction affairs, 229
Intrauterine devices (IUDs), 332,
  555, 556, 560
Intrinsic marriages, 222
  choosing a, 237–238
In vitro fertilization (IVF),
  332–333, 567–568
  pain in treatment, 336
Involuntary infertility. *See*
  Infertility
Involuntary stable singles, 156
Involuntary temporary singles,
  156
IQs
  of only-children, 317
  sperm banks and, 333
  in stepfamilies, 525
Iron John (Bly), 76
Irving method tubal ligation, 557
I-statements, use of, 256
IUDs, 332, 555, 556, 560
IVF. *See* In vitro fertilization (IVF)

Jealousy, 230
Jellies, contraceptive, 561–562
Jewish persons
  geographic segregation,
    188–189
  infertility treatment for, 336
Jobs. *See* Working
Johnson, Lyndon, 29
Joint marital counseling, 580
The Joy Luck Club (Tan), 366
The Joy of Sex (Comfort), 138
Juvenile delinquency, 438

Keogh plans, 583, 584
Kin-scription, 41
Kinship, 220–222. *See also*
  Extended family; Families
  marriage and, 222
Kin-time, 41
Kin-work, 41
Kitchen-sink fight, 252–253
Knowledgeable choice, 16–17
Koop, C. Everett, 137
Kwanzaa, 43

Labia majoria, 541
Labia minora, 541
Labor, 553
Laboratory observation, 44–45
Labor force, 391. *See also*
  Working women
  divorce and women in,
    470–471

hours per week spent in, 393
male dominance in, 59–60
Labor unions, flexible scheduling
  and, 421
La familia, 221
Lagon, Patrick, 233
Laissez-faire discipline, 372
Laparoscopy techniques, 557
Late-term abortions, 329
Latex condoms, 131–132
Latin persons. *See* Hispanics
Least interest, principle of. *See*
  Principle of least interest
Leaving home, reasons for,
  149–151
Leftover embryos, 333–334
Legal custody, 499–502
Legal divorce, 477
Legal issues
  of abortion, 328
  of cohabitation, 162, 164–166
  custody of children, 497
  divorce and, 471
  of homosexual couples,
    162–163
  of interracial marriages, 193
  of marriage agreements, 249
  of pregnancy, 320–321
  relationship violence and,
    294–295
  of remarriage, 524
  reproductive technology and,
    334–335
  of same-sex marriage, 232–233
  for stepfamilies, 524
Legitimate needs, 85
Legitimate power, 271
  cultural context of, 275
Leisure hours, 409
Lesbians
  abuse among, 293–294
  case studies of, 46
  comparing gay male and
    lesbian sexual behaviors, 115
  cross-sex friendships and, 231
  enhanced relationship quality
    of, 163
  marriage between, 232–235
  parenting by, 355–357
  power and, 279
  public views on gay issues, 112
  sexuality and, 111–114
Leveling, 255–256
Liability insurance. *See* Insurance
Liberals, 28
Life course solution, 422–423
Life insurance, 165
Life satisfaction
  among adults, 170
  of single persons, 168–175

Two-earner couples, 392, 400–402
 ambivalence and, 428
 conflict and, 427–428
 elder care and, 420
 empathy with partners, 428–429
 equitable balance, striking a, 429
 family leave, 420
 feelings, sharing of, 428
 flexible scheduling, 420–421
 gender strategy and, 425
 geography and, 415–416
 juggling family and work, 412–416
 life course solution, 422–423
 love and appreciation in, 429
 mixed couples, 426
 Mommy Track, 421–422
 parenting, effect on, 349
 reinforcing cycle, 408
 relationship and, 424–429
 role conflict and, 412
 second shift in, 406
 self-employment, 401
 social policies and, 416–424
 traditional families and, 409
Two-location families, 416
Two-person single careers, 395
Two-stage marriage, 204–205
Typical families, 40–41

Ultrasound fetal monitoring, 551–552
Umbilical cord, 549
Unconscious decisions, 16
Unemployment
 parenting and, 359
 rates, 11
Uniform child custody acts, 499
Unitarian Universalist Association, 232
United Nations Fourth World Conference on Women, 58
United We Stand America, 137
Unmarried couples. *See also* Cohabitation
 as domestic partners, 9
 as families, 5–6
 increase in households, 7
Unpaid family work, 402–408
 juggling employment and, 408–416
 reinforcing cycle, 408
Upper class families, 363
 parenting, 362–363
Upwardly mobile black men, 176

Up with Down's Syndrome Foundation, 340
Urethra
 of female, 541
 of male, 544
Uterus, 541
Utilitarian marriage, 222
 choosing a, 237–238

Vacations, marriage agreement provisions for, 242
Vagina, 541
 orgasm and, 575–576
Vaginismus, 576
Vaginitis, 571
Value maturity, 214
Values. *See also* Family values; Social influences
 attraction and, 196
 heterogamy and, 195–196
 individualistic values, 20
 knowledgeable choices and, 17
Values stage of relationship, 197
Vasectomy, 555, 557
Velásquez, Nydia, 58
Verbal skills, 64
Vertical family, 458
Vestibule, 541
Video dating, 190
Vietnamese families, 221–222
Vietnam War, 11
Vimule cap, 556
Violence, 39, 171, 286–289. *See also* Child abuse and neglect; Sexual abuse and harassment; Wife abuse
 among homosexual couples, 293–294
 annual incidence rates for, 288
 in courtship process, 204
 curbing relationship violence, 294–295
 divorce and, 473
 elder abuse, 299–300
 against homosexuals, 113
 husband abuse, 293
 between intimates, 286
 mutual violence between spouses, 293
 between siblings, 371
 stress and, 438
 three-phase cycle of, 289
Virtual kin, 220
Visitation, 496
 stepparents' rights, 524, 533
Vital marriages, 224
Voluntary stable singles, 156
Voluntary temporary singles, 156, 172

Volunteer participants in experiments, 45
Vulva, 541

Wage withholding for child support, 485
War and Peace (Tolstoy), 261
War on Poverty, 29
Weak families, 453
Webster v. Reproductive Health Services, 328
Welfare, child support and, 485
Wheel of love, 98–99
White-color workers, 15
Whitehead, Mary Beth, 334
White House Conference on Families, 29
Widowed persons, 446–447. *See also* Remarriage
 numbers of, 148
 self-esteem of, 171
Wife abuse, 289–292
 child abuse and, 298
 culture and, 291–292
 reasons for, 289–290
 risk factors, 290
 shelters for abused women, 292
Wills, cohabitation and, 165
Winning and conflict, 258
Women. *See also* Feminism; Homemakers; Mothers
 caring and, 86–87
 in college, 70
 double standard, 118–119
 expressive character traits, 52
 femininities, 54
 as full-time homemakers, 394–397
 gender expectations and culture, 53–54
 genital structures, 541–542
 HIV/AIDS and, 129
 in politics, 57–58
 schools, girls in, 68–70
 second shift, 405–407
 sex ratio and, 147
 sexual pleasure and, 125
 sharing men, 163, 166–168
 single men compared to, 172
 stalled revolution, 406
Women's culture, 76
Working, 397–399. *See also* Income; Labor force; Unemployment
 dependent family members, caring for, 402–403
 divorce and, 470–471, 473

family-friendly workplace policies, 423–424
 flexible scheduling, 420–421
 homemakers, women as, 394–397
 loss of job, 438
 Mommy Track, 421–422
 overtime, 409
 providers, husbands as, 392–394
 role conflict and, 412
 second shift, 405–407
 speed-up of, 409
 traditional model, 391–397
 unpaid family work, 402–408
 unpaid family work and, 408–416
Working class families, 15
 black men's friendships, 176
 extended families in, 458
 parenting in, 361–362
Working poor, 360
Working women, 397–399
 future for, 399
 postponement of parenthood, 316–317
 power of, 274
 Pregnancy Discrimination Act, 312
 as role models, 409
 sexually segregated jobs, 398
 types of jobs, 397–399
World War I, 397
World War II
 family disorganization and, 34
 family life and, 11
 working women and, 397

You Just Don't Understand (Tannen), 255, 260
Young adults
 divorce of parents and, 490, 503
 living with parents, 8
 parents of, 30, 379–380
Young Fathers' Program, 325

Zablocki v. Redhail, 510
Zero-parent children, 365–368
ZIFT (zygote intrafallopian transfer), 568
Zoning laws, cohabitation and, 164
Zygote, 549
Zygote intrafallopian transfer (ZIFT), 568

TO THE OWNER OF THIS BOOK:

We hope that you have enjoyed *Marriages and Families,* Sixth Edition, as much as we enjoyed writing it. We would like to know as much about your experience as you would care to offer. Only through your comments and those of others can we learn how to make this a better text for future readers.

School _____ Your instructor's name _____

1. What did you like the most about *Marriages and Families,* Sixth Edition? _____

_____

_____

_____

2. Do you have any recommendations for ways to improve the next edition of this text? _____

_____

_____

_____

3. In the space below or in a separate letter, please write any other comments you have about the book. (For example, were any chapters or concepts particularly difficult?) We'd be delighted to hear from you! _____

_____

_____

_____

_____

_____

_____

_____

_____

_____

_____

_____

Optional:

Your name _____ Date _____

May Wadsworth quote you, either in promotion for *Marriages and Families* or in future publishing ventures?

Yes ☐   No ☐

Thanks!

FOLD HERE

CUT PAGE OUT

FOLD HERE

NO POSTAGE
NECESSARY
IF MAILED
IN THE
UNITED STATES

BUSINESS REPLY MAIL

FIRST CLASS   PERMIT NO. 34   BELMONT, CA

Postage will be paid by addressee

Mary Ann Lamanna / Agnes Riedmann
Wadsworth Publishing Company
10 Davis Drive
Belmont, CA 94002